3/15/16

Fred,

In the Bell System years our paths crossed frequently, and converged at least three or four times. Under your guiding hand I learned a lot that went into this book, which would not have happened except for several events that you had a hand in. I've always appreciated our association and your integrity that helped make PUB a great place to spend the first half of my working life.

Gary

THE IRWIN HANDBOOK OF TELECOMMUNICATIONS

FIFTH EDITION

THE IRWIN HANDBOOK OF TELECOMMUNICATIONS

JAMES HARRY GREEN

McGraw-Hill
New York San Francisco Washington, D.C. Auckland Bogotá
Caracas Lisbon London Madrid Mexico City Milan
Montreal New Delhi San Juan Singapore
Sydney Tokyo Toronto

The McGraw-Hill Companies

1 2 3 4 5 6 7 8 9 0 DOC/DOC 0 9 8 7 6 5

ISBN 0-07-145222-2

Printed and bound by R.R. Donnelley & Sons Company.

This book is printed on recycled, acid-free paper containing a minimum of 50% recycled, de-inked fiber.

Library of Congress Cataloging-in-Publication Data

Green, James H. (James Harry)
The Irwin handbook of telecommunications / by James Harry Green.— 5th ed.
p. cm.
ISBN 0-07-145222-2 (hardcover : alk. paper)
1. Telecommunication systems—United States. 2. Business enterprises—United States—Communication systems. I. Title.

TK5102.3.U6G74 2005
621.382—dc22 2005005849

CONTENTS

Chapter 9

Structured Cabling Systems 139

Chapter 10

Local Area Networks 155

PART TWO

SWITCHING SYSTEMS

Chapter 12

Chapter 13

Chapter 21

Wireless Communications Systems 365

Chapter 22

Video Systems 387

PART FOUR

CUSTOMER PREMISE SYSTEMS

LIST OF FIGURES

LIST OF TABLES

INTRODUCTION TO THE 5TH EDITION

When this volume reaches the bookstores, it will have been in publication for more than two decades. That is but a fraction of telecommunications' lifespan, which started with the invention of the telegraph in the mid-nineteenth century. Despite the enormous amount of technical progress telecommunications saw in its first hundred and fifty years, more pivotal changes have come about in the past thirty years. The telecommunications system has been fundamentally reshaped, not so much by technology as by political processes that have resulted in a veritable explosion of companies vying for advantage. For many companies implosion may be a more apt term than explosion. Thirty years ago when the U.S. Department of Justice was launching its antitrust case that was to break up the Bell System, AT&T was the dominant company in telecommunications and one of the largest in the world. Now, except for its name, which survives in the products of its acquirers, the company has vanished from the telecommunications landscape.

AT&T's post-divestiture history illustrates the hazards of trying to predict the direction of technology. With the best think tank and some of the top scientists in the world, AT&T bet that its future would flourish in long distance, equipment manufacturing, and computers. Now long distance has become a commodity with waning profitability. Learning that its competitors were reluctant to purchase equipment with the AT&T logo attached, it spun off its manufacturing arm, Western Electric, into Lucent Technologies. Lucent, in turn, divested its customer equipment manufacturing and sales arm into Avaya. AT&T was correct about the future of computers, but after a hostile takeover of NCR Corporation, it held it for six years and returned it to independence as a publicly-traded company. It bet heavily on wireless and cable television with acquisitions of McCaw Cellular and TCI, but spun off or sold those entities.

To AT&T's management, which was faced in the 1980s with the prospects of losing an anti-trust case, its regulated operating companies looked like losers. Divesting them into seven regional operating companies looked like its best alternative. The divestiture was good for the shareowners, particularly those that were fortunate enough to sell their stock at the right time. With the exception of US West, which allowed itself to be acquired by Qwest, a company with huge indebtedness, the Bell Operating companies have thrived; so much so that SBC in 2005 acquired its former parent.

One major milestone in the post-divestiture years was the Telecommunications Act of 1996. As Chapter 1 explains, this legislation was designed to allow competition into the local network, which the divestiture decree had left intact.

Many interests inside the industry view the 1996 act as misguided, and as this book is written, Congress is considering a rewrite. It will be interesting to watch the progress of this legislation because many powerful companies have a vested interest in its shape.

The Telecommunications Act states its purpose in the introduction:

> "To promote competition and reduce regulation in order to secure lower prices and higher quality services for American telecommunications consumers and encourage the rapid deployment of new telecommunications technologies."

Whether the country has progressed much toward these goals depends on your viewpoint. From the average residential consumer's standpoint, it has done little but raise prices. Competition in the local network has occurred, but most residential users have the same telephone company they had in 1996 and they are paying more for local service. The Bell companies have shed most of their restrictions on providing long distance, so consumers are enjoying costs that are about 10% of pre-divestiture costs. For large long distance users, competition has been a boon, but the tradeoff has been higher local service costs and a profusion of confusing taxes, fees, and surcharges. Moreover, in most of the country the old Bell monopolies appear to be resurrecting as two giants, SBC and Verizon, recreate the old AT&T structure, albeit this time without manufacturing.

One way this landscape is markedly different from the old staid Bell System is its dynamism. Alliances are fluid and failures are copious. Large players such as MCI-WorldCom, Global Crossing, Williams Communications Group, GST Telecommunications, Inc., Winstar Communications, and Rhythms NetConnections, Inc. have sought bankruptcy protection in the past few years. Besides the mergers that formed SBC and Verizon, cellular carriers have combined and recombined. Cingular Wireless, 60% owned by SBC, (Bell South owns the rest) merged with AT&T Wireless, forming the largest cellular company in the United States, a dubious distinction that may persist only until Verizon acquires more wireless carriers. Meanwhile, Nextel and Sprint merged to form the third-largest wireless company. The two largest telecom manufacturers, Nortel and Lucent, have struggled to stay afloat and have seen their market capitalization dwindle to a fraction of its previous values. Their shareowners can take small comfort from the fact that they have plenty of company along with those who were unfortunate enough to be owners or debtors of MCI and Qwest, among others.

In this shifting environment, it is difficult to keep track of the players, let alone their services and products. The purpose of this book is to present an overview of how this complex system fits together as part of the worldwide telecommunications network. We discuss how each of the telecommunications technologies works in language that anyone with an interest in the subject can understand. Obviously, some complex concepts have been simplified. Each of these chapters would easily fill several volumes for a in-depth treatment. For those who want more detail, the bibliography lists many excellent books on the subject.

This book is intended to be an introduction to telecommunications. No previous technical knowledge is assumed, so it starts with the basics and builds to a level that a technically informed professional needs to understand. The audience is assumed to be involved in enterprise telecom management, but many other professions will find it equally useful. Sales and marketing people, product managers, and even system developers should find it useful in seeing how their products fit into the whole. Several colleges and universities have used previous editions as a text. An instructor's manual for this edition is available through normal university channels.

The book is divided into five parts, as were previous editions, corresponding to major divisions in telecommunications equipment. Chapter One is an introduction to voice and data. The remainder of Part One is devoted to concepts that are common to the industry. In Part One, we discuss voice and data fundamentals, pulse code modulation, outside plant, structured wiring, access technologies, local area network principles, and the other building blocks of telecommunication networks.

Part Two covers switching. The part begins with a discussion of signaling, including new protocols Session Initiation Protocol (SIP) and ENUM, which are new since the last edition, and hold considerable promise for the future. A chapter on the public switched telephone network follows, discussing how it works and the quality requirements that IP must achieve to support voice. Two chapters follow to explain in overview how local and toll switches and integrated services digital network (ISDN) function. Circuit switching has been at the heart of the telephone industry for more than a century and retains stability and service quality that packet technologies cannot yet provide. We devote a chapter to it. Part Two ends with a discussion of softswitches, which are a new generation of IP switches that serve advanced IP networks.

Part Three covers transmission equipment. Separate chapters discuss the fundamental technologies of fiber optics, microwave radio, satellite transmission, cellular and PCS radio systems, wireless, and video. Fiber lies at the heart of the telecommunications infrastructure and is arguably the most important development in the industry's history. It displaced long-haul microwave, but that technology is becoming more important than ever with an emphasis on communications mobility. Customer demand is fueling a host of new wireless services and protocols that operate in the microwave bands and are receiving a great deal of attention. Video is also becoming a vital Internet access service, and more. The new hybrid fiber-coaxial cable architecture enables cable to compete with the conventional telephone system.

Part Four discusses customer premise equipment. As with the public telephone network, customer premise switching is evolving to IP. We begin this part with a discussion of station equipment, followed by a chapter that discusses features that customer premise switching equipment supports. Chapters follow on conventional digital switching and the newer IP switching. We next discuss automatic media distribution systems, which are evolving from the older automatic call distribution systems. These respond to customer demands for contact alternatives besides the telephone. Other chapters discuss voice processing, electronic messaging, and facsimile.

Part Five pulls together the building blocks we have discussed in the earlier chapters into completed and functioning telecommunications networks. This part illustrates the tremendous variety of alternatives that are available and discusses how and where they are applied. We begin this part with the discussion of enterprise networks, which is a blanket term covering the networks organizations use to link the enterprise. Following that, other chapters cover metropolitan area networks, wide area data networks, frame relay, asynchronous transfer mode, and IP data networks. The IP chapter discusses multi-protocol label switching (MPLS), which is evolving into a platform for handling multimedia applications over IP networks. We discuss testing and network management systems and how they are evolving to enable humans to cope with the increasing complexity of modern networks. The final chapter in the book looks ahead a few years with a view of where telecommunications technology is headed.

The telecommunications industry is fraught with jargon and acronyms. It is impossible to avoid these and repeatedly spelling them out would be tedious. Accordingly, the principal acronyms are spelled out at least once in the text and explained in context. The Acronym Dictionary in Appendix A lists most of these and the Glossary in Appendix B defines them. Telecommunications glossaries are plentiful and more exhaustive explanations of terms can be found by entering them in an Internet search engine. The book ends with a bibliography of relevant books on this fascinating subject.

PART ONE

Introduction

We are at the threshold of a compelling age in which boundless information lies at our fingertips and it is clear that the future belongs to those who know how to use it. Most users give little thought to how information flows from its sources to their screens, printers, and files, but it all happens through telecommunications, which is the subject of this book. This introductory part covers concepts that are an essential prerequisite for understanding how these marvelous technologies work.

The first chapter addresses questions that would undoubtedly puzzle anyone who is not familiar with telecommunications history. A newcomer viewing the intricacies of the interlocking organizations that deliver and control services would ask, "Why in the world do they do it that way?" This chapter explains that things are not shaped as they are because it is the best way to design a telecommunications system. Politics have played a more important role than technology, particularly in the last decade or two.

The entire industrialized world is in the throes of a major transition. In most of the world except for North America, the government owned the telecommunications system; in North America it was a regulated monopoly. Now a confluence of powerful factors is reshaping the way we communicate and work. Regulation is gradually giving way to competition under the resolute direction of governments that are unwilling to leave control of such vital services to the guidance of Adam Smith's "invisible hand."

After the short history in the first chapter, we look in more detail at the building blocks of voice and data systems. We review the structure of the public switched telephone network, which shares the underlying infrastructure with data networks. We offer an overview of data technology including protocols, which are the language that computers use to communicate, while insulating users from their complexity. Chapter 5 discusses pulse-code modulation, which forms the basis for the worldwide digital network. Subsequent chapters provide an overview of outside plant, structured wiring, power systems, and other elements that are common to all systems. One chapter is devoted to the basics of local area networks, which are the link between the desktop computer, the office file and print servers, and the wide area networks including the Internet.

At the conclusion of this part, you should have a good understanding of the elements of telecommunications systems and be prepared to understand the details that the subsequent chapters discuss in more depth.

CHAPTER 1

A Brief History of Telecommunications

The turn of the new millennium was a grand time to be an entrepreneur in the telecommunications industry. The economic wisdom of the past was yielding to a widespread belief in a new marketplace that did not conform to historical verities. Venture capitalists lavished money on Internet-related business plans that, as it turned out, had scant hope of success. Hundreds of new companies materialized, filing initial public offerings even before demand for their product was validated. Investors gobbled up stocks in companies that had histories measured in mere months.

The euphoria began to collapse around the middle of 2000 in a period that is now known variously as The Telecom Winter or The Technology Bubble. It turned out that much of the new economy was built on an illusion. Stocks sold for prices that bore little relationship to profitability. Earnings were inflated by accounting artifices, and purchases were financed on credit, often provided by the manufacturers that produced the equipment. The market for technology stocks collapsed and companies closed their doors. Although the period of readjustment was a wrenching experience for those that endured it, the gloom is now fading. In the longer term telecommunications continues to transform the world in ways that few could have predicted a decade ago.

For the first century following invention of the telephone, the industry clicked along with steady, but unspectacular change. Voice was the dominant application on a network that was designed with little capability for carrying data. When mainframe computers came on the scene in the 1960s, the public switched telephone network (PSTN) was the only medium available for connecting remote terminals to a central computer. Data communications equipment and protocols were designed to fit within the technical constraints of voice circuits.

Toward the end of telecom's first century, several forces more or less coincided to compel more change in two decades than the industry had seen in the previous 100 years. The telephone industry was a staid monopoly, most of it owned and governed by AT&T, and regulated by state and federal government agencies. In 1975 the Department of Justice (DOJ), responding to pressures from a variety of interests, filed suit to dissolve the monopoly. The litigation stretched over 7 years, during which period microprocessors were transforming the computer industry and fiber-optic technology was at the threshold of a massive development surge. The world's telecommunications networks were migrating from analog to digital and the design and manufacturing techniques used in computers could be adapted easily into telecommunications products. The DOJ's lawsuit ended abruptly in 1982 when AT&T agreed that it would yield to government demands that it divest its local operating companies and begin the transition to a competitive telecommunications system.

At the time this transformation was taking place, data transmission comprised an increasing amount of traffic on the PSTN. Early data applications involved mostly human-to-machine communication. Remote offices needed to access the databases that had become the mainstay of large businesses. Ordinary telephone circuits were more than adequate for the purpose because a single voice circuit could support more than a dozen people keyboarding simultaneously. Desktop computers changed all that.

Inside the offices of the era, local area networks (LANs) began to link desktop computers. LANs communicate at high speeds that far exceed the capabilities of a telephone circuit. Desktop computers replaced the dumb terminals of early mainframes, and data networks evolved into networks of interconnected LANs. LAN-to-LAN communications required increasing amounts of bandwidth that the existing services and protocols could not support.

Technology came to the rescue. At the time of the Bell System's breakup, fiber optics was leading-edge technology. In less than 10 years most of the microwave systems and undersea cables that had served the world for decades were replaced with fiber. Data devices now had a new transmission medium that enabled them to communicate in their natural digital form. No longer was it necessary to convert a digital signal to analog so it could travel over a voice circuit. Companies built fiber optic networks with enormous amounts of capacity, far outstripping demand. Prices fell and new techniques for data communications came into play.

While this transformation was occurring, a consortium of government, commercial, and educational organizations was working on new methods for computer communications, resulting in a network known as ARPANET. Heretofore, voice and data networks had been proprietary. AT&T's Bell Laboratories designed most of the hardware that connected telephones and data devices to the network. IBM, as the dominant computer manufacturer, designed the high-level data protocols. Other computer manufacturers developed proprietary protocols of their own. Although the various companies and agencies of the world recognized the need for international standards, the market was slow to change.

ARPANET, as it turned out, was the necessary catalyst for change. Its protocols were developed in an open environment. When it evolved into the Internet and the World Wide Web brought the benefits to the public, proprietary standards began to give way to open standards. New protocols and services were produced to meet the demand for increased amounts of data bandwidth, and still the evolution continues under the driving forces of deregulation, competition, technology, open standards, and the emerging Internet.

A SHORT HISTORY OF THE BELL SYSTEM

No organization in the world has influenced telecommunications to the degree that the Bell System did. The Bell System traced its roots back to Alexander Graham Bell's invention of the telephone in 1876, and survived for nearly 110 years. For more than six decades prior to the 1984 dissolution of the Bell System, AT&T operated as a regulated monopoly, but it was not always so. Following expiration of Bell's patents in 1893, the early years of the telephone's existence were characterized by freewheeling and open competition, at least in the United States.

In most of the world the post office ran the telephone system as a government monopoly. In the United States, AT&T, which was incorporated in 1885, rose to dominate the industry by purchasing small telephone companies and acquiring franchise rights to provide service in most metropolitan areas of the United States. They concentrated on the profitable core areas of cities and left much of the sparsely populated rural areas to other companies. By the turn of the twentieth century, some 4000 independent telephone companies representing about 40% of the telephones in the country served the United States. AT&T owned the rest.

AT&T's expansionary tactics attracted the attention of antitrust authorities. In 1913, under pressure from the Department of Justice, AT&T signed an agreement known as the Kingsbury Commitment. In this agreement, AT&T promised to cease acquiring smaller independent telephone companies and to interconnect with other companies to form a national network. The ownership pattern was established by then and the result was a patchwork of telephone companies. The larger cities had an AT&T-owned company in the core and smaller independent companies in the surrounding rural areas.

The Principle of Regulated Monopoly

In 1917, the United States entered World War I. In the following year the federal government nationalized the nation's telephone system, keeping it under government control for 1 year. In the process of returning to private ownership, Theodore N. Vail, AT&T's president at the time, proposed the principle of regulated monopoly as an alternative to government ownership. The argument suggested that certain utilities such as power, water, and telephone are natural monopolies.

Competition for these services, the reasoning went, was not in the public interest because the duplication of facilities was expensive and redundant access to the public right-of-way was disruptive.

Regulation placed several aspects of telephone service under government control. Regulated companies filed tariffs with the regulatory commissions, which in most states was a public utilities commission (PUC). Tariffs published the terms and conditions of service including prices and service areas. The public needed to know nothing about the technical aspects because the telephone companies owned and maintained everything associated with the service. The interstate portion of telephone service was then under the jurisdiction of the Interstate Commerce Commission (ICC). ICC regulation continued until 1934, when Congress passed the first telecommunications act. Among other provisions, this act then established the Federal Communications Commission (FCC) and gave it responsibility for regulating interstate communication services. In addition, the act charged the FCC with regulating use of the airwaves.

Through the years of regulated monopoly, the local exchange carriers (LECs) and the utility commissions agreed to certain principles. One of these that still survives is the principle of *universal service*. Recognizing that the value of every telephone is enhanced by the number of other telephones it can call, the objective of keeping service ubiquitous and affordable guides both technical and regulatory decisions at all levels. Of particular interest was, and still is, how to assure that rural and high-cost service areas are furnished affordable service, which means that rates are lower than the cost of providing service.

PUCs and LECs worked together in wary collaboration to hold down telephone rates through a series of initiatives. One was reducing labor and maintenance expenses with a steady flow of technical improvements. Dial systems replaced manual switchboards. Cable and multiplexing equipment replaced open wire transmission systems. Microwave radio provided transcontinental toll service at a cost far less than the trouble-prone wire and coaxial cable it replaced. These innovations required capital investment, which in turn increased depreciation expenses and mandated rates that were sufficient to pay a reasonable return.

Telephone Rate Subsidization

Despite the technical progress, rural telephone systems were, and still are, more expensive to construct per subscriber than urban systems and they are less profitable because of the lower density. Rates, however, are generally based on the number of phones in the local calling area. This tends to shift the higher rates to the urban areas, which is the reverse of the cost structure. The result is that urban telephones subsidize rural telephones.

Another way of keeping basic rates down is to load costs on discretionary services. Where rates are still regulated, the PUCs and LECs hold public hearings in which they negotiate an authorized return on invested capital. After the return on investment is approved, the commissions and LECs decide how to allocate the consequent revenue requirement to various services while keeping basic residential rates as low as feasible. Business rates are often pegged at three or four times residential rates even though the cost of providing service is not much different. Rates for services such as trunk hunting and later push-button dialing are set well in excess of the cost of providing them. When the LEC owned the telephones, extensions, colored telephones, push-button dials, and even coiled handset cords commanded premium prices, contributing revenues to the LEC and subsidizing local rates.

The discretionary service that bore the bulk of the subsidy in the pre-competitive market was long-distance. AT&T's Long Lines division carried practically all interstate and international long-distance calls in the United States. The subsidization method was subtle. The FCC prescribed a rate of return on invested capital for Long Lines. The LECs, both Bell and independent, billed their customers for the long-distance calls. The LECs contributed their interstate revenue to a settlement pot that Long Lines administered. At the end of the settlement period, each company withdrew from the pot enough money to repay its cost of providing service plus a return on its invested capital at the same rate as the FCC prescribed for AT&T. If revenues were insufficient to support earnings at the authorized level, the carriers lived with the shortfall until the FCC authorized a rate increase. Fortunately, increased efficiency kept rates relatively stable. Telephone rates increased at a rate much lower than advances in the cost of living, and over the years, long-distance rates dropped because of increasing efficiency. However, they were then much higher than they are today.

One of the subtle elements in the long-distance subsidy was the manner by which the LECs established their cost of providing long-distance service. Most of their plant investment was involved in providing both local and toll service. All of their equipment, including local switching systems, outside plant, private branch exchanges (PBXs), and telephone instruments, was included in plant investment at the time. The question was how the costs were allocated between local and toll. The LECs conducted studies to determine the percentage of usage of various elements of plant investment for toll calls. The PUCs had an incentive to attribute as much of the investment as possible to interstate toll because the LECs recovered those costs from the toll settlement pot. Toll revenue reduced the LECs' total revenue requirement, which in turn helped hold local rates down. It also, incidentally, helped to impede innovation. Under the regulatory environment, equipment was often left in service long after it had become technologically obsolete because it helped minimize depreciation expense.

This environment served the nation until the late 1960s, when several events occurred that foreshadowed the eventual collapse of the Bell System. Up to that point, virtually all telephone innovation emanated from Bell Laboratories. The Bell System was a vertically integrated entity in which R&D, manufacturing, installation, and the provision of all phases of customer service were under unified Bell control. The technical standards that were set at the time were Bell standards and were not subject to outside review.

AT&T's organization had a great deal to do with future events that reshaped the industry. The Company owned either all or a majority of the stock of its operating telephone companies, of which there were 22 at the time of divestiture. These companies, known as Bell Operating Companies (BOCs), were responsible for providing telephone service to the customers and most intrastate long distance. AT&T's Long Lines subsidiary provided interstate long-distance service between the companies. Another subsidiary, Western Electric Corporation, manufactured most of the telephone equipment that the BOCs and Long Lines used. Figure 1-1 shows the organization of the vertically integrated Bell System at the time of divestiture.

AT&T and Western Electric collectively owned Bell Telephone Laboratories. Bell Labs had two major roles. Western Electric's portion designed the products that Western manufactured. It was funded through equipment sales to the Bell companies. The other half of Bell Labs was funded through license fees that the BOCs paid to AT&T. This portion performed fundamental research and developed methods of managing and maintaining telephone plant. Ironically, one of Bell Laboratories' most significant discoveries, the transistor, made AT&T's dissolution technically feasible.

Although the concept of digital communications was developed in 1941, it was little more than a curiosity at the time because it relied on vacuum tubes, which were unreliable power and space hogs. The transistor, which was patented in 1948, offered a superior alternative to the vacuum tube, but the exclusive right to the patents placed an enormous amount of market power in the hands of one company. This, among several other complaints, drew the attention of the antitrust authorities.

FIGURE 1-1

Simplified Organization of the Pre-divestiture Bell System

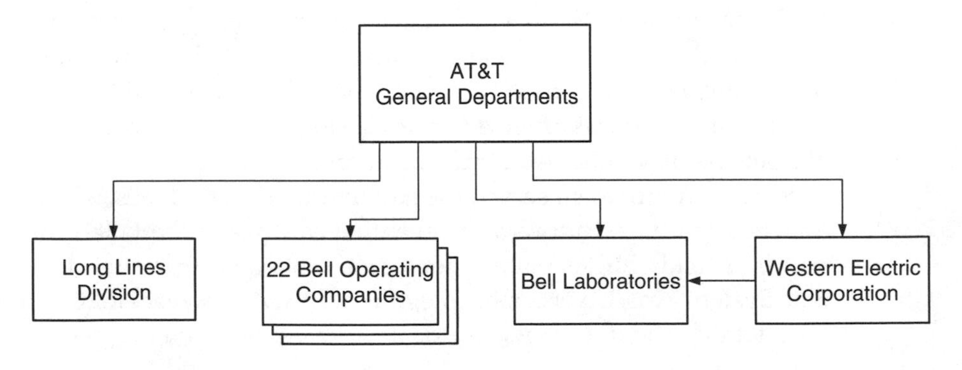

The First Antitrust Case

In 1949, 1 year after the invention of the transistor, the DOJ filed an antitrust suit against AT&T. The suit was resolved with a consent decree in 1955. That decree, known as the Final Judgment, required AT&T to license its inventions, including the transistor, to outsiders. Except for certain sales to the government, Western Electric was confined to producing telecommunications equipment for the Bell System and many of its product lines of the time were spun off or sold.

This situation existed for the next 20 years until 1975, when the Department of Justice filed another antitrust case against the Bell System. That case ended with the Modified Final Judgment (MFJ), which was the decree that dissolved the Bell System. The issues behind that litigation explain how the telephone industry acquired its current shape.

Interconnection

The Bell monopoly on telephone service had gradually been crumbling at the time the DOJ filed the suit. The first chink in the Bell armor was the 1968 Carterphone case. Carterphone was a device that connected a privately owned radio dispatch service to the PSTN. Interconnection to the PSTN was technically uncomplicated. Ham radio operators had been doing it for years, albeit without the approval of the LECs. Citing its tariff prohibition against so-called "foreign attachments," the BOC involved, Southwestern Bell, demanded the removal of the Carterphone device. The issue was appealed to the FCC, which ruled that the foreign attachment provision in the tariff was not justified. The FCC ordered the Bell companies to permit foreign attachments through a protective coupling arrangement (PCA) that was designed to prevent foreign attachments from harming the network or its technicians.

The PCA persisted until 1976, when the FCC adopted its Part 68 regulations. These rules establish a process for registering all devices that attach to the PSTN, including those belonging to the LECs.

Long-Distance Restrictions

The Bell monopoly was further weakened in court battles between AT&T and MCI. In 1969, Microwave Communications, Inc. was granted an FCC license to provide point-to-point communications over its private microwave system between Chicago and St. Louis. AT&T argued, unsuccessfully as it turned out, that their tariffs granted them the exclusive right to provide such circuits. Their main argument was that as a regulated common carrier they were required to serve all comers. A competitor that had no such service obligation could cut rates, capture the most profitable routes, and leave the remainder for the regulated carrier. The result, AT&T argued, would be an increase in rates for those remaining customers. The argument failed to convince the courts. AT&T lost the case and MCI remained in the private line business over their microwave network.

MCI's customers began to request that the company permit them to dial up connections in the cities in which their private networks terminated. MCI obliged by obtaining ordinary telephone lines from the BOCs and permitting their customers to dial connections across them. AT&T appealed to the FCC that MCI was encroaching on its monopoly of message telephone service. Moreover, AT&T argued, the main reason MCI could cut its rates was that they were not required to subsidize local service. The FCC sided with AT&T in the case, but MCI appealed to the courts, which reversed the FCC. The FCC and the Bell System then instituted a method of compensating the LECs for local exchange access through two different methods called Feature Group A and Feature Group B. Feature Group A is a service that is technically identical to an ordinary telephone line, except that usage is metered both incoming and outgoing.

With MCI's success, other interexchange carriers (IXCs) entered the market. They complained that the technical connection between AT&T and the LECs was discriminatory. The transmission quality of feature group services was inferior to AT&T's and their subscribers had to dial an access code and personal identification number plus the terminating telephone number. This required dialing an additional 12 or so digits for every telephone call compared to AT&T's customers, whose lines were automatically identified in the LEC's central office. The competitive carriers demanded access service equal to that the LECs provided for themselves and Long Lines. The LEC central offices, however, were not designed for competitive access and the carriers had little incentive to provide service that cut into their long-distance revenues.

Equipment Manufacturing

Manufacturers complained to the DOJ that the BOCs had little incentive to purchase telecommunications equipment from companies other than AT&T's internal manufacturer, Western Electric. The vertically integrated Bell System with its design and manufacture of telecommunications equipment represented a market that was difficult for outsiders to penetrate. The BOCs purchased Western Electric equipment without competitive procurement, leaving outside manufacturers a market that was mostly confined to the independent telephone companies.

The Second Antitrust Case

In this milieu, the Department of Justice filed another antitrust case against the Bell System, seeking to break AT&T's monopoly. By this time, solid-state electronics had developed to the point that design and manufacturing techniques used for computers were equally applicable to telecommunications equipment. The markets for equipment sales and provision of long-distance service were enormous and outside companies were eager to get a share. AT&T mounted a vigorous defense, but in early 1982 they surprised the world by acceding to dissolution.

The terms of the decree, which were a modification of the 1955 case, transformed telecommunications throughout the world. The decree permitted AT&T to keep its manufacturing and Long Lines subsidiaries and it was relieved of an earlier restriction that prohibited the company from manufacturing and marketing computers. The local telephone business was divided into seven regional companies known as regional Bell Operating Companies (RBOCs). Under the decree, the RBOCs retained their monopoly on local service, but were prohibited from providing long distance service except within a narrowly defined region known as a local transport area (LATA). The LATAs correspond roughly to a standard metropolitan statistical area as defined by the Census Bureau. The LATA boundaries were established by a political process that was intended to maximize long-distance competition by restricting the area the RBOCs could serve with long-distance. The RBOCs were required to reconfigure their switching systems to permit subscribers to choose a primary interexchange carrier (PIC). When a subscriber dialed the digit 1 long-distance access code, the call would switch through the PIC.

Under the MFJ, Bell Laboratories was divided into two parts. AT&T kept the Bell Laboratories name along with the portion that performed product-oriented research for Western Electric. The RBOCs retained the other half under joint ownership and named it Bell Communications Research (BellCore).

Another important aspect of the decree was replacement of the long-distance subsidy. The FCC decided that a system of access charges would replace the previous division of revenue process. Access charges consist of two elements. Subscriber line access charges are a fixed per-line cost that the LECs bill their subscribers. The charges started low, but have increased through the years, with businesses paying a higher fee than residences. The LECs often use billing terminology that makes it appear as if subscriber line access charges are a government fee, but in fact, they are simply a convenient way of mandating nation-wide local service rate increases without the need to work through every PUC.

The other portion of exchange access charges comes from a usage fee that the LECs levy on the IXCs. The LECs provide ports for terminating long-distance calls into their networks and bill the IXCs at an FCC-authorized cost per minute. The IXCs, in turn, include the cost in their long-distance rates. This structure motivates the IXCs to bypass the LECs by extending direct lines to their larger customers. The industry refers to this as *dedicated access*, whereas terminating through the LEC is referred to as *switched access*.

Figure 1-2 shows the cost elements of the two different types of long-distance access. The LEC usually provides the dedicated access line. It routes through the central office, but it is not switched.

In the years since the breakup of the Bell System, access charges have been the subject of much controversy and political maneuvering. The FCC has changed the formula several times, but access charges remain a combination of fixed and variable charges. Usage costs have steadily declined while the fixed fees have increased. Subscriber line access charges for both businesses and residences have increased to

FIGURE 1-2

IXC Access to the Subscriber

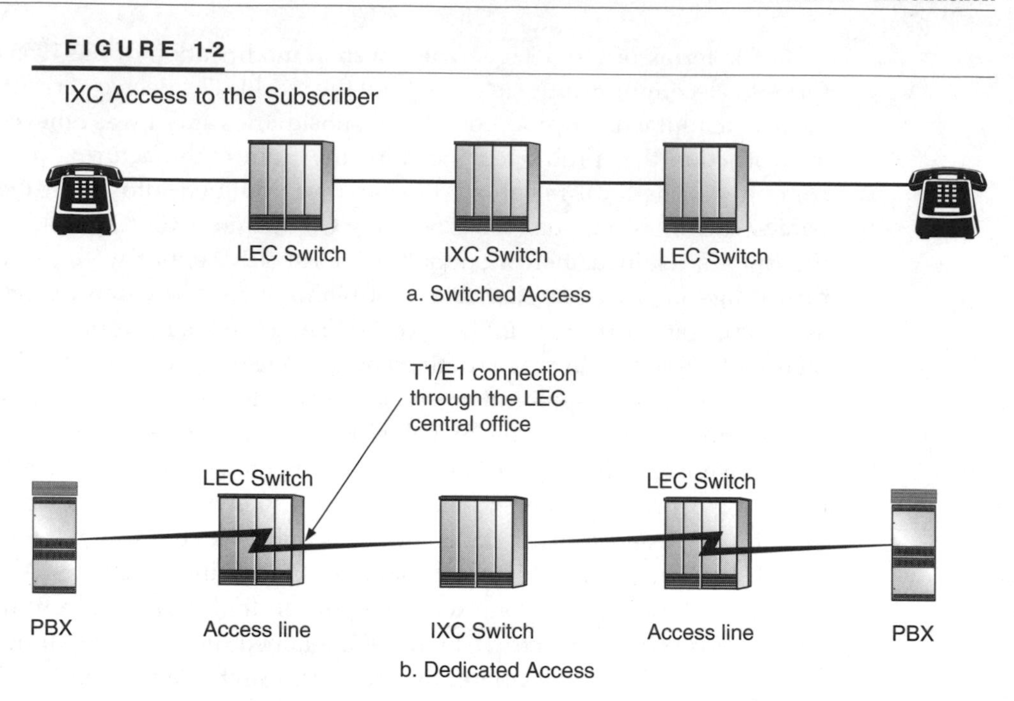

about $9 and $5 dollars per line, respectively. In addition, the FCC in 2001 permitted to the LECs to bill the IXCs a monthly fixed fee for each presubscribed line. This charge is known as the presubscribed interexchange carrier charge (PICC), otherwise known as the pixie charge. Most IXCs also flow this cost to their users.

The Telecommunications Act of 1996

If the dissatisfaction of all parties is evidence of an equitable court case, then the decree that broke up the Bell System was a rousing success. The IXCs were unhappy with access charges, the LECs chafed under restrictions that prevented them from providing interexchange services to their customers, and would-be competitors complained about the local service monopoly. The market for local service in the United States is estimated at about $90 billion per year and this remained in the hands of regulated companies that were forgetting their common heritage and increasingly viewing one another as competitors.

Congress stepped into the melee in the first half of the 1990s, and after interminable hearings, produced the Telecommunications Act of 1996. The act has many important provisions, not only for telephone companies, but also for the broadcast industry, and it is commended to anyone who wants to understand the telecommunications environment in the United States today. The most important aspect for our purposes is the opening of the local exchange network to competition. The act required the RBOCs to allow competitors to connect to their local exchange networks at technically feasible interface points and to allow competitors to share facilities such as conduits, pole lines, and central office buildings on a space-available basis. Competitors were permitted to lease facilities such as cable pairs and switch ports as *unbundled network elements* (UNEs).

In exchange for giving up their monopoly, the RBOCs were permitted to enter the interexchange long-distance market after they demonstrated that their networks were open to competition. Today, the RBOCs have been granted permission to enter the long-distance market in all states.

The Universal Service Fund

It was clear that the benefits of competition would have undesirable side effects. Competitive local exchange carriers (CLECs) would concentrate on capturing the large and most profitable customers, leaving the less profitable smaller and rural customers to the incumbent local exchange carriers (ILECs). As profitable services migrated to competitors, the ILECs' costs would not drop accordingly, prices would rise, and rural subscribers would be adversely affected.

The act directs the FCC to establish a *universal service fund* (USF) and methods of collecting money to fund it. Congress decreed that certain telecommunications expenditures of schools, libraries, and rural hospitals would also be subsidized by the fund, a process that goes by the name of E-rate. The USF is levied on the gross receipts of telecommunications carriers, most of which pass the costs directly to their customers.

The Aftermath of the MFJ

The competitive telecommunications arena has been an interesting experience, although it is too early to foresee the outcome. Some significant shifts have occurred in the two plus decades since divestiture. AT&T, once one of the largest companies in the world, is a shadow of its former self. Through acquisitions, it made forays into computers, cellular, and cable businesses only later to divest them. It spun its manufacturing business off into Lucent Technologies, which in turn split off customer premise equipment manufacturing and sales into a separate company, Avaya. Lucent retained the Bell Laboratories name, but that organization is greatly reduced in scope as Lucent has struggled to recover from the telecom winter. AT&T today is being acquired by SBC Corporation.

The seven RBOCs have gradually combined, leaving only one, Bell South, in its original shape. Southwestern Bell (southern Midwest) merged with Pacific Bell (California) and Ameritech (northern Midwest) to form SBC Corporation. In 2005 SBC acquired its former parent, AT&T, which had shrunk to a shell of its former self. Bell Atlantic (middle Atlantic states) merged with NYNEX (New York and New England) and then subsequently with GTE to form Verizon. These mergers give both Verizon and SBC a nation-wide footprint. US West allowed itself to be acquired by Qwest, a company that was then riding the dot-com boom of the 1990s.

The company is now laden with heavy debt and has a market capitalization that is a fraction of its earlier peak. The RBOCs sold BellCore, the name of which is now Telcordia. It is interesting to note that the RBOCs, which many viewed as the stodgy and unexciting portion of the telecommunications industry have, with the exception of Qwest, surpassed their former parent.

Most of the objectives of the antitrust case have been achieved. The long-distance market is no longer regulated and tariffs for domestic long-distance have been eliminated. Long-distance prices are a small fraction of predivestiture days. The result has been huge savings for all but small and uninformed long-distance users. The result has been much less favorable for local service users. By the time the various fees, access charges, and surcharges are added to the basic service, costs have more than doubled since divestiture.

The competitive local market is far less mature and more difficult than long-distance. Anticipating opportunities to participate in a huge market, hundreds of companies jumped into the local exchange business and invested huge sums that evaporated into bankruptcy. In the rapture of the dot-com boom, investors pumped millions into companies that discovered that building a local telephone network is neither easy nor inexpensive.

The Telecommunications Act requires number portability among LECs. It also requires interconnection at technically feasible interface points and at prices based on the cost of service. Arguments have raged through the regulatory agencies, the FCC, and the courts about the appropriate method of calculating costs and setting prices for UNEs. The ILECs claim that the revenues are less than their cost of providing service, and the CLECs contend that the prices are too high. The RBOCs complained in 2004 to the courts that they were being forced to lease facilities to CLECs below cost. The courts sided with the RBOCs and the FCC declined to appeal to the U.S. Supreme Court. Consequently, RBOCs have raised their UNE prices. It seems clear that the most successful CLEC businesses will rely on ILEC facilities to the least degree possible.

The competitive local market is a changing field with a future that is difficult to foresee, but competition is alive in the local exchange. Larger businesses now have a choice of several carriers in most metropolitan areas. Most residences and small businesses have little if any choice of carriers because the cost of providing service is too high to be profitable. That situation may change in the future with some of the technologies we will discuss in this book. Meanwhile, another revolution is occurring; one that was but dimly seen of the time of the antitrust activity: the Internet.

A SHORT HISTORY OF THE INTERNET

The Internet's impact on life in the twenty-first century is difficult to exaggerate, and its major effects have scarcely been felt. Today's Internet traces its origins back to a 1968 project in which the Advance Research Projects Agency (ARPA) awarded

Bolt, Beranek, and Newman (BBN) a contract to develop a network to support information exchange among computers. At that time, computers were mainframes with far less processing power than today's desktop computers, but they were the information repositories for government, commercial, and educational institutions. The value of information, the world had learned, is enhanced by its accessibility over a network.

The telecommunications network of that time was largely analog. Digital multiplexing was in its infancy and confined to metropolitan areas. Therefore, computers had to communicate over voice circuits that were transported over cable or microwave radio. By today's standards, they were noisy with narrow bandwidth; adequate for voice, but unsatisfactory for data transfer, which requires error-free communication.

ARPANET was designed from the start as a packet network, which has the advantage of interleaving multiple sessions in the same data stream. The idea of transmitting data in packets was far preferable to sending a continuous flow. Since there was no way to prevent data errors, the solution was to resend errored data. Sending it in small segments was much more efficient than resending the entire message.

The early applications of ARPANET were e-mail and file transfer. The computers at that time were text oriented and the procedures of these two applications required learning a command language that was not congenial for general use. The Internet throughout the 1970s and 1980s was in the developmental stages and successfully proved the viability of the techniques. Transport Control Protocol (TCP) was developed and then split to separate Internet Protocol (IP) to form TCP/IP, which is the mainstay of data communications today.

Computers shrank in size while their processing power grew. In the network, microwave radio systems gave way to fiber optics, digital circuits replaced analog, and higher speed applications became feasible. As desktop computers became prevalent, the public began to adopt computer communications and applications such as bulletin boards and public databases such as AOL and CompuServe came on the market. BBN, which had designed the original ARPANET, introduced the first nationwide packet network, Telenet. Applications such as USENET appeared on the Internet and a variety of e-mail programs emerged. Those on the Internet used Simple Mail Transfer Protocol (SMTP), while those on other networks used a variety of protocols, most of which interoperated only with difficulty. Text-oriented protocols such as File Transfer Protocol (FTP) and TELNET were developed for the Internet. Servers running applications such as Gopher became information repositories and search tools such as Archie and Veronica allowed users to find and download information from the growing network of host computers.

In 1984, the Defense Department separated DARPANET, as the network had become, into a separate military network, and the nonmilitary functions became the National Science Foundation Network. Bandwidths were increased, and the network grew rapidly. In 1993, NSF formed the Internet Network Information

Center (InterNIC) to handle domain name registrations. In 1995, NSF announced that it would no longer allow direct access to its backbone, effectively turning the Internet over to private enterprise.

Throughout the 1980s and 1990s, the value of the Internet was amply demonstrated, but its arcane command language limited its usefulness. Also, access to the Internet required dedicated circuits that most individual users could not justify. That situation changed with the development of graphics-oriented hypertext transfer protocol (HTTP) and the World Wide Web. Internet service providers (ISPs) provided both direct and dial-up channels to connect users to the network. Web browsers eliminated the need to understand the command language and the Internet burgeoned.

Today's Internet is a collection of interconnected independent networks. Barriers to entry are practically nonexistent, so as a result there are thousands of ISPs. A tier-1 ISP has global routing tables. Some tier-1 ISPs serve customers directly but others primarily provide transit traffic to lower level ISPs. Smaller ISPs provide direct access to customers and purchase bandwidth from backbone carriers. If an ISP has sufficient traffic to justify direct connection with another ISP, the two form a peering relationship in which they agree to provide direct access to each other's customers. The motivation for peering relationships is to reduce the cost of transit traffic, which is expensive since it may have a global reach. ISPs also reduce the amount of traffic by providing caching servers to store the most frequently accessed Web pages.

If an ISP cannot complete a connection locally, it hands the traffic to a regional or tier-1 carrier. These carriers meet for exchange of traffic at a network access point (NAP) or at regional Internet exchanges (IEs). The NAPs and IEs have a secure physical facility. They provide collocation space, protected power, and a high-speed backbone to carry traffic between networks. ISPs may connect at these points to exchange traffic with their peers or they may use point-to-point circuits between themselves.

When the Internet started, the world had a single unified network that was designed for voice, and data networks were formed from its elements. Now the trend is back to a single network, but this time it is a data network with voice as one of its applications. The Internet is designed to be fast and cheap, but it is a best-effort network and is not designed for the degree of reliability required for commercial-grade telephony. It could be made so, but at a cost. In other words, the Internet can be cheap or it can be high quality, but not both. Nevertheless, the transition to a converged network is underway. This transition has hardly begun and its long-term effects are not yet understood. The only thing certain is that people need to communicate, markets will continue to expand to meet the demand, and the evolution is far from over.

TELECOMMUNICATIONS STANDARDS

One aspect of the Internet that is not apparent to lay people is its effect on standards. Before about 1980, most telecommunications standards were proprietary, emerging from the laboratories of large companies such as AT&T and IBM.

The International Telecommunication Union (ITU), which is an agency of the United Nations, promulgated international standards but they had little effect in the United States. ITU often followed the lead of American companies, adopting standards that were almost, but not quite like U.S. standards. Before the world was interlaced with undersea fiber-optic cables, the lack of uniformity made little difference because the North American telecommunications systems connected with European systems through gateways that made the differences irrelevant.

The divergence of European and North American systems had several causes. One was the lack of clear channels for international cooperation. Another resulted from the slowness with which international standards progressed from proposal to adoption. Companies can develop proprietary standards in a fraction of the time required for international adoption. IBM produced its Systems Network Architecture (SNA) with its own resources while the International Standards Organization (ISO) later developed its Open Systems Interconnect (OSI) architecture that is similar, but not identical. AT&T, eager to replace signaling systems that were vulnerable to toll fraud, developed its common channel interoffice signaling (CCIS) and applied it several years ahead of the ITU standard, which subsequently evolved into Signaling System 7 (SS7).

From the first, the Internet Engineering Task Force (IETF) developed its standards through an open process. The ITU also uses an open process, but most of its standards are focused on telephone common carriers while the IETF is data-oriented. The ITU operates under the auspices of multiple governments, and is therefore less agile than the IETF, which has close ties to education and industry. No one is compelled to use IETF standards, but the fact that they are needed to communicate over Internet serves as an irresistible lure to use them for general data communications.

Proprietary standards work to the advantage of the developer, but they have distinct disadvantages for users. For one thing, no proprietary protocol meets all needs. SNA, for example, is designed for mainframe computers that are used by companies with large, centralized networks. Although it can be adapted to LANs, it is not optimized for inter-LAN communications. The major problem with proprietary protocols is the lack of interoperability with other protocols, which results in an inability to purchase products from other manufacturers. One of the hottest technologies on the market today is wireless LANs. The technologies have been well known for years, but until the Institute of Electrical and Electronic Engineers (IEEE) developed 802.11 standards, the proprietary networks were expensive and rarely applied.

Unfortunately, the need for standards and the need for technical progress sometimes conflict because standards cannot be set until the technology has been validated in practice, and the only way to prove the technology is through extensive use. Therefore, when it comes to the time to set a standard, a large base of installed equipment has been designed to proprietary standards. The procedures of many standards-setting organizations preclude their adopting proprietary standards, even if the manufacturer is willing to make them public. Competing developers are represented on the standards-setting bodies and resist the adoption of proprietary standards. As a result, even after standards are adopted, a considerable amount of equipment exists that does not conform to the standard.

A good example of this is in the Ethernet switch standards that apply power over station cable. Without such power, voice over IP (VoIP) telephones must be locally powered, and will therefore be inoperative during power failures unless they are individually equipped with an uninterruptible power supply. The methods of accomplishing power over Ethernet (PoE) were proprietary until IEEE released the 802.3af standards. Now any compatible VoIP telephone can work over another manufacturer's Ethernet switch.

How Standards Are Developed

The field of players in the standards process is vast, and sometimes not closely coordinated. Standards progress from conception to adoption in four key stages:

1. Conceptualization
2. Development
3. Influence
4. Promulgation

In the United States, just about anyone can conceptualize the need for a standard. Before development begins, however, some recognized body must accept responsibility for the task. For example, LAN standards could have been developed by any of several different organizations; IEEE undertook the task. Likewise, Internet standards are the purview of the IETF. Both organizations have open membership and do their work through working groups.

Participation in working groups is largely voluntary, and is usually funded by the standards influencers, which are the companies or associations with the most to gain or lose. Governmental organizations also wield influence in the standards process. Companies with vast market power also fill the role of standards influencers. Often, their influence is enhanced because they have already demonstrated the technical feasibility of the standard in practice.

The standards promulgators are the agencies that can accredit standards and produce the rules and regulations for enforcing them. In the United States, ANSI is the chief organization that accredits standards developed by other organizations such as IEEE and Electronic Industries Alliance (EIA).

Standards Organizations

Many organizations and associations are involved in the standards process. The following is a brief description of the role of the most influential of these.

International Telecommunications Union (ITU)

ITU was formed in 1865 to promote the mutual compatibility of the communications systems that were then emerging. The ITU, now a United Nations-sponsored organization to which most of the world's countries and over 500 private members belong, promulgates international standards. It operates through three sectors: ITU-T (telecommunications), ITU-R (radio), and ITU-D (development). ITU does its work through study groups that work in 4-year time increments. After a 4-year session, the study groups present their work to a plenary assembly for approval. Plenary assemblies coincide with leap years.

In some countries where a state-owned agency operates the telecommunications system, ITU recommendations bear the force of law. In other countries compliance is largely voluntary, although the motivation to accept standards is so great that most proprietary systems do not survive long on the market once a standard has been adopted.

American National Standards Institute (ANSI)

ANSI is a private, nonprofit standards body in the United States that administers and coordinates the U.S. standardization and conformity assessment system. Its goal is to enhance U.S. competitiveness in the world economy by promoting voluntary adherence to standards. ANSI does not develop standards, but instead provides the venue through which entities can cooperate to agree on standards. It represents about 200 companies and more than 270 accredited standards developers.

The ANSI X.3 committees handle information-related standards; T1 committees handle telecommunications standards. Both consumers and manufacturers are represented on ANSI committees, and cooperating trade groups that follow ANSI procedures do much of the work. IEEE and EIA are two prominent organizations that develop standards through ANSI.

International Standards Organization (ISO)

ISO is an association of standard-setting organizations from the various nations that participate in the process. Unlike representatives to the ITU, which are governmental delegations, ISO operates as a bridge between government and private organizations. In the United States, ANSI is the ISO representative and advisor to the State Department on ITU standards. The most familiar ISO standards in the telecommunications industry are the standards that support the Open Systems Interconnect (OSI) model.

Internet Engineering Task Force (IETF)

The IETF has many roots in the academic community, which has a tradition of open communication and peer review. Adherence to its standards is voluntary, but it is the means by which the Internet operates and any network connected to the Internet can do so only by following its protocols.

The standardization process starts with an Internet Draft, which is open for comments. Drafts may progress on the standards track as request for comments (RFC), or they may be allowed to expire. When the RFC has been thoroughly reviewed and tested, it may be adopted as "Internet Standard," and given a number STDxxx. It still, however, retains the RFC designation. The process is described in RFC 2026, which is available on the IETF's Web page.

Industry and Professional Associations

Three of the most important industry and professional associations from a telecommunications standards standpoint are IEEE, EIA, and the Telecommunications Industry Association (TIA). EIA has produced many standards that are important to the telecommunications industry. For example, most data terminal devices use the RS-232 interface standard in their interconnection with circuit equipment. More recently, EIA collaborated with TIA to produce commercial building telecommunications wiring standards, which we discuss in Chapter 9.

The IEEE is a professional association that has had an important effect on standards activities such as the LAN standards that its 802 committee developed. These standards, discussed in more detail in Chapters 6 and 10, use the framework of ISO's Open Systems Interconnect model and ITU-T protocols to develop local network alternatives. IEEE also produces software engineering standards as well as many standards that affect the power industry.

Other associations include the United States Telephone Association, Corporation on Open Systems, an industry group that is promoting open architectures, Computer and Business Equipment Manufacturing Association, Internet Security Alliance, Exchange Carrier's Association, Open Software Forum, and in Europe, the Standards Promotion and Applications Group.

The transition from proprietary to open standards has had an enormous salutary effect on the telecommunications industry, particularly from the consumer's standpoint. The need to rely on a single manufacturer as the source of telecommunications products and services is gradually diminishing. With more choices and interoperability assured, users can purchase with growing confidence that they are able to change suppliers if the source is no longer reliable.

Dozens of telecommunications products have been affected by this change. To name a few, facsimile, wireless, LANs, routers, Ethernet switches, cell phones, multiplexers, and modems that were once proprietary now adhere to standards that diminish the risks and lower the cost for the user.

CHAPTER 2

Introduction to the Public Switched Telephone Network

This chapter and the next present an overview of voice and data networks. For the most part, voice and data ride separate networks, although both use the same fiber-optic physical medium. The PSTN is a circuit-switched network. Users set up worldwide connections by dialing a series of digits. The network switches together a series of circuits and assigns them for the exclusive use of the session. When one of the parties hangs up, the circuits return to a common pool. The data counterpart of the PSTN is the Internet, which is composed of broadband facilities that multiple users time-share. Data networks are beginning to carry voice traffic, but the PSTN carries the vast majority of voice calls today and will continue to do so for at least the next decade. It also carries a significant amount of data traffic, mostly in the form of fax transmissions and dial-up Internet access.

The major strengths of the PSTN lie in its ubiquity and ease of use. It is a tribute to the inventive genius of the thousands of scientists, engineers, and technicians who design, install, and maintain the many parts of the PSTN that its complexity is so well hidden that users do not have to know any more than how to dial and answer telephone calls. In reality, however, the network is composed of countless circuits, switching systems, radio and fiber-optic systems, signaling devices, and telephone instruments that interoperate as part of a coordinated whole, even though they are owned by many unrelated companies. Every day, the industry adds new devices to the network, and they must function with other devices that were designed and installed more than 30 years ago.

The word "network" is an ambiguous term. It can describe the relationship of a group of broadcasting stations or a social fabric that binds people of similar interests. In telecommunications the usage is more specific, but the precise meaning of the term must be derived from context, and this can be confusing to those outside the industry. In its broadest sense a telecommunications network is a combination of all the circuits and equipment that enable users to communicate. All of the switching apparatus, trunks, subscriber lines, and auxiliary equipment that support communications are classified as elements of the PSTN. In the strictest sense of the word, the direct interconnection of two telephones meets the definition of a network, but this is a restrictive kind of private network because it lacks accessibility. In a broader sense, public voice networks support connectivity, which is the capability of reaching other stations anywhere in the world without the need for complex addressing.

To comprehend the systems discussed in this book, it is necessary to understand how they fit as part of a coordinated whole. This chapter discusses the major elements of the PSTN without explaining how or why they work. Many of these building blocks are used by both voice and data systems. Many terms are introduced and explained in detail later in the book. You are cautioned that, despite its origins in scientific disciplines, the vocabulary of telecommunications is frequently ambiguous, and the meaning of many of its terms must be taken from context. When we use technical terms, we define them in context to illustrate their meaning.

In this chapter, we discuss several elements that are common to both voice and data networks. The backbone that carries both voice and data rides on fiber-optic cables for the most part. Lightwave, which is the subject of Chapter 17, provides the huge amounts of bandwidth that transport the world's voice and data traffic. Inside this backbone both voice and data move as digital signals—zeros and ones that find their way from source to destination through this maze.

Just as the human body can be viewed either as a unit or as an assembly of systems such as the digestive, respiratory, and circulatory, the telecommunications network can be visualized as a whole, but better understood as interactive systems. We will first discuss the major systems in broad terms, and in subsequent chapters examine them in more detail.

THE MAJOR TELECOMMUNICATIONS SYSTEMS

Figure 2-1 shows the major classes of telecommunications equipment and how they fit together to form a communications network. Unlike the systems of the human body, the systems in telecommunications are not tightly bound. Each element in Figure 2-1 is largely autonomous with the systems exchanging signals across defined interfaces.

Customer Premise Equipment (CPE)

Located on the subscriber's premises, *station* or *terminal* equipment is the only part of the telecommunications system the users normally contact directly. It includes the telephone instrument itself and the premise wiring that connects to the LEC's demarcation point. For our purposes, station equipment also includes other apparatuses such as PBXs, multiline key telephone equipment used to select, hold, and conference calls, videoconferencing equipment, faxes, and other devices that subscribers normally own and users operate. CPE has two primary functions.

FIGURE 2-1

The Major Classes of Telecommunications Systems

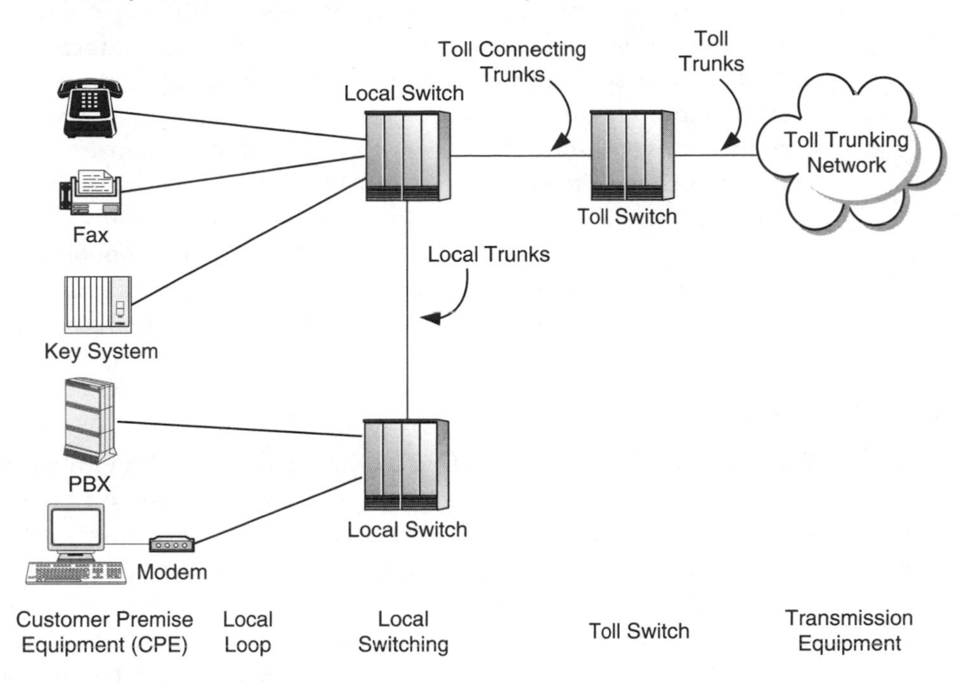

It is used for intra-organizational communication within a narrow range, usually a building or campus, and it connects to private or common carrier facilities for communication over a wider range.

Subscriber Loop Plant

The *subscriber loop*, also known as the *local loop*, consists of the wire, poles, terminals, conduit, and other fixtures and that connect CPE to the LEC's central office. Before deregulation, LECs had a monopoly on the local loop. Now alternatives are becoming available. For some services it is possible to transport information over cable television, and in other cases to use facilities such as microwave radio or fiber optics to bypass LEC equipment. A class of common carrier known as a competitive access provider (CAP) provides such service in most major metropolitan areas. CAPs and LECs as well as some CATV companies offer service over fiber-optic-based facilities. Both voice and data equipment use the same types of loop facilities.

Local Switching Systems

The objective of the telecommunications system is to interconnect users for the duration of a session. Either the user dials these connections, or the connections are wired in the LEC's central office and remain connected until the service is discontinued. This latter kind of circuit is called *private line* or *dedicated*. The industry uses the term *provisioning* to refer to the process of setting up dedicated circuits for customers and internal facilities to interconnect its central offices.

The local switching office (often called an *end office* or *class 5 office*) is the termination point for local loops. Loops used for switched services are wired to computer-driven switching systems that can switch them to other loops or to *trunks*, which are circuits to other local or long distance switching offices. Loops used for private line services are directly wired to other loops or to trunks to distant central offices. Until recently, local switching was a monopoly service of the LECs, but the Telecommunications Act of 1996 opens local switching to CLECs. The traditional telephone company is known as the incumbent LEC or ILEC.

Interoffice Trunks

Because of the huge concentrations of wire that converge in local switching offices, the number of users that a single central office can serve has practical limits. Beyond a certain limit, ordinary telephones will not operate without electronic amplification, which is costly. Therefore, in major metropolitan areas, the LECs place multiple central offices according to population density, and interconnect them with *local trunks*. A switching system and its associated trunks and subscriber loops is known as a *wire center*. Central offices exchange signals and establish talking connections over these trunks to set up paths corresponding to telephone numbers dialed by the users.

Tandem Switching Offices

In a region with multiple central offices, some of which belong to CLECs or small independent telephone companies, it is too expensive to provision direct trunk groups between every pair of offices. The left side of Figure 2-2 shows a fully meshed network in which each central office is connected to every other central office with direct trunks. Direct trunks are effective when the traffic volume is sufficient to justify them, but they have a major drawback. If the trunk group fails, subscribers in the affected central offices would be unable to reach one another. The telephone network in larger cities is therefore engineered with *tandem switches* as shown in the right side of the figure. A tandem switch provides a measure of service protection because calls have alternate paths to reach the destination. Where the traffic volume requires at least 24 trunks, central offices are provisioned with direct trunks, which are known as *high-usage* trunks. If fewer trunks are needed, or in case of overflow, traffic switches through the tandem. The alternate path is known as the *final* trunk route. Service providers engineer final routes to a higher standard than high-usage trunks because callers encountering a busy on the final route have no alternative but to hang up and try again.

FIGURE 2-2

Tandem Switching

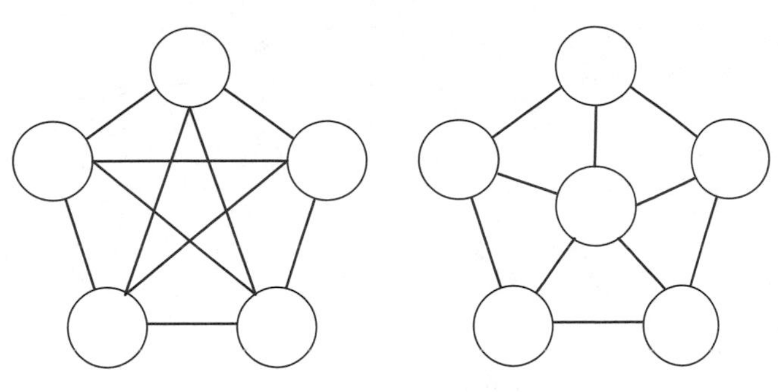

In addition to local tandem switches, RBOCs provide *LATA tandem* switches to provide IXCs with access to all of the end offices in a LATA. IXCs choose whether to run trunks directly to the end office or connect them through the tandems. IXCs also deploy tandem switches of their own. A traffic engineering process determines whether to run direct trunks or switch them through tandems. Traffic engineers make the decisions based on cost, service level, and reliability.

A special type of tandem switch known as a *gateway* interconnects the telephone networks of different countries when their networks are incompatible. Gateways are also used to connect calls carried over the IP network to the PSTN.

Private enterprise and governmental organizations also employ tandem switches to connect networks of dedicated circuits. The circuits connecting these private switches are known as *tie lines*. These private networks, known as *electronic tandem networks* (ETNs), are engineered using the same principles as public networks. They are used for internal connections and for connection to the PSTN at interconnection points known as "hop-off" points. If a PSTN call is connected at the origin, it is called "head-end hop-off." If the private network carries the call as far as it can and then connects to the PSTN, it is referred to as "tail-end hop-off."

Interexchange or Toll Trunks

Telephone service areas are divided into *exchanges*, which are geographical areas roughly approximating a metropolitan area in which the LEC provides a uniform collection of services and rates. Exchanges may be further subdivided into wire centers, which are geographical areas that describe the boundaries served by a particular central office. Figure 2-3 shows the layout of a typical telephone service area. Interexchange trunks that connect offices within the local calling area are known as *extended area service* (EAS) trunks, while those that connect outside the local calling area are called *toll* trunks. Trunks that connect the LEC to an IXC are

FIGURE 2-3

Telephone Service Areas

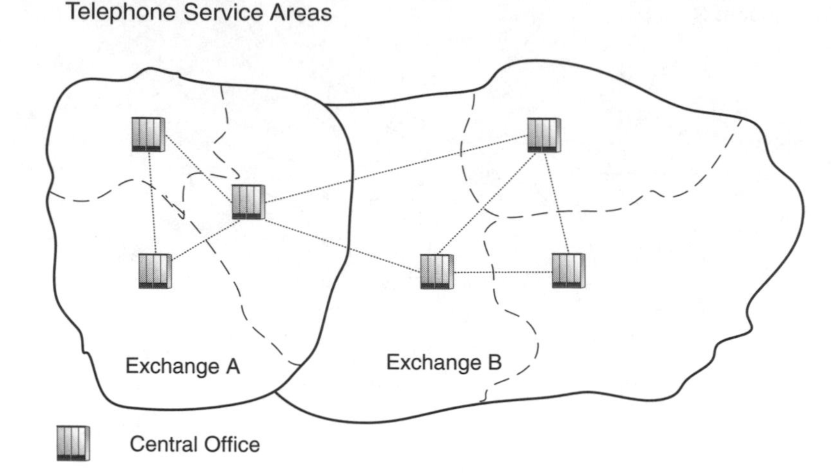

called *inter-LATA connecting trunks*. A service area such as this would typically include several other switches, which are not shown for reasons of clarity. The ILEC would usually provide at least one local tandem switch. They provide a LATA tandem in each LATA, and IXCs often install toll tandem switches in each LATA. The exchange would also include CLEC switches. IXCs and CLECs usually serve a much larger geographical area from a single switch than the ILECs do.

TRANSMISSION SYSTEMS

The process of transporting information in any form including voice, video, and data between users is called *transmission* in the telecommunications vernacular. The earliest transmission took place over open wire suspended on poles equipped with crossarms and insulators, but that technique has now all but disappeared. For short ranges, some trunks are carried on pairs of copper wire, but most trunks now ride on fiber optics. For longer ranges, interoffice and interexchange trunks are transported primarily over fiber optics, with microwave radio and satellites used to a limited degree. *Multiplexing* equipment divides these backbone transmission facilities into voice channels.

Fundamentals of Multiplexing

The telecommunications industry has always invested in research and development, much of it directed toward methods of superimposing an increasing amount of information on a single transmission medium. Multiplexing is the process of placing multiple voice or data channels over one facility. The basic building block of the telephone network is the *voice grade* channel occupying 300 to 3300 Hz of bandwidth. *Bandwidth* is defined as the information-carrying capacity of a telecommunications facility. The bandwidth of a voice-grade channel is far less than the bandwidth of high fidelity music systems that typically reproduce 30 to 20,000 Hz, but for telephone sessions voice grade bandwidth is sufficient. For the first six or seven decades of telephony, open wire or multiple-pair cables were the primary transmission media. At first, each pair of wires carried one voice channel. However, these media (known as *facilities* in telecommunications terminology) have enough bandwidth to carry multiple channels.

Frequency Division Multiplexing (FDM)

In the telephone industry, the term *carrier* is used to describe the application of multiple voice channels to a transmission medium. With the development of the vacuum tube, carrier telephony became feasible. The earliest carrier systems increased the capacity of a pair of wires by FDM. Broadcast radio is an example of FDM in action, except that it does not allow two-way communications. In an FDM carrier system, each channel is assigned a transmitter–receiver pair or *modem* (a term derived from the words MOdulator and DEModulator). These units operate at low power levels, are connected to cable rather than an antenna, and therefore do not radiate for any appreciable distance. FDM carrier systems were once the backbone of the telecommunications network, but digital multiplexing systems have now replaced them.

Time-Division Multiplexing (TDM)

Voice circuits originate and terminate as analog signals, but virtually all communication today is carried over digital circuits. Either the telephone instrument or the interface circuits in the central office digitize the analog signal. With rare exceptions, once a connection is digitized, it is unlikely to be converted back to analog until it reaches the destination. The transition from analog to digital transmission has resulted in a significant quality improvement. Noise has dropped to insignificant levels and the volume of the voice signal is no longer a function of circuit length.

A digital voice circuit, derived through a process called *pulse-code modulation* (PCM), occupies 64 Kbps of bandwidth. Multiple digital circuits are combined into wider bandwidths through TDM. A TDM signal is conceptually like a time-shared computer, which enables multiple users to share processing power. The capacity of the digital transmission medium is so great compared to the needs of a voice channel that each channel can have access to the full bandwidth long enough to transmit one octet, which is 8 bits. The time slot recurs 8000 times per second, with users unaware that others are sharing the medium. Chapter 5 explains how PCM works. The theory of PCM is not new. An IT&T scientist in England conceived it in 1938, although the concept was not feasible until the development of solid-state electronics. When large-scale integration became available, the economics shifted in favor of digital transmission.

Higher Order Multiplexing

The basic building block of digital multiplexing systems in North America is known as a *digroup*, which comprises 24 voice channels. A digroup occupies 1.544 Mbps of bandwidth. The system is known as T1 carrier and the signal is known as a DS-1, with the DS standing for "digital signal." The European TDM hierarchy is different than the North American system. It uses 64 Kbps of bandwidth for the voice channel, but its carrier system, known as E1 has 2.048 Mbps of bandwidth. E1 supports 30 voice plus two signaling channels.

At the T1/E1 level, the bandwidth can be divided into voice channels in voice CPE such as a PBX or key system or it can be converted to analog or digital channels using a device known as a *channel bank*. Channel banks are used when the CPE is not TDM compatible, which many smaller key systems are not. The entire bandwidth of a nonchannelized T1/E1 can also be applied directly to a data device such as a router.

Both the North American and European systems multiplex the bandwidth up to higher levels than T1/E1. Channels can be interleaved all the way up to the capacity of a fiber-optic pair, which today can carry more than 600,000 voice channels. By using *dense wavelength division multiplexing* (DWDM), the capacity can be multiplied more than 100 times. Chapter 17 discusses the higher order multiplexing protocol, which is known as SONET/SDH.

Analog and Digital Transmission

When people speak into a telephone instrument, the voice actuates a transmitter to cause current flowing in the line to vary proportionately or *analog*ous to the changes in sound pressure. Because people speak and hear in analog, there was, until a few decades ago, little reason to convert an analog signal to digital. Now, digital transmission has replaced analog for four primary reasons:

- Digital equipment is less expensive to manufacture.
- An increasing amount of communication takes place between computers, which are inherently digital.
- Digital transmission provides higher quality.
- Digital signals can be compressed to occupy a fraction of the bandwidth of an analog signal.

The higher quality of digital signals results from the difference in the methods of amplifying the signal. In analog transmission, an audio amplifier known as a *repeater* boosts the signal, together with any noise on the line. With digital transmission, *regenerators* detect the incoming bit stream and create an entirely new signal that is identical to the original. The result is a channel that is practically noise free.

SWITCHING SYSTEMS

A telecommunications system must provide connectivity, or the capability of accessing other users wherever they are located. Telephone switching systems connect lines to other lines or to trunks that are then switched together to build up a connection to the destination. Switching systems have undergone five distinct generations, with each bringing enhanced features that were either costly or impossible with their predecessors.

Early Switching Systems

The earliest switching systems were manually operated. Telephone lines and trunks were terminated on jacks. Operators interconnected lines by inserting plug-equipped cords into the jacks. The connection remained in place until the parties signaled that they were finished and the operators pulled down the connection. In 1891, a Kansas City undertaker named Almon B. Strowger, convinced that operators were listening to his conversations, patented an electromechanical switch that made connections without the need for operators. This system, called step-by-step, progressively builds a connection through devices that advance from one stage to another in step with dial pulses. Although the system is obsolete now, Strowger systems were common until late in the twentieth century.

The process of setting up a telephone call involves a series of functions such as lifting the receiver, connecting dial tone to the line, sending the called number to the central office, signaling the distant telephone, ringing the line, and so forth. In the Strowger system, all of the equipment involved in setting up the call remained connected for the entire session. More important, it was also inflexible. If a call encountered blockage, the system was incapable of rerouting it to a different path. It could only signal the user to hang up and try again. The human operators had a big advantage in this respect. They could manage several calls simultaneously, and choose alternate routes if the primary was blocked. Considering the fact that the switch was effectively just connecting two cable pairs, to solution was to use an intelligent device to set up the connection and drop off.

Common Control Switching

The second generation of switching used *common control* equipment to set up the talking path through a simple switching fabric known as crossbar. The brain of a crossbar switch was a *marker*, which was an electromechanical device. The marker selected the path through the central office by directing the operation of other single-purpose devices that were used for one stage of the setup and then released. One such device, still used in electronic switches, is a *register*, which sends dial tone to the user and prepares to receive digits. When the required digits are received, the register relays them to a device known as a *translator* that determines from the dialed digits which route the call should take. At this point, the register's work is done and it drops off and prepares to receive another call. At each step in the process, common control equipment is brought into play long enough to do its function. Common control equipment is more complex than directly controlled switching, but it is faster and more efficient.

Computer-Controlled Switching Systems

Markers in crossbar systems were effectively computers with wired logic. In the mid-1960s, the age of the stored program control (SPC) central office was born. The No. 1 ESS, some of which are in use today, is a processor-driven third-generation system using a reed relay switching fabric. Although the control is a digital computer, the switching fabric of the No. 1 ESS is analog.

Western Electric, which has evolved into Lucent Technologies through a series of reorganizations, manufactured the No. 1 ESS. The equipment was not widely used outside the United States, but other manufacturers produced similar systems. Today, analog switching systems of all types are obsolete. New systems are either TDM digital or softswitches.

Digital Switching Systems

When large-scale integrated circuits were perfected in the 1970s, it became technically feasible to develop a TDM switching fabric to replace the analog relays of No. 1 ESS. Current central office technology has a digital switching fabric controlled by a central processor. Most modern switching equipment, ranging from small PBXs to large toll tandem switches that can handle thousands of trunks, uses TDM technology. As we will explore in more detail in Chapter 14, digital switches match efficiently with digital transmission systems. By contrast, an analog switching system must convert signals to digital to send them across the digital transmission network.

Softswitching Systems

The next generation switching system is the *softswitch*. The central processor generally runs on a commercial operating system such as Microsoft Windows or Linux. Voice is digitized in the telephone set and converted to IP so that the output of the telephone is a stream of IP packets. The endpoints on the network have IP addresses, either instead of or in addition to regular telephone numbers. The switching fabric, typically a high-end Ethernet switch, connects IP packets between the source and destination. The transition from TDM to IP switching will be gradual because of the large investment the LECs have in conventional digital switches.

Numbering Systems

The PSTN numbering system is an ITU-T standard, E.164, which is uniform throughout the world. E.164 addresses consist of a country code, an area code, and a telephone number that is generally associated with a particular central office;

the exception being when the number is relocated to another switch through number portability. The exact structure of the numbering plan varies with country, but the pattern is compatible worldwide. In this section, we review briefly how telephone numbering operates.

The telephone numbering system is hierarchical and geographically oriented. The highest tier in the hierarchy is the *country code*. Below that, countries are assigned a second tier code that is known in North America as an area code or *numbering plan area* (NPA). The NPA is divided into central office areas that are also unique within certain limits. Until 1995 a 0 or 1 as the second digit identified the NPA, but the pool of available numbers was nearing exhaust, so on January 1, 1995 the new North American Numbering Plan was introduced. This plan permits any digit from 0 to 9 in the second position. Area codes were once unique within a geographical area, but in some NPAs *overlay area codes* have been assigned. With an overlay area code, two or more NPAs cover the same geographical area. Overlay area codes require callers to dial the area code to make a local call.

The next level is the *central office code* or *prefix*, abbreviated as NXX. In most parts of the world a central office code is a unique geographical area. In the United States, however, the Telecommunications Act of 1996 required LECs to support number portability as a way of promoting competition in the local network. Within limits, users are free to change carriers and retain their previous number. This means the NXX is no longer tied to a geographical location. Number portability is confined to an ILEC's rate area, which is generally equivalent to an exchange. The number portability rules are partly prescribed by law, and partly by company policy. For example, nothing in the FCC rules requires an ILEC to port numbers for one of its customers from one wire center to another. Some ILECs do permit this, however, for customer convenience. Anyone planning to move and retain the telephone number is well advised to verify that portability is permitted.

The lowest tier on the PSTN switching hierarchy is the *end office*, also known as a class 5 office. The latter designation referred to the lowest level of a five-level hierarchy that AT&T used before divestiture. Under that plan a call could theoretically be composed of as many as nine links switched in tandem. Now, long-distance networks tend to be flat, although most long-distance connections route through one or more tandem offices. Switching systems contain routing algorithms to enable them to select the appropriate group of trunks, always attempting first to connect to a distant central office over a high usage trunk. Calls progress from system to system until the terminating office is reached. Systems route calls by exchanging signals over an external data network known as SS7.

COMPETITIVE LOCAL EXCHANGE SERVICE

The Telecommunications Act of 1996 opened the local exchange network in the United States to competition. Since then, the results have been somewhat checkered and few CLECs have been profitable, but unquestionably, the competitive local network is here to stay. CLEC services contrast to ILEC service in several ways. For one, CLECs are under no obligation to serve all comers. They are usually unregulated and usually require their customers to subscribe to a minimum quantity of lines. Twelve-line minimums are typical for CLECs that own their switching equipment. CLECs that resell ILEC service may have a lower requirement.

The act requires the ILECs to offer service to the CLECs at any technically feasible interconnection point. This requires the ILECs to provide access to their cable pairs, trunking networks, and switching systems. These facilities are known as unbundled network elements (UNEs). A UNE may be a standalone element such as a cable pair or it may be a group of UNEs. The latter arrangement, called UNE-P, may provide the equivalent of a business or residence line, a basic rate ISDN line, or even a Centrex system. UNE-P (the "P" stands for platform) has been controversial throughout the industry, with the ILECs urging the FCC to eliminate the requirement. At the time this is written, the FCC still requires ILECs to provide UNE-P, but they have authorized significant price increases.

CLECs frequently need a cable pair or fiber UNE to reach their customers. Some CLECs have fiber networks that enable them to reach large users directly, but few have access to copper cable except through the ILECs. Since cable pairs terminate in central offices, access to cable requires collocation with the ILEC or space in a nearby building that has wire connections to the ILEC central office. The act requires the ILECs to permit collocation on a space-available basis. In most cities nearby space has been converted to so-called "equipment hotels," which lease space to CLECs and other service providers.

A CLEC may have one central switch and reach other central offices over its own or the ILEC's network, or the switch may be distributed by means of remote switching units or subscriber line carrier. Chapter 14 discusses both of these methods. In any event, the CLEC typically has a limited number of central offices in each exchange and relies on interconnection to the ILEC for access to the cable pairs. Otherwise, switching and transmission equipment that the CLEC employs may be identical to the ILEC's equipment.

The ILECs sell service to the CLECs at wholesale rates, and generally the CLECs offer services at lower rates than the ILECs. The basic cost of the service may be only one of the differences, however. Regulatory commissions permit, but do not mandate charges such as subscriber line access and number portability. CLECs may choose not to levy these charges. They are also free to make their service attractive by offering other inducements such as reduced-cost long distance and enhanced features such as caller ID, often bundled with unrelated services such as Internet access.

Industry Requirements and Processes

The telephone industry requires a complex set of procedures to ensure that changes are communicated among companies and that all participants keep their numbering and routing tables updated accordingly. When the local network was noncompetitive, the LECs handled most of the details internally. The RBOCs collectively owned BellCore, which set many of the standards and procedures and the RBOCs funneled the necessary information to the IXCs and independent telephone companies. A competitive network, however, requires that sources independent of the carriers be established to handle the various databases. CLECs negotiate interconnection agreements with the ILECs, which require a usage forecast and assist the CLECs in the planning process.

Each company needs an operating company number, which the National Exchange Carrier Association (NECA) assigns. They also need an access customer name abbreviation and a revenue accounting code number, which Telcordia assigns. The North American Numbering Plan Administration (NANPA) assigns NPAs and NXXs.

Each carrier requires an SS7 provider. The larger carriers provide their own signaling facilities while the smaller carriers contract with an outside firm. The entire call setup and number portability process requires the use of databases that are accessed through SS7. For example, the line information database (LIDB) provides telephone line numbers used by operator services systems to process and bill services such as calling card, third number, and collect calls. LIDB is also used for calling card validation, fraud prevention, and service restrictions. The 800 database identifies which IXC is the carrier for toll-free calls, and the internetwork calling name service (ICNAM) database associates calling names and numbers to support caller ID. The Number Portability Administration Center (NPAC) provides the database that identifies which switch serves a dialed number. Carriers also must support emergency calling through E-911 in North America.

The Alliance for Telecommunications Industry Solutions (ATIS) provides the standard ordering forms and codes for requesting UNEs and interconnection to ILEC loops and trunks. Telcordia produces the Local Exchange Routing Guide, whereby carriers keep updated with new NPA/NXX combinations. They also produce identifiers by which carriers encode their locations, trunks, and equipment.

APPLICATION ISSUES

Telephone services, both local and long distance, represent a significant expenditure for most enterprises. With minor exceptions, the service offerings of most carriers, both local and toll, are identical. It is not safe to assume, however, that the differences between the carriers are insignificant and that services can be selected based only on price.

Telecommunications services are not commodities. Because of differences in architecture, some services may be more vulnerable to failure than others. An IXC with its toll switch in a different city than the service delivery point is marginally more exposed to failure than one in the same metropolitan area. Self-healing networks can reduce the vulnerability, but an access circuit a mile long is less exposed than one 100 miles long. By the same token, an ILEC with multiple switches in a metropolitan area is less vulnerable to total failure than a CLEC that covers the entire area with a single switch. The difference in cost may be worth the risk, but that is a judgment that should be made only after understanding the differences.

Other differences are not apparent at first glance. A good example is Centrex, which is a service offering that provides PBX-like functions from a central office. A user such as a bank with multiple branches may find it costly to obtain Centrex in a multicentral office metropolitan area, whereas a CLEC with a single switch can provide Centrex throughout the area. ILECs are usually regulated to some degree and are prevented from negotiating costs and terms and conditions of services that are published in their tariffs. They can, however, negotiate special contracts with larger customers.

The issues in purchasing products and services are beyond the scope of this book. They are covered in a companion volume to this book, *The Irwin Handbook of Telecommunications Management*.

This chapter has covered voice networks in overview. In the next chapter, we will examine data networks in overview and discuss differences between voice and data. It is important to recognize that although voice and data are separate at this time, they are converging, at least to some degree, down to the desktop level.

One cannot help being awed by the intricacy of the PSTN, which is effectively a time-shared computer with millions of terminals distributed worldwide. The complexity is evident from this brief overview, but it becomes even more impressive as the details emerge. The marvel is that the system can cover such a vast geographical area, can be administered by hundreds of thousands of workers, contain countless pieces of electrical apparatus, and still function as reliably as it does. As we discuss these elements in detail, the techniques that create this high-quality service will become easier to understand.

CHAPTER 3

Introduction to Data Communications

In the minds of most people, data communication is a recent phenomenon associated with computer communications, but data messaging predated the telephone by several decades. Samuel F.B. Morse invented the first data communication device, the telegraph, in 1844. The telegraph and its successor the teletypewriter played an important role in communications for the first century of their existence. These devices were the principal means of sending messages over long distances until computers appeared.

Computers began as text-based systems that used punched cards for input and magnetic tape for storage. At first, the mainframe computer was confined to a self-contained processing center. Video display terminals (VDTs) gradually replaced keypunch machines as the principal input device. VDTs are known as "dumb" terminals because they have no processing power of their own. It all resides in the host computer. Output devices of the time were punched cards, tape, or high-speed line printers.

The methods of remote computer communications were primitive at first, and the means of cramming information into voice channels was little understood. One of the earliest applications involving computer communications was the U.S. Air Force's Semi-Automatic Ground Environment System (SAGES). SAGES integrated data from radar stations into control centers that later formed the basis of the air traffic control system. Development of SAGES began in 1957, a scant dozen years after development of ENIAC, which was the ancestor of today's computer systems.

As computer applications evolved from centralized to distributed, enterprises needed to move terminals to outlying offices to support applications such as reservations and banking. Computer ports could easily be extended to distant terminals with a modem running over a telephone circuit. Remote printers, successors to the teletypewriter, had printing speeds in the order of 100 words per minute, more than the speed of most typists. One hundred wpm is only about 80 bps, and a telephone circuit supports 2400 bps with the simplest of modems. It was not cost-effective to use a fraction of the capacity of a voice circuit for a single VDT. Moreover, 1 bit in error could render a message stream unusable. The industry needed a reliable means of sharing circuits and detecting and correcting errors.

Two methods of solving this problem evolved. In one solution, IBM's Binary Synchronous Communications (BSC), a remote controller communicated with the host computer. The host sent short polling messages to determine if any of the VDTs attached to the controller had traffic to send. If so, the controller sent a block of information with an error-detection block attached. If the host received the block error-free, it returned an acknowledgment. If not, the controller resent the block until the host acknowledged receipt. This method provided reliable communications, but since the transmitting end could not send a block until the previous one was acknowledged, it was slow. IBM upgraded BSC to SNA in later years. SNA permits the sender to launch multiple packets without waiting for acknowledgment. The service that supports BSC and SNA is called *multidrop*, a name derived from the fact that multiple controllers share a single circuit back to the host.

The second method, which most non-IBM computers used, was a data multiplexer. A multiplexer connected to terminal ports on the host computer and communicated with a matching unit that supported terminals at the distant end. The multiplexer pair had two functions: one, they subdivided a voice channel into multiple data streams, and two, they carried on an error detection and correction dialogue to ensure data integrity.

Both of these methods are still used with variations, but only to a limited extent because servers have replaced most mainframes and minicomputers. As long as the keyboard remained the primary input device, a voice circuit was more than adequate. Applications, however, evolved into file transfers and local area network (LAN) interconnection, both of which involve computer-to-computer communications. The mismatch in speed with which a hard disk can deliver information and a modem can transmit it puts a severe damper on efficiency. LANs move data within the office at much higher speeds. The earliest LANs were in the order of 1 Mbps, but these soon increased to 10 and 100 Mbps. As organizations interconnected their LANs and distributed information over wide areas, the bandwidth of an analog circuit became a severe choke point. Users, having no idea where information was stored, noted the vast difference in response time between a local file delivered over a high-speed LAN and one delivered over a voice-grade wide area network (WAN). Distributed applications needed a better match between LAN and WAN speeds.

The common carriers' first solution was to provide digital circuits. The basic digital circuit in North America supports 56 Kbps of bandwidth. This is more than twice as fast as an analog circuit, but a LAN is many times faster. The digital 56 Kbps circuit was barely a waypoint on the trek to higher bandwidths.

The carriers' difficulty lay in the availability of transmission resources. Fiber optics was in its infancy in 1980 and long-haul circuits were transported over analog microwave radio. Digital microwave was available, but its error rate limited it to a span of about 500 miles (800 km). The restrictions of the analog voice network remained until fiber optics relieved the bottleneck.

Another major change in computer applications was the transition from text-based information to graphics. Computers had been migrating toward graphics for several years, but the Web accelerated the change. Desktop computers replaced dumb terminals as the I/O device, and graphics-based operating systems replaced command-driven systems. File sizes grew by orders of magnitude and with the growth of desktop computing and the need for Internet access, network managers had little choice but to expand bandwidths to meet the demand.

This brings us to the current state of data communications, which is mixed, but moving toward a unified network. Desktop computers have replaced most of the older VDTs. Multiplexers have largely been supplanted by devices that interconnect LANs, and the older data networks have evolved into networks of interconnected LANs. To begin the discussion of data networks, an understanding of some key terminology is in order. We start the chapter with a discussion of *protocols*, which data devices use to communicate with one another. Following that, we discuss networks in overview, focusing on LANs, which are private data networks that operate at high speed over a narrow range and WANs, which have a global reach. Then we review the common carrier services that subscribers use to interconnect their LANs so users can access resources from the desktop without regard to where they are located.

PROTOCOL TERMINOLOGY AND FUNCTIONS

Protocols are specialized computer programs that either exist in software or in firmware chips. Most desktop computers contain both varieties. The Ethernet LAN protocol, which comes with all new computers, is encoded into firmware. For users to access the Internet they must load the software protocol, Transport Control Protocol/Internet Protocol (TCP/IP), which is provided with the computer's operating system.

Most protocols are written in modules or layers, each of which has defined functions and interfaces. A modular structure allows developers to write application software without concern about how the program operates over the wide area. If a function or specification changes, it is not necessary to change the entire protocol stack; only the affected layer and its interfaces with other layers are changed. Many international standards are based on the OSI model, which is an ISO structure that defines the functions of each layer.

LAN standards are a good example of the efficacy of a layered protocol. LANs operate at the first two layers of the OSI model: physical and datalink. Network interface cards (NICs) connect computers to the physical layer, which is twisted pair wire, fiber optics, coaxial cable, or wireless. Network operating systems use higher layer protocols that talk to any NIC. The NIC manufacturers provide software drivers to enable the functions in their cards to communicate with the network operating system across a protocol known as the logical link layer (LLC), which is a subset of the datalink layer.

Protocol Functions

Diplomatic protocols prescribe a complex set of rules that dictate the way people interact in international relationships. Diplomatic protocols prescribe such customs as who is seated next to whom, how officials of different ranks are addressed, and what kind of response is appropriate to another's statement. The protocol metaphor is apt for data because devices must behave according to defined procedures. In a layered protocol these functions are usually assigned to one layer, but the rules regarding this are not rigid. For example, every Ethernet NIC has a unique and exclusive protocol and address known as media access control (MAC) embedded in firmware. The MAC address identifies the station within its own LAN segment, but to communicate across a WAN a higher level address is required, the most prevalent of which is IP.

A fundamental purpose of protocols is to establish a session across a network. A session begins when a user logs on to a distant computer, and ends with log-off. The protocol authenticates the parties before permitting communication to begin. Data networks handle sessions in two distinct ways: *connectionless* and *connection-oriented*. In a connection-oriented protocol, the devices establish a physical or logical connection across the network. The connection is set up at the start of the session and remains for its duration. The connection can be circuit switched as it is in the telephone network, or it can be a *virtual* connection that is defined in a software path and shares bandwidth with other sessions. In a connection-oriented session the packets contain a path identifier, but they do not need to carry the address of either the sender or the receiver after the path is set up.

In a connectionless session, a single packet known as a *datagram* is launched into the network and delivered to the distant end based on its address. The postal service is an example of a connectionless operation. The user does not know or care how a letter gets to its destination provided the post office delivers it to the addressee intact and in a timely manner. LANs are connectionless. A station launches a stream of information onto the network. All stations copy the message but retain it only if they are the addressee. Connectionless operation requires that each packet or frame contains the address of the sending and receiving stations.

The following points discuss the generic functions of protocols. The application and the communications medium dictate whether a specific function is needed. Within a LAN, which has a short and relatively secure transmission medium, stations can communicate without many of the protocols that a session needs across the more hostile environment of a WAN.

- *Communications control*: Protocols can be classified as peer-to-peer or master–slave. In the latter case, the master controls the functioning of the datalink and controls data transfer between the host and its endpoints. All communication between slave stations goes through the master. Peer-to-peer protocols do not use a controller. Devices can communicate directly with one another.
- *Error detection and correction*: Protocols check for errors, acknowledge correctly received data blocks, and send repeat requests when blocks contain an error. Sophisticated protocols can acknowledge multiple packets using one of two procedures. A *selective repeat* acknowledgment enables the receiving device to request the sending end to resend specific packets. In the *go-back-n* method the receiver instructs the sender to resend an errored packet and all subsequent packets.
- *Link management*: After the session is set up, the protocol controls the traffic flow and data integrity across the datalink.
- *Setting session variables:* The protocol determines such variables as network login and authentication and whether the session will be *half-duplex*, meaning information flows in one direction at a time or *full duplex*, meaning information can flow in both directions simultaneously.
- *Synchronizing*: At the start of a session data devices exchange signals to determine such factors as the data transfer rate and whether they will use compression or encryption. Modems exchange signals to determine the highest speed at which they can exchange data and fall back to a lower speed if the circuit will not support the maximum.
- *Addressing*: Every session requires an address to set up a connection if the protocol is connection oriented or to route packets if it is connectionless. Not all protocols contain addresses. Many of them rely on higher or lower layers for addressing.
- *Routing*: In data networks that have multiple routes to the destination, the protocol determines the appropriate route based on conditions such as cost, congestion, distance, and type of facility.
- *Data segmenting and reassembly*: The protocol segments a continuous data stream from the source into frames, cells, or packets as appropriate. The term *protocol data unit* (PDU) refers to the unit into which the data stream is divided. The PDU is equipped with header and trailer records for transmission over the network. Headers and trailers contain extra or *overhead* bits that contain information the protocol uses for routing and error checking. At the distant end, the protocol strips the overhead and reassembles the data stream for delivery to the receiver.
- *Data formatting*: The bit stream may require conditioning before transmission and restoration after reception. For example, conditioning could include encryption or compression.

- *Supervision*: The protocol establishes a connection, determines how the session will be started and ended, which end will control termination of the session, and how billing will be handled.
- *Flow control*: Protocols protect networks from congestion by sending signals to the source to halt or limit traffic flow.
- *Failure recovery*: If the session terminates unexpectedly, the protocol sets markers and enables recovery without starting over.
- *Sequencing*: If data blocks arrive out of their original sequence, the protocol delivers them to the receiving device in the correct order.

DATA NETWORKS

Unlike voice sessions, which have predictable and uniform bandwidth requirements, data sessions have unique requirements that depend on the characteristics of the application. Users have many network alternatives to match the needs of data sessions. Data sessions have much different characteristics than voice. For one thing, they are lengthy. Contrasted to voice sessions, which typically average 3 or 4 min, data sessions extend from log-on to log-off, which may be days or months apart. During the session the user may download a file, which puts a momentary high demand on the network, but then suspend any further activity while working on the file or attending to other business. Meanwhile, the network makes its bandwidth available to other sessions that have noncoincident peaks. Here are examples of typical data sessions:

- *Inquiry–response*: Client-server applications such as Web browsing are characterized by a small transmission upstream to a server, which may result in a large downstream transmission.
- *Remote data entry*: Access to a bank ATM machine is an example. Transmissions are short, and may be approximately symmetrical, but data flows in only one direction at a time.
- *Database backup*: A computer may back up its hard disk to a storage area network. Data flow is heavily weighted in the upstream direction.
- *LAN interconnection*: This application has symmetrical data flow on the average, but it is characterized by bursts of heavy flow.

The bursty nature of many data applications means that fixed-bandwidth circuits impose limitations because the larger the file, the longer it takes to transfer the data. Most data applications work best in an environment that provides a variable amount of bandwidth, expanding to meet short-term needs, and during low-usage periods, making it available to other sessions. Also, circuit setup time, which is relatively insignificant to a voice session, is lengthy in data terms. Many data applications need a fast response from the other end. If the devices had to carry out a lengthy handshaking routine each time they needed access, the application would be handicapped.

Network Access Methods

If multiple devices share the transmission medium, some means of allocating access is needed. Circuit switching can be used, and for some applications such as dial-up Internet access and occasional connections to remote access servers and computer ports it is effective. For many data applications, however, the setup time is too long, the bandwidth is too narrow, and dial-up access costs too much. Most data applications need a full-time connection to the network.

One common method of allocating access is *polling* in which a device at the head end of the network sends short messages to each node in turn, asking if it has traffic to send. IBM's BSC and SNA use this method.

A third method is *contention* access, which has the characteristics of a party line. Devices listen to the network to see if it is busy and defer their transmissions until it is idle. Ethernet uses this method. This method works well if all devices can hear all of the transmissions on the network. If they cannot, however, transmissions may collide, necessitating the devices to recognize the collision and retransmit. Collisions may be reduced by permitting devices to transmit only during their allocated time slots.

Another method of regulating access is by using an intelligent device such as a router or switch to accept packets from the sources, buffer them briefly, and release them in a disciplined fashion that does not exceed the capacity of the transmission medium. The TCP/IP protocol operates in this way. IP is a connectionless protocol that allows packets to traverse the packet as individual PDUs known as *datagrams*. A datagram is an independent, unacknowledged packet, which can arrive out of sequence, with an error, or not at all. TCP, as we will discuss in Chapter 6, provides the stability that teams with the unreliable IP to form the protocol that most of the world uses.

Data Network Addressing

A major difference between the PSTN and data networks lies in the addressing method. A voice station address is the familiar E.164 telephone number, which is unique for each station throughout the world. Data addresses are also unique, but they are designed for machines to read. Humans would find it difficult to work with an address such as 00-05-E9-3F-88-4A, which is expressed in a form known as *hexadecimal*. All computers use the binary form of addresses—hexadecimal, which consists of a series of 4 bits, is just a convenient way to make it less daunting to humans. Hexadecimal has a number base of 16. It uses the digits 0 to 9 plus letters A to F as symbols. For example, binary 1111 equals hex F.

Data protocols operate with a variety of addressing methods. For example, the address mentioned above is typical of an Ethernet address, a unique address that is permanently encoded in the card. The computer that contains the card can be named to make the address easy for humans to use, but this raises a complication. Any computer can be connected to an Ethernet port anywhere in the world, and its owner could request access to the Internet or a private network. The computer name and Ethernet address must be correlated somehow. The Address Resolution Protocol (ARP), which runs on every LAN, translates the computer name to its MAC address.

The Internet has a completely different addressing scheme. IP addresses are expressed in decimal form such as 23.128.55.3. Network users would find this address complex, but network technicians use it every day, and often must know its binary equivalent. For users, IP addresses are expressed as a uniform resource listing (URL) such as username@company.com. Domain name service (DNS) translates between the URL and the IP address. Every computer that connects to the Internet must inform the network through its setup routine of the URL of its DNS. We will discuss this in more detail in Chapter 6. For now it is sufficient to understand that many different protocols have unique addressing methods and where they must interoperate, some method of address resolution is provided.

Local Area Networks

A LAN is a high-speed privately owned data network that links computers and peripherals over a short range. The primary motivation for LANs arose from a need to share peripherals such as printers, and to access files that are shared in servers. The industry proposed several proprietary LAN solutions, but none caught on until the IEEE 802 committee standardized two LAN protocols. One, 802.3, is an offshoot of Ethernet, a proprietary product that Xerox developed and offered to the committee. The 802.3 standard that IEEE adopted is similar to Ethernet, but not compatible. Nevertheless, 802.3 is universally known by the Ethernet name. IBM promoted a different standard, which became 802.5 token ring. An 802.5 network uses a token-passing protocol that is a variation on polling. Before a station can transmit, it must capture a software token that circulates around the ring. The vast majority of LANs use Ethernet.

LANs form the foundation of nearly every network today. Computers come equipped with Ethernet ports, which enable them to connect directly into an office network. The Ethernet port connects via high-quality twisted-pair wire to a shared hub or, more frequently, to a switch that connects ports long enough to pass a data frame. Within the range of an Ethernet segment, which is nominally 100 m from station to hub, data travels at 100 Mbps or more. A variety of high-speed methods, the most common of which is a fiber-optic link where it is practical, or a WAN can expand segment diameters where common carrier facilities are used.

Wide Area Networks

When LANs must extend outside the boundaries of privately owned facilities, they normally use common carrier facilities. *Facility* is a generic term that describes the combination of local loops and long-haul circuits that make up a connection between subscribers' endpoints. A variety of services and associated protocols are available for connecting LANs. The industry typically shows the WAN as a cloud as in Figure 3-1. The LAN owner is concerned with how to feed the service provider's devices at the network edge, but has little knowledge of how the data flows inside the WAN. In this figure, routers at each customer location connect LANs to the WAN. The industry has a sub-category of the WAN called a metropolitan area network (MAN). The major difference between the two is in the span of the network. A MAN typically does not extends outside a metropolitan area. We can confine the discussion to WANs because the services and protocols are essentially identical. The distinctions between the MAN and WAN are blurred and the differences are unimportant for now. We briefly discuss the types of facilities available for WANs here, and will elaborate on them in later chapters.

Common carrier WANs facilities can be classed as one of seven types: point-to-point, multidrop, circuit switched, message switched, packet switched, frame switched, or cell switched. Any one or a combination of these can serve as the WAN cloud.

Point-to-Point Circuits

A point-to-point circuit is a dedicated private line that is directly wired between two endpoints. Analog voice-grade circuits are available, but rarely used today. Much higher bandwidths are used for data transport. Digital circuits of 56 Kbps in North America or 64 Kbps in Europe are available, with bandwidths extending as high as 40 Gbps. Some carriers provide "dark" fiber, which subscribers can illuminate with their own equipment and use up to its capacity.

FIGURE 3-1

LANs Connected with a Wide-Area Network

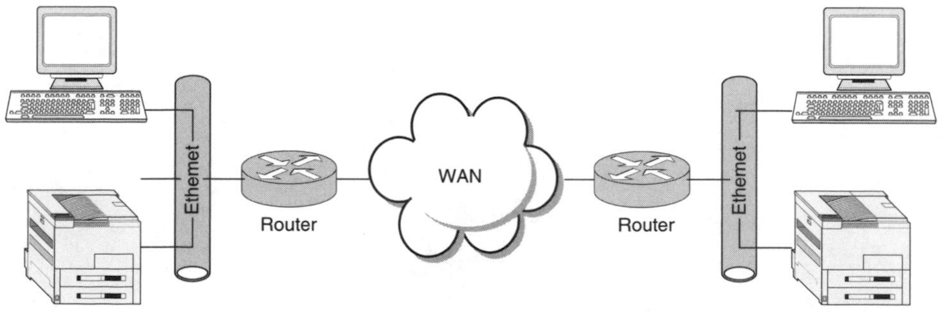

Point-to-point circuits can be terminated on a variety of different devices; data circuits usually terminate on routers. Carriers typically price dedicated or private line facilities in three segments: originating loop, *interoffice channel* (IOC), and terminating loop.

Multidrop Circuits

The simplest point-to-point circuits connect directly between two endpoints that have exclusive access to the circuit. If the endpoints do not have enough traffic to use the capacity of a circuit, multiple circuits can be bridged in the central office in a configuration similar to Figure 3-2. Legacy data networks such as IBM's SNA have such a topology. A front-end processor polls controllers in round-robin fashion. If a terminal attached to the controller has traffic to send, the controller forwards it. Otherwise, the controller responds that it has no traffic. The central unit also sends output messages to the controllers when it has downstream traffic. Circuit time is consumed with polling messages and negative responses, adding to the overhead. A multidrop circuit is an effective way of sharing capacity in a distributed network where no single station needs more than a fraction of the circuit time. Automatic teller machines often use this type of network.

FIGURE 3-2

A Multidrop Data Network

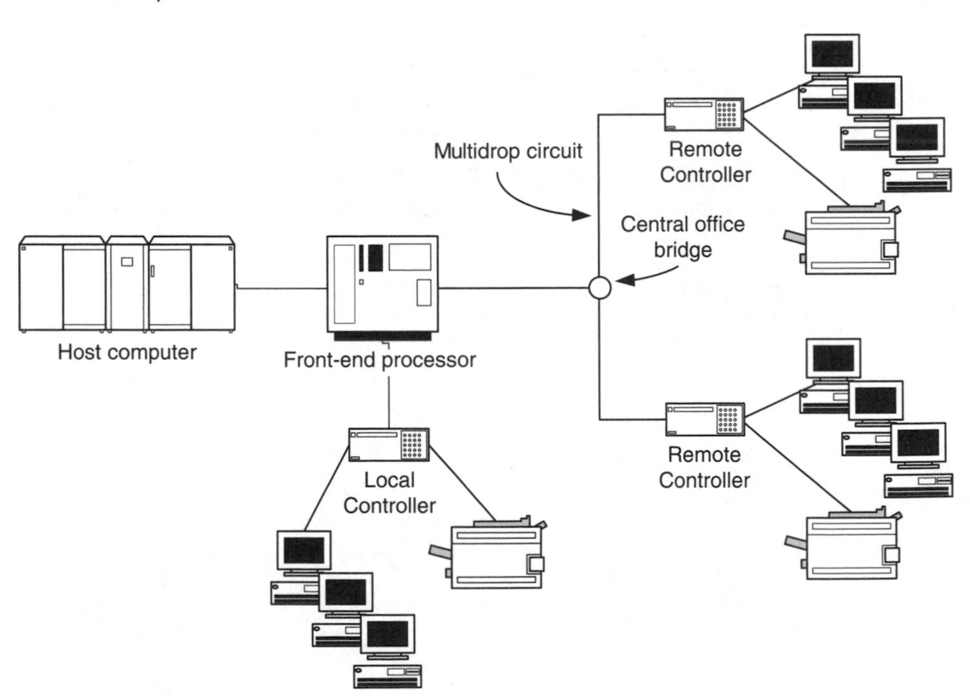

Circuit Switching

In a circuit-switched network, stations connect to a central switch in a star configuration. Fax machines use the PSTN almost exclusively. Other data devices may employ it as a matter of convenience; otherwise, it is rarely used for data except for access to a WAN or LAN. Circuit switching has a major advantage compared to other sharing methods in that the endpoints have exclusive use of the link to the central switch, which reduces security problems. The usage-sensitive nature of circuit switching and its bandwidth restrictions limit its use in data communications. Also, circuit switching wastes bandwidth. When a station has no data to send, the circuit's bandwidth cannot be allocated to another station.

Message Switching

Message switching networks are sometimes called *store-and-forward*. Stations connect to a computer that accepts messages, stores them, and delivers them to their destination. The storage turnaround time can be either immediate for interactive applications or the message may be delayed for forwarding when circuits are idle, rates are lower, or a busy device becomes available. Data networks used message switching in the past, but it is not common today except for electronic mail, the essence of which is store-and-forward.

Packet Switching

A packet network consists of a collection of nodes connected with links as shown in Figure 3-3. Subscriber stations connect at the network edge with either dedicated or dial-up access circuits. The interface between the network and the subscriber station is the X.25 packet switching protocol. Data travels across the network in packets, which consist of address and control headers, a data payload block, and an error-checking trailer. Each node checks incoming packets for errors, and forwards them to the next node only after they are error-free. Inside the network, the service provider determines how to interconnect the nodes and how much bandwidth to provide. Stations communicate across a packet network with virtual circuits. A *virtual circuit* is a connection between endpoints defined in software over a path that is shared with other stations. Packet networks offer both *permanent virtual circuits* (PVCs) and *switched virtual circuits* (SVCs). The carrier provisions a PVC. An SVC is a temporary circuit set up for the duration of a session.

Common carrier packet networks are known generically as public data network (PDNs). Users access the network over either dedicated or dial-up connections. Before the Web became popular, many public databases allowed their users to dial local connections into a PDN node to set up a connection to the database provider.

Frame Switching

Packet networks were popular in the 1970s when physical circuits were analog, error rates were high, and most applications were text-based. When fiber optics

FIGURE 3-3

A Packet Network

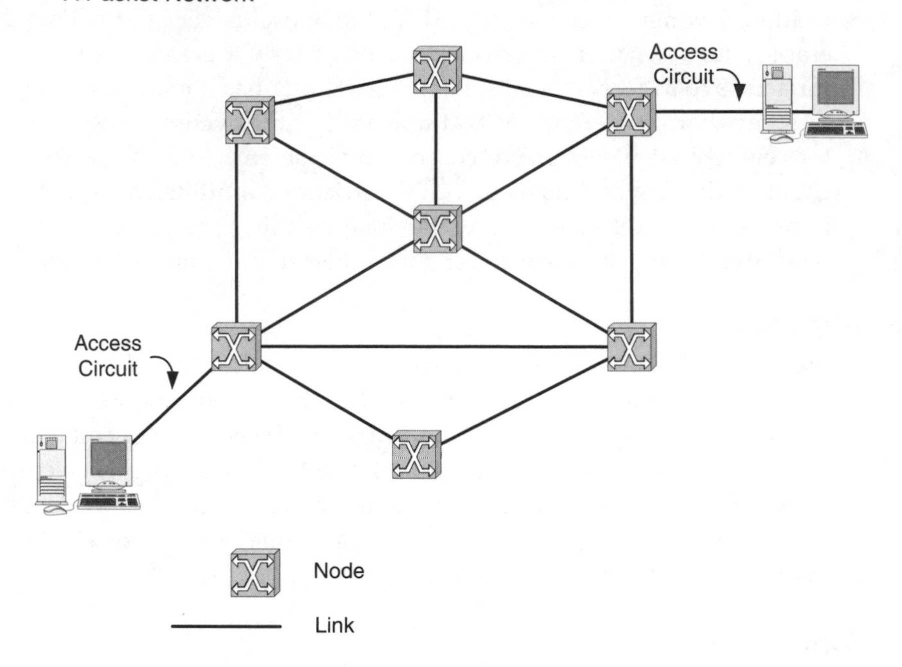

became available, errors dropped to a fraction of their previous level. The error rate was so low, in the order of one error or less per billion bits, that link-by-link error checking was a waste of time and processor resources. The frame relay protocol was the next phase of packet network development. The architecture is similar to that of a PDN, except that dial-up access is rare. If a frame relay node discovers an error, it discards the packet and leaves it to the endpoints, which are usually routers, to arrange retransmission.

The method of charging on frame relay networks is also different. Packet networks typically base their charges on kilopackets transmitted. By contrast, frame relay costs are based on bandwidth. The subscriber pays the cost of the access circuit, the bandwidth of the access port, and the cost of the PVC. For each PVC, the customer chooses a *committed information rate* (CIR). The CIR is the minimum rate the network guarantees to carry. The frame relay protocol allows the data rate to burst up to the speed of the access port if capacity is available. A user could, for example, purchase a 64 Kbps access circuit and port, but choose a lower CIR, such as 16 Kbps. The network would guarantee to carry at least 16 Kbps. If capacity was available, the network would permit the subscriber to send bursts up to 64 Kbps. If capacity is not available, the network can mark frames as discard eligible, carry them if capacity permits, but discard them to protect itself from overload.

Another major departure from the packet network is in the bandwidth offered. Packet networks typically support dedicated or dial-up analog access. Most PDNs had a maximum access bandwidth of 56 Kbps. Frame relay always uses digital access. The minimum access circuit speed is 56 Kbps, and bandwidths of up to 45 Mbps are available.

A major advantage of frame relay is that it offers the security of a dedicated network. A customer's packets are never accessible on another customer's premises. Circuits are shared only in the backbone, and all backbone nodes are confined to common carrier premises. Frame relay is discussed in more detail in Chapter 34.

Cell Switching

Cell relay, the predominate example of which is asynchronous transfer mode (ATM), is a combination multiplexing and switching protocol. The architecture of the network is similar to a packet network except the protocol slices data streams into short cells, 48 octets in ATM, and forwards them across the network with a short header, which is five octets in ATM.

ATM is used in both private and public networks. Unlike other data protocols, cell relay is designed to be used in both voice and data networks. ATM circuits form much of the basis for the Internet and for transporting voice across common carrier networks. It is also commonly used as the access protocol for digital subscriber line (DSL).

IP Networks

A growing tendency in the industry today is to use IP networks. Architecturally, an IP network is similar to the packet-switched network in Figure 3-3. In fact, it is a packet network, but with some important differences compared to a PDN or frame relay. The nodes in a packet network are usually routers, which are special-purpose computers that are optimized for data communications. Routers are available with a variety of input ports, including frame relay and ATM as well as dedicated lines. Since IP is a connectionless protocol, packets can be sent to any station that has an IP address. This is both an advantage and a disadvantage. The main advantage is that it is not necessary for the service provider to define a PVC. The major disadvantage, however, is that an IP network is exposed to all the security threats that plague the Internet.

The Internet service provider (ISP) is the Internet's counterpart of the LEC in the PSTN. ISPs serve the customers and provide access into networks of regional or backbone carriers. Customers may connect to the ISP with either dedicated or dial-up service. Dedicated service may be a point-to-point digital line, but smaller users use either cable access from a CATV provider or DSL from a LEC. The ISPs may choose to connect themselves through private peering arrangements, or they may connect through a network access point (NAP) or a metropolitan area exchange (MAE). NAPs are currently located in New York, Miami, Chicago, San Francisco, and Washington, DC, with large carriers providing the hosting facilities.

MAEs are located in most major cities worldwide. The main requirement for an MAE is collocation space in a building that is secured against all of the usual hazards such as fire, sabotage, and power failure, plus the availability of broadband fiber facilities with protected bandwidth. Figure 3-4 illustrates how the Internet is configured.

Routers attempt to find the most effective path for packets to reach the destination. They do their work in two stages. The first is to consult their routing tables, which may be extensive databases of the ways to reach a destination address. If the destination cannot be reached over a route contained in the table, the router forwards the packet to another router that is closer to the NAP or a peering point. Packet forwarding is the second stage of this two-step process.

IP addresses can be either public or private. If public, the address is directly reachable over the public Internet, opening the network to security concerns. The first step in insulating the LAN from the Internet is *network address translation* (NAT). As its name implies, NAT translates the public address space to private addresses behind a *firewall*. The private address space cannot be reached directly from the public Internet. A firewall is either a hardware device or a software applet that enables the network administrator to restrict who can access the internal network. One reality of security is that convenience and protection are usually diametrically opposed. The best firewalls do not necessarily block viruses and worms because holes in the firewall must be opened to allow

FIGURE 3-4

Internet Configuration

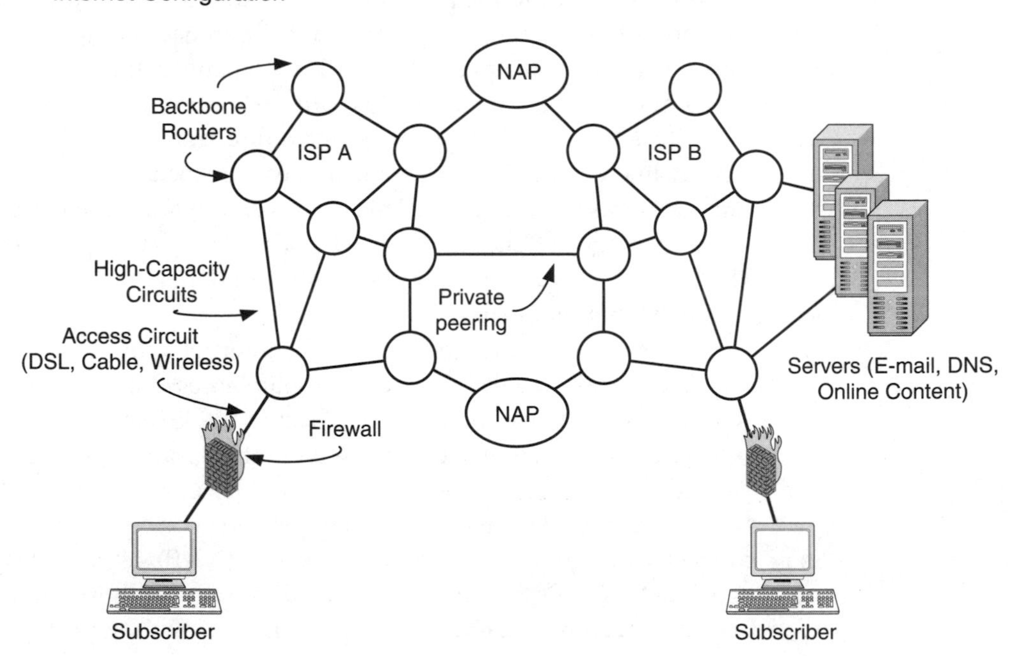

e-mail and others must be opened for voice and other applications. To derive a private network over IP, users can form virtual private networks (VPNs) by "tunneling" through the Internet. A VPN tunnel uses encryption and authentication to ensure that even if communications are intercepted, the intruder will not be able to read them. IP networks and security will be addressed further in Chapters 10 and 36.

Network Hierarchies

Large data networks are usually hierarchical in form. ISO defines terminology for three levels of network entities: *end system* (ES), *intermediate system* (IS) or *area*, and *autonomous system* (AS). An ES is a device such as a terminal, personal computer, or printer that does not perform traffic forwarding or routing functions. ESs are typically connected to an IS, which is a device such as a router, switch, or bridge that routes and forwards data packets. There are two types of IS: intradomain and interdomain. An AS, also known as a *domain*, is a collection of networks under common administrative control. An intradomain IS communicates within a single AS, while an interdomain IS communicates between ASs. An AS may be subdivided into areas, which is a logical grouping of network segments. Figure 3-5 illustrates the concept. The devices connecting the areas are generically known as gateways; the gateway function is usually contained in a router.

FIGURE 3-5

Hierarchy of Networks

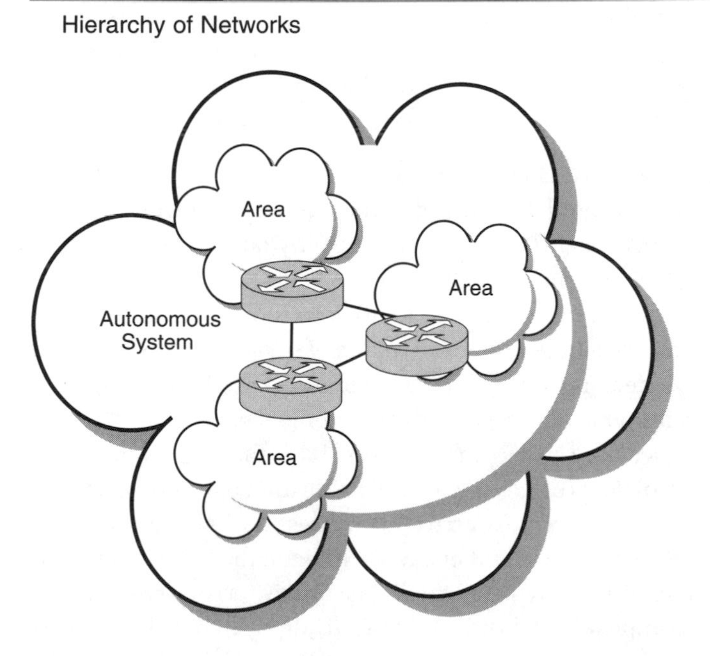

Data Network Performance Measures

A crucial question for every network manager is how much bandwidth is enough. Bandwidth is cheap in a LAN, but in the MAN and WAN, where it is leased from a common carrier, it is costly. Insufficient bandwidth results in lost productivity as users wait for the network to respond. *Response time* is defined as the interval between the time the user presses a key or clicks a button to launch a demand and the screen begins to fill. Response time can usually be improved by increasing bandwidth, but at a higher service cost.

The most effective data networks make bandwidth available to meet the demand. Since traffic demands come in peaks that do not coincide, the bandwidth is allocated to multiple sessions. *Throughput*, which is defined as the number of data bits correctly transferred per unit of time, and reliability are increased by using mesh networks. The nodes in data networks are either computers or routers that have the intelligence to move traffic to another route if the primary route is congested. The links in a full-mesh network connect every node to every other node. Full-mesh connectivity between nodes is usually unnecessary. Providing several paths to and from each node offers high reliability as well as the ability to avoid congestion, but at a higher cost.

Every network has a bandwidth ceiling that is set by some choke point, which is usually the bandwidth of the access circuit. If the access circuit is uncongested, it can devote its entire bandwidth to the application that requests service. In a lightly loaded circuit, the response time is easy to predict—it is a function of the throughput and the size of the file. If several applications are using the access circuit simultaneously, each receives some portion of the bandwidth. How much they get depends on the router programming. Some packets are more important than others. Perhaps the importance depends on who is using the network, or maybe it depends on the nature of the application. This issue becomes more important as data and voice attempt to share the access circuit.

Differences in Quality Requirements

The PSTN has evolved to the point that quality is rarely a problem for voice. Quality is subjective, and carriers measure it by taking controlled opinion surveys. The standard of reference is the so-called "toll-quality" circuit, which earns a mean opinion score of 4.4 on a scale of 1 to 5. For the Internet to approach toll quality, several measures known as *quality of service* or QoS must be applied. Users do not always demand toll quality as evidenced by the fact that they accept lower quality on cell phones in favor of the convenience.

Data, voice, and video can and do share the same circuits, but they assess quality in completely different ways. Reliable data communications must be free of errors; that is, every bit must arrive at the destination exactly as it left the source. Any deviation constitutes a bit error. Microwave radio has a much higher bit-error rate than fiber optics, which usually is nearly error free, but no transmission medium is completely devoid of errors. Data systems detect errors and correct them by retransmitting the errored information. Voice sessions, by contrast, can tolerate a bit-error rate that would render a circuit unusable for data. Error correction is essential for data, but it is a waste of time and resources for voice. Delay-sensitive applications such as voice and video make no attempt to detect and correct errors and dropped packets.

TCP/IP protects itself from congestion in a variety of ways, but the input buffers of routers may fill to capacity before the router can forward packets to the next hop. When this happens, the protocol permits the router to discard packets. Voice can tolerate a certain level of packet loss, but interactive video is highly sensitive to lost packets. The percentage of lost packets is an important measure of network quality.

Delay or *latency*, which is the time required for the signal to travel from source to destination, is also a critical quality measurement for voice and interactive video. If the latency in a voice session is excessive, the participants tend to talk over each other and natural conversation becomes difficult. Voice connections ideally have a maximum of 150 to 200 milliseconds (ms) of delay. Satellite circuits, most of which have been replaced by fiber optics, have a delay of about 250 ms. Although people can talk over a satellite circuit, it can be disconcerting and carriers avoid it if terrestrial circuits are available. To the data user, delay is relatively unimportant. If the response time is half a second longer than normal, it may hardly be noticeable.

Noise and *loss* are two phenomena that have an adverse affect on both data and voice. They are characteristic of analog circuits, but affect digital circuits to a minor degree. Loss refers to the drop in volume from one end of a circuit to the other. When loss is excessive, a voice user has difficulty hearing and the difficulty increases with a higher noise level. Data are more tolerant of loss than voice. Since data signals terminate in electronic apparatus, they can be amplified to the level the equipment requires, but noise is amplified with it. Noise is important to data only if it causes errors. To both voice and data, the signal-to-noise ratio is the important factor.

Echo is another quality impairment that is important to both voice and data. Echo occurs when a signal hits an irregularity that reflects a portion of the signal back to the source. Excessive echo can render a circuit unusable for both voice and data. It is generally resolved by using *echo cancellers*, which are devices that look at the forward signal, compare it to the reverse signal, and cancel the reverse path if the signals match.

Differences in Session Requirements

The nature of a voice session is much different than a data session. Voice sessions begin with setting up a connection, which is exclusive for the duration of the call, and then tearing the connection down. When a data user dials up a connection, such as remote access to an ISP or a corporate network, the session is, for all practical purposes, a voice session, with the modem emulating the human voice. Dial-up sessions are acceptable for some data applications, including Internet access, but dial-up is usually an expedient that is used only when dedicated access is unavailable.

Voice sessions are usually user to user, extending to multiple users in the case of a conference call, but easily confined to fixed bandwidth. The nature of a voice session leaves half the bandwidth of the circuit unused. In addition, during pauses in conversation, no information flows in either direction. Data sessions can make use of this idle bandwidth, but it is important that long data packets not block short voice packets.

DATA APPLICATION ISSUES

This chapter has touched on most of the concepts needed for a broad understanding of how data networks function and how their quality is measured. Data and voice both ride over the same physical fiber-optic backbone, but at the circuit level they are mostly separate. That, however, is changing. The trend is toward convergence of all media, including voice, data, video, graphics, and multimedia applications over a combined network. Before that transition can occur, several issues remain to be resolved, so the networks will remain separate for the next several years.

Today, frame relay is the most popular data service. Frame relay also carries a limited amount of voice and video. The main reason for its popularity lies in its simplicity and security. A frame relay network is provisioned within common carrier facilities, and is fully as secure as a network composed of dedicated point-to-point lines. It is economical, and offers bandwidth on demand up to the capacity of the input port.

The major frame relay carriers also offer IP networks, and are recommending that their customers move their service from frame relay to IP VPNs. For the customer, an IP network has a major attraction. The typical enterprise is widely distributed and international frame relay is expensive. Most sites already have dedicated Internet access. If the same circuit can be used for access to both the Internet and internal enterprise network, the cost may be significantly less. A transition to an IP network would be easy to justify except for security and privacy concerns. To implement an IP VPN, the customer either must set up and manage appliances that provide the tunnels or turn the task over to the carrier. Either approach raises the cost to the point that much of the expected saving evaporates.

Like the telephone network, the IP network is a complex and intricate collection of circuits, routers, and servers. The Internet is most effective for those that have full-time connections to it, but dial-up access through the PSTN is an alternative for those that do not.

For carriers the choices are point-to-point circuits, MPLS, or ATM. For enterprise networks, the choices are point-to-point circuits, frame relay, or IP, with the latter gaining in popularity. The three services have a narrow range of overlap in their applications and differences that must be clearly understood.

Access Channel

The service providers usually obtain the access channel from the customer premises to the carrier's *point-of-presence* (POP) from the LEC or a CAP. The access channel cost within an LEC's wire center is based on bandwidth. If the POP is in a different wire center than the point of service delivery, the access channel may also be distance sensitive, meaning the longer it is the more it costs. Therefore, the location of the service provider's POP is important. To be price competitive, the carrier may select its POP for rating purposes at a different location than the point of connection to its backbone network. Nevertheless, the cost of the access channel is apt to vary widely among carriers and types of service. IP networks can use DSL or cable for the access channel, which offers inexpensive access that carriers can provision in short intervals.

Pricing Structure

Unlike frame relay and IP, which are not distance sensitive, the cost of point-to-point circuits varies with the length of the IOC. The cost of all three alternatives varies with the amount of bandwidth. IP is connectionless, so packets can be routed to any endpoint. Frame relay has a PVC charge, which defines the endpoints that can communicate across the network, although stations that lack a PVC can still communicate through the central site. In addition, a CIR charge applies to each PVC. Multiple PVCs to the same location may share access channels.

Bandwidth

Point-to-point bandwidth can be obtained up to the capacity of a fiber-optic wavelength or lambda. Frame relay providers usually offer a maximum bandwidth of DS-3.

Security

From a security standpoint, a dedicated channel offers the maximum security. Frame relay is a close second. With both dedicated circuits and frame relay, the circuit is accessible only on the carrier's premises. Security is the weakest point in an IP network and is the primary factor weighing against its use in sensitive applications. Although IP can theoretically be made as secure as frame relay or dedicated circuits, IP is vulnerable to such hazards as denial-of-service attacks that cannot affect frame relay or dedicated circuits. Some applications require firewall holes that destructive viruses and worms can penetrate. IP networks can be secured, but to preserve security levels they require more administrative attention than frame relay.

Connectivity

A major advantage of IP is its connectivity. Any IP address that is part of the public addressing structure can be reached from any other location. This is a major advantage compared to frame relay, which offers the equivalent of a dedicated connection. Frame relay PVCs are provisioned using an identifier known as a datalink connection identifier (DLCI), which defines the circuit. A node on a frame relay network can communicate with another node only by defining a PVC to that node or by communicating through another node.

Service Level Agreements (SLAs)

The three services are much different in the ability of the carriers to provide tight SLAs. SLAs are generally quoted on the basis of factors listed below.

Availability

Availability is defined as a percentage of time the network is available for service. For example, a network with 1 h of the outage per year has an availability of 99.989%. This is calculated with the formula:

$$\text{Percent availability} = 1 - (\text{outage hours}/8760\ \text{h/year})$$

Availability is normally quite high for all three types of facility. Point-to-point circuits are not subject to degradation during overload periods as are IP and frame relay. IP reliability is somewhat less than frame relay, which is not subject to hacking attacks.

Order Intervals

The user is concerned only with obtaining the access channel in an IP network. Once the channel is up, the endpoint can connect anywhere in the world without carrier intervention. Access channels can be provisioned quickly by using DSL or cable. If a dedicated access channel is required, the order interval is the same as for dedicated or frame relay circuits, but a major advantage of IP is the speed with which it can be provided.

The longest intervals can be expected with point-to-point circuits, which must have two access circuits and an IOC provisioned, often by multiple carriers. To some degree the same is true of frame relay, except that the service provider can provision the DLCI quickly, and can make changes to the port speed in a day or two.

Technical Variables

The tightest limits on such variables as noise, bit-error rate, latency, and packet loss can be obtained with point-to-point circuits, and to a marginally lower degree, with frame relay. Some frame relay providers offer classes of service with latency and packet loss limits that are tight enough to support VoIP.

The ability of ISPs to offer SLAs on technical variables depends on the degree of control they have over the network. Large ISPs may have their own backbone, and therefore be able to control the network, but smaller ISPs may have little or no control after they hand off packets to a backbone provider. In any case, latency and packet loss will be higher on an IP network than frame relay.

CHAPTER 4

Data Communications Principles

Not long ago, desktop computers were rare, and the Internet was the province of a handful of intellectuals in universities, government agencies, and large companies. Most people had no idea what a modem was or why they would ever need one. Now, desktop computers are selling for well under $1000, hard drive capacities run into the multi-gigabytes, the applications are almost unlimited, and the Internet has attracted millions to the data communications world.

Telecommunications and the computer are partners in a marriage that has changed the way people store, access, and use information. In the heyday of the mainframe, databases were centralized and a specialized program was required to access them. That program was an application that ran on the mainframe, which transmitted information over a structured network to the terminal that requested it. The terminal could modify the record only to the extent that the program permitted. The mainframe is quite efficient at what it does, but its applications are limited. No computer of the mainframe era has applications such as word processing, spreadsheet, or e-mail that approach the effectiveness of those on desktop computers.

The merger of mainframe and telecommunications makes possible many applications including automatic teller machines, airline reservation systems, and credit card verification networks. These use text-based clients, and work fine on dumb terminals. More effective, however, are graphics-based applications that enable the user to merge images and multi-color graphics with text. These require desktop processing power, graphic displays, and storage capacity that outstrip the capabilities of computers and networks of a few years ago.

This chapter is the second of several that deal with data communications. Chapter 3 was an overview intended to put data communications in perspective. That chapter omitted details and used terminology that is difficult to grasp without a deeper understanding of how the major components of a data network function down at the bit level.

As we discussed in Chapter 3, data protocols function in layers. The lowest level is the physical layer, in which bits move across a transmission medium, which is a pair of wires, a coaxial cable, a radio channel, or an optical fiber pair. Before computers came on the scene, data communications used only the physical network. Two teletypewriters (or two desktop computers with their serial ports connected) can exchange data over two pairs of wires, but with limitations. Disturbances from a variety of sources can mutilate bits and render the message useless or worse if the error passes undetected.

This chapter expands on Chapter 3 with explanations of how data devices ensure error-free transport across telecommunications media that are subject to a variety of disturbances. This chapter is hardware-oriented, with enough discussion of protocols to prepare you for a discussion of pulse-code modulation in Chapter 5 and a more detailed protocol discussion in Chapter 6.

DATA COMMUNICATIONS FUNDAMENTALS

Data communications is an endless quest for perfection. Billions of dollars move among financial institutions throughout the world, software traverses the Internet, and goods and services move through electronic document interchange. Billions of transactions are carried daily without as much as one misplaced bit. Binary digits, from which the word *bit* is derived, are either right or wrong and a single bit in error can convert a file to rubbish. Some data applications can tolerate an occasional error, but most require absolute integrity regardless of whether the transmission medium is the finest fiber-optic cable or a deteriorated rural wire line running through a swamp.

The devices that originate and receive data are called *data terminal equipment* (DTE). These can range from computers to simple receive-only terminals or printers. DTE connects to the telecommunications network through *data communications equipment* (DCE), which converts the DTE's output to a signal suitable for the transmission medium. DCE ranges from line drivers to complex modems and multiplexers.

The basic information element that a computer processes is the bit, which is represented by the two digits 0 and 1. Processors manipulate data in groups of eight bits known as *bytes* or *octets*. To make binary digits easier for humans to manipulate, octets are often split into groups of four bits and represented as the hexadecimal digits 0 to 9 and A to F. Inside the computer data travels over parallel paths. Parallel transmission is suitable for short distances to peripherals such as printers, but for communications over a range of more than a few feet, the eight parallel bits are converted to serial as Figure 4-1 shows. This serial bit stream is coupled to telecommunications circuits through some type of DCE.

FIGURE 4-1

Parallel to Serial Conversion

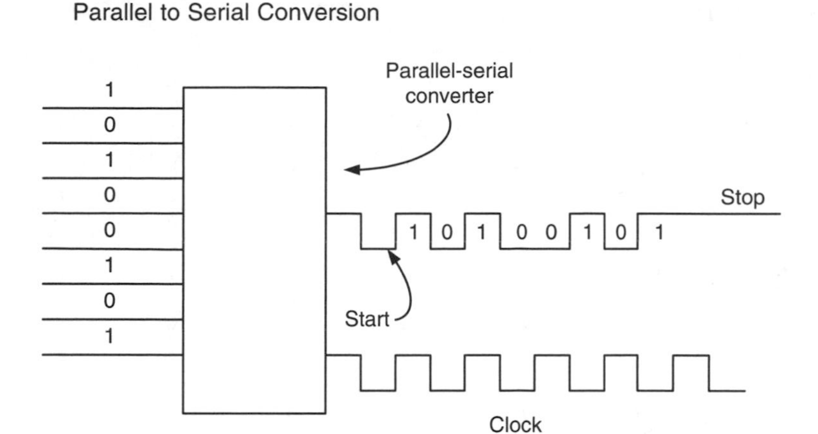

Coding

The number of characters that binary numbers can encode depends on the number of bits in the code. Early teletypewriters used a five-bit code called Baudot that had a capacity of 2^5 or 32 characters. Today's telecommunications device for the deaf (TDD) uses this code. Thirty-two characters are not enough to send a full range of upper and lower case plus special characters. In the Baudot code, shift characters trigger the machines between upper and lower cases. The receiving device continues in its current mode until it receives a shift character. If it misses a shift, the device continues to print and the transmission is garbled.

To overcome this 32-character limitation, teletypewriters evolved to a seven-level code known as the American Standard Code for Information Interchange (ASCII). This code, which Table 4-1 shows, provides 2^7 or 128 combinations. Although the ASCII code uses seven bits for characters, eight bits are transmitted. The eighth bit is used for error detection as described later.

Several other codes are used for data communications. Many IBM devices use Extended Binary Coded Decimal Interchange Code (EBCDIC), which Table 4-2 shows. EBCDIC is an eight-bit code, allowing the full 256 characters to be encoded. Neither EBCDIC nor ASCII can begin to represent all of the characters and technical symbols in the world's languages. Unicode is a 16-bit code that provides a platform for encoding any symbol into binary. Not only are all current and archaic languages encoded, but the standard also includes musical symbols and geometric shapes. Code compatibility between machines is essential. Because EBCDIC, ASCII, and Unicode are widely used, some applications will require code conversion. Most Web browsers and many application programs support Unicode and can convert between the codes.

TABLE 4-1

American Standard Code for Information Interchange (ASCII)

$b_7\ b_6\ b_5$ $b_4\ b_3\ b_2\ b_1$	000	001	010	011	100	101	110	111
0000	NUL	DLE	SP	0	@	P	‘	p
0001	SOH	DC1	!	1	A	Q	a	q
0010	STX	DC2	“	2	B	R	b	r
0011	ETX	DC3	#	3	C	S	c	s
0100	EOT	DC4	$	4	D	T	d	t
0101	ENQ	NAK	%	5	E	U	e	u
0110	ACK	SYN	&	6	F	V	f	v
0111	BEL	ETB	‘	7	G	W	g	w
1000	BS	CAN	(	8	H	X	h	x
1001	HT	EM	)	9	I	Y	i	y
1010	LF	SUB	*	:	J	Z	j	z
1011	VT	ESC	+	;	K	[	k	{
1100	FF	FS	,	<	L	\	l	|
1101	CR	GS	-	=	M	]	m	}
1110	SO	RS	.	>	N	^	n	~
1111	SI	US	/	?	O	_	o	DEL

Data Communication Speeds

Data communications speeds are measured in bits per second (bps) or some multiple: kilobits (Kbps), megabits (Mbps), gigabits (Gbps), and terabits (Tbps). It is easy to confuse this with storage and file sizes, which are always quoted in bytes. As a matter of convention, bits are shown in lower case and bytes in upper. For example, 1 Mbps is megabits and 1 MBps is megabytes.

Early data applications were limited by the speed at which an operator could type or toggle a telegraph key. When punched paper tape teletypewriters were developed, the operator could type off-line, and then send at the full speed of the device. Teletypewriters using the Baudot code around the time of World War II ran at 60 words per minute, which was roughly 30 bps. Later ASCII machines upped the speed to 100 wpm, still a fraction of the data-carrying capacity of a voice circuit. Analog multiplexers at the time subdivided voice channels so they could carry multiple telegraph or teletypewriter signals.

The telephone channel bandwidth of 300 to 3300 Hz imposes an upper limit on data transmission speeds. Many people use bit rate and baud rate interchangeably to express the data-carrying capacity of a circuit, but they are not technically synonymous. Bit rate is the number of bits per second the channel can carry. Baud rate is the number of cycles or symbol changes per second the channel can support. The bandwidth of a voice channel is limited to 2400 baud, but higher bit rates are transmitted by encoding more than one bit per baud. A 19,200-bps modem, for example, encodes eight bits per baud. The latest version of high-speed

TABLE 4-2

Extended Binary Coded Decimal Interchange Code (EBCDIC)

Bits 8 7 6 5 \ 4 3 2 1	0000	0001	0010	0011	0100	0101	0110	0111	1000	1001	1010	1011	1100	1101	1110	1111
0 0 0 0	NUL	SOH	STX	ETX	PF	HT	LC	DEL			SMM	VT	FF	CR	SO	SS
0 0 0 1	DLE	DC_1	DC_2	DC_3	RES	NL	BS	IL	CAN	EM	CC		IFS	IGS	IRS	IUS
0 0 1 0	DS	SOS	FS		BYP	LF	EOB	PRE			SM			ENQ	ACK	BEL
0 0 1 1			SYN		PN	RS	UC	EOT					DC_4	NAK		SUB
0 1 0 0	SP										“	.	<	(	+	|
0 1 0 1	&										!	$	*	)	;	¬
0 1 1 0	-	/										,	%		>	?
0 1 1 1										\	:	#	@	‘	=	
1 0 0 0		a	b	c	d	e	f	g	h	i						
1 0 0 1		j	k	l	m	n	o	p	q	r						
1 0 1 0			s	t	u	v	w	x	y	z						
1 0 1 1																
1 1 0 0		A	B	C	D	E	F	G	H	I						
1 1 0 1		J	K	L	M	N	O	P	Q	R						
1 1 1 0			S	T	U	V	W	X	Y	Z						
1 1 1 1	0	1	2	3	4	5	6	7	8	9						

PF	Punch Off	UC	Upper Case	PN	Punch On	PRE	Prefix (ESC)
HT	Horizontal Tab	RES	Restore	EOT	End of Transmission	RS	Reader Stop
LC	Lower Case	NL	New Line	BYP	Bypass	SM	Start Message
DEL	Delete	BS	Backspace	LF	Line Feed	Others	Same as ASCII
SP	Space	IL	Idle	EOB	End of Block		

modems uses somewhat higher baud rates than 2400 to achieve downstream speeds as high as 56 Kbps. This speed requires one end of the connection to have digital connectivity to the central office.

Modulation Methods

A data signal leaves the serial interface of the DTE as a series of *baseband* voltage pulses as Figure 4-2 shows. Baseband means that the varying voltage level from the DTE is impressed without modulation directly on the transmission medium. Baseband pulses can be transmitted over limited distances across copper wire from a computer's serial interface. RS-232, or more accurately EIA-232, is the standard that most computers support. The standard limits the cable length to 50 ft (15 m), although many users operate it successfully over longer distances.

The length limitations of EIA-232 result from its use of unbalanced transmit and receive leads, each of which is a single wire, sharing a common return path. A balanced transmission medium such as EIA-423 can transmit over much greater distances because the transmit and receive paths are separate. EIA-423 has a length limit of 4000 ft (1200 m). EIA-232 can operate over longer distances by using a limited distance modem or a line driver that matches a short serial interface to a balanced cable pair. A balanced transmission medium is inherently less susceptible to noise.

For transmission over voice-grade channels, a modem modulates data pulses into a combination of analog tones and amplitude and phase changes that fit within the channel pass band. The digital signal modulates the *frequency*, the *amplitude*, or the *phase* of an audio signal as Figure 4-2 shows. High-speed modems use all three methods. Amplitude modulation by itself is the least-used method because it is susceptible to noise-generated errors. It is frequently used, however,

FIGURE 4-2

Data Modulation Methods

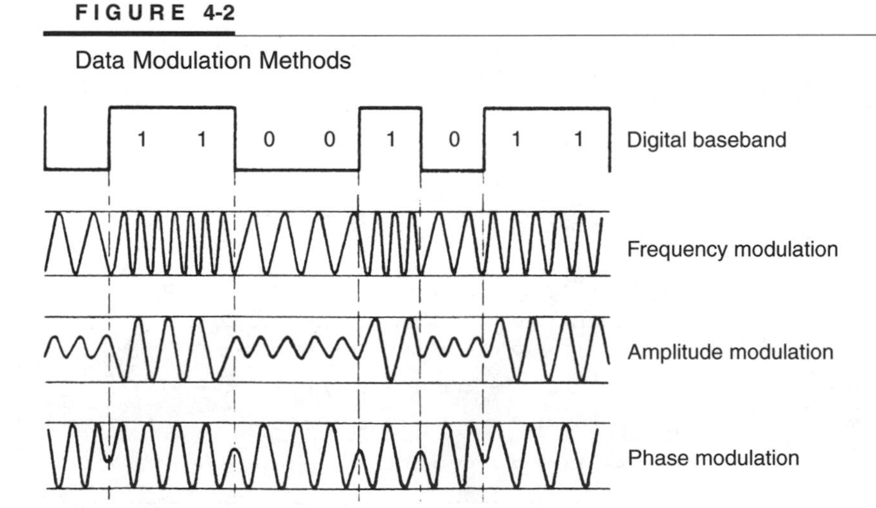

in conjunction with frequency and phase changes. Frequency modulation is an inexpensive method used with low-speed modems. To reach speeds of more than 300 bps, phase shift modems are employed.

Quadrature Amplitude Modulation (QAM)

Modems use increasingly complex modulation methods for encoding multiple bits per baud to reach speeds approaching the theoretical limit of a voice grade circuit. Since an analog channel is nominally limited to 2400 baud, to send 9600 bps, e.g., four bits per baud must be encoded. This yields 2^4 or 16 combinations that each symbol can represent. High-speed modems use a method known as QAM. In QAM, two carrier tones combine in quadrature to produce the modem's output signal. The receiving end demodulates the quadrature signal to recover the transmitted signal. In 16 QAM, each symbol carries one of 16 signal combinations. As Figure 4-3 shows, any combination of four bits can be encoded into a particular pair of X–Y plot points, each of which represents a phase and amplitude combination that corresponds to a 4-bit sequence. The four bits can be expressed as hexadecimal in 2^4 or 16 combinations. This two-dimensional diagram is called a *signal constellation*.

The receiving modem demodulates the signal to determine what pair of X–Y coordinates was transmitted, and the four-bit signal combination passes from the modem to the DTE. If line noise or phase jitter affect the signal, the received point will be displaced from its ideal location, so the modem must make a best guess

FIGURE 4-3

Signal Constellation in a 16-Bit (2^4) QAM Signal

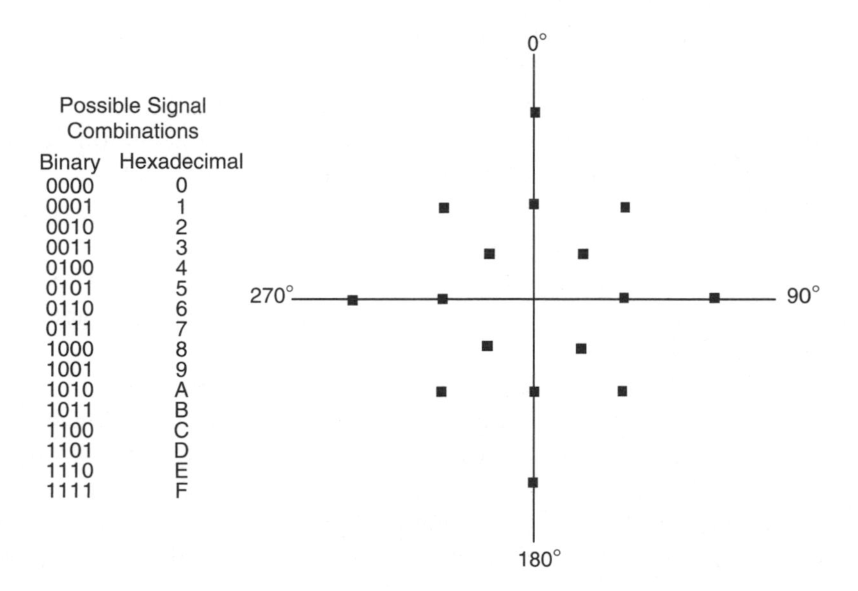

as to which plot point was transmitted. If the signal is displaced far enough, the receiver guesses wrong, and the resulting signal is in error.

Even higher rates can be modulated, with each additional bit doubling the number of signal points. A 64-QAM signal encodes 2^6 bits per symbol, a 128-QAM signal results in 2^7 combinations, and a 256-QAM signal results in 2^8 combinations, bringing the signal points closer together and increasing the susceptibility of the modem to impairments. DSL modems use QAM to modulate a broadband data signal above the voice circuit on a telephone cable pair.

Trellis-Coded Modulation (TCM)

TCM is an encoding method that makes each symbol dependent on adjacent symbols. In a 14,400-bps modem, for example, data is presented to a TCM modulator in six-bit groups. Two of the six bits are separated from the signal, and a code bit is added. The resulting signal is two groups: one three-bit and one four-bit. These combine into 2^7 bits, which are mapped into a signal point and selected from a 128-point signal constellation. Since only six of the seven bits are required to transmit the original signal, not all 128 points are needed to transmit the signal, and only certain patterns of signal points are defined as valid. If an error causes an invalid pattern at the receiver, the decoder selects the most likely valid sequence and forwards it to the DTE. If it guesses wrong, the DTE's error-correction mechanism arranges for retransmission. TCM reduces the signal's susceptibility to line impairments and reduces the number of retransmissions.

Full- and Half-Duplex Mode

Full-duplex data systems transmit in both directions simultaneously. Half-duplex systems transmit in only one direction at a time; the channel reverses for transmission in the other direction. Half-duplex is used only on dedicated line facilities where the application is inherently half-duplex. A good example is an automatic teller machine where the user interacts with a central computer to make deposits and withdrawals with each end of the transmission sending a short message that identifies the user and actuates the transaction.

LECs provide dedicated analog circuits as either two-wire or four-wire, but almost all data private lines use four-wire facilities. LECs also offer dedicated digital circuits. A 56-Kbps digital channel (64 Kbps in Europe) provides a four-wire digital channel. This service uses a bipolar modulation method, which is discussed in Chapter 5.

The LECs' and IXCs' inter-office facilities are inherently four-wire. To provide end-to-end four-wire facilities, the LEC assigns two cable pairs in the local loop. Two-wire facilities can support full-duplex operation by using modems that separate the two directions of transmission. *Split channel* modems provide the equivalent of four-wire operation by dividing the voice channel into two segments, one for transmit and one for receive. Dial-up modems support 2400-bps full-duplex communication over two-wire circuits using the ITU V.22 *bis* modulation method, 9600 bps using V.32, 14,400 bps with V.32 *bis*, 33,600 bps with V.34, and 56 Kbps with V.92 modulation.

Synchronizing Methods

All data communications channels require synchronization to keep the sending and receiving ends in step. The signal on a baseband data communications channel is a series of rapid voltage changes, and synchronization enables the receiving terminal to determine which pulse is the first bit in a character.

The simplest synchronizing method is *asynchronous*, sometimes called stop–start synchronization. Asynchronous signals, illustrated in Figure 4-4, are in the one or *mark* state when no characters are being transmitted. A character begins with a start bit at the zero or *space* level followed by eight data bits and a stop bit at the one level. The terms mark and space originated in telegraphy and extend to teletypewriters. A teletypewriter needs line current to hold it closed when it is not receiving characters. Some asynchronous terminals also use a current loop over ranges greater than the EIA-232 serial standard supports. Current loops have largely disappeared from public networks because they generate noise and the ILECs cannot guarantee that circuits will be assigned to metallic cable.

Asynchronous signals are transmitted in a character mode, i.e., each character is individually synchronized and unrelated to any other character in the transmission. One drawback of asynchronous communication is the extra two overhead bits per character that carry no information. Asynchronous communication also lacks the ability to correct errors. Asynchronous has a major advantage of being a simple and universal standard. Nearly every desktop computer has a serial port that can be attached to a modem for communications wherever a telephone can be found. Modems overcome much of the asynchronous deficiency by implementing an error detection and correction dialogue.

Protocols intended for LANs and WANs use *synchronous* or block mode protocol. Figure 4-5 shows a High-Level Datalink Control (HDLC) synchronous frame. The PDU structure is different for other datalink protocols such as Ethernet, but the principles are the same. An information block is sandwiched

FIGURE 4-4

Asynchronous Transmission

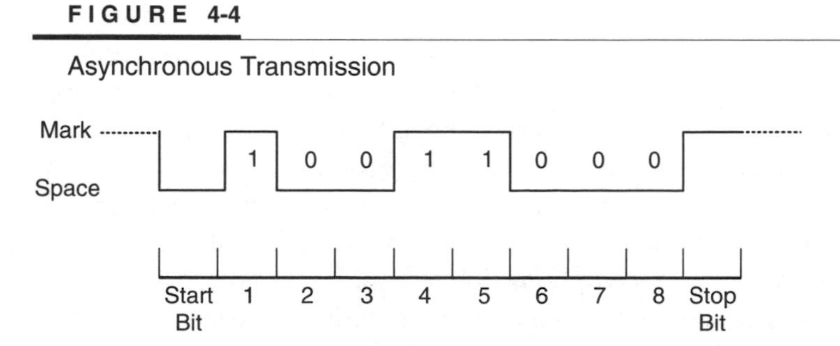

FIGURE 4-5

HDLC Frame

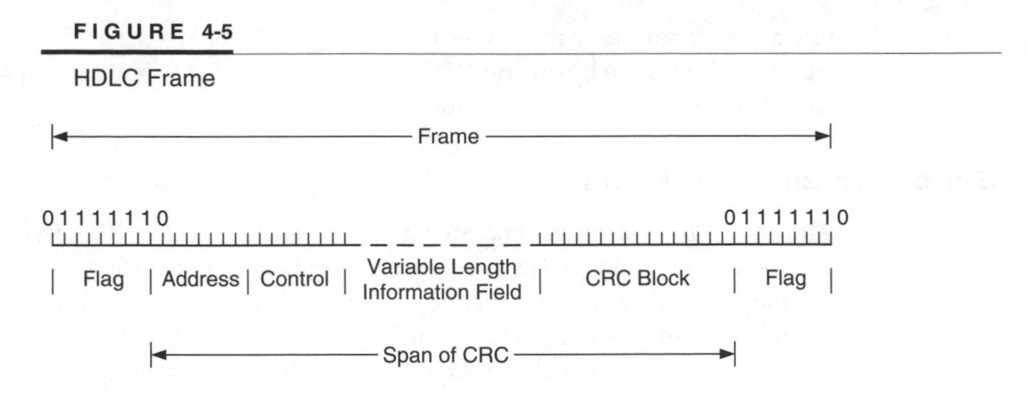

between header and trailer records. The header contains addressing and control information, and the trailer handles error detection and correction as explained in the next section. A starting flag contains a unique bit pattern that prepares the receiving device to receive the frame. The header and trailer lengths are set in the protocol and the control octet contains, among other things, the length of the data block. The network administrator adjusts the length of the block to fit the needs of the application and the characteristics of the network.

Error Detection and Correction

Errors occur in all data communications circuits. Where the transmission is text that people will interpret, a few errors can be tolerated because the meaning can be derived from context. Teletypewriters have no error-correction capability, but they use other techniques to flag errors and users can interpret the message from context. Data applications are not tolerant of errors, but voice and video can accept a high rate of errors with imperceptible effect. This section discusses causes, detection, and correction of data communication errors.

Causes of Data Errors

The type of transmission medium and the modulation method have the greatest effect on the error rate. Any analog transmission medium is subject to external noise, which affects the amplitude of the signal. Atmospheric conditions, such as lightning that cause static bursts, and noise induced from external sources such as power lines all cause errors. Digital circuits carried on fiber optics are immune to these influences, but digital radio is susceptible to these as well as signal fades. Fiber-optic systems exhibit an infinitesimal error rate until something fails in the electronics and the system switches to a standby channel. Technicians probably cause the bulk of data errors. Any circuit is subject to errors during maintenance activities and external damage or interruption by vandalism. Even LANs within a single building are subject to occasional interruptions due to equipment failure or human error. The best error mitigation program is a design that reduces the susceptibility of the service to errors. Nevertheless, errors are inevitable and corrective measures are essential.

Parity Checking

The simplest way of detecting errors is *parity checking*, or *vertical redundancy checking* (VRC), a technique used on asynchronous circuits, particularly on teletypewriters. In the ASCII code set, the eighth bit is reserved for parity. Parity is set as odd or even, referring to the number of 1 bits in the character. As Figure 4-6 shows, DTE adds an extra bit, if necessary, to cause each character to match the parity established for the network.

Most asynchronous terminals can be set to send and receive odd, even, or no parity. When a parity error occurs, some terminals can register an alarm, but for the most part parity is useless in computers. Parity has two drawbacks: there is no way to tell what the original character should have been and, worse, if an even number of error occurs, parity checking will not even detect that there was an error. In today's data networks, parity is irrelevant, although it is part of modem setup strings.

Echo Checking

The receiving computer may echo the received characters back to the sending end. This technique, called *echo checking*, is suitable for detecting some errors in keyboarded text. The typist sees an unexpected echoed character, backspaces, and retypes it. An error in an echoed character is as likely to have occurred on the return trip as in the original transmission, so the receiver may have the correct character but the transmitter believes it was received in error. Although echo checking is ineffective in machine-to-machine communications, some computers still use it. You may inadvertently set up your modem to display characters locally and they are also echoed from the distant end. In this case double characters appear on the screen.

Cyclical Redundancy Checking (CRC)

Synchronous data networks use CRC to detect and correct errors. The sending DCE processes the bits in each frame against a complex polynomial that always results in a remainder. The remainder is entered in an error check block following the data block. The receiving DCE recalculates the CRC field against the header and data block and compares it to the received CRC. If the two match, the frame is acknowledged; otherwise the protocol returns a message that instructs the sender to retransmit. The sender must, therefore, buffer PDUs until it receives

FIGURE 4-6

Character Parity

	Bit 8	Bit 7	Bit 6	Bit 5	Bit 4	Bit 3	Bit 2	Bit 1
ASCII a		1	1	0	0	0	0	1
Odd Parity	1	1	1	0	0	0	0	1
Even Parity	0	1	1	0	0	0	0	1

an acknowledgement. The probability of an undetected error with CRC is so slight that it can be considered error-free. Synchronous datalink protocols acknowledge which frames have been received correctly through a process that Chapter 6 describes. If a PDU is not acknowledged before the protocol times out, the sending end retransmits it. This can result in duplicate PDUs, so the protocol must detect these and kill the duplicate.

One bit in error in a frame is fully as detrimental as a long string of errors. Most carriers quote the bit-error rate (BER) or error-free seconds (EFS) in their SLAs. Errors often come in groups and since one bit-error destroys the frame, a high BER may be a somewhat deceptive quality measurement. A better measurement of datalink quality is the block error rate (BLER), which is calculated by dividing the number of errored blocks or frames received over a period by the total number of blocks transmitted. A device such as a front-end processor or a protocol analyzer can compute BLER.

Forward Error Correction (FEC)

When the BLER of a circuit is excessive, throughput may be reduced to an unacceptable level. The longer the data block, the worse the problem because of the amount of data that must be retransmitted. FEC can help bring the error rate down to a manageable level. In FEC systems, an encoder on the transmitting end processes the incoming signal and generates redundant code bits. The transmitted signal contains both the original information bits plus the redundant bits. At the receiving end, the redundant bits are regenerated from the information bits and compared with the redundant bits in the received signal. When a discrepancy occurs, the FEC circuitry on the receiving end uses the redundant bits to generate the most likely bit combination and passes it to the DTE. Although FEC is fallible, it reduces the BLER and the number of retransmissions.

Throughput

One critical measure of a data communication circuit is its throughput, defined as the number of information bits correctly transferred per unit of time. Although it would be theoretically possible for the throughput of a data channel to approach its maximum bit rate, in practice this can never be realized because of overhead bits and the retransmission of errored PDUs. The following are the primary factors that limit the throughput of a data channel:

- *Modem speed*. The faster the modem the less the time taken to transmit a block of data.
- *Half- or full-duplex mode*. With other factors equal on a private line circuit, full-duplex circuits have greater throughput because the modems do not have to reverse between transmitting and receiving.
- *Error rate*. The higher the error rate, the more the retransmissions and the lower the throughput.

- *Protocol.* Different protocols have different overhead bits and error-handling methods. Also, some protocols cannot be transmitted over a particular network and must be encapsulated in another protocol, which increases overhead.
- *Size of data block.* If the error rate is high, short data blocks are more efficient because the retransmission time is high. If the error rate is low, long data blocks are more efficient. The shorter the data block, the more significant the header and trailer as a percentage of PDU length. When the data block is too long, each error necessitates retransmitting considerable data. Optimum block length is a balance between time consumed in overheads and in error retransmission.
- *Propagation speed.* This factor is the time required for data to traverse the circuit. It depends on the length of the circuit and the type of transmission medium. Satellite circuits have the greatest delay.

The network administrator optimizes the throughput of a data channel by reaching a balance between the above variables.

DATA COMMUNICATIONS EQUIPMENT

An effective data communications network is a compromise involving many variables. The nature of data transmission varies so greatly with the application that designs are often empirically determined. The network designer arrives at the most economical balance of performance and cost, evaluating equipment alternatives as discussed in this section.

Terminals

The dumb terminal of the past is giving way to the PC and a variety of hand-held and wireless devices that communicate with special hosts. Dumb terminals still have their uses. For example, PBXs, routers, multiplexers, and other such devices are equipped with EIA-232 ports so they can be configured from a terminal. Since a serial port is a standard feature of most desktop computers, it is simple for a computer to emulate an asynchronous terminal. Telecommunications software ranges in features from simple dumb terminal emulation to full-featured intelligent terminal applications. In the latter category, a desktop computer can upload and download files from and to its own disk, select and search for files on the host, and even interact with the host without a human attendant.

Modems

Since the early 1980s, modems have undergone a dramatic evolution. To discuss modems, it is useful to classify them as dial-up and private line. In the dial-up category, modems have almost become a commodity. They are manufactured to international standards, and nearly every computer contains one. The interface between DTE and the modem is standardized in most countries, with the predominant interfaces being the EIA-232, EIA-449, and ITU V.35. EIA and ITU standards specify the functions of the interface circuits but do not specify the physical characteristics of the interface connector. Connectors have been adopted by convention, e.g., the DB-25 connector has become a de facto standard for the EIA-232 interface. Not all 25 pins of the DB-25 are necessary in most applications, so many products use the nine-pin DB-9 connector.

Dial-Up Modems

Like other telecommunications products, modems have steadily become faster, cheaper, and smarter, with V.92 modems being the modern norm. Modem setup was once somewhat tricky, but most devices are now self-configuring except for special terminal emulation functions. Dial-up modems either plug into a desktop computer expansion slot or are self-contained devices that plug into the computer's serial port. Modems support an error correction protocol and implement V.44 data compression. Most also support fax.

The switched telephone network carries a considerable share of asynchronous data communication. Therefore, many modem features are designed to emulate a telephone set. The most sophisticated modems, in combination with a software package in an intelligent terminal, are capable of fully unattended operation. Modems designed for unattended, and many designed for attended, operation include these features:

- dial tone recognition
- automatic tone and dial pulse dialing
- monitoring call progress tones such as busy and reorder
- automatic answer
- call termination

Dial-up modems operate in a full-duplex mode. When two modems connect, they go through an elaborate exchange of signals to determine the features the other modem supports. Such features as error correction and compression are examined. High-speed modems test the line to determine the highest speed with which they can communicate and fall back to that speed.

The V.92 Standard

An ordinary telephone circuit is designed to support voice communications and has inherent characteristics that limit its bandwidth. For years, engineers believed that 33.6 Kbps was the maximum speed that a voice-grade circuit could carry. With the popularity of the Internet, companies began seeking ways to increase modem speeds. Engineers reasoned that in virtually every telephone connection, most of the circuit is digital and that only the local loop from the central office to the user's premise is analog.

As we will discuss in Chapter 5, every time a circuit undergoes an analog-to-digital conversion, a bit of the quality is lost. If the connection could be digital all the way except for the loop on the modem user's end of the circuit, only one analog conversion would take place, and the majority of the connection would be digital. Several companies began experimenting with an approach to increase modem speed. Compatible modem protocols would be used at each end of the connection, but the portion of the circuit from the central office to the ISP would be digital.

Several proprietary protocols came on the market before ITU-T approved the V.90 standard in 1998. The V.92 standard followed and improved on V.90 with innovations such as V.44 compression and a quick-connect procedure. Although 56 Kbps is possible, it is not always achieved. Poor phone-line quality limits speed and one end of the connection must be digital. Note that V.92 modems are asymmetric. They download at speeds up to 56 Kbps, but are limited to 33.6 Kbps in the upstream direction.

Private Line Modems

Analog private lines are rapidly becoming outdated, and private line modems are replaced by their digital equivalents, so little additional development work is being conducted. Different manufacturers use proprietary formats to encode the signal, compress data, and more important, communicate network management information. Private line modems can be classed as synchronous or asynchronous, half- or full-duplex, and two- or four-wire, with the latter being the most common. Circuit throughput can be improved by using data compression. With data compression and adaptive equalization, it is possible to operate at 19.2 Kbps or higher over voice-grade lines.

Special Purpose Modems

The market offers many modems that fulfill specialized requirements. This section discusses some of the equipment that is available:

- *Alarm reporting modem.* This class of modem has connections for accepting and relaying alarms from external devices. It may also monitor the ASCII bit stream of a channel looking for particular bit patterns. When alarms occur, the modem dials a predefined number.
- *DSL and cable modems.* These devices, discussed in more detail in Chapter 8, are used for access, primarily to the Internet.
- *Dial-backup modems.* A dial-backup modem contains circuitry to restore a failed leased line over a dial-up line. The restoral may be automatically initiated on failure of the dedicated line. The modem may simulate a four-wire private line over a single dial-up line, or two dial-up lines may be required.

- *Fiber-optic modems*. Where noise and interference are a problem, fiber-optic modems can provide high bandwidth at a moderate cost. Operating over one fiber pair, these modems couple directly to the fiber-optic cable.
- *Limited distance modems*. Many LECs offer limited distance circuits, which are essentially a bare nonloaded cable pair between two points within the same wire center. LDMs are inexpensive modems operating at speeds of up to 19.2 Kbps. Where LDM capability is available, the modems are significantly less expensive than long-haul 19.2-Kbps modems.

Data Service Units/Channel Service Unit (DSU/CSU)

A DSU/CSU connects DTE to a digital circuit. It provides signal conditioning and testing points for digital circuits. For example, the bit stream from a data device is generally a unipolar signal, which must be converted to a bipolar signal for transmission on a digital circuit. The CSU/DSU does the conversion, and provides a loop-back point for the carrier to make out-of-service tests on the circuit. Operating at 56 and 64 Kbps, DSUs are full-duplex devices. They are available for both point-to-point and multidrop lines.

Multiplexers and Concentrators

A data multiplexer subdivides a voice-grade line so it can support multiple sessions, usually from dumb terminals. Multiplexers come in two varieties. A standard data multiplexer carves the line into multiple channels. For example, the 2400-baud capacity of a voice-grade circuit could be divided into 16 channels of 150 baud each, with more capacity than a typist can use. The nature of many data applications is such that the terminals are idle a great deal of the time. With a straight multiplexer the idle time slots are wasted. A *statistical multiplexer* is able to make use of this time by assigning time slots as necessary to meet the demand. A typical statmux might provide 32 time slots on a single voice-grade line.

A *concentrator* is similar to a multiplexer, except that it is a single-ended device. At the terminal end, devices connect to the concentrator exactly as they would connect to a multiplexer, and the concentrator connects to the facility. At the host end, the facility connects into the host or front-end processor. A concentrator matches the characteristics of the host processor.

The primary application for multiplexers is in data networks that use asynchronous terminals. Since many of these devices cannot be addressed and have no error correction capability, they are of limited use by themselves in remote locations. The multiplexer provides end-to-end error checking and correction and circuit sharing to support multiple terminals. Although multiplexers are still available, they are being displaced by local area networks linked over digital circuits.

LAN Equipment

Much of the hardware we have discussed in this section is of limited applicability in today's network because it is obsolete, replaced by LANs and equipment for interconnecting them. We will go into these devices in considerably more detail in subsequent chapters, but we discuss them here briefly to complete the equipment picture and to prepare for the protocol discussion that follows in Chapter 6.

Hubs

The earliest LAN segments used coaxial cable as a transmission medium. Coax is bulky and unwieldy and is now obsolete for LAN use. Modern networks use unshielded twisted-pair (UTP) wire that is carefully designed and constructed to support data communications up to 1 Gbps over distances of 100 m. See Chapter 9 for additional information on the wiring infrastructure.

At first, stations were connected together with a central multi-port hub. Many hubs still exist, but they are being phased out because they have a major drawback. All of the stations connected to the segment share the bandwidth and contend with one another for access. When stations attempt to transmit simultaneously, they collide, their transmissions are mutilated, and they must retransmit the frame. When the traffic reaches the point of excessive collisions, throughput drops off and the LAN must be broken into smaller segments. This is accomplished by means of a *bridge*.

LAN Bridges

A LAN bridge is a two-port device that interconnects two segments. Its method of operation is simple. By listening to traffic on the network, it learns which MAC addresses belong to which segment and builds a table. If the sending and receiving addresses are on the same segment the bridge ignores the frame. If the addressee is on the other segment, it lets the frame across the bridge. If the bridge does not have the addressee in its table, it broadcasts a query on both ports. Bridging has further limitations that we discuss in later chapters, among which is its two-port limit. The solution is to eliminate both bridges and hubs by using a *switch*.

Ethernet Switches

An Ethernet switch is, in effect, a multi-port bridge. Each station is assigned to a port, so potential collisions are eliminated. The switch learns the station's MAC address on each port and connects the sending and receiving ports long enough to pass a frame. Some LANs share switch ports with hubs, but the cost of switches is so low that sharing ports is usually more trouble than it is worth. Switches are effective devices that nearly every LAN uses, but they have a drawback of their own: they cannot handle more than one route between the originating and terminating ports. For this, we need a router and a different addressing scheme. Switches operate on the MAC address, but packet flows require an IP address.

Routers

Routers are the workhorses of the Internet. They are specialized computers that connect a user's LAN to the Internet, and within an IP network they consult routing tables to determine which of their alternative routes is the most effective one to carry a packet. They are more expensive than switches, and although they can be used in a LAN, their more common use is in the WAN.

DATA COMMUNICATIONS APPLICATION ISSUES

Several clear trends are shaping data communications networks. The most significant is a decline in proprietary protocols in favor of TCP/IP. The shift to open protocols gives users more vendor choices, which, in turn, lowers costs. Terminals are disappearing as centralized databases move from mainframes to servers. The dumb terminal of the past is now a desktop computer that has internal processing power. As a result, popular office applications run on the desktop machine and the database runs on a server. Client software on the desktop computer communicates with the database. In some cases the client is proprietary, while in others it is a Web browser. All of this renders the central computer–dumb terminal combination obsolescent. Even operating systems are succumbing to the trend toward openness as Linux migrates to the desktop and increases in popularity as a server operating system. Manufacturers and developers, losing their proprietary advantages, must distinguish themselves by providing additional features, services, and improved setup routines.

The trend away from mainframes and terminals drives a similar transition in the network. The applications for conventional packet switching and message switching have shrunk to insignificance. Point-to-point circuits are still the choice for many enterprise networks, but where in the past multiplexers were employed to subdivide dedicated bandwidth, now bandwidths are increasing to support the demands of the desktop and server environment.

Much of the discussion in this chapter has revolved around analog transmission and modems. Although private line modems are fading from significance, dial-up modems are still widely used. They are included with every laptop and most desktop computers, and are used as backup for data private lines. By far the bulk of data traffic uses digital transmission, the subject of the next chapter.

CHAPTER 5

Pulse-Code Modulation

All voice sessions begin and end as an analog signal in a telephone handset, but between central offices the transport is digital. The analog portion of the connection may be no longer than the length of the handset cord because PBXs and ISDN convert the signal to digital in the telephone set. Plain old telephone service, known in the industry as POTS, is analog all the way to the central office. If the switching system is digital its line interface digitizes the voice. If the system has an analog switching matrix, the voice is converted in the trunk circuit and travels to a distant central office as a digital signal. The process for digitizing an analog signal is known as pulse-code modulation (PCM).

Alec Reeves, an ITT scientist in England, patented PCM in 1938. Although the system was theoretically possible then, it was not technically feasible. Pulse generating and amplifying circuits required vacuum tubes and their size and power consumption consigned PCM to the shelf for another 20 years. With advances in solid-state technology, PCM became commercially feasible in the 1960s. Large-scale integration continued the reduction in cost, size, and power consumption to the point that today even PCs can digitize voice.

Digital technology has replaced analog in all modern switching and transmission systems. Central offices, PBXs, and most key systems use digital technology. The advantages are substantial. Analog circuits do not lend themselves well to integrated circuitry, whereas digital circuits use the same manufacturing techniques that have resulted in dramatic cost reductions in all electronic devices. Digital switches can interface digital multiplexing systems directly without the need for an analog-to-digital conversion. Digital circuits are also less susceptible to noise. In analog circuits, noise is additive, increasing with system length, but in digital carrier the signal is regenerated at each repeater. Over a properly engineered system, the signal arrives at the receiving terminal with unimpaired quality and a negligible error rate. Furthermore, while analog signals occupy a fixed amount of bandwidth, digital signals can be compressed into a fraction of their original bandwidth.

This chapter explains the process for converting analog signals to digital. We discuss T1/E1 carrier. This system forms the foundation for the digital hierarchy SONET/SDH, which interconnects the world over the fiber-optic backbone. The chapter concludes with a look at voice compression and some of the equipment that carriers and enterprise networks use to configure and manage digital systems.

DIGITAL CARRIER TECHNOLOGY

The basic digital multiplexing system is known as T1 carrier in North America and E1 in Europe. The T carrier name came from the Bell System's carrier nomenclature, which assigned letters to each successive model that Bell Labs designed. The T1 carrier system consists of 24 channels, each of which occupies a bandwidth of 64 Kbps. T1 samples a voice signal, converts it to an 8-bit coded digital signal, and interleaves it with 23 other voice channels. The E1 system uses the same sampling method, but applies 32 channels to the line, of which the first channel maintains synchronization and passes control information and the 16th channel is used for signaling. Although this discussion refers to voice channels, devices such as routers use nonchannelized bandwidth. T1/E1 multiplexers can channelize part of the bit stream and leave the remainder unchannelized for data.

The initial T1/E1 application was on copper cable pairs with regenerators spaced at intervals of 6000 ft (1800 m). This spacing coincided with the spacing of load coils, which are applied to voice circuits to improve transmission. When digital carrier became available, cable between wire centers was deloaded, load coil cases were removed, and regenerators were installed in the space. As the popularity of digital carrier grew, digital radio entered the market to further the transition from analog. Fiber optics, with its higher quality and lower space requirements, provided the final impetus for the transition from analog to digital.

Analog To Digital Conversion

A five-step process converts analog to digital:

- Sampling
- Quantizing
- Encoding
- Companding
- Framing

According to Nyquist's Theorem, if an analog signal is sampled at a rate twice its highest frequency, the samples contain enough information to reconstruct the original signal. Communication channels filter the voice to a nominal bandwidth of 4000 Hz, so twice the highest frequency is a sampling rate of 8000 times per second.

FIGURE 5-1

Sampling and Quantizing

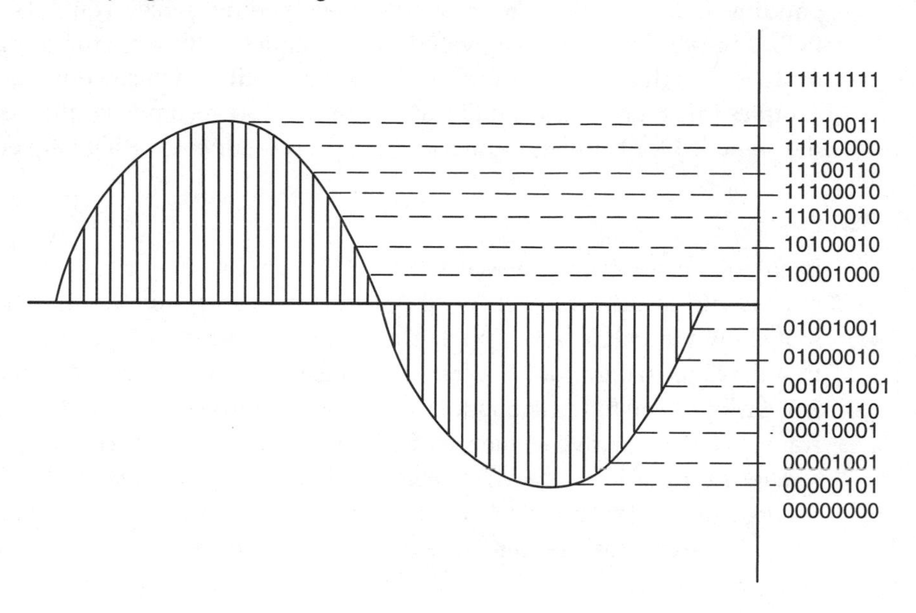

A process called *quantizing* scales the amplitude of each sample to one of 2^8 or 256 discrete steps. Figure 5-1 illustrates the process. The output of the encoder is a stream of octets, each representing the magnitude of a single sample.

The quantizing steps, as shown in Figure 5-2*a*, do not precisely represent the original waveform. The error is audible as quantizing noise, which is present only when a signal is being transmitted. Quantizing noise is more audible with low-amplitude signals than with high. To compensate, the encoder divides low-level signals into more steps and high-level signals into fewer steps as shown

FIGURE 5-2

Companding in a PCM Channel

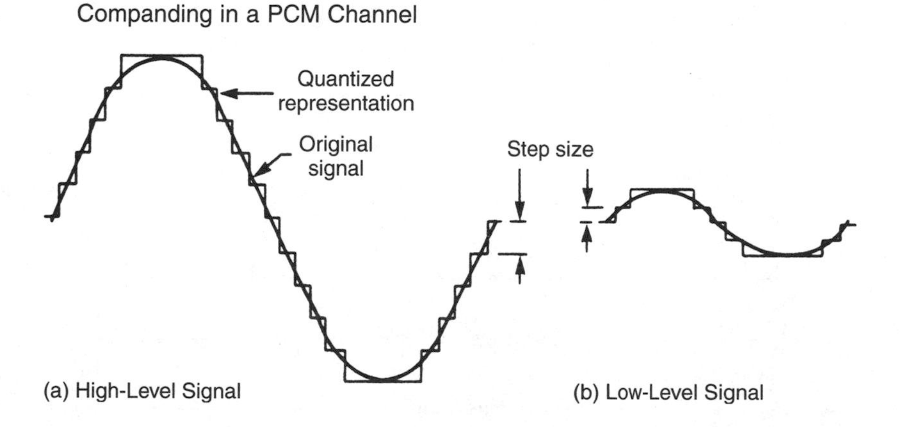

in Figure 5-2*b*. The decoder expands the signal back to its original form. The compression and expansion process is called *companding*. In the United States, companding follows a formula known as μ-law coding, which converts 14-bit linear PCM samples into 8-bit compressed PCM samples. In Europe, the companding formula is a slightly different form called A-law coding, which converts 13-bit PCM samples into 8-bit compressed PCM samples. ITU-T recommendation G.711 defines the standard for both algorithms and the process for converting between them.

Framing

Each voice channel generates a bit rate of 64 Kbps (8000 samples per second × 8 bits per sample). The 24 channels in the North American system result in the frame format shown in Figure 5-3. A single framing bit is added to generate a 193-bit frame that is 125 μs in duration. The frame repeats 8000 times per second for a line rate of 1.544 Mbps. The framing bits follow a fixed pattern of zeros and ones through 12 frames. This repetitive sequence of 12 frames is called a *superframe* (SF). This system of multiplexing is known as *byte-interleaved* or *synchronous* TDM. Each channel occupies a constant position in the transmission frame known as a *time slot*.

E1 carrier has the same 64 Kbps channel, but multiplexes 32 rather than 24 channels for a 2.048 Mbps bit rate. Of the 32 channels, 30 are used for information channels, one channel is used for frame alignment, and one for signaling. The separate signaling channel enables E1 to provide a full 64 Kbps of bandwidth per channel. In T1, the least significant bit is sometimes used for signaling, as we discuss in the next section.

Companding and bit-rate differences make North American and European digital carrier systems incompatible. This incompatibility was of little consequence when the systems were developed in the 1960s because they were not interconnected. With the development of undersea fiber-optic cable, however, the issue of end-to-end connectivity of incompatible systems became an issue. This became one of the driving forces behind the synchronous optical network (SONET), which is discussed later in this chapter. In Europe, SONET is known as the synchronous digital hierarchy (SDH).

FIGURE 5-3

T1 Frame

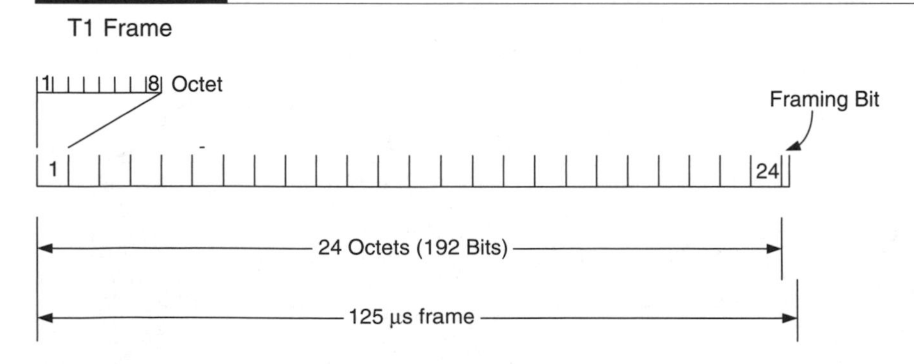

Bit-Robbed Signaling

The original application of T1 was for interoffice trunking in metropolitan areas. Since signaling is a binary function, it is feasible to use a portion of the T carrier signal to convey the on-hook or off-hook status of the channel. See Chapter 12 for further discussion of how signaling systems work. SF framing, also called D4 framing, uses the least significant bit in every sixth frame for signaling. The technique is known as *bit robbing* and the bits are called the A and B bits. The A bit is robbed from the sixth frame and B bit is robbed from the 12th frame. The distortion resulting from bit robbing has no effect on voice or modem data, but the forced bit error prevents the use of all eight bits for digital data. Therefore, digital devices use only seven of the eight bits, which results in a bandwidth of 56 Kbps. Special dataport channel units provide access to the usable bandwidth.

Extended Superframe (ESF)

Several factors make SF framing obsolete. For data circuits the ability to use only seven of the eight bits per channel reduces the capacity by 12 percent. Although analog voice signals are unimpaired by bit-robbed signaling, ISDN requires 64 Kbps channels. Moreover, local switching systems now use external SS7 signaling almost exclusively, which means they no longer need link-by-link signaling. ANSI standard T1.403, introduces a format, known as *extended superframe* (ESF), which provides 64 Kbps clear channels. Under ESF, the 8000 bps framing signal, also called the Fe channel, is given added functions. To detect errors, 2 Kbps is used for 6-bit CRC. A 4 Kbps facility datalink (FDL) is used for end-to-end diagnostics, network control, and maintenance functions such as forcing loopback of a failed channel. The remaining 2 Kbps are used for framing and signaling.

The CRC code detects, but does not correct, errors. The CRC code is calculated at the source and then again at a terminal or intermediate point. If an error is detected, the equipment can flag the fault before a hard failure occurs. The receiving equipment calculates the performance of the facility from the CRC results and stores it or sends the information back to the originating equipment over the FDL.

DIGITAL TRANSMISSION FACILITIES

The basic digital transmission facility is a T1/E1 line, which operates over twisted-pair wire and terminates in an office repeater at each end. The function of the office repeater is to match the impedance of the line to the equipment, to feed power to the repeaters, and to maintain line synchronization. The line repeaters, which are spaced every 6000 ft (1800 m), regenerate the incoming pulses and feed them to the next repeater. When T1/E1 first went in operation, most interoffice trunking was carried on loaded copper cable. The load coils were placed with 3000-foot end sections facing the central offices, and intermediate load coils at 6000-foot spacing. If an end section was less than 3000 ft, the line was built out electrically to the required length. Loading, as discussed in Chapter 7, improves voice quality up to 4 KHz but cuts off higher frequencies. Most copper cables in the interoffice trunking network have subsequently been replaced with fiber optics, eliminating repeaters.

T1/E1 is also used extensively for customer premise applications. Most large users connect their PBXs to the IXC with T1/E1. ISPs connect to their larger users with direct circuits of T1/E1 or higher bandwidths. T1/E1 is used as the access medium for IP and frame relay data networks, and point-to-point T1/E1s are used for LAN-to-LAN connections, voice and data private lines, and a host of similar applications. At the customer end, the circuit terminates in a device known as a *channel service unit* (CSU).

The T1/E1 Carrier System

Figure 5-4 is a block diagram of a T1/E1 carrier system. At the central office the line may terminate in a digital switch as shown in the figure, in a channel bank if the switch is analog, in a multiplexer, or it may be connected straight through the central office to another location. The line usually terminates on jacks in a *digital signal crossconnect* (DSX) panel. The DSX jack is a test point that enables rapid patching between lines. Figure 5-5 is a photograph of a working DSX-1 jack field. T1/E1 systems to customer premises may terminate on a variety of devices. PBXs, routers, and T1/E1 multiplexers or integrated access devices (IADs) are typical. In the central office, the line is often extended to an IXC, a CLEC, or to another customer location. In such applications the line is terminated in a multiplexer or a *digital crossconnect system* (DCS), which is a high-speed electronic switch.

Office Repeaters and Channel Service Units

An office repeater terminates the T1/E1 line at each central office. In the receiving direction, the office repeater regenerates the signal. Its transmit function is to

FIGURE 5-4

T1 Carrier Block Diagram

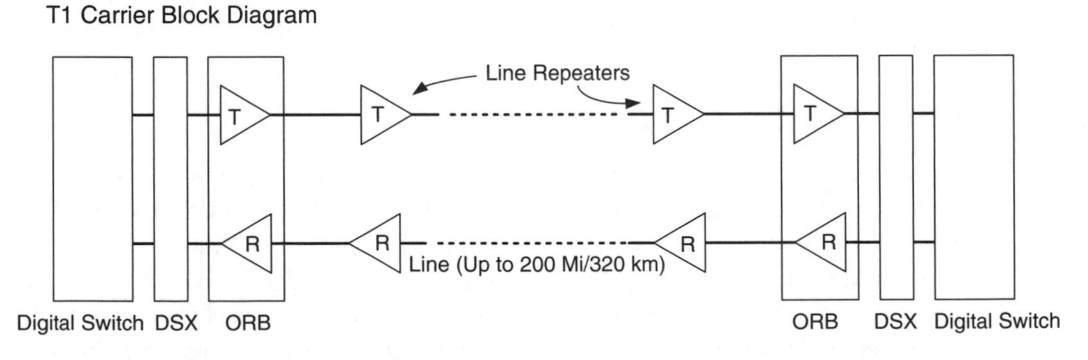

FIGURE 5-5

DSX-1 Panel (Photo by Author)

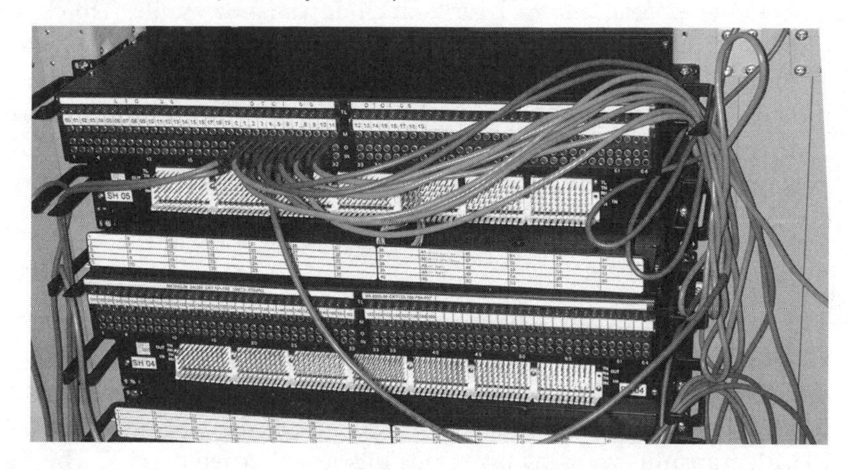

couple the bipolar signal to the line, feed power to the line repeaters, and ensure that the line has the required density of ones bits. A CSU, which is similar to an office repeater, terminates the far end of a T1/E1 line feeding customer premises. The CSU fulfills the following functions:

- Terminates the circuit, including lightning and surge protection
- Regenerates the signal
- Loops the digital signal back to the originating end on command
- Monitors the incoming line for bipolar violations or CRC errors
- Generates a signal to maintain synchronization on the line if the terminal equipment fails
- Maintains the ones density requirement of the line
- Provides signal lamps and line jacks for testing and monitoring
- Provides line build-out (LBO) if necessary

Channel Banks

A T1 channel bank consists of 24 channels called a *digroup*. Some manufacturers package two digroups in a 48-channel framework, which is often known as a D4 channel bank. The 24 or 48 channels share a common power supply and other common equipment. Most channel banks have a rack-mounted metal framework with backplane wiring designed to accept plug-in common equipment and channel units.

A variety of plug-in channel units are available to provide special transmission functions. Numerous signaling options are also available. For example, foreign exchange (FX) channels are often used to connect analog devices to the central office over a T1/E1 line. At the FX subscriber end the channel bank provides dial tone and ringing and detects off-hook circuit seizures. A mate unit at the office end relays the on/off-hook signal from the subscriber and detects dial tone from the central office, relaying it to the subscriber end.

Other special channel types include dataport units, which accept direct digital input from data devices, and program channel units, which replace two or more voice channel units. Program units use the added bit streams to accommodate a wider channel for use by radio and television stations, wired music companies, and other applications that require a wide-band audio. Program channels with 5 KHz bandwidth replace two-voice channels, and 15 KHz units replace six-voice channels.

T Carrier Lines

T1/E1 carrier lines can operate on twisted-pair wire for about 200 miles (320 km). Circuits that long are rare because most private and common carrier applications are deployed over digital microwave or fiber-optic facilities. A T1 line signal is bipolar, as shown in Figure 5-6. A bipolar signal, also called *alternate mark inversion* (AMI), transmits zeros as 0 V. Ones bits are alternately ±3 V. The bipolar signal offers two benefits. First, the line-signaling rate is only half the rate of 1.544 Mbps, because in the worst case of a signal composed of all 1s, the signal would alternate at 772 Kbps. The second advantage is the ability of a bipolar signal to detect line errors in SF frames. If interference or a failing repeater adds or subtracts ones bits,

FIGURE 5-6

T1 Carrier Line Signals and Faults

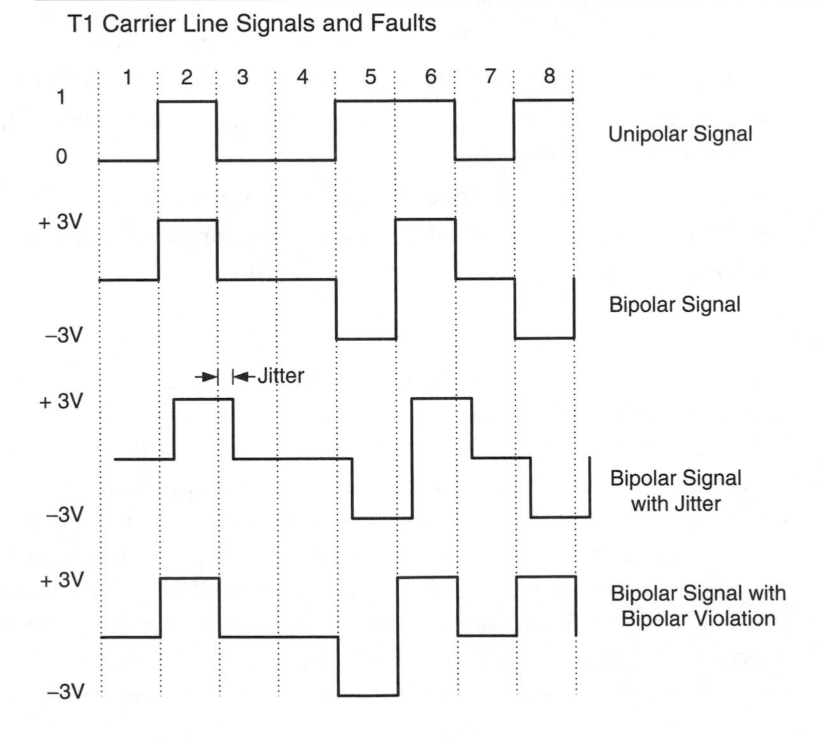

a bipolar violation results, which indicates a fault. A bipolar violation occurs when two ones bits of the same polarity arrive in sequence. Most terminating equipment have a BPV light that flashes if violations are occurring.

T1/E1 relies on ones bits to maintain synchronization. A digital carrier system must be able to support any bit pattern, including long strings of zeros. The SF specification permits a maximum of 15 consecutive zeros, in which case the CSU stuffs in a bit pattern with sufficient ones to maintain synchronization. ESF systems replace the straight bipolar signal with a coding scheme known as *bipolar with 8-zero substitution* (B8ZS). In this method, any string of eight zeros is replaced with an intentional bipolar violation at the fourth and seventh bits. The CSU detects the BPV and replaces it with a string of eight zeros. The B8ZS coding scheme is not compatible with the earlier T-carrier lines, but most modern regenerators, office repeaters, and CSUs are ESF-compatible.

Line Repeaters

Line repeaters are mounted in apparatus cases, which are watertight for mounting on poles and in manholes. Repeaters analyze each bit in the received signal, determine whether it is a zero or a one, regenerate the pulse, and put it in the output time slot. Repeaters are needed only on copper cable. Fiber-optic lines can easily span the distance between central offices without regeneration.

Incoming pulses are received in one of three states—plus, minus, or zero. If the incoming pulse exceeds the plus or minus threshold, the repeater generates a one output pulse. Otherwise it registers a zero. Phase deviations in the pulse, which are additive along a T-carrier line, are known as *jitter*. Excessive jitter, as illustrated in Figure 5-6, can cause errors in data signals.

The transmit and receive paths of T1/E1 signals must be isolated to prevent crosstalk coupling. If excessive crosstalk occurs between the high-level pulses of a repeater's output and the low-level received pulses of adjacent repeaters, errors will result. The transmit and receive paths are isolated by assigning them to separate cables or to partitions within a specially screened cable. In larger cables, groups of pairs can be shielded by physical separation.

T1/E1 Carrier Synchronization

T1/E1 signals are synchronized by extracting timing pulses from the incoming bit stream. The repeaters detect the transition of each ones bit and use this as a clock signal to keep the system synchronized. To understand why synchronization is important, consider the timing diagram in Figure 5-7. If the clock of one terminal device runs slightly faster or slower than the clock at the other end, a point will be reached where an extra bit will be inserted or a bit will be lost, and the frame will lose synchronization momentarily. This condition is known as a *slip*.

Slips have a negligible effect on voice circuits, but framing loss causes data errors. To prevent slips and consequent data errors, the entire digital network is

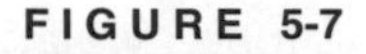

FIGURE 5-7

T1 Carrier Line Slip

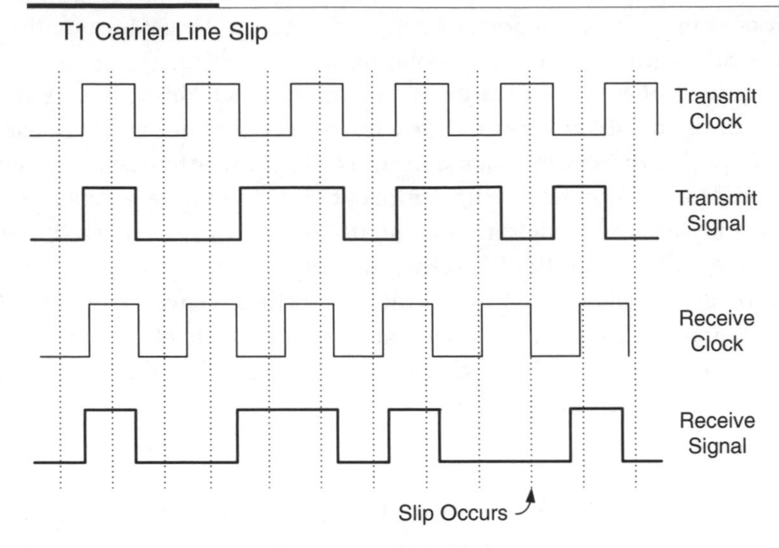

synchronized. In a point-to-point T1/E1 circuit, one device is configured as the master and the other is a slave. A precise clocking rate is not important if the two stand alone because the slave extracts clocking from the ones bits in the received signal to maintain synchronization. When the network is connected to a common carrier network, as with a long distance service, it is vital that the entire network slave from the common carrier because the carrier is tied to a higher level in the national clocking structure.

Nationally, digital network synchronization is maintained through a four-level hierarchy of clocks. ANSI standard T1.101 defines the four levels, which are known as Stratum 1 through Stratum 4 timing levels. Stratum 1 clocks drive accurate oscillators using cesium and rubidium clocks with a maximum drift of 1×10^{-11}. Stratum 2 clocks, which are used in common carriers' toll centers, are slightly less accurate—1×10^{-10} in the short term—but if synchronization with the Stratum 1 clock is lost it will still maintain an acceptable amount of stability. Stratum 3 and 4 are less stable and generally depend on synchronization from higher levels to maintain an acceptable degree of performance. Private network equipment generally employs Stratum 3 or 4 clocking.

THE DIGITAL SIGNAL HIERARCHY

To this point we have discussed carrier systems, with only a brief mention that T1/E1 is but one level in a hierarchy of digital signals. The lowest level in the hierarchy is DS-0, which is a 64 Kbps signal in both systems. The digital signal level can be multiplexed up to much higher levels. Bell Laboratories defined the original DS hierarchy for North America. The plan was to apply digital signals to metallic circuits—twisted-pair wires for DS-0 and DS-1, and coaxial cable for DS-3 (45 Mbps) and DS-4 (274 Mbps). The development of fiber optics intervened, however, providing bandwidths so far in excess of DS-4 that the hierarchy became obsolete.

The European hierarchy, which was developed independently from North American standards, begins by multiplexing 32 DS-0s into an E1 line at 2.048 Mbps. The E1 line is also known as CEPT1. The North American and European hierarchies do not mesh comfortably despite their use of the same sampling rate. SONET/SDH, which is discussed later in this section, resolves this incompatibility. Table 5-1 shows the signal levels and bit rates of the North American and European digital hierarchies.

As we have discussed, a DS-1 signal can be applied directly to a T1 carrier line. The next level in the hierarchy is DS-2, which consists of four DS-1s. DS-2 was intended for twisted-pair cable, but is rarely used because fiber optics makes it obsolete. It is important to understand DS-2, however, because it uses a bit-by-bit interleaving process that leaves the four DS-1s asynchronous with respect to each other. The next level in the hierarchy is DS-3, which multiplexes seven DS-2s into a 44.736 Mbps bit stream. The bit rate of the 28 underlying DS-1s is 43.232 Mbps; the extra bits are used for framing and overhead. The DS-1s cannot be extracted from a DS-2 or DS-3 by their bit position. The only method of dropping bandwidth at intermediate points is with multiplexers.

The bandwidth of even the earliest fiber-optic cables far exceeded the DS-4 bit rate, and there was no standard for higher rates. Consequently, manufacturers developed proprietary multiplexing methods based on DS-3 multiples. The fiber transmission systems required additional overheads for synchronization, alarming, and order wires. These proprietary systems were not end-to-end compatible. The lack of compatibility plus the need for international connectivity led to the development of the SONET/SDH hierarchy.

TABLE 5-1

Digital Signal Hierarchy

North American Hierarchy			European Hierarchy		
Signal	Bit Rate	Channels	Signal	Digital Bit Rate	Channels
DS-0	64 Kbps	1 DS-0	64-Kbps	64 Kbps	1 64-Kbps
DS-1	1.544 Mbps	24 DS-0s	E1	2.048 Mbps	1 E1
DS-2	6.312 Mbps	4 DS-1s	E2	8.45 Mbps	4 E1s
DS-3	44.736 Mbps	28 DS-1s	E3	34 Mbps	16 E1s
DS-4	274 MBps	168 DS-1s	E4	144 Mbps	64 E1s

Synchronous Optical Network/Synchronous Digital Hierarchy (SONET/SDH)

SONET/SDH is a hierarchy of optical standards that has replaced the previous digital signal hierarchy above DS-3. The hierarchy has no defined limit; it can be extended as fiber-optic bandwidth increases. Today, the hierarchy is defined to STS/OC-768, which is approximately 40 Gbps. Table 5-2 shows the levels of SONET/SDH that are currently defined. STS/OC-768 is at the outer reaches of fiber capabilities and is not yet widely deployed.

The lowest SONET level is STS-1, which has a line rate of 51.840 Mbps. STS stands for synchronous transport signal. It defines the electrical characteristics of a SONET signal. The optical signal is characterized by corresponding optical carrier (OC) levels. Although the STS and OC designations are distinctly different, the industry tends to refer to SONET by its OC identification. The SDH levels are known as synchronous transport module (STM). As the table shows, an STM-1 and an OC-3 have the same line rate of 155.520 Mbps. As a result, European and North American signals are compatible at STS-3 and above.

In addition to resolving mid-span meet and international compatibility issues, SONET/SDH brings several other benefits. A major advantage lies in the fact that it is a synchronous system. As we discussed earlier, T1 is synchronous, but as it is multiplexed up to T-3, it becomes bit-interleaved. As a result, multiplexers are required to extract a single DS-0 or DS-1 from the DS-3 bit stream. SONET/SDH is byte synchronous up to the highest multiplexing level. Individual DS-1s and DS-0s can be extracted from the bit stream with inexpensive add-drop multiplexers. SONET/SDH offers the following advantages:

- ◆ It offers bandwidths more commensurate with today's fiber-optic systems than the older digital signal hierarchy.
- ◆ It merges North American and European hierarchies at higher rates.
- ◆ It offers multi-vendor interoperability over fiber-optic systems, replacing proprietary frame overheads that prevent end-to-end interoperability.

TABLE 5-2

SONET/SDH Hierarchy

SONET Signal	Line Rate (Mbps)	SONET Capacity	SDH Signal	SDH Capacity
STS/OC-1	51.840	28 DS-1s or 1 DS-3	STM-0	21 E1s
STS/OC-3	155.520	84 DS-1s or 3 DS-3s	STM-1	63 E1s or 1 E4
STS/OC-12	622.080	336 DS-1s or 12 DS-3s	STM-4	252 E1s or 4 E4s
STS/OC-48	2488.320	1344 DS-1s or 48 DS-3s	STM-16	1008 E1s or 16 E4s
STS/OC-192	9953.280	5376 DS-1s or 192 DS-3s	STM-64	4032 E1s or 64 E4s
STS/OC-768	39,813.120	21,504 DS-1s or 768 DS-3s	STM-256	16,128 E1s or 256 E4s

- It offers centralized end-to-end network management and performance monitoring.
- It permits add–drop capability with inexpensive multiplexers.

SONET Framing

SONET starts with the basic DS-0 signal: an 8-bit octet repeated every 125 ms, and multiplexes it upward to an STS-1 frame. Figure 5-8 shows an STS-1 frame, which consists of nine rows and 90 columns. Note that each row has three octets of overhead plus 87 octets of payload. Nine rows times 90 octets per row equals a single frame of 810 octets, which, when repeated 8000 times per second, comprises the 51.840 Mbps STS-1 signal. Multiple STS-1 signals interleave to form the higher levels. At each level of the hierarchy, the payload and overhead octets are exact multiples of STS-1. For example, an STS-3 signal has 270-octet rows, of which nine octets are overhead, and 261 octets are payload.

For payloads below STS-1, SONET defines synchronous formats known as *virtual tributaries* (VTs). DS-1s, for example, are mapped into VTs of 1.728 Mbps, and E1s into VTs of 2.304 Mbps. To compensate for frequency and phase variations between either STS or VT payloads, SONET/SDH uses a concept known as *pointers* to align the payload and minimize jitter.

Figure 5-9 shows graphically how VTs and STSs are multiplexed to higher orders. Tributaries such as DS-1 and E1 are mapped into VTs by adding path overhead and justification bits. VTs and STSs are aligned by including pointers in the path overhead to locate the first byte of the information payload. To map

FIGURE 5-8

SONET STS-1 Frame

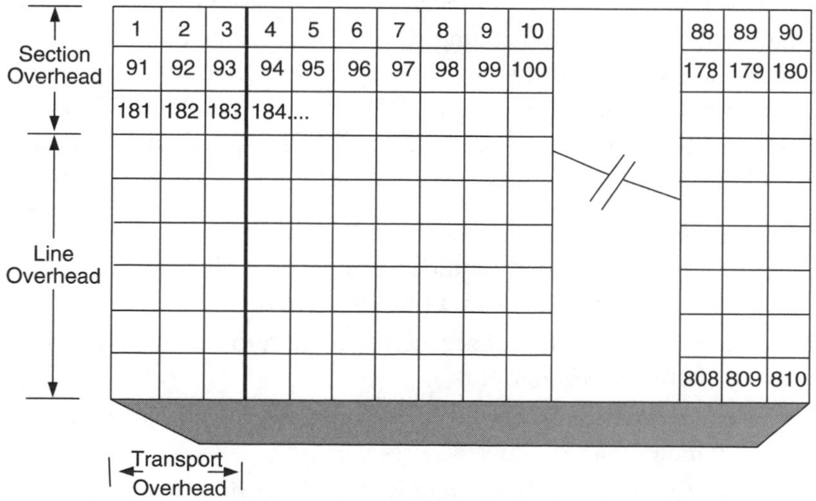

Section Overhead
Line Overhead
Transport Overhead
1 2 3 4 5 6 7 8 9 10 ... 88 89 90
91 92 93 94 95 96 97 98 99 100 ... 178 179 180
181 182 183 184...
808 809 810

FIGURE 5-9

SONET/SDH Multiplexing Hierarchy

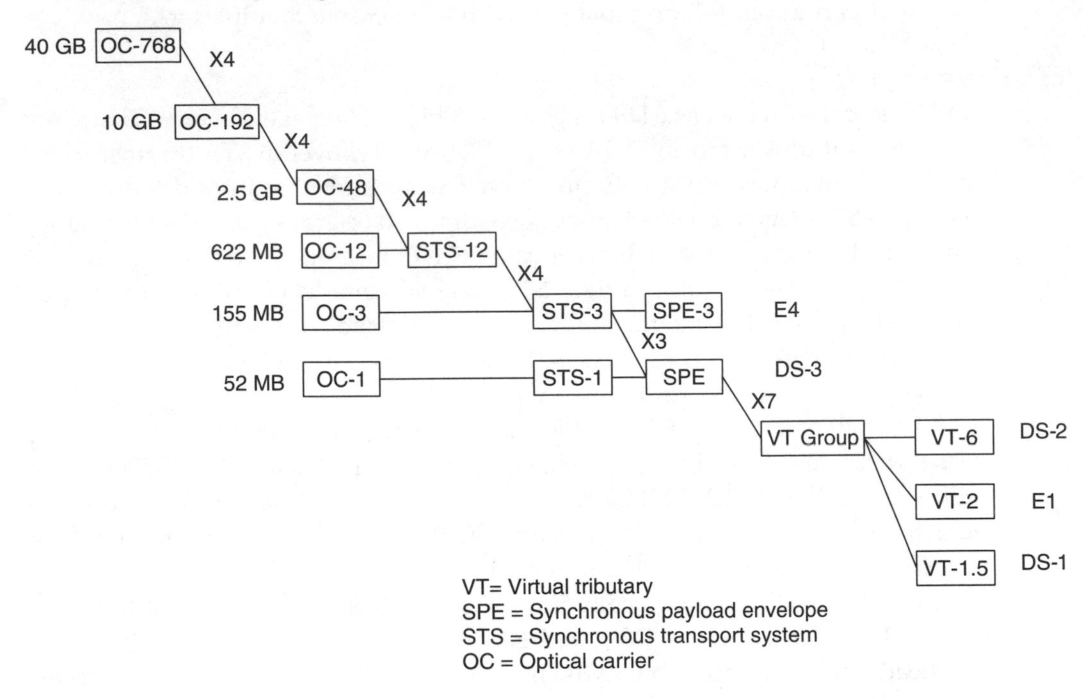

the tributary signals into the STS-1 frame, spare capacity in the path overhead is filled with stuffing bits to fill out the frame.

Overheads

The overhead bits in SONET/SDH frames are used for operations, administration, maintenance, and provisioning (OAM & P) in four levels: path, line, and section overhead signals plus the photonic layer. Note that these four levels are contained within the physical layer of the OSI model. SONET/SDH is a physical layer protocol that moves bits across a fixed path. It does not have any switching or routing capability. The four levels of the SONET/SDH signal are:

- *Photonic*, which is concerned with the optical transmission path. Physically, the photonic layer consists of lightwave terminating equipment, lasers, and photonic diodes.
- *Section*, which is concerned with monitoring and administration between regenerators.
- *Line*, which is concerned with monitoring and administration across maintenance spans.
- *Path*, which is concerned with end-to-end transmission.

Section overhead carries framing and error-monitoring signals that all devices on the line can monitor and process. Line overhead extends between central offices. It carries performance information plus automatic protection switching and control information. Path overhead provides performance and error information plus control signaling between the endpoints.

LECs and IXCs use SONET/SDH to connect digital services between central offices and directly to end-users. Enterprise networks also employ SONET/SDH over private and public fiber-optic systems. SONET/SDH is often deployed in a ring configuration with two fiber-optic pairs carrying service in opposite directions. The multiplexers can detect a loss of signal and loop back over the protection pair, providing automatic restoration with only a small loss of data. The ring topology is discussed in more detail in Chapter 17, Optical Networking.

DIGITAL CROSSCONNECT EQUIPMENT

SONET/SDH facilitates *grooming*, which is the process of combining and configuring digital signals to optimize use of the bandwidth. Circuits can be groomed through physical DSX panels or through electrical DCSs.

Digital Crossconnect Panel

Physical crossconnect positions are known as DSX panels. A DSX-1 operates at DS-1 and a DSX-3 at DS-3. Figure 5-5 shows a DSX-1. The physical transmit and receive legs of the digital circuit are wired through jacks in this panel to provide test access and to enable the equipment to be patched to another T1/E1 carrier line. Patching is used to rearrange circuits and to connect service to a spare line when repeaters fail. The ability to patch to spare lines is a vital part of service restoration in common-carrier networks. In private networks the DSX panel is useful for temporary facility rearrangement. For example, a T1/E1 facility might be used for a voice tie line during working hours and patched to a high speed multiplexer for after-hours data transfer.

Digital Crossconnect System

DCS, functioning at the electrical level, is an integral part of the digital backbone. It is the electronic alternative to physical crossconnects. DCS provides the capability of connecting and rearranging DS-0s, DS-1s, and DS-3s in software. Bandwidth is dropped and added and connected as if it were a wired connection. DCS connections are provisioned as opposed to being made on the fly. Once a crossconnect is made, it remains in service until further activity rearranges it. Figure 5-10 shows a Lucent Digital Access Crossconnect System.

As an example of DCS application, suppose an IXC has elected to centralize its switching for a region in a particular city. To be cost-competitive in the access

FIGURE 5-10

Lucent Digital Access Crossconnect System (Photo by Author)

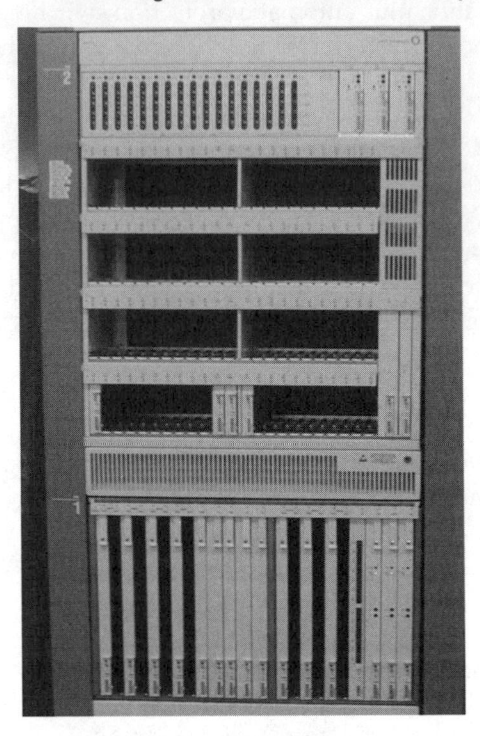

facilities with other IXCs, it establishes rating POPs in other cities. The IXC uses a DCS to combine T1/E1 access circuits from customers in the originating city to form a broadband signal that can be combined over a digital backbone to its switch. Crossconnects are also possible at the optical level, as discussed in Chapter 17.

The DCS system offers a high degree of flexibility and eliminates much of the labor associated with circuit rearrangements. Routing changes can be centrally controlled and even provisioned through a computerized operational support system (OSS). DCS also plays a key role in service restoration. When a major switch node fails, a DCS system can quickly reroute traffic around the point of failure. It also improves facility utilization by making use of idle capacity.

Most large LECs offer DCS as a service. Since the T1/E1 lines extend through the LEC's central office anyway, a subscriber can combine them with a centralized DCS. For example, an ISP could use the LEC's DCS to combine DS-1 access circuits into a DS-3 or OC-3 backbone. They could also be connected to different backbones to achieve diversity. The principal issue in using a public versus a private DCS is the degree of control the LEC provides over the DCS facility. If the LEC allows the network operator full control over circuit configuration, the use of private versus common carrier DCS is primarily an economic issue.

VOICE COMPRESSION

PCM, known in ITU terms as G.711, is an efficient encoding algorithm that provides excellent fidelity and clarity for a voice signal. Voice quality is based on the subjective rating of a large universe of samples in which users rate connections of varying quality. ITU-T recommendation G.114 covers voice quality standards. ITU has standardized two recommendations for measuring voice quality. Recommendation P.800 covers a mean opinion score (MOS) of voice quality. Voice samples are played over transmission media with various levels of loss, noise, and clarity to a group of users who score the connections on a scale of 1 to 5 with 5 high. A score of 4 or higher is defined as toll quality. P.861, perceptual speech quality measurement (PSQM) is a mechanized process using a computer to derive scores that correlate to MOS scores. PSQM is intended for the circuit-switched network and is not particularly reliable for some impairments that may arise on an IP network. Table 5-3 shows the MOSs of six compression standards. Compression and decompression algorithms require processing time. The time is insignificant on circuit-switched networks, but added to other delays in an IP network, the standards with the greatest compression may add an unacceptable amount of delay. The column on accumulation delay compares the standards. This section discusses two of the most common compression algorithms.

Adaptive Differential Pulse-Code Modulation (ADPCM)

An ADPCM transcoder doubles the capacity of a T1 line by encoding at 32 Kbps rather than 64 Kbps. ADPCM uses the same sampling rate as PCM, but instead of

TABLE 5-3

Quality Ratings of Voice Compression Standards

Compression Standard	Name	Payload Bit Rate	Accumulation Delay	Mean Opinion Score
G.711	Pulse-code modulation (PCM)	64 Kbps	0.75 ms	4.4
G.726	Adaptive differential pulse-code modulation (ADPCM)	16 to 40 Kbps	1.0 ms	4.2
G.728	Low-delay code excited linear prediction (LDCELP)	16 Kbps	2.5 ms	4.2
G.729A	Conjugate structure algebraic code linear prediction (CSACELP)	8 Kbps	10 ms	4.2
G.723	Multipulse maximum likelihood quantization (MPMLQ)	6.3 Kbps	30 ms	3.9
G.723	Algebraic code excited linear prediction (ACELP)	5.3 Kbps	30 ms	3.5

quantizing the entire voice signal, it quantizes only the changes between samples. A circuit known as an adaptive predictor examines the incoming bit stream and predicts the value of the next sample. ADPCM encodes the difference between the actual sample and the predicted sample into 16 levels. The encoder adapts to the speed of change in the difference signal: fast for speech-like signals and slow for data signals.

ADPCM, which ITU-T recommendation G.726 covers, can sample at various rates:

- Five bits = 40 Kbps
- Four bits = 32 Kbps
- Three bits = 24 Kbps
- Two bits = 16 Kbps

The most common implementation of ADPCM is 32 Kbps, which provides voice quality that is almost equal to full PCM.

Code Excited Linear Prediction (CELP)

CELP technology provides higher compression ratios than many of the other algorithms while maintaining satisfactory voice quality. Numerous standard and proprietary variations of CELP are in the market. ITU-T recommendation G.729A covers conjugate structure algebraic CELP. G.728 covers low-delay CELP. G.723.1, which is part of H.323 videoconferencing over IP, also includes CELP options at 5.3 and 6.3 Kbps. In addition, manufacturers have proprietary versions of CELP.

CELP operates in a manner similar to ADPCM in that the algorithms rely on analyzing the relationship between multiple voice samples. A "codebook" at each end stores samples, which are compared to find a match. The matched sample and deviation are transmitted across the circuit and filtered with a model of the speaker's voice. CELP is optimized for voice. Data signals are transmitted at a lower rate, generally about 1200 bps.

One compression technique is silence compression. During normal speech activity, regular CELP coding is used. During silent periods a silence insertion descriptor (SID) is transmitted at a lower rate. The SID includes a comfort noise generator so the circuit sounds alive.

PCM APPLICATION ISSUES

Several initiatives are underway that will change the shape of public networks in the future, but for now, most voice and data traffic rides on the TDM infrastructure at bandwidths up to the capacity of SONET/SDH. In developed countries the fiber-optic backbone extends down to the central office level in virtually every major metropolitan area. SONET/SDH networks are usually deployed in a ring topology. Inherent in many multiplexers is the ability to detect a fiber break and heal the network with minimal interruption.

Where the need for bandwidth is sufficient to bring fiber to the building level, bandwidth can be delivered to equipment rooms in the same increments as in the metropolitan area. Copper wire predominates in the local loop and to the desktop, but the vision of unlimited bandwidth wherever someone is willing to pay for it is not far in the future. Meanwhile, the technology we have described in this chapter is the default method of building the physical layer. This section discusses applications for the digital network.

Carrier Backbones

The carriers are the major bandwidth consumers. At the fiber level, bandwidth is increasingly sold in lambdas, which is an entire optical wavelength. Dense wave division multiplexing, which is discussed in Chapter 17, applies multiple light wavelengths to the fiber. Each wavelength can be subdivided with SONET/SDH or fed directly with a protocol such as IP or even Ethernet. Some classes of carriers specialize in wholesaling bandwidth to other carriers that subdivide and sell it to users. A point that is often overlooked is the fact that fiber bandwidth is not necessarily the main concern of end users. The chief issue is service, and bandwidth becomes a service by overlaying an OAM&P process. The efficacy of that process must be considered in selecting service.

On the data level, the service provider is the ISP or the frame relay or ATM provider. Many LECs and IXCs offer those services in addition to other providers that are not associated with any carrier. On the voice level the provider is the IXC and the LEC. All of these use fiber bandwidth, usually at DS-3 or higher. The LECs' trunking network offers virtually unlimited capacity at any OC level with fiber connectivity to CLECs and independent telephone companies. Bandwidth is configured through DCS systems and multiplexers, terminating on digital switches at any level from DS-1 to OCx.

Enterprise Network Access and Backbones

LECs and IXCs gain access to their larger customers over digital circuits in the metropolitan network and the local loop. Between the customer and the central office the facilities may be either copper or fiber. The SONET backbone provides the bandwidth, and the ways of applying it are varied. Figure 5-11 shows some alternatives. Bandwidth can be separated with voice and data occupying separate segments or both voice and data packets can be fed at the source and use an integrated pathway to the destination. Frame relay and ATM both use T1/E1 and higher for access circuits. Larger organizations use T-3, DS-3, or higher levels for Internet access.

Private organizations also use the digital network to interface routers and voice and data switches. Routers are available with electrical and optical

FIGURE 5-11

Customer Locations Connected through SONET Rings

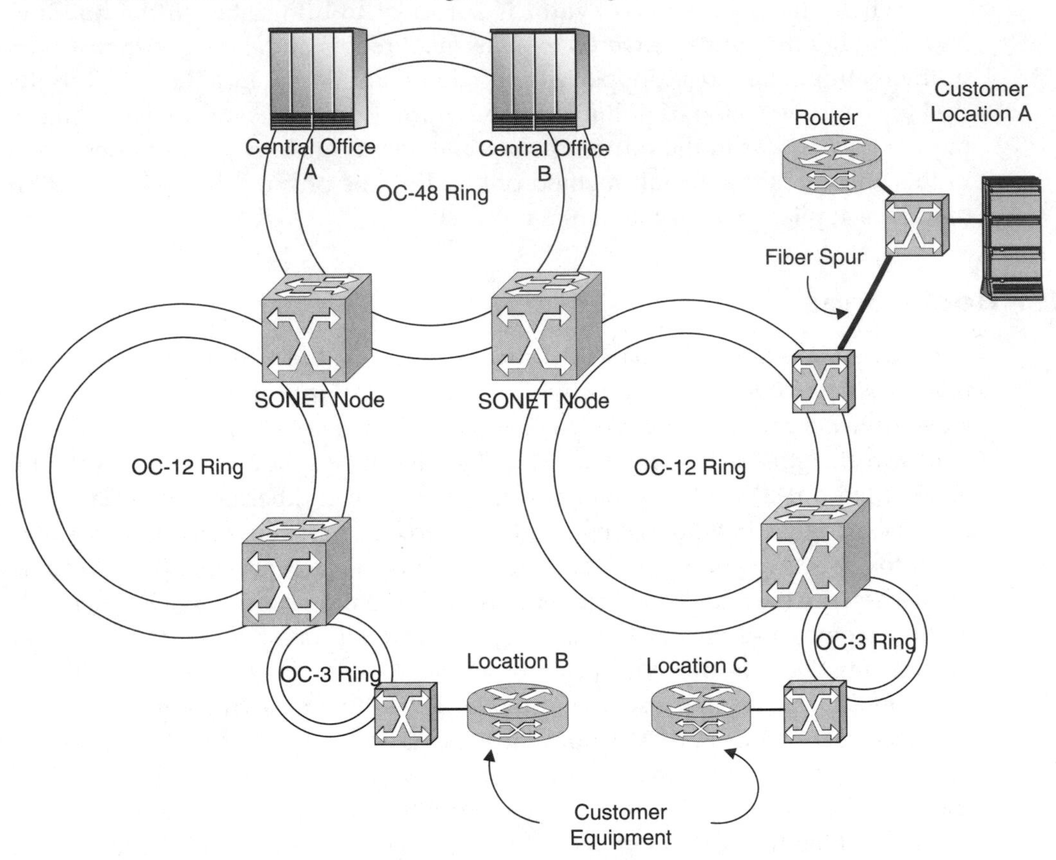

interfaces to use SONET/SDH bandwidths or direct connection to fiber optics. PBXs interface T1/E1 for connections to the PSTN and Ethernet for VoIP connections.

The digital hierarchy discussed in this chapter is the essence of the current TDM network. SONET/SDH is expensive and inflexible, so there is a movement to do away with it at the physical layer in favor of direct IP or Ethernet connection. That will probably occur gradually. In the meantime, however, a significant portion of the network rides over the SONET/SDH backbone at the physical level. These bandwidths are the primary means the enterprise network has of connecting to the public networks. DS-1s are widely deployed, and larger enterprises connect with DS-3 or OC levels. Large central office switches connect to their peers with optical facilities and large backbone routers have optical interfaces. In all such cases, however, switching and routing are done at the electrical level for these functions are in their infancy at the optical level.

CHAPTER 6

Data Communications Protocols

The data industry borrowed the protocol term, an apt metaphor, from the world of diplomacy. In the realm of international affairs, diplomats are concerned with such details as the syntax for expressions, the order of precedence, the form of greeting, and other conventions that puzzle most of us. The same is true in communications networks. Human interaction over the network is possible only after the devices they are using have negotiated the details of the session. The voice network has protocols that are familiar to everyone: dial tone, ringing and busy signals, reorder, and recorded announcements. These are the human-to-machine communications involved in setting up a call. In the data world, the prompts and mouse clicks are the equivalent. Behind the scenes, the machine-to-machine communications are invisible to the user. The protocols select the circuits and set up the connections without the users' needing to understand anything about the internal working of the network.

Addressing and signaling are different in voice and data networks. Data users communicate through applications such as e-mail and Web surfing which are simple to operate because the protocols shield the user from the complexity. The application designer likewise does not have to worry about what happens inside the network. Designers can program to a known set of interfaces and the protocols take over from there.

In Chapter 3, we discussed the main functions that data protocols perform. This chapter expands on that with more detailed information. The chapter starts with a discussion of the OSI model, which is the basis for many practical protocols. Next, we discuss Ethernet, otherwise known as IEEE 802.3, a protocol that runs on most LANs. Then we consider a related protocol, TCP/IP, which runs on top of LAN protocols and operates the Internet and most private networks in the world today.

THE OPEN SYSTEMS INTERCONNECT MODEL

Figure 6-1 shows the seven layers and the PDUs each uses for communicating with its peer across the network. Controlling communications in layers adds overhead because each layer communicates with its peer through header records, but layered protocols are easier to administer than single-layer protocols and provide greater opportunity for standardization. Although protocols are complex, the functions in each layer are modularized so system designers write their code to known application program interfaces (APIs). Layered control offers an opportunity for standardization and interconnection between the proprietary architectures of different manufacturers. The seven OSI layers are defined below. Table 6-1 lists the OSI layers and some of the standards that apply to each layer.

FIGURE 6-1

OSI Reference Model

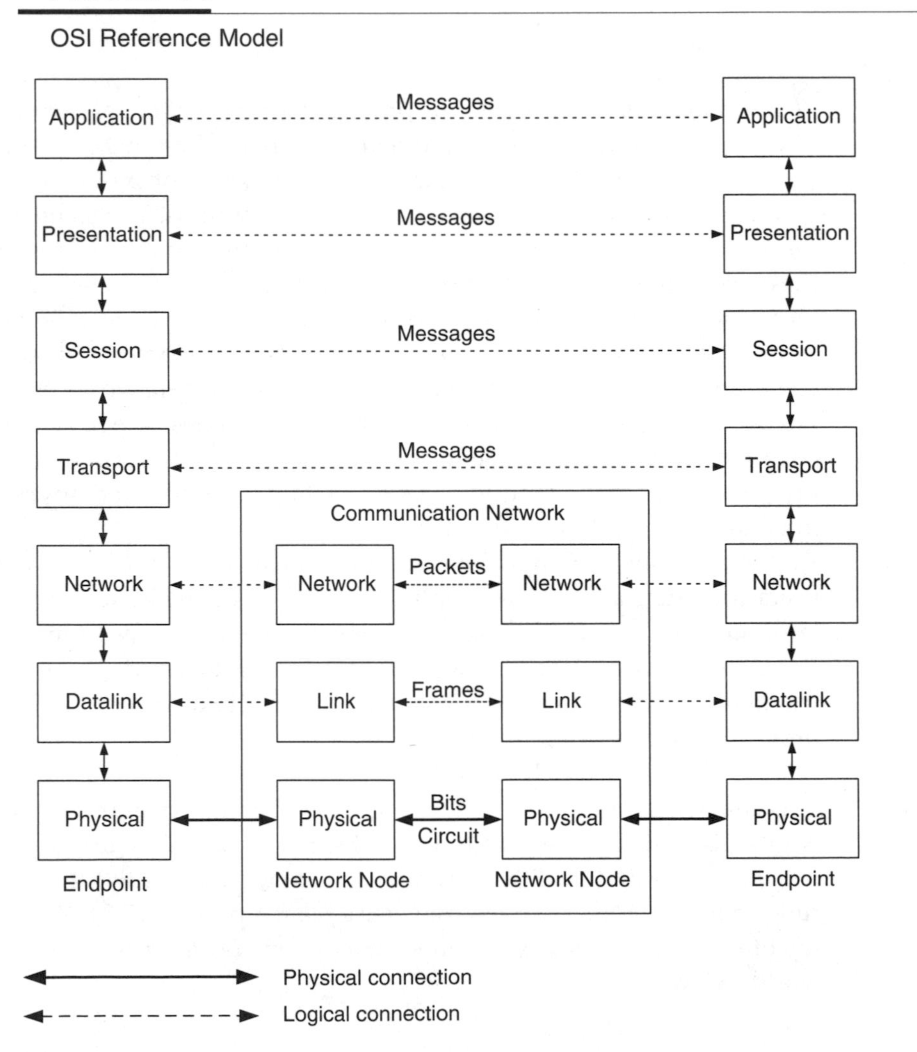

TABLE 6-1

Representative ISO, ITU-T, and IETF Protocol Standards at Their Corresponding OSI Layers

	Layer	Common Standards
1	Physical	EIA-232, EIA-422, V.35
2	Datalink	High-Level Data-Link Control (HDLC), Balanced Link Access Procedure (LAPB), Designated Link Access Procedure (LAPD)
3	Network	X.25 Packet Level Protocol, Internet Protocol (IP), ISO 8473 Connectionless Network Protocol (CLNP)
4	Transport	Transport Control Protocol (TCP), User Datagram Protocol (UDP), ISO 8473 Connectionless Transport Service
5	Session	ISO 8306 and 8037, ITU-T X.215 and X.225
6	Presentation	ISO 8822 and 8823, ITU-T X.216 and X.226
7	Application	Virtual Terminal Protocol, X.400 Message-Handling Service (MHS), X.500 Directory Service, X.700 Common Management Information Protocol (CMIP), File Transfer, Access, and Management (FTAM)

Although the OSI model is a structure for developing standards, it is not in itself a complete networking protocol. Its purpose is to establish a framework that developers can use to devise protocols that can be used independent of manufacturers' proprietary protocols. Most applications today communicate with TCP/IP, which has become a de facto international standard. Although TCP/IP does not conform exactly to the OSI model in the upper layers, its modular design and multitude of coordinating protocols adhere to the same principles as OSI. Virtually all IP networks run on top of Ethernet, which implements OSI's physical and datalink layers.

Layer 1—Physical

Layer 1 transfers bits across a circuit, which can be any transmission medium including wire, fiber optics, coaxial cable, or wireless. The physical layer contains the rules for the transmission of bits between endpoints and standardizes pin connections, modulation methods, and multiplexing over the physical medium. Two devices can communicate using only the physical layer. For example, the serial ports of two computers can be connected through an adapter known as a null modem, which connects the transmitting data and signaling leads of each computer to the receiving leads of the other. SONET/SDH, which we discussed in the previous chapter, is a family of coordinating layer 1 protocols.

Layer 2—Datalink

Datalink protocols provide link integrity to transmit frames of data between endpoints. The protocol accepts bits from the physical layer, assembles them into a frame, and calculates CRC. If a network protocol is used, the link protocol passes the data block with control and CRC blocks attached. As Figure 4-5 shows, flags of a specific bit pattern mark the beginning and ending of the frame. A header contains address and control information, followed by an information field and a trailer containing CRC bits for error correction. The datalink protocol at the receiving end calculates the CRC of the received frame. It acknowledges correctly received frames and requests retransmission if necessary.

The principal international standard, HDLC, has numerous subsets, of which Balanced Link Access Procedure (LAPB) and Designated Link Access Procedure (LAPD) are common. The former is used in packet switched data networks, and the latter is the access protocol for ISDN. Frame relay is a layer 2 protocol that transmits HDLC frames across a network path that is defined in software.

Layer 3—Network

The network layer routes packets between endpoints on a network. A packet is a frame with a header that contains addressing and other information. For example, the IP header, shown in Figure 6-2, contains a time-to-live (TTL) field. If a packet is not delivered before the time expires, it is killed. The type of service (TOS) field informs downstream applications of what type of information the packet contains.

The network layer can be either connectionless or connection oriented. A connectionless protocol such as IP relies on a higher level protocol to assure data delivery and integrity. A connection-oriented protocol such as X.25 sets up a connection across the network before packet exchange can begin.

FIGURE 6-2

IP Packet Structure

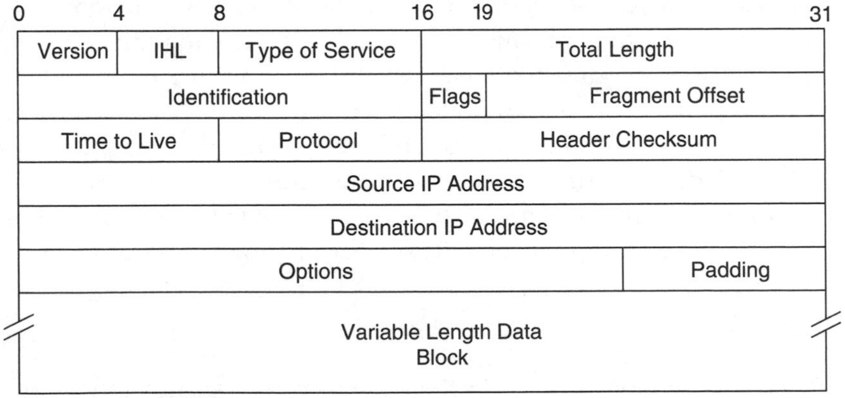

<table>
<tr><td>0</td><td>4</td><td>8</td><td>16</td><td>19</td><td>31</td></tr>
<tr><td>Version</td><td>IHL</td><td>Type of Service</td><td colspan="3">Total Length</td></tr>
<tr><td colspan="3">Identification</td><td>Flags</td><td colspan="2">Fragment Offset</td></tr>
<tr><td colspan="2">Time to Live</td><td>Protocol</td><td colspan="3">Header Checksum</td></tr>
<tr><td colspan="6">Source IP Address</td></tr>
<tr><td colspan="6">Destination IP Address</td></tr>
<tr><td colspan="5">Options</td><td>Padding</td></tr>
<tr><td colspan="6">Variable Length Data
Block</td></tr>
</table>

IHL = IP Header Length

Layer 4—Transport

The transport layer assures integrity between endpoints. Transport protocols establish and terminate connections, segment data into manageable PDUs, and reassemble them at the receiving end. Layer 4 is responsible for flow control, sequencing, and end-to-end error correction. TCP, which is discussed later in this chapter, is the most widely used transport layer protocol, although it is not an OSI protocol. UDP is a connectionless transport protocol that is used by VoIP, Simple Network Management Protocol (SNMP), and other applications that do not require data integrity.

Layer 5—Session

Session service is an optional function that may be embedded in the application as opposed to employing a separate protocol. The session layer establishes and terminates connections between users of session services and synchronizes data transfer between them. It negotiates the use of session layer tokens, which it requires users to have before they can communicate. It provides synchronization points in the data being transferred so that if the session is interrupted, the applications may be able to recover without retransmitting everything.

Layer 6—Presentation

This layer provides the syntax for the session. For example, it might translate between ASCII and EBDIC if the two systems use a different format. If data compression and encryption are used, they communicate through this layer.

Layer 7—Application

The application layer is the interface between the network and the application running on the computer. Examples of application layer functions now in use are ITU-T's X.400 Electronic Mail Protocol and its companion X.500 Directory Services Protocol. Message Handling System (MHS) is an important protocol for enabling X.400 e-mail systems to communicate. ISO's File Transfer, Access, and Management (FTAM) is a protocol for managing and manipulating files across a network. Other protocols include Virtual Terminal (VT), which provides a standard terminal interface, and Electronic Document Interchange (EDI), which uses the MHS platform for transferring electronic documents across networks.

Most vendors at one time or another agreed to support OSI, which is ITU-T recommendation X.200, in a quest for international standardization. It forms the basis for Ethernet, which is discussed in the next section, but with variations. At the higher layers, protocols such as X.400 and X.500 are applied today, but many of the layer 3 to 6 protocols are displaced by TCP/IP, which is more agile and better suited to intermachine communications.

ETHERNET: IEEE 802.3

When IEEE began developing LAN standards in 1980, no one could foresee the impact that desktop computers would have, but the need for a simple, high-speed network to support office applications and manufacturing was then evident. Standards work began with modest objectives of sharing facilities: files, expensive peripherals such as printers and plotters, and software applications. It was not easy to foresee additional requirements that have since materialized, including access to mainframe computers and connectivity to WANs. It was not clear then that processing power and storage would become so inexpensive that servers would far outnumber mainframes and that bandwidth requirements for LANs would increase accordingly.

Ethernet, otherwise known as IEEE 802.3, is the dominant LAN protocol. Nearly every computer sold today has an Ethernet adapter and most Internet sessions originate and terminate in an Ethernet connection. The protocol originated in Xerox's Palo Alto Research Center. Xerox and their development partners Intel and Digital Equipment Co. offered Ethernet to the IEEE as the LAN standard. The committee adopted the principles, but with enough variation that Ethernet and 802.3 are not compatible. The term Ethernet has stuck, however, and in keeping with industry practice, we use it synonymously with 802.3 in this book. Although it is only one of three IEEE LAN standards, Ethernet is the foundation of the vast majority of LANs.

Ethernet has its roots in the Aloha protocol, which the University of Hawaii devised for data communications over radio between Oahu and the outlying islands. Although all islands could communicate with Oahu, they could not necessarily hear transmissions from other islands. If two stations transmitted simultaneously, their signals collided, were mutilated, and had to be retransmitted. The Aloha protocol uses a contention access method. A station with traffic to send listens to determine if the network is idle, and if so, sends a frame of information. Stations can detect other transmissions in a wired network, but a finite time known as the *collision window* is required for a pulse to traverse the length of the medium. As Figure 6-3 shows, a node at one end of a LAN may begin to transmit without detecting that a node at the other end of the network has also begun to transmit, so the frames collide. Any station detecting the collision transmits a jamming signal. The stations cease transmitting and back off a random time before attempting to re-access the network.

The generic name for the Ethernet protocol is Carrier Sense Multiple Access with Collision Detection (CSMA/CD). A contention network can be visualized as a large telephone party line with no central control. Nodes contend for access under control of the MAC protocol that is embedded in their NICs. When a node has traffic to send, its NIC listens to the network and if it is idle sends a frame. Stations connected to the same cable belong to a *collision domain* in which they are susceptible to colliding with other transmissions. As we mentioned in Chapter 4,

FIGURE 6-3

Collision in a Contention LAN

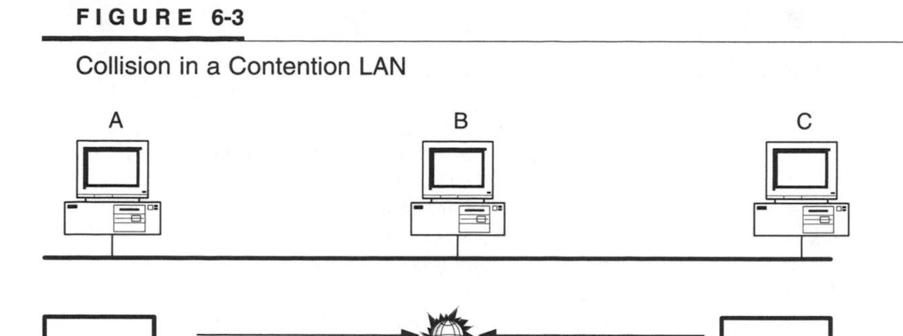

Frames Collide

bridges, switches, and routers can segment LANs to reduce the size of or eliminate collision domains.

The 802.3 frame, shown in Figure 6-4, must always be at least 64 octets long, and no more than 1518 octets in length, of which 1500 octets is payload and the rest is overhead. The minimum frame length is required to support collision detection. If the frames are too short, a collision could escape undetected because the frame could have completed transmission before the collision could be reported and retransmission initiated. The frame includes a variable length pad field to build out the frame length if necessary.

The original Ethernet protocol was intended for RG-8 coaxial cable operating at a data speed of 10 Mbps. The IEEE standard has been revised to support a variety of transmission media. Its speed has been scaled to 10 Gbps and its reach has expanded. Table 6-2 lists some of the vast number of standards that have

FIGURE 6-4

Ethernet Frame

Preamble (7)
Start Frame Delimiter (1)
Destination Address (6)
Source Address (6)
Length/Type (2)
Client Data (variable)
Padding (variable)
Frame Check Sequence (4)

Order of Transmission

Figures in parentheses show the number of octets

TABLE 6-2

Selected IEEE 802 Local Area Network Standards

Standard	Description
802.1	Overview document containing the reference model, tutorial, and glossary
802.1b	Standard for LAN/WAN management
802.1p	Specification for LAN traffic prioritization
802.1q	Virtual bridged LANs
802.2	Logical link control (LLC)
802.3	Contention bus standard 10-Base-5 (Thicknet)
802.3a	Contention bus standard 10-Base-2 (Thin net)
802.3b	Broadband contention bus standard 10-Broad-36
802.3d	Fiber optic inter-repeater link (FOIRL)
802.3e	Contention bus standard 10-Base-T
802.3i	Twisted pair standard 10-Base-T
802.3j	Contention bus standard for fiber optics 10-Base-F
802.3u	100 Mbps contention bus standard 100-Base-T
802.3x	Full duplex Ethernet
802.3z	Gigabit Ethernet
802.3ab	Gigabit Ethernet over Category 5 UTP
802.3ae	10 Gigabit Ethernet
802.3af	Power over Ethernet
802.3ah	Metropolitan Ethernet
802.3ak	10 Gigabit Ethernet over Copper 10GBASE-CX4
802.4	Token bus standard
802.5	Token ring standard
802.5b	Token ring standard 4 Mbps over unshielded twisted pair
802.5f	Token ring standard 16 Mbps operation
802.6	Metropolitan area network DQDB
802.7	Broadband LAN recommended practices
802.8	Fiber optic contention network practices
802.9a	Integrated voice and data LAN
802.10	Interoperable LAN security
802.11	Wireless LAN standard
802.12	Contention bus standard 100VG AnyLAN

Note: This tables lists only the major standards in the 802 series. Each standard has numerous sub-parts, many of which are omitted for clarity. IEEE standards are available online at http://standards.ieee.org/.

been built around the 802.3 protocol. Ethernet clearly illustrates the benefits of the standards approach. Before the 802 committee completed its work, several proprietary LAN protocols had been developed. These were sound methods of fulfilling the need for an office network, but none was able to develop enough momentum to endure. Once the industry was assured of enough volume to justify LAN developments around 802.3, the market thrived.

A LAN, operating at the first two layers of the OSI model, is a complete network, but it needs higher layer protocols to make it operational. These protocols reside in the network operating system (NOS), which contains the functions that application programs must have to communicate with servers, printers, and other shared peripherals. Chapter 10 discusses NOSs in more detail.

TRANSMISSION CONTROL PROTOCOL/INTERNET PROTOCOL (TCP/IP)

In the 1970s, the Department of Defense commissioned the development of the TCP/IP suite of protocols to provide interoperability among computers. The protocols emerged from research that spanned three decades. ARPANET, as it was called in its early days, was a loosely confederated collection of networks operated by colleges, universities, and defense-related companies and agencies.

In late 1989, the original ARPANET evolved into a network that subsequently became the Internet. Internet is a collection of independent networks that are interconnected to act as a coordinated unit. Commercial companies operate the networks, but each is independent and no single body has overall control. The IETF controls the structure of the protocols, which compensate for the unreliability of the underlying IP network and insulate users from the need to understand the network's physical and addressing architecture.

The original Internet had four primary purposes:

- To provide electronic mail service to the users
- To support file transfer between hosts
- To permit users to log on remote computers
- To provide users with access to information databases

A TCP/IP Internet fits in a three-layer framework atop the physical and datalink layers as shown in Figure 6-5, which compares TCP/IP to the OSI model. The TCP/IP and OSI protocols do not correspond directly, and the boundaries between layers are not identical, but they share the layered concept and have corresponding functions. The application services layer defines the interface to the TCP/IP network.

The primary application layer protocols are SMTP for e-mail, FTP for file transfers, terminal emulation protocol (TELNET) to enable users to run commands on a remote computer over the network, and HTTP to support database access through Web browsers.

TCP/IP is a suite of protocols that coordinate to support an Internet, which is a collection of network segments. Figure 6-6 illustrates broadly how a TCP/IP network functions. The protocol stack runs on hosts that may be part of the same segment or independent of each other. We assume that both hosts are connected to a LAN, and that a router interfaces the LAN to a wide area network. A single carrier may provide the WAN, or as the figure shows, the WANs may be separate and interconnected with routers. Before any data can be transferred, TCP sets up a connection with its peer at the receiving endpoint. The application, receiving an indication that the connection is established and TCP is ready, sends a data stream across the interface. TCP segments the data stream, appends a transport header, and hands it to IP. The IP portion of the protocol appends a datagram header, which contains the receiver's IP address, and hands a packet stream to the router. The router consults its

FIGURE 6-5

TCP/IP Protocols Compared to OSI Model

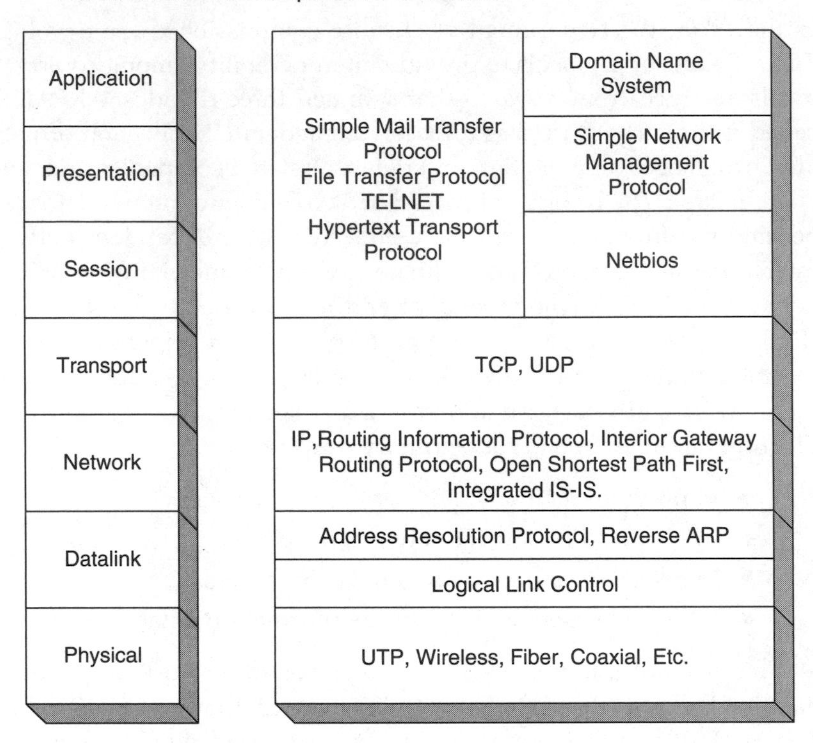

FIGURE 6-6

A TCP/IP Network

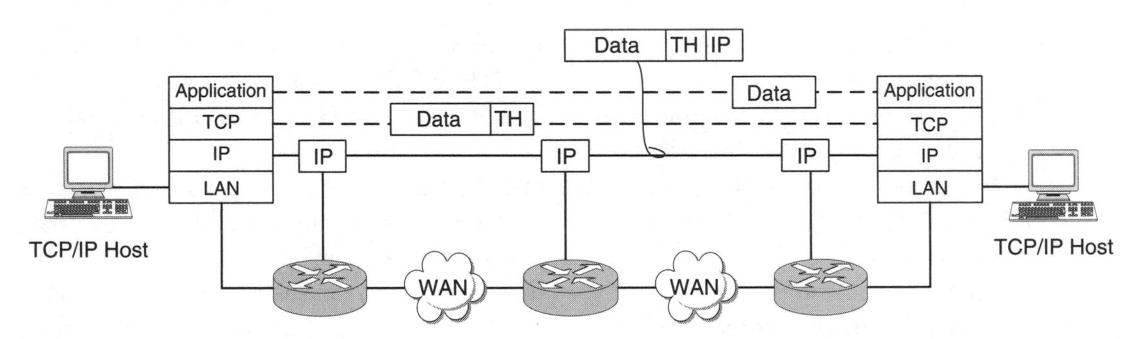

tables, decides which route to use, and forwards the packet stream to the next router in sequence until they reach the destination. TCP at the receiving host checks for errors, acknowledges the packets, strips the headers, and hands the data stream to the application. This section examines the principal protocols in more detail.

Transmission Control Protocol (TCP)

TCP is a connection-oriented guaranteed-delivery protocol that interoperates with IP to deliver packets across a network. The transmitting and sending ends of the circuit exchange setup packets to establish a connection before information packets begin to flow. TCP/IP is independent of the physical medium. The protocol makes use of the physical and datalink layers on the LAN. Between the LANs, the WAN protocols handle packet transfer. TCP accepts data streams from the application program, breaks them into datagram size, and hands them to IP for routing across the network. TCP on the receiving end checks for errors, reassembles the original data stream, and passes it to the application. TCP handles packet sequencing, flow control, retransmission of errored or missing packets, and deletion of duplicate packets. TCP can be used with a variety of packet level protocols, but it is usually paired with IP.

TCP can support either go back-n or selective repeat packet acknowledgment. Under go back-n, when the receiving end acknowledges packets, it sends the sequence number of the next packet it is prepared to accept, which communicates to the sender the number of the last correctly received packet. The sender transmits the next packet in sequence plus all subsequent packets, whether they were in error or not. Under selective acknowledgment, the receiver acknowledges specific packets. The acknowledgment process provides a convenient method of flow control known as "sliding-window." If the receiving end has plenty of space in its buffers, it sends a window message to the sender indicating how many packets it is prepared to receive before it sends an acknowledgment. When the buffer begins to overflow, it reduces the size of the window in its next acknowledgment. In this way, the protocol keeps the sending end from overrunning its buffer space and keeps the network operating at maximum throughput. At the worst extreme, the receiver could acknowledge packets one at a time, which would slow data transfer to a minimum.

The sending TCP sets a retransmit timer for each packet. This timer specifies how long to wait for acknowledgment before assuming the packet failed to reach the destination. When the timer expires, the sending host resends the packet and waits twice as long as the time set in the preceding window. Each time an acknowledgment is not received, the retransmit timer doubles. This process may result in duplicate packets, so the receiving TCP host must be prepared to discover and discard duplicates.

TCP delivers packets to specific *ports* or *sockets* depending on the application. Both TCP and UDP, which is discussed next, use 16-bit values called port numbers to identify the TOS requested by the application. Certain applications normally use well-known ports. For example, FTP uses port 21, HTTP uses port 80, and TELNET uses port 25. The port identification, which is contained in the TCP header, aids the receiving application in identifying the type of data being received.

User Datagram Protocol (UDP)

Error correction delays packet flow. To a data application, delay is either invisible or appears as a longer response time, which may be annoying but does not affect integrity. Voice and interactive video are less tolerant of delays. Too much delay results in users talking over each other, making it difficult to communicate. Furthermore, errors are inconsequential to voice and video. TCP, therefore, is not an appropriate protocol for these time-sensitive applications. Voice and video, as well as other protocols such as SNMP, use UDP for the transport protocol. UDP is a connectionless protocol that contains no provision for error correction, packet sequencing, or retransmission.

A UDP datagram has considerably less overhead than TCP. The header contains only the source and destination ports, the length of the data block, and a checksum for verifying the header. Voice and video packets are normally quite small. To minimize transmission time, it is desirable to keep the transport header as short as possible.

Internet Protocol (IP)

Routers at IP nodes usually have multiple paths for delivering packets. When a packet arrives, a router reads the destination address, consults its internal routing table to determine the appropriate route, and forwards the datagram over the physical network to the destination. Routing tables are either static or dynamic. Static tables are set in the router's software and remain until they are reconfigured. Static tables are appropriate when the router has a limited number of choices, but they are inflexible and unable to adapt to changing circumstances. Routers with dynamic tables exchange messages to keep each other updated with network changes such as circuit failures and route reconfiguration.

An IP datagram contains a header that is 20 octets or more in length plus a data field that can be up to 65,535 octets long. The maximum length permitted by a network is called its *maximum transfer unit* (MTU). For example, Ethernet has an MTU of 1500 octets. When it is necessary to send an IP datagram across a network with a short MTU, IP divides the datagram into fragments and reassembles it at the destination, a process called *fragmentation*.

IP is an unreliable best effort protocol. Each node along the network attempts to forward packets to the next node, but if it is unsuccessful, it is permitted to discard the packet. Routers attempt to forward messages to the destination over the shortest path, but there is no assurance that a packet is always heading toward its destination, so packets can arrive out of sequence. Every packet contains a TTL counter that has a maximum value of 255 s. Each router that forwards a datagram decrements the TTL by one. If a router receives a packet with an expired TTL timer, it discards the packet. This process prevents undelivered packets from traveling the Internet forever. A protocol known as Internet Control

FIGURE 6-7

E-Mail Application Using TCP/IP over Ethernet

Layer					Unit
Application				The data block contains the message text	Data
Transport			TCP	The data block contains the message text	Packet
Network		IP	TCP	The data block contains the message text	Packet
Datalink	Ethernet	IP	TCP	The data block contains the message text	Frame

Message Protocol (ICMP) reports errors and messages regarding datagram delivery and alerts the sender when a destination is unreachable.

To clarify the way these protocols interact, assume an e-mail application such as Microsoft Outlook originates an e-mail message and sends it to an e-mail server across Ethernet. As shown in Figure 6-7, the application sends a data stream to TCP, which chops it into packets, appends a TCP header, and sends it to IP for routing. IP appends a header containing the sending and receiving addresses, TOS, and other such information, and hands it to the Ethernet LLC. The LLC appends a header containing, among other things, the receiving MAC address, and arranges for transmission across the physical medium. If the sending driver does not have the MAC address of the receiver in its cache memory, it sends an Address Resolution Protocol (ARP) request, which is discussed later, to determine it. The receiving MAC, recognizing its address, strips the Ethernet header and passes the packet to IP, and on up to the mail server through TCP.

IP Addresses

Every device connected to a TCP/IP network requires at least one unique IP address. The address consists of four octets expressed in decimal format and separated by dots, which makes it easier to read than in binary form. A sample address would be 188.12.2.1. The addresses identify the network to which the device attaches as well as the address of the device itself. IP addresses are composed of three parts: class, network portion, and host. The addresses are classed as A, B, C, and D to accommodate the needs of networks of different sizes. Class E addresses are experimental, and are not assigned at this point. Figure 6-8 shows how the 32-bit structure is divided between networks and hosts. The actual quantities of networks and hosts are less than the full range because of certain reserved addresses. For example, binary addresses of all ones are broadcast to all stations in a network.

Address classes can be recognized from the digits in the first octet. Table 6-3 shows the four classes of address, the address ranges, and the binary digits that start the address range. The address classes are not as rigid as they would appear

FIGURE 6-8

IP Address Classes

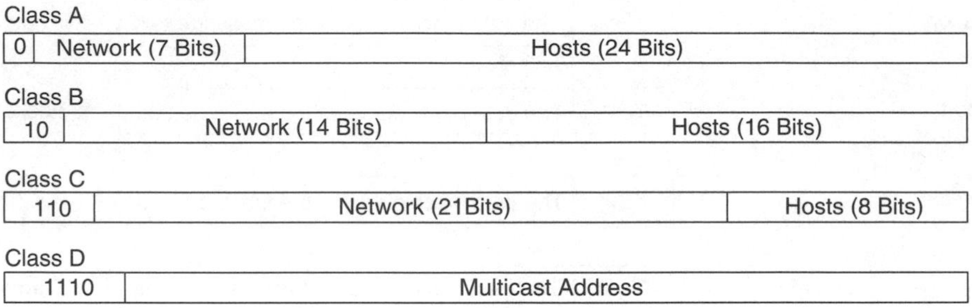

Class A

0	Network (7 Bits)	Hosts (24 Bits)

Class B

10	Network (14 Bits)	Hosts (16 Bits)

Class C

110	Network (21Bits)	Hosts (8 Bits)

Class D

1110	Multicast Address

TABLE 6-3

IP Address Ranges

Address Class	Start Address	Finish Address	Binary First Digits
A	0.0.0.0	127.255.255.255	0
B	128.0.0.0	191.255.255.255	10
C	192.0.0.0	223.255.255.255	110
D	224.0.0.0	239.255.255.255	1110
E	240.0.0.0	255.255.255.255	1111

to be because of the use of subnet masks, which borrow bits from the host portion of the address and uses them as a subnet. Subnet masks have all ones bits in all fields of the IP address except for the host field. Also, classless interdomain routing (CIDR) is a technique for making more efficient use of IP numbers. CIDR uses a prefix to the IP address to indicate how many bits are used for the network prefix.

The Internet Network Information Center (InterNIC) assigns IP addresses. Of the four segments of IP addresses, InterNIC assigns the first segment in class A addresses, the first two segments in class B addresses, and the first three in class C addresses. The network administrator assigns the lower address levels. Class A addresses are issued to only organizations large enough to make efficient use of the lower level addressing structure.

The IP address architecture yields 2^{32}, or a potential of 4,294,967,296 addresses. This would seem to be more than enough addresses to satisfy the world's requirements for many years, except that some addresses are reserved for special purposes. Moreover, it is impractical to use all the addresses within a range because they are assigned to entities that retain some numbers for growth and flexibility in address assignment. Addresses are also being consumed by connecting unconventional devices to the Internet. For example, a refrigerator could be connected to a home network and programmed to report its internal temperature to a monitoring station.

Such uses will eventually exhaust the address supply, necessitating an updated version known as IPv6. IPv6 provides 126-bit addresses, which should be plenty to handle all conceivable devices that might connect to the Internet. IPv6 has been slow to catch on. Part of the reason is that many devices are incompatible with the expanded address and will have to be replaced. Perhaps more important, network address translation (NAT) has slowed the exhaustion of numbers.

Network Address Translation

NAT permits an entity to use internal addresses that are not accessible from the public Internet. The process usually runs on a router and translates between the internal and external addresses. NAT is analogous to a hotel that has one public address, but keeps the identities of its guests private except when the guest grants permission for connection to the outside world. The IETF in RFC 1918 designated certain blocks of addresses for use behind NAT. These blocks are: 10.0.0.0 through 10.255.255.255, 172.16.0.0 through 172.31.255.255, and 192.168.0.0 through 192.168.255.255. Users cannot register these address blocks and they are frequently used for private networking plans. Not only does NAT conserve IP addresses, it also improves security because outsiders cannot reach an internal address directly. Networks that will never connect to the Internet are free to use any IP addresses they like, but any organization doing so should be aware that a future Internet connection will likely require changing existing addresses.

Address Resolution Protocol

Devices that connect to IP from a LAN have dual addresses—the NIC's MAC address and an IP address. The MAC address is permanent since it is encoded into the NIC, but the IP address may change when the device moves to a different LAN segment. The network needs a method of correlating IP addresses to physical addresses. The problem is further complicated with MAC addresses, which are 48 bits long and cannot be encoded into the 32-bit IP address. ARP maps IP addresses to hardware addresses. When a router needs to know how to reach a LAN station, it broadcasts an ARP request and the station with that IP address responds with its MAC address. The device that sent the ARP message retains the address mapping in cache memory to avoid loading the network with repeated ARP messages.

A special case occurs with devices such as diskless workstations that are unable to store their IP addresses permanently. Before they can operate they must determine their IP address when all they know is their physical address. This is done through Reverse Address Resolution Protocol (RARP). At startup, such stations broadcast their physical address, asking a server to respond with their IP address. Once this initial transaction is completed, the workstation is able to respond to ARP messages.

Dynamic Address Assignment

In large networks with many devices, IP address assignment becomes a time-consuming chore. Each device must have an IP address, and if the device moves to a different subnet, the address changes. Furthermore, devices such as laptop computers may move to different subnets several times per day, needing a different IP address each time. Static IP address administration is inflexible, time consuming, and runs the risk of assigning duplicate addresses.

TCP/IP specifies a protocol known as Bootp, which enables diskless PCs to request an IP address assignment from a server. This eliminates the need to visit each node to enter an IP address, but Bootp has a major drawback: once it assigns an address, it has no way of recovering it if it is not used. Dynamic Host Configuration Protocol (DHCP) maintains a pool of IP addresses. When a PC logs into a server, if it does not have an IP address, DHCP assigns one from its pool. The address has a definite expiration time so that addresses can remain inactive only for a limited time. DHCP saves administrative time, and conserves IP addresses that may be in short supply.

Routing

The early Internet defined two classes of device, hosts and gateways. A host was any device directly connected to the network that used or produced information. A gateway was a computer that routed information among hosts. The early gateways were standalone computers, but today routers perform the gateway function. Routing involves two functions: route calculation and packet forwarding. Routers contain tables and logic to determine the most effective route for reaching a host. Routes are either static or dynamic in nature. Static routes are table mappings that are programmed into the router and remain until they are changed. Static routers are satisfactory for simple networks with predictable traffic.

Dynamic routing protocols adapt to network changes such as circuit failures and automatically select the best routes. Routing protocols select the best path for packets to travel by the use of metrics such as cost, bandwidth, and route congestion. Routing algorithms populate the routing tables with information about the topology of the network and the best way to reach the next hop. Routers transmit messages among themselves to communicate the status of their links and to update their tables. A link failure, for example, would trigger messages alerting other routers to the change in status. The process of optimizing routes by exchanging update messages is known as *convergence*, which is not to be confused with voice–data integration.

Intermediate systems have the capability to forward packets between subnetworks. Routers that lack this capability are known as end systems. IS routers are further classified as those that can communicate only within domains and those that communicate both within and between domains or autonomous systems. These are known as intradomain ISs and interdomain ISs, respectively. Autonomous systems operate under the same administrative authority and control.

The most popular interior routing protocols are Routing Information Protocol (RIP), Interior Gateway Routing Protocol (IGRP), Open Shortest Path First (OSPF), and Integrated IS-IS. RIP is used in small networks, but it lacks the sophistication of the other protocols which are used in larger networks. The two primary exterior routing protocols are External Gateway Protocol (EGP) and Border Gateway Protocol (BGP). EGP is an early protocol that is unsatisfactory for the Internet today, and is being phased out in favor of the current version, which is BGP-4.

Router algorithms are classified as *link-state* or *distance vector*. Link-state algorithms flood all routers in the network with information about the state of the router's links and each router builds its tables with a picture of the network. Routers using distance-vector algorithms send their routing tables only to their neighbors. The more sophisticated the algorithm, the more the route can be based on multiple metrics such as delay, bandwidth, path length, reliability, cost, and traffic load.

Routers form the first line of defense against outside intrusion. Through *access control lists*, routers can reject packets from particular IP addresses or limit access to ranges of addresses. They can also filter packets based on port number. Network administrators often reject packets addressed to ports that are not needed or are known to admit certain types of hacking attacks. Routers are discussed in more detail in Chapter 36, IP Networks.

Domain Name Server (DNS)

IP addresses are easy for machines to handle, but they are difficult for people to remember. Therefore, Internet destinations have a mnemonic address known as a uniform resource listing (URL) which takes the form of name@organization.suffix (e.g. John.Jones@McGraw-Hill.com). A complex network of servers known as DNSs translates from URLs to IP addresses. Every station on a network is programmed with the IP address of its DNS.

At the top of the DNS hierarchy is the *route node*, which provides pointers to the *top-level domain* (TLD) servers. The TLDs are divided into three categories:

- Generic: e.g., .com, .net, .org
- Country code: e.g., .us, .au, .uk
- Chartered: e.g., .gov, .mil, .edu

Chartered domains have specific admission requirements and country code domains are limited to the countries involved. Generic domains can be issued to users in any country.

DNS servers are assigned zones throughout the Internet and positioned at each level of the hierarchy. The master server for a particular zone is the definitive data source for that zone. The DNS protocol propagates information downward to servers within the hierarchy. The servers use a lookup process known as *resolving*. The TLD servers always know how to reach the lower level domains within their purview. Second-level domains are issued to large organizations such as ISPs, and these may host third and fourth level domains down to small companies and individuals. Caching servers retain data on frequently accessed domains such as Amazon.com so that DNS requests are handled at the lowest level feasible.

DNS management and administration is a complex process involving international relationships. In the United States, Internet Corporation for Assigned Names and Numbers (ICANN) manages DNS. Another company, VeriSign, handles domain name registrations through several other companies. The policy in the United States is to keep the process competitive, but other countries do not necessarily handle domain registration that way.

APPLICATION ISSUES

This discussion has touched on only the most common protocols. The OSI model is the foundation for the first two protocol layers and serves nearly every session that is used today. In the wide area, every common carrier uses SONET/SDH to some degree at the physical layer, its bit structure riding above the fiber-optic infrastructure. The vast majority of LANs use Ethernet, and as it evolves from a hubbed to a switched structure its limitations disappear. Ethernet extends into the metropolitan area, with some tendency toward using it in the wide area directly over fiber-optic wavelengths, eliminating the SONET/SDH structure.

Above the first two layers, the impetus has switched to IP with one of its transport protocols, TCP or UDP. Much of the development of the ISO-ITU protocols happened concurrently with the development of ARPANET and its evolution into Internet. TCP/IP has become such a universal protocol, thanks to its inherent simplicity, that it has forestalled additional growth of OSI and much of it has become irrelevant. It was not easy to foresee at the time these systems were under development, but the computer market changed so that the architecture drifted away from OSI's top-down structure to TCP/IP's peer-to-peer model. OSI is closely aligned to IBM's SNA, using the same seven-layer structure, but with somewhat different layer definitions. TCP/IP was designed for computer-to-computer communications. When the applications shifted from text based to graphics, the network model had to change. Were it not for TCP/IP, the OSI model would undoubtedly have adapted, but the advent of the public Internet was such a powerful force that adopting TCP/IP, which runs on nearly any computer, was a natural choice.

Ethernet also has emerged as a clear winner in the LAN, to the point that token ring and its high-speed variation FDDI are also becoming moribund. When the 802 committee was doing its initial work, IBM contended that Ethernet with its contention protocol was not sufficiently robust to support applications that required predictable timing. Token ring, IBM's alternative, lagged Ethernet development long enough that by the time it arrived the rest of the computer world had settled on Ethernet. When switched Ethernet eliminated contention and speeds increased to first 100 Mbps, then to 1 Gbps, the robustness argument disappeared.

Many network services have been built around the structure of the ISO physical and link layer protocols. Frame relay, ATM, Switched Multimegabit Data Service (SMDS), Transparent LAN Service (TLS), and Metropolitan Ethernet service are examples of the variety of common carrier services that build on the lower levels of OSI. Beyond that, public packet switching service, SS7 signaling network, ISDN, and a host of other services use the higher layers as well.

CHAPTER 7

Outside Plant

Outside plant is the collection of cables, poles, conduit, and fiber optics that interconnect central offices and connect the central office to the subscribers' premises. The local loop is the most expensive and the least technically complex portion of the entire telecommunications system. Wide-bandwidth signals travel across the country in ribbons of fiber-optic cable, are digitally routed and switched, but finally must be converted to analog and piped to the customer over a pair of wires that may cut off any frequency higher than 4 kHz. Even though the local loop is less than ideal, together with its supporting conduit it is arguably the most valuable asset the ILECs have, and one that cannot easily be duplicated or replaced by their CLEC competitors. Furthermore, technologies such as DSL, which Chapter 8 discusses, increase the value of the local loop.

The local loop is often referred to as the "last mile," and it consists largely of twisted-pair copper wire enclosed in cables that are routed through conduit, buried in the ground, or mounted on poles to reach the end user. Except for metropolitan areas where many buildings are served by fiber optics, the local loop has not changed much over the years. Insulation has improved, cable sheaths have evolved from lead to nonmetallic, and improved splicing techniques have increased cable-splicing productivity. Much of the outside plant that once was aerial has migrated underground, which makes it less vulnerable to damage. Otherwise, cable today is technically little changed from that placed a century ago.

As the telephone network developed, central offices were linked with high-quality copper cables that were enclosed in an underground conduit. Manholes were placed to provide splice points and to hold load coils, which were placed at 6000-ft (1800 m) intervals to improve voice frequency response. As conduits filled to capacity, the next step was to remove load coils and install analog carrier, which typically multiplexed 12 channels on two copper pairs. In the 1960s digital carrier began to replace analog, doubling the capacity of the cable. Twenty years later, fiber optics changed the character of interoffice trunking completely. Its small size and enormous capacity has now all but eliminated copper cable in the plant between central offices.

Eventually, the local loop will undoubtedly migrate to fiber optics. The large ILECs, SBC and Verizon, have announced plans to place fiber loops gradually over an indefinite period. For now, copper cable will serve the majority of subscribers for several reasons. First is the matter of economics. Fiber-optic cable has enormous bandwidth, but it must be multiplexed, and multiplexing equipment is expensive for the subscriber's premises and requires local powering. Today's copper cable carries power to the customer, and with backup batteries in the central office, telephone service is effectively immune to commercial power failures.

Conversion to fiber is also deterred by the magnitude and cost of the task. LECs install fiber as needed to meet service demands, but the benefits are insufficient to justify extending it to millions of residences and small businesses. Competition, however, is changing the mix. Cable providers have fiber to network nodes now, and are inspiring the ILECs to follow suit. Many observers believe that video-on-demand will drive the conversion to fiber in the local loop, but so far the service has not caught on. The industry has even given fiber-in-the-loop technology the name FITL, but it has yet to make a significant impact except for feeds to large businesses.

This chapter discusses how outside plant is designed and constructed. Our focus is LEC plant, but the principles are equally applicable to private organizations that use copper cable to link their buildings. Interbuilding fiber applications are somewhat more complex, and are discussed in Chapter 17.

OUTSIDE PLANT TECHNOLOGY

Outside plant (OSP), diagrammed in Figure 7-1, links the central office to subscribers' premises. *Feeder cables* enclosed in conduit connect the central office to office buildings or neighborhood centers. Feeder pairs are terminated directly on terminals in office buildings, usually in enough quantity to provide Centrex service, which requires a pair per station. In neighborhood centers feeder cables are cross-connected to smaller *distribution cables*, which are placed along streets and alleys to provide connections to homes and smaller commercial buildings. Terminals provide access to the cable pairs, which are connected to the subscribers' premises with *drop wire*.

Supporting Structures

Cable is classified according to its supporting structure:

- *Aerial* cable supported on pole lines
- *Underground* cable enclosed in buried conduit
- *Buried* cable placed directly in the ground

Aerial cable is lashed to a galvanized metallic strand known as a *messenger*, which is attached to the poles. Self-supporting aerial cable contains an internal

FIGURE 7-1

Major Components of Outside Plant

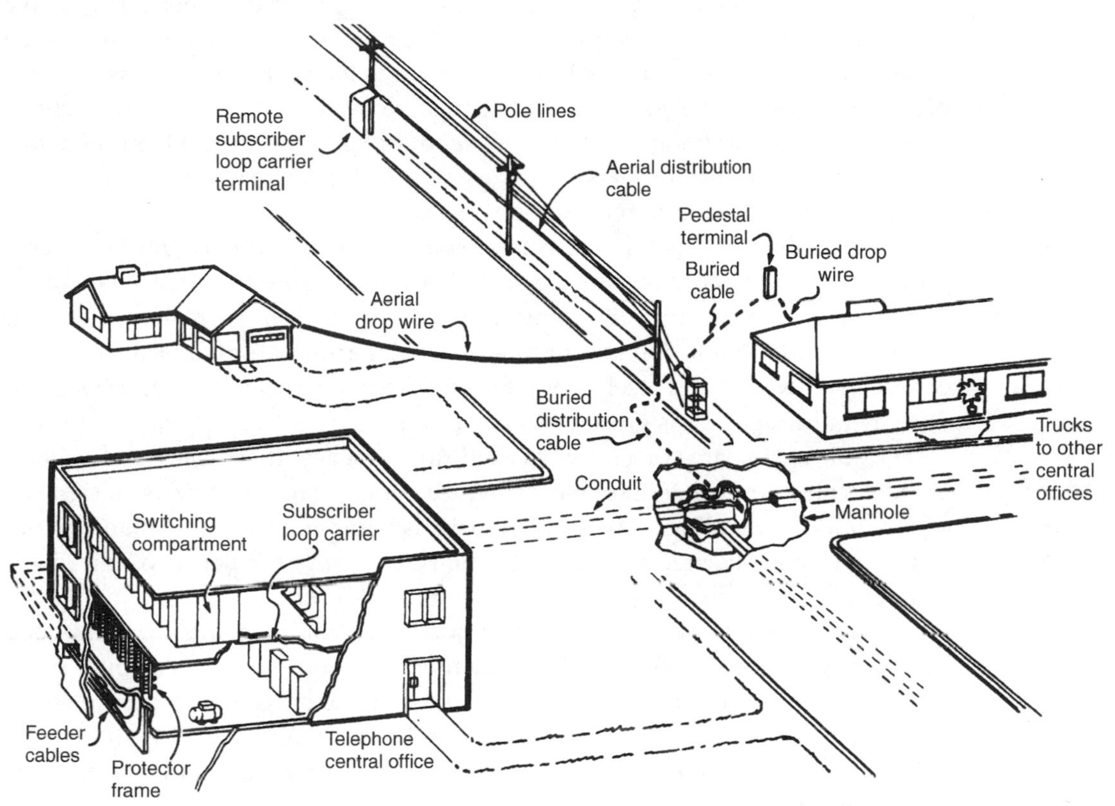

messenger. Down guys and anchors are placed at bends and at the ends of cable runs to relieve strain on the poles. Aerial cable is unsightly and vulnerable to storm damage, but it is less expensive than placing cable underground.

Underground cable is placed in conduit. Cost aside, conduit is always the preferred method of placing cable under ground. Where future additions and rearrangements will be required, LECs provide empty ducts for expansion. Manholes are placed at intervals for splice points and to house T-carrier repeaters and load coils.

Direct burial is often less expensive than conduit. Buried cable is either placed in an open trench or plowed with a tractor-drawn implement that feeds the cable underground through a guide in the plow blade.

Cable Characteristics

Twisted-pair cables are classified by the wire gauge, sheath material, protective outer jacketing, and the number of pairs contained within the sheath. Sizes range from one- or two-pair drop wire to 3600-pair cable used for central office building entrance. Cable size is limited to the size of conduit ducts, which are up to 4 in. (10.5 cm) in internal diameter. The sheath diameter that a conduit will hold depends on wire gauge and the number of pairs. Cables of larger sizes, such as 2400 and 3600 pairs, are used primarily for entrance into telephone central offices. Wire gauges of 26, 24, 22, and 19 AWG are used in loop plant. Cost considerations dictate the use of the smallest wire gauge feasible, consistent with technical requirements. Finer gauges are used close to the central office to feed the largest concentrations of users. Coarser gauges are used at greater distances from the central office as needed to reduce loop resistance.

Cable sheath materials are predominantly high-durability plastics such as polyethylene and polyvinyl chloride (PVC). Cable sheaths are constructed to protect against damage from lightning, moisture, induction, corrosion, rocks, and rodents. In addition to the sheath material, submarine cables are protected with coverings of Kevlar and steel armor. Besides the outer sheath, cables are shielded with metallic tape that is grounded on each end.

Cable pairs are precisely twisted to preserve their electrical balance. Unbalanced pairs are vulnerable to induced noise, so the twist is designed to ensure that the coupling between cable pairs is minimized. Cable is manufactured in layers of pairs twisted around a common axis, with each pair in a unit given a different twist length. These units are called *complements*.

Cable pairs are color coded within 50-pair complements. A color-coded binder is wrapped around the pairs to identify each complement. At splice points, the corresponding binder groups and pairs are spliced together to ensure end-to-end pair identity and continuity. Cables can be manually spliced with compression sleeves or ordered from the factory cut to the required length and equipped with connectors.

Splicing quality is an important factor in preserving cable pair balance. Older cables are insulated with paper and were spliced by twisting the wires together. These older splices are often a source of imbalance and noise because of insulation breakdown and splice deterioration. To prevent crosstalk, it is also important to avoid splitting cable pairs. A split occurs when a wire from one pair is spliced to a corresponding wire in another pair. Although electrical continuity exists between the two cable ends, an imbalance between pairs exists, and crosstalk may result.

Cable splices and terminations are placed in aerial or aboveground closures such as the ones shown in Figure 7-2. Cables must be manufactured and spliced to prevent water from entering the sheath because moisture inside the cable is the most frequent cause of noise and crosstalk.

Loop Resistance Design

Outside plant engineers select the wire gauge to achieve an objective loop resistance commensurate with the characteristics of the switching system. All telephone switching systems, including central office switches, PBXs, and key telephone

FIGURE 7-2

A Pedestal Terminal (Photo by Author)

systems are designed to support a maximum loop resistance range. The loop resistance includes the following elements:

- Battery feed resistance of the switching system (usually 400 Ω)
- Central office wiring (nominally 10 Ω)
- Cable pair resistance (variable to achieve the design objective of the central office or PBX)
- Drop wire resistance (nominally 25 Ω)
- Station set resistance (nominally 400 Ω)

Central office switches can signal over loop resistance ranges in the order of 1300 to 1500 Ω. If the fixed elements listed above total 835 Ω, that leaves some 500 Ω for cable pair resistance. This method of design is called *resistance design*. Most PBXs have less range, typically 400 to 800 Ω. The range can be extended with subscriber carrier or range extension devices, which boost the nominal –48-V central office battery to –72 V. The range limitation of subscriber loops depends on the current required to operate PBX trunk circuits, dual tone multi-frequency (DTMF) dials, and the telephone transmitter. Designers aim for loop current between 23 and 50 mA.

A further consideration in selecting cable is the capacitance of the pair, expressed in microfarads (μF) per mile (1.6 km). Ordinary subscriber loop cable has a high capacitance of 0.083 μF/mile. Low capacitance cable, used for trunks because of its improved frequency response, has a capacitance of 0.062 μF/mile. Special 25-gauge cable used for T carrier has a capacitance of 0.039 μF/mile.

Special types of cable are used for cable television, closed circuit video, LANs, and other applications. Some types of cable are constructed with internal screens to isolate the transmitting and receiving pairs of a T-carrier system.

Feeder and Distribution Cable

Cable plant is divided into two categories in the local loop—feeder and distribution. Feeder cables extend from the central office to a serving area. Main feeders are large backbone cables that exit the central office and are routed, usually

FIGURE 7-3

Feeder and Distribution Service Areas

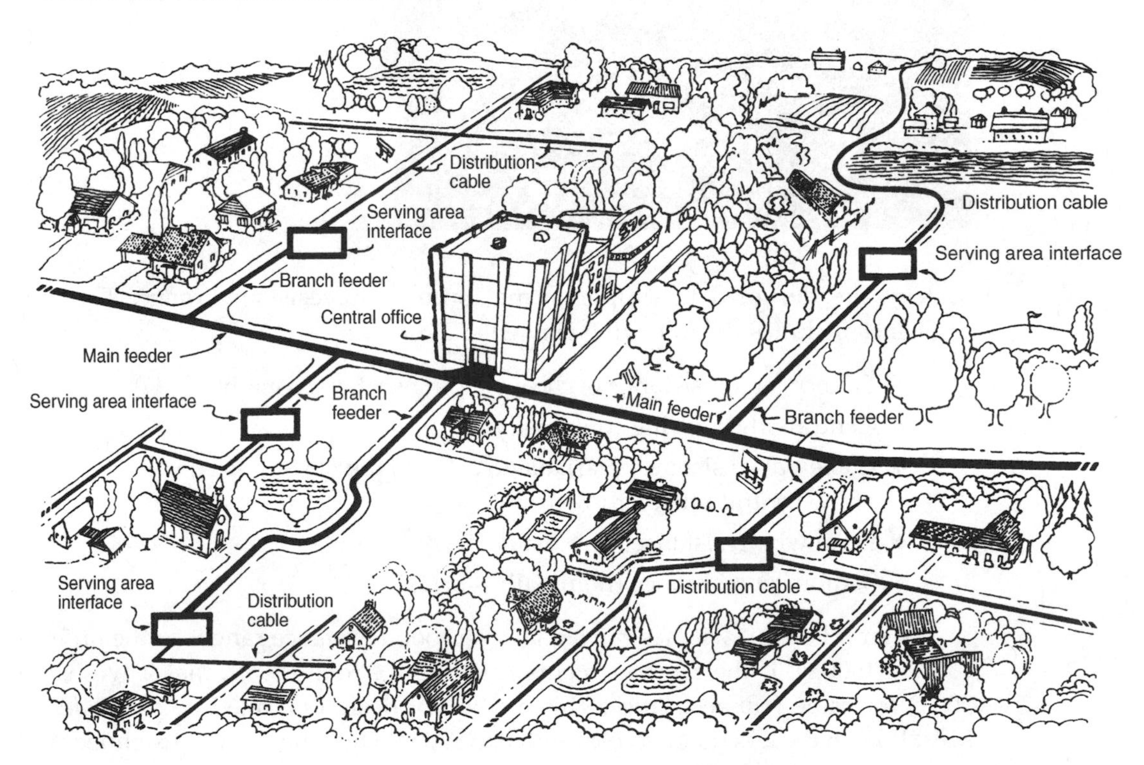

through conduit, to intermediate branching points. Branch feeders are smaller cables that route pairs from the main feeders to a serving area. Distribution cable extends from a serving area interface to the users' premises. Figure 7-3 shows the plan of a typical serving area.

Where 25 pairs or more are needed in a building, the LEC splices distribution cable directly into the building. Otherwise, the feeder cable may terminate in a crossconnect cabinet similar to the one shown in Figure 7-4, which serves as a junction point between the feeder and distribution cable. Terminals provide access to cable pairs, either on binding posts or direct connection. Terminals may be mounted on the ground in pedestals, in buildings, on aerial cable messengers, or underground. Aerial or buried drop wire runs from the terminal to a protector at the user's premises.

To the degree feasible, feeder and distribution cables are designed to avoid *bridged tap,* which is any portion of the cable pair that is not in the direct path between the user and the central office. Bridged tap has the electrical effect of connecting a capacitor across the pair and impairs the frequency response of the

FIGURE 7-4

Serving Area Interface Terminal (Photo by Author)

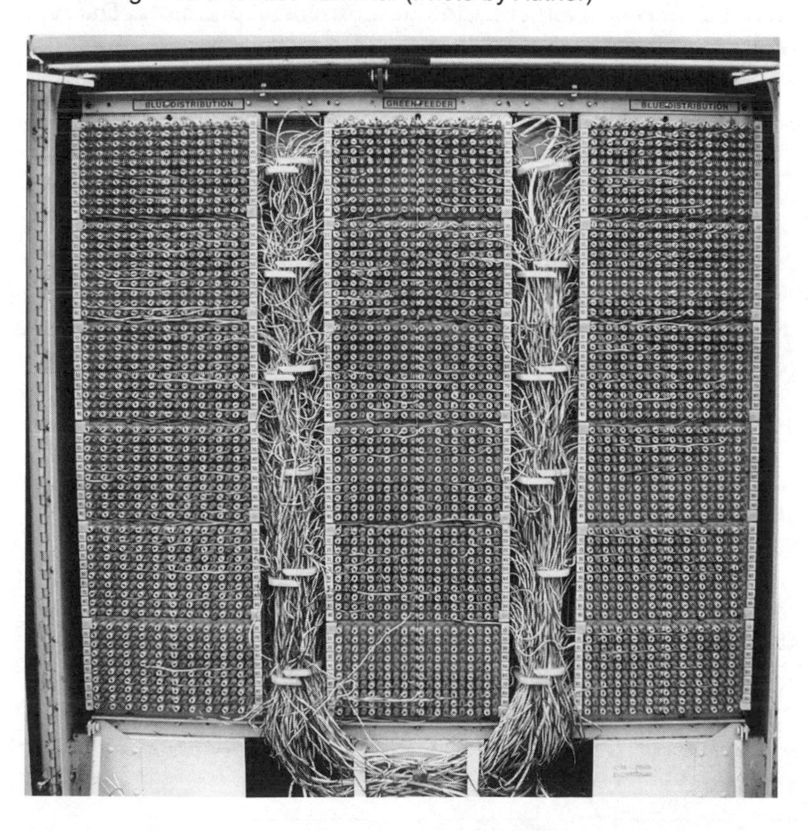

circuit. Bridged tap is inherent in the way drop wire is often connected to the subscriber's premises. The drop wire is bridged across the distribution cable pair, and the unterminated pair continues on to the end of the cable. Bridged tap often must be removed (i.e. the pair cut off beyond the drop) before the loop can be used for DSL or high-speed modems.

The frequency response of long subscriber loops is improved by *loading*. Load coils are small inductors wound on a powdered iron core as shown in Figure 7-5. They are normally placed at 6000-ft (1800 m) intervals on loops longer than 18,000 ft (5.5 km). Load coils are contained in weatherproof cases that are mounted on poles or in manholes. Load coils flatten the frequency response in the voice band, but attenuate higher frequencies. They must be removed from cable pairs that support high-frequency applications such as T carrier and DSL.

ELECTRICAL PROTECTION

Whenever communications conductors enter a building from an environment that can be exposed to a foreign source of electricity, it is essential that electrical protection be used. Protection is required for two purposes—to prevent injury or death to personnel and to prevent damage to equipment in case of lightning strike or cross with electrical power. Common carriers protect both ends of cables between their central offices and the subscribers' premises. The type of protection provided is designed to prevent injury or death to personnel, but may not be sufficient to prevent damage to delicate telecommunications and computer equipment. Interbuilding cables on customer premises may require protection, and the carrier will supply it only if it owns the cable. LECs normally wire a building to their point of demarcation and leave it to the subscriber to place any cable needed beyond that point.

Protection requirements are based on the National Electrical Safety Code. ANSI T1.316-2002, *Electrical Protection of Telecommunications Outside Plant*, and much of the overview in this section is more stringent than the code.

FIGURE 7-5

Toroidal Load Coil

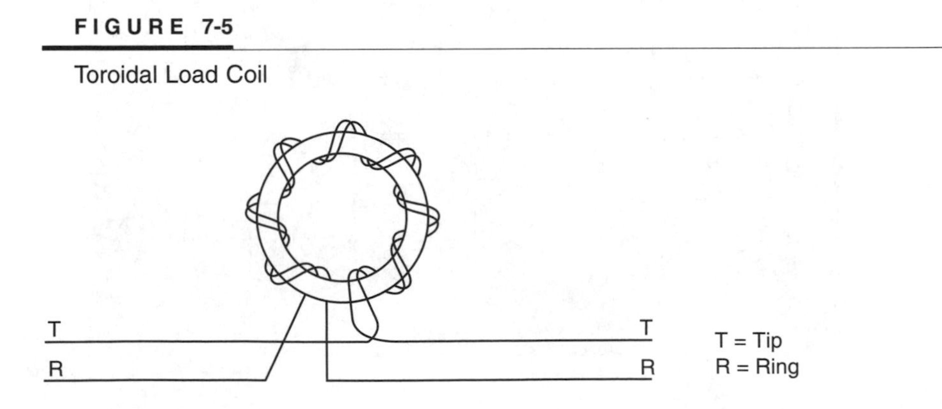

Determining Exposure

The first issue in determining protection requirements is whether the cable is considered exposed. An exposed cable is one that is subject to any of the following hazards:

- Contact with any power circuit operating at 300 V rms (root mean square) or more from ground
- Lightning strike
- Induction from a 60 Hz source that results in a potential of 300 V rms or more
- Power faults that cause the ground potential to rise above 300 V rms

All aerial cables are considered exposed. Even though a short section of aerial cable may not be in proximity to power at the time it is constructed, aerial power may be added later and expose the cable. A cable is considered to be exposed if any pairs within are exposed. All buried and underground cables should be considered exposed unless one or more of the following conditions exist:

- The region experiences five or fewer thunderstorm days per year and the earth resistivity is less than 100 m Ω.
- A buried interbuilding cable is shorter than 140 ft (43 m) and has a shield that is grounded on both ends.
- A cable is totally within a cone of protection because of its proximity to buildings or other structures that are grounded.

Zone of Protection

Structures extend a zone of protection that diverts lightning strikes and shields the cable from damage. As Figure 7-6 shows, if a mast or building is at least 25 ft (7.6 m) high, it offers lightning protection to objects within a radius equal to the mast height. If the structure is 50 ft (15.2 m) high, it protects within a radius of twice the mast height. To illustrate, assume that a cable runs between two buildings, each of which is 50 ft high. Each building extends a zone of protection of 100 ft, which means that a buried cable 200 ft long would not be considered exposed to lightning.

For structures higher than 50 ft, the zone of protection concept can be pictured with the "rolling ball" concept. Visualize a ball 300 ft (91 m) in diameter rolled up against the side of the structure as in Figure 7-7. The zone of protection is shown as the shaded area. Note that the zone of protection applies only to lightning, not to power exposures. If a cable rises above the elevation of surrounding terrain such as on a hilltop or tower it should be considered exposed to lightning even though it is in an area that would otherwise be excluded by the earth resistivity and thunderstorm frequency requirements.

Normally, ground is considered to be at zero potential, but current flow from a power fault or lightning strike can cause the ground potential to rise. Induction occurs when power lines and telephone cables operate over parallel routes. Under

FIGURE 7-6

Cone of Protection Provided by Vertical Grounded Conductors

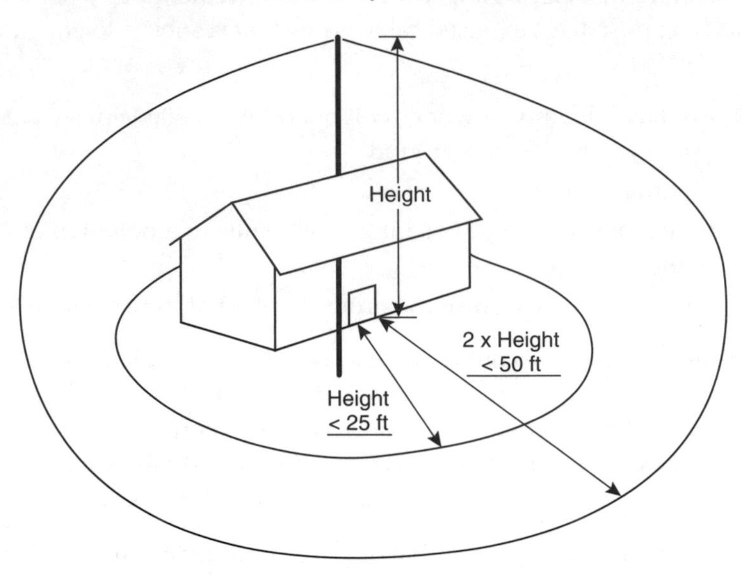

normal conditions the magnitude of the induction is not so great as to constitute a hazard, but when telecommunications lines are unbalanced or when a fault occurs in the power line, the amount of induced voltage can rise to hazardous levels. This is not a concern in most private networks, but designers should be alert to the possibility of induction whenever power and telecommunications circuits share the same route, though they may not share a pole line.

FIGURE 7-7

Protected Zone from Lightning Strike Using Rolling Ball Model for Structures Greater than 50 ft Tall

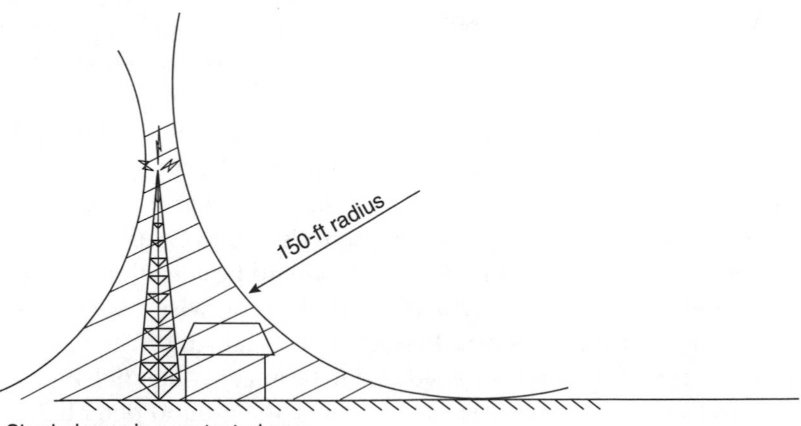

Shaded area is a protected zone

Although a circuit is protected to eliminate hazards to users, the equipment attached to it may be sensitive to foreign voltages. Since the equipment provider is in the best position to know of this sensitivity, all requests for proposal and purchase orders should require the vendor to specify the level of protection required.

Protection Methods

Personnel and equipment can be protected from the hazards of unexpected contact with a foreign source of electrical potential by the following methods:

- Insulating telecommunications apparatus
- Shielding communications cables
- Grounding equipment
- Opening affected circuits
- Separating electrical and telecommunications circuits

This section discusses how each preventive measure is applied to telecommunications circuits.

Insulating Telecommunications Apparatus

The first line of defense against contact with foreign voltages is insulation. Polyethylene, which is the insulation used with most copper cables, has a conductor-to-conductor breakdown value of from 1000 to 4000 V. Although this is enough to guard against high voltages, a lightning strike or power cross may destroy the insulation, so it alone is not enough to satisfy electrical protection requirements.

Shielding Communications Cables

Cables can be shielded from lightning strikes by placing a grounded conductor above the cable so it intercepts the lightning strike. A grounded shield wire can be placed above aerial cable to attract the lightning strike to itself. Shield wires can also be buried above a communications cable. If there is enough separation to prevent arcing between the shield and the cable, this method is effective.

Grounding Equipment

An important principle of electrical protection is to provide a low-impedance path to ground for foreign voltage. Both carbon and gas tube protectors, which are illustrated in Figure 7-8 operate on the principle of draining the foreign voltage to ground.

The simplest form of protector is the carbon block. One side of the carbon block is connected to a common path to ground. It is essential that the ground path be a known earth ground. In most buildings the grounding point for the power entrance is suitable. A metallic cold-water pipe may be a satisfactory ground, but only if the entire water system is metallic. To ensure an effective water-pipe ground, the pipe should be bonded to the power ground with a copper wire of at least #6 AWG.

The conductors are connected to the other side of the carbon block. When voltage rises to a high enough level to arc across the gap, current flows, the block

FIGURE 7-8

Station and Central Office Protection Equipment

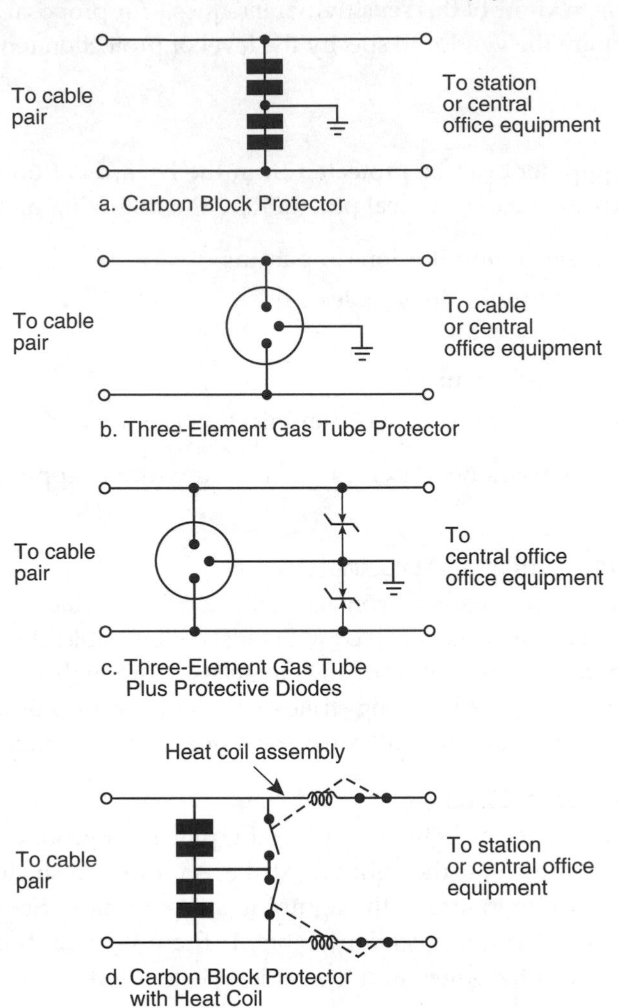

a. Carbon Block Protector

b. Three-Element Gas Tube Protector

c. Three-Element Gas Tube Plus Protective Diodes

d. Carbon Block Protector with Heat Coil

fuses, and the cable pair is shunted to ground. When a carbon block protector is activated, it is destroyed and must be replaced.

A gas tube protector connected between the communications conductors and ground is a better, but more expensive alternative. Like the carbon block protector, its purpose is to provide a low-impedance path to ground for foreign voltage. The electrodes of the gas tube are contained in a glass envelope that is filled with an inert gas. When the breakdown voltage is reached, the gas ionizes and current flows until the voltage is removed. When the voltage is removed, the tube restores itself. Although gas tubes are more expensive than carbon blocks, the self-restoring effect may repay the additional cost. They are particularly effective in sensitive apparatus that is easily damaged by relatively low voltages.

Another type of protector is the *heat coil*, which is a spring-loaded device that grounds conductors when it operates. Heat coils protect against sneak currents, which are currents that flow from voltages that are too low to activate a carbon block or gas tube protector. The heating effect of the sneak current is sufficient to melt a low-melting-point metal that keeps the electrodes separated. When the metal is melted, the spring forces the electrodes together and the circuit is grounded until the heat coil is replaced. Heat coils are normally installed on central office protector frames.

At the user's end of the circuit, protectors range from a simple single-pair device to multiple-pair protected terminals. Although station protectors are adequate to prevent injury to users, they are often inadequate to prevent damage to delicate electronic equipment.

Opening Affected Circuits

Everyone is familiar with the next method of protecting circuits and equipment: a fuse or circuit breaker. If the communications conductors are opened before they enter the building or before they reach the protected equipment, current cannot flow and damage the equipment or reach the operator. The LECs often install a fuse cable between its distribution cable and the building entrance. A fuse cable is a short length of fine-gauge cable, usually 27-gauge. In the case of sustained high current the finer gauge fuse cable will open before the protected cable.

A fuse, by its nature, takes time to operate. Current flow during lightning strikes tends to be short, lasting less time than it takes to open the cable. Therefore, fuse cables are effective against power crosses, but not against lightning.

Separating Electrical and Telecommunications Circuits

Another method of protecting from accidental cross with electrical power is adequate spacing. Buried power and telecommunications cables share a joint trench to many buildings. The minimum acceptable separation between power and telecommunications circuits in a joint trench is 1 ft; more separation gives an additional measure of protection. The sharing of a joint trench with at least the minimum separation does not, of itself, create an exposure condition.

DIGITAL LOOP CARRIER (DLC)

DLC is increasingly used to deliver telecommunications service to concentrations of subscribers. The DLC unit is fed from the central office with either fiber optics or copper cable. The feed from the DLC to the subscriber is usually copper cable. DLCs are available in two configurations. One alternative uses matching central office and remote units with analog interfaces to the central office line circuits. The other uses a standard interface known as GR-303 in the central office. The GR-303 interface matches the remote unit to any compatible digital switch. The degree to which the DLC increases the channel-carrying capacity of a cable pair is known as its *pair gain*. For example, a 24-channel DLC operating over separate transmit and receive pairs has a pair gain of 22.

DLC provides better transmission quality than standard cable facilities. The transmission loss is fixed, normally at 5 dB or less, and channel noise is lower than cable pairs. The remote terminal is contained in a pole-mounted cabinet or in an environmentally controlled vault that is mounted either above or below ground. The enclosure provides space for multiple racks of equipment, battery backup, alarms, and other facilities to provide reliability that is equivalent to cable pairs. The feed from the central office is often a self-healing fiber ring. In addition to POTS service, the complex may be equipped for DSL. A variety of plug-in units are available to support services such as ISDN and DSL to subscriber locations that are outside the cable-pair range for these services.

OUTSIDE PLANT APPLICATION ISSUES

The focus of this chapter has been on LECs' outside plant applications, but the same principles apply to private cable systems. Private campus networks in the past have had little choice but to use copper cable to link buildings because fiber-optic equipment was too expensive for voice applications. Today, VoIP and fiber-connected remote switching units often make it feasible to eliminate copper cable between buildings except for alarm systems. Wireless also offers an alternative to both copper and fiber in small locations. Nevertheless, copper cabling will survive on campuses for many decades to come.

Any new cable construction should include plans for both fiber and copper cable. For telephone systems the break-even point between running copper cable directly to the PBX as opposed to using a fiber remote depends on the number of stations. For a few stations, perhaps 20 or fewer, copper cable is usually less expensive and more reliable than a separate remote unit. Fiber will be part of almost any plan, either initially or in the future, so where fiber is run anyway, larger concentrations of stations can be connected to the main PBX with fiber. One major advantage of fiber is the fact that electrical protection is not needed.

Cable can be buried directly in a campus environment, but conduit should be used if possible. Conduit not only offers physical protection, it also facilitates future rearrangements. Flexible conduit can be plowed underground. That method is often used with small conduits intended for fiber alone. Copper cable in sizes above 25 pairs requires conduit too large and inflexible to be plowed. Trenching is usually more expensive and disruptive than plowing, but it is easy to place additional empty conduits while the trench is open. If the conduit is trenched, there is usually little reason to use than the full 4-in. size. Copper cable and fiber can share the conduit if it is large enough. In a shared conduit the fiber should be protected by installing an inner duct, which is a 1-in. flexible sub-conduit.

An investment in cable is far more durable than the apparatus that uses it. Fiber, copper cable, and conduit have service lives measured in decades. With the right design, the service life of cables connecting between buildings should approach the lives of the buildings themselves.

CHAPTER 8

Access Technologies

If the world were ideal from a telecommunications standpoint, fiber optic access with multimegabit bandwidth into the Internet would be available to every business and home at an affordable price. Although that day will eventually arrive, for now residences and small businesses have to make do with what is available. Most larger businesses have fiber optic feeds in to their buildings and plenty of bandwidth. They connect to their ISPs with T1/E1, DS-3, or some higher multiple. Small business and residences use one of the technologies described in this chapter: DSL, cable, or wireless. The bandwidths of these alternatives do not approach that of fiber optics, and the prices are affordable for households of moderate income, but beyond the reach of low-income households. In the United States well under half of the households have full-time Internet access despite agreement of government agencies that it is a national priority.

The importance of access technologies cannot be overstated. The Internet is a crucial force in accelerating social and economic development. It has the potential to provide jobs, improve productivity, and create new services and industries. It can expand democracy by giving people direct access to government information and services. It facilitates teleworking, distance learning, telemedicine—the list of potential applications is almost endless. The lack of broadband access creates a chasm increasingly known as the "digital divide." When broadband access is limited to only a few, class distinctions widen as one group of citizens has at its fingertips access to services and information that others lack.

Although the technologies discussed in this chapter are usually considered for Internet access, nothing prevents their use for access to other services such as frame relay. Many service providers offer VoIP telephone service as well. This chapter explains how these three principal access technologies and their many variations work.

DIGITAL SUBSCRIBER LINE (DSL)

DSL operates over the ILECs' copper wire local loop. Although the loop is not ideal for broadband communications, it is widely available, inexpensive, and does the job effectively. Figure 8-1 shows the layout of DSL service. A specialized modem at the subscriber's premises connects over the cable pair to a *DSL access multiplexer* (DSLAM) in the central office. Some varieties of DSL derive a data channel over a POTS line and support both voice and data, while other types are exclusively data. At the central office, the DSLAM routes the telephone channel to the local switch and feeds the data portion into the backbone network, where it routes to the ISP. The backbone network is typically frame relay, ATM, or IP.

DSL Standards

DSL comes in an alphabet soup of options going by the names of HDSL, SDSL, IDSL, ADSL, RADSL, and VDSL, collectively described by the term xDSL. Each of these has different characteristics and fits a different niche in the marketplace. To add to the confusion, two different modulation methods, discrete multitone (DMT) and carrierless amplitude phase (CAP) are used. Table 8-1 lists the types of DSL and their approximate transmission range.

Early DSL service required installation by a specialist, but this was expensive and time consuming. To compete with cable providers, the Universal ADSL Working Group was formed in 1998 and charged with developing a better standard. The principal objective was to make it easy for subscribers to purchase self-configuring modems and eliminate the need to dispatch technicians to set them up. The resulting protocol, known as G.lite, can usually be installed without

FIGURE 8-1

DSL Configuration

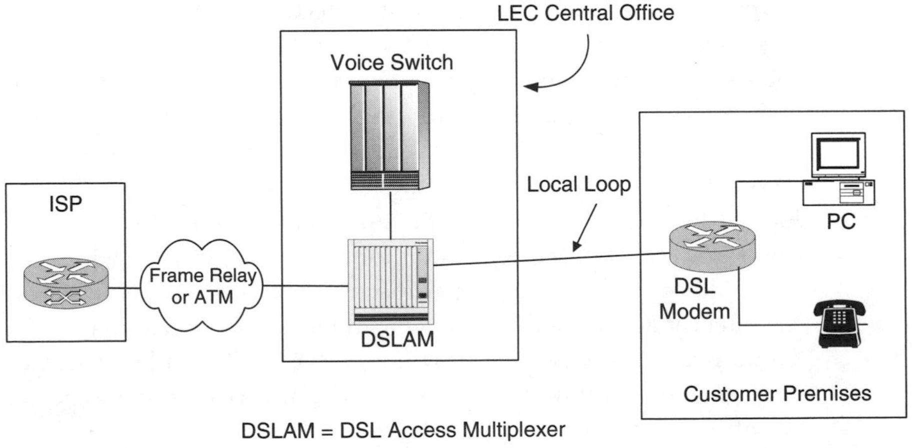

TABLE 8-1

DSL Operating Parameters

Type of DSL	Acronym	Upstream Bandwidth	Downstream Bandwidth	Range in feet(meters)
Asymmetric Digital Subscriber Line	ADSL	16 to 640 Kbps	1.5 to 9 Mbps	18,000 (5500)
High Bit-Rate Digital Subscriber Line	HDSL	1.544 or 2.048 Mbps	1.544 or 2.048 Mbps	12,000 (4000)
ISDN Digital Subscriber Line	IDSL	144 Kbps	144 Kbps	18,000 (5500)
Single-Pair Digital Subscriber Line	SDSL	1.544 or 2.048 Mbps	1.544 or 2.048 Mbps	12,000 (4000)
Splitterless DSL	G.Lite	16 to 640 Kbps	1.5 to 6 Mbps	18,000 (5500)
Very High Bit-Rate Digital Subscriber Line	VDSL	1.5 to 2.3 Mbps	13 to 52 Mbps	1000 to 4500 (330 to 1500)

difficulty. Bridged tap may be a problem, however, because the cable records do not always show whether it exists, so the ILEC may have to condition the cable pair to make DSL work.

DSL also has range limits that make it unavailable to some telephone subscribers. Since the ILECs normally load any cable pairs greater than 18,000 ft (5500 m), this forms the maximum loop length for DSL. It can operate over longer loops, but transmission will be affected on the telephone circuit. ILECs may resolve the loop-length issue by installing neighborhood gateways that are fed by fiber optics. Specialized DLCs can be obtained with the DSLAM function built into the system. Neighborhood gateways permit the LECs to keep the loop length within a limited range of 1000 to 2000 ft. (330 to 660 m). Very high bit-rate DSL (VDSL), which is capable of handling video and high-speed data signals, can operate over these short loops.

DSL Technology

DSL can be categorized by two primary criteria: the line coding and modulation methods. The methods are similar to those used in high-speed modems as we discussed in Chapter 4.

QAM Coding

Early 8-QAM modem systems used eight phase and two amplitude levels to transmit four bits over each symbol. By increasing this to 16-QAM, 12 phase changes are combined with two amplitude levels. Amplitude changes occur on only four of the phase changes, so both a phase change and an amplitude change would be valid in some symbols but not in others. Trellis coding defines the way in which signal transitions are allowed to occur. Signal transitions that do not follow the expected pattern are detected as errors and may be corrected with FEC. By using finer phase and amplitude changes, QAM can be extended to six bits per symbol (64-QAM), and eight bits per symbol (256-QAM).

Modulation Methods

DSL modems use either of two modulation methods: DMT and CAP. DMT, which is ANSI standard T1.413, separates the data spectrum into 256 channels called *bins*, each of which is 4 KHz wide. A simple splitter separates voice from the data signal. Data at rates up to 60 Kbps per bin are modulated onto each channel using QAM. The modems detect any impaired channels and spread the data to unimpaired channels. DMT is inherently rate-adaptive, adjusting the data rate to the maximum that the medium will support.

In the other method, CAP, the two directions of transmission are separated on different frequency segments and modulated with QAM. At the receiving end, a simple filter separates voice and data. The modems use adaptive equalization to compensate to some degree for the effects of splices, bridged tap, and gauge changes. CAP is a proprietary method, but despite its lack of standard status, it has a large following.

Spectral Compatibility

The degree to which DSL signals can coexist in the same cable is a matter of concern to the ILECs. The limiting factor is near-end crosstalk, which is most severe near the central office because the cable density is highest there. DSL systems use the lowest frequencies for the upstream direction to compensate for the fact that crosstalk is less severe at lower than at higher frequencies. To control crosstalk, the LECs limit the number of pairs that are assigned to DSL in each cable complement.

DSL Characteristics and Applications

Each of the DSL types listed in Table 8-1 has a particular function. This section briefly discusses the characteristics and applications of each type.

Asymmetric Digital Subscriber Line (ADSL)

ADSL is the most common type and is most attractive for casual users. It supports bandwidth as high as 1.5 Mbps downstream over a range of up to 18,000 ft. Upstream the rate drops to between 64 and 640 Kbps. The service is acceptable for Web surfing, which is inherently asymmetric. A splitter at the subscriber's premises separates the voice and data signals and terminates the line with constant impedance. The splitter has to isolate the data signal from sharp impedance changes that result from users' lifting the handset or unplugging telephone sets.

G.lite eliminates the splitter by retraining itself rapidly in the face of these sudden impedance changes.

Very High Bit-Rate Digital Subscriber Line (VDSL)

VDSL was developed to support video or high-speed data. The wire length is kept short by locating an access node near subscribers and feeding both voice and data to the node over fiber, using wire in the last span. Rates as high as 52 Mbps can be supported over a 1000-ft (330 m) range, but the rate drops to about 13 Mbps beyond 4500 ft. (1500 m).

High-Speed Digital Subscriber Line (HDSL)

ILECs use HDSL to provide T1/E1 service over two twisted pair wires up to 12,000 ft (4000 m) long. This service is not sold to subscribers as DSL service, but as T1/E1 point-to-point service.

Single-Pair Digital Subscriber Line (SDSL)

SDSL, also known as symmetric DSL, provides T1/E1 service on a single cable pair and derives a POTS line under the data signal. It transmits and receives in the same band of frequencies using an echo-canceling protocol.

ISDN Digital Subscriber Line (IDSL)

IDSL provides 128 Kbps of bandwidth using the same technique as BRI. This method has little application because its bandwidth cannot compete effectively with other DSL technologies.

CABLE ACCESS

Cable systems were originally constructed with one-way amplifiers, which made them unsuitable for access service. Most cable operators have now upgraded the distribution cable to two-way, and the backbones to fiber optics. The industry refers to this as hybrid fiber-coax (HFC). Fiber trunk cables are routed from the service provider's central location, which is known as the *headend*, to a neighborhood center where the signal is converted to coaxial cable and routed to the customer's location. Today, cable is the most prevalent method of providing residential Internet access. Chapter 22 discusses the architecture of cable systems in more detail.

When the first two-way systems went into operation, the industry lacked standards, so early cable systems were constructed using proprietary protocols. IEEE organized the 802.14 committee to develop a standard, but the cable industry developed a method known as data over cable system interface specifications (DOCSIS). In areas where the cable provider offers service meeting DOCSIS specifications, cable modems can virtually be self-configuring. Unlike DSL, which is generally open to any ISP, most cable systems are restricted to the cable provider's ISP. The FCC exempts cable from the open access requirements of DSL. In 2005 the U.S. Supreme Court upheld the FCC.

Cable Access Technology

Cable operators inject information and entertainment channels at the headend. Standard television channels are 6 MHz wide, and with QAM modulation, one channel can carry 27 Mbps of downstream data. The service provider decides how many subscribers will share this bandwidth. The cable operator uses DHCP to furnish IP addresses from a central pool. The subscriber can couple directly to a cable modem. If more than one computer is connected over the modem, each computer gets a separate address. Some cable providers limit the subscriber to one IP address. Others do not limit the number of addresses, but charge for each simultaneous address the customer uses. Most subscribers connect a router to the cable modem and use NAT ahead of private IP addresses to make one address suffice. The asymmetric nature of Internet access fits well into the cable transmission scheme with its limited upstream and abundant downstream bandwidth.

Downstream packet flow is a simple matter. Each station listens for its IP address and ignores the other packets on its segment. Upstream is more difficult because the medium is shared and the protocol must have some method of granting access to stations with traffic to send. Any station on a segment can receive any packet, so the protocol encrypts the packets for privacy using RSA, which is a public key algorithm.

Data over Cable System Interface Specification

CableLabs, Inc., a consortium of equipment manufacturers developed DOCSIS as a standard method of providing Internet access over a 6 MHz RF channel. In Europe, DOCSIS is known as a Euro-DOCSIS, and is derived from the U.S. version. Although DOCSIS is designed for cable, with an RF front end and modifications it can also be used in wireless Multipoint, Multichannel Distribution Service (MMDS), and Local Multipoint Distribution Service (LMDS), which we discuss in the next section.

The cable modem (CM) is a frequency-agile transceiver that is tuned to upstream and downstream channels. A splitter on the customer's premises separates the data and television portions of the signal. The CMs are all tuned to the same channel and use their IP address to pick their packets from the stream. The CM usually connects to the customer's router with Ethernet.

When a CM is first turned on, it begins scanning the cable for a downstream data channel. From this it learns the frequency of the upstream channel. It then broadcasts its presence to a *cable modem termination service* (CMTS). The CMTS identifies the CM by its MAC address, validates its right to use the service, obtains an IP address from the DHCP server and sends the IP address to the CM. With this initial handshake completed, the CM is prepared to communicate over the cable. If the user replaces the CM, it must be reregistered with the CMTS.

Besides the CMTS, the headend includes a variety of specialized servers such as a DHCP server, an authorization server, and a time-of-day server. A DNS server correlates URLs to IP addresses and a Trivial File Transfer Protocol (TFTP) server is provided to facilitate file transfer. In addition, most service providers offer e-mail and have servers to cache popular Web pages to reduce the amount of bandwidth required in the Internet backbone.

DOCSIS uses a frame format similar to Ethernet in both the upstream and downstream directions. The LLC is standard IEEE 802.2 protocol. The rest of the datalink layer has two other sublayers in addition to the LLC: link security and MAC. The link security sub layer includes three security protocols: Baseline Privacy Interface (BPI), Security System Interface (SSI), and Removable Security Module Interface (RSMI). The BPI encrypts data traffic between the user's modem and the CMTS. Figure 8-2 compares the DOCSIS architecture with the OSI model.

The MAC layer includes collision detection and retransmission, error detection and recovery, and procedures for registering modems. It also performs ranging, which enables the CMTS to evaluate the time delay to each cable modem and allocate the upstream timeslots accordingly.

The downstream channel, which is standardized as ITU J.83, is either 64 QAM or 256 QAM. The downstream payload of 64 QAM is approximately 27 Mbps. Eight amplitude levels are used to modulate the carriers and six bits of data are transmitted at a time. The payload of 256 QAM is 39 Mbps using 16 amplitude levels and transmitting eight bits of data at a time. The bandwidth

FIGURE 8-2

DOCSIS Compared to OSI

<table>
<tr><th>OSI</th><th colspan="2">DOCSIS</th></tr>
<tr><td>Higher Layers</td><td>Applications</td><td rowspan="3">DOCSIS Control Messages</td></tr>
<tr><td>Transport</td><td>TCP/UDP</td></tr>
<tr><td>Network</td><td>IP</td></tr>
<tr><td>Datalink</td><td colspan="2">IEEE 802.2
Link Security (BPI, SSI, RMSI)
Media Access Control</td></tr>
<tr><td rowspan="2">Physical</td><td>Upstream</td><td>Downstream</td></tr>
<tr><td>TDMA
5-40 MHz
QPSK/16-QAM</td><td>MPEG-2
54-850 MHz
64/256-QAM</td></tr>
</table>

BPI = Baseline Privacy Interface
SSI = Security System Interface
RMSI = Removable Security Module Interface

of the RF signal is from 180 KHz to 6.4 MHz with the data rate varying from a low of 320 Kbps to a high of 20.5 Mbps.

All users on a coaxial segment share the aggregate bandwidth, so the actual throughput any user will experience is lower and varies with the amount of activity. The service provider can maintain its service level by regulating the number of subscribers that share a channel. The upstream direction uses either *quadrature phase shift keying* (QPSK) or 16 QAM. Although 16 QAM has twice the data rate of QPSK, the latter is more tolerant of interference. The upstream direction is of much lower speed than the downstream channel, having a typical bandwidth of 300 Kbps to 1 Mbps. The upstream channel uses TDMA to enable users in a cable segment to share the channel. The stations cannot hear each others' transmissions, so the protocol provides for collision detection and retransmission.

Downstream data is encapsulated in MPEG-2 frames. MPEG stands for the Motion Picture Experts Group, which developed a standard method of transmitting compressed digital video. MPEG-2 provides studio-quality video, including support for HDTV. It allows multiple channels to be multiplexed into a single data stream.

DOCSIS specifies Reed-Solomon FEC as a means of improving error performance. FEC adds redundant bits to the bit stream and sends it along with the information bits. If errors occur the decoder attempts to correct them before the bit stream is presented to the application, with the final error check done at the receiving apparatus.

WIRELESS ACCESS

Wireless is the only broadband access method available to many users. Cruise ships and airlines offer their passengers telephone and Internet access over satellites. Emergency vehicles and mobile services all rely on radio waves for access to voice and data networks. Also, many subscribers, particularly in rural areas, are beyond the range of DSL and cable is unavailable. The industry provides a variety of wireless alternatives that we discuss briefly in this chapter, and in more detail in Part 3.

The FCC regulates the use of the airwaves. Most of the radio spectrum is licensed, meaning an FCC license must be granted for its use. The FCC in the United States and regulatory agencies in most other countries also allocate some unlicensed spectrum for certain applications. Licensed spectrum is the most reliable from the user's standpoint because the licensee must conduct interference studies before the license is granted. Satellites, cellular, and some services that use the acronym *wireless local loop* (WLL) use licensed spectrum. Unlicensed spectrum, which the 802.11 protocols use, is ubiquitous, but open to a variety of uses and the user has no assurance against interference.

Wireless has much in common with cable in that the medium is shared. As with cable, the downstream direction is straightforward because the receiving device is set up to respond to its address and ignore the others. The upstream direction is more difficult because multiple contenders are vying for access to the same bandwidth. Upstream sharing takes place by one of three multiple access methods: frequency division (FDMA), time division (TDMA), or code division (CDMA).

FDMA is the least favored method because each user is assigned a frequency slot. If the user does not transmit at a particular instant, the slot is wasted. Analog cellular, which uses a calling channel to assign users to a frequency pair, is an example of FDMA. TDMA divides the upstream bandwidth into time slots. When stations can hear one another as they do with Ethernet, they defer transmissions until no other station is transmitting. When they cannot hear one another, as is the case with satellites, cable, and cellular, access is regulated by assigning time slots from a master station.

CDMA, also known as *direct sequence spread spectrum* (DSSS), shares a wide band of frequencies. A code embedded in the transmission enables the receiver to pick the desired signal from the jumble. DSSS is analogous to a crowded room filled with people speaking many different languages. Each listener is able to focus on his or her native language and pick out that conversation from a cacophony of voices.

Radio signals are convenient for mobile and portable applications, but the path is unreliable because of the ultra-high frequencies involved. Line-of-sight between transmitter and receiver is required for reliable communication, and even when the signal is strong, multipath fading may attenuate it. Fading occurs when a signal is reflected from some object. The reflected path is somewhat longer and arrives at the receiving antenna slightly out of phase with the direct path. Obstructions in the path also attenuate radio signals.

WiFi 802.11

The FCC allocates unlicensed frequencies in the industrial, scientific, and medical (ISM) and unlicensed national information infrastructure (UNII) bands. The frequency ranges are 2.4 to 2.4835 GHz and 5.1 to 5.825 GHz. Collectively, the LAN services that ride these bands are known as wireless fidelity (WiFi). In the 2.4 GHz band, users have two protocol choices. The first, designated as 802.11b, has a maximum data rate of 11 Mbps. The other, 802.11g operates at 54 Mbps. In the 5 GHz band, 802.11a operates at a maximum data rate of 54 Mbps. The 2.4 GHz band is shared with a variety of devices such as wireless LANs, microwave ovens, and cordless telephones, so interference can be a limiting factor. The FCC requires ISM devices to use spread spectrum modulation with a maximum of 1 W of power. The throughput achieved will be much lower than the quoted data rate because of protocol overheads that are discussed in more detail in Chapter 21.

The main advantage of WiFi is its ubiquitous coverage. Airports, coffee shops, and even some street corners are covered by wireless signals. The protocols provide wired equivalent privacy (WEP), which prevents users from accessing the signal unless they have the necessary permission, but many wireless signals are open. In other cases a service provider restricts access to paid subscribers by redirecting users to a logon application. WiFi is discussed in more detail in Chapter 21.

WiMax 802.16

The WiMax protocols are not approved at the time this is written. IEEE is working on the protocols and expects them to be completed in 2007. Meanwhile, several proprietary protocols using WiMax techniques are currently on the market. WiMax is discussed in more detail in Chapter 21. It is intended for both fixed and mobile applications and is expected to have a range of around 10 miles (16 km). The frequency spectrum is not allocated yet, but it will be in the microwave range. Briefly, WiMax operates by exploiting multipath fading, which is normally considered an impairment to microwave transmission. WiMax broadcasts the signal in small segments over separate channels, using orthogonal frequency division multiplexing (OFDM), which is a technique similar to DMT modulation in DSL. The signal bounces off multiple obstructions and is combined in the receiver to reconstruct the original signal. WiMax is intended for service providers to bypass the wired local loop for access and point-to-point services in the metropolitan area.

Satellite Access

Satellite service is the only method of providing access from airlines and ships at sea. It can also be used from terrestrial locations, but the cost of the upstream channel is prohibitive for most residential users. Where two-way service is not feasible, telephone service can be used for the upstream direction and satellite downstream. The main drawback to satellite access besides the upstream channel issue is latency. The delay of a satellite channel is detrimental to some applications. For example, the delay is too great to support VoIP.

Local Multipoint Distribution Services

LMDS is intended to bypass the copper local loop with an economical service that supports telephone, Internet access, and video. WLL providers bid for right to use frequencies in the 28 to 31 GHz range known as the A block with 1150 MHz of bandwidth and the B block with 150 MHz of bandwidth. The service provider places a hub in the center of a serving area. Subscriber locations are equipped with small rooftop antennas and transceivers as shown in Figure 8-3. The base station contains a variety of servers to support the services offered. Upstream access is allocated with TDMA using PSK or QAM modulation.

At the high microwave frequencies used, the signal path is subject to disruption from rainfall and fading. An acceptable link of as much as 8.5 miles (14 km) may be achievable in some climates, while in heavy rainfall areas the maximum link may drop to 1.5 miles (2.5 km).

FIGURE 8-3

Local Multipoint Distribution Service Architecture

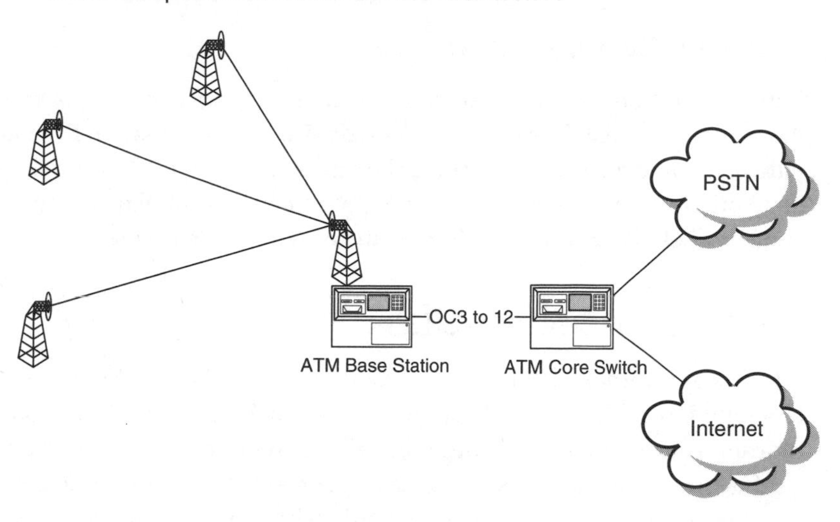

Multipoint Multifrequency Distribution Service (MMDS)

MMDS was originally intended as a wireless replacement for cable television with 200 MHz of spectrum allocated between 2.5 and 2.7 GHz. The architecture of MMDS is similar to LMDS, using head-end equipment like that used in a CATV system. For video reception, a set-top converter demodulates the incoming signal to the frequency of conventional television channels. The MMDS signal is transmitted from an omni-directional antenna. Repeaters may be used to extend the range or to fill dead areas caused by shadows in the coverage area.

MMDS was initially intended for one-way video but it is now authorized for two-way service, making it suitable for Internet access. Typical service offerings provide downstream transmission rates of 1 Mbps or higher, scaleable up to 10 Mbps, and upstream speeds up to 512 Kbps, which makes it competitive with DSL and cable access. The MMDS spectrum is shared with instructional television fixed service (ITFS), which is intended for distance learning video. The 6 MHz video channels can be modulated with data signals using the same concepts and in many cases the same hardware as cable modems.

Cellular Digital Packet Data (CDPD)

CDPD, which is discussed in more detail in Chapter 20, is a packet overlay on cellular sites. The technology uses idle analog cellular channel capacity to support TCP/IP sessions at speeds up to 19.2 Kbps. The main advantage of CDPD is its coverage, which is available wherever the service provider has converted the cell site. Unlike other access methods, CDPD charges are usage sensitive. While it is a suitable service for e-mail or remote access to a data network, the slow speed and the per kilopacket charge make it a poor choice for Internet access.

Third Generation Cellular (3G)

Third generation (3G) cellular is intended to support Internet access from a cell phone or equivalent. The most common device is a combination PDA and cell phone with a small keyboard to overcome the limitations of the 12-button keypad. Despite the restrictions of the keypad and the tiny screen, 3G cellular will form the basis for a wide variety of new applications, some of which we discuss in Chapter 20.

ACCESS APPLICATION ISSUES

In terms of affordable full-time Internet access, the United States lags much of the developed world. Although about three-fourths of American households have Internet access, current estimates are that it will be as late as 2008 before broadband development reaches as high as 50 percent. This is in feeble contrast to Asian countries such as Japan and South Korea where broadband access is approaching 80 percent.

The choice of access technology depends to a large degree on what is available. The only universally available choice is dial-up, which is a poor alternative for full-time access. Even ISDN with its 128 Kbps of bandwidth is insufficient for many users and if the service is usage sensitive, the cost may be prohibitive. Only the casual residence user and smallest businesses will find dial-up access acceptable. For a business with multiple employees who must have access to e-mail, a full-time connection is essential. For large businesses that host Web servers, T1/E1 and higher multiples are the only acceptable alternative. In between are residences and countless smaller businesses that cannot justify the cost of T1/E1. The choice between DSL, cable, and wireless may boil down to the question of what is available.

Regulatory Issues

In the United States, Congress and the FCC have a major role to play in broadband development and the outcome is difficult to predict. Both bodies have announced that broadband development is a major goal, but it is still fraught with uncertainty. The large ILECs, SBC, and Verizon have announced plans to build FITL following the FCC's determination that they will not be required to share it with competitors. Left to their own volition, the ILECs are likely to build fiber where it is the most profitable, starting with upscale neighborhoods. This, however, is counter to the goals of municipalities that regulate the use of the public right-of-way. Their goals are often to bring affordable access to the least developed neighborhoods. The states have the additional goal of bringing broadband access to rural areas, where the low density makes it the least cost-effective. Meanwhile, some municipalities take matters in their own hands and build fiber optics or WiFi access, often against the objections of the LECs. Major conflicts are likely to erupt over these issues in the next few years.

Choice of Medium

Cable access is available in localities in which the cable operator has upgraded the facility to two-way and provides access service over the cable. This excludes many small communities and rural areas, although cable access is becoming more widely available outside large metropolitan areas. Cable is often not available in office buildings, many of which are not wired for cable, but this is also changing as cable providers recognize the scope of the small business market and prepare to offer telephone service.

DSL also has availability problems. First is the question of loop length and bridged tap, which precludes DSL for around 40 percent of the telephone subscribers. A certain critical mass of subscribers is needed to make it economical for the service provider to install DSLAMs and the connecting data network. Many rural communities do not have enough potential subscribers to make it worthwhile for the LECs to provide DSL.

This leaves wireless as an alternative. Terrestrial wireless depends on having line-of-sight, which precludes it for a significant number of potential subscribers. Furthermore, it is a shared medium and therefore subject to the same kind of bandwidth limitations that cable experiences. Satellite technology circumvents line-of-sight restrictions, but radio uplinks are expensive, so many satellite services use the telephone line for the upstream direction.

Wireless has a huge potential, but the spectrum is limited. Furthermore, licensed spectrum, which is the most effective for the service providers, is costly. During the market bubble of the late 20th Century, several service providers spent enormous amounts of money to acquire licensed spectrum. The service demand failed to materialize and companies went bankrupt. From a service provider's standpoint, WiMax is currently considered the application with the most potential. At this writing, several companies have announced plans to offer WiMax service, but they have two choices: use proprietary protocols or wait for IEEE to complete its work. If the experience with WiFi is any indication, the market will not develop until open protocols are complete.

Power companies are beginning to consider using their distribution lines to provide Internet access. Fiber optics would be used to bypass the high-tension lines and the signal would be applied as an RF carrier to the distribution plant. This technique has several problems that must be overcome. First is bypass of transformers, which block any RF signal. This problem is not technically challenging, but the second issue of RF radiation is formidable. Applying high-speed signals to the power line will undoubtedly result in radiation that has the potential of interfering with other services. The American Radio Relay League, in particular, has raised questions with the FCC about interference. Power line access is merely a vision at present, but it is an option to watch.

Over-Subscription

The principle of over-subscription is built into every telecommunications network. The available capacity is always engineered at less than total aggregate demand because the service provider knows that during normal conditions, users will not all require service simultaneously. During abnormal conditions, which are caused by out-of-the-ordinary events such as storms, disasters, political disturbances, and such circumstances, any network will be blocked or provide longer response times than normal.

The objective of subscribers with always-on access is to obtain acceptable performance in the access channel. DSL subscribers have unblocked service in the channel up to the DSLAM, but the channel is shared beyond that to the ISP. Cable is shared in the coaxial segment. The cable provider can regulate the number of households that a coaxial segment serves. Wireless and cable are similar in their response to loads. Both have contention in the upstream access channel and between the service provider's headend and the Internet. The Internet, in particular, has numerous chokepoints and it is impossible for the user to determine where the bandwidth restriction actually is.

The issue of oversubscription is not critical for Internet access because most users can tolerate additional delay without adverse impact. Telecommuting subscribers, however, may not be able to tolerate oversubscription because their applications may require predictable delay. Unfortunately, about the only way to be certain is to try it.

Voice and Video Service

Internet access is asymmetric in nature, so the limited upstream bandwidth of cable, wireless, and DSL are appropriate. Once voice and video are added to the equation, however, a symmetric channel with QoS capability is required. Some cable and wireless providers offer a total package of telephone, entertainment, and Internet access over their medium. Some DSL providers offer voice over DSL (VoDSL), using a single unbundled cable pair to provide packetized voice channels that are multiplexed with data packets.

CHAPTER 9

Structured Cabling Systems

A structured cabling system is the foundation of any LAN. During the life of a typical office building, people will move several times, computers will be updated every three or four years, and servers and printers will be changed for newer and faster models. A properly designed cabling system will sustain these changes without interrupting operations, and will support future applications that cannot be predicted at the time of installation.

When terminals were the method of accessing the corporate mainframe, the physical medium was coax, twinax, or shielded wire. These are not only hard to install but also more difficult to manage. With these rigid wiring systems, when users changed location, the old cable was hard to remove. It was easier to abandon the old cable and install a new one. As a result, ducts and conduits became clogged with dead wire, which then became impossible to remove. Until the late 1980s when AT&T demonstrated that properly manufactured unshielded twisted pair wire (UTP) had plenty of bandwidth to support data, most data manufacturers insisted on shielding. The LAN medium was either thick or thin coax, which were paragons of inflexibility. Telephone cable was unshielded, but each vendor had its own installation standards so telephone wiring was an uncoordinated concoction of wire sizes ranging from 2-pair to 25-pair cable.

In the early 1990s the EIA and TIA collaborated to develop standards for a structured cabling system. The initial standards covered cable and connector performance standards to ensure data rates of at least 100 Mbps could be sustained over distances of up to 100 m. These standards, the current versions of which are listed in Table 9-1, have been updated several times to further tighten the performance criteria. Structured wiring standards are among the most important in the world because they ensure that cabling and equipment will be compatible. Structured wiring has evolved out of the office and into homes, many of which are being wired to support Internet access from living areas and smart appliances that require Ethernet connectivity.

TABLE 9-1

TIA/EIA Cabling Standards

Standard	Description
TIA/EIA-568-B Series	Commercial building telecommunications cabling standard
TIA/EIA-569-A	Commercial building standards for telecommunications pathways and spaces
TIA/EIA-569-A-6	Commercial building standards for telecommunications pathways and spaces for multi-tenant buildings
TIA/EIA-606	Administration standard for commercial telecommunications infrastructure
TIA/EIA-570-A	Residential and light commercial telecommunications cabling standard

The primary advantages of using UTP over other media are cost and ease of installation. Wire is readily available from many vendors, and trained personnel can install it with simple and inexpensive hand tools. Wire is manufactured in multiple twisted-pair cable, both shielded and unshielded, and in flat ribbons for under-carpet installation. It is durable and with its sheath intact it is impervious to weather.

Unfortunately, the standards have not eliminated uncertainty for those who plan cabling systems because the target keeps moving. When the standards were first developed, 100 Mbps was the outer range of data transmission speeds in the local network. Fast Ethernet was 10 times the speed of Ethernet, and it was difficult to imagine why higher speeds to the desktop would be needed. If selected users needed higher speeds someday, designers reasoned, fiber optics would be the answer. Since then, 100 Mbps Ethernet has become the default, and the price of gigabit Ethernet NICs and switch ports are dropping to the point that they are only marginally more expensive than 10/100 Ethernet. It is not clear that a gigabit is needed to the desktop, but the wiring standards have expanded to support it and the industry will undoubtedly evolve in that direction.

This chapter presents a summary of structured wiring principles and standards. The standard documents are much more detailed and should be consulted for specifics.

OVERVIEW OF STRUCTURED WIRING

Structured wiring brings numerous benefits to the owner:

- The system is modular and easy to expand.
- Compatibility with equipment of different manufacturers is assured.
- Moves, adds, and changes of voice and data stations are facilitated.
- Reliability is improved by providing cabling that is virtually fault free.

- Problem detection and isolation are enhanced by providing a standardized layout and documentation.
- Network segmentation is facilitated by locating stations around wiring closets that limit the length of wire runs.

Although structured wiring does not appear technically complex on the surface, it must be installed to exacting standards and tested for compliance. Structured wiring, particularly category 6 and above, is a complete system that includes wire, jacks, and patch cords. Unless these are installed and terminated in accordance with the manufacturer's instructions, the wiring system may fail to meet the specifications. Installation considerations include maintaining sufficient distance from sources of electromagnetic interference (EMI), controlling bend radius, and preserving twist in terminating the wire on patch panels and wall outlets.

Elements of a Structured Wiring System

A structured wiring system is composed of one or more building blocks. Figure 9-1 shows the elements, using a campus wiring system as an example.

FIGURE 9-1

Campus Wiring Topology

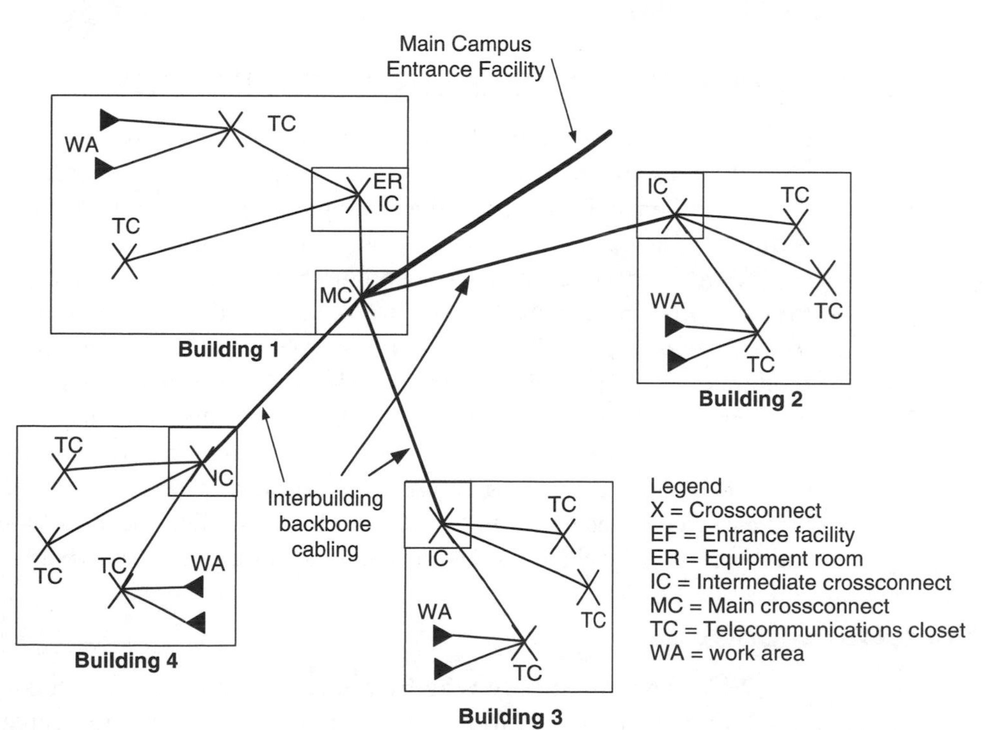

- *Main crossconnect* (MC) is the point at which the LEC's facilities enter the customer's premises. The MC includes the LEC's demarcation point.
- *Equipment rooms* (ER) are building areas intended to house telecommunications equipment. Equipment rooms may also include the functions of a telecommunications closet.
- *Intermediate crossconnect* (IC) is a point for interconnecting backbone cables.
- *Telecommunications closets* (TC) are rooms in which wiring is terminated. A building may have multiple telecommunications closets.
- *Backbone cable* (also known as riser cable) connects ERs to TCs.
- *Work area* (WA) is the user's workspace. Telecommunications outlets (TOs) are located in the work area. TOs are usually called jacks.
- *Horizontal wiring* runs from the equipment room or telecommunications closet to the work area.

Any transmission facility offers a tradeoff between bandwidth and distance. With other factors equal, the greater the bandwidth requirement, the shorter the distance over which wire can support it. Structured wiring standards specify that the horizontal wiring *link* is limited to a maximum of 90 m. With an allowance of a maximum of 10 m for patch cords, the total end-to-end *channel* length is 100 m. Within this distance, the specifications of a properly manufactured and installed wiring system will support high-bandwidth applications within the limits of the wiring category.

TIA/EIA STRUCTURED CABLING STANDARDS

Structured cabling standards are specified by TIA/EIA 568 standards in the United States and ISO/IEC-11801 Generic Customer Premises Cabling in most of the rest of the world. Although the principles are similar, much of the terminology is different, and TIA/EIA terms are used in this chapter.

Wire is installed in a star configuration with horizontal wiring extending from the work area to the TC or ER with no splices or bridged tap. The standard recognizes four-pair 100-Ω category 5E UTP and 62.5/125-μm or 50/125-μm multimode fiber-optic cable. The standard calls for at least two four-pair UTPs to each work area; additional media are optional. Coaxial cable is no longer recommended, but it can be installed to support video. In most offices video will be carried over UTP, either as IP video or by using baluns. Balun is a contraction of balanced–unbalanced. It is a device to convert coax to balanced UTP. Buildings such as hotels will prefer coax over UTP for video because its bandwidth supports more channels.

Wire Categories

TIA/EIA standards specify wire by category. The lowest is category 3, which is suitable for voice and 10 Mbps LAN cabling. Category 3 and category 4, which

TABLE 9-2

Characteristics of Unshielded Twisted Pair Wire

Parameter	Category 5E	Category 6	Proposed Category 7
Frequency range	1–100 MHz	1–250 MHz	1–600 MHz
Channel Insertion Loss	24 dB	35.9 dB	54.1 dB
Channel NEXT	30.1 dB	33.1 dB	51 dB
Power-sum NEXT	27.1 dB	30.2 dB	48 dB
ACR pair-to-pair	6.1 dB	− 2.8 dB	− 3.1 dB
Power-sum ACR	3.1 dB	− 5.7 dB	− 6.1 dB
ELFEXT pair-to-pair	17.4 dB	15.3 dB	
Power-sum ELFEXT	14.4 dB	12.3 dB	
Channel return loss	10.0 dB	8.0 dB	8.7 dB

Note: These parameters are not official and are subject to change. TIA/EIA specifications can be obtained from Global Engineering Documents at 1-800-854-7179 or www.global.ihs.com.

was intended for 16 Mbps token ring LANs, are still available, but rarely installed because the incremental cost of Category 5 is minimal. Category 5 is likewise considered as legacy cable today, although it is satisfactory for 100 Mbps operation. The characteristics of the most recent standard, category 5E, are shown in Table 9-2 along with category 6 and category 7. The latter is not an approved standard, so many of its characteristics are provisional and subject to change. Category 6 is intended for gigabit Ethernet. The next sections discuss the meaning of the wire parameters outlined in the table. The standards specify both link and channel standards as well as permanent link standards, which are normally those to which the link is tested.

Shielded vs. Unshielded Wire

The bandwidth of UTP may be insufficient for some applications. For these, the standard recognizes fiber-optic cable and shielded twisted-pair wire. The emerging category 7 standard uses shielded wire. Shielding is a metallic braid or layer of foil surrounding all or some of the conductors in the cable. Users often intuitively assume that STP is superior to UTP, but for properly designed applications through category 6, it is not. The purpose of the shield is to reduce EMI, which it does by attenuating the electrical energy radiated from the cable and minimizing energy coupled from outside sources. There are two theories on how shielding operates: the field effect or the circuit effect. The field effect theory holds that the shield reflects and absorbs the interfering waves. The circuit effect assumes that interfering signals generate a secondary field in the shield, canceling the original field. It is difficult to connect the shield to ground in some types of connectors; so most shielded cables include a drain wire, which is in direct contact with the shield throughout the cable length.

The key to avoiding the need for shielding lies in proper balance of the transmission medium. Balanced apparatus applies its signal across the two wires of a pair, or in some applications across the wires of two or four pairs. If an interfering signal is induced into the cable, the two wires of a pair will receive an equal voltage if they are twisted so they have equal exposure to the interfering signal. If the voltage on both wires is equal, the receiving device does not detect a voltage difference across the pair, so the interference has negligible effect. In practice, balance is not perfect, but the circuit has enough margin that it does not have to be.

Many organizations choose shielded wire for one of two convincing reasons: the equipment manufacturer specifies it, or it seems the most conservative approach. Both are rational reasons for using shielded wiring, but the cost is significantly more than unshielded. It is more difficult to install and terminate, and it has higher loss.

To sum up the shielded vs. unshielded question, most administrators should consult an authority on their particular installations before making the decision. A properly installed category 5E or category 6 UTP system will support gigabit Ethernet, and this should provide plenty of bandwidth for the future. If the equipment specifies shielded wire, then the manufacturer's recommendations should be followed.

Cable Installation Testing

Cable installations must be tested to demonstrate compliance with the specifications. Cable test sets must be designed to test all of the parameters of the specification and must show the margin, which is the difference between the measured attenuation and the maximum allowed by the specification. Test sets are usually automated and print out the results of each measurement.

Insertion Loss

Insertion loss is a measure of attenuation, which is the loss of signal strength over the span of the wire. If the received signal is too weak, the equipment may interpret the bits incorrectly. Attenuation is measured in decibels (dB) of signal loss. Cable testers measure attenuation by injecting signals of varying frequency and known power level at one end of the cable and measuring the received level at the other end.

Near-End Crosstalk (NEXT)

Crosstalk refers to the amount of coupling between adjacent wire pairs. When current flows through wires, some of the signal is induced into adjacent conductors. NEXT is measured at the transmitting end, where the outgoing signal is at the highest level and the received signal is at the lowest because it has been attenuated by the loss in the cable. NEXT is minimized by controlling the balance of the wire during manufacture and installation. If wire is twisted so that each pair has equal exposure to the other pairs, current flows cancel each other.

NEXT is measured by injecting a signal on each of the four pairs of UTP cable and measuring the signal level induced into each of the other three pairs. The amount of crosstalk coupling is a function of the wire, patch cords, and the TOs. Crosstalk is kept to a minimum in the wire by carefully controlling the length and tightness of twist in terminating the wire in the outlets and patch panel.

Power Sum NEXT

Power sum refers to the summation of interfering signals from all adjacent cable pairs. Full-duplex transmission such as that used in gigabit and 10G Ethernet uses all four pairs of wire. In contrast to NEXT measurements, which measure the crosstalk from a single pair, power sum NEXT measurements measure the total amount of crosstalk from all the other pairs. Power sum measurements are made on four-pair cables, and on larger cables that aggregate multiple four-pair wires.

Attenuation-to-Crosstalk Ratio (ACR)

ACR, also known as the signal-to-noise ratio, is calculated by dividing attenuation by NEXT. A high ACR means the received signal has a comfortable margin over the noise level. A high ACR enables the receiver to distinguish between wanted and interfering signals. Power sum ACR is used in category 5E and 6 cables.

Far-End Crosstalk (FEXT)

FEXT measures crosstalk at the receiving end of a cable. It is an important measurement on cables that must support protocols such as gigabit Ethernet that use all four pairs. FEXT is measured by injecting a signal on one pair and measuring it at the far end of the cable.

Equal-Level FEXT (ELFEXT)

ELFEXT measures the amount of signal coupling from a transmitter at the near end into the far end of another pair with the effects of attenuation removed. ELFEXT eliminates the effect of cable length on the FEXT measurement. It is required pair-to-pair and power sum for gigabit Ethernet cable certification.

Return Loss

When a signal reaches the receiving end of a circuit, if the impedance of the cable and the connecting device have a mismatch in impedance, some of the signal is reflected to the source. Return loss is the ratio expressed in decibels of the power of the outgoing signal to the power of the returned signal. The higher the return loss, the better the match, and therefore, the more complete the power transfer.

Backbone Cable

Backbone cable can be any combination of single-mode fiber, multimode fiber, or twisted-pair wire. Table 9-3 shows the maximum lengths of each medium

TABLE 9-3

Backbone Cable Specifications

Type of Medium	Maximum Range
Single-mode fiber	3000 m/9840 ft
50/125 μm or 62.5/125 μm multimode fiber	2000 m/6560 ft
Twisted-pair copper <5 MHz	800 m/2625 ft

permitted by the standard between the MC and the next level in the cabling hierarchy, which may be the equipment room or telecommunications closet.

The backbone cable can be connected through an IC point, but the standards recommend no more than two hierarchical levels of crossconnects.

Connectors and Patch Panels

The standards specify the TOs in the work area and patch panels or punch-down blocks in the telecommunications closets and equipment rooms. The standard does not specify the physical jack style, and a variety of configurations are available from different manufacturers. Data wire is terminated on RJ-45 jacks on both ends. The standard recommends two wiring patterns for jacks and patch panels as shown in Table 9-4: T-568A and T-568B. Although either standard is acceptable, T-568B is most commonly applied to office cabling.

Categories 5, 5E, and 6 wire all use the same type of wiring connectors. The connectors for each level are manufactured to progressively higher specifications,

TABLE 9-4

TIA/EIA 568 Jack Wiring Standards

Pair	Pin	Color
T568A		
1	4–5	BL/W-BL
2	3–6	W-O/O
3	1–2	W-G/G
4	7–8	W-BR/BR
T568B		
1	4–5	BL/W-BL
2	1–2	W-O/O
3	3–6	W-G/G
4	7–8	W-BR/BR

Color Codes: W, White; BL, Blue; O, Orange; BR, Brown.

so they are not interchangeable. Category 7, if it is approved, will use shielded wire and a different type of connector.

Two methods are available for terminating wire in the telecommunications closets and equipment rooms: punch-down connectors and patch panels. The latter are preferred for data wiring, and the former for conventional voice. Data wiring usually connects to switch ports that are equipped with RJ-45 jacks, and these connect to wire runs with patch cords. Data wire should always be terminated on patch panels. It is difficult to make jumper wire connections and retain category 5E compliance. Jumper wire is preferred for voice wiring, on the other hand, because it is less expensive than patch panels. PBX ports and riser cable are usually terminated on 66-type or 110-type punch-down blocks. With the proper kind of administration system, the crossconnections can remain in place after the service is disconnected.

Equipment Room

A telecommunications equipment room of some sort is essential for every office that has its own telephone system, which is usually centrally located. The room should be centrally located if possible to reduce the length of wire runs. It must be well lighted with sufficient aisle space on all sides of the equipment to enable maintenance personnel to access it. The room must be located above any flooding danger and free of overhead pipes that could burst and flood the equipment. Backboard space sufficient to terminate cables, ports, and wall-mounted apparatus is essential. Access to the room must be controlled with locked doors and, if appropriate, alarm system.

Telecommunications Pathways and Spaces

The industry offers numerous methods of installing and concealing telecommunications cables. In older buildings designers could not foresee the impact of telecommunications, so wiring methods are often a compromise between utility and aesthetics. In a new building, the choice of wiring method is generally inherent in the design of the building and should be considered carefully.

TIA/EIA 569 standards cover commercial building telecommunications pathways and spaces. Figure 9-2 shows the elements of building pathways as described in the standard. The pathways include the conduits, floor cells, raceways, and cable troughs that support the backbone and horizontal cables. They also support entrance cables and pathways for the grounding systems.

Fire codes also regulate pathway choices. In any building that uses air plenum space for wiring, the cable insulation must be plenum rated. Plenum cable uses a smokeless sheath such as Teflon to minimize toxic fumes in case of fire. Non-plenum wire installed in plenum spaces must be enclosed in conduit. Building codes also limit the length of outdoor cable sheath brought into the

FIGURE 9-2

Typical Commercial Building Wiring Topology

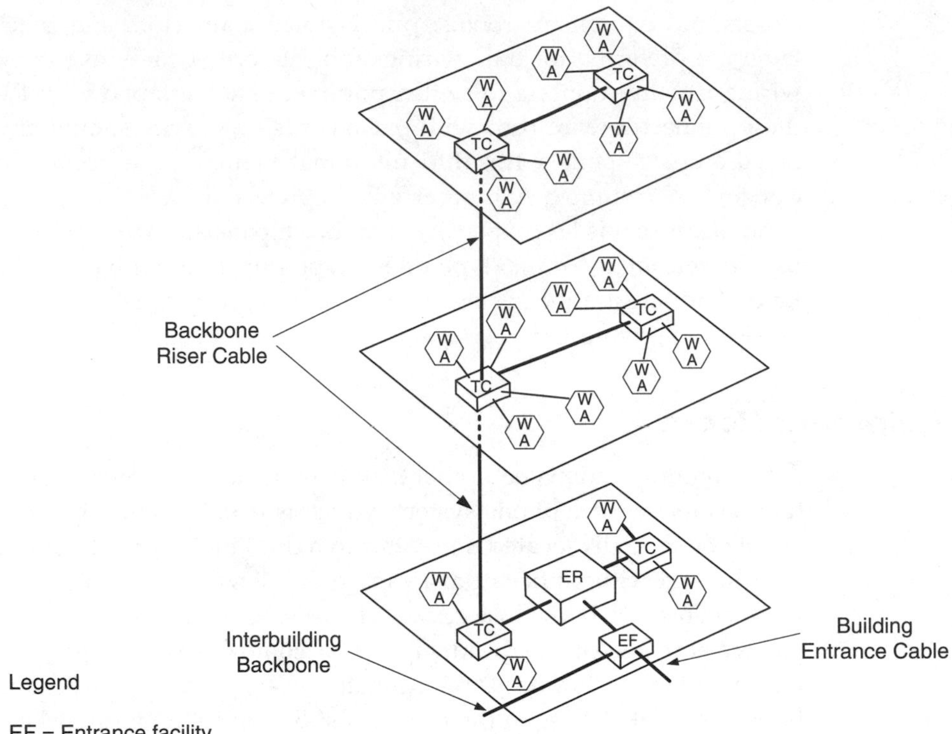

building to 50 ft unless the cable is in metallic conduit. For that reason, as well as for ease of rearrangement, the pathway provided from the point of entrance to the MC should always be conduit if possible. Otherwise it may be necessary to splice the entrance cable to a cable with an indoor sheath.

Conduit also facilitates placing the backbone cable. Campus backbone cable can be buried directly, but the major cost of installation is opening a trench and conduit adds only a small increment to the total cost. Fiber should be placed in an inner duct if conduit space permits or if not, in a separate conduit. It is also good practice to place empty conduit while a trench is open.

Horizontal cable is usually installed in four-pair increments. Twenty-five pair cable is available and can be used effectively where multiple TOs are clustered and provided the cable passes the power sum specifications. Horizontal cable routes from the TC to the work area by one of the pathways that are discussed next.

Suspended Ceilings

The area above a suspended ceiling is an excellent space for concealing horizontal wire. Wire must be fastened to the building structure, and not laid on the ceiling grid. Wire is brought into the work area through conduit placed in the walls and stubbed into the ceiling, run in surface-mounted raceway, fished through walls, or placed in telepower poles.

Floor Cells

Many buildings are constructed with under-floor cells for power and telecommunications. Access to the cell is by core drilling. TOs are mounted in monuments that fasten to the floor. Floor cells add to the building's structural integrity, but they are expensive and moves, adds, and changes are costly and disruptive.

Raised Flooring

Some buildings are equipped with raised flooring built on a metal framework. Cable is placed under the floor and brought into the work area through holes in the floor tiles. Raised flooring is an excellent method of concealing wiring, but it is too costly for most office areas.

Cable Trays and Raceways

Cable trays are an effective way of routing large quantities of wiring along both horizontal and vertical surfaces. Cable trays are particularly effective in warehouses, factories, and other areas where appearance is not important. Enclosed raceways can be surface-mounted to conceal wire. Surface-mounted raceway comes in a variety of colors to blend with the decor.

Telecommunications Closets

Telecommunications closets must be placed with close attention to maximum horizontal and backbone wire lengths. Good practice limits the number of telecommunications closets while restricting the wire length to 90 m. Telecommunications closets must be built large enough to provide adequate wall space for terminating horizontal and backbone cables. Many telecommunications closets also contain hubs, routers, and other active equipments. Swing-out relay racks are available where space does not permit installing a floor-mounted rack. As with equipment rooms, telecommunications closets need to be well lighted and secure. Telephone wiring is normally terminated on punch-down connectors. Data wire is terminated in patch panels as shown in Figure 9-3 and connected to the horizontal wiring with patch cords.

Work Area

Telephone instruments and data workstations are installed in the work area. Horizontal wiring terminates in modular jacks and connects to voice and data apparatus with patch cords.

FIGURE 9-3

Station Wiring Topology

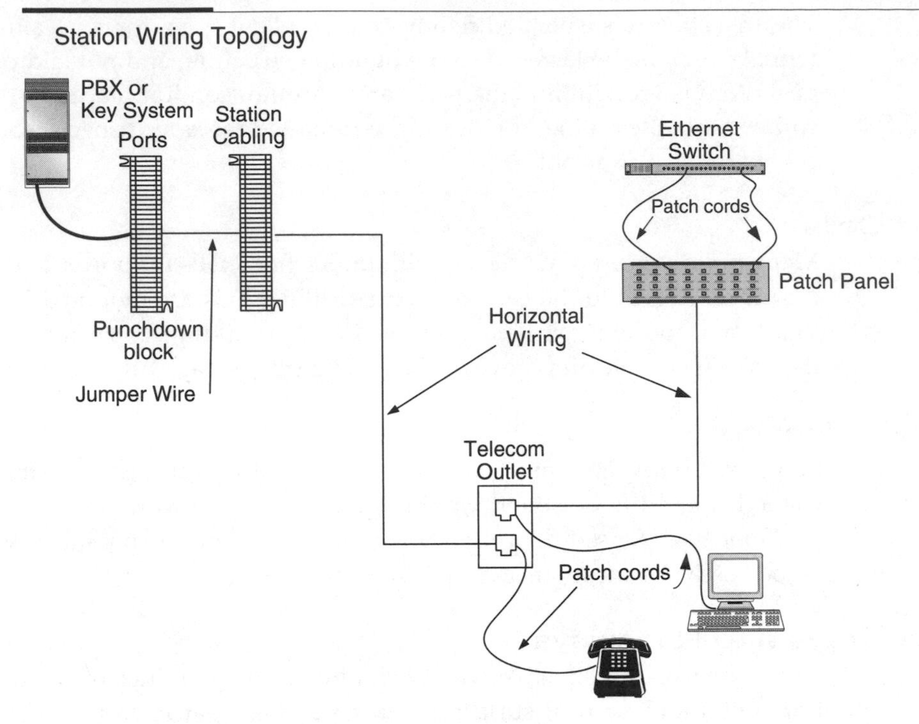

Wire Administration

TIA/EIA 606 covers administration of telecommunications wire. To make it easy to administer, wire and terminations must be numbered, labeled, color-coded, and designated in a manner that is easy to follow. Before the standard was developed most wire installation companies had a preferred method of installation and undocumented wiring systems were common. In the standard, separate designations are used for backbone, horizontal, and grounding paths. Poorly documented wiring systems cause technicians to spend unnecessary time in toning and tracing.

The TIA/EIA 606 standard specifies the types of records that are to be kept. Records include required and optional linkages and identifiers. Identifiers are the designations that are assigned to items of plant such as cables, manholes, conduits, and bonding and grounding locations. Linkages are the logical connections between identifiers and their corresponding records.

Bonding and Grounding

Modern electronic equipment is sensitive to improper grounding. Many cases of erratic operation, unexplained noise, and periodic equipment failure can be traced to inadequate grounding. TIA/EIA 607 describes bonding and grounding requirements in commercial buildings. The standard specifies a single grounding point called the telecommunications main grounding busbar (TMGB), with all equipment room and telecommunications closet grounds brought back to this point. The TMGB is bonded securely to the building's electrical power ground and to the building's metal framework.

Each TC and ER must contain a telecommunications grounding busbar (TGB) that is connected to the TMGB over the telecommunications bonding backbone (TBB). The purpose of the TBB is to equalize, as far as possible, the ground potential of the various building locations housing telecommunications equipment. The gauge is appropriate to the size and structure of the building. The minimum size is #6 AWG, but it may be as large as #3/0 AWG. Each TGB also is bonded to the metal building frame with a minimum of #6 AWG wire. It is important that architectural drawings specify the installation of these grounding and bonding elements when a building is constructed, and that an electrical engineer specifies the ground wire size.

NATIONAL ELECTRICAL CODE® (NEC)

The NEC specifies fire resistance standards for communications cables to protect people and property from fire hazards. The code, product of the National Fire Protection Association, addresses the methods of limiting the hazards of cable-initiated fires and cable-carried fires. The code requires that communications and signaling wires and cables in a building be listed as suitable for the purpose. The following summarizes the code requirements.

Article 770 Optical Fiber Cables

This article lists three categories of optical fiber cable:

- OFNP—Cables capable of being installed in ducts and plenums without the use of conduit.
- OFNR—Optical fiber riser cables.
- OFNS—Specified for general-purpose use other than plenums and risers.

Article 800 Communications

This article lists six categories of copper communications cables:

- CM cables for general purpose use except plenums and risers
- CMP cables for use in plenums
- CMR cables for use in risers
- CMX cables for residential and for use in raceways
- CMUC cables for undercarpet use
- MP cables for multipurpose use (must satisfy requirements for CM, CMP, and CMR)

Any cable used for telecommunications must be tested and listed as meeting the fire resistance, mechanical, and electrical standards of the testing laboratory. The code requires that jumper wire be fire resistant. It also requires that equipment intended to be electrically connected to a telecommunications network is listed for the purpose. The NEC specifies requirements for separation between communications and power conductors, and states what types of cables can share raceways and closures.

STRUCTURED CABLE APPLICATION ISSUES

Structured wiring standards have resolved many issues for the enterprise telecommunications manager, but plenty of choices remain. One of the most perplexing is the question of how many cable runs to install. The standard calls for two runs per work area, but VoIP vendors tout as a major advantage the ability to support both voice and data on the same wire run. If an IP PBX is being installed, is it safe to install only one run per work area? Contractors will tell you that it is a lot cheaper to install wire while the office is unoccupied and the ceiling tiles are not in place.

What category of wire should be installed? Any building owner or network manager wants to avoid replacing the wiring in the future, but the higher the wire category, the more it costs. Overkill in wire may add something in the way of peace of mind, but if it is never used to its capacity it will do little to enhance telecommunications performance.

A major consideration is providing adequate structure for pathways between the communications closets and the work area and the equipment rooms. Conduits are great for future rearrangements, but they cost more. Plenty of riser cable should be provided for telephone service unless VoIP is being planned. Four riser pairs per 100 square feet is a reasonable rule of thumb. This section discusses some of the factors that should be considered in acquiring a new wiring system.

Universal Voice and Data Wiring

Many designers recommend installing the same category of wire for both voice and data, reasoning that the universality of the wire provides the flexibility of using voice or data on either cable. This strategy is complicated, however, by the method of terminating wire in the telecommunications closet or the equipment room. Data cabling must be terminated on patch jacks because patching is the only way to ensure channel performance. Voice cabling can be installed on patch jacks, but if a conventional PBX is used, punch-down blocks are cheaper and easier to use. If wire is terminated on punch-down blocks, it cannot be tested to the category standards. If the telephone system is conventional now, but will eventually be IP, then the voice cabling will probably be unused, so it may not pay to use high-quality wire for voice. However, a high percentage of the cost of the cabling system is in installation and subsequent administration, which means it is unwise to cut costs by reducing wire quality.

Cost Differential

A major factor is the cost differential between wire categories. The difference in cost between category 5E and category 6 lies mostly in the material cost—the installation labor, which is a large part of the cost, is about the same. As the cost differential between wire grades drops, the reason for installing a lesser grade of wire disappears. As a result, categories 3, 4, and 5 are rarely used today. When the category 7 standard is approved, it will undoubtedly be significantly more expensive than category 6, so managers will want to consider it carefully before deciding to install it, particularly in preference to fiber.

Future Gigabit and 10G Ethernet

The choice of wire is further complicated by the direction of the data industry toward higher and higher speeds to the desktop. The point will eventually be reached where further increases in bandwidth to the desktop are of no value. In fact, 10 Mbps to the desktop is more than enough now for all but a small fraction of users. Workers who use the network for Web surfing and standard office applications such as word processing, presentation, e-mail, and spreadsheet do not put a heavy load on the network. If the trend continues toward decreasing cost of NIC cards and switches, gigabit Ethernet to the desktop will become standard, so installing category 6 wire now may future-proof the wiring system.

Stability of the Office Layout

A properly designed and installed structured wiring system can and should have a life of 10 years or more. In buildings with fixed walls the telecommunications wiring should last as long as the electrical wiring and it should be installed with the same degree of forethought. The end of the wiring plan's life is apt to be caused by factors other than obsolescence.

Stability of the office layout is an important factor. Many offices use open space and modular furniture that is meant to be rearranged. The telecommunications wire is installed in troughs in the cubicles. When the cubicles are rearranged, it is usually necessary to pull out the old wire and replace it. If the building has open space and modular furniture and the company rearranges offices with some frequency, the life of the station wire tends to be short. If so, it does not pay to buy a higher grade of cable than is currently needed.

An important issue is the accessibility of the cables for future rearrangement. In an office with fixed walls the life of the cable plant is much longer and it may be advantageous to buy category 6 wire even though category 5E is adequate.

Another factor is who is paying for the wire. Tenant companies probably will not value beyond the duration of the lease. The building owner, on the other hand, can increase the value of the floor space by providing high-quality structured cabling and pathways.

Fiber to the Desktop

Another question managers often ask is whether to place fiber for future use even though it is not needed now. The answer to that depends on how difficult it is to add it in the future. It is risky to place fiber now for future use. So many different grades and types of fiber are available that it is nearly impossible to foresee what future equipment might require. If fiber to the desktop will potentially be needed in the future and if it would be expensive to install, consideration should be given to placing conduit. Furthermore, as cable improves, the need for fiber decreases.

Manufacturer's Extended Performance Warranty

Cabling systems are often installed with the expectation of future applications, but they are used on LANs that do not begin to test the ultimate capabilities of the wire. An extended performance warranty, which most of the major wiring manufacturers offer, is a good way to address these issues. If the system fails to support the application for which it is designed, the manufacturer (not the vendor) stands behind the installation. Since the manufacturer offers the warranty, it survives the potential demise of the vendor that installed it. Without a performance warranty you are left with the manufacturer's materials warranty, which is of limited value considering that labor is usually more than half the cost of an installation.

CHAPTER 10

Local Area Networks

LANs are the glue that binds the office. They are the means of sharing files and common resources such as printers. They provide access to wide area networks, host computers, and servers. They are the foundation of office automation with its productivity-enhancing features such as electronic mail, scheduling, information warehouses, and collaborative work on shared applications. By definition, a LAN is a network dedicated to a single organization, limited in range, and connected by a common communication technology. This definition has important implications. Because the network is private, it can be specialized for a function. Security is less critical than in wide area networks, and because range is limited, a LAN can operate inexpensively at high speeds. LANs started out as narrow-range systems, but the enterprise network of today has evolved into a network of interconnected LANs.

LOCAL AREA NETWORK TECHNOLOGY

A LAN has the following characteristics:

- High speed that permits users to transfer data seamlessly across the network.
- A restricted range. UTP LAN segments have a maximum radius of 100 m.
- An access protocol that permits nodes to share a common transmission medium.
- A network operating system (NOS) that permits workstations to address resources as if they were directly attached.

A LAN operates at all seven layers of the OSI protocol stack. Personal computers, printers, and other devices connect directly to the transmission medium through a NIC, which operates at the datalink level. The NIC has the following functions:

- Provides a physical interface to the transmission medium.
- Monitors the busy/idle status of the network.
- Buffers the speed of the attached device to the speed of the network.
- Assembles the transmitted data stream into frames for transmission on the network and restores the frames into a data stream at the receiving end.
- Recovers from errors and collisions that result from simultaneous transmissions or other disruptions.

LANs can be classified according to four criteria:

- Topology
- Access method
- Modulation method
- Transmission medium

Topology

Network topology is the pattern of interconnection of the network nodes. LANs use the same topologies as wide area and metropolitan networks: star, bus, ring, and branching tree as shown in Figure 10-1. The original Ethernet began as a bus with stations tapped into coaxial cable at intervals. When the preferred medium changed to UTP, Ethernet evolved to a star topology with the wire legs connected through a hub. Although the topology changed, a hubbed network still operates as a bus with every station hearing all traffic to the network. Most Ethernet LANs use a switch instead of a hub, which provides each wire leg with a nonshared path. Ring topologies are part of the LAN standards, but are less common. The IEEE 802.5 token ring protocol and fiber distributed data interface (FDDI) both use a ring topology. The 802.4 token bus protocol is wired as a bus, but the tokens flow over a logical ring. IEEE 802.3b broadband Ethernet operates over a video cable system using its branching tree topology. Although broadband Ethernet is still standard, it is rarely used in LANs.

Access Methods

All networks need some method of regulating access to the transmission medium. LAN access methods are classified as contention or noncontention. CSMA/CD, which we discussed in Chapter 6, is a statistical access method. It relies on the probability that its nodes will get enough share of the bandwidth to send their traffic in a timely manner. During heavy load periods access may be delayed, so CSMA/CD is not satisfactory for applications that rely on predictable network access. Contention

FIGURE 10-1

Network Topologies

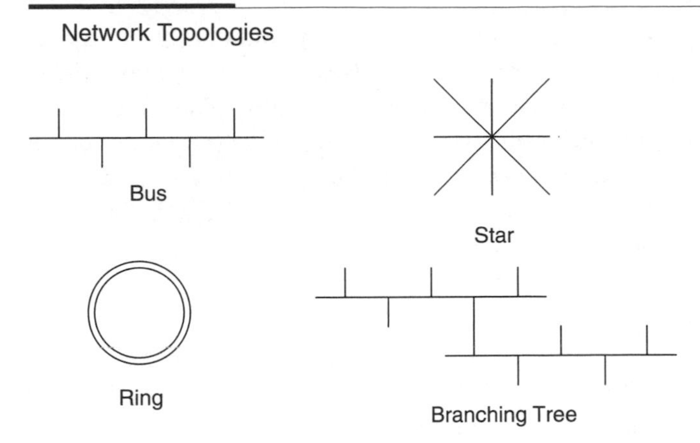

access is simple in the receiving direction. Each node reads the destination address, copying the frame if it is the addressee or otherwise ignoring it. In switched Ethernet the CSMA/CD protocol is still active, but the switch eliminates collisions.

Token passing is a deterministic access method that assures each station of equal access to the bandwidth. A token is a unique combination of bits that circulate through the network following a predetermined route. If a node has traffic equal to or higher in priority than the indicator in the token, it captures the token, sends its traffic, and replaces the token on the network. The advantages of control are purchased at the price of greater complexity. One of the nodes in a token network must be equipped to initiate recovery action if the token is lost or mutilated, which can occur if a node fails or loses power at the time it possesses the token. Other functions required of the control node include the removal of persistently circulating frames, removal of duplicate tokens, control of priority, and addition and removal of nodes. All nodes contain the logic to assume control if necessary.

In 802.5 token rings, frames advance around a physical ring with the token always following the same route. A station must possess the token before it can send traffic. On incoming traffic the station copies the frame if it is the addressee; otherwise it repeats the traffic to its downstream neighbor. The sending station removes the frame and passes the token downstream to the next station. Although token ring networks are wired as a ring, the wire is configured as a star. Two pairs of wire from each station terminate in a multinode access unit (MAU) in the center of the network. All traffic passes in and out of the MAU on its trip around the ring. The MAU bypasses a port if its node is inoperative.

In the 802.4 token bus network, frames are broadcast to all nodes simultaneously, but only the node with the token can transmit, which eliminates collisions. The token can be passed to any other node without any regard to its physical position on the network. Token bus networks are used primarily in manufacturing automation protocol (MAP) networks, and have no office application.

Modulation Methods

LANs use one of two methods of modulating data on the transmission medium. In the first, known as *baseband*, light pulses or square wave pulses of DC voltage are applied directly to the transmission medium. The other method, known as *broadband*, divides the transmission medium into frequency bands or channels. Each broadband channel can be multiplexed to carry data, voice, or video. This method is common in cable access, but broadband LANs have practically disappeared on customer premises.

Transmission Media

All of the transmission media used in other networks—twisted pair wire, coaxial cable, fiber optics, radio, and infrared—are employed in LANs. Radio and infrared are used for wireless applications and are discussed in Chapter 21. Table 10-1 lists the shorthand designations for the various media that IEEE has standardized for LANs.

Coaxial Cable

The original Ethernet used thick RG-8 coaxial cable equipped with a string of transceivers that tapped into the coax at prescribed intervals. The original thick coax alternative was unwieldy because the cable had to be routed past each workstation and tail cables run to the transceivers. This medium, which is obsolete, is known as 10-Base-5. The 10 stands for 10 Mbps, Base means baseband, and 5 means a segment length of 500 m. A subsequent variation known as 10-Base-2 or thin Ethernet uses RG-58 cable. RG-58 coax is easier to handle than RG-8, but it introduced additional drawbacks. The segment length is restricted to 185 m, which seems like plenty until you consider how much cable is needed to loop it past each workstation, while concealing it in walls and ceilings. Furthermore, the stations are daisy-chained together. If the chain is broken, it kills every downstream workstation. Coaxial cable and transceivers have been relegated to the museum shelf as far as LANs are concerned.

Twisted Pair Wire

Twisted pair wire was once considered to lack the bandwidth and noise immunity required for reliable LAN operation until UTP that is manufactured to tight specifications entered the market. With the approval of EIA/TIA 568 standards, UTP has become the medium of choice. Properly installed category 5E and category 6 wires have plenty of bandwidth to support gigabit Ethernet. The shorthand designation of Base-T prefixed by the speed in megabits per second is used for UTP LANs.

Fiber Optics

Fiber-optic cable with its wide bandwidth can support data speeds far higher than most LANs need and it is somewhat more expensive, but it brings advantages

TABLE 10-1

IEEE 802 Local Area Network Standards

802.1	Overview document containing the reference model, tutorial, and glossary
802.1b	Standard for LAN/WAN management
802.1p	Specification for LAN traffic prioritization
802.1q	Virtual bridged LANs
802.2	Logical link control
802.3	Contention bus standard 10-Base-5 (Thicknet)
802.3a	Contention bus standard 10-Base-2 (Thin net)
802.3b	Broadband contention bus standard 10-Broad-36
802.3d	Fiber optic inter-repeater link (FOIRL)
802.3e	Contention bus standard 1-Base-5 (Starlan)
802.3i	Twisted pair standard 10-Base-T
802.3j	Contention bus standard for fiber optics 10-Base-F
802.3u	100 mb/s contention bus standard 100-Base-T
802.3x	Full duplex Ethernet
802.3z	Gigabit Ethernet
802.3ab	Gigabit Ethernet over Category 5 UTP
802.3af	Power over Ethernet
802.3ak	10 Gigabit Ethernet over Copper 10GBASE-CX4
802.4	Token bus standard
802.5	Token ring standard
802.5b	Token ring standard 4 Mbps over unshielded twisted pair
802.f	Token ring standard 16 Mbps operation
802.6	Metropolitan area network DQDB
802.7	Broadband LAN recommended practices
802.8	Fiber optic contention network practices
802.9a	Integrated voice and data LAN
802.10	Interoperable LAN Security
802.11	Wireless LAN standard
802.12	Contention bus standard 100VG AnyLAN

Note: This table lists only the major standards in the 802 series. Each standard has numerous subparts, many of which are omitted for clarity.

such as immunity to NEXT and external noise sources. It has one major drawback as a substitute for UTP in that it cannot provide power to a VoIP telephone, which means such instruments must be locally powered. Fiber can be used as a link between the workstation and its hub, or as a medium to connect hubs and switches. It is the main transmission medium for 10G Ethernet, which can also operate for short distances on twinaxial wire. Ethernet can operate over a fiber-optic repeater link (FOIRL). The 10/100-Base-F standard allows a fiber link to be used on short wavelength fiber for up to 2 km between repeaters. The standard defines three types of fiber segments as shown in Figure 10-2:

- 10/100-Base-FL: fiber link segment
- 10/100-Base-FP: star topology using a passive optical coupler at the hub
- 10/100-Base-FB: fiber optic synchronous backbone link

FIGURE 10-2

Fiber-Optic Links

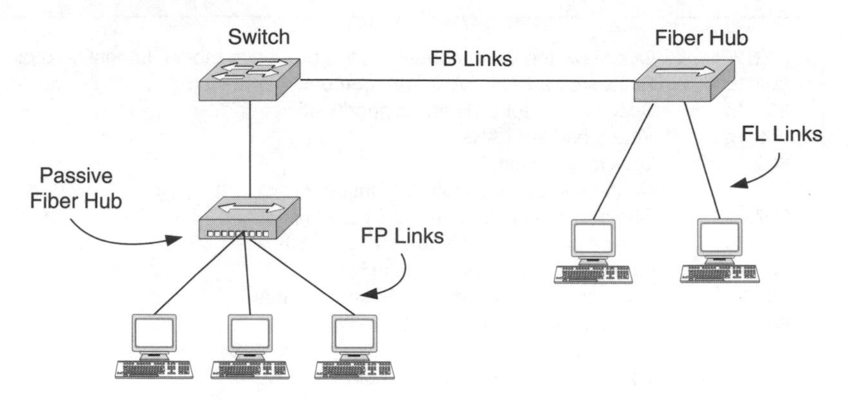

ETHERNET STANDARDS

Although today's Ethernet can trace its ancestry back to the original Xerox protocol, not much else about it is recognizable today. UTP and, to a lesser degree, fiber optics have displaced the original coaxial medium. The original 10 Mbps speed has increased by three orders of magnitude and there is no indication that it will not progress further. This section discusses some of the variations of 802.3.

Figure 10-3 shows the basic structure of the 802 protocol stack, which applies to token ring and token bus as well as Ethernet. In the original implementations of Ethernet the attachment unit interface (AUI) was a separate connector. Today, all of the components are included in the NIC and the medium dependent interface is either an RJ-45 connector for UTP or a fiber-optic connector for fiber media.

Fast Ethernet

For the first few years of LAN operations, the 10 Mbps speed of standard Ethernet was plenty for most users. As applications became more complex, and particularly as voice and video began to operate over the LAN, more bandwidth was needed. IEEE standardized several alternatives for various media, as Table 10-2 shows, and fast Ethernet rapidly caught on in the market. The prevalent protocol is 100-Base-TX, which is a fast version of 10-Base-T and uses the same protocol and frame structure. The 100-Base-T4 protocol obtains high speed on category 3 wiring by using all four pairs. The 100VG AnyLAN protocol also uses all four pairs of category 3 wire, but it does not use the standard Ethernet protocol. Instead, the hubs poll the stations to see if they have traffic to send, which eliminates collisions. These two alternatives have had little impact because most offices have upgraded their wiring to category 5 or higher. Today most NICs, hubs, and switches are 100-Base-TX and operate at either 10 or 100 Mbps. If the medium supports the higher speed, it is used; if not, the NIC and port downshift to 10 Mbps.

FIGURE 10-3

IEEE 802 Protocol Stack

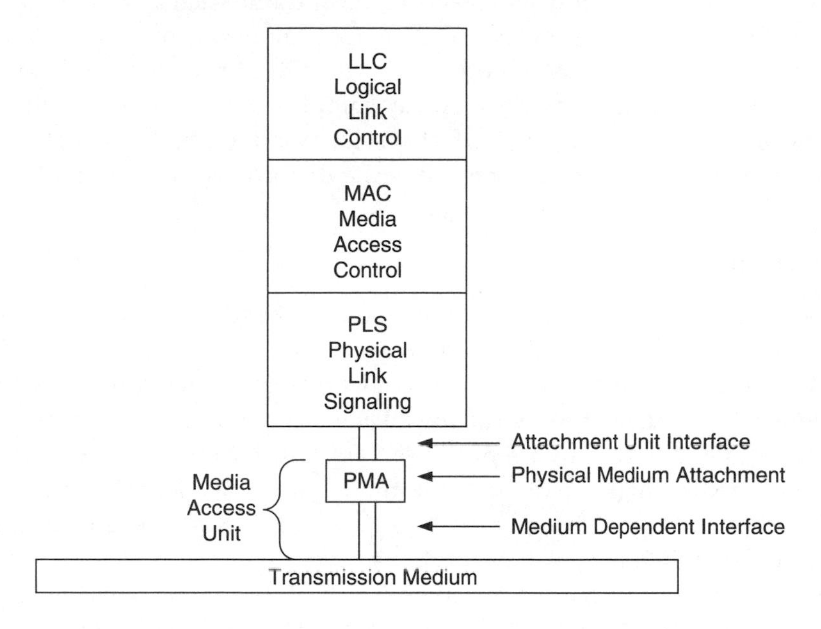

TABLE 10-2

Ethernet Media Standards

Designation	Medium	Segment length
10-Base-2	RG-58 coax	185 m
10-Base-5	RG-8 coax	500 m
10-Base-FL	850 nm fiber	≤2 km
10-Base-T	≥Cat 5 UTP	100 m
100-Base-FX	1300 nm fiber	≤2 km
100-Base-SX	850 nm fiber	≤300 m
100-Base-TX	≥Cat 5 UTP	100 m
100-Base-T4	≥Cat 3 UTP	100 m
100VG-AnyLAN	≥Cat 5 UTP	100 m
1000-Base-LX	1300 nm fiber	≤10 km
1000-Base-SX	850 nm fiber	≤500 m
1000-Base-T	≥Cat 5E UTP	100 m

Note: Segment lengths indicated in the table above vary depending on the grade of fiber. Refer to the IEEE standards for details. IEEE standards are available online at http://standards.ieee.org/.

LAN Traffic Prioritization

The standard Ethernet protocol provides best-effort service, which is not good enough for carrying time-sensitive frames such as voice and video. The 802.1p specification gives layer 2 switches the ability to prioritize traffic at the MAC layer.

To implement 802.1p, switches must be capable of grouping incoming traffic into separate classes. The specification defines up to eight classes of traffic. The highest level is typically assigned to critical traffic such as routing table updates. The next levels are for delay-sensitive applications such as voice and interactive video. Managers can assign other priority levels for traffic that is more business critical—for example, when its delay might adversely affect productivity. If no priority level is set, the best-effort service is automatically assigned. If switches have inadequate buffer capacity, they can discard low-priority traffic.

Gigabit Ethernet

Fast Ethernet had scarcely made its mark on the user market before the industry began working on a modification to increase the speed by a factor of 10. The gigabit Ethernet frame structure is nearly identical to the 100 Mbps and 10 Mbps versions. The common frame structure means that relatively inexpensive components can be employed to link users and servers. Gigabit Ethernet operates over a variety of media: single-mode fiber, multimode fiber, coaxial cable, and UTP. Gigabit Ethernet uses a coding method known as 8B/10B. It operates in a full-duplex mode for switch-to-switch connections.

Although the main application for gigabit Ethernet is in the backbone, it is increasingly being deployed to the desktop. The need for this kind of speed to the desktop is rarely demonstrable, but the history of fast Ethernet is repeating itself. The price of gigabit cards and switch ports is dropping to the point that the premium over fast Ethernet is small. Today, most desktop computers are delivered with fast Ethernet cards, but gigabit Ethernet will likely become the norm in the future.

10 Gigabit Ethernet

A major reason Ethernet has endured is its ability to scale and adapt to ever-higher speeds. The 802.3ae standard for the latest iteration, 10G Ethernet, was approved in 2002. It uses the same frame structure, MAC, and protocol stack as the other versions of Ethernet, which means that it fits seamlessly into the current network infrastructure. It is also designed to interoperate with SONET. Aside from a 10-fold increase in speed, the main difference between 10G and other versions is that it does not currently operate over UTP, and it operates only in a full-duplex mode. The collision detection protocol is therefore unnecessary in this version.

The specification defines both LAN and WAN interfaces. The protocol is intended for use in metropolitan networks as well as in the enterprise LAN backbone. The protocol operates over both single- and multimode fiber. Over single-mode it can extend up to 40 km, which enables its use to interconnect LANs across the WAN without reframing or protocol conversion.

Power over Ethernet (PoE)

VoIP products are designed to use an Ethernet interface and to be compatible with existing LANs. Conventional Ethernet switches do not deliver power over the wire, so telephone instruments must be locally powered. During power failures, IP telephones are inoperative unless the entire office electrical system is backed up.

Pending development of a standard, VoIP manufacturers either used proprietary switches or power was injected in the station cable with a midspan device that connected between the switch and the telephone. Standard 10/100 Mbps Ethernet uses only two of the four UTP pairs, so power can be fed over the vacant pairs. This solution is acceptable, but it does not work with full-duplex gigabit Ethernet, which uses all four pairs. Also, if the power is short-circuited, the power feed must be interrupted to eliminate the potential of fire.

The 802.3af standard addresses these concerns. Both midspan and through-the-switch powering are included in the standard. To permit the use of all four pairs for data, PoE furnishes power through the center tap of a transformer that feeds 48 V power over two pairs. Power is cut off if the telephone is unplugged or if the wires are shorted. If the wrong device is plugged into the outlet, the power will not damage it.

Given the maximum resistance of a UTP run, the highest amount of current that can flow is 350 milliamps. This limits the power drain of endpoints to 15.4 W. The PoE standard defines four classes of endpoints, according to how much power they draw:

- Class 0: (the default, when a device's class is unknown) 15.4 W
- Class 1: 4 W
- Class 2: 7 W
- Class 3: 15.4 W

In addition to powering IP telephones, PoE is useful for powering 802.11 wireless access points, which are often placed in ceilings where power is not readily available.

LAN ARCHITECTURES

Ethernet would never have grown so dramatically were it not for structured wiring using UTP in a star topology. Wires in a hubbed arrangement run from each node to a central hub, which is a nonintelligent device that receives a signal on one port and repeats it to every other port. As a repeater, a hub effectively doubles the diameter of the network, but every station is a member of a collision domain. All workstations in the domain share the total bandwidth, which is typically about 60 percent or less of the data speed. A hubbed network cannot reach the full wire speed because as the network approaches capacity, collisions increase and throughput falls off. In an ordinary office with a limited number of workstations, a hubbed network provides enough bandwidth because the portion allocated to each station is still more than most of them need. Many networks grow to the point, however, that collisions increase and throughput drops to an unacceptable level. When this happens, the solution is to segment the network with a bridge.

Bridging and Switching

A bridge, as shown in Figure 10-4, is a two-port device that contains a table listing the MAC addresses of all the stations on each segment. If the sending and receiving stations are both on the same segment, the bridge ignores the frame. If the receiving station is on the opposite segment, the bridge repeats the frame to the other segment.

Segmentation with bridges solves the collision problem up to a point, but when the network outgrows the capacities of two segments and is segmented again, some traffic must transit two or more bridges to reach its destination. If a segment must support transit traffic, the advantages of segmentation are lost. Adding more bridges eventually becomes self-defeating. The solution is a multiport bridge in the center of the four segments so that traffic has to make only one

FIGURE 10-4

LAN Segmented with a Bridge

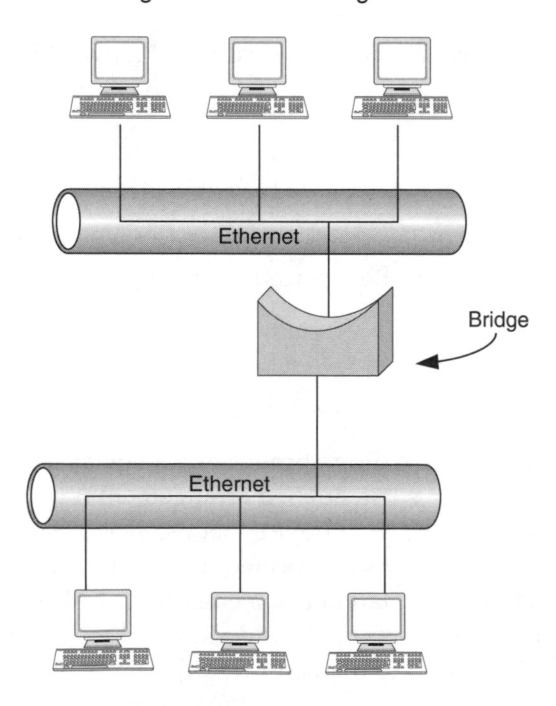

hop to reach the destination. This is the function of an Ethernet switch. Figure 10-5 shows a four-segment network connected with a switch. When the switch recognizes that the traffic is destined for a segment different from the originating segment, it switches the two segments together long enough to pass a frame.

This alternative still leaves stations hubbed together and still subject to collisions. This is solved by making the switch large enough to eliminate hubs. If each station has the exclusive use of a switch port, it becomes a single-station segment. Each station has full wire speed of the protocol—typically 100 Mbps. Moreover, since the station no longer has to listen to the receive pair to detect collisions, it can operate in full duplex. Not all applications can profit from FDX, but applications such as interactive video operate more effectively. FDX effectively doubles the speed of the network, i.e., 10 Mbps Ethernet becomes 20 Mbps and 100 Mbps becomes 200 Mbps in equivalent throughput.

Switching Technology

Ethernet switches have three different methods for connecting the transmitting and receiving ports. A *cut-through* switch makes the connection as soon as it reads the MAC address of the addressee. If the frame is invalid because it is too short (i.e., a *runt* frame) or perhaps mutilated by collision, a cut-through switch forwards it anyway. A *store-and-forward* switch reads the entire frame and checks the CRC before forwarding it to the output port. A store-and-forward switch is somewhat slower than a cut-through switch, but it sends only valid frames to the other segment. The third type is a *fragment-free* switch, which forwards or not, based on

FIGURE 10-5

LAN Segmented with a Switch

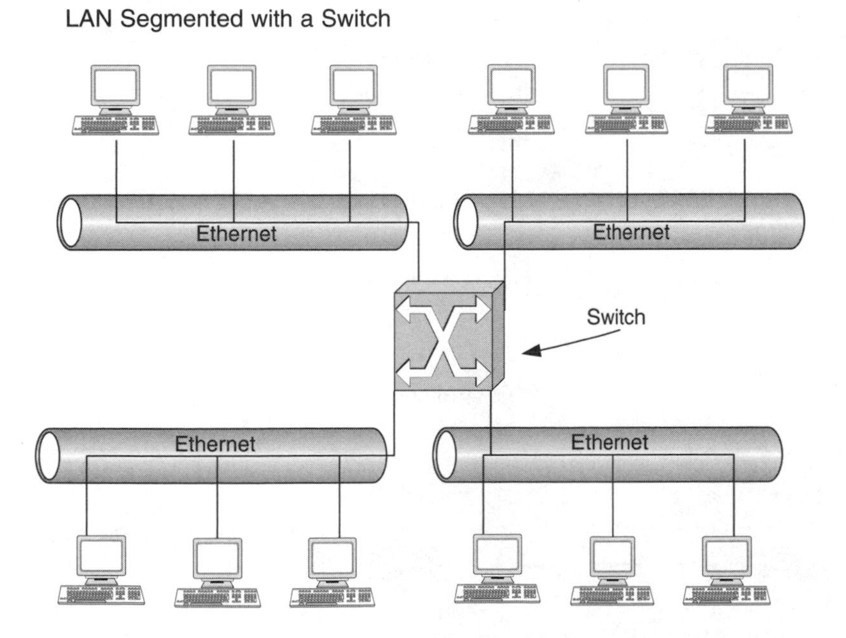

the first 64 bytes of the frame. This reduces runts and mutilated frames since the majority of collisions occur at the start of the frame.

Switches are further classified by their architecture. A matrix switch has an internal grid for making connections between input and output ports. A shared-memory switch stores frames in an internal buffer and gates them in turn to the output port. A shared-bus switch has an internal transmission path that all of the ports share on a time-division basis. Whatever the architecture of the switch, it is essential that it has sufficient capacity in its backplane to handle the packet flow. The total potential demand on the switch is the product of the port speed and the number of ports. The raw performance of the switch is designated as the number of frames per second it can switch in and out.

Switches are also classified according to their position in the LAN. Work group switches serve the users with direct connections and these feed into *core switches* as Figure 10-6 shows. It is less critical that work group switches be able to support the total potential demand because not all stations transmit simultaneously. Core switches must have a much greater capacity than work group switches because they receive packet streams from the work group switches and any packet loss degrades network performance.

Traffic in a LAN flows in short bursts, so switches have internal buffering to handle the peaks, but they also need some means of protecting themselves against buffer overflow. One method is flow control—the switch sends a message

FIGURE 10-6

Ethernet Switch Hierarchy

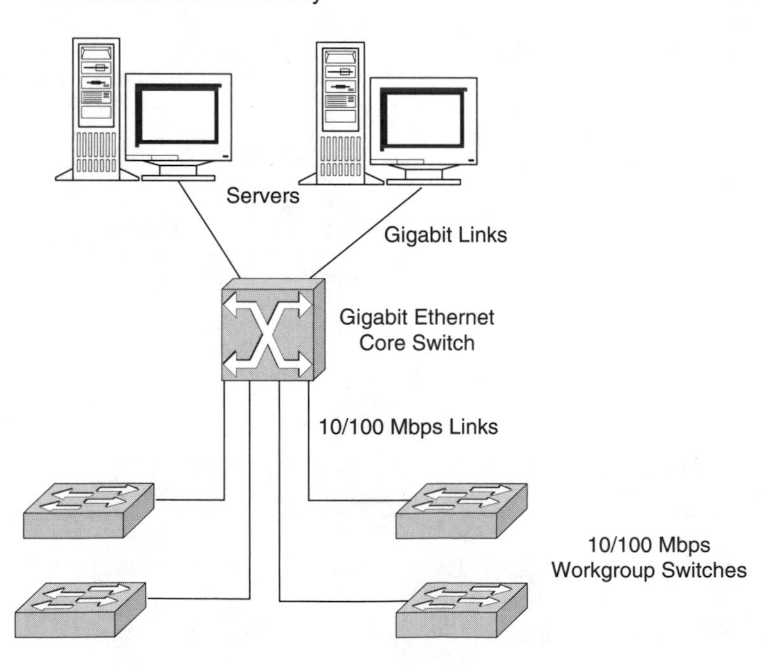

to the sending station to stop transmitting. The other method is to discard packets. Both of these can be avoided by providing sufficient buffering capacity.

Switches are self-configuring devices. When a switch is first connected to the network, it learns the MAC address of the first station that transmits a frame to another node. Since the switch does not know which port the receiving station is connected to, it broadcasts the frame to all of its other ports. When the receiving port acknowledges the frame, the switch adds its MAC address to its table and forwards the acknowledgement to the sending station. As the switch gradually populates its tables, if it knows the destination port, it no longer needs to send frames to any other port. To prevent old MAC addresses from remaining in its tables forever, the switch has an aging process to delete addresses that are not used within a specified interval.

Layer 3 Switches

A switch, like a bridge, operates at the datalink layer of the OSI model. Although most LANs use IP addresses, a layer 2 switch reads only the MAC address and ignores the IP address. Routers are considerably more expensive than switches and they are also slower. Routers do their packet forwarding in microprocessor-based engines, while switches operate through application-specific integrated circuit (ASIC) hardware. The performance of layer 3 switches is often quoted in terms of millions of packets per second (pps), while routers have packet throughput in the range of 100,000 pps to more than one million pps.

A layer 3 switch is essentially a router that does its switching in hardware instead of software. The principal difference is that routers have larger routing tables and interface cards to enable them to interface WAN protocols such as frame relay or a T1 connection to an ISP. Switches are fast and accurate at what they do, but they primarily relay Ethernet connections to other switches or to end users. These connections can be either fiber or copper, depending on the configuration of the switch. Routers are discussed in more detail in Chapter 36, IP Networks.

Spanning Tree

Since bridges, and therefore layer 2 switches, have only one route for reaching a station, they lack the robustness of alternate routing. In a three-bridge network such as that shown in Figure 10-7, it is advantageous to provide more than one route for Station A to reach Station B in case the primary route fails. Although this increases reliability, the bridges, lacking routing capability are incapable of determining which route to use. With switched networks it is easy to inadvertently create multiple paths to a destination, which the switch cannot handle. The solution is to run the *spanning tree algorithm* on the switches. Spanning tree ensures that even if more than one route is available, only one is active at a time. If the active route fails, the spanning tree protocol alters the table in the switch so it uses the other route. Inactive routes are disabled to prevent packets from looping.

FIGURE 10-7

Spanning Tree Protocol

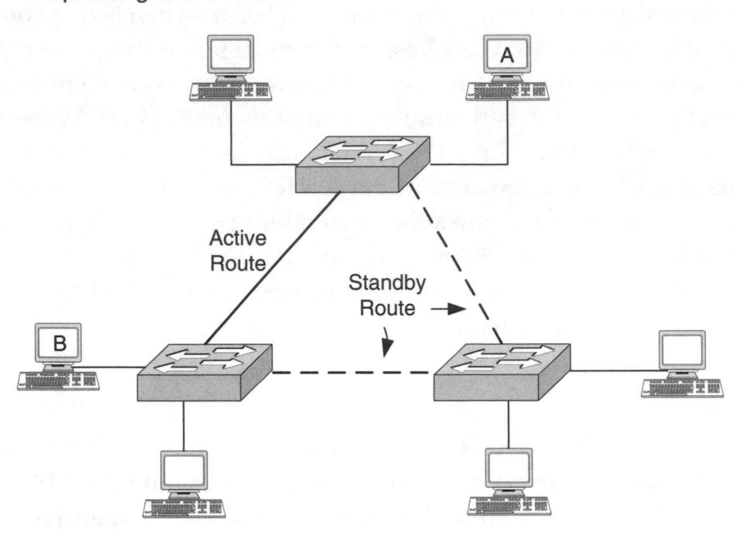

Virtual LAN (VLAN)

The theory of LAN development is that a network starts as one or more large segments. As the network grows, it is segmented by bridging or routing. As long as the users on the segments remain in the same physical proximity, this method of segmentation works well. If a user moves to a location that is served by a different segment, the benefits of segmentation may be lost because traffic from that node flows over both segments as well as the backbone. The solution is a VLAN, which is a collection of stations that are grouped together logically even though they are physically separate. The grouping is some common attribute such as belonging to the same organization. The main benefit of a VLAN is reducing the amount of broadcast and backbone traffic that traverses the entire LAN.

The membership of the VLAN is contained in switch tables. The administrator accesses the switch directly or through TELNET and uploads the name, domain name, and port assignments of stations belonging to the VLAN. The broadcast domain is then based on something other than physical location of the station. VLANs are also set up to improve security by confining traffic flow to members. Personnel who are assigned to special projects or are part of the same organization are often assigned to VLANs. Switches communicate configuration information among themselves with the VLAN trunking protocol.

NETWORK OPERATING SYSTEMS

By themselves, the 802 networks implement only the first two layers of the OSI model, and are incapable of complete communications. The higher protocol layers, which are needed to complete the network, are implemented in the NOS. The following are the primary functions of the NOS:

- Capture calls from the application program to the computer's operating system and convert them to NOS calls.
- Manage shared disks in the file server to maximize the efficiency of information transfer between the network and the server.
- Manage security on the network, permitting users to offer their files for shared access and restrict them from unauthorized access.
- Designate and manage shared resources as if they were directly attached to the users' personal computers.
- Provide tools with which the network administrator can manage the network.

The concepts of *redirection* and *virtual resources* are important to understand any NOS. In the process of logging onto the network, the desktop device's operating system interacts with the NOS to define virtual disk drives and printer ports. Virtual drives and ports act as if they were installed as hardware, but they actually exist on a server. For example, a PC might be equipped with one hardware line printer port, defined as LPT1. The NOS allows the administrator to identify additional printers that are attached to other devices or directly to the network. When the application program sends a document to the designated printer, the NOS intercepts the data stream and redirects it to the network printer based on its MAC or IP address.

File service works in the same manner. NOSs define drive designations as a path to the file server's disk and underlying directories. The file server then behaves as if it was directly attached to the station. When the application program makes a file call, the NOS redirects it over the network to the file server.

NOSs can be classified as peer-to-peer or server-centric. A peer network enables all users to offer their resources for sharing. A server network concentrates the resources in one or more servers. Nothing precludes the use of dedicated servers on a network of devices all running a peer operating system. Any station can be designated as a server, offering its resources as print or file services, to other stations on the network without actually running a server operating system. Network attached storage is also an inexpensive substitute for a central server.

Peer-to-Peer Networks

Peer networks enable users to share directories, disk space, printers, or other resources that are attached to their computers. If a directory is marked for sharing other users have access to the files it contains. If a printer is marked for sharing, print jobs are spooled from the network to that printer. Peer network capability is inherent in Microsoft's Windows and Apple desktop operating systems.

Peer NOSs are usable in small offices that do not have regular network management or a great concern for security. Users must, however, be trained to observe network discipline. For example, if users re-boot or turn off their computers while other nodes are accessing a file or using a printer, the data or print job are likely to be corrupted. Normal file shutdown procedures inform users when others are attached to their computer, so if procedures are followed the results are satisfactory.

Enforcing security on a peer network may be difficult. If all files are stored on a server, the network administrator can prevent users from accessing files by marking the directory as unshared. With some peer networks the entire directory is either sharable or not, in which case it is impossible to allow some users rights to access files while denying them to others.

Network statistics are generally not provided by a peer NOS. Lacking a file server and routine backups, file backup may be haphazard. Without a dedicated server, it is also difficult to be sure where the latest version of files is stored because users can copy a file from another user's disk, modify it, and save it to their own hard disk. Although some companies operate peer networks successfully with numerous stations, most networks with more than a handful of stations will find it preferable to use a server.

Server-Centric Network Operating Systems

The server-centric network has functions such as file, print, fax, modem pool, and remote access centralized in one or more servers. A server NOS has significantly more administrative features than a peer NOS and provides statistics that the administrator can use to assess performance. The operating system runs in a robust processor that provides the capabilities of mapping virtual drives and virtual printers on their systems through the NOS. The server supports a universal naming convention such as \\server_name\path_name that allows users to address resources if the network manager grants them permission. Instead of installing software on every machine, the NOS may permit the administrator to control the use of licensed software by entering the number of licenses the company has. Users can load the software from the server up to the number of licenses.

LAN APPLICATION ISSUES

Several issues in addition to the hardware and NOS must be considered when deciding which product to purchase. These issues are discussed in this section.

Security

LAN security is perhaps the most important issue confronting network managers. Denial-of-service attacks, worms, and viruses have compromised many networks and are a constant threat. Stand-alone LAN segments are a low-security risk, but when they are connected to the Internet, security becomes a major issue. Not all risks are external. LAN managers must cope with internal attempts to access unauthorized files or attack the network for vengeance or just the challenge of doing it. Managers must provide adequate firewall protection to keep intruders and harmful files out of the network. Critical files and communications must be encrypted to preserve privacy. LAN NOSs are capable of fending off most internal attacks. Fundamental to security is keeping all physical access points accessible only to authorized personnel.

Fault Tolerance

A fault-tolerant LAN can handle irregularities in network operation while allowing users to continue processing or, at worst, to terminate their operation without losing data. Network irregularities fall into three categories, each of which may require a different recovery strategy: faults, which are software or hardware defects; errors, which are incorrect data; and failures, which are breakdowns of network components. Users do not distinguish between these three categories; their concern is the continued operation of the network.

Fault-tolerant LANs employ a combination of hardware and software, usually consisting of the following elements:

- An uninterruptible power supply to protect against AC power failure or at least permit a graceful shutdown.
- A duplicate file allocation table (FAT) kept on the disk.
- Software that logs what users were doing at the time of a crash and saves their files to help bring a failed system back to the starting point.
- Redundant hardware, such as mirroring servers, duplicated bridges, redundant cabling, and multiple disk drives that mirror each other so the second drive takes over when the first fails. RAID (redundant array of independent disks) technology is even more fault-tolerant because data is striped across multiple drives, any of which can fail without data loss.
- Applications software that stores information between saves.
- Careful backup procedures and management of backed-up files.

A key to fault tolerance is a network management system that alerts the administrator to impending problems by detecting minor faults before they develop into major problems.

Network Management

Larger networks can often profit from a network management network system. Components installed on the network should be SNMP-compatible. Chapter 39 provides information about network management systems.

Internetworking

Most multisite companies connect their LANs with an Internet. Internetworking raises several issues that affect the choice of NOS. Most LANs require an IP addressing system, with NAT and DHCP operating on a server or router. Some peer protocols do not support TCP/IP, some older protocols cannot be routed, and some network protocols are not supported by current routers. Refer to Chapter 36 for further information on internetworking.

Infrastructure Planning

Companies that occupy multiple floors of a building or have multiple departments will probably segment the LAN with switches and connect to the Internet through routers. The architecture of the system should be planned in advance to be certain that the infrastructure has been built to accommodate the necessary hardware. Planning is especially important when a structured wiring system is being installed. It is always less expensive to install additional capacity in the horizontal and backbone wiring paths at the time of initial wiring than it is to add it later.

The minimum of two horizontal wiring paths specified by EAI/TIA 568 is not enough for many companies. It is difficult to look far enough in the future to predict bandwidth requirements, but fiber optics in the horizontal distribution system may be needed. Installing empty conduit at the time of twisted pair installation saves the labor cost of adding it later.

Backbone Architecture

LAN architecture is flexible and it can often be rearranged by changing the equipment if the proper infrastructure is in place. In planning a LAN, determine whether LAN workgroup switches will connect to a gigabit switch in a collapsed backbone. The choice affects the number of fibers in the backbone. Consider whether the company needs a VLAN architecture.

Network Speed and Throughput

Throughput, or rate of transfer of information between nodes, is an important issue with LANs. In most cases, 100 Mbps Ethernet will be used in the workgroup and workgroup switches will connect to the backbone at either 100 Mbps or over gigabit backbones. Many organizations will consider using gigabit Ethernet in LAN segments as the price of NICs and switches continues to drop. The aggregate amount of traffic the segment imposes on the backbone must be taken into consideration. Although the network itself may have a high throughput, communication may be slowed because of characteristics of the file server.

CHAPTER 11

Network Equipment

The market offers a dazzling array of telecommunications equipment. Enterprise networks will not apply some devices directly because they are intended for carrier use, but many products are common to both carriers and private networks and the criteria for selecting them are common to both. This chapter discusses evaluations of some of the devices that we have discussed in the first part of the book with an emphasis on LAN and transmission products. The primary criteria for choosing products are discussed in broad terms. We also discuss the common equipment, which is the apparatus such as power equipment, distributing frames, and cable racking that apply to multiple classes of equipment. Additional evaluation considerations can be found in product-oriented chapters.

GENERAL NETWORK EQUIPMENT EVALUATION CRITERIA

Certain factors are common to the evaluation of most telecommunications equipment. The amount of effort worth spending on evaluation depends on the consequences of a poor decision. Some consequences are easy to determine. High maintenance and repair costs can be tracked, but the cost of lost productivity and missed opportunities often far exceed the cost of the device. The more severe are the consequences, the more it pays to research the product exhaustively.

The main problem in evaluating equipment is how to get accurate data. Some of the factors discussed in this section can be obtained from manufacturers' specifications, but some knowledge of how the manufacturer measures its benchmarks is needed because the differences may be subtle and only a thorough hands-on evaluation will pick them up. Experience of other users is often a good guide, provided you are able to get user evaluations from independent sources. This section briefly discusses factors that should be considered in comparing products.

Operations, Administration, Maintenance, and Provisioning (OAM&P)

OAM&P support is the most important differentiator between many products. Enterprises have procedures in place for OAM&P support, and the degree to which the product fits with existing procedures is important. Remote monitoring and provisioning capability are critical at sites that do not have local technicians. Other factors include alarms, remote diagnostics and trouble-shooting capability, statistical information collection, remote backup, and restoral of the database; all of these facilities improve service response time and reduce travel costs.

Network management capability is a closely related issue. Some equipment has a proprietary interface and operates through a custom terminal. A major objective of most network managers is to operate any device over a universal terminal using SNMP. Consider how the product is accessed. Is it over a dumb terminal interface using a proprietary command language? Does the manufacturer provide a GUI to make it easy to administer the device as opposed to learning a command language? What kind of training does the manufacturer provide?

Adherence to Network Equipment Building System (NEBS)

Most central office equipment manufacturers in the United States follow the Telcordia NEBS guidelines in manufacturing their equipment. NEBS specifies standard relay rack and bay dimensions, environmental requirements including temperature, humidity, fire, earthquake, acoustical, electrical safety, grounding, and a host of other recommendations. Manufacturers that are providing equipment to common carriers usually adhere to NEBS standards because the carriers demand compliance. NEBS compliance is also important for enterprise networks because it assures the buyer that the equipment has undergone the required survivability testing. NEBS not only discusses equipment, but also covers how building space must be conditioned to support telecommunications equipment. It describes three levels of assurance:

- Level 1 is for office class environments and is intended for noncritical systems.
- Level 2 is described as failure-tolerant services that need to be reasonably fault-free, but not completely bulletproof.
- Level 3 is for carrier-class environments where systems are critical and failures cannot be tolerated.

Failure Rate

For years, the primary criterion used in evaluating telecommunications products was mean time between failures (MTBF). Today, MTBF is becoming almost irrelevant as a product differentiator. For one thing hardware failures are rare except for electromechanical apparatus. MTBF figures are expressed as a function of hardware, but software is the essence of many products and software failures are impossible to predict using MTBF evaluation techniques. A third reason for the diminishing importance of MTBF is the fact that it is impractical for users to get an independent validation. The manufacturer can claim an MTBF that is difficult to validate and not worth the effort unless you are purchasing the item in large quantity. More important is the incidence of out-of-box failure. Once the product has successfully operated for the first 30 days, failures are rare.

Warranty

With all else equal, it is better if the warranty is longer. A key issue is whether the dealer or the manufacturer is providing the warranty. The manufacturer's warranty covers defects in materials and may be of limited value since equipment failure rate is usually low. With products that have a high installation component, the manufacturer's warranty alone is of marginal value. A good example is structured wiring. If a category 6 wiring plan fails to support gigabit Ethernet, it is a small comfort to have the manufacturer deliver a load of new wire and jacks.

Configuration Backup

Most apparatus has configuration information that is stored in RAM or flash memory. RAM is flexible but can be lost during power failures. If RAM memory is used, a backup method of restoring the configuration, such as booting from disk or tape, should be provided. Does the product provide some means of offloading the configuration to magnetic media? Is special apparatus required? In case of power failure, does the system self-restore? How long does it take?

Scalability

Many telecommunications products are modular in configuration, which means that the purchase of plug-ins can be deferred until they are needed. The main concern is the upper size limit of the device. When the capacity limit is reached, the product has to be replaced.

Power Considerations

Power drain for some components is so small as to be insignificant, but for devices that operate 24 × 7, low power consumption can be a deciding factor. The type of power is also crucial. Forty-eight-volt power is standard for most central office apparatus and is common in large enterprise installations because a central 48-V string eliminates the need for multiple UPSs.

Floor Space Requirements

As with electric power, the floor space occupied by equipment is an ongoing cost for the life of the product. If space is at a premium, then high-density products are preferred. High-density components mean more economical use of floor space, but plug-ins are more expensive and utilization is harder to control. For example, if a PBX line port supports 32 ports, when the 33rd port is needed, a card is purchased, but 31 ports are vacant.

Vendor and Manufacturer Stability

Telecommunications is littered with the remains of many failed companies. When the vendor goes out of business, another dealer can usually be found if the manufacturer is stable, but when the manufacturer goes out of business, the result is orphaned equipment. How long has the manufacturer been in business? If the vendor fails, how easy is it to find another vendor? What support does the manufacturer provide to its vendors? For example, does it provide a technical support hotline?

Documentation

Documentation has long been one of the weak spots in the IT industry, including both telecommunications and computers. The trend is away from printed manuals and toward information provided online or on CD-ROM. Both of these are advantageous to the manufacturer and the user because it is easier to keep them updated, but this is no guarantee of documentation quality.

LOCAL AREA NETWORK EQUIPMENT

The main components of a LAN are NIC, cabling plant, switches, and servers. NICs are essentially commodities. They are included with nearly any device intended for Ethernet connection and they are self-configuring. Cabling has been covered in Chapter 9, which leaves switches and servers as the primary LAN components. This section discusses evaluations of switches and network operating systems (NOSs).

Ethernet Switches

Ethernet switches are the most critical component of LANs. They can be broadly categorized into workgroup and backbone or core switches, depending on their position in the office architecture. A workgroup switch has multiple ports to feed LAN segments, which may consist of a single station or multiple stations if the port feeds a hub. The size of the workgroup switches varies with the size of the organization and the configuration depends on how the servers are distributed.

An organization of moderate size with centralized servers is likely to choose a collapsed backbone in which multiple workgroup switches feed into a backbone switch that distributes the traffic to the appropriate servers.

The primary factor in choosing switches is the maximum forwarding rate the switch can support without packet loss. Packet loss requires recovery at the transport layer, which can increase user response time. Retransmitted packets also add to the traffic load, so the most effective switches can handle close to 100 percent of the theoretical load.

Switch forwarding rate is a function of packet size. Short packets, which are typical of voice and interactive video, impose a higher demand on the switch than large packets. LANs have a packet distribution that usually consists of many large packets and, depending on the applications, either few or a significant number of small packets. Packet loss is most apt to occur during congestion periods when multiple input ports are contending for a limited number output ports. The criteria listed in the next sections are important in evaluating switches.

Switch Configuration

Ethernet switches are available in expandable, stackable, modular, or fixed configuration. Modular switches have a multislot chassis that accepts cards or blades of various speeds and port quantities. They are scalable and therefore best suited for larger workgroup and backbone applications. Stackable switches are also scalable in that multiple units can be interconnected to increase the port count. Fixed switches come with a set or expandable number of ports. They are most appropriate for small workgroups. Backbone switches usually have some number of gigabit ports in addition to fast Ethernet. All switches used in a distributed environment should have fiber-optic connections. Switches are also available in managed and nonmanaged configurations.

Can the switch deliver wire speed across the projected distribution of frame sizes? Can it prioritize traffic based on 802.1p?

Quantity of MAC Addresses Per Port

Switches must have sufficient memory to support address tables up to the number of MAC addresses that can be stored per port. Switches with a small number of MAC addresses per port cannot be used in backbone configurations that interconnect shared segments. In the latter configuration, the switch must store the MAC address of each station that resides on that port and requires enough memory to store them.

Fault Tolerance

Fault tolerance is important in offices where a large number of users would lose service in case of switch failure. Switches with high port count or those intended for backbone applications require fault-tolerant features such as redundant power supplies, and in the case of modular switches, hot-swappable modules.

Packet Filtering

Store-and-forward layer 3 switches may filter packets based on source or destination address or protocol type. Switches may be capable of filtering out broadcast and multicast packets that exceed a defined threshold. Packet filtering is most important for backbone applications, but many workgroup switches filter unwanted traffic between the workgroup and the backbone.

Spanning Tree Support

Most Ethernet switches include support for the IEEE 802.1d spanning tree protocol, so managers can install switches in parallel, while avoiding active loops within the network. How quickly can the switch reconfigure on failure of the primary route? Does it support 802.1w rapid reconfiguration?

Port Mirroring

Port mirroring allows traffic from any port to be repeated on a designated port, so a single protocol analyzer or remote monitoring probe connected to that port could be used to analyze the traffic.

VLAN Support

VLAN capability allows the manager to define logical groups of users without regard to their physical location. VLAN confines broadcast traffic within the boundaries of the VLAN and eases the burden of moves, adds, and changes. Can the switch properly tag and distinguish traffic among multiple VLANs?

Network Operating Systems

Server-centric NOSs enable a network administrator to exercise controls such as the ones listed in this section. These are included in any NOS, but competing products have different methods of implementing them.

- *Access control.* The administrator can control who has rights to access files and directories. Typical methods of administering rights include:
  - Who is authorized to access what files and with what level of rights?
  - What shared resources are available and to what group of users?
  - How is security managed to prevent unauthorized access and alteration of files?
  - What administrative tools are available to evaluate network performance, control access to the network, and add, move, and change nodes?
  - What other networks are accessible and how access is controlled?
  - What maintenance, statistical, and trouble-shooting tools are available to alert managers to impending problems and assist in restoral when they occur?
  - What utility programs are made available to the users and the administrator?

- *Rights administration.* The administrator can grant or revoke rights to directories or individual files. Files and directories can be hidden to conceal their presence from users who lack the appropriate rights.
- *User groups.* Users can be assigned to groups that have a common set of interests. Members can send messages, the supervisor can assign rights to selected files and directories, and users can share exclusive use of printers.
- *Security administration.* The administrator can assign and control passwords, user IDs, and other means of regulating security. The administrator can force nontrivial passwords of a particular length and can force users to change passwords as needed.
- *Login administration.* Login procedures can be established for every user. These define virtual drives, set up port redirection for printers, send messages to the users, and with other commands allow the administrator to otherwise customize configurations for users and groups.
- *Printer administration.* The network administrator can identify stations as printer control stations. These stations can delete or move print jobs in queue, establish or change print priorities, pause the printer to change forms, and other such tasks concerned with printer management.
- *Administrative tools.* Many LAN managers underestimate the amount of administrative effort it takes to keep a network operating. The network administrator is defined, in software, as the person who has the right to modify and control the rights of other users. In most networks, the rights approximate the following list, which is ordered by the degree of control the right provides, with the highest rights listed first. Each level has the rights of succeeding but not of preceding levels of authority:
  - *Hidden.* The presence of the file or directory is concealed from unauthorized users.
  - *Parental.* Has full control over any file in the directory.
  - *Private.* Only specified users can read files.
  - *Modify.* Can change any file in the directory.
  - *Read/write.* Can read any file or write any file in the directory but cannot modify existing files.
  - *Read only.* Can read files in the directory but cannot add or change a file.
- *Directory administration.* The operating system permits managers to create a tree-structured directory structure on the server. Files are contained in directories and subdirectories; the network administrator can regulate rights at any level from a file up to the entire directory.
- *Statistical information.* The operating system provides statistical information that enables the administrator to isolate trouble and reconfigure

the network as necessary. For example, the administrator needs to know the volume of traffic on the network and the frequency with which users encounter delays. Contention networks should show the frequency of collision. The operating system should inform the administrator of factors such as the number of cache hits that affect the operating efficiency of the file server.

TRANSMISSION EQUIPMENT

This section covers some of the most common transmission devices that are prevalent in enterprise networks. Individual chapters in Part III discuss other products in more detail.

Channel Service Unit/Data Service Units (CSU/DSU)

In North America the CSU is part of customer premise equipment; in most of the rest of the world it is part of the network and the LEC provides it. Technically, the CSU and DSU functions are separate, but as a practical matter they are nearly always included in the same device and the separation of functions is not relevant. The CSU may be a separate device that connects between the T1/E1 line and the apparatus, but increasingly it is integral to the apparatus. The CSU should support both SF and ESF T1 lines, and when used on the latter, it should provide for retaining performance data such as slips and framing errors in storage for diagnosing line troubles. Ideally, the information is provided over SNMP. Features such as the following diagnostics and trouble-shooting add value to the CSU:

- *Loopback capability*. The CSU can be looped on its WAN side or DTE side to isolate trouble. Frame relay CSUs may provide for looping back PVCs.
- *Statistical information*. ANSI T1.403 standards establish the reporting capabilities of the CSU. A CSU that is not ESF-compatible may not be compatible with T1.403 standards, and may or may not be capable of reporting line irregularities such as clock synchronization and framing errors.
- *Add-drop capability*. The CSU can bring out some quantity of DS0s to separate ports. One to four V.35 ports are typical.
- *Automatic configuration*. The CSU detects circuit characteristics such as line coding and framing and configures itself accordingly.
- *Frame relay configuration*. The CSU has a discovery mode that determines critical factors such as PVC and local management information (LMI) and verifies network integrity automatically.
- *Protocol analysis*. CSU allows manager to set traps and analyze data at the IP layer.

Integrated Access Devices (IAD)

IADs do not lend themselves to easy classification because manufacturers can build access devices in any configuration and label them as an IAD. The simplest form is essentially a channel bank, which is a device that subdivides T1/E1 lines into DS0 channels, but has no intelligence. Some products support data and special services as well as voice channels. With different plug-in modules, a T1/E1 line can be divided into any combination of channelized voice and nonchannelized data within the total bandwidth of the line. Some systems offer channel units that can be reconfigured remotely.

The primary issue in selecting an IAD is the signaling capability of the channel units. FXO and FXS channel units are used for the office and subscriber ends of a POTS line. They make the signaling conversion so that a DS0 line emulates a metallic connection. E&M channel units are used for terminating devices that have two-wire or four-wire analog voice paths and use analog central office signaling. A typical application is tie lines between PBXs and/or key systems. E&M signaling is discussed in more detail in Chapter 12. Other channel units provide for DS0 data connections, wide-band audio, DSL, and even routers in some product lines.

T1/E1 Multiplexers

The next level up on the scale from a channel bank may be defined as a T1/E1 multiplexer or IAD, depending on the manufacturer's designation. Voice channels are provided as per the discussion above. Any bandwidth not required for voice can be used for data. The data module typically has a DSX-1 output using an RJ-45 connector. It can be set up as Nx64 with *N* being the amount of unchannelized bandwidth provided for data.

Some products can be networked to form an integrated voice and data network with alternate routing capability and sophisticated network management. Some multiplexers have the capability of monitoring multiple points in the network, reporting malfunctions, and keeping the network manager supplied with usage and performance information from all the nodes. Another vital feature of many multiplexers is their ability to reroute circuits during failure or congestion. Two different systems are used to keep data flowing. A table-based system is composed of fixed routing tables that instruct the multiplexer how to act in the face of a failure. A parameterized or rule-based system develops a global view of the network and responds flexibly.

Voice over DSL

Another type of IAD provides integrated voice and data over a DSL line. The device typically packetizes a fixed number of voice channels. The voice is prioritized and routed to a specialized DSLAM, which routes data packets toward the Internet and voice packets to analog ports for connection to a TDM switch or as an IP voice stream to a softswitch.

Add-Drop Multiplexers (ADM)

An ADM subdivides a T1/E1 line into two or more portions. An ADM normally has a straight-through T1/E1 connection that can be terminated in a PBX. It also has one or more V.35 connections that can be cabled to another device such as a router. One frequent use of an ADM is to split a T1/E1 line access to the IXC so that part of the line is used for access to the IXC's long-distance network and the remainder to a data network such as frame relay or, in some cases, to the Internet. Most CSUs are available with add-drop capability. Since the T1/E1 line must be terminated in a CSU anyway, this is an inexpensive method of obtaining add-drop capability.

Digital Crossconnect Systems (DCS)

A DCS has a crossconnect matrix that allows input ports to be switched to selected output ports. DCS systems are available in a wide variety of sizes and configurations. Based on the configuration of their switching matrix, DCSs come in three types: narrowband, wideband, and broadband. Narrowband units switch at the DS0 level, wideband at the VT1/DS1 level, and broadband at the STS-n/DS3 level. A DCS operates at the electrical level, but larger systems have optical interfaces and convert between OC and STS levels. DCS systems are not optical crossconnects, however. Even though bandwidth may enter and leave the device at the optical level, it is converted and crossconnected at the electrical level.

DCSs are classified by the DS levels they support. For example, a 3/1/0 DCS is capable of grooming and time-slot interchange of T3 and T1 services down to the DS0 level. Systems are usually modular, which allows the owner to defer the purchase of port hardware until needed. The switching matrix should be completely nonblocking. Also important is the ability to upgrade the operating system without interrupting service.

A key factor in evaluating DCS systems is how the crossconnections are mapped. In some systems the mapping is entered from a terminal and remains fixed, but in other systems it may be remapped under software control. For example, a DCS could be set up with one pattern for normal working hours and another pattern at night.

A DCS may function as an M1-3 multiplexer, i.e., it separates a DS3 into its constituent DS1 channels. An M1-3 multiplexer is a less expensive way of gaining access to the DS1 channels if the terminating equipment is collocated with the multiplexer. The DCS is vital if some of the DS0s or DS1s must be remapped and connected to the outgoing bandwidth. With conventional multiplexers, channels would have to be connected back-to-back.

Other important uses of a DCS that affect the choice of product may include these:

- *Inverse multiplexing*. The unit has the capability of aggregating low bandwidth channels into a higher bandwidth channel.
- *Fractional T1 grooming*. The unit can accept FT1s from multiple sources and aggregate them into higher bandwidth.

- *T1 to E1 conversion.* System can convert between the European and North American systems including the A-law to μ-law signaling conversion.
- *Type of interface ports.* Some systems include a combination of T1/E1, T-3/E3, HDSL, and other types of ports.
- *Trouble isolation.* Some products allow loopback of the channels and may have built-in bit-error rate tester for line evaluation.

COMMON EQUIPMENT

Common equipment supports all classes of telecommunications. Common equipment, as covered in this chapter, includes:

- relay racks and cabinets;
- distributing frames;
- alarm and control equipment;
- power equipment.

Relay Racks and Cabinets

PBXs, ACDs, routers, hubs, patch panels, and other telecommunications equipment can be mounted in either cabinets or relay racks. In cabinetized equipment the interbay cabling is often contained within the cabinet. In rack-mounted equipment the cabling is external and is supported by overhead cable troughs or run through raceways in the floor. Figure 11-1 shows a typical overhead racking configuration. Because of the quantities of cables involved and the need for physical separation in some cables, overhead racking is the most common method in both central offices and large PBXs. Cables can be run through closed cable trays or in open ladder racks. The latter are generally less expensive than cable trays and permit airflow around the cables. For example, in Figure 11-1 a conduit to contain fiber-optic cable is mounted below the metal framework.

To control noise and crosstalk in telecommunications equipment the manufacturer's specifications must be followed for the type and layout of cabling. As with outside plant cable, the twist in interbay cable controls crosstalk and prevents unwanted coupling between circuits. In addition, cables often must be run in separate troughs that are segregated by signal level and kept physically separated so that signals from high-level cables cannot crosstalk into low-level cables. For some types of cable, shielding is required to further reduce the possibility of crosstalk.

In central offices and PBXs alike, many critical leads have maximum lengths that cannot be exceeded. If lead lengths are exceeded signal loss between components may be excessive, signals may be distorted, or in high-speed buses timing

FIGURE 11-1

Central Office Racking Layout (Photo by Author)

may be affected by propagation delay. Manufacturer's specifications must be followed rigorously with respect to lead length.

Distributing Frames

Temporary connections or those requiring rearrangement terminate on crossconnect blocks mounted in distributing frames such as the one shown in Figure 11-2. Distributing frames also provide an access point for testing cable and equipment. The size and structure of a distributing frame are dictated by the quantity of circuits to be connected. Cabling to the central office equipment routes through openings at the top of the frame, fastens to vertical members, and turns under a metal shelf or mounting bracket that supports the crossconnect blocks. The crossconnect blocks are multiple metallic terminals mounted in an insulating material and fastened to the distributing frame. Equipment and lead identity is stenciled on the blocks.

Large PBXs may use the same hardware as central offices or they may use a wall-mounted backboard that holds 66-type blocks or 110 connectors as shown in Figure 11-3. Wall-mounted frames are satisfactory in small installations, but in PBXs with more than about 1000 stations the frame grows too large to be administered efficiently. Jumpers are long and wiring trough congestion becomes a problem. To relieve jumper congestion, hardware is available to mount wiring blocks on double-sided freestanding frames.

FIGURE 11-2

Central Office Distributing Frame (Photo by Author)

FIGURE 11-3

110-Type Connector Field (Photo by Author)

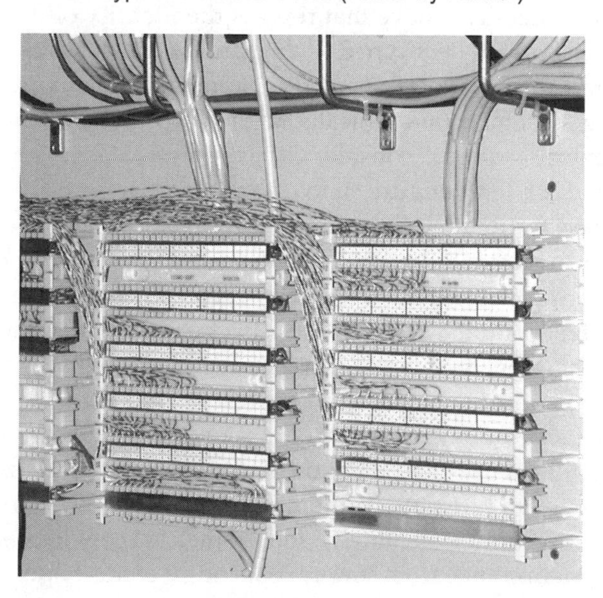

Protector Frames

Incoming circuits that are exposed to power or lightning are terminated on protector frames. The protector module forms the connection between the cable pair and the attached equipment. As discussed in Chapter 7, if excessive current flows in the line the protector opens the circuit to the central office equipment and grounds the conductors. If excessive voltage strikes the line, carbon blocks inside the protector module arc across to ground the circuit. Modules are manufactured with gas tubes where these are needed to protect vulnerable central office equipment such as digital switches.

Alarm and Control Equipment

Most telecommunications equipment includes integral alarms with external contacts. The extent and type of alarming varies with the manufacturer, but generally alarms draw attention to equipment that is marginal or failed and direct the technician to the defective equipment. Equipment alarms light an alarm lamp on the chassis and operate external contacts that are used for remoting the alarm and for operating external audible and visual alarms. Most central offices contain an office alarm system to aid in locating failed equipment. Alarms are segregated into major and minor categories to show the seriousness of the trouble; different tones sound to alert maintenance personnel to the alarm class and location. Besides audible alarms, aisle pilots and bay alarm lamps guide maintenance personnel to the room, equipment row, bay, and the specific equipment in trouble.

In offices designed for unattended operation, network management equipment transmits the alarms to a distant center over telecommunications circuits. The alarm remote is generally a slave that reports the identity of the alarm point. The central system is typically equipped with a processor and database that pinpoint the trouble and may diagnose the cause. Some equipment, including most electronic switching systems, communicates with a proprietary remote that provides the equivalent of the local console. Other remote alarms report building status such as open door, temperature, smoke, and fire alarms.

Offices designed for unattended operation frequently include control apparatus for sending orders from a distant location over a data circuit. For example, microwave and fiber-optic equipment usually have control systems that enable technicians to transfer working equipment to a backup channel. Offices equipped with emergency generators frequently are arranged for engine start and shutdown and transfer to and from commercial power.

The more extensive private telecommunications networks use central office techniques for reporting alarms and diagnosing trouble. Most PBX manufacturers support their systems with a remote maintenance and testing system that enables technicians to diagnose trouble remotely, and, sometimes, switch around failed apparatus. Alarm systems range from simply reporting a contact opening or closure over a circuit to more elaborate systems that report values to a remote center, support remote diagnostics, and maintain a trouble clearance database.

Power Equipment

Equipment in most central offices and many enterprise networks operates from –48 V DC, which is the typical voltage supplied by central office charging and battery plants. PBXs operate either on –48 V or on commercial AC. Microwave equipment usually works on –24 V DC in radio stations and –48 V DC in central offices. Some central office equipment operate from alternating current (AC) and requires a DC to AC converter known as an *inverter* to provide an uninterrupted source during power failures. Commercial uninterruptible power supply (UPS) equipment, which we will discuss later, contains a built-in battery supply to keep AC-operated equipment functioning through power failures. The difference between an inverter and a UPS is that the former operates off a central battery supply and the latter has its own batteries.

Most central offices and PBXs in hospitals and other organizations that cannot tolerate system failures have an emergency generator to carry the load and keep the batteries charged during prolonged power outages. The emergency generator connects to the charging equipment through a power transfer circuit that cuts off commercial power while the generator is online.

Batteries and Charging Equipment

Storage batteries use technology similar to automobile batteries. Lead acid and nickel–cadmium cells are common and some equipment use batteries with solid electrolyte. Power is distributed from the battery plant to the equipment over bus bars, which are conductors large enough to carry the total load. To minimize the amount of voltage drop, batteries are installed as close to the equipment as possible. Under normal operation the –48 V bus is actually floated from the charger at –52 V. Most central office equipment can tolerate a drop to 44 V or less without affecting equipment operation.

A storage battery has three principal elements: positive and negative plates, electrolyte, and the case. The plates, which are made of a metallic substance such as lead or nickel–cadmium, are suspended in electrolyte. In contrast to automobiles, which normally have the negative pole grounded, in telecommunications equipment the positive pole is grounded to prevent electrolysis.

Temperature has a significant effect on battery life and capacity. Battery capacity is highest during moderately warm temperatures, but as temperature increases, battery life is shortened. Conversely, cooler temperatures extend battery life, but below freezing, capacity is reduced. To maintain the best balance between capacity and life, batteries should operate at approximately room temperature. Obviously, this is impractical in remote locations that lack heat and air conditioning.

In such locations, batteries should be positioned to minimize their exposure to temperature extremes.

Storage batteries are evaluated by their capacity, usually stated in ampere-hours, type of plate material, and electrolyte. Central office batteries are usually strings of individual cells, each having a nominal voltage of 2.17 V. The cells must be in leak-proof and crack-proof cases, preferably with a sealed electrolyte. Manufacturers also specify batteries by their expected service life. Long-life central office batteries have sufficient plate material to last for up to 20 years with proper maintenance.

The length of time the equipment can operate under power failure conditions depends on the current drain and the capacity of the batteries. For example, if the equipment draws 10 A and the battery string can supply 100 A-h, the equipment could operate for 10 h under power failure conditions.

Uninterruptible Power Supplies (UPS)

Commercial AC power is occasionally subject to the following irregularities:

- *Blackouts*: total failures of commercial power.
- *Brownouts*: reductions in voltage.
- *Surges*: momentary voltage changes.
- *Transients*: momentary open-circuit conditions.
- *Spikes*: sharp pulses of high voltage that rapidly rise and decay.
- *Frequency variations*. momentary or prolonged deviations from the nominal 60-Hz power line frequency.

Power irregularities can damage equipment or interrupt service. The severity of the problem varies with locale and season. In some localities outages happen so infrequently that an occasional failure can be tolerated. In other parts of the country outages are a regular occurrence, particularly in bad weather.

Power-line conditioning equipment removes spikes, transients, and surges. Blackouts and brownouts require some form of backup power. Equipment such as computers, tape and disk drives, many PBXs, and most key telephone systems operate from commercial AC and should be protected by a UPS.

UPS capacities range from enough to enable a device such as a file server to shut down gracefully to ones with enough capacity to operate computers, a PBX, and auxiliary equipment such as modems, multiplexers, and voice mail through a prolonged power outage. If the UPS does not have enough capacity to operate through an outage, the protected device should be triggered to shut itself down. Some UPS supplies can dial a pager to notify the administrator of the power outage. Some supplies are SNMP-compatible, so they can be monitored and controlled from a network management system.

UPS supplies are classified as static or rotary. Static supplies are less complex and are available in three general types: offline, online, and line interactive.

FIGURE 11-4

An Offline UPS

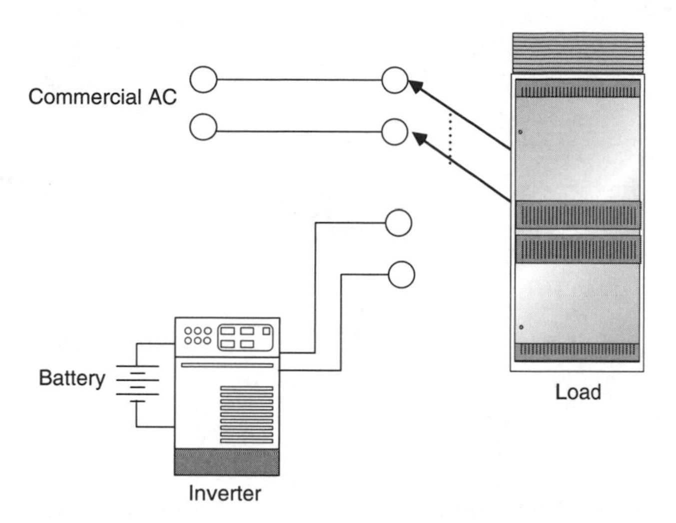

Offline UPS

An offline supply, sometimes known as standby power source or standby UPS, monitors the power line and switches the load to its internal inverter when the power falls outside limits. Figure 11-4 shows schematically how this type of supply works. The inverter converts DC to AC. It connects permanently to a storage battery that charges from the AC source. The AC source carries the load and on failure, the load switches to the output of the inverter. A short break in power occurs when the load transfers. The break runs from 5 to 20 ms, which is short enough to keep most apparatus working. During brownout conditions, however, the switching time may be longer. Some offline UPSs have ferrite core transformers that provide a flywheel effect to keep power supplied to the load long enough to prevent any interruption. Many inexpensive offline UPS supplies lack line conditioning, have no frequency regulation, and provide limited or no surge and spike protection.

Online UPS

An online UPS, which is shown in Figure 11-5, supplies power continuously to the protected apparatus. The commercial power source keeps the UPS battery charged. When the power fails, the inverter continues to function without a break because the inverter supplies power directly. The charging apparatus keeps the equipment completely isolated from power line irregularities. Some types of online supplies have a dangerous flaw in that if the UPS itself fails, the load is isolated from the commercial source. When selecting equipment of this type, be certain that the unit has bypass circuitry to connect the protected equipment directly to the commercial source in case of UPS failure.

FIGURE 11-5

An Online UPS

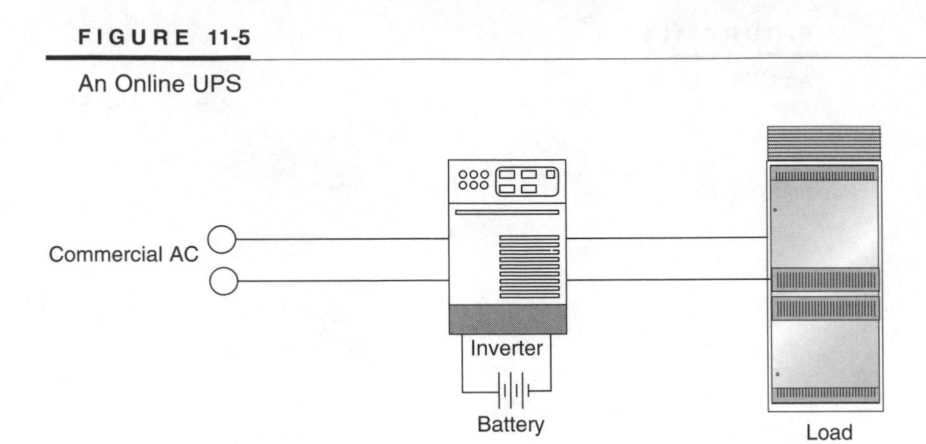

Line-Interactive UPS

A line-interactive UPS has some of the characteristics of both an online and an offline UPS. The protected equipment is powered from commercial power during normal conditions, but the UPS has circuits that can boost or reduce the voltage from the source. This technique provides voltage regulation, which the offline supply lacks. If the commercial supply fails, the UPS increases the amount of voltage boost in a time interval approaching that of an online supply.

Rotary UPS

Static UPSs are usually sufficient except in critical applications that cannot tolerate any interruption. A rotary UPS uses a motor generator with a flywheel to absorb any momentary interruptions. Power to critical loads comes directly from the generator, which is driven by commercial AC under normal conditions, and switches to an auxiliary engine during failures. The flywheel keeps the generator producing output while the engine starts. Some rotary supplies have the generator and auxiliary engine on the same shaft. When a power failure occurs, the engine starts and when it comes up to speed, a clutch automatically engages.

UPS Evaluation

UPS manufacturers quote the capacity in KVA (kilovolt-amps). In DC systems, power is the product of volts and amps, e.g., 120 V × 10 A = 1200 W. In AC systems, however, voltage and current can be out of phase, so supplies are quoted with a *power factor*, which expresses a ratio between the theoretical and actual power-delivering capacity of the system. In the above example, if the UPS has a power factor of 0.8, then the supply must deliver 1500 W. Usually, it is wise to purchase a supply with plenty of extra capacity. This will result in a larger battery and the supply can carry the load for a longer period. This is particularly important as the batteries age and lose capacity.

The *crest factor ratio* is another important criterion. This factor evaluates the supply's capability of handling load peaks. Technically, it is the ratio between the nonrepetitive peak load the supply can provide and the linear root mean square (RMS) load it supplies. The ratio should be at least 2.5; the higher the better.

The output wave shape is often important. Commercial AC is a pure sine wave, and the more effective UPS supplies also furnish sine wave output. Less-expensive supplies provide square wave output, which could affect the operation of the power supply in the protected equipment. The supply should retain its wave shape throughout the rated power drain.

The amount of voltage regulation and the backup time are two more important factors in evaluating a UPS. The supply should maintain voltage within ±3 percent under its rated capacity. The amount of backup time is determined by comparing the power drain of the protected equipment to the power furnished by the UPS and how long the equipment is to be protected. This factor is a function of battery capacity. A key telephone system or PBX normally is protected through the longest expected power outage. It may be necessary to power a file server or computer only long enough to allow a graceful shut down since desktop devices are down anyway.

Emergency Engine Generator

Common carriers and organizations such as hospitals and public safety organizations that require continuity of telecommunications service must use an auxiliary engine generator to furnish power. The generator connects through a transfer switch to the charging equipment. If a generator is available, less battery capacity is needed because the battery must furnish power only until the generator starts. Generators must be operated under load periodically to ensure that they are operable in emergency conditions. Diesel fuel tends to deteriorate over time. Many manufacturers provide natural gas or propane-powered engines, the fuel for which is more reliable and may comply with local air-quality requirements.

PART TWO

Switching Systems

Switching lies at the heart of every telephone session. The PSTN uses circuit switching, which sets up a call by connecting circuits end-to-end. At the edge of the PSTN most circuits are analog and connect to stations over copper wire. Larger businesses, which are those needing more than about 15 lines, often use digital circuits at the edge, and within the core of the PSTN all circuits are digital. Most IP PBX sessions are switched at the edge in an Ethernet switch. If the call is completed over IP, it is switched or routed in the core and switched again at the edge. Most VoIP calls hop off the IP network to ride the PSTN for at least part of the distance. Whether the call is circuit- or packet-switched, the physical infrastructure of the core rides on fiber optics.

The hybrid nature of the voice network evolves toward IP, although not rapidly. The PSTN has reached maturity and development that remains is mostly fine-tuning. This is not to say that the PSTN is perfect, but that further improvements are most likely to develop on IP. The reason for this is inherent in the nature of the two networks, or more precisely, the endpoints. The PSTN developed from the start as a vehicle for providing efficient telephone service to the masses. To keep the cost as low as possible, the telephone was intended to be cheap, simple, and dumb. As such, it has withstood the test of time despite the fact that the 12-key dial pad imposes severe limitations on the scope of features it can provide. The endpoint in an IP network is assumed to be intelligent and programmable. It therefore has considerably more flexibility because it can communicate with servers that can be programmed with unlimited features.

Circuit switches have long had similar architectures regardless of the manufacturer. A central processor controls a hardware switching fabric that is connected to subscriber lines on one side and trunks on the other. The products on the market have gravitated toward a uniform set of features that are tightly integrated with basic call processing. The processors are proprietary and operate under control of a generic program that is likewise proprietary and integrated with call processing. With the onset of VoIP, a new concept known as a *softswitch* is coming on the market. The hardware platform is a server using industry-standard components. The operating system is commercial and call processing outside the basic functions is likely to be customized in packages produced by a third-party developer.

The purpose of this part of the book is to explore and understand the ramifications of different alternatives for switching systems and the signaling systems by which they interoperate. We begin with an explanation of signaling systems, which are the nerve center of the PSTN. Following that are three chapters dealing with switching. Chapter 13 discusses the PSTN, its architecture, its features, and the ways in which it is evolving. Chapter 14 covers circuit switching and the architecture of the hardware. Chapter 15 discusses softswitches and VoIP. Throughout this part we examine QoS and how it is achieved in circuit- and packet-switched networks.

CHAPTER 12

Signaling Systems

The objectives of any telephone network are to establish a connection between endpoints, to monitor the circuit while it is in use, to disconnect it when the users hang up, and to compile information for billing the call. Users need to know nothing of how the network establishes the connection; they merely enter the destination address (the E.164 number) and let the system select the route. These functions require a unified signaling system.

An ongoing discussion among network designers is whether intelligence should be internal to the network, external, or contained in the edge devices. The Internet designers have more-or-less settled on keeping the core of the Internet simple in favor of making it fast and flexible. The core devices switch and route packets with little or no concern about what they contain. This keeps the network operating at high speed, but it detracts from its ability to distinguish traffic classes and prioritize them accordingly.

For years, the PSTN had dumb endpoints (telephones) and concentrated its intelligence on the switches. The first switches were manual, and offered the advantages of human control. The early switches that replaced manual switchboards were merely slaves to the dial pull. If the switch encountered blockage, it lacked the ability to try another route. Common control switches came along around the mid-20th Century and brought intelligence with them—first in the form of electromechanical control, and then full stored-program control. The interconnecting trunk network operated under switch control and had no intelligence. Signaling was *in-band*, meaning that signals set up the call link-by-link over the talking path.

In-band signaling has several drawbacks. For one thing, every circuit requires a dedicated signaling device, the only purpose of which is to convert tones into DC signals. Toll offices of the period contained racks and racks of single-frequency (SF) signaling sets. Furthermore, conventional signaling lacks look-ahead capability. A circuit can be set up cross-country or internationally, only to encounter a busy signal, wasting expensive circuit capacity. Another drawback is setup time. When circuits are built up with progressive signals over the circuits, the time required to set up the call is much longer than it takes to direct call setup end-to-end from a central computer. Most important to the service providers is the fact that in-band signaling is susceptible to fraud. Toll thieves used devices that emulated signaling tones and fooled the billing apparatus into thinking that the call was not completed. The solution to so-called blue box fraud was a separate external signaling network. That network, SS7 is the primary topic of this chapter. SS7 is designed for the circuit-switched network and does not lend itself well to VoIP. Several other initiatives we discuss in this chapter are either available or under development to support call control over an IP network.

SS7 is both the brain and the nerve center of the PSTN. The intelligence remaining in the switch relates primarily to internal call control and feature operation, with translation and routing functions offloaded to SS7. Phone numbers once were glued to a particular switch in a particular wire center, but no longer. Number portability demands that numbers be independent of switch location. Before a call can be completed, the originating switch must determine where to send it based on a regional database that is accessed over SS7. Cell phones rely on SS7 to communicate information about roaming subscribers. Special signaling features such as caller ID and last-call return, which fall under the category of Custom Local Area Signaling Services (CLASS), require a separate signaling network because it is impractical to send high-speed data messages in the bandwidth confines of a voice channel.

CLASS notwithstanding, signaling services are still quite rudimentary compared to where they will be in the future. The current industry buzzword is "presence," which means you are able to advertise to the network your location and preferred means of contact. The manual switchboard was an early form of presence that its replacements cannot emulate. Users are not unanimous in their belief that universal accessibility is a good thing, but without question, the presence capability in the PSTN is lacking. At least we have gone beyond the busy signal, which in the past meant the call had reached a dead end. Now calls can overflow to voice mail, which is a near-universal capability. If a PBX supplies the voice mail, the caller can escape to a coverage position, but if that position is not staffed, the caller lands back in voice mail again with no clue about how to reach the callee.

Call forwarding is a crude form of presence. With CLASS you can screen which calls are forwarded, but the network cannot inform a caller that you are out of the office and best reached with instant messaging or e-mail. You can do that with your voice mail greeting, but users are notoriously careless about keeping their greetings current. The IETF's solution is Session Initiation Protocol (SIP). This protocol has considerable promise, but it is immature and will remain so for several years. SIP will eventually replace H.323, which is an ITU-T signaling protocol that most VoIP products rely on today. H.323 is a stable protocol, but its shortcomings are the source of inspiration for SIP designers. The IETF and ITU-T are working together on another protocol, Megaco, which fulfills a similar function.

PSTN SIGNALING TECHNOLOGY

Signaling systems can be classified by the method of exchanging signals: direct current (DC), tone, bit-robbed, and common channel. POTS service uses a combination of DC and tone signaling in the local loop. The familiar dual-tone multi-frequency (DTMF) signals have almost completely replaced the rotary telephone dial, which is a DC device that operates by opening and closing the loop in step with the dial pulses. Multifrequency (MF) tone signaling was used on all long distance circuits in the past, but except for a few minor applications, SS7 has replaced it. Bit-robbed T-carrier signaling is an in-band system that is used in both trunks and loops. Bit-robbed signaling is disappearing, but it will persist in North America for some years.

Signals can be grouped by the four functions they perform: Supervising, addressing, alerting, and call progress.

- *Supervising* is monitoring the status of a line or circuit to determine if it is busy, idle, or is requesting service. The term is derived from the manual switchboard from which operators literally supervised the call. Switchboards displayed supervisory signals by an illuminated lamp on the key shelf. In the network, supervisory signals are conveyed as voltage levels on signaling leads or the on hook/off hook status of signaling tones or bits.
- *Addressing* is the process of transmitting route and destination signals over the network. Addressing signals include dial pulses, tone pulses or data pulses over loops, trunks, and signaling networks.
- *Alerting* informs the callee of an incoming call. Alerting signals are either audible or visual.
- *Call-progress* signals such as busy and reorder tones inform the user of the status of the call setup process.

To illustrate these four functions, consider what happens when you lift a telephone handset to place a call. Lifting the handset places a DC short-circuit on the line and current begins to flow. The line circuit in the central office detects the current flow and responds by returning dial tone and attaching a register to receive the called number. This process is called *loop signaling* or *loop start*. Although the line is short-circuited for DC, a transformer in the telephone instrument isolates the DTMF tones and voice signal from the short. As you press the DTMF buttons on the telephone, the central office registers the address, sets up the path, and rings the called telephone. If the called station is busy the central office sends a busy signal. If all circuits are busy it sends a *reorder* or fast busy tone.

The type of signaling used in the loop depends on the class of service. POTS lines use loop signaling. PBXs that connect to the PSTN with analog trunks use a variation known as *ground-start signaling*. Because of glare, which we discuss later, ordinary loop-start signaling is unsatisfactory for PBXs. Most LECs offer T1 trunking for PBXs. This type of trunking uses DTMF signaling, or in North America, it may use bit-robbed signaling.

ISDN lines use external signal channels to support functions that cannot be carried in-band. For example, with analog signaling systems, when a station is busy, the signal stops at the end office unless the user has chosen the call-waiting feature, which interrupts the call in progress. ISDN signals with data messages over a separate out-of-band signaling channel. If a call arrives while the line is busy, the network can send a message that displays on the callee's telephone. The user can choose to interrupt the original session or not, depending on who is calling.

In-Band Signaling

Using the general call state model shown in Figure 12-1, we next consider the differences between in-band and common channel signaling methods. Although the trunk signaling method discussed below is obsolete, it illustrates the signaling functions. The line signals still operate as the model shows.

In the idle state, subscriber loops have battery on the ring side of the line and an open circuit on the tip. No loop current flows in this state. Signaling equipment associated with the trunks indicates on-hook or idle status, either with a signaling tone or with digital trunk signaling bits. The local central office continually scans the subscriber line and trunk circuits to detect any change in their busy/idle status. When station A lifts its receiver off hook, current flows in the subscriber loop, signaling the local central office of A's intention to place a call. The central office responds by marking the calling line busy and returns dial tone to the calling party. Dial tone is one of several call-progress signals that telephone equipment uses to communicate with the calling party. It conveys the readiness of the central office to receive addressing signals.

Station A transmits digits to the central office using either DTMF or dial pulses. The system registers the digits and checks its translation tables, determining that the call is not local. The switch must determine which IXC handles A's long distance calls and which trunk route connects to that IXC. From the address, the system determines that the call must connect to tandem B, which belongs to the preferred IXC. The local switch at A checks the busy/idle status of trunks to B, and seizes an idle trunk. If no trunks are idle, switch A returns a reorder tone to station A. When the switch seizes an idle trunk, the caller at A hears nothing except, perhaps, a faint click. If central office A has local automatic message accounting (LAMA) equipment, it registers an initial entry to identify the calling and called number, and to prepare the LAMA equipment to record the details

FIGURE 12-1

General Call State Model

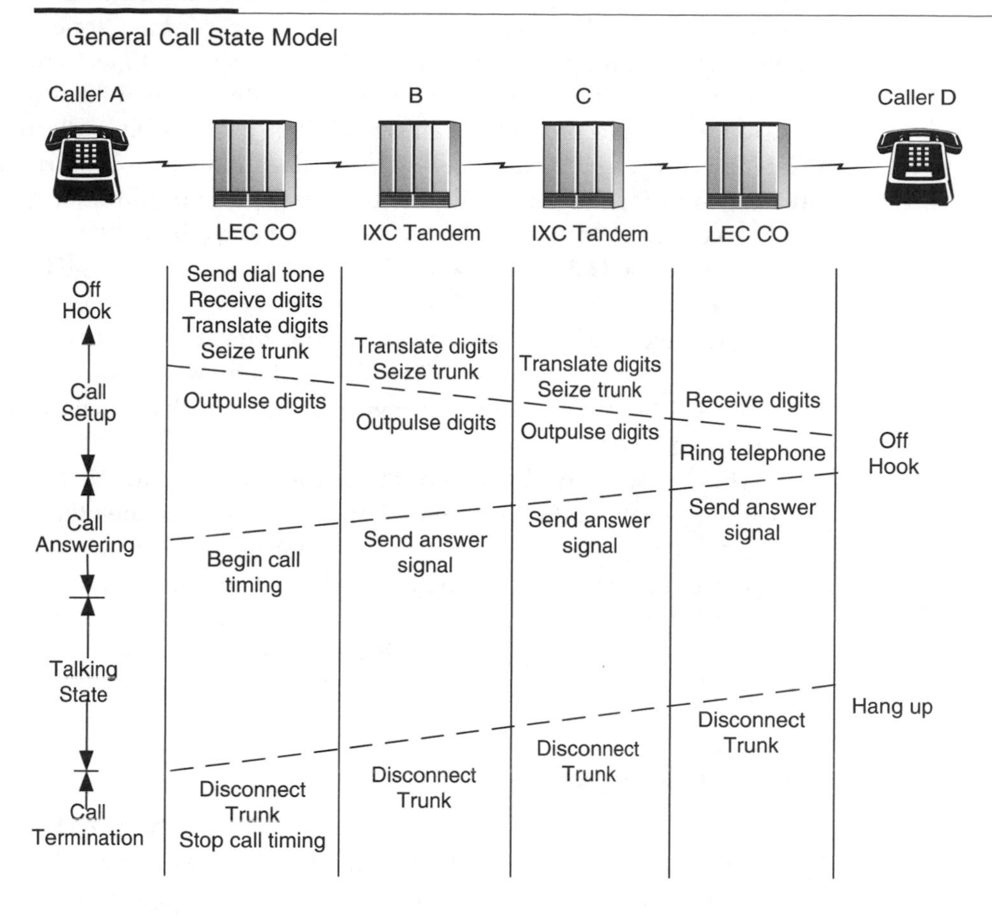

of the call when D answers. If the office does not have LAMA it sends the initial entry to an AMA office over a centralized automatic message accounting (CAMA) trunk.

The trunk seizure removes the tone from the channel or changes the signaling bits to show the change in status. Tandem office B, detecting the status change, returns a signal, usually a momentary interruption in the signaling tone toward A. This signal, called a wink, signifies that B is ready to receive digits. Detecting the wink, central office A sends its addressing pulses toward B. These are MF pulses, which are conveyed by coding digits with combinations of two out of five frequencies. Tandem office B continues to send an on-hook tone toward office A, and will do so until station D answers. At this point, office A has completed the originating functions, and awaits the completion of the call.

Tandem B translates the digits and picks a route to Tandem C, selects an idle trunk, and seizes it. Tandem C, detecting the seizure, sends a start signal to B, and prepares to receive digits. B detects the start signal and sends the digits forward.

Tandem C repeats the process to local central office D. Office D tests the called station for its busy/idle status, and if busy, returns a busy tone over the voice channel. The calling party, recognizing the call progress tone, hangs up and the switches take the connection down. The originating switch adds no completion entry to the AMA record, so the call is not billed. If station D is idle, office D sends a 20 Hz alerting signal to ring the bell in D's telephone. It also returns audible ring (another call progress tone) over the transmission path to the originating party. The line continues to ring until D answers, A hangs up, or the equipment times out. When station D answers, central office D detects the change in status as line current begins to flow. This trips the ringing signal and stops audible ringing. Office D changes the signaling status toward C from on-hook to off-hook by interrupting the SF tone. C sends the off-hook signal to B, which transmits the off-hook signal to A. The AMA equipment registers call completion, indicating the time that charging begins.

When either party hangs up, the change in line current indicates a status change to its central office, which forwards the change to the other end by restoring the SF tone. Office A registers a terminating entry in the AMA equipment to stop charging. The SF tones are restored to all circuits to show circuit status as idle. All equipment then is prepared to accept another call. Today, T-carrier and common-channel signaling have obsoleted SF signaling except for a few specialized applications.

Call Setup with Common Channel Signaling

Consider now how common channel signaling handles the call. Figure 12-2 shows the architecture of the network, the details of which are discussed in a later section. A separate data communications network connects through an interface known as a *service switching point* (SSP) in each central office. SSPs are software applications running in the central offices, and are linked via data circuits to *signal transfer points* (STPs), which are routers for signaling messages. STPs connect to *service control points* (SCPs), which contain servers with databases of network information that the network nodes can access.

Referring to the call setup model in Figure 12-1, assume that the stations are basic rate ISDN (BRI). A BRI station at A sends a setup message over its D (data) channel to local central office A. The message includes the address of the terminating station plus information about the call, such as the type of call. Call types can be data, voice, video, etc. Central office A selects the appropriate combination of B (bearer) channels based on the call type, and sends a data message to its STP. The STP sends a data message to the SCP's database to retrieve information about the originating and terminating stations. For example, the user's class of service indicates whether the called and/or calling stations are members of a virtual voice network, and whether the connection is switched or dedicated. The STP selects a route to the destination, allocates circuits, and sends connect messages to the

FIGURE 12-2

Signaling System No. 7 Architecture

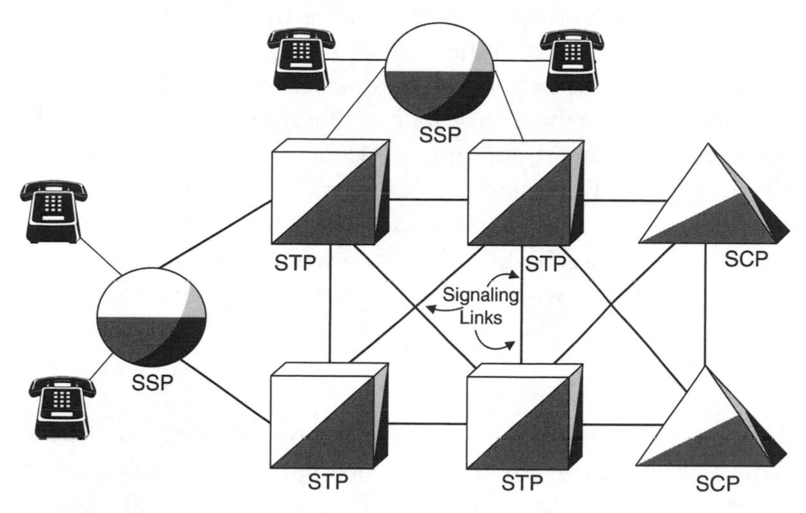

switches. Each stage of the call: ringing, connect, and disconnect is signaled with a data message.

Q.931 Connection Control Protocol

ISDN uses the Q.931 network layer protocol for call control. The central office stores information about the subscriber's services and features in its service profile ID (SPID). The central office and CPE communicate over the D channel with a variety of message types such as setup, suspend, call proceeding, disconnect, and corresponding acknowledgment messages. The Q.931 messages begin with a single octet protocol discriminator, followed by an octet that contains the length of the call reference, which is a field that distinguishes between the calls being managed over the D channel. This is followed by the message type and information elements that are needed by the message type. For example, the setup message contains the 10-digit calling and called party numbers.

A call consists of a message exchange between the caller and the central office switch. The caller sends a setup message to the switch, which returns a call proceeding message and a setup message to the callee. If the callee responds to the setup it rings the phone and sends an alerting message to the caller. When the callee answers, the switch sends a connect message, which the caller acknowledges. Q.931 emulates the steps that the caller, callee, and switch take in completing a call with in-band signaling.

E&M Signaling

By long-standing convention, signaling on interoffice circuits uses two leads designated as the E or recEive and the M or transMit leads for conveying signals. External signaling sets and the built-in signaling of T carrier channels use E&M signaling to communicate status to attached central office equipment. Signaling equipment converts the binary state of line signals, (tone on or off for analog and zero or one for digital equipment), to actuate the E&M leads. There are five different types of E&M signaling interfaces, but in the most common type the M lead is grounded when on-hook. Applying –48 volt battery to the M lead indicates an off-hook seizure. The E lead is open when on-hook; the signaling set applies ground to the E lead when it receives an off-hook signal from the distant end.

Glare

All of us have had the experience of picking up the telephone handset to place a call and finding that the line is already connected to an incoming call, but the bell has not rung. This condition, known as *glare*, occurs when both ends of a circuit are simultaneously seized and the callee picks up the handset during the silent period between ringing intervals. Glare is easy to resolve in ordinary telephone circuits (the parties both say hello), but it creates a problem in trunks. One way of preventing glare on trunks is by using one-way signaling. On small trunk groups the use of one-way trunks is inefficient from a traffic-carrying standpoint. To accommodate two-way trunks, signaling and switching systems must prevent glare or resolve it when it occurs. In the worst case, when glare occurs the equipment is unable to complete the connection, the circuit times out, and the user receives reorder.

ISDN circuits are not subject to glare, but other types of loops are. PBXs cannot use loop-start two-way central office trunks because the ringing signal occurs at 6-s intervals. For up to 6 s the PBX would be blinded to the possibility of an incoming trunk seizure. It could seize a circuit for outgoing traffic, unaware that a call had been connected from the other end. Ground start lines provide an immediate trunk seizure signal toward the PBX. When either end seizes the line for an outgoing call, it grounds the tip side to signal the seizure to the other end.

SIGNALING SYSTEM 7

ITU-T has adopted SS7 as the world standard common-channel signaling system. SS7 uses a separate packet switched network to pass call setup, charging, and supervision information. It also can access the carrier's database to obtain account information such as features and points served on a virtual network. The purposes of SS7 can be summarized as follows:

- *Improves call management*. The system handles call setup and disconnect. SS7 handles end-to-end supervision and timing.

- *Enhances network management.* SS7 is responsible for routing and congestion control, functions that previously each switch handled internally. It provides status information to network elements and collects performance information.
- *Separates network control from hardware.* With other signaling systems, network control is embedded in the underlying hardware. SS7 control is independent of the circuits and hardware.
- *Supports user database.* SS7 provides database information for functions such as number portability, virtual networks, enhanced services, and caller identification and verification.
- *Handles addressing and supervision.* Signaling protocols carry line status, calling and called numbers, credit card numbers, and other such information through the network.
- *Supports 800 number portability.* When an 800 number is dialed, an SS7 message to a central database returns the identity of the IXC.
- *Supports local number portability.* An advanced intelligent network (AIN) feature in local central offices uses SS7 to relay to the originating switch the identity of the LEC that handles the number.

With in-band signaling, audible signals are passed over the channel that will be used for talking, so each circuit is automatically tested for continuity. With common channel signaling this test is not possible; therefore, the equipment makes a separate path assurance test before connecting the call.

Signaling System 7 Architecture

Figure 12-2 shows the architecture of SS7. Each carrier obtains its own signaling network; the networks are interconnected to enable carriers to interoperate. Each node has an address known as a *point code.* When the SSP receives a service request from an end office, it sends a service request to its SCP and suspends call processing until it receives a reply. The STPs are geographically dispersed redundant nodes that are interconnected over diverse paths. STPs are deployed in pairs so that the failure of one system will not affect call processing. STPs pass the call setup request to an SCP over direct circuits or by relaying it through another STP.

The SCP is a high-speed database that is deployed in pairs, each member having duplicates of the database. The database holds circuit and routing information, and for customers that are connected through a virtual voice network, the database contains customer information such as class of service, restrictions, and whether the access line is switched or dedicated. The SCP accepts the query from the STP, retrieves the information from the database, and returns the response on the network. The response generally takes the same route as the original inquiry.

Signaling System 7 Protocol

SS7 uses a layered protocol that resembles the OSI model, but has four layers instead of seven. Figure 12-3 shows the protocol stack compared to the OSI model. The first three layers are called the message transfer part (MTP). The MTP is a datagram service, which means that it relays unacknowledged packets. The MTP has three layers, which form a network similar to X.25 and IP. The functions of these layers are:

- The *signaling datalink* is the physical layer. It is a full duplex connection that provides physical links between network nodes.
- The *signaling link layer* is a datalink that has three functions: flow control, error correction, and delivery of packets in the proper sequence.
- The *signaling network layer* routes messages from source to destination and from the lower levels to the user part of the protocol. Its routing tables enable it to handle link failures, and to route messages based on their logical address.

FIGURE 12-3

SS7 Protocol Architecture Compared to OSI Model

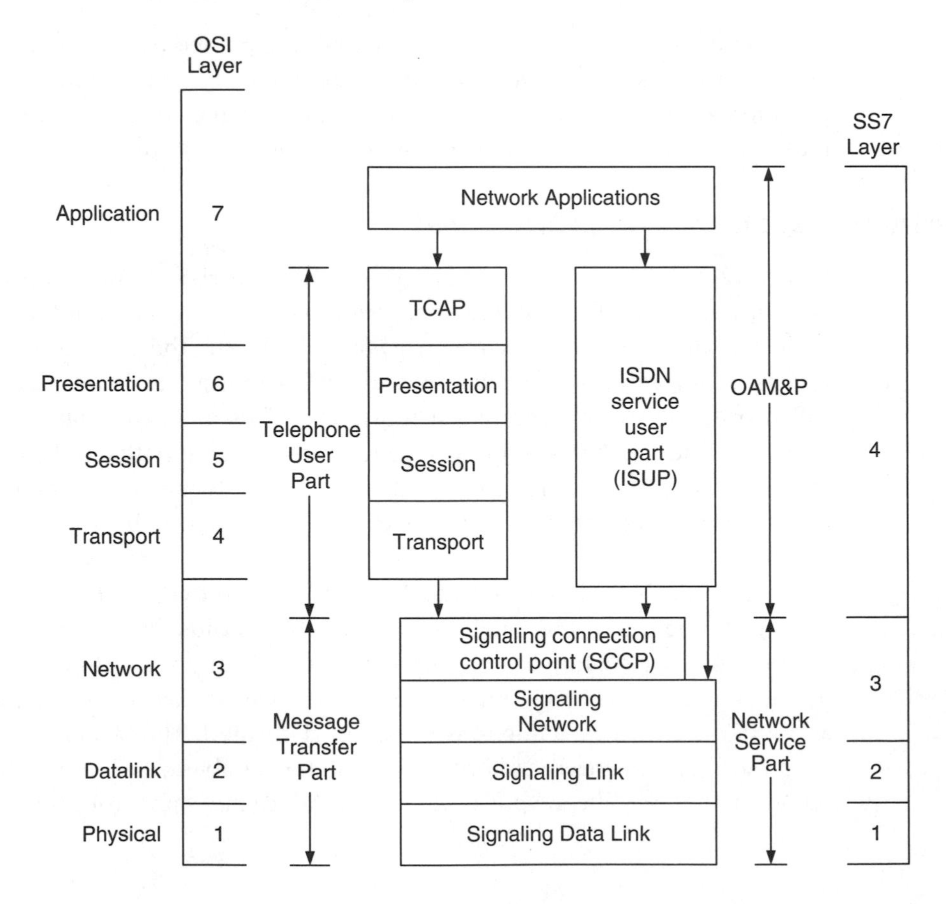

The fourth layer is called the *signaling connection control* part (SCCP). It is responsible for addressing requests to the appropriate application, and for determining the status of the application. An application, for example, might be a toll-free service request. The ISDN Service User Part (ISUP) relays messages to ISDN users. The user in this context refers to the interface with the end user's equipment, and not to the user itself. The ISUP handles call setup, accounting and charging, and circuit supervision for ISDN connections.

Custom Local Area Signaling Services

Most LECs offer a suite of CLASS services. CLASS services bring to residence and small business users features that are available in most PBXs. These additional features improve service for the users and generate revenue for the LECs. Most CLASS features depend on SS7 for communication between central offices, and some operate only over ISDN. Some of services require special telephone sets or external adapters to use them.

Initially, many of the CLASS services were assumed to be possible only with ISDN. With the large number of analog telephones in operation, the LECs had substantial motivation to make CLASS features available to analog subscribers. The result is analog display services interface (ADSI) protocol, which supports features such as calling line ID on an analog line. An ADSI-compatible telephone can receive a short modem message from the central office between the first and second rings. The central office relays caller ID over this connection.

The following lists the principal CLASS features with a short description of what they do.

- *Anonymous caller rejection* allows the subscriber to reject calls from callers who have blocked identification of their telephone number. Rejected calls are sent to a recorded announcement.
- *Automatic callback* lets the subscriber know when the called line is available, and automatically redials the called number unless canceled by the caller.
- *Automatic recall* enables a callee to return a call without knowing the calling number. If the number is busy, the user is informed with an announcement and can elect to continue with the call or drop it.
- *Calling name delivery* sends the listed directory name of the calling party to a display on an ISDN or ADSI telephone set after the first ringing cycle.
- *Calling number delivery* delivers the calling number to the called party. The number can be displayed on an ISDN or ADSI telephone set after the first ringing cycle.
- *Calling number delivery* blocking enables a calling party to block transmission of calling party identification. Blocking can be implemented on a per-call or per-line basis. To communicate with lines that have

anonymous caller rejection, the user can dial a code to unblock the feature for that call.

- *Customer-originated trace* allows a user to initiate a trace on the last call placed. The service is useful for tracing harassing or obscene calls. The calling number is not delivered to the customer, but is entered on a log at the LEC or a law enforcement agency.
- *Distinctive ringing/call waiting* announces calls from a selective list with a special ringing tone. If the user subscribes to call waiting, the call waiting tone is also distinctive.
- *Selective call acceptance* allows users to enter a list of numbers from which calls will be accepted. Calls from numbers not on the list route to an announcement and are rejected.
- *Selective call forwarding* allows users to enter a call-screening list. Calls from stations on the list are forwarded, but calls from other stations ring only at the dialed number.
- *Selective call rejection* allows users to create a screening list. Calls from these stations route to an announcement that states that the called party is not accepting calls. Calls from other stations will be routed through normally.
- *Visual message-waiting indication.* Subscriber is notified of voice-mail message by indicator lamp, message display, or interrupted dial tone on a nondisplay set.

Signaling Systems and Calling Line Identification

Calling line identification (CLID) is often confused with automatic number identification (ANI). The difference between the two is rooted in how signaling is handled in local central offices. This section discusses the differences between the two, and explains when the delivered number may be identical in either case. The distinction is important to anyone who plans to capture calling numbers, whether for personal purposes or for business purposes such as routing calls in a call center.

ANI identifies the originating party's billing number, and the user cannot block it. The ANI process sends the calling party's billing telephone number to the recording office. The purpose of ANI is to bill long distance calls. The IXC may deliver it to the called party for identification but that is not its primary purpose. The ANI number may be the directory number of the line that originates a call, but often the ANI is another number assigned for billing purposes. This is often the case with large businesses that have all charges billed to a single telephone number. The same business may have numerous trunks, each of which the LEC sends to the IXC as CLID. If the call comes from a PBX equipped with ISDN trunks, the station number may show as the CLID if the PBX is programmed to send station numbers to the central office, and if the central office is programmed to send it forward.

IXCs forward ANI to some accounts if the customer asks for, and sometimes pays extra for the service. This feature enables the customer to use ANI for such purposes as calling up a computer screen from the database with an incoming call. Originating callers can block CLID, but they cannot prevent the IXC from forwarding the ANI over a toll-free trunk to the called subscriber. Note that ANI is provided only over T1 lines and not over regular telephone lines, which are used for switched access toll-free service.

Signaling methods are different for ANI and CLID. The LEC can pass ANI to the IXC via in-band signaling over CAMA trunks, which use MF signaling. They can also pass ANI over an SS7 datalink, and can pass it to the customer over primary-rate interface (PRI) trunks. CLID is passed between LEC offices only by SS7. It identifies the calling line's directory number, which may be the same as the billing number, or may be nonpublished. The two numbers are usually the same for single line residences, but different for multiline businesses. An SS7-equipped central office retains the CLID of both the originator and terminating party for later use if either party initiates subsequent calls between them using the special CLASS features. CLID is not used for billing.

CLID is usually transmitted on most calls carried on the LECs ' networks. Where SS7 signaling has not been installed, CLID is not transmitted because it cannot be sent over MF trunks. For CLID to work reliably, the LECs must have agreed to send line identification between their offices. LECs that are equipped for SS7 usually forward CLID to the IXCs, but the IXCs and cellular/PCS carriers sometimes do not forward CLID. Therefore, CLID is not received on every call. Also, CLID is not transmitted for users who have subscribed to nonpublished number service. These calls usually are identified with a message that indicates the line is a private number.

IP SIGNALING

The PSTN and IP networks have major differences in signaling. Either network handles voice calls internally using protocols that are mature and well understood, but connections between the two networks require a gateway to convert the protocols. Today the bulk of the calling is from the IP network to the PSTN since the vast majority of telephone calls are handled over the PSTN. The traffic mix is changing, however, and the need for seamless traffic flow between IP and PSTN is becoming essential. It is especially critical that IP-to-IP calls avoid the use of the PSTN link if possible because it involves conversion from digital to analog back to digital again.

Part of the difficulty lies in the bodies that control the standards: ITU-T for the PSTN and IETF for the Internet. The PSTN is intended to be under tight control of diverse carriers, while the Internet is intended to be simple, fast, open, and under loose control. Numerous protocols are either in place or under development to transform the Internet to carrier-class stability. At the time this book is written, however, the obstacles preventing the Internet from being equivalent to the PSTN as a consumer choice are still formidable.

The H.323 Protocol Family

H.323 is the primary ITU-T recommendation for signaling over an IP network. We will cover the protocol suite, which includes video and audio codecs, compression methods, and multipoint communication in more detail in Chapter 16. The signaling standards for H.323 are H.225 Call Control Protocol and H.245 Control Protocol for Multimedia Communication. An important difference between IP and PSTN endpoints (the standard calls them terminals) is the need in IP to determine the endpoint's capabilities. As a dumb device, a telephone has well-known standard capabilities and limitations in contrast with an intelligent IP endpoint that may have data and video capabilities.

H.245 negotiates channel usage and capabilities. Its sister protocols, H.225, and supplemental protocols in the H.450 series interact to exchange messages among endpoints using ASN.1 syntax. Messages include the following functions:

- Determines which endpoint is the master and which is the slave.
- Determines the multimedia capabilities of endpoints.
- Opens and closes logical channels for exchange of information.
- Actuates supplemental services such as transfer, hold, call park, etc.
- Handles multiport conferences.

H.323 endpoints can communicate as peers, or the network can include a device known as a *gatekeeper* to manage traffic flow. If a gatekeeper is used, it controls a process known as registration, admission, and status (RAS). The network may also include a gateway to interface the packet network with the circuit-switched voice network. The gateway translates H.225 messages on the IP network to Q.931 messages toward the PSTN.

H.225 sets up connections between H.323 endpoints by exchanging messages on the call-signaling channel. The call-signaling channel can be set up between endpoints or between an endpoint and the gatekeeper. It uses Q.931 signaling over TCP/IP to transport the messages. H.225.0/RAS protocol is used for registration, admission control, and status between endpoints and gatekeepers. A RAS signaling channel carries messages between an endpoint and a gatekeeper. The signaling process is essentially the same as circuit-switched devices use except that the endpoints have intelligence, whereas the brains of a POTS telephone are in the central office.

H.323 is a large and flexible protocol with many options, which can be a drawback. Even though the standard is open, device interoperability is a concern, particularly over the Internet. Unless the products have been designed to interoperate, the probability is high that devices manufactured by different companies will not communicate. H.323's primary drawbacks are bulkiness, complexity, and lack of scalability. Much of the complexity is required to support multimedia conferencing and much of it is overkill for voice. It is difficult to work through firewalls and NAT and it is slow because of the number of messages required for call setup. Furthermore, it is not well integrated with SS7. Nevertheless, it is the only complete protocol family for implementing VoIP and it has a lot of momentum behind it.

Media Gateway Control (Megaco)

Megaco is an abbreviation for Media Gateway Control Protocol. Megaco is being developed through the joint efforts of the IETF and ITU-T and carries the dual label of Megaco and H.248. Megaco controls the interaction of a media gateway (MG), which converts between voice and IP, and a media gateway controller (MGC). The industry refers to the MGC as a softswitch. We will cover the concept in more detail in Chapter 16. SIP, which is discussed later, places call control in the endpoints. Megaco places it in a central controller.

STREAM CONTROL TRANSMISSION PROTOCOL (SCTP)

SCTP is a transport protocol that was originally developed to transport SS7 signaling messages over a connectionless packet network. The purpose is to enable services such as VoIP, ISDN, and ATM to use packet networks for signaling. While that is still its main purpose, it supports transport services to a broader range of applications. It provides the same transport services as TCP such as packet acknowledgment, sequencing, and data fragmentation, but it has superior congestion-avoidance capabilities and resistance to denial-of-service attacks.

Another way SCTP differs from TCP is its multihoming capability. This allows SCTP endpoints to have multiple IP addresses, making them reachable over multiple paths. A session between two SCTP endpoints is known as an *association*. An association begins when a call is initiated and remains active until terminated through a shutdown process. SCTP packets consist of a 12-octet header containing source and destination port number, a verification tag that identifies the association, and a checksum. Following that is an 8-octet field identifying the data contained in trailer records known as *chunks*.

Chunks are variable length records containing user data or a variety of control messages. When a call is being initiated, the setup data is bundled into a control chunk known as a "cookie," which contains message authentication codes that are used to prevent denial-of-service attacks. SCTP packets flow to the destination through a primary path. The protocol monitors the path using "heartbeat" chunks. If the primary address repeatedly drops chunks, or if the session experiences path failure or congestion, it can steer the flow to an alternate IP address. Since the protocol can redirect traffic to a separate IP address, the protocol is practically immune to physical network failures. The association can also use multistreaming, in which it sends data over multiple SCTP streams. If one stream is lost, the other streams within the association are not affected.

SESSION INITIATION PROTOCOL

The main benefits of VoIP lie in its ability to offer users choices and flexibility that the PSTN cannot provide. In the PSTN, users' choices are limited to the features that the service provider elects to offer and these are often limiting. Consider, for example, the issue of accessibility. Today's user typically has many methods of being contacted: telephone, fax, mobile, e-mail, pager, and so on, each of which has a unique address. The instrument associated with the address has certain capabilities but the caller has no way of knowing what they are. This leads to ineffective communication attempts because the caller has no idea where the callee is located nor which communications medium has the best chance of succeeding.

The development cycles in the PSTN for new features are long and users have limited control of the features. The call-waiting feature is a good example. On an analog line, subscribers can turn it off and on. When call waiting is on, it can interfere with modems so the user must manually disable it before the session starts. In a voice session, the user can choose to answer an incoming call or let it ring, either of which may convey an undesirable impression to the other party. Some LECs offer whisper call waiting in which the second caller's ID is announced so that the other party does not hear it, but this interrupts the call. Call waiting is inflexible because of the inherent need for signaling over the voice channel. ISDN offers out-of-band signaling, but it is not widely enough used to entice many programmers to develop software to handle calls in accordance with user preference.

The answer, IETF believes, lies in SIP, which is a simple, lightweight signaling protocol that is intended for multimedia applications over an IP network. Although SIP is a modular component of IP telephony, it can operate over any network. Its purpose is to establish, change, and disconnect sessions between endpoints. It has significant differences from other signaling protocols, which are intended for centralized networks such as the PSTN. Instead of communicating via data messages, SIP uses text messages between endpoints and servers that perform such functions as keeping track of a mobile station's whereabouts, its communication capabilities, and the user's preferred means of being contacted (or not). SIP addresses are a URL with SIP: appended to the front of it; for example, SIP: fname_lname@mcgraw-hill.com. Unlike the PSTN in which the station is associated with a physical location, a SIP station can be located anywhere and the network endeavors to find it to establish a session.

A conventional voice network concentrates intelligence in the network. Routing, status of the station and feature capabilities are all contained in the switch and the station is a slave. A SIP station has enough intelligence that the network needs only routing information, which resides in a server. A SIP telephone, therefore, has many capabilities that a conventional telephone lacks. Although a conventional telephone might have a display, the intelligence associated with the display is contained in the network. A SIP telephone, by contrast, could access an external database to look up the caller ID, even displaying a photo it retrieved from file. In contrast to the PSTN with its centralized architecture, SIP uses the multiple paths of IP, resulting in a robust signaling network with no single points-of-failure.

SIP is not a standalone protocol. It uses other IETF protocols to perform selected functions. A multimedia session might be announced using Session Announcement Protocol (SAP). Session Description Protocol (SDP) provides the details of the session including invitations, terminal requirements such as video and whiteboarding, bandwidth requirements, contact information, and so on. Internet standard URLs are used for addressing. Telephone Routing over IP (TRIP) provides routing information in a manner that is analogous to BGP-4 in a data session. RSVP reserves the bandwidth and a protocol known as Diameter provides authentication, accounting, and authorization (AAA).

SIP Functions

Presence is a critical issue to many people. They must be accessible to some types of calls and prefer to be inaccessible to others. In a conventional telephone system such as a PBX, the user has selected features such as call forwarding to improve accessibility, but the feature has severe limitations. If a callee might be at one of four locations, the phone could ring enough times at each one sequentially before reaching a destination, which might turn out to be voice mail. Locating a mobile user without human intervention is difficult or impossible. Consequently, a high percentage of calls are not completed.

SIP supports mobility by registering the user's location and capabilities. For example, a multiuser conference might involve participants, some of whom have video capability and some who do not. For some participants that capability depends on where they are at the time of the conference. With SIP, users' preference files can inform the application that is setting up the conference of location, capabilities, and preferences on how to be reached, find their current address, and connect them if available. With the appropriate application software, the entire session could be set up at a mutually convenient time without involving the multiple phone calls and e-mails that would otherwise be required. SIP and its associated protocols provide the definitions of the call setup and tear down. Participants can make updates while the call is in progress by setting additional SIP messages.

SIP Architecture

The SIP architecture and main components are shown in Figure 12-4. The SIP client resides in the endpoint. Its user agent contains the station's feature information. The client registers its location and availability with the location server. When the user's location changes, this information is forwarded to the redirect server, which sends call setup requests to the appropriate server. The redirect server does not initiate

FIGURE 12-4

SIP Architecture

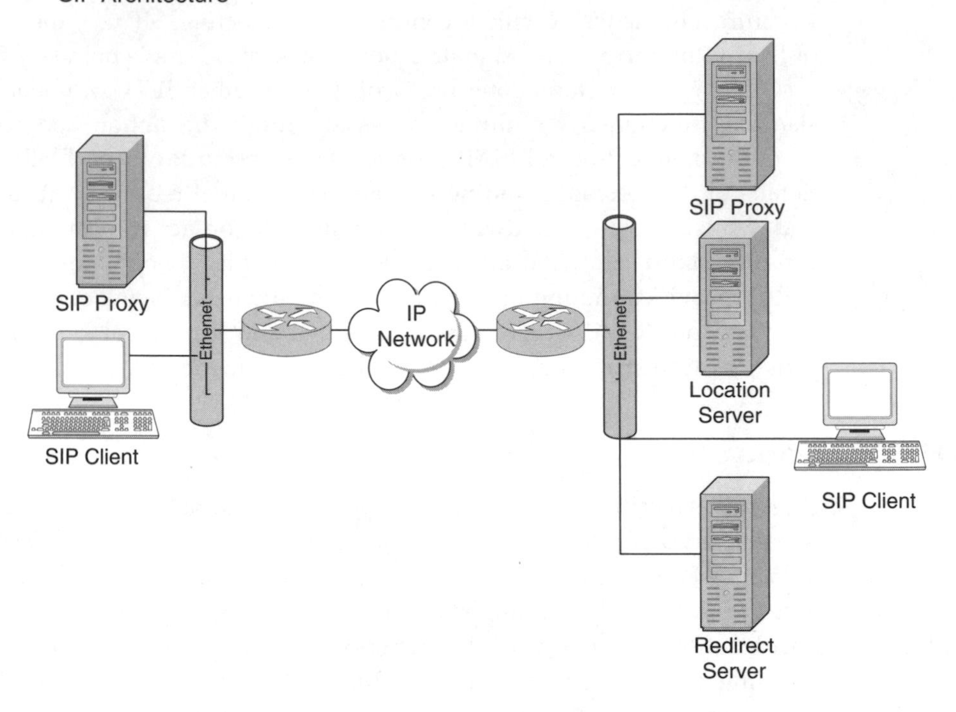

requests or accept calls. The SIP proxy server is a rendezvous point at which callees are available. It determines where to route signaling information and retains the user's signaling preferences. The proxy server may try a single destination, or through a process known as forking, it may try several destinations simultaneously.

Figure 12-5 shows an example of call setup over a SIP-enabled network. The caller initiates a session with an Invite request asking the callee to join the session. The caller's invitation travels over the LAN to the local SIP proxy, which relays the request to the distant SIP proxy. If the callee is not reachable on that network, the callee's proxy redirects the call to the network were the callee is reachable. Servers retain no information about the user's current state, but upon discovering the state such as busy on another call, the server takes action based on the user's preference. Choices might include rules such as call secretary first, and if secretary is unavailable, send call to voice mail unless it is spouse and in that case play a voice message that indicates the callee will be working late tonight. The rules could get complicated and could even have some unexpected results. The user controls the options, typically with a setup screen accessed with a browser.

SIP Applications

SIP facilitates potential applications that make VoIP attractive. Here is a representative list of possibilities.

FIGURE 12-5

SIP Call setup

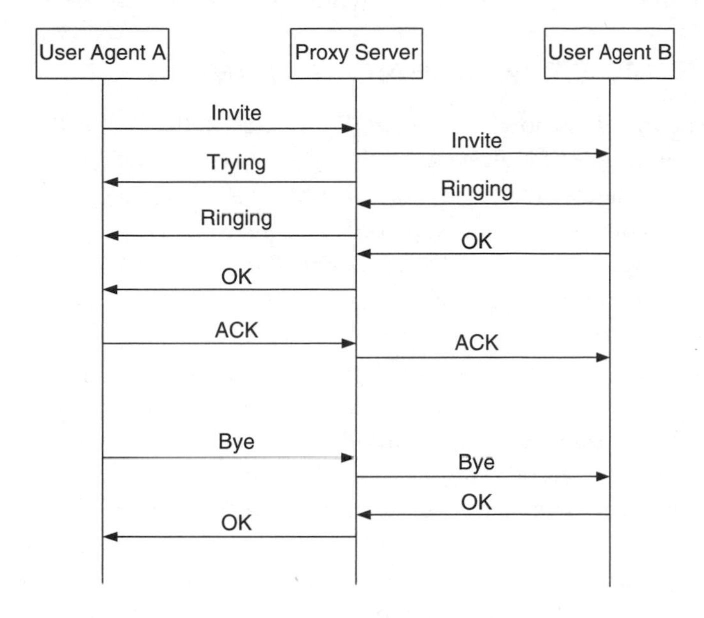

- Reach callees through multiple methods such as through a SIP address embedded in a Web page, voice link in an e-mail message, or an e-mail message translated to voice.
- Selective call redirection based on variables such as user's calendar, who is calling, or where they are calling from (e.g., a local call might be sent to voice mail, but an overseas call redirected).
- Collaborative working over the Internet on files or shared applications.
- Selective notification and alerting to voice-mail messages, stock market results, and other significant events.
- Automatic conference setup based on participants' availability obtained from their electronic calendars.
- Fax delivery over the Internet.
- Remote and collaborative whiteboarding over the Internet.

Decentralizing call control opens the door to some interesting applications that are otherwise either difficult or impossible. SIP is an incomplete and immature protocol at the time of this writing, but it is gathering considerable momentum. Most of the companies that support VoIP will eventually abandon their proprietary or H.323 signaling implementations in favor of SIP. Once this happens, the imaginations of countless developers will produce some interesting applications that will transform the way people work and communicate.

Until SIP becomes a reality, investments in proprietary IP phones should be scrutinized. Many manufacturers offer enhanced features on their IP phones, but

SIP is likely to spawn a new generation of applications that have not been conceived yet. Prospective purchasers are well advised to inquire whether today's proprietary IP phones can be inexpensively upgraded to SIP.

ELECTRONIC NUMBER MAPPING (ENUM)

A major issue in VoIP is how to reach an IP phone from the PSTN. The process in the other direction is not difficult because the calling party knows the E.164 address of the callee and a gateway makes the connection, but a caller from the PSTN is unlikely to know the IP address of a VoIP station. Therefore, to go from PSTN to IP requires a gateway that does the E.164-to-IP translation. Today, most VoIP providers do their own translation and have exchange agreements with competing vendors, but as VoIP grows, that process will become unwieldy. ENUM maps E.164 numbers to a uniform resource indicator (URI). An Internet URL is the most common form of URI.

ENUM runs on servers that retrieve an indicator known as a NAPTR (naming authority pointer), which provides the instructions for reaching a particular node. The server can be thought of as a large directory assistance database that lists multiple ways of reaching users such as fax, mobile, e-mail address, phone number, voicemail, and even their public encryption key. ENUM intends to use the DNS structure to retrieve NAPTRs. The domain.arpa is proposed as the domain name. The E.164 telephone number is mapped in reverse order using a dot between each digit and the.arpa suffix. For example, McGraw-Hill's main telephone number including country code is 01-212-904-2000. In ENUM terms this would be expressed as 0.0.0.2.4.0.9.2.1.2.1.0.arpa.

ENUM proponents must resolve many issues before the protocol is adopted. The chief issue is who controls it. ITU controls PSTN country codes, which are the top-level of the E.164 hierarchy. Each country administers its own internal numbering plan. In the United States, the FCC has overall responsibility and Neustar, Inc. is the North American Plan Administrator. State commissions are involved in NPA boundaries and issues such as overlay area codes. The DNS system is more open, but international interests are gradually demanding more control over how it operates.

Security is also a vital issue with ENUM. Essential servers must be insulated from the attacks that have frequently infected servers and stations on the Internet. The responses from ENUM servers must be guaranteed valid and correct, which means it must be impossible for an intruder to intercept and alter messages to and responses from an ENUM server.

Other issues include how to handle nonpublished numbers, emergency calls, slamming, number portability, and keeping databases synchronized. The international flavor of ENUM assures that the process of resolving these issues and making it publicly accessible will be lengthy.

CHAPTER 13

The PSTN and Quality Requirements

Circuit switching is the essence of today's PSTN. With it, users can reach any telephone anywhere in the world, including cell phones simply by dialing a series of digits. SS7 directs the switches to connect circuits from source to destination and return them to common pools when the session ends. The PSTN is effectively a vast and highly distributed time-shared computer. Countless switches interoperate seamlessly despite the fact that multiple and competing companies own the hardware and serve the customers. The telephone is so easy to use that nearly every preschooler knows how to place and answer calls. Irregularities such as noisy and low-volume connections and fast-busy tones that signify ineffective call attempts are rare.

Circuit switching has evolved through five different generations: manual, direct-controlled electromechanical, common-controlled electromechanical, processor-controlled analog, and processor-controlled digital or TDM switches. The PSTN is currently evolving toward a hybrid combination of circuit and packet switching and this chapter discusses that transition. The first three generations of switches have effectively disappeared. Analog switching is diminishing as digital switches replace them, and the sixth generation of softswitches is beginning to appear.

Local switching is the most complex part of the PSTN because of the wide variety of functions and features that must be activated through the telephone, an endpoint with no intelligence. Although manufacturers continue to improve their TDM switches, the technology has reached maturity and most new development will focus on VoIP and softswitches. Nevertheless, TDM switches will be around for many years because they satisfy the needs of most users. They are in place and working, and although softswitches claim to be significantly less expensive, the economic benefits of replacing them will come about slowly. Moreover, circuit switches inherently deliver service quality and universal connectivity that VoIP is struggling to emulate.

The telephone industry recognized the limitations of the PSTN several decades ago and devised ISDN as the solution. ISDN retains circuit switching while it offers expanded feature capability by separating signaling from the talking path. It also offers end-to-end digital connectivity, but that has proved to be of minimal value to most users. Primary rate ISDN has fulfilled its promise for large businesses and is widely applied, but basic rate, which is intended for residence and small business service, has had practically no impact in North America. Nor is it likely to in the future because the industry has turned its face toward VoIP.

Despite the enthusiasm VoIP is generating, circuit switching and POTS will be around for many years because of something ISDN advocates discovered: POTS fulfills the telephone needs of the vast majority of subscribers. Furthermore, as this chapter explains, quality of service (QoS) in circuit switching and VoIP are distinctly different. The Internet can become as ubiquitous as the telephone network, but is designed as a best-effort network, which is inconsistent with commercial grade telephone service. In this chapter, we examine the PSTN and its characteristics and discuss what changes must occur for the transition into the sixth generation of switching.

THE ARCHITECTURE OF THE PSTN

The PSTN is a hierarchical network that consists of two types of switches: end offices and tandem switches. Before the breakup of the Bell System, the LECs owned the end offices and the first one or two levels of the tandem switch network, and AT&T's Long Lines owned the higher-order toll tandems. The network was designed with a five-level switch hierarchy that connected circuits end-to-end. With the increasing power and intelligence of switching systems, that hierarchy has evolved to a flatter structure. The lowest level in the hierarchy is the end office, which was then, as now, called a class 5 office. Telephone subscribers receive their dial tone and all elements of local exchange service from a class 5 office.

The objective in circuit-switched network design is to determine the most economical configuration of circuits and switches based on peak and average traffic load, required grade of service, and OAM&P costs. It is practical to connect a limited number of switch nodes with direct trunks as Figure 13-1 shows, but as the number of nodes increases, the number of circuit groups soon becomes unwieldy. To control costs and simplify trunking, the carriers interconnect the nodes with tandem switches. In such a network direct trunk groups are provisioned between switch pairs that have enough traffic to justify them; otherwise traffic flows through the tandem. The direct trunks are called *high-usage trunks*. The switches always use the direct trunk group as the first routing choice because it is the most economical since it avoids tandem switching costs. When the direct trunks are all busy, the switch routes the traffic to the tandem over a group of *final trunks*. When callers encounter blockage on the final trunk group, the switch returns reorder and the caller must hang up and redial.

Tandem switching provides service protection in addition to reducing the number of direct trunk groups required. If a high-usage trunk group fails,

FIGURE 13-1

Direct and Tandem Trunking

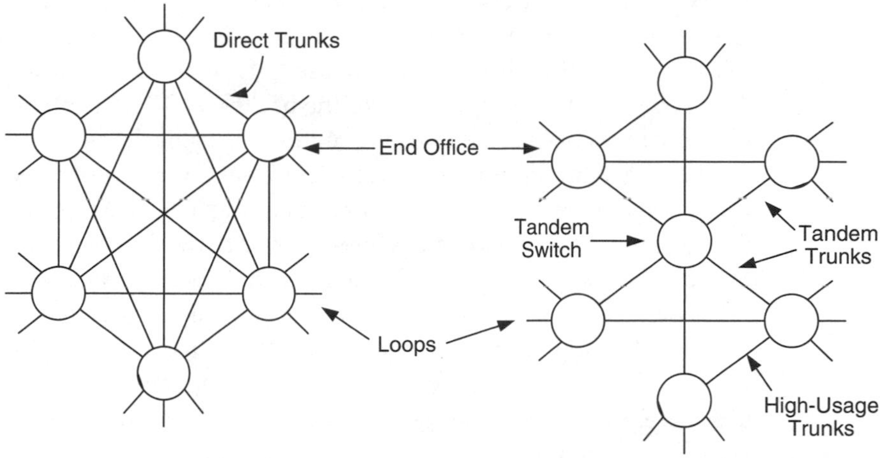

the traffic can be sent through the tandem. IXCs and LECs alike configure their networks according to the principle of providing direct trunk groups where feasible and overflowing through one or more tandems. Traffic engineers collect data about the peak and average traffic flow to particular destinations. All switches register traffic flow by trunk group in hundreds of call seconds (CCS). CCS is the product of the number of calls and the average call-holding time in the trunk group divided by 100. For example, a group of trunks that carried 30 calls in an hour, each with an average holding time of 180 seconds would have a total load of 54 CCS.

Traffic engineers consult tables or computer programs that project the number of trunks needed to carry the traffic within a specified blockage objective during the busiest hour. High-usage trunks are deliberately designed to overflow to final groups during traffic peaks. The final groups are designed to a higher standard because it is in these groups that the carrier delivers the true grade of service. Carriers measure trunking grade of service by percentage of blockage. A P.01 grade of service, e.g., means 1 percent of the calls will be blocked using the Poisson traffic formula. Traffic tables and the method of using them are provided in the companion volume to this book, *The Irwin Handbook of Telecommunications Management*.

Telecommunications networks exist in an environment that is continually changing. Service demands are not constant—they vary by time of day, day of the week, and season of the year. Demand is continually evolving in response to changing calling habits and business conditions. Competition and new technology have a substantial effect on cost and demand. Also, network design is always a compromise that seeks to balance the utilization of facilities with the provision of satisfactory service. Constant data collection and attention to traffic loads are needed to keep the core network at a satisfactory grade of service.

VoIP in the PSTN Core

As we have seen, trunks between class 5 switches are segregated into multiple trunk groups and trunking complexity increases as the network grows. The unit of circuit-switched trunk growth is typically DS-1. If the load is high enough to justify at least a DS-1, direct trunks are used; if not, the traffic flows through a tandem. To derive direct trunk groups over the backbone SONET infrastructure, either tandem switches or DCS systems are required at critical junction points. To keep traffic flowing without blockage, traffic engineers must collect and monitor data and provision circuits in anticipation of a changing load. This cost of traffic engineering is one of the drawbacks of a circuit-switched network. A second and equally important drawback is the length of time the provisioning process requires.

Broadband loops will eventually replace copper cable, but at the present state of development, circuit switching from the end office to the subscriber is unavoidable. In the core of a circuit-switched network, carriers must provide alternate routes to make service robust. This requires expensive toll tandem switches at frequent intervals. If the network edge is circuit switched, but converted to IP in the core, large routers can route traffic away from congestion automatically, and consequently reduce the need for high-usage trunks, DCS systems, and large toll tandem switches.

The concept of traffic engineering is different in a circuit-switched network than in an IP network. In circuit switching, traffic engineers forecast load in terms of circuit usage to destinations, measure usage, and prepare orders to provision the circuits and switching capacity in time to meet the forecast demand. In TDM networks each trunk group is separately engineered based on its peak traffic load. Traffic engineering in an IP network has the same objectives of forecasting and measuring traffic, but in a packet network the backbone bandwidth is shared so the engineering in the core is concerned with total packet flow, not with circuit usage. If voice traffic is converted to IP and routed in the core, the packet network, in effect, becomes a large tandem switching system.

Several major IXCs are transitioning their voice backbone to IP, replacing TDM switches with a routed core. The edge switches require enough bandwidth to connect their traffic to the core, but trunks are no longer provisioned based on switch-to-switch traffic. Moreover, the total traffic load is reduced because of IP compression. If silence suppression is used the load is further reduced. More than half of a typical voice session is idle time because of silent periods and the half-duplex nature of a voice call.

Call Admission Control (CAC)

In a combined voice and data enterprise network, CAC must be applied to voice traffic in a different manner than data traffic, which is self-policing. In a data flow, if a burst of traffic oversubscribes the bandwidth, the protocols may relieve the congestion by queuing, buffering, or dropping packets, delaying the traffic until the network is prepared to carry it. This method of relieving congestion cannot be used on real-time traffic, which is sensitive to delay and packet loss. If a circuit-switched network cannot carry the traffic, it denies admission. Dial tone is delayed when the switch is overloaded and reorder is applied when the outgoing trunk groups are congested. CAC in a combined voice–data network keeps voice traffic off the network if it cannot be carried. For example, if two switches are connected by an IP trunk group, the CAC function must determine whether adding another call to the network will begin to cause packet loss, and reroute the next calls over the PSTN or return reorder.

Voice Call Limiting

CAC mechanisms must operate under rules that are designed to avoid admitting additional calls to a congested network. One simple method is to limit the number of voice calls on an IP trunk group based on the administrator's knowledge of the number of calls the group can handle. While this method can avoid congestion, it does not take into account the dynamically changing data load, which may allow many more calls to be carried at times. This method is similar to that used on circuit-switched networks where the number of direct trunks is limited and the call overflows to a higher-cost trunk group or is blocked.

Measurement-Based CAC

Measurement-based mechanisms use probes to measure the delay and packet loss. For example, the controller could ping the destination and keep the traffic off if the response time fell outside a defined range. A ping is not a reliable method of measurement, however, because it does not resemble a voice packet. Furthermore, any QoS mechanisms in the path will not recognize the ping as a voice packet and give it priority treatment. Ping is, at best, a crude measurement of the network characteristics. More effective is a specialized probe that measures the calculated planning impairment factor recommended in ITU-T G.113. The main advantage of a probe is its ability to measure any network independent of ownership and resources in the routers.

Resource-Based CAC

A third mechanism is resource-based. The CAC controller either reserves the necessary resources or calculates the resources needed and compares them to those available. Resource measurement techniques determine the bandwidth available at every link and node and use RSVP to reserve the bandwidth. This method requires the resources on the network to support the necessary protocols, so it is not practical across networks with multiple ownership. Reservation systems are the only ones that can guarantee the availability of bandwidth for the duration of the call. All other mechanisms are just a one-time snapshot of whether the bandwidth is available or not before admitting the call.

The amount of bandwidth needed is a function of several factors such as the voice compression protocol, header size, and whether header compression is used. RSVP reserves 80 Kbps for G.711 compression. For highly compressed protocols such as G.723.1 (5.3 Kbps) it reserves 22 Kbps of bandwidth per call.

QUALITY OF SERVICE

As we discussed earlier, one measure of service quality in the PSTN is the percentage of blocked calls. Also important as a measure of service quality is transmission performance, which consists of three variables:

Loss, which is the drop in volume between the sending and receiving ends of a connection;

Noise, which is the presence of any unwanted signal such as hum, static, or crosstalk;

Echo, which is the return of the talker's voice over the receiving channel with a noticeable delay.

In analog networks, loss and noise are major problems because the network amplifies the voice signal to compensate for loss, and noise is amplified with it. The digital network nearly eliminates loss and noise problems because the signal is regenerated at intervals, so the only place loss and noise can impair a connection is in the copper local loop. Circuits that are extended to customers' premises over T1/E1 using either copper or fiber optics are virtually free of noise and loss.

Echo is present any time a connection undergoes a two-wire-to-four-wire conversion, which happens every time a two-wire loop is used. The circuit that does the conversion is known as a *hybrid* and it always has imperfections that reflect some of the outgoing signal back to the source. Although echo is present on every telephone call, it is not objectionable on local and short-range toll calls because the delay is so slight that it sounds like *sidetone*, which is the feedback that makes a telephone instrument sound alive. Terrestrial TDM circuits ride on fiber optics, which has a propagation velocity approaching the speed of light. Echo suppression is not needed until the delay approaches 50 ms. Delay is inherent in every VoIP session, however, so echo suppression is always required.

Transmission quality on a voice connection is subjective. Talkers range from those who speak in a whisper to those who hardly need a telephone within a block or two. Listeners range from those with acute sensitivity to the hearing-impaired. Circuit noise is tolerable if the signal is loud, but the lower the volume, the more the noise interferes with communication. The industry does not attempt to satisfy all of the users all of the time, but instead relies on subjective measures of so-called "toll-grade" transmission quality.

Voice quality is based on users' subjective ratings of a large universe of connections of varying quality. In a TDM network the standard of reference is a 64 Kbps connection with a fixed amount of loss and noise. ITU-T recommendation G.114 covers voice quality standards. ITU-T recommends two standards for measuring voice quality. Recommendation P.800 covers a method for deriving mean opinion score (MOS) of voice quality. Voice samples are played over transmission media with various levels of quality to a group of users who rate

TABLE 13-1

Quality Ratings of Voice Compression Standards

Compression Standard	Name	Payload Bit Rate (Kbps)	Accumulation Delay (ms)	Mean Opinion Score
G.711	Pulse-code modulation (PCM)	64	0.75	4.4
G.726	Adaptive differential pulse-code modulation (ADPCM)	16 to 40	1.0	4.2
G.728	Low-delay code-excited linear prediction (LDCELP)	16	3 to 5	4.2
G.729A	Conjugate structure algebraic code-excited linear prediction (CSACELP)	8	10	4.2
G.723	Multipulse maximum likelihood quantization (MPMLQ)	6.3	30	3.9
G.723	Algebraic code-excited linear prediction (ACELP)	5.3	30	3.5

the connections on a scale of 1 to 5 with 5 high. A score of 4 or higher is defined as toll quality. P.861, perceptual speech quality measurement (PSQM) is a mechanized process using a computer to derive scores that correlate to MOS scores. PSQM is intended for the circuit-switched network and is not valid for some types of IP network impairments. For example, a VoIP signal can be compressed to a fraction of the 64 Kbps bandwidth and the more the compression, the lower the voice quality.

Table 13-1 shows ITU-T ratings of various types of voice compression. ADPCM, which is ITU recommendation G.726, shows some slight loss of quality compared to an uncompressed 64 Kbps voice signal (G.711), which receives a score of 4.4. Code-excited linear prediction provides quality approximating that of a cellular telephone. The ratings in this table are on circuits that have no additional impairments such as packet loss. This table also shows the accumulation delay of the coder, which adds to the other delays that are inherent to IP sessions.

VoIP Impairments

In addition to the impairments that affect TDM networks, VoIP is subject to delay, delay jitter, and packet loss. In addition, data compression may reduce voice intelligibility. ITU-T recommendation G.113 defines an impairment/calculated planning impairment factor (ICPIF), which compiles delay and packet loss into a value that can be used to assess user reaction to the network. Any factor of 20 or lower is considered to be adequate for packet voice. This section discusses how these impairments arise, and what can be done to cure them.

Delay or Latency

Latency is the delay from the time a voice signal leaves the transmitter till it arrives at the receiver—called "mouth-to-ear" latency. Every connection has a certain amount of delay. In TDM networks three factors cause delay. The first is the time required to sample and quantize the voice signal. Next are end-to-end propagation delays, which are a function of the speed of the transmission medium. The velocity of a radio signal in free space propagates at a little less than 300,000 km/s. In fiber the signal propagates at about 205,000 km/s. Finally, each switching stage delays the signal a small amount. Altogether, the one-way delay in a transcontinental TDM connection is in the order of 40 ms or less, with propagation delay the main factor.

ITU-T recommends that one way end-to-end delay be held to 150 ms or less for a voice connection. Delays up to 400 ms can be tolerated depending on user expectations. For example, a 400 ms delay would be unreasonable on a short call, but a user might find it acceptable on an intercontinental call to a remote area. When delay exceeds about 400 ms, the connection is practically unusable. When delay exceeds a threshold of 150 to 200 ms users attempt to talk over each other and talkers miss the short feedback responses that listeners make. Satellite circuits have a delay of about 250 ms and although they are usable for voice sessions, some users find them disconcerting.

VoIP is subject to the same digitizing and propagation delays that affect TDM. It also has three other sources of delay: compression, framing, and routing. Compression is not required in VoIP, but the improvement in bandwidth utilization is one of the primary motivations for using VoIP, particularly in enterprise networks. As Table 13-1 shows, the more the signal is compressed the less intelligible it is and the greater the delay.

VoIP segments the digital voice stream into frames containing about 20 ms of voice for a rate of 50 packets per second (pps). Framing delay is the time it takes to fill a frame. At 8000 samples per second, a 20 ms frame of uncompressed voice would hold 160 samples, or 1280 bits. If the voice is compressed to 8 Kbps, the frame would hold 160 bits. A larger frame can be used, but it takes longer to fill, which adds to latency. Quality is also adversely affected if a packet is discarded—the longer is the frame the more difficult it is to interpolate lost data.

The IP protocol stack, which typically runs in the gateway, packetizes the frames. A Real Time Protocol (RTP) header precedes each voice packet. RTP contains timing information that the jitter buffer on the receiving end uses to remove variations in packet spacing. The IP header contains the source and destination IP addresses, the IP port number, packet sequence number, and other protocol information needed to transport the data. The RTP header is 16 octets long, but it can be compressed. It is added to a UDP header of 8 octets and an IP header of 20 octets for a total of 44 octets of header to carry 160 bits or 20 octets of payload. The inefficiency of this proportion of overhead defeats much of the purpose of using VoIP in the first place, so some systems enclose more than one frame of data in a packet. At 50 pps × 64 octets per packet, the data rate is about 25.6 Kbps per channel.

Packets are queued at each router in the path. The delay depends on the router's processing power, the load on the network, and the speed of the access circuit. The method the router uses to prioritize voice packets also affects delay. As packets traverse the network they are subject to propagation delay, which may be insignificant compared to the time required for each router to check its table before forwarding the packet. MPLS shortcuts the routing process, but the total delay is proportionate to the number of routers the packets encounter on the way to the receiver.

Once the packets arrive at the receiver, the process of queuing and compressing must be reversed. The router recognizes the packet as a VoIP packet, and sends it to the compression and decoding device. Packets may be further delayed at the receiver, particularly if they have to go through a firewall.

Jitter

As packets travel the network, they are subject to conditions that cause them to arrive at uneven intervals, a condition known as jitter. Jitter is caused by variations in congestion, queuing time at routers, and packets taking diverse paths. It can be corrected within limits by buffering packets to absorb the slack and releasing them at regular intervals. In most products the jitter buffer can be configured, which can be a delicate balancing act. If the buffer is too small, distortion will be audible and if it is too large, delay will be excessive.

Compensating for jitter is not just a simple matter of storing and releasing packets. The buffer must compensate for silence suppression and must check for lost and out-of-sequence packets. RTP sets a silence suppression bit in the header immediately following a period of silence. The jitter buffer detects the silence suppression bit and ignores the gap between the packets spanning each side of the silent period.

If packets are lost, the arrival time between two packets will appear to be excessive even in the absence of jitter. The jitter calculation checks the sequence number to detect packet loss and takes this into consideration. Packets arriving out of sequence have a similar effect.

Packet Loss

The third variable affecting quality is packet loss. In the face of congestion, routers can dump their buffers, leaving it to the application to recover. Bit errors are inconsequential in a voice session, so VoIP runs under UDP, which does not correct errors. If a packet is discarded, the gateway predicts what the bits would have been and interpolates bits to bridge the gap. Although error correction is unnecessary for voice it is required for call control signaling protocols such as H.225, so these operate under TCP.

A voice session can survive with a certain amount of packet loss. The amount of distortion resulting from lost packets is a function of packet length. If a packet contains 20 ms of speech, a lost packet would contain 1/50th of a second of speech or a fraction of a syllable, which is hardly noticeable. From a quality standpoint, the nature of packet loss is important. If lost packets are randomly spaced, the effect on speech is slight compared to the loss of a group of packets, which might occur if a router flushes its buffers. With other factors being equal, a VoIP network can tolerate from 2 percent to 5 percent packet loss without a significant effect on intelligibility. If packet loss and delay are both excessive, the signal may be so distorted as to render the session unusable.

End-to-End Delay Budget

The total delays in a VoIP network are the sum of a variety of fixed and variable factors that are listed in Table 13-2. Since we are ideally aiming for an end-to-end delay of 150 ms, and 200 ms in the worst case, the variable or controllable factors must be evaluated in the network to ensure that they are within expectations.

Several of the variables affecting delay are controllable. First is the choice of coder. Table 13-1 shows that the greater the compression the more latency the coder adds to the mix. The decompression process is approximately the same regardless of coder chosen. The network access and egress delays are a function of the bandwidth chosen for the access network, i.e., a T1/E1 access pipe moves bits 24 to 30 times faster than a 64 Kbps circuit. The delay in the network depends on the carrier's service level. Delay is a function of the distance the packets must travel plus the latency introduced by the routers and switches the packets traverse. Latency will naturally be greater in intercontinental networks than in short domestic routes. An approximation of the latency of the physical network can be computed by dividing the route length in kilometers by 205,000 km/s, the approximate propagation speed of fiber optics. For example, a 4000 km route would have a delay of 4000/205,000 = 19.5 ms.

The interprocessing delays that result from queuing the packets are affected to some degree by the speed of the router, but the primary factor is the scheduling algorithm chosen, which we discuss in the next section.

TABLE 13-2

VoIP Delays

Factor	Range of Delays (ms)
Voice coding	20 to 45
Ingress interprocess	10 to 15
Network access	0.25 to 7
Network delay	20 to 50
Network egress	0.25 to 7
Egress interprocess	10 to 20
Jitter buffer	10 to 30
Decompression	10
Total	80.5 to 184

QUALITY OF SERVICE PROTOCOLS

Most quality problems resulting from congestion can be solved by providing plenty of bandwidth. Carriers that have their own fiber optic capacity can often provide enough bandwidth so that delays resulting from router queuing and packet discard are held to a minimum. Some carriers even use the Internet for part of their network, but to do so they must monitor quality and switch the traffic to a more stable platform if congestion begins to develop.

The main factor in carrier networks is the use of QoS protocols to provide the stability and precedents that isochronous traffic needs. This is not feasible in the Internet because of the diverse ownership and lack of controls over the router protocols. Several protocols are available to ensure that the network reserves enough bandwidth and boosts time-sensitive traffic along without unacceptable delay. Although the QoS protocols discussed in this section are not technically part of the PSTN, they are discussed here because they are used with VoIP sessions that must emulate the PSTN. Chapters 36 and 37 provide additional details.

Reservation Protocol (RSVP)

RSVP is not a routing protocol; it is a signaling protocol that works through the routing protocol to set up the path. An application running under RSVP sends a reservation request forward through the IP network using RSVP messages following the path set up by the routing protocol. RSVP communicates with admission and policy control modules in the node. Admission control determines whether the node has resources to provide the requested QoS. Policy control determines whether the user has permission to request the reservation. If the checks succeed, RSVP sets a QoS class in its packet classifier.

Each node in the path stores path-state indicators containing the IP address of the previous node. The path state is used to return the reservation confirmation over the same pathway using an RESV message. The RESV message provides receivers with information about the path and traffic load so they can reserve the bandwidth. If the path changes during the session, RSVP adapts to the change.

Each node along the path must support RSVP and have the bandwidth available to reserve for the session. If either of these fails, the reservation request fails. If the session becomes inactive, RSVP sends keep-alive packets to confirm that the session is still alive. When these stop, the reservation times out and the resources are returned to the pool.

RSVP emulates a circuit-switched connection by providing controlled latency and sufficient bandwidth to support the session. It is inefficient for short flows because the setup time becomes a significant part of the total session time. It is also reliable only in a network of known characteristics such as a carrier's IP network. Three classes of service are defined. Guaranteed service provides bandwidth of known delay characteristics and no lost packets. Controlled load service is an intermediate service providing a better class of service than the third category, which is best effort, the service class now provided by the Internet. RSVP falls under a category the IETF defines as Integrated Services (IntServ).

Differentiated Services (DiffServ)

DiffServ takes a different approach than IntServ by identifying time-sensitive packets at the application and placing class-of-service information in the IP header. The type-of-service (ToS) field in the IPv4 header marks packets to receive a particular forwarding treatment at each node; i.e., to forward it immediately or delay it in favor of a higher priority packet.

With DiffServ, it is not necessary to exchange QoS requirements between the source and destination, which reduces the setup time compared to RSVP. RSVP's setup time can be a substantial part of the total in short flows. DiffServ does not guarantee a level of service. It is designed to ensure that services requiring explicit QoS are identified and classified as priority traffic. It is less complex than RSVP and scales better, but it lacks admission control, so too much high-priority traffic can swamp the network.

Multiprotocol Label Switching (MPLS)

MPLS short cuts the routing process by mapping IP addresses to simple fixed length labels, which are numbers that identify a data flow. A sequence of labels and links is called a *label-switched path* (LSP) also known as an *MPLS tunnel*. MPLS supports a variety of QoS functions including bandwidth reservation, prioritization, traffic engineering, traffic shaping, and traffic policing on almost any type of interface.

Routing is simplified because it happens only once at the edge of the network in the *label edge router* (LER). Routers in the path read the label instead of the packet header and route the call along a path that is defined by the label. Traffic is assigned to a *forwarding equivalence class* (FEC), which is a group of packets to the same destination and which share the same QoS requirements. An MPLS-compatible router is called a *label switched router* or LSR. Packets enter and leave the core network through ingress and egress LERs. The ingress LER inserts the label ahead of the IP header. As packets flow through the network, each MPLS device checks its label tables or label information base to determine how to forward the traffic.

MPLS is at the heart of most VoIP carrier networks and in combination with RSVP it is capable of delivering QoS that is comparable to circuit switching, provided every node in the path supports the protocols. The only way this can be assured is for the carrier to control the network design. This limits the degree of network interoperability that the PSTN provides, and is contrary to the free-wheeling nature of the Internet.

Router Queue Scheduling Protocols

The method routers use to process their queues is an important factor in determining latency. In the absence of a prioritization scheme, routers process packets first-come-first-served, which is unsuitable for time-sensitive packets unless the load is light. Most routers are capable of prioritizing traffic using a variety of techniques.

Priority queuing is the technique in which the traffic is divided into various priorities and assigned to queues, which the router processes in priority order. All the traffic in the highest priority queue is processed ahead of traffic in the second priority queue, and so on. The problem with this approach is that since voice and video must be given highest priority, when the load is high they can preempt data and lengthen the response time. *Custom queuing* alleviates this somewhat by dividing the bandwidth on a percentage basis as part of the router setup process. The queues are processed in round robin fashion, but the only prioritization given to voice and video is the percentage of bandwidth they are allocated.

Weighted fair queuing (WFQ) divides the bandwidth among all the traffic flows entering the queue. All flows by default have a precedence of zero, and are given a weighting of the precedence plus one. If five flows of equal priority were flowing into the router, each would get 20 percent of the bandwidth. If one of these were a voice flow, it might be given a precedence of, say, 5. This would mean that the voice flow would receive a precedence of 6 (5+1) and the other four flows would have a precedence of 1. The total weighting would be 10, so the voice flow would receive 60 percent of the bandwidth, and the other four flows would each receive 10 percent. Priority classifications include source and destination addresses, protocol, and session identifier based on the three IP precedence bits in the packet header. Routers may use a combination of priority queuing and WFQ with voice applied to the priority queue and data applications distributed among queues using WFQ.

Weighted random early detection (WRED) is a method of determining which packets are dropped when the router must discard traffic. In the absence of any prioritization, the last packets in are dropped first. WRED predicts the need to drop packets, and begins dropping low-priority packets to prevent congestion rather than waiting until congestion occurs. When WRED needs to drop TCP/IP packets, it drops all of them within a TCP window because if one packet within the window is dropped, all of them must be retransmitted anyway.

Classifying Traffic

A final QoS issue is how to classify the traffic or as some call it "coloring" the packets so they are forwarded to the appropriate queue in the switch or routers. The network devices then must apply the appropriate scheduling techniques for unloading their queues. The traffic class may be defined by the type of user, by port number, or by the applications such as voice, video, or e-mail. This classification is ideally handled by a policy management system, but policy management is largely proprietary and will remain so until standards are developed. The other alternative is to have the application dictate the appropriate class at the origin. The 802.1p Ethernet priority bits define a variety of parameters for providing QoS including such variables as frame error rate, loss, sequencing, duplication, data unit size, user priority, delay, and throughput.

The 802.1p protocol provides a total of eight traffic classes and eight user priorities. The highest class is network control, which is given precedence because of its role in controlling the network. Voice and video are given controlled delay, which is next highest. From there the classes descend to the lowest, background class, which is suitable for such applications as games. The standard is implemented in network interface cards in response to priority requested by the application. In order to get end-to-end QoS, all the network elements have to support the standard.

The advantage of using 802.1p to classify traffic is simplicity. Switches can read the information and forward frames with the prioritization intact. The drawback is that the information is stripped off when the frame passes through a router. This argues for the second method, which is setting the ToS field in the packet. This works in some cases, but it does not work with layer 2 switches, which do not read packet headers.

Essential Characteristics of the PSTN

The PSTN has certain characteristics that all circuit-switched class 5 and above central offices must support. This section provides an overview of how the PSTN functions, both from a technical and a regulatory standpoint.

The North American Numbering Plan

The North American Numbering Plan is a subset of ITU E.165. Neustar, Inc. is the North American Numbering Plan Administrator (NANPA). NANPA assigns area codes, which are also known as NPA codes, for all of the countries included in the North American plan. It also assigns central office codes in the United States.

Area codes have the form NXX, where N is any digit from 2 through 9 and X is any digit from 0 through 9. There are 800 possible combinations under this format, but some combinations are not available or have been reserved for special purposes. Codes with the format N11 are called service codes and are not used as area codes. These are used for such services as emergency calling (911), directory assistance (411), and travel information (511). Codes where the second and third digits are the same are called easily recognizable codes. These designate special services such as the series 800, 888, 877, etc., that are used for toll-free calling.

Until recently an NPA always had an exclusive geographic boundary. As more and more prefixes were assigned to support CLECs, cellular and paging providers, and large companies that requested exclusive prefixes, NPAs were split into smaller areas and numbers assigned to the new NPA were changed. To forestall number changes, some state regulatory commissions elected to use *overlay area codes*. An overlay area code follows approximately, but not necessarily exactly, the same boundaries as the original NPA. Overlay area codes require the subscribers to dial 10 digits to complete local calls. In a non-overlay NPA, calls are completed with seven digits.

Central office codes, also known as NXXs or prefixes, are digits 4, 5, and 6 of a 10-digit telephone number. NANPA assigns central office codes using guidelines from state and federal authorities. Several factors are leading to the exhaust of available numbers, so NXXs are assigned within strict rules, particularly where the available numbers in an area code are in danger of being exhausted. Among the factors contributing to number usage is the assignment of prefixes to entities that cannot use all the numbers. For example, many small communities have a switch that serves less than 1000 telephones and the rest of the prefix is unused. To make better use of prefixes, the industry has introduced *thousands-block pooling*, which subdivides the prefix into blocks of 1000 numbers.

A major difference between the PSTN and an IP network is the fact that a PSTN number is associated with a physical location. Numbers can be ported or forwarded, but they are reached by a traceable path. Even with cell phones, which can roam anywhere the service provider supports, the number is associated with a switch that has a fixed location. This inflexibility is a disadvantage from a mobility standpoint, but it allows accurate dispatch of emergency vehicles for wired locations. Cellular emergency calls are a separate issue, which we discuss in Chapter 20.

Local Number Portability (LNP)

The Telecommunications Act of 1996 mandates that customers be permitted to change service providers without changing their telephone number. Within the bounds of a rate area, subscribers can change their geographic location or can change between classes of service such as from POTS to Centrex without a number change.

To implement LNP, each wireline switch is identified with a 10-digit *location routing number* (LRN). The LRN for a particular switch is a "native" NPA-NXX that serves as a network address for that switch. Carriers query the LNP database to determine the LRN corresponding to the dialed telephone number. The originating switch routes the call to the terminating switch based on the LRN. Figure 13-2 illustrates how the process works. An LNP database serves the territory covered by each of the seven original RBOCs. After the caller dials the telephone number, the originating switch sends a query over SS7 to the LNP database through its STP. The SCP that hosts the LNP returns a message containing the LRN of the recipient switch. The originating switch sends the call to the recipient switch, which completes the call.

Access to the Local Exchange Network

Until AT&T's divestiture in 1984, the telephone network in the United States was designed for single ownership in each exchange. In some countries it still is, but those countries that have deregulated long distance provide access to multiple IXCs. This section discusses the architectures of local and long distance telephone

FIGURE 13-2

Local Number Portability

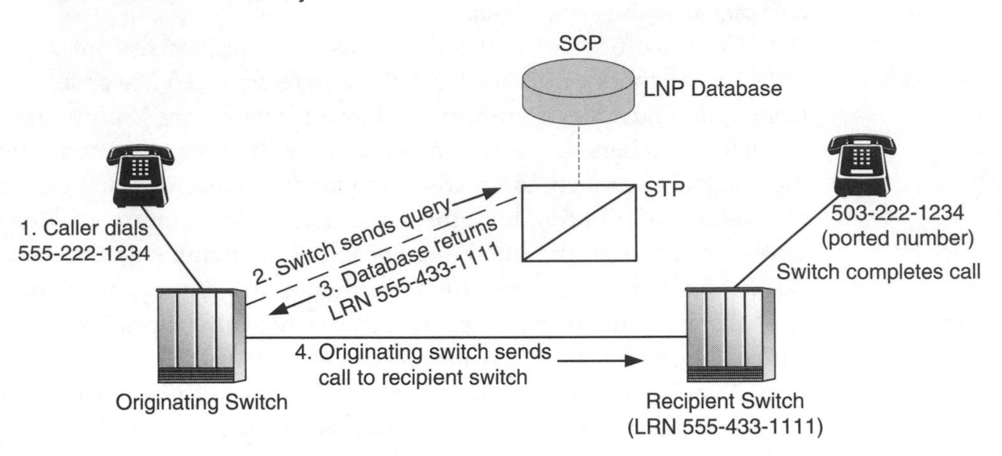

networks, how the IXCs obtain local access, and the way IXCs sometimes bypass the local networks to provide service directly to end users.

Local Access Transport Areas (LATAs)

The terms of the agreement between AT&T and the Department of Justice prohibited the RBOCs from transporting long-distance traffic outside LATAs. The Telecommunications Act of 1996 provides for the eventual lifting of this restriction as local networks are opened to competition. At the time of this writing, the restrictions prohibiting the RBOCs from carrying inter-LATA toll have been lifted in most states, but the LATA boundaries remain. RBOCs that have been relieved of the inter-LATA restrictions provide long-distance service through a subsidiary. LATAs correspond roughly to Standard Metropolitan Statistical Areas defined by the Office of Management and Budget.

Equal access requires intelligence in the class 5 office to route the call to the trunk group serving the primary IXC. Subscribers register their choice of PIC with their serving LEC. Separate PICs are provided for inter-LATA and intra-LATA traffic. IXCs select the method of accessing the LECs. They can connect to the LEC with direct trunks, or they can gain access to any exchange in the LATA through an access tandem. Every end office in the LATA, including CLEC offices, connects to the LATA tandem with direct trunk groups. Access through the tandem carries a usage charge, so IXCs use direct trunks to the end office where they are cost-effective.

Maintenance and Administrative Features

Central offices include many features to monitor the system's health. These features enable the system to respond automatically to abnormal conditions through a local or remote maintenance and control center (MCC). The features in this section are essential for any carrier switch. Most enterprise network switches include the same features, although often to a less sophisticated degree.

Fault Detection and Correction

The central processor continually monitors all peripheral equipment to detect irregularities. When a peripheral fails to respond correctly, the processor signals an alarm condition to the MCC and switches to a duplicated element if one is provided. The MCC interfaces maintenance personnel to the fault-detecting routines of the generic program. At the MCC, the central processor communicates its actions with messages on a CRT, a printer, or both. Depending on the degree of sophistication in the program, the system may register the fault indication or may narrow the source of the fault down to a list of suspected circuits. For unattended operation the MCC transmits fault information to a control center over a datalink.

The processor also monitors its own operation through built-in diagnostic routines. If it detects irregularities, the on-line processor calls in the standby and goes off-line. All such actions to obtain a working configuration of equipment can be initiated manually from the MCC or from a remote center. The ultimate maintenance action, which can be caused by an inadvertently damaged database or a program loop, is a restart. A restart is usually done only manually and as a last resort because it involves total loss of calls in progress and loss of recent changes to the database.

Communications Assistance for Law Enforcement Act (CALEA)

CALEA requires all wireline, cellular, and broadband carriers to assist law enforcement personnel who are acting under court order. The act requires a carrier to ensure that its equipment, facilities, or services are capable of enabling the government to intercept wire and electronic communications. Carriers must provide access to reasonably available call-identifying information and deliver intercepted communications and information to law enforcement agencies.

Circuit-switched networks have little difficulty in complying with CALEA requirements, but VoIP is another matter. Calls placed over residential broadband and calls between SIP endpoints can flow directly between endpoints in RTP packets. The softswitch controlling the packet flow does not have access to the voice packets, which makes it difficult, if not impossible, to comply with wiretapping requests. The FCC permits packet-mode providers to request exemption from CALEA requirements, but as VoIP gains popularity, it is difficult to see how a disparity between circuit-switched and packet-switched requirements can survive.

Emergency Calling

In North America, the E-911 system is designed to ensure that emergency vehicles from the correct jurisdiction are sent to an address from which a call originates, regardless of whether the caller is able to communicate with the emergency services operator. In most jurisdictions, the LEC boundaries do not align with the boundaries of the emergency service agencies and the agencies' boundaries do not align with one another. Furthermore, city and county boundaries are also no guide to which public safety answering point (PSAP) may handle an emergency call. Therefore, the E-911 system must have a database that correlates a calling number with the appropriate responding agency.

Figure 13-3 shows the principal components in an E-911 system. Each locality has one or more central offices designated to route emergency calls to the correct PSAP. This switch is known as the *selective router* (SR). It may be known by another name such as the E-911 tandem or the E-911 control office. Every end office in the serving area connects to the SR with a trunk group for handling emergency calls.

When a subscriber places a 911 call, the end office sends the voice call and the calling number to the SR. The SR looks up the emergency service number (ESN), which identifies the serving PSAP. The call is switched to the PSAP, which interrogates the 911 database for the address of the calling telephone number. The database returns the name, address, location, and the appropriate emergency responders for that address. The process of automatically displaying the caller's number, address, and supplementary emergency information is known as *automatic location identification* (ALI). The 911 database is often known as the ANI/ALI database

Providing accurate location information on emergency calls is a major issue for any organization that has private switches distributed over a wide area while sharing a common group of trunks. Unless other measures are taken, the host

FIGURE 13-3

E-911 Diagram

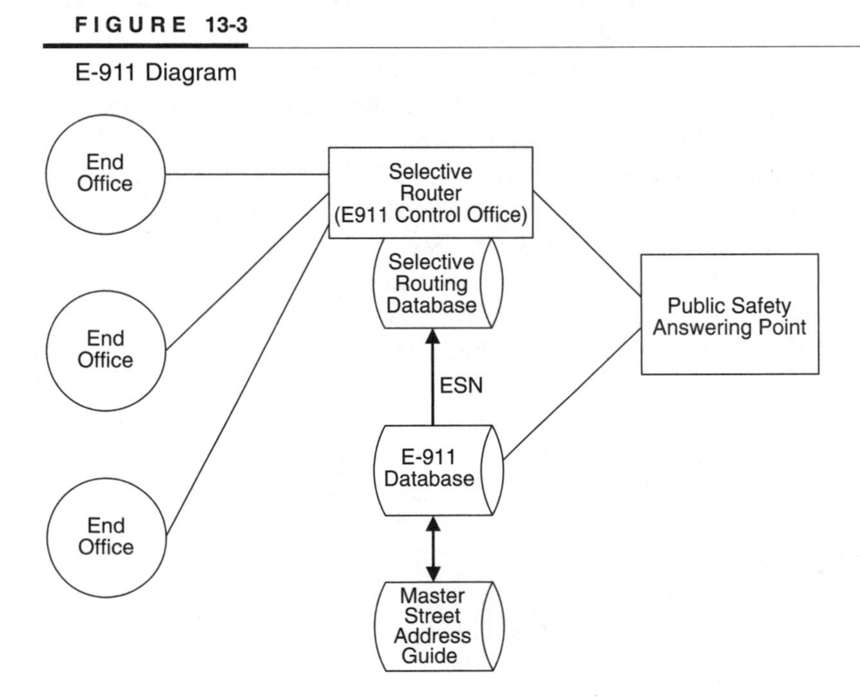

switch location will display in the PSAP and emergency vehicles will be dispatched to the wrong address. The process of supporting private systems to the E-911 system is known as *private switch automatic line identification* (PS/ALI). PS/ALI identification of the source of a 911 call can be accomplished by several different methods:

- A POTS line is connected to the CPE switch. The automatic route selection in the switch selects that line for 911 calls.
- A CAMA trunk can be connected to the serving central office. The CAMA trunk contains the calling station identity.
- If the PBX connects to its serving CO over PRI, the CO can receive the calling station identification and pass it to the SR.

VoIP raises critical issues with respect to E-911 because the phone is mobile and the IP address has no fixed relationship to the location. The service provider maintains a default location address for the serving trunk group, but VoIP enables users to transport their phones to other jurisdictions, even to other area codes. The issues are similar to cell phones, but the solutions proposed there involve either the use of GPS or the triangulation from cell sites. GPS is technically a suitable method for VoIP, but requires an unobstructed view of at least three satellites, which is not reliable.

The most practical solution relies on the fact that every phone is connected to a switch port that has a fixed geographical location. If a VoIP phone dials 911, the call controller communicates with a server that registers the physical location of the phone. The call switches to the PSAP through the SR while a gateway communicates location information through the PS/ALI process. The process for handling this is not standardized at this writing.

Essential Service and Overload Control

Switching systems are designed for traffic loads that occur on the busiest normal business days of the year. Occasionally, peaks occur that are higher than normal. Heavy calling loads can occur during unusual storms, political disorders, and catastrophes. During these peaks, the switching system may be overloaded to the point that service is delayed or denied to large numbers of users. Central offices include *line load control* circuitry that denies service to nonessential users so that users essential to public safety and health can continue to place calls. Nonessential users are assigned to two groups. When line load control operates, one group is denied service while the other group is permitted to dial. The two groups are periodically reversed to give equal access to both nonessential groups.

Many local offices have network management provisions to prevent overloads. For example, *dynamic overload control* automatically changes routing tables to reroute traffic when the primary route is overloaded. Code conversion allows the system to block traffic temporarily to a congested central office code.

This feature enables the blocked system to take recovery action without being overwhelmed by ineffective attempts from a distant central office. These provisions are circuit-related and not applicable to VoIP except in the case of gateways, which are subject to the overload conditions on their circuit side.

Telecommunications Service Priority

The Telecommunications Service Priority (TSP) Program provides national security and emergency preparedness (NS/EP) users with priority authorization of telecommunications services that are vital to coordinating and responding to crises such as hurricanes, floods, earthquakes, and other natural or human-caused disasters. During such a disaster, telecommunications service providers may become overwhelmed with restoration requests. The TSP Program provides service vendors with an FCC mandate for prioritizing service requests by identifying those services critical to NS/EP. A telecommunications service with a TSP assignment is assured of receiving full attention by the service vendor before a non-TSP service.

Trunk Maintenance Features

Local central offices have varying degrees of trunk maintenance capability. The system monitors trunk connections in progress to detect momentary interruptions or failures to connect. The system registers the failure and enters it on a trunk irregularity report, which technicians use to pattern trunk trouble. The system marks defective trunks out of service and lists them on a trunk out-of-service list. When all trunks in a carrier system fail, the system detects a carrier group alarm, marks the trunks out of service in memory, and through its alarm system reports the failure to the MCC.

Line Maintenance Features

Switching systems contain circuits to detect irregularities in station equipment and outside plant. Like trunk tests, these are made on a routine or per-call basis. On each call, many systems monitor the line for excessive external voltage (foreign EMF), which suggests cable trouble. Line insulation tests (LIT) are made routinely during low-usage periods to detect incipient trouble. The LIT progresses through lines in the office on a pre-programmed sequence and measures them for foreign voltage or low insulation resistance, which is low resistance between the tip and ring or from each side of the pair to ground. These tests detect outside plant troubles such as wet cable, and terminal, drop wire, and protector problems.

Switching systems must deal with *permanent signals*, which are caused by a telephone receiver off hook, cable trouble, or a defective station protector. Any of these irregularities place a short circuit on the line, and the line circuit attaches the line to a register, which furnishes dial tone and prepares to accept digits. The system must be capable of detecting and dealing with receivers off hook and multiple short circuits that result from cable trouble. Such lines are marked out of service temporarily so they do not tie up common equipment in the central office.

One method of resolving permanent signals is to connect the line to a series of tones and recordings such as a recording that asks the caller to hang up the line. Then after a suitable interval it may be connected to a progressively louder series of tones to attract the caller's attention.

Configuration Management

When the office is initially configured, its software retains a map of equipped and spare facilities. When the office is upgraded the configuration management system informs the switch of the presence of new equipment. Daily service order activity modifies the software to add new subscribers, delete disconnected ones, or change features. The configuration management system enables technicians to enter order activity through a terminal or through file transfer.

As discussed in the next section, the system samples call activity and reports traffic information such as call counts, call completions, and line, trunk, and feature usage. The system monitors CPU activity and reports grade of service measurements such as dial tone delay and common equipment usage.

Traffic-Measuring Equipment

Circuit switches are engineered based on usage. Usage information is based on the number of times a trunk or line circuit is seized and the average holding time of each attempt. Software registers measure the number of times the circuit is seized and elapsed time that the circuit is busy. The registers are periodically unloaded to a processing center for summary and analysis. Class 5 offices that have integrated VoIP and separate VoIP gateways must also collect usage information toward the PSTN because it is needed to determine the bandwidth of the connecting trunks. Traffic-measuring equipment can provide a variety of other data for administering the central office. For example, database information is provided to update the availability of vacant lines and trunks.

LOCAL CENTRAL OFFICE SERVICE FEATURES

All central offices provide certain basic features. This section discusses some of the most common features and explains their purpose.

Automatic Number Identification (ANI) and Automatic Message Accounting (AMA)

ANI automatically identifies the calling party for billing purposes. On POTS lines the number is identified from the cable pair. In ISDN and VoIP lines call setup messages identify the calling number. AMA equipment interrogates ANI equipment to determine the identity of the calling party. AMA equipment is classified as local AMA (LAMA), in which call collection is done in the local central office, or centralized AMA (CAMA). A CAMA office connects calling subscribers to a center where call details are recorded for all subtending central offices. AMA equipment records call details at each stage of a connection. The calling and called party numbers are registered initially. An answer entry registers the time of connection, and the terminating entry registers the time of disconnect. The storage medium is tape, disk, or solid-state memory. Call records are sent to distant data processing centers over a datalink.

Local Measured Service (LMS)

Many telephone companies base their local service rates on usage. Flat rate calling is often available, with LMS as an optional service class. With the LMS class of service, calls inside the local calling area are billed by the number of calls, time of day, duration of call, and distance between parties. The method is similar to long distance billing except that calls may be bulk-billed with no individual call detail.

Centrex

LECs furnish a PBX-like service through equipment located in the central office. The switching equipment is a partition in the end office called a *Centrex common block*. Centrex features allow direct inward dialing (DID) to a telephone number and direct outward dialing (DOD) from a number. For calls into the Centrex, the service is equivalent to individual line service. Outgoing calls require users to dial an access code such as 9, after which the central office returns the second dial tone.

Stations within a common block can have an extended suite of features. Internal calls can be abbreviated to four or five digits instead of the seven digits required for ordinary calls. Centrex users have access to all of the calling features of the central office, which are listed in a later section, as well as most of the features of a PBX such as call pickup, call hold, and class-of-service restrictions. An attendant console located on the customer's premises links to the central office over a separate circuit. These features are discussed in Chapter 24, Customer Premise Switching System Features.

Centrex brings several advantages to the customer. Reliability is high since the LEC provides backup power and the stability of a switch that is more reliable than most PBXs. The customer avoids the expense of powering the switching equipment. The LEC handles problems of growth and obsolescence and the customer is relieved of many administrative tasks associated with switch ownership. A major advantage is the ability to link telephones in multiple locations with a single switch.

Offsetting this to some degree is the difficulty of feature activation in analog Centrex. Telephone sets are POTS sets operating over cable pairs and the user activates features with a dial code. For example, call pickup might be *8. Users tend to forget the codes, and the features are consequently under-utilized. Some switch manufacturers offer digital Centrex with TDM feature telephones similar to those provided with PBXs. Feature telephones have displays, which improves usability, and functions such as call pickup and transfer are activated by pushing a button. Digital Centrex is limited, however, by the distance from the central office. Feature telephones have a range limitation much less than POTS phones and will not work outside the home central office.

IP telephony offers the possibility of rejuvenating the LECs' Centrex offerings. A media gateway (MG) can be placed on the subscriber's premises. If the subscriber has multiple locations, a separate MG can be placed on each, and the entire complex can be served from a single controller. This technology is described in more detail in Chapter 16. IP telephones overcome the range restrictions of TDM telephones and offer the benefits of expanded displays.

Routing to Service Facilities

All central offices provide access to service facilities such as operator and repair service. All local switching systems also provide access to call progress tones such as busy, reorder, vacant number tone, etc., and recorded announcements for intercepted numbers and permanent signals. Some systems also provide access to local testing facilities.

Call-Processing Features

Local central offices provide a variety of service features that enhance call processing. The user signals the central office equipment with a momentary on-hook flash on a POTS line or a data message on IP or ISDN. On a POTS line the system responds to the flash with stutter dial tone to show that it is ready to receive instructions. Many call-processing features that once were reserved for business use are now becoming popular with residential subscribers. As residential users increasingly subscribe to second lines and custom calling features, such services as call pickup and do-not-disturb are becoming popular. The following is a brief description of the principal features provided by most end offices in addition to the CLASS features discussed in Chapter 12:

Call forwarding enables the user to forward incoming calls to another telephone number. While this feature is activated, calls to the user's number route automatically to the target telephone number. When the user of a forwarded line picks up the telephone to place a call, the system sends stutter dial tone as a reminder that the line is forwarded.

Call forward remote access permits the user to activate or deactivate call forwarding from a remote location.

Call transfer allows the user to add a third party to a call as with three-way calling. One party can then hang up, leaving the other two parties in conversation.

Call waiting sends a tone to signal a user on a busy line of another incoming call. The user can place the original call on hold and talk to the waiting call by pressing the switch hook. Some LECs offer "whisper" call waiting so the other party is not aware of the interrupting call. A *cancel call-waiting* feature enables users to disable the service temporarily to avoid having a modem call interrupted by call waiting.

Distinctive ringing allows the LEC to assign as many as four separate directory numbers to a single line. By using an adapter that responds to the unique ringing pattern, the user can route calls to a separate telephone, fax machine, modem, or other analog device. Most fax machines respond to the distinctive ring without an adapter.

Gab line enables callers to dial a number to which multiple callers can be connected simultaneously. Some LECs use the gab line, which is popular with teenagers, as a revenue-generating feature.

Hot line automatically places a call to a directory number when the handset is lifted.

Multiline hunt connects incoming calls to the first idle line in a group.

Speed calling enables the user to dial other numbers with a one- or two-digit number. Most telephones also provide speed calling without involving the central office.

Three-way calling allows the user to add a third party to the conversation by momentarily holding the first party while a third number is dialed.

Voice mail allows callers to leave messages when the phone is busy or unanswered and alerts the user to waiting messages.

Warm line is similar to hot line except that the predetermined number is not dialed until after an elapsed time. The service is useful for handicapped people who may be unable, at times, to dial the phone.

CHAPTER 14

Circuit Switching Systems

Circuit switches serve the PSTN as end offices and tandems, and the enterprise network as PBXs and key systems. End offices are designed around the characteristics of the local network as it has been since the early years of the telephone. The central office switch nests in the center of a serving area or wire center. Feeder cable radiates out to neighborhood centers where it connects to distribution cable that terminates on a protector at the subscriber's premises. This model serves most of wireline telephones in the world, but it is evolving as LECs replace copper cable with fiber optics. Every metropolitan area in the United States with enough business development to support high-bandwidth service is now served with fiber. New residential areas are also likely to be served with a fiber feeder and the line terminations extended over copper. The length of the copper loops is growing shorter to enable the ILECs to serve their subscribers with higher DSL speeds.

In one sense of the word the local network is becoming less centralized as the line terminations move closer to the subscribers, but in another sense it is becoming more centralized as smaller central offices, known as community dial offices, are replaced with remote switching units homing on large central switches. In the final analysis, circuit switching is inherently centralized because it relies on centralized intelligence and is analogous in many ways to a mainframe computer.

Circuit switching is evolving on both the line and the trunk sides of the switch. As discussed in Chapter 13, the trend is away from tandem switching and toward routing in the PSTN core, with the edge of the network remaining circuit switched. LECs are testing the softswitching model at the edge with limited trials and occasional central office upgrades, but the evolution will extend over several years.

Fiber optics has had an even more dramatic effect in the trunking network than on the line side. Analog trunks have practically disappeared. The unit of trunking is no longer the individual trunk; it is the DS-1 and increasingly DS-3 or OC-3. The LECs have plenty of fiber in the ground and the cost of expanding it is insignificant, so it is often effective to over-provision trunks and save on administrative costs. This factor impedes the growth of IP trunking because of the lack of a compelling economic justification.

Competition is another significant factor in the evolution of the local network. The CLEC industry has undergone a dramatic shakeout in the past few years. Many have merged and more yet have vanished, but the signs are that the CLEC industry is beginning to grasp the nature of the competitive local market, although shifting regulatory decisions have added to the uncertainty.

Some CLECs provide service through their own class 5 switches, but many use unbundled network elements from the ILECs. That option is disappearing, however, as the FCC ordered in 2004 that ILECs have no obligation to furnish unbundled access to mass market local switching. Switch-based CLECs tend to serve an entire metropolitan area from a single switch, often using the ILEC's loop plant to reach their customers. CLECs typically focus on business customers that require enough lines, usually 12 or more, to justify T1/E1 access. This is changing the structure of the local network away from the bare cable pair and toward more T1/E1 access in the loop.

CIRCUIT SWITCHING ARCHITECTURE

All digital circuit-switching systems include the elements shown in Figure 14-1. Our focus in this chapter is on central office switches, but PBX architectures are similar. The principal elements are

- A TDM *switching fabric* that connects paths between input and output line and trunk ports.
- A *central controller* that directs the connection of paths through the switching network. The central control often offloads part of its load to peripheral processors.
- *Databases* that store the system configuration and trunk and subscriber line features.
- *Line ports* that interface outside plant for connection to users. All local and PBX switching systems include line ports; tandem switches may have only a few specialized line ports.
- *Trunk ports* that interface interoffice trunks, service circuits, and testing equipment.
- *Service circuits* that provide call progress signals such as dial tone, ringing, and busy tones.
- *Common equipment* such as battery plants, power supplies, testing equipment, and distributing frames.

Manufacturers approach switch design from different perspectives, but the outcome is nearly identical in the products used in North America. At least to

FIGURE 14-1

Local Central Office Configuration

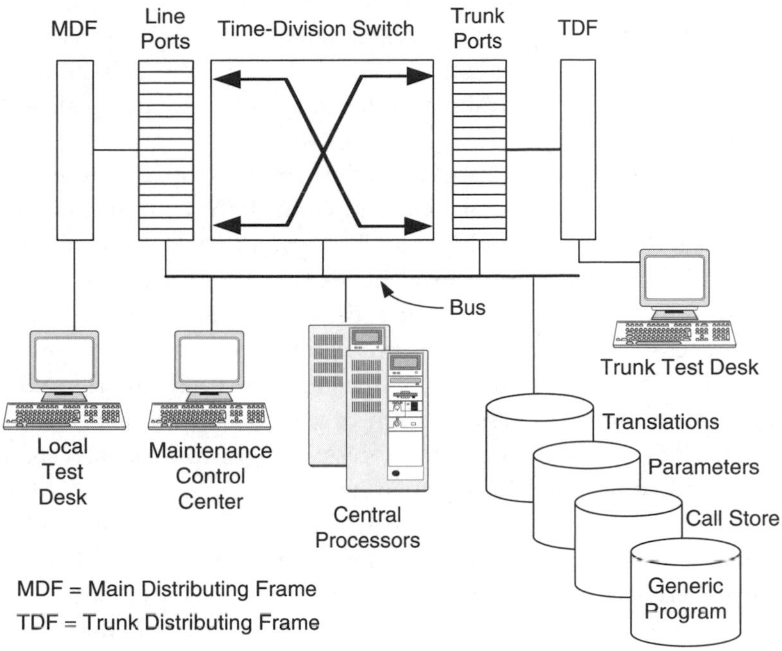

MDF = Main Distributing Frame
TDF = Trunk Distributing Frame

some degree, this is because the RBOCs are by far the largest purchasers of switching systems, and their requirements evolved from common ownership. Through their jointly owned company Bellcore (now under independent ownership as Telcordia) the RBOCs produced a specification known as the Local Switching System Generic Requirements (LSSGR). This series of documents defines the functionality of the PSTN in North America. Equipment manufacturers develop their products to adhere to the features and service standards of the LSSGR. Most CLECs and independent telephone companies purchase products developed to the LSSGR, so as a result the differences among products are nearly invisible to subscribers.

The end office is evolving to support IP trunking. Some manufacturers have IP enabled their end office products. The switch has TDM switching fabric, but has an IP trunking interface. If the end office is not IP enabled, then a separate gateway is required. In either case, the switch typically uses SIP, Megaco, or H.323 signaling on the IP side of the network.

The software in most digital central offices is modular. Features such as ISDN and Centrex are usually optional, which allows the manufacturer to reduce the price in residential and rural areas where the demand for these features may be low. Many CLASS features require optional software, and voice mail is always an add-on, often from a third-party manufacturer.

Switching System Control

Circuit switches may distribute control and some call-processing functions to peripheral units, but ultimate control is centralized. The central controller monitors the health of peripherals, collects traffic data, provides access to the database, and other functions common to an operating system as well as providing basic call processing. The central controller in a digital switch is always redundant to provide high reliability.

Switching systems are designed as central control, multiprocessing, or distributed control. In a central control system, all the call processing is concentrated in a single location. A multiprocessing system has two or more processors that share call-processing functions. The sharing divides the call-processing load or assigns one set of functions to one processor and another to its mate. For example, one processor could handle call processing and the other maintenance. If either failed, the survivor could assume the entire load. The distributed method of control uses multiple processors, each of which handles a designated part of the switch. For example, a separate processor might control each shelf or frame with the central control monitoring their health.

The software that drives electronic switching systems falls into four categories. Some manufacturers use different terminology and there may be architectural differences between systems, but the functions reside in every digital switch. The *operating system* ties the elements of the switch together, takes care of input and output functions, and supervises the general health of the system. Closely tied to the operating system is the call-processing software or *generic program*. The generic program contains the logic for activating the features of the switch and maps the connections through the switching fabric.

The third type of software is the *parameters*. The parameter database contains the types, quantities, and addresses of the major hardware components. The processor controls these components, mapping their busy/idle status, setting up paths through the switching fabric, monitoring them while calls are in process, and releasing the path when the call ends.

The fourth software element is the *translations*, which are a database of port assignments and specify the features that are assigned to each port. Each trunk and station in a switching system is assigned through the translation tables, and the features associated with that port are defined in the tables. For example, trunks are translated as to location, signaling, and type. Stations are translated as to location, restrictions, and features.

Changes to a switching system's database of line, trunk, and parameter translations are accepted only after the system makes internal checks to ensure the accuracy of the input record and to assure that the integrity of existing records will not be damaged. Updates of the database are allowed only from authorized input devices, and then limited with password control and authentication to ensure that only qualified personnel may access the files. A copy of the database also is kept off-line in disk storage so it can be reinserted if the primary file is damaged or destroyed. The manner of assuring database integrity varies with the manufacturer and is an important consideration in evaluating local switching systems.

Switching Networks

Nearly all currently manufactured systems use a PCM switching fabric, which switches calls in 64 Kbps increments. The basic function of the switching fabric is to provide paths between the input and output ports. The network design objective is to provide enough paths to avoid blocking users, while keeping costs under control. A switch that contains fewer paths than terminations is called a *blocking switch* because not all users can be served simultaneously.

A blocking switch has concentration in the line ports, but not in the trunk ports. Line circuits are installed in line groups that share links to the switching fabric. A switch with a concentration ratio of 4:1 would have enough links that 25 percent of the ports in a line group could be connected simultaneously. The switch administrator determines the number of links required based on traffic volume.

A nonblocking switch does not ensure that users will never encounter blockage because trunking is always designed to some level of blockage. Also, if the system is configured without enough common equipment, users will encounter delays that they will interpret as blockage. For example, too few digit receivers results in slow dial tone. Most LECs provide plenty of digit receivers to handle normal peak loads, but during disaster conditions the receivers may become overloaded.

Switches are also subject to processor overloads, which can result in a variety of call-processing delays. Switches are rated by the number of *busy hour call attempts* (BHCA), which is a factor that specifies the number of calls the system can handle during load peaks. The term call attempt may bear little relationship to the actual number of calls handled by the system. Not only are call originations and terminations counted as call attempts, but accesses to features such as call pickup, call transfer, and call waiting, which require attention by the central processor, are included.

Modern switching networks are wired in stages as shown in Figure 14-2. Each stage consists of a switching matrix that connects input links to output links. Links, which are often called *junctors*, are wired between switching stages to provide a possible path from any input port to any output port. The network shown in Figure 14-2 is blocking because it has twice as many input ports as output paths.

LECs engineer the line concentration ratio to support the traffic intensity of the wire center. In the past LECs have found heavy daytime usage by businesses and light usage for residential subscribers, peaking in the early evening hours. Business and residence load peaks could be expected to balance each other to some degree. With the popularity of the Internet, however, the LECs have found that traditional traffic engineering assumptions are no longer valid. Residential

FIGURE 14-2

A Switching Network with 2:1 Concentration

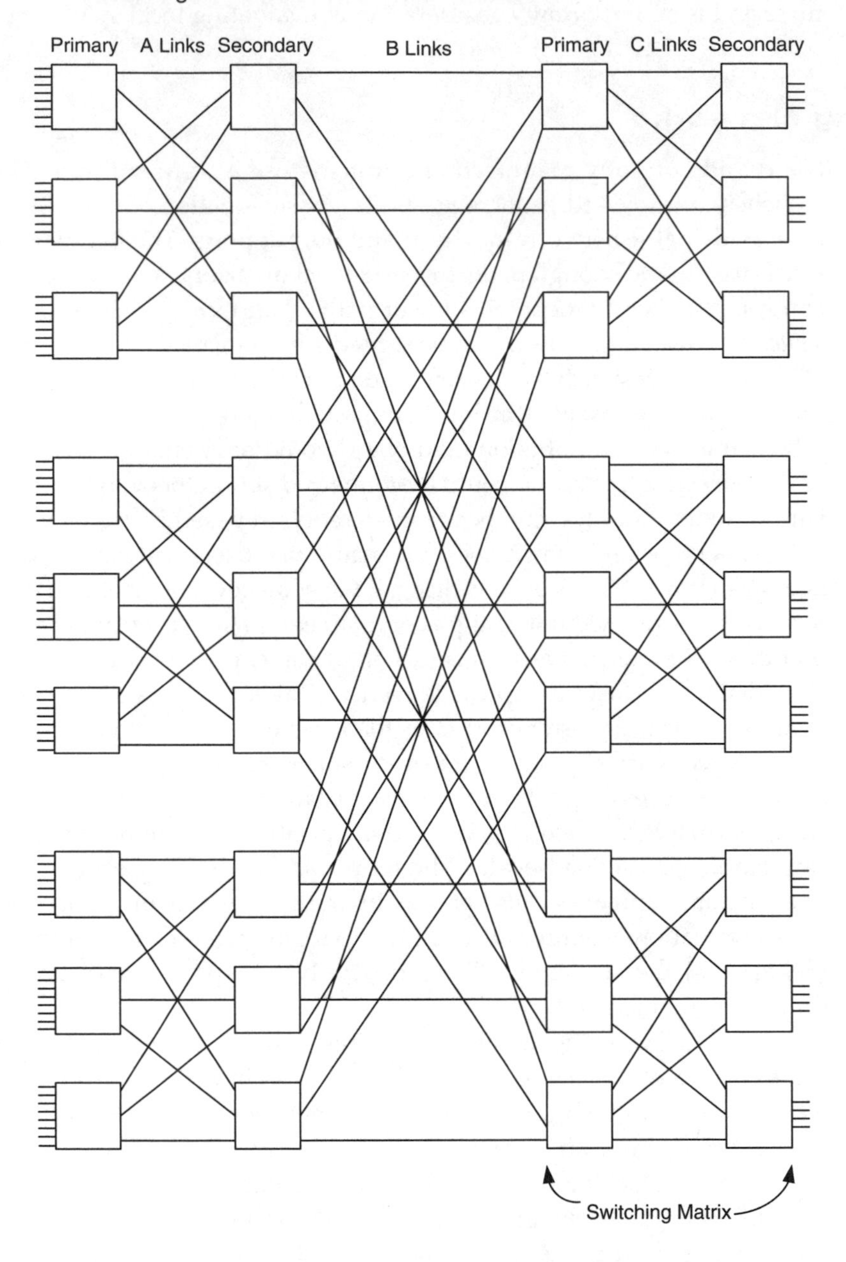

subscribers may obtain a second line for Internet access and instead of an average call length of 4 min, the average may be 30 min or more. This results in unexpectedly high usage on trunks as well as the switching network. As broadband Internet access increases in popularity, voice traffic patterns are returning to normal.

PCM Switching Networks

The capacity of a switching network is a function of the number of switching stages it has. The switching matrix is physically limited by the number of terminations it can support. To avoid blocking, the controller must have multiple choices of paths through the network. These paths are obtained through multiple switching stages so that each stage has enough choices that the probability of blocking is reduced to a level consistent with the grade-of-service objectives. Note in Figure 14-2 that each switching stage has paths to diverse modules in the next stage.

In earlier switch generations, the switching network elements were analog and the paths through the switch were physical wires. Some of these switches still survive, and are equivalent to digital switches except that the line circuits are completely analog, where in digital switches the line circuits include A-to-D conversions. Internally, PCM networks connect a digitally encoded signal over 8-bit parallel paths.

The switch modules are of two types: time switches and space switches. The objective of TDM switching is to connect a channel that enters the switch on one time slot to an outgoing time slot that is spatially separate. A switching stage, therefore, must either do a time slot interchange or make a physical switch to another time division module. The network is rated by the sequence of switching stages. For example, a TSST network would have time-division switches in the first and last stages and space switches in the center stages.

Time division switching is implemented in a time slot interchange element as shown in Figure 14-3. In this case, Station A is assigned to time slot 3 and Station B to time slot 1. A TSI receives digital pulses during one time slot, stores them for one processor cycle, and releases them during the proper time slot in the next cycle. For two stations to talk through the switch the controller assigns them to the same time slot.

If the buses were physically separated so they were unable to reach the same time slot interchange, they would need a space-switching stage, which is accomplished by gating the buses together at the proper instant.

FIGURE 14-3

Time Slot Interchange

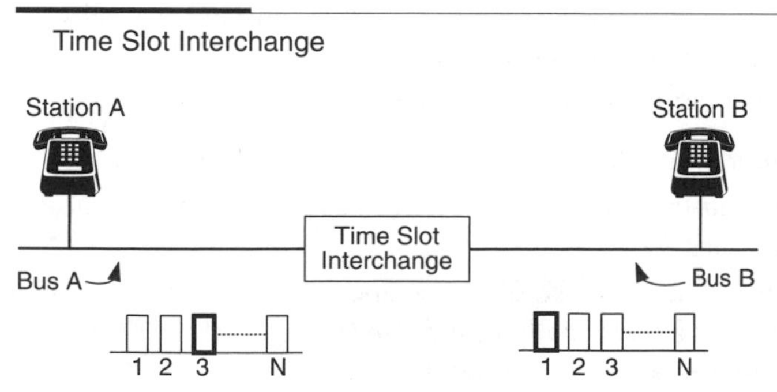

FIGURE 14-4

Space Division Switch

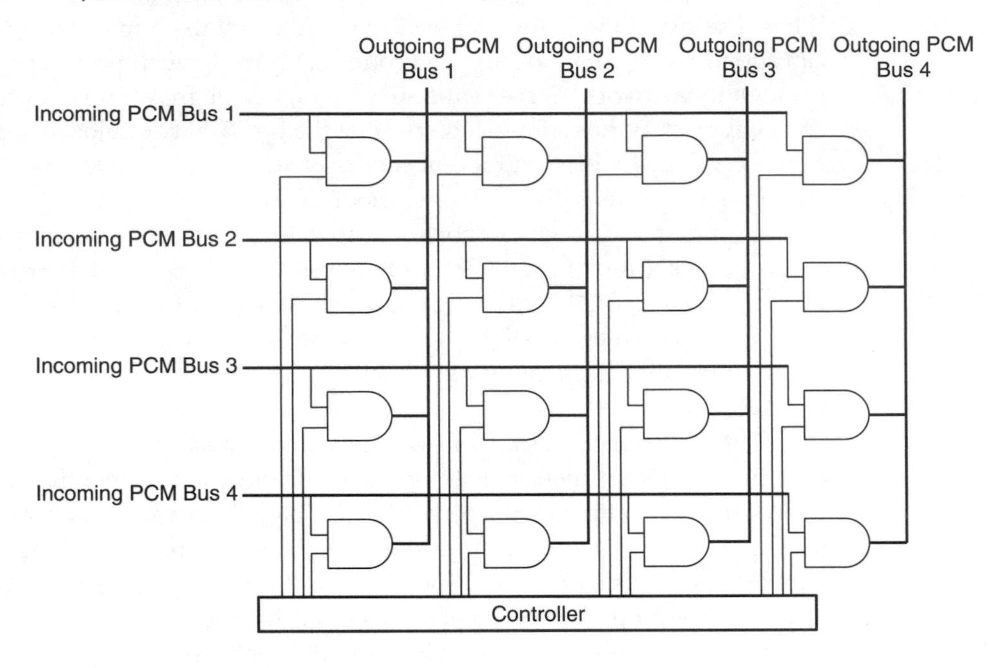

Figure 14-4 shows conceptually how a space-division switch operates. Signals arriving on a TDM bus are applied to one input of an AND gate. The controller applies a signal to the other input of the gate associated with the selected path, which sends the pulses on the outgoing bus. Practical switching networks contain a combination of space- and time-division switches. For example, Figure 14-5 shows a four-stage time–space–space–time (TSST) network.

Call Processing

Most digital switches use similar techniques for processing calls. The following is a short description of a typical call process, although different manufacturers have different terminology for the elements. The discussion assumes that the call originates and terminates in the same central office and does not consider number portability. The principal processing elements are

- *Scanners* to detect changes in states of lines and trunks.
- *Signal distributors* to transmit signals from scanners to call-processing programs.
- *Registers* that furnish dial tone and accept and register dialed digits. Registers are normally dial pulse (rotary dial) or DTMF, and they reside on the trunk side of the switch.

FIGURE 14-5

Time–Space–Space–Time Digital Switching Network

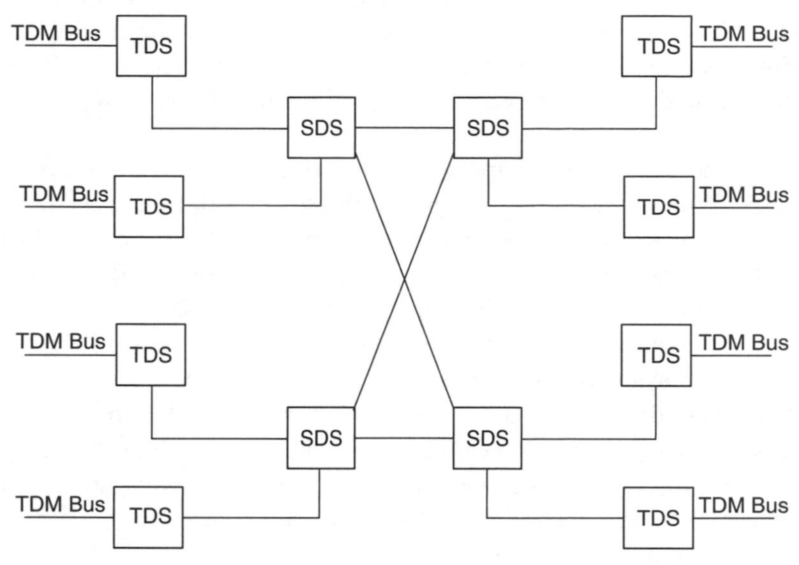

- *Generic program,* which is the call-processing instructions contained in the program store.
- *Call store* to provide temporary scratch-pad memory that is used to store the details of calls in progress.
- *Data store,* which stores the line translations.
- *Time slots,* which, in a digital switch, are units of time reserved for the parties to share during a session.
- *Switching fabric* to make connections between lines and trunks.

The calling party initiates a call by lifting the receiver from the switch hook, which causes current to flow in the line. A line scanner detects the change in the state of the line and sends a signal through a signal distributor to the generic program, which marks the line busy in call store memory and consults the data store about features and options available to the line. The processor marks a path through the network, assigns a register for the call, and reserves it a time slot.

The processor connects the line to the register, which sends dial tone toward the line. The register receives tones or dial pulses and stores them. When digit reception is complete, the processor consults the data store about the address of the called number. It marks the called line and reserves time slots through the network—one for the originating party and one for the terminating party.

The processor connects the called number to a ringing source. The calling number is connected to an audible-ringing source. The terminating line is monitored for *ring trip*, which is a short circuit on the line that indicates that the called party has answered. When the called party answers, the ringing and audible ring signals are removed, and the path that was reserved through the network is connected. If the line is metered, an initial connect entry is registered in the AMA equipment.

The connection is supervised for an indication that either party has gone on-hook. When either party hangs up, the scanner detects the change in state and notifies the processor. The processor restores the circuits, marks the time slots idle, and restores the call store registers. The line status is changed in memory from busy to idle, and the terminating entry is made in AMA.

Calls outside the same central office are handled similarly, except that two switching systems are involved. SS7 selects the path between the offices and receives notification of call answer so it can inform the offices to connect the call through. During call processing, many events other than those described above may occur. For example, one party may flash the switch hook, which recalls the processor to send second dial tone and to be prepared for another process such as conferencing.

Central Processor

The central processor in digital switches is similar to that used in commercial computers, but with some important differences. First, the SPC processor is not only fault tolerant; it is almost fail safe. Although digital switches can fail, their design objective is no more than 1 h outage in 20 years, which is an outage tolerance several orders of magnitude better than that of most computers. Central office processors have extensive trouble detection modules that are designed to ensure reliability.

A second difference is in the nature of the processing task. Call processing is input/output intensive, with little requirement for arithmetic operations compared to commercial computers. Where a commercial computer is overseeing several dozen peripherals, a central office processor is managing tens of thousands of individual terminals, any of which can spring to life at any time and demand service within a second or less.

The processor operates under the direction of a generic program, which contains the call-processing details. Features can be added by replacing the generic program with a new issue. The call-processing steps are closely integrated with the hardware in most systems, and although the processor might be programmed in a high-level language, the logic is not readily portable from one processor to another.

The processor is also involved in collecting statistical information for administering trunk groups and common equipment. The information is usually dumped to an external processor that assists by finding the busy hour and calculating trunk requirements. The processor is also constantly looking for circuit and system irregularities. It is designed to survive abnormal conditions by shedding low-priority tasks such as service order processing.

Synchronization

Networks of digital switches must be closely synchronized to prevent transmission errors. If two interconnected systems do not have a common synchronizing source, their clocks will run at slightly different rates. This means that occasionally the receiving end will miss a bit or will sample the same bit twice, either of which results in a bit error, a condition discussed in Chapter 5 as a slip. Slips have little impact on voice transmission, but they are detrimental to data.

Circuit-switched networks are kept in synchronization by slaving switches from a highly stable master clock known as the basic standard reference frequency (BSRF). Each central office has its own clock that can run freely with a certain degree of stability. The highest level clock below the BSRF is a Stratum 1 clock, which has an accuracy of at least 1×10^{-11}. Three lower levels of clocking offer progressively less accuracy and lower cost. Timing passes down from each higher class office to the offices that home on it. This type of synchronization is called *isochronous*. When clocking is lost from a higher level office, lower level devices run freely. The greater the differences in clocking between two such devices, the higher the number of slips that will occur.

Line and Trunk Circuits

All switching systems are equipped with circuits to interface the switching network to stations, trunks, and service circuits such as tone and ringing supplies. In some systems, these circuits are external devices. In other cases, they are integral to the switching equipment.

Line Circuit Functions

In a digital central office, line circuits have seven basic functions that can be remembered with the acronym BORSCHT. Analog central office line circuits require five of the seven functions because they have two-wire switching networks, the hybrid and coding functions are omitted. The BORSCHT functions are

- *Battery* feeds from the office to the line to operate station transmitters and DTMF dials.
- *Over voltage* protection is provided to protect the line circuit from damaging external voltages that can occur during the time that it takes the protector to operate.
- *Ringing* connects from a central ringing supply to operate the telephone bell.

- *Supervision* refers to monitoring the on-hook/off-hook status of the line.
- *Coding* converts the analog signal to a PCM bit stream in digital line circuits.
- *Hybrids* in digital line circuits convert between the four-wire switching fabric and the two-wire cable pair.
- *Testing* access is provided so an external test system can access the cable pair for trouble isolation.

In PBXs and digital central offices, line circuits reside on plug-in modules. Because much of the cost of the system is embedded in the line circuits, shelves are installed, but to defer the investment, line modules are added only as needed. Analog line circuits are permanently wired in frames that are connected to a distributing frame for crossconnection to the cable pairs. The battery feed is different between circuit-switched and VoIP systems. Circuit-switching systems feed power to the telephone over the cable pair. This is not possible with VoIP phones in the local network since they have no metallic connection to the central office.

Line Equipment Features

Line switch frames in digital switches are constructed modularly with several line modules concentrated into a smaller number of links to the switching network. Two different architectures are employed in the line interface. In the codec-per-line architecture, each line module contains a separate analog-to-digital converter. The 64 Kbps bit streams from multiple modules combine in a multiplexer into a high-speed bit stream and route to the switching matrix. In the shared-codec architecture, the output of the analog line circuits is switched to a group of shared codecs. The former method, while more expensive, reduces the service impact of a single codec failure. The line module also provides testing access to the local test desk through a relay that transfers the cable pair to a testing circuit.

Line modules are produced in different varieties to provide features required by special types of subscriber equipment. The simplest module provides POTS features to analog telephone sets. Business-class line modules provide features for ground start. Switches equipped for digital Centrex require digital line modules that communicate with proprietary telephone sets. Message-waiting line modules are available to light lamps or provide stutter dial tone on telephones associated with LEC-provided voice mail. Switches offering ISDN service require line modules supporting T and U basic rate interfaces.

LECs offer both line-side and trunk-side T1/E1 connections to their switches and it is important to understand the differences. A line-side T1/E1 port emulates a telephone set. It uses loop signaling and furnishes dial tone toward the subscriber line. A typical application is providing T1/E1 service to a PBX or key system that has analog ports. The line can be terminated in a channel bank that has channel units corresponding to the central office side of the connection. A foreign exchange subscriber (FXS) channel has a T1 signaling interface toward the central office and provides ringing and dial tone toward the station.

A trunk-side T1 signals switch to switch with E&M or D-channel signaling. Channels can be set up as DID, one-way out, one-way in, or two-way, depending on the LEC's service offerings and the capabilities of the CPE switch. Two-way DID is a popular service for PBXs and key systems that are not PRI compatible because it offers many of the same features as PRI.

Trunk Circuits

Trunk circuits interface the signaling protocols of interoffice trunks to the internal protocols of the switching system. Most offices connect trunking under control of SS7. Digital switches normally provide for both analog and digital trunks, although analog trunks are disappearing from service. Trunk circuits may interface at the DS-1, DS-3, or OCx level, depending on the manufacturer. The trunk circuits are on cards that plug into modules that contain the trunk controllers.

Most switches have the capability of automatic trunk testing, which allows technicians to schedule the system to run trunk maintenance tests. Trunk tests are of two types: functional and diagnostic. Functional tests check signaling capabilities by connecting to responders in the distant central office. Diagnostic tests are internal to the switch, and check that the trunk hardware is operating properly.

All tandem switching systems, most end offices, and some PBXs include trunk testing circuitry to make transmission and supervision measurements on central office trunks. These circuits vary from 1004 Hz tone supplies that can be dialed from telephones served by the switching system to trunk testing circuits that enable two-way transmission and supervision measurements on trunks.

Service Circuits

All types of switching systems require circuits that are used temporarily in routing and establishing connections. These applications are briefly discussed so readers will understand how the services are obtained and applied in all types of switching systems.

Ringing and Call Progress Tone Supplies

All switching systems that interface end users require 20 Hz ringing supplies generating approximately 90 V to ring telephone bells. In addition, switching systems require audible ringing supplies, busy tones operating at 60 interruptions per minute (ipm), and reorder tones operating at 120 ipm. In digital switching systems, these tones are generally created in firmware.

Recorded Announcements

Recorded announcements provide explicit information to the user when calls cannot be completed and tone signals are insufficient to explain the cause. For example, calls to disconnected numbers are connected to recorded announcements. When a transfer of calls is required, the system routes the incoming call to an intercept operator; otherwise, calls to nonworking numbers route to a recorder. Announcements are also used on long-distance circuits to indicate temporary circuit or equipment overloads. Often these are preceded by a three-tone code called special identification tones (SIT) so automatic service observing equipment can collect statistics on ineffective dialing attempts.

Permanent Signal Tones

A permanent signal occurs when a line circuit is off-hook because of trouble or because the user has left the receiver off-hook. Permanent signals in trunk circuits occur because of equipment malfunctions or maintenance actions. Most switching systems use combinations of loud tones and recorded announcements to alert the user to hang up the phone. Permanent trunk signals are indicated by interrupting the supervisory signal at 120 ipm, which flashes the supervision lights attached to E&M leads in some signaling apparatus.

Testing Circuits

Central offices include a suite of features that are used for subscriber line testing. Cable pairs are subject to a variety of ravages from weather and dig-up and the central office is designed to protect itself from damage and assist in isolating trouble. Switches have a series of test lines that technicians can use to test loop transmission. For example, most offices have a milliwatt test line that attaches a 0 dBm level tone to the line from which the test number is dialed. A technician can measure the 1,004 Hz loss of the line with a transmission test set. A quiet-line termination is used for noise tests, and a multitone line tests for loss at various frequencies in the voice band.

Most switches provide connectivity for a local test desk. Technicians in a test center can dial a connection that enables them to look at the DC characteristics of the line. The test desk measures foreign voltages, noise, open and shorted cable pairs, and other such irregularities, often by comparing the line impedance with that of a good line. Foreign or unexpected voltages are usually the result of cable trouble. A cable pair normally has –48 V battery connected to the ring side of the line. The local test desk can remove the battery and if it finds that voltage is still present on the cable pair, it is generally a sign that the pair is crossed with another pair—often caused by wet cable.

It should be evident that much of the complexity in a local central office is the result of its interface with copper cable pairs on the subscriber side. As the network evolves toward fiber optics, the degree of complexity will diminish, but for now, it is inherent in the structure of the local exchange network.

DISTRIBUTED SWITCHING

Planning for service expansion is a difficult proposition in fast-growing areas. New business or residential developments can spring up rapidly, putting a strain on switching and outside plant facilities. Instead of bringing the lines to the central office with more copper or fiber outside plant facilities, it is often more economical to move the front end of the switch nearer to the users. Distributed switching reduces the number of circuits needed between the central office and the users and transmission is improved because the loop is shortened and linked to the central office with low- or zero-loss trunks. Also, features such as voice mail can be provided from a centralized unit to communities where it might otherwise not be cost effective. Digital switches offer at least two methods of distributing line interfaces closer to the subscriber: a remote switch unit (RSU) and digital loop carrier (DLC).

Remote Switch Unit (RSU)

An RSU essentially extends the line circuits and sometimes trunk circuits closer to the subscriber. An RSU supports a full complement of subscriber line options. Some RSUs also support local trunking, which is important if the RSU serves a distant community because much calling is local and running the connection back to the host is not economical. Most switches have a maximum range over which they can host RSUs, but it is generally in the range of hundreds of miles.

The umbilical back to the host unit is critical. It must be sized to support the traffic load. If the RSU has local trunking, the load may be low, but if all trunking is terminated on the host, the traffic on the umbilical will be high. Some RSUs have *intra-calling* capability, which enables them to switch station-to-station calls internally within the RSU. This feature is important in RSUs that serve a community because the intra-calling load is apt to be high and it is wasteful of resources if the calls must loop through the host, using two channels per session.

A related feature is standalone capability or survivability, which enables the RSU to function if the host or the umbilical should fail. The RSU typically has a minimal database, so feature functionality is apt to be marginal when it operates in a standalone mode. The administrator is also concerned about how the database is kept updated. In the worst case, separate entries are required in the host and the remote. Double-entry translations are undesirable because of the risk of discrepancies. The most effective systems keep the remote databases automatically synchronized from the host.

To minimize the possibility of umbilical failure, the LECs feed remotes over a self-healing facility, usually a SONET ring or some other form of automatic route protection. The remote is mounted in a controlled environment that provides the same protective features that the central office enjoys: backup power, environmental controls, and testing and administrative capabilities.

A remote line module (RLM) is similar to an RSU except that it has no intra-calling capability. Calls between two stations within the module require two central office links. If the link to the host fails, calling within an RSU is still possible, but an RLM is dead without the umbilical.

Digital Loop Carrier (DLC)

DLC is an alternate form of remote. DLCs come in both single- and double-ended configurations. A double-ended DLC has matching units in the field and the central office. The central office end connects to analog line ports in the switch and effectively extends the ports to the field over T1/E1 lines. Small DLCs may serve 24 or 30 channels over a single T1/E1 line. For larger groups of subscribers the unit may offer concentration. A typical configuration might serve 96 subscribers over two T1 lines, which provides two-to-one concentration.

The other form of DLC is single- ended using an interface known as GR 303 in the central office. The GR 303 interface is a T1 connection into the switch, interfacing a DLC unit in the field. GR 303 is an industry standard that can be used for other purposes besides DLC. For example, a Lucent Technologies product known as i-Merge provides a VoIP connection for digital central offices through GR 303.

DLCs and RSUs provide comparable services and may have much the same features, including intra-calling and standalone capability. DLC is preferred in smaller line sizes because the RSU is generally designed for larger line concentrations. DLCs generally do not require the environmental controls of an RSU, which is designed for a central office environment. DLCs are available in pole-mounted configurations and some products mount in closures that are designed to be compatible with the product. Battery backup is a standard feature and some units may be solar powered. The line density is usually not great enough to justify the cost of an emergency generator.

TANDEM SWITCHING

Tandem switching capability is an optional feature of most digital switches. Most class 5 switching products can function as a combination end office and tandem switch, but where the application is pure trunk switching, a switch designed for the purpose is often used. The industry has three classes of tandems: local, LATA access, and toll. Although the functions are unique, the hardware may be the same with a different software load.

LECs employ local tandems in most metropolitan areas to terminate calls within the local calling area. Any end office in the local area is reachable through the local tandem. ILECs, independent telephone companies, and CLECs all use the local tandem as an overflow from direct end-office trunks or as a means of reaching other end offices if the traffic load is not enough to justify direct trunks.

The RBOCs provide LATA tandems, which are trunk switches that extend to every end office in the LATA. Every ILEC and CLEC switch is reachable through the LATA tandem, either as the only trunk group, or as overflow from direct trunk groups from the toll tandems. LATA tandem trunks provide a path for the IXCs to use to terminate switched access calls to their subscribers. All switch-based IXCs terminate trunks on the LATA tandem and pay a per-minute cost for the traffic.

The third class of tandem is a toll tandem. Class 5 offices can also do toll switching, but toll tandems have limited line-side features and provide a rich set of trunk features that end offices do not always support. Because of their heavy reliance on trunk quality, all tandem switches require trunk test positions. Trunks are switched through the switching network to the test position for making continuity, transmission, and supervision tests. In a network of multiple tandem switches, technicians communicate between test positions by dialing a special access code, usually 101. This connects the two test positions so technicians can talk and test over the trunk. Several different codes are employed in the long-distance trunks of the IXCs, LECs, and some private networks for testing as shown in Table 14-1. These test lines can be used for both manual and automatic tests.

To test direct trunks between two offices automatically, computer-controlled test equipment dials a remote office test line (ROTL) over a trunk. The ROTL seizes an outgoing trunk and dials the test line number over the trunk. A responder at the distant office interacts with the ROTL and the responder at the near-end office to make two-way transmission, noise, and supervision measurements. The test system registers the test results, automatically takes defective trunks out of service at the switching system, and marks them for maintenance action.

Operator Service Positions

Local exchange and interexchange carriers require operator service positions with some of their tandem switches. Most carriers that provide operator services centralize the function with one operator center serving multiple tandems. Operator service functions include intercept, directory assistance, and toll and assistance.

TABLE 14-1

Trunk Maintenance Test Lines

100	Balance	Provides off-hook supervision and terminates the trunk in its characteristic impedance for balance and noise testing
101	Communications	Provides talking path between test positions to support trunk tests
102	Milliwatt	Provides a 1004-Hz signal at 0 dBm for one-way loss measurements
103	Supervision	Provides connection to signaling circuit for testing trunk supervision
104	Transmission	Provides terminations and circuitry for two-way loss and one-way noise tests
105	Automatic transmission	Provides access to a responder to allow two-way loss and noise tests from an office equipped with an ROTL and responder
107	Data transmission	Provides access to a test circuit for one-way voice and data testing. Enables measurements of P/AR, gain/slope, C-notched noise, jitter, impulse noise, and various other circuit quality tests.

The toll and assistance function helps callers complete collect, third number, and credit card calls. Intercept is the function of assisting customers with calls to disconnected numbers. Directory assistance provides telephone number lookup for callers.

The objective of intercept systems is to complete as many calls as possible with an interactive voice response unit (IVR), which Chapter 28 discusses in more detail. IVR systems read out new numbers if one is available on disconnected numbers and number changes. If the caller remains on the line the system connects the call to an operator service position. Operators look up the number and transfer the caller to a voice announcement unit to read out the number.

An important feature of operator centers is the ability to recognize customers that require hotel class-of-service treatment. Hotels and similar residential units such as dormitories and hospitals often provide DID to the rooms. To prevent the occupant from accepting collect or third party calls to the room, the customer requests a trunk class from the LEC that identifies it as a hotel. The operator service position accepts only paid outgoing calls from this class.

Figure 14-6 shows the components of an operator service center and a tandem switch system. Incoming and outgoing digital trunks terminate on the switching network. A central processor controls call processing. The system directly switches calls that do not require operator assistance. If the caller dials an assistance code, the processor routes the call through the switching network to the front end of the center, which is an automatic call distributor (ACD).

The ACD delivers calls to the appropriate position. The position controller receives calls from the ACD and supplies the circuitry to enable the operator to communicate with the various subsystems. Within limits, the larger the work group, the more efficiently positions can be staffed. See Chapter 27 for further information on the functions and method of operation of an ACD.

Tandem Switch Features

As the tandem network converts to IP, unique features remain that in most cases work more effectively with circuit switching. IXCs offer customers direct connections to their tandem switches. Customers that have enough traffic to justify the cost of a T1/E1 access line to the IXC receive reduced rates because the IXC does not have to pay switched access charges to the LEC. Consequently, the switch must support features that are intended for end-customer operations.

User-Dialed Billing Arrangements

Special networks that register and verify charging information handle most coin and calling card calls. Calling card calls are verified over a datalink to a centralized database. The customer dials the desired number, and then when signaled by the system, dials a card number. The system sends a message to the database to verify the validity of the card number, after which it switches the call.

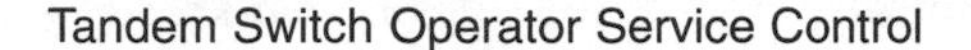

FIGURE 14-6

Tandem Switch Operator Service Control

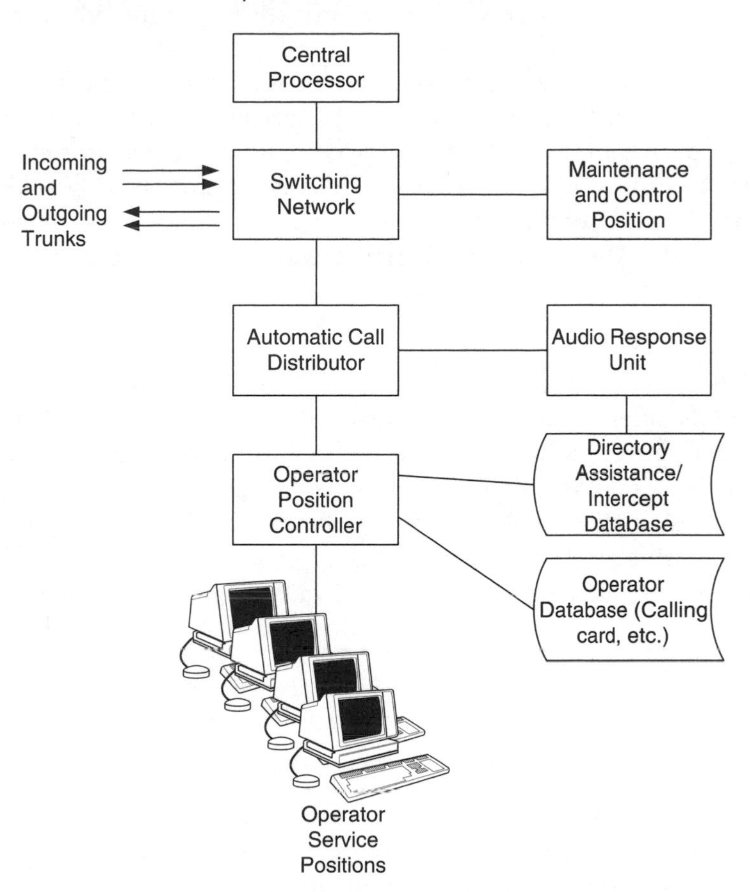

Coin calls are connected to a circuit that computes the rate, informs the user of the charge with a voice announcement unit, registers the values of the coins deposited, and connects the call. The equipment monitors the conversation time and reconnects the voice announcement unit when the caller is required to make an additional coin deposit.

Authorization Codes

Tandem switches provide authorization codes to identify subscribers, determine their class of service and features, and to collect call details by specific user within the subscriber's organization. Most carriers offer both verified and nonverified authorization codes. The switch accepts a verified authorization code only if it is listed in its database. This feature prevents unauthorized individuals from placing long-distance calls without a validated code. Most switches offer as many as seven digits for each authorization code, which makes it difficult to select a valid code by chance. An accounting code may also be required in addition to the authorization code. The system may also accept and verify bankcards and credit cards offered by the LECs and IXCs. The authorization is handled by a datalink to an external center.

A nonverified authorization code relies on the subscriber to assign code numbers; the switch will accept any valid digits. Subscribers often use nonverified codes for such purposes as distributing costs among departments and to clients, but unlike verified codes, unverified codes do not offer security against unauthorized use. With either type of code the switch sends a tone to prompt the caller to enter the code. After an interval of unsuccessful or no attempts to enter the code, the switch terminates the call.

Virtual Networks

Some IXCs offer virtual private networks to their customers, a service that requires support from the tandem switch. A virtual private voice network is one that operates as if it is composed of switched private lines, but which, in reality, is derived by shared use of the carrier's switched facilities. Voice VPNs are not to be confused with data VPNs, a concept that is explained in Chapter 31. The database for a virtual private network is contained in the IXC's SCP database.
A virtual private network handles calls in three manners:

- Dedicated access line to dedicated access line
- Dedicated access line to switched access line
- Switched access line to switched access line

A dedicated-to-dedicated call bypasses the LEC and avoids access charges in both the originating and the terminating direction. The dedicated-to-switched call eliminates the access charge in the originating, but not the terminating direction. For switched-to-switched access line calls, the virtual private network handles calls like regular long-distance calls except for features and restrictions.

Virtual private networks also emulate many features of electronic tandem networks. For example, users can dial on-net stations with an abbreviated dialing plan. Virtual private networks offer a full restriction range such as blocking calls to overseas locations, selected area codes, central office codes, or even selected station numbers. If the virtual network is used in conjunction with account codes, calls can be restricted for certain station numbers. For example, a company could allow its accounting personnel to call the accounting department in another branch, but calls to all other numbers could be blocked.

Call Reorigination

This feature enables subscribers to place multiple calls through the network without redialing the carrier and reentering the authorization code. Manual call reorigination requires the caller to dial a digit such a * or # to place another call, after which dial tone returns to the user, who can continue to dial calls. Automatic call reorigination returns dial tone automatically if the caller remains off-hook after the called party hangs up.

Accounting Information

Tandem switches provide message accounting information to allocate communications costs to the users. This information is either provided in machine-readable format for separate processing or processed by the carrier's accounting office to assemble completed message detail. IXCs distinguish themselves by the type of billing information they provide, but billing is handled by an external accounting center. The switch's role is to provide the raw call details.

Network Management Control Center

Most tandem switching systems provide either a centralized or localized network management control center (NMCC) to administer service and performance on private and public networks. The center collects all network management information in a single location. Typical NMCC functions include

- Manual and automatic trunk testing using apparatus and techniques discussed above
- Statistical compilation and analysis to determine loads and service levels and to determine when circuit types and quantities should be changed
- System performance monitoring, including diagnostics and maintenance control of all system features and circuits
- Alarm surveillance to detect and diagnose troubles and determine status of switching and trunking equipment
- Performance and status logging to monitor and log system history, including records of trouble and out-of-service conditions

Call Progress Tones and Recorded Announcements

Every tandem switch must be capable of inserting recorded announcements or tones to prompt callers on system usage or to inform them when the call has reached an unexpected termination. Callers are familiar with the most common progress tones such as busy, fast busy (reorder), dial tone, ringing, and audible ringing.

ADVANCED INTELLIGENT NETWORK (AIN)

LECs have long been hampered by the inability to react quickly to customer demands for specialized services. Until AIN the switching services the LECs offered were restricted to those provided by the particular central office manufacturer. If an LEC uses central offices from more than one manufacturer, the service offerings may lack uniformity across the serving area. New service offerings require the manufacturer to upgrade the generic program, the scheduling of which may be unacceptable in this day of fast-moving competition.

AIN provides services based on customized logic located at centralized nodes in the network. Central offices use the SS7 network to communicate with the central nodes. The SSP in the end office communicates over SS7 links with the central database in the SCP. The SSP determines when a call requires an AIN service based on variables such as the digits dialed or the class of service. This process is called *triggering*. When the SSP is triggered, it sends a message to the SCP asking for instructions. The SCP responds with a message that tells the switch how to handle the call. The features can be provided via datalink from a central point to all compliant switches in the LEC's network. This process is the means by which LEC switches implement number portability.

IP INTEGRATION

The digital central office is undergoing a transformation into a combination digital switch and softswitch. The new architecture retains the conventional line and trunk ports, but adds line and trunk gateways to interface the VoIP world. The system so configured can therefore serve existing customers with no change, but can add VoIP when the demand arises. Conceptually such a system is a hybrid of the circuit-switching architecture discussed in this chapter and the softswitch that is discussed in Chapter 16.

The major advantage this configuration brings, aside from avoiding obsolescence, is the ability to serve a wide area from one central office. IP modules can be placed in large customer concentrations to provide Centrex service or to relieve outside plant congestion. Neighborhood centers can be served with a choice of conventional or IP services with a remote connected to the host central office over IP riding on a fiber backbone. Few existing central offices have been retrofitted to the new architecture, but it provides a logical way for LECs to preserve their existing investment while meeting competition from VoIP providers.

CHAPTER 15

The Integrated Services Digital Network

The PSTN has taken shape over more than a century of progressive technical advancements. Before the advent of reliable computers, PSTN features were constrained by the limits of relays and wired logic that were complex, inflexible, and expensive to maintain. The first departure from electromechanical switches was Western Electric's No. 1 ESS, the first of which was installed in 1965. That system had to meet a rigid set of requirements of interoperability with existing systems and the need to preserve investment in outside plant, station equipment, and interoffice trunking. The No. 1 ESS was crude by today's standards, but it launched a transition into the intelligent network.

Except for agreements on the use of electromagnetic spectrum, international standards were weak and ineffective in the United States in the mid-20th century. Network transmission and data protocol standards were, to a large degree, dictated by AT&T and IBM, respectively, resulting from their market power. Both companies worked with international bodies to devise mutual standards, but the results were often incompatible. The OSI model is derived from IBM's SNA, but SNA is a working network architecture, while OSI is not. AT&T, unwilling to wait for international bodies to agree on digital carrier and common channel signaling standards, moved forward with their own standards. SS7 has since replaced AT&T's common channel interoffice signaling, but the digital hierarchies remained incompatible until SONET/SDH standards brought them together at higher levels.

In this context, the Consultative Committee on International Telephone and Telegraph (CCITT), which has since evolved into ITU-T, began working on modernization of the PSTN in the 1970s, but the original ISDN guidelines were not produced until 1984 in CCITT recommendation I.120, and this was far from complete. Computers at the time were primitive and costly and 64 Kbps of bandwidth was more than enough for almost any application that was then foreseen. One possible exception was large file transfers, but at that time a 100 MB disk was considered huge. By the time the recommendations matured, data requirements had changed so dramatically that ISDN was largely irrelevant for data.

ISDN was developed to resolve what was then perceived to be deficiencies in the PSTN that hampered both voice and data transmission. One was the need for multiple interfaces depending on the type of service. Service delivery to customer premises was on copper cable pairs and different interfaces were needed for various types of service. ISDN was the only technology at the time that could offer end-to-end digital connectivity, which data needed.

In the vision of the time, ISDN would breach the transmission speed limits of the PSTN for connectivity to value-added networks, which were then accessed over low-speed modems. It would provide flexibility for developers to bring new products to market and it promised interoperability between products of different vendors so that any ISDN telephone could operate on any vendor's switch. For voice users, a separate signaling channel would give the users insights into the nature of incoming calls, even when the line was active. It would enable distinctive treatment of the call, depending on who was calling.

ISDN might have fulfilled its potential if it had reached the market earlier and had been priced attractively and promoted actively, but none of these happened in North America. The standards were slow to develop and when they did, products were not interoperable. Some IXCs used proprietary ISDN protocols to deliver services their customers were requesting. For example, calling number delivery was a desirable service to many call centers. Some IXCs delivered caller ID in-band and others used a proprietary ISDN. Wags joked that the acronym stood for "innovations subscribers don't need," or "I still don't know."

ISDN requires digital switching. Many of the LECs' No. 1 ESS switches had to be replaced to offer the service and the LECs did not perceive that the demand existed. For existing digital switches, some LECs were slow to invest in the upgrade to ISDN until they were certain that the demand existed. Many LECs charged such a premium for the service that customers had no incentive to subscribe except for dial-up video conferencing. Perhaps the crowning blow to basic-rate ISDN, which is aimed at residences and small businesses, is the availability of DSL and cable Internet access. In comparison to the speeds available with these services, ISDN is too slow and too costly.

Primary-rate ISDN is a viable service offering. Where it is priced competitively with analog or non-ISDN digital PBX trunks, PRI offers many advantages and is usually chosen over other alternatives. Trunks can be used interchangeably between incoming, outgoing, and DID, which increases trunk efficiency. The ability to place end-to-end digital calls anywhere in the world is far from assured, however, because of analog local loops.

Several key applications generate demand for ISDN. For business subscribers, ISDN is an effective medium for videoconferencing. ISDN lines are effective for backing up private networks such as frame relay. Where the LEC tariffs are reasonable, ISDN trunking offers key advantages for PBX owners. Another major trend is telecommuting. As more incentives develop for workers to spend at least part of the week working from home, ISDN can be an economical way of retaining a link to the office at reasonable speed, although here also it is losing the battle to IP.

ISDN TECHNOLOGY

ISDN lines come in two varieties, BRI and PRI. BRI consists of two 64 Kbps B (bearer) channels plus one 16 Kbps D (data) channel. In ISDN shorthand the interface is called 2B+D. The D channel carries signaling between the central office switch and terminating equipment, which could be a telephone set, personal computer, video conferencing system, router, or similar device. The B channels are used for so-called bearer services such as voice, data, or video. Through a process known as *bonding*, the two channels of a BRI can be tied together to provide 128 Kbps of bandwidth. Most LECs offer BRI as a line-side option from local switches, and most major PBXs support BRI line cards. A special station set is required for ISDN, or as shown in Figure 15-1, ordinary station equipment can be installed behind a terminal adapter (TA). The station equipment can be a telephone set, a card in a key telephone system, a card in a PC, or a direct interface into a Group 4 fax machine or a video codec. Unlike a POTS telephone, an ISDN instrument has intelligence that enables it to perform selected functions in concert with the central office.

FIGURE 15-1

ISDN Architecture

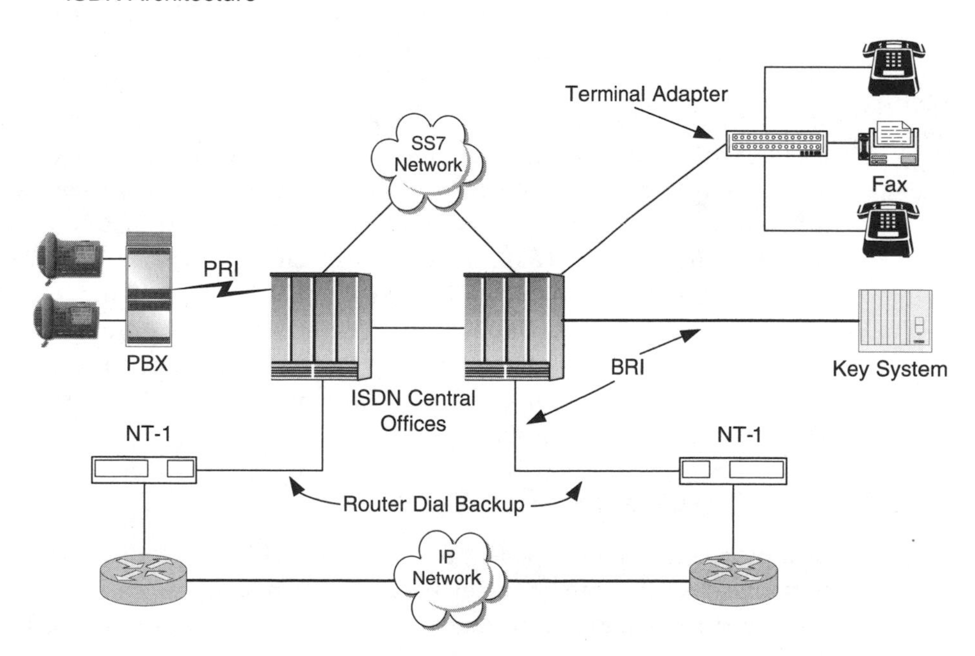

A PRI interface is a 24/32 channel T1/E1 circuit with one or two channels reserved for signaling. A PRI can connect to any device designed to be PRI compatible, including PBXs, routers, T1/E1 multiplexers, central office switches, tandem switches, or computers. The equipment can select 64 Kbps channels singly, in pairs, or with some systems, in an N X 64 configuration. N X 64 means contiguous bandwidth equal to N quantity of 64 Kbps channels. For example, 384 Kbps is a popular bandwidth for conference-quality video, and is defined as an HO channel. The LECs provide dial tone over PRIs, and the major IXCs provide direct customer access over PRIs. The separate data channel allows communication between the IXC's switch and the PBX so channels can be switched at will to outgoing, 800, 900, video, data, or any other such service that can be delivered over T1/E1.

The objectives of ISDN are

- To provide end-to-end digital connectivity
- To gain the economies of digital transmission, switching, and signaling
- To provide users with direct control over their telecommunications services
- To provide a universal network interface for voice and data

ISDN ARCHITECTURE

ISDN network architecture is based on ITU-T standards, with standards in the United States largely driven by Telcordia. ISDN standards are formulated on the first three layers of the OSI Model. The standards specify how information is encoded and how supplementary services such as calling features are provided.

In the United States, the two major switch manufacturers, Nortel and Lucent Technologies, developed their own ISDN protocols initially. This first edition was known as NI-1, which defines the most commonly needed set of features for regular telephone service plus several Centrex features. The main objectives of NI-1 are terminal portability and switch interoperability. Terminal portability enables users to move to a new location and have their telephone work without concern about the type of central office system. Switch interoperability enables switches of different manufacture to communicate over SS-7. Unfortunately, these objectives were not all realized because the switches were not completely compatible.

A subsequent version, NI-2, expands on NI-1 by defining the PRI standard and expanding BRI capabilities. In NI-1, some BRI features were proprietary. In NI-2, they are standardized. Features such as D-channel backup and *nonfacility associated signaling* are introduced. NFAS permits multiple PRIs to share D channels. The current version of ISDN is ISDN 2000.

BRI Network Terminations

BRI intends to simplify the POTS network interfaces, but it introduces a language and interfaces that can be somewhat bewildering. Figure 15-2 illustrates the interfaces and the equipment designations. In most of the world the LEC provides the interface termination from the local central office, but in the United States

FIGURE 15-2

BRI Interfaces

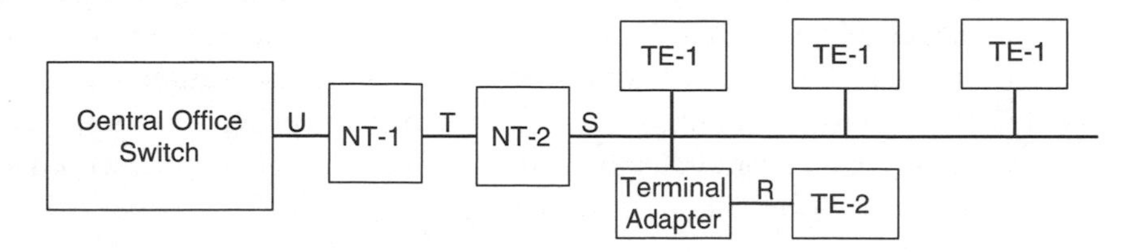

the customer must furnish a device known as the network termination 1 (NT1). The NT1 communicates with the central office over a two-wire U interface and provides a four-wire S/T interface toward the customer. The S/T interface provides separate transmit and receive paths for attached devices.

Many devices have the NT1 built in. The NT1 handles physical layer functions, which include these:

- Separation of the 144 Kbps line signal into its constituent B and D channels
- Termination and conversion between two-wire and four-wire interfaces
- Monitoring of performance and maintenance functions
- Provision of loop timing

An NT-2 is an optional terminator connecting between the NT1 and ISDN-compatible devices. A terminator that includes both NT1 and NT-2 functions is known as an NT12. The NT-2 allows up to eight devices to share a passive bus. Only two of the devices can be active at one time, but any of them can be addressed through the NT-2. The D channel can be used for data communications from other devices on the bus. The protocol provides a method for allowing multiple devices to contend for the signaling channel. The NT-2, which may be built into a PBX, multiplexer, local area network, or terminal controller, performs the datalink and network layer functions.

Terminal equipment that is BRI compatible can plug directly into an NT1 interface or across a shared bus. In ISDN terminology compatible equipment is known as TE-1. Equipment that is not BRI compatible is known as TE-2 equipment, and must plug into a TA. TAs are required for non-ISDN voice terminals and equipment with non-ISDN interfaces.

ISDN standards define five points of demarcation:

- R interface is a link between non-ISDN equipment and an ISDN TA.
- S interface connects ISDN terminals to NT-2 and NT12 devices.
- T interface connects NT-2 and NT1 devices.
- U interface connects NT1 and NT12 devices to the public network.
- V interface, located in the ISDN node, separates the line termination equipment from the exchange termination equipment.

The ISDN access line replaces the local loop of pre-ISDN services. It is a single twisted-pair metallic line with a maximum length of 6500 m. An NT1 and the ISDN switch communicate over a full-duplex connection by using 2B1Q (2 binary, 1 quaternary) signal with echo cancellation. The transmitting power of the four-wire input to the NT1 splits between the line and an equalizing network in the NT1. An electronic filter determines whether a line signal is original data or an echo caused by a mismatch between the line and the network. Echoes are canceled out so only the original signal remains.

Service Profile Identifiers (SPIDs)

SPIDs are an optional feature that identifies the services and features the central office switch provides. SPIDs are configured at device initialization. The SPID format is usually the 10-digit phone number of the ISDN line, plus a four-digit number that identifies features on the line. The SPID varies with LEC, central office equipment type, and generic features the switch supports. The SPID can be the most confusing aspect of ordering a BRI line although some devices are self-configuring.

PRI Network Terminations

ISDN PRI is a standardized architecture for the interface between CPE and the PSTN. The PRI protocol is in three layers:

- Layer 1 (DS-1): Electrical, synchronization, and framing
- Layer 2 (LAPD): Packetization, error detection, and flow control
- Layer 3 (Q.931): Call control messages

In North America a PRI is 23B+D, while in the European system it is 30B+2D. B channels can be bonded to provide H channels, which provide higher bandwidth. The H channel standards are

- H0 = 384 Kbps (6 B channels)
- H10 = 1472 Kbps (23 B channels)
- H11 = 1536 Kbps (24 B channels)
- H12 = 1920 Kbps (30 B channels)—International (E1) only

PRI offers features that are valuable to many customers. Depending on the carrier's service offering, any of the CLASS features discussed in Chapter 12 can be provided. PRI service offers the following advantages compared to using analog trunks:

- Hardware costs are reduced in the PBX. In most PBXs, a PRI card costs about the same as an eight-port analog trunk card, and uses one-third the number of slots.
- ISDN trunks with their call-by-call service selection provide listed directory number, outgoing, DID, and other services without the need for

special trunk groups. Depending on the size of the system, trunk requirements can be reduced substantially.

- Transmission performance is enhanced. Digital trunks can be operated with no loss and imperceptible noise. By contrast, the loss and noise of analog trunks increase with the distance from the central office.
- ISDN trunks can be provided over self-healing networks, reducing vulnerability to outage.
- Trunks can be added up to the capacity of a PRI with no wiring work or hardware additions on the customer's premises.

The following is a brief description of some of the more popular features:

- *Call-by-call service selection*: Without call by call, a CPE switch must be equipped with multiple specialized trunk groups. For example, separate groups might be required for DID, two-way CO trunks, outgoing local trunks, foreign exchange trunks, toll-free lines, and tie trunks to a distant PBX. With call by call, any B channel can be designated for any purpose. The CPE switch and the central office negotiate the call and select the channel or channels to assign it to.
- *D-channel backup*: A B channel can be selected as backup to the D channel to prevent service failure in case the signaling channel fails.
- *Nonfacility associated signaling*: NFAS enables multiple PRIs to share one or more D channels.
- *Release link trunk*: This feature is similar to antitromboning. If a user on a CPE switch receives a call from the central office and forwards or transfers it to another node on the PSTN, the two switches recognize that connection to the CPE switch is no longer required and drops the trunks.
- *Trunk antitromboning*: Tromboning refers to a situation in which two B channels associated with the same D channel are being used in parallel. For example, if a user in a CPE switch receives an incoming call, and conferences on a third party on the PSTN, two B channels are tied up. If the two switches are equipped for antitromboning, they release one of the B channels.
- *Two B-channel transfer*: This feature enables a user to transfer two B channels to another termination.
- *Wideband dial-up*: This feature permits dialing a connection using multiple B channels.

ISDN APPLICATION ISSUES

The primary issues surrounding ISDN are cost and availability. Where the service is both available and cost effective, managers will find little reason not to use it.

Cost must be examined closely before reaching a conclusion. Even if the cost is greater than an equivalent number of analog lines, the two-way nature of ISDN means that fewer trunks will be needed for the same grade of service.

Anyone who is selecting a new PBX, or even a key system, should consider ISDN compatibility. Applications are emerging that require the digital connectivity of ISDN, and as the service becomes common, applications such as these will develop:

- Video conferencing, primarily desktop devices
- High-speed access to remote databases such as the various on-line services and Internet
- High-speed remote access to LANs
- High-speed image applications such as Group 4 facsimile
- Second line in the home applications driven by the development of telecommuting
- Dial backup for routers to restore leased line failures

The same list of applications can also apply to Internet, so to some degree IP and ISDN compete to provide the same services.

Many LECs offer attractive ISDN tariff rates where facilities permit. Note, however, that in many cases only one switching system in a wire center is equipped for ISDN, which means that a line cannot be moved between switches without a number change. Some LECs permit customers to avoid the number change by using number portability. The major IXCs provide PRI as an alternative to the T1/E1 connections over which they offer bulk outgoing and incoming services. With call-by-call service selection the channels can be allotted to any service, which increases utilization and reduces costs.

CHAPTER 16

Softswitches

Local service providers find it difficult to differentiate their offerings because the features in most class 5 switches are so similar. There are several reasons for this. One is the standardization that the industry has always practiced. For decades, Bell Laboratories and Western Electric were the industry vanguards and the other manufacturers tended to follow their lead. Another reason was the transition from electromechanical to electronic switching. It is easy to implement new features in electronic switches, but introducing them too rapidly would hasten the demise of electromechanical systems that still had years of service life.

Class 5 switches have always been inflexible and features have evolved slowly. Feature activation is deeply ingrained in the generic programs of the circuit switches of the world. New features reach the market only after arduous testing. By design, the architectures are closed and do not lend themselves readily to the changing milieu of VoIP. This has led to a new switching architecture known as the softswitch. The softswitch is to the circuit switch as the client-server computer model is to the mainframe; decentralized, open, and offering varied features, not only from the manufacturer, but also through third-party developers.

Softswitches generated a lot of enthusiasm in the late 1990s as the CLEC boom took root. TDM switches are expensive and do not scale well in lower line sizes. They also lack multimedia support and CLECs' business plans usually assume delivery of a service package distinguished from ILECs' services. CLECs can collocate with the ILEC wire centers, but it is difficult to reach the subscriber density needed to justify the cost of a digital switch, even with the use of remotes. Under the telecommunications act, CLECs could use ILEC switching as an unbundled network element, but that means the ILEC switch controls the features, and even that option looks bleak in the face of FCC actions. The only way a CLEC can differentiate its service on anything but cost is to own the switch. The softswitch fits the CLEC business model because its front-end elements, the media gateways (MGs) can be collocated in a small space with the ILEC central office and connected throughout the metropolitan area to a central controller.

The technology collapse at the turn of the century arrested softswitch development. Bankruptcy rates in the CLEC industry were high. Manufacturers saw their markets collapse as their customers disappeared, leaving behind an abundance of used equipment that was sold at bargain prices. The market is beginning to revive as this is written and it extends well beyond the CLEC industry. The IXCs are converting the core of the PSTN to VoIP and ILECs are making trial forays into softswitching at the edge. Among other expectations the ILECs have is rejuvenation of their Centrex service offerings, which have suffered from lack of feature flexibility. With intelligent endpoints, a softswitch can provide a common user interface that makes it easier for users to activate features that are difficult to control from a 12-key pad. There is an old joke about how many telephone experts it takes to set up a conference call: something that everyone stumbles over.

SOFTSWITCH ELEMENTS

Service providers have varied motivations for installing a softswitch. Selling points are reduced capital costs, reduced operating expense, opportunities to increase revenue, and the convergence of voice and data onto a single packet network that carries both the signaling and the media streams. The softswitch and presence-related protocols offer the ability to integrate telephony with other modes of communication such as instant messaging.

To some degree the softswitch model is based on providing improved subscriber services, but more to the point is a fundamental change in local service architecture. The local loop is the choke point in the network and its viability is being challenged by cable, wireless, and the telephone companies themselves. Service providers can use a softswitch to deliver services without taking a big leap into the unknown with an expensive TDM switch. Ongoing costs are also lower. A softswitch requires much less floor space and consumes less power and air conditioning. Bear in mind, however, that the softswitch rarely furnishes power to the user, so telephone service will not survive power outages unless the service is deliberately designed for survivability. Some savings result from bypassing the ILECs' access network and reducing toll costs, but as long-distance costs drop, the saving from bypass is not as great as it was several years ago.

The main motivation for the softswitch is to support multimedia over IP, enabling carriers to create new services. Most softswitches provide open APIs and are designed to interoperate with application servers from different manufacturers. The industry expects that third-party developers, in conjunction with new protocols such as SIP, will invigorate the market with a flurry of innovation. The value proposition for VoIP is revenue-generating services that enhance productivity through improved scheduling and collaboration. The next sections briefly discuss some of the potential features that are difficult or impossible with conventional switches.

Presence Services

As we discussed briefly in Chapter 14, much of the expected innovation falls under the category of presence, a concept that is familiar to instant messaging users. With new services that an IP network facilitates, presence includes registration and monitoring of the location and busy–idle status of any endpoint including VoIP clients, wireless phones, POTS phones, IP softphones, and multimedia clients. The presence engine can be linked to the online calendar to convey the callee's status and availability. The system can treat calls different if the callee is on vacation, traveling out of the office, at home sick, or in a meeting.

The presence engine provides services to both end users and to network services. Users could communicate their presence to a level as granular as they chose. For example, a user could register with the presence engine as not only out of the office, but in a hotel room and connected to the Internet. A variety of network elements such as office phone, cell phone, e-mail, online calendar, and laptop PC could be programmed to provide presence information to a server without the need for the user to take any action. The presence engine facilitates applications such as find-me–follow-me service. For example, a call to a user's office might be forwarded to a cell phone if the cell phone is turned on and idle and the availability database indicates that the user is not in a meeting. The user can update the status dynamically by actions such as turning the cell phone on or off or posting a new appointment to the calendar.

Enhanced Collaboration Services

Today, multimedia conferencing requires a variety of devices and the question arises as what capabilities the endpoints have and how they can be linked. Some endpoints might have video capability and be available on IP, others only over BRI, and others audio only. *Service blending* allows participants to request the network to establish a conference call based on the identities and capabilities of the participants. For example, some users might not have video capabilities but everyone else does. If a conferee is on another session, the network could announce that a session is waiting and extend the announcement across multiple media, not just the telephone.

Setting up a conference today usually requires many phone calls or e-mails, particularly when the conferees are in different organizations and their calendars are on a firewall-protected network. Meeting times and locations must be determined, people must advise whether they will participate in person, by telephone, or by video, and they may need to communicate a fax number for receiving documents. The conference chair must distribute time schedule, agenda, call-in numbers, and passwords to off-site participants. When the conference starts, call-in participants must identify themselves with a session number and password, and in many cases the chair must challenge new entrants to identify themselves as they join the conference. When someone drops out, the participants may hear a tone, but have no idea who left. The process consumes valuable time. With enhanced conferencing, the application can consult calendars and meeting room schedules and distribute the conference announcement over the network. Invitees can accept, decline, or suggest an alternate time by whatever medium is convenient to them.

Some dial-in TDM conference bridges facilitate part of the process, such as identifying participants as they connect to or leave a conference, but the process is primitive compared to the possibilities. In today's audio conferences the conferees are asked to introduce themselves, but people who do not regularly work together usually have a difficult time recognizing voices. With enhancements, the network could authenticate the conferees by a variety of techniques, challenging people for a password only if they have not been identified by network authentication. In a multimedia conference, the network could identify who is speaking by displaying his or her name on participants' computer screens.

Communications could be enhanced by distributing media such as text, pictures, and video clips while the conference is in session without the need to ask everyone how best to receive them. For spontaneous conferences, the system could find the participants, specify their method of access to contents and services, display presence information for the other participants, and link information such as charts, video, audio clips, and so forth. If some invitees are unable to attend, they may send personal avatars to record sessions or to present prerecorded comments.

Many of these features are available on a TDM network today, but using them requires a manual effort and interrupts the conference. For example, if the conference chair needs to send a document in today's conferences, the off-site conferees must all be questioned whether they can receive it by e-mail or fax, what addresses or numbers to send it to, and other questions that waste time. These are all issues that the presence engine can determine and handle without intervention.

Personal Assistant Services

Personal productivity can be enhanced in many ways with the advanced services that multimedia over IP facilitates. Intelligent agents in the presence engine can learn the users' preferences by observing their behavior. They can learn who the user communicates with and give these priority treatment. A simple example is the way many e-mail packages work today. They determine who the user communicates with frequently and, after two or three keystrokes in the address box, bring up a short list of potential addressees. The ability to participate in office activity becomes independent of location. For example, employees might want to watch a corporate announcement on video from the convenience of their home television, office PC, or from a wireless handset while traveling. An intelligent agent could deliver those options.

An intelligent agent works for individual users across all communication media including e-mail, voice mail, instant messaging, and desk and cellular telephones. The agent could prioritize incoming calls on the basis of the caller's identity, which it receives from the network. For example, e-mails might be routed to a high-priority folder or a particular caller might receive priority treatment in the call-waiting indication. High-priority sessions could be rerouted automatically, and lower priority sessions sent to an alternate destination such as voice mail. Multiple voice mail greetings could be triggered based on the system's knowledge of the user's presence. Today, call pickup works only if the members of a pickup group can hear the phone ring. With presence-enabled call pickup, the members of a pickup group could be alerted to answer the phone without even being on the same system.

These services require many applications that do not exist today. To be effective, the services need close interaction between private networks such as an IP PBX, private IP networks, and public networks including the PSTN and the Internet. With a converged network as the platform, service providers can offer a bundled set of services that share a common user interface and seamless internetworking.

IP Centrex

The ILECs find softswitches intriguing because of the possibility that they may rejuvenate the Centrex market. Centrex has always suffered from drawbacks that caused many prospective customers to shy away from it. Centrex is constrained by the characteristics of the local loop, which prevents the use of feature telephones on customer premises. One option is to use analog telephones, but these require users to dial codes to access features. To gain pushbutton access to features, users often put CPE switches; PBXs and key systems, across the Centrex line. In effect, the subscriber pays twice for the features, once in the Centrex and once in the CPE switch. The LECs' answer is ISDN Centrex, but as explained in Chapter 15, ISDN has never fulfilled its potential in North America.

Moreover, every Centrex station requires a separate cable pair, which requires a considerable amount of outside plant compared with a PBX, which serves multiple customers with T1/E1 line access. If the subscriber has enough lines to justify a remote, the LEC can install one on the customer's premises, but the issue of the analog telephone or ISDN phones remains. With IP Centrex most of these objections are solved. Large locations can be served with a MG and smaller locations with IP telephones. Besides reducing the number of cable pairs needed, a major benefit of IP Centrex is the availability of IP feature phones. IP telephones support features that equal or exceed those of proprietary PBX phones and with advanced signaling protocols, the features discussed above can be provided.

SOFTSWITCH ARCHITECTURES

The architecture of a softswitch is flexible and varies with the manufacturer. Figure 16-1 shows a representative softswitch architecture. MGs interface

FIGURE 16-1
Softswitch Architecture

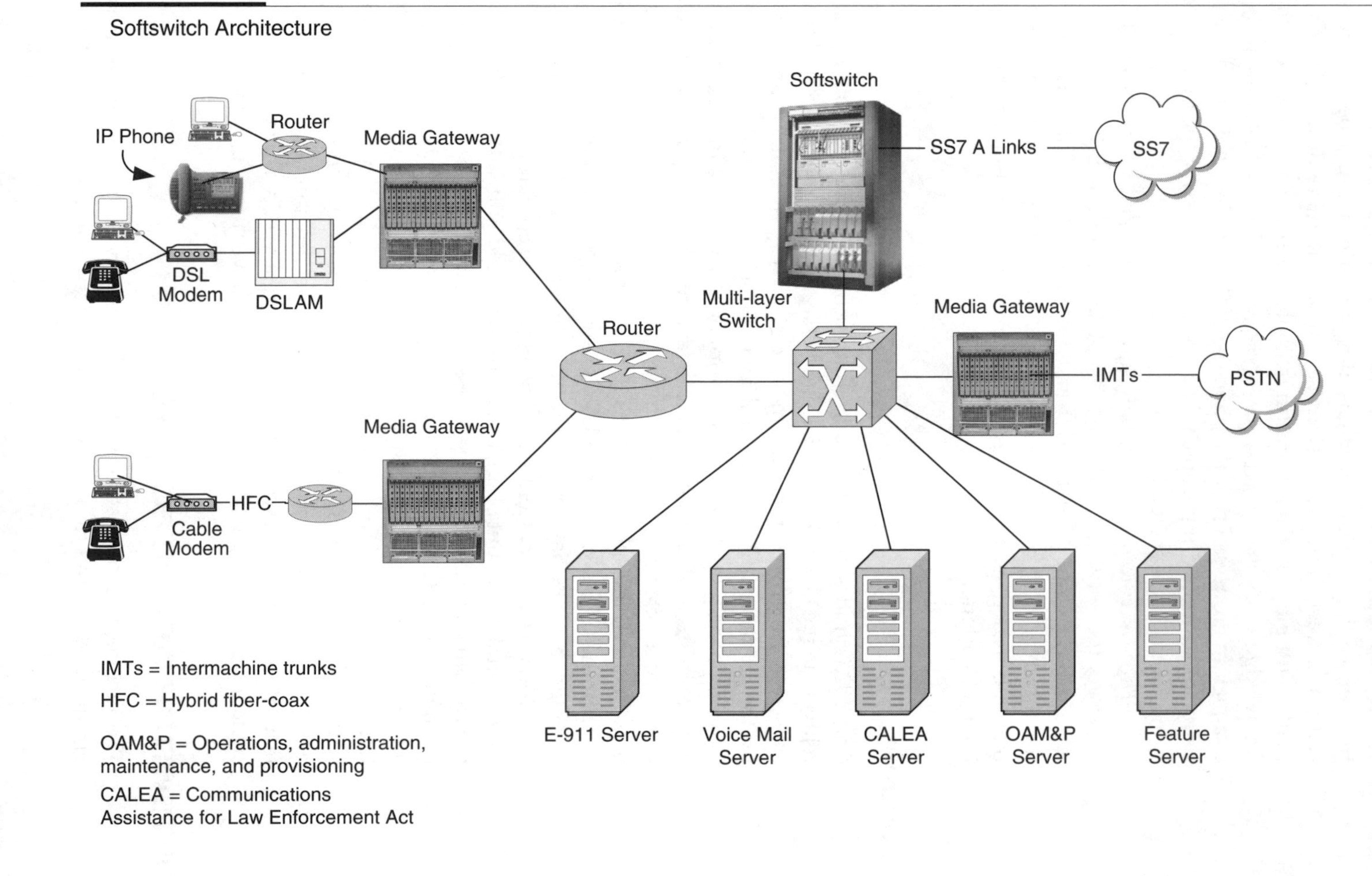

the complex to line-side and trunk-side interfaces. The line side connects to a variety of devices through conventional cable pairs, DSL, or cable. The switch is intended to be fully distributed. The processor resides in a server that can be located anywhere that has connectivity with sufficient bandwidth to the MGs. A circuit switch, by contrast, can distribute its line units close to the customer in a remote switching unit and can support digital loop carriers that offer a similar configuration, but the similarity ends at the line and trunk interfaces. The linkage between a circuit switch and its satellites is TDM lines and trunks or analog lines. The softswitch may connect to circuit switches with TDM trunks, but to other softswitches it employs IP over a circuit with appropriate bandwidth.

Media Gateway

In a softswitch network stations connect to an MG. The MG is equivalent to the line-circuit frame in a circuit-switched network. It handles supervision, supplies ringing and dial tone, plays tones, opens ports, and establishes and releases connections under the control of a media gateway controller (MGC). All of the call-processing elements of a circuit switch are required in the softswitch, although they may reside in separate servers. The MG has minimal intelligence except for the possibility that it may have standalone capability. The MG can be configured to play audio or video messages and perform enhanced services such as interactive voice response and multimedia conferencing. MG capabilities can include processing audio, video, and T.120 data streams alone or in any combination. Standalone capability is important in case of link failure to the MGC.

An MG resides at the edge of the network. It can be placed on customer premises if the number of lines is sufficient to justify the cost; otherwise, it can be collocated with the MGC or placed in an environmentally controlled enclosure in a neighborhood center. In that respect, it is no different than an RSU in a TDM switch. The MG, however, has capability for media conversion: for example, IP phones can talk to POTS or ISDN phones through the MG. The MG contains a digital signal processor to provide echo suppression. Line-to-line POTS calls are not converted to IP, but calls outside the MG are converted to IP and remain IP until they reach the destination gateway. It is important that once a voice signal has been packetized that it not be converted to analog and then packetized again because of the additive delay and distortion. MGs can be equipped with a variety of interfaces such as POTS, IP, ISDN, DLC, and Ethernet.

Media Gateway Controller

A benefit of VoIP technology is that networks can be built with either a centralized or a distributed architecture, allowing service providers to build networks with both simplified management and endpoint innovation. Centralized architectures use an MGC, also known as a call agent or softswitch, to control MGs.

The MGC has the same basic call-handling functions as a TDM switch except it is designed to operate in a client-server mode with the MG as the client. The main difference between the MGC and a TDM switch is that the MGC is built with APIs to enable developers to write applications that are not feasible with a conventional switch. Circuit switches have AIN capability, which provides some ability to write applications, but the amount of control is limited. The addition of new features is slow because feature operation is bound to basic call processing, which is in turn tied to the hardware. Extensive testing of any new feature is required to ensure that basic call processing has not been impaired. The softswitch has the advantage of being designed from the ground up with APIs, but the same restrictions on basic call processing still apply. The manufacturer must control the generic program, or if it is open, revisions to the code must be permitted only after extensive testing.

MEDIA GATEWAY CONTROL PROTOCOLS

A major selling point of the softswitch is its use of standard protocols, which provide vendor independence and interoperability. At the time this book is written, the protocols are in transition, so much of the softswitch operation is proprietary. Protocols such as SIP and Megaco/H.248 are still in the draft or transitional stage. Many vendors purport to support them, but until the standards are accepted, interoperability is not assured. Nevertheless, the basic softswitch structure is flexible enough that it can be readily adapted when standards are complete.

The H.323 family of standards is an ITU-T recommendation for handling multimedia over a packet network. H.323 is the default protocol that most VoIP and video applications use today. SIP is often erroneously viewed as a replacement for H.323, but it is not. H.323 is a set of standards that contain all of the elements of a complete multimedia network. The protocols not only handle signaling, but also media conversion and transport. SIP, by contrast, calls on other IETF protocols such as URLs for addressing, DNS for service location, and TRIP for call routing to handle functions that are contained within the H.323 protocol family.

Another major point of departure between SIP and H.323 is in the degree of centralization. Centralized systems are easier to manage, provision, and control, but to some degree centralization suppresses feature development. Intelligence in a decentralized network is distributed between the controller and the endpoints. H.323 and SIP operate between peer clients while MGCP and Megaco operate in a master–slave environment. SIP, in keeping with the orientation of the IETF, is a decentralized protocol and assumes that the endpoints have intelligence. In a SIP phone the call-control signaling runs directly on the end device. H.323 is a centralized protocol, which fits the orientation of ITU-T and in many ways aligns with a circuit-switched network. The slave devices are gateways, IADs, and phones. The Megaco control model aligns with traditional telephony, while SIP's origins and therefore its structure are closely allied to the open and decentralized Internet.

Intelligence consists of call state, calling features, routing, provisioning, billing, and other aspects of call handling. Call-control devices are called gatekeepers in H.323 and proxy or redirect servers in SIP.

Centralization has both advantages and disadvantages. On the positive side, it is easier to implement features in a centralized environment because they run in a server and not in a multitude of clients. Distributed endpoints such as those SIP uses can deliver a wider variety of services because the endpoints are intelligent and can interact without a central controller. As the industry matures, it is most likely that devices will support all of the protocol families rather than concentrating on one alternative.

This section discusses the two principal standards for controlling multimedia sessions over data networks. We covered SIP in the signaling chapter because it is primarily a signaling protocol.

The H.323 Protocol Family

The primary protocol for connecting between packet and voice networks is H.323, an ITU standard for multimedia communications over a packet network. The original protocol was introduced in 1996 to communicate over LANs that had no QoS capability. It is designed to support interoperability between devices of different manufacture. For the most part, the objective is achieved; however, the protocol has many options and plug-and-play interoperability is not assured. The protocols are independent of the type of network and the platform they operate on. They are intended to provide multipoint and multicast services and a rudimentary form of bandwidth management. Table 16-1 lists the protocols that are included within the recommendation.

Figure 16-2 shows the software architecture of an H.323 network. The left side of the stack is concerned with audio and video input and output, while the right side is concerned with signaling and control. The H.323 protocols run on top of the network layer interface. H.323 is independent of the type of packet network, although the most frequent network is IP using the TCP for portions of circuit setup and signaling where reliable communications are required. The unreliable UDP is normally used for transporting voice and video packets after the session is established. Support for G.711 voice, which is the conventional 64 Kbps PCM protocol, is required by the standard. Compression protocols are optional, but are used by most VoIP applications to gain greater bandwidth efficiency.

The complete network consists of four elements: terminals, gateways, gatekeeper, and multipoint control unit (MCU). The only required elements are the terminals, which can be any devices capable of supporting the protocol stack. IP telephones, videophones, desktop computers, and similar devices can function as H.323 terminals and can carry on a session using URLs for addresses. Voice support is mandatory for terminals; video and data support are optional.

TABLE 16-1

The H.323 Protocol Family

Recommendation	Title
Audio codecs	
G.711	Pulse code modulation audio codec 56/64 Kbps
G.722	7 kHz audio coding at 48/56/64 Kbps
G.723.1	Dual rate speech coders for 5.3 and 6.3 kb/s
G.728	Speech codec for 16 kb/s using code excited linear prediction
G.729	Speech codec for 8/13 kb/s using conjugate-structure algebraic-code-excite linear-prediction
Video codecs	
H.261	Video codes for P × 64 Kbps
H.263	Video coding for <64 Kbps communication
Call control	
H.225	Call control protocol
H.235	Security protocol
H.245	Media control protocol
Real-time transport	
RTP/RTCP	Transport and delivery of real-time audio and video
Data conferencing	
T.120	Real time data conferencing protocol
Supplementary services	
H.540.1	Generic functional protocol for the support of supplementary services in H.323
H.450.2 and 450.3	Call transfer and call diversion supplementary services for H.323

FIGURE 16-2

H.323 Protocol Stack

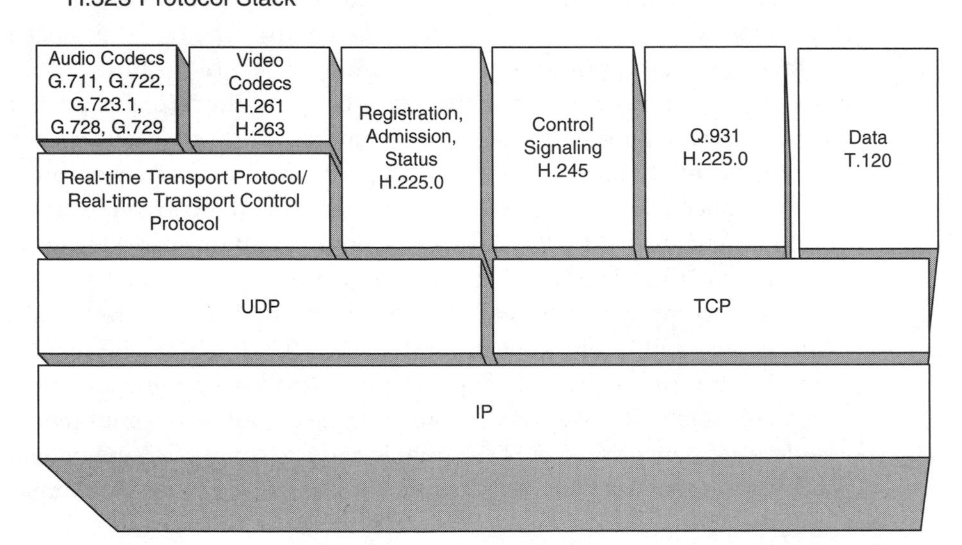

Gateway

To communicate outside the bounds of an IP network, a gateway is required. The gateway is composed of two elements: an MG and an MGC. For calls from the PSTN to a packet network, the incoming signal may be either digital or analog. If the signal is analog, the MG digitizes it. The MG compresses and packetizes a voice signal into packets containing about 20 ms of voice. From a packet network to the PSTN, the gateway collects packets, buffers them to reduce jitter, decompresses the voice signals, and connects them to voice circuits as a stream of PCM samples.

The MG is a protocol converter that talks H.323 toward the IP network and the protocol of the other network, which is likely to be PSTN. It performs the signaling conversion between the packet protocol H.225 and external protocols such as SS7 and Q.931. The MG also performs the media conversions such as converting between different compression protocols. Figure 16-2 shows the compression standards that are available. The gateway is not called into operation on internal IP network calls. Its only function is to convert protocols and media to non-H.323 networks. Gateways are a potential choke point in the network, so they must be sized to prevent blockage. A VoIP network could have a gateway with as few as one voice port or it could contain several gateways, each with multiple ports. The ports can be analog, connecting directly to an analog telephone as, for example, an off-premise extension, or the ports can be T1/E1 for connection to a PBX or tandem switch.

Gatekeeper

The second optional feature is the gatekeeper, which is a centralized controller. The gatekeeper can be set up to act as a virtual switch with all of the resulting controls of a circuit switch. The network can function without the gatekeeper, but if one is present, it has certain mandatory functions. The first is address translation between IP addresses and E.164 numbers. The second function, admission control, determines whether devices are authorized to connect. The function is similar to station restrictions in a circuit-switched PBX. The third function is bandwidth control. The administrator can set thresholds for the number of concurrent sessions. When a threshold is reached, the gatekeeper denies admission to additional sessions. These functions fall under the heading of *zone management*. A zone in an H.323 network is similar to a domain in an IP network. The zone consists of all the endpoints; terminals and gateways, for which the gatekeeper has registration authority.

Admission control is not the same thing as congestion control. The gatekeeper does not have the capability of looking at QoS indicators such as lost packets and delay and determining from that whether quality has fallen below acceptable levels. Many VoIP networks function with a form of congestion control, but it is proprietary and not a part of the H.323 specification. Also, bandwidth management should not be confused with true QoS, which is necessary when packets must span multiple networks.

In addition to the mandatory functions, a gatekeeper can also include several optional features. One is signal message routing. The gatekeeper can permit the endpoints to signal independently, but the gatekeeper can be configured to control the routing of signaling messages. Lacking this capability, the endpoint knows only the route to the gatekeeper and a single route to other destinations. The gatekeeper can handle call authorization, imposing additional restrictions such as time of day or restrictions on features. In addition to bandwidth control, it can support bandwidth management. For example, an endpoint could request additional bandwidth and the gatekeeper determines whether to grant the request or not. The gatekeeper can also handle management functions such as accounting, billing, charging, handling the directory, managing SNMP functions, and collecting usage information. The gatekeeper must distinguish between traffic types. For example, a gatekeeper might let data packets flow between the LAN segment and the WAN, but treat voice packets as a chargeable call.

If the gatekeeper is present, the network cannot operate without it. Therefore, multiple gatekeepers with load-balancing capability may be required and the servers in which they reside should have redundant processors and hot-swappable cards. If the gateway in an IP network fails, the effect is similar to a trunk group failure in a circuit-switched network. Internal communications are still possible, but the network is cut off from the outside world.

Multipoint Control Unit

The MCU is a complex part of H.323 and is the main reason the protocol is bulky. It is brought into play whenever the session consists of more than two endpoints. The MCU negotiates session variables such as what codecs will be used. If format conversion is required, such as converting NTSC video to one of the European formats, the MCU does the conversion. It negotiates the session bandwidth with the gatekeeper and takes care of ancillary functions such as audio mixing, video switching, and data distribution. For the latter, the H.323 protocol suite includes T.120, which is used for data transmission in a multimedia session.

The H.323 specification deals with functions, not hardware configuration. The gatekeeper and MCU may be built into the gateway or they can reside in separate servers.

H.225 Registration, Admission, and Status (RAS)

The RAS protocol carries admission control, bandwidth management, and status determination messages. It also carries messages that associate an endpoint's address with its signaling channel transport address. RAS is used only when a gatekeeper exists. Its functions include gatekeeper discovery, which endpoints use to identify their gatekeeper. When the gatekeeper exists, all endpoints within a zone must register so the gatekeeper knows the aliases and transport addresses of all the endpoints in its zone.

Calls are set up in H.323 by exchanging RAS messages. The H.225 protocol operates between endpoints and gatekeepers over a RAS signaling channel, which runs under UDP. Endpoints negotiate with the gatekeeper for admission to a zone using H.225. The gatekeeper either permits the stations to signal directly or requires them to communicate through the gatekeeper. Another function of the H.225 protocol is call signaling. Setup messages are carried over TCP/IP, either directly between endpoints or through the gatekeeper. Endpoints use H.245 control signaling to communicate their capabilities to support various media streams. H.245 negotiates bit rates and codec selection based on user preference and the codecs that are installed in the endpoint devices. Before the endpoints can communicate, they must negotiate compatibility through the H.245 protocol, after which the protocol opens logical channels that can be used for audio, video, and data communications.

The output of the codec is applied to the transport network through Real-Time Transport Protocol. RTP provides end-to-end delivery services for data with real-time characteristics, such as interactive audio and video. RTP works in conjunction with its counterpart from the signaling side Real-Time Transport Control Protocol (RTCP)—RTP carries the media and RTCP carries control and status information and monitors QoS. These two protocols handle such functions as timing, sequence numbering, time stamping, source identification, and identification of the type of payload as audio or video. RTP supports media transport to multiple destinations using multicast distribution if the underlying network supports it. RTP does not provide any QoS services and does not guarantee delivery or packet sequencing.

If video is supported on the network, video coding protocols H.261 and H.263 are both available but only H.261 is mandatory. These protocols support bandwidths in multiples of from 1 to 30 times 64 Kbps. Note that 30 is the full bearer channel bandwidth of E1, which supports near-broadcast quality video.

Resource availability indication (RAI) is a RAS message that the gateway sends to the gatekeeper to inform it of the gateway's ability to handle more calls. The gateway calculates whether it has the necessary bandwidth; the method of making the calculation is not part of the RAI protocol.

In the absence of a gatekeeper, endpoints may communicate directly, but if a gatekeeper is involved, the session must initiate through the gatekeeper. The originating endpoint sends a message to the gatekeeper to initiate a session to the called endpoint, which could be another station or a gateway. If the station were restricted from placing such a call, the gatekeeper would deny permission; otherwise, the gatekeeper checks the busy–idle status of the called endpoint. If it is available, gatekeeper notifies the calling endpoint to proceed. The calling endpoint sends a message to the called endpoint, which wakes up, sends a message, and receives an acknowledgment from the gatekeeper. Then the two endpoints begin a station-to-station handshake that results in establishing a session. When one of the parties hangs up, the session is torn down.

The process sounds involved, but it is essentially the way circuit-switched devices operate. The main difference is that stations in a circuit-switched network have no intelligence. They are slaves to the intelligence in the central processor. The endpoints in a VoIP network have varying degrees of intelligence, but most are capable of addressing other endpoints.

T.120 Real-Time Data Conferencing Protocol

T.120 is an ITU-T standard for real-time multipoint data communications. It is often used in conjunction with H.323 for mixing voice and video with data communications for such applications as desktop video and data conferencing, electronic whiteboard, and the like. It can be used for multipoint data in real time and it is designed to permit interoperability between multiple vendors. T.120 supports a wide range of transport options including the PSTN, ISDN, TCP/IP, and other packet switched networks. T.120 can connect multiple endpoints together to form a domain, the nodes of which can connect to an MCU for multipoint conferencing. The protocol provides for both reliable and unreliable data delivery as determined by the application.

The T.120 protocol provides for both network interoperability and application interoperability. The former provides the capability of setting up conferences, providing reliable transport, adding and dropping users, and other such functions the application needs. While the primary application is conferencing, many other applications are available with T.120. Network games and simulations and distance learning are a few of the applications that are possible by marrying T.120 to H.323.

H.323 Summary

Since its introduction in 1996, H.323 has undergone a continuous stream of version releases to add functions to support a full-scale VoIP network. The current version at this writing is version 5, but more work remains to be done. Version 2 added an alternate gatekeeper for service protection and enabled the gateway to relay its resource availability to the gatekeeper. Also, fast-connect messages with a limited number of steps were introduced in version 2. Subsequent releases added features such as call transfer, call pickup, and caller ID to emulate the most common features that PBXs and class 5 offices provide. Version 3 introduced address resolution and pricing exchange between administrative domains. Earlier releases had difficulty maintaining lip synchronization between voice and video, so version 4 allows for multiplexing both voice and video in the same media stream.

One drawback to earlier versions of H.323 was its inability to add new features without upgrading to a new release. Version 5 introduced generic extensibility framework (GEF), which allows new features to be added without changing the basic software. Most new features that came in with version 5 such as number portability, call prioritization, and QoS monitoring and reporting have been implemented using GEF.

Media Gateway Control (Megaco)

Megaco/H.248 Gateway Control Protocol is the joint effort of ITU-T and the IETF. Megaco/H.248 defines a centralized architecture for supporting multimedia applications, including VoIP. The protocol is an extension of MGCP, which was an earlier IETF control protocol. MGCP is used in several operational products, but it does not reflect true industry consensus. It is incomplete and less efficient than Megaco; for example, it offers limited support for networks other than the PSTN, whereas Megaco can be used on other networks such as ATM.

Megaco/H.248 is a single approach to controlling all gateway applications including PSTN trunking gateways, analog line and phone interfaces, Internet telephones, ATM interfaces, announcement servers, and similar devices. It is not tied to any peer-level call control, so it can be used with either SIP or H.323. MGCs communicate peer-to-peer with each other using H.323 or SIP and with MGs using Megaco/H.248. The signaling logic is located in MGCs and media logic is located in MGs. A single MGC can control multiple MGs, which can be optimized for their function. MGs have limited complexity and are therefore lower in cost. Feature changes are implemented in the controller, which makes them easier to implement than if they were distributed to the endpoints. Figure 16-3 shows the architecture of Megaco/H.248. The MGC layer implements call-control intelligence and features such as forward, conference, transfer, hold, etc. It implements peer level protocols to interact with other MGCs and PSTN signaling such as SS7.

FIGURE 16-3

Megaco/H.248 Gateway Control Architecture

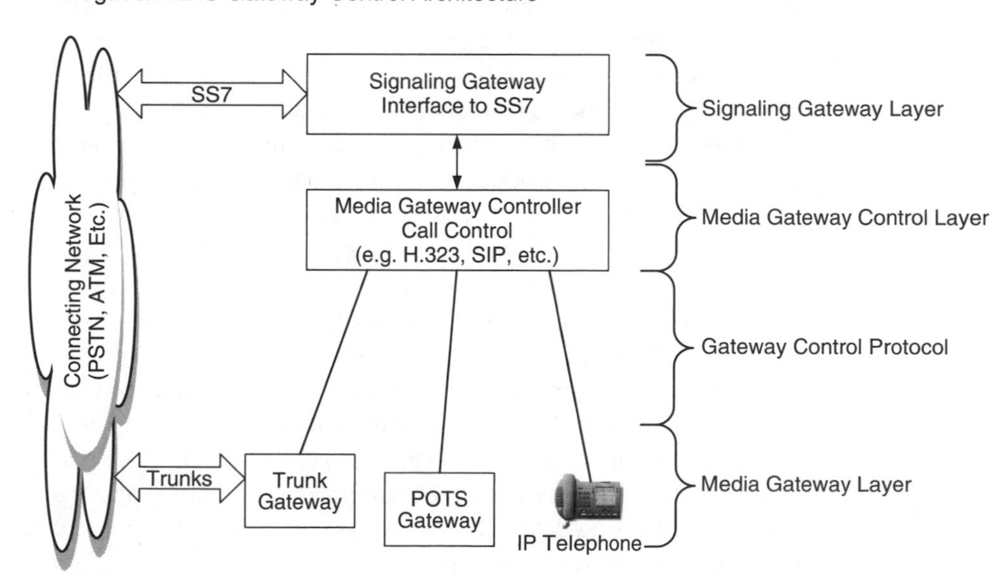

Megaco defines two basic elements: terminations and contexts. Terminations are of two types, physical and ephemeral. *Physical terminations* are ports such as analog or ISDN lines or IP connections to SIP phones that are provisioned on the gateway. *Ephemeral terminations* are setup in the call process and exist only for the duration of their use. For example, a flow of RTP packets is an ephemeral termination.

A context is an association of media flows among a group of terminations. A simple session might have only two terminations per context, but a conference call could have multiple terminations. The MG, under MGC control, creates and names a context when the first termination is added, and releases contexts when the session ends. The MGC can prioritize contexts for emergency calls or to control routing under heavy load conditions. A termination may have multiple streams, in which case the context is a multistream context. A video session, for example, has both video and audio streams and could have a data stream as well. Terminations that are not associated with any other terminations, such as idle lines, are known as *null terminations*. One property of the MG is the number of terminations in a context. For example, an MG that allowed only point-to-point connections would allow a maximum of two terminations per context.

Megaco/H.248 uses commands to control signals, terminations, and contexts. The command structure is simple and easily grouped together into transactions, which reduces messaging overhead. The protocol has only seven simple commands: Add, Subtract, Move, Modify, Notify, Audit, and ServiceChange. For example, the Add command adds a termination to a context and the Subtract command removes it. When the last termination is subtracted, the context releases. The Move command moves terminations between contexts. The MG uses the Notify command to inform the MGC of events such as an endpoint going on-hook. Groups of commands form transactions, which are a series of commands that are executed in sequence.

Megaco defines *packages*, which are sets of properties for terminations with uniform characteristics. For example, an analog loop has a package with events defined for on/off-hook status, ringing, and statistics. The MG may keep statistics such as lost packets and operational measurements and report these to the MGC.

A call session consists of a series of messages. When a caller lifts the receiver off hook, the MG recognizes the signal and sends a Notify message to the MGC. The database in the MGC performs the function of admission control; i.e., is the user recognized and authorized? The MGC instructs the MG to send dial tone and to collect the dialed digits. When dialing is complete, the MG forwards the called number to the MGC for routing instructions. The context could be established with another station controlled by the same MG or established across conventional TDM or an IP packet trunk to a distant MG or to the PSTN. The MGC instructs the MG serving the called party to begin ringing the telephone and the originating MG to return audible ring. When the session is established, the MGC begins timing the call. When either party hangs up, the stations return to idle status in the database.

These functions are almost identical with the functions that a conventional end office performs in setting up and managing a session. The difference is that the switching function, which is carried on in the MG, is distributed throughout the serving area and controlled from a central location. In a conventional switch, processors are associated with each switch and management of an interswitch session involves two processors. The softswitch uses a server platform in a redundant configuration running a commercial operating system. The servers can be distributed to different locations for additional service protection, and compared to data processing servers they are hardened to improve reliability.

Softswitch Call Control Summary

To summarize the foregoing discussion, it should be apparent that the principal issues are centralization versus decentralization and master/slave versus peer-to-peer communication. The traditional telephone carriers have favored the centralized and master/slave model, while the Internet has evolved from a decentralized peer model. The question is whether one of these will prevail. The long view is impossible to predict, but in the short term it is clear that both models will remain in service.

Another question for the future is whether H.323 will be a short-lived protocol that is destined to be replaced by SIP and its related protocols. That may happen eventually, but in the meantime, H.323 has a considerable amount of momentum and exists in millions of devices. The protocol stack runs on every PC that supports Microsoft's Net Meeting. Every major manufacturer of VoIP equipment supports H.323, and elements of it such as RTP and RTCP are used by SIP and Megaco/H.248. Moreover, H.323 provides the centralization and stability that common carriers favor. As new protocols develop, multimedia systems likely will support them all.

PART THREE

Transmission Technologies

If we define telecommunications as the movement of information, transmission technologies are the means by which it is moved. The technology that makes intercontinental communication possible is beyond the comprehension of most users and it is a tribute to the genius of its designers that users do not need to understand the technology to use it. Two major trends in transmission are clear. One is the migration of everything possible to fiber optics. This has been going on for nearly three decades and the developed world has reached the point that everything feasible to migrate to fiber has been completed, except for taking it down to homes and small offices, and that transition is underway. Now fiber owns the long haul and metropolitan networks except where the user station is mobile, and that belongs to the second technology: wireless in one of its many forms.

If it is difficult to imagine beaming information through a tiny strand of glass not as thick as a human hair, encoded by a laser the size of a grain of salt, and moving it halfway around the world at nearly the speed of light, then wireless technology is just as amazing. The difference is that nearly everyone alive today grew up in the age of radio and television and they take the technologies for granted. To anyone who understands the complexities, however, the new wireless technologies are nothing short of astounding. When you consider that the cell phone not only contains RF circuitry, but also a complex control system that keeps the system aware of its whereabouts, responds to calls, and hops to assigned radio channels, it is amazing that so much can be packed into such a small package. Add to that a digital camera, a Web browser, a global positioning locator for emergency calls, and a cost brought down to give-away levels and the result is almost unbelievable.

Most people do not realize what an unstable medium the radio spectrum is or how scarce it is. The laws of physics limit the amount of spectrum, but technology keeps finding ways to use it more effectively. Everyone knows that microwave communications requires line-of-sight, but researchers are finding ways to use multipath reflections, which have always been considered an impairment, to bend, or rather reflect, the radio waves around obstacles.

As with other telecommunications technologies, transmission is profoundly affected by the Internet and the migration to IP. Here, wireless has plenty of room to improve. Cellular supports IP services, but despite VoIP's potential advantages of improved spectrum utilization the standards are still several years in the future.

Although fiber and wireless have the spotlight at the moment, it would be premature to write the epitaph for copper cable and coax. Although both have disappeared from the long-haul network and nearly so from the metropolitan network, coax is alive and well for video transmission and copper is preferred inside buildings. Wireless offers in-building mobility, but it cannot match the bandwidth, stability, and security of wired Ethernet.

This part reviews both wired and wireless technologies. The part begins with chapters on optical networking, microwave, and satellite. Next, we look at mobile, cellular, and PCS radio, wireless voice and data systems, and conclude by examining video systems.

CHAPTER 17

Optical Networking

Occasionally, a development comes along that revolutionizes an industry. Such is the case with fiber optics. The importance of fiber optics to the world economy cannot be overstated. Today, every major industrialized center is linked to the rest of the world with fiber-optic cable, and the result is high-quality and low-cost bandwidth. Fiber serves at the building level in commercial and industrial centers, freeing them from the restrictions of copper cable and enabling broadband communications that connects entities with high-speed pipelines. Without fiber the Internet would be relegated to its origins of slow-speed connections between text-based computers. Internet, telecommuting, video conferencing, distance learning, and a host of other productivity-enhancing applications ride the fiber backbone. Fiber is the default method of linking LANs and it is a shortsighted organization that installs a campus backbone without providing fiber optics between the buildings.

The conversion to fiber in the public backbone network is complete. The bandwidth restrictions to residences are beginning to disappear as fiber moves into the distribution network. Cable companies serve their customers with hybrid fiber–coax (HFC) and LECs are replacing their copper feeder cables with fiber to neighborhood centers. Eventually, fiber will be brought to the proximity of most households and the capability of end-to-end digital connectivity will be realized, but that day is still in the future.

Fiber optics arrived at an opportune time in telecommunications history, providing unlimited bandwidth and interference-free communications in a world that was rapidly exhausting the microwave spectrum. Fiber provides such high quality that it matters not whether the endpoints of a session are next door or half a world apart. The cable is fabricated from silicon, the most abundant substance on Earth, and in terms of energy consumption the electronics are far more efficient than the technologies they replaced. Best of all, once the cable is in place, it can be expanded to many times its original capacity by upgrading the electronics.

Today carriers routinely deploy fiber with bandwidths that would have been unthinkable a few years ago. OC-192, which is roughly 10 Gbps, is common and OC-768 at 40 Gbps is beginning to appear. DWDM divides the fiber into multiple channels, each on a unique wavelength capable of carrying the bandwidth of the base fiber. Lightwave amplifiers increase the span between regenerator points and optical cross-connects make it possible to route light streams without converting them to electrical signals. This chapter discusses how these hair-thin optical waveguides are manufactured and deployed into the worldwide optical infrastructure.

LIGHTWAVE TECHNOLOGY

The use of light for communication is an idea that has been around for more than a century. Alexander Graham Bell, in the first known lightwave application, received a patent for his "Photophone" in 1880. The Photophone focused a light beam from the sun, modulated it with voice, and radiated its free space to a nearby receiver. The system reportedly worked, but free-space light radiation has several disadvantages that the devices available at that time could not overcome. Like many other ideas this one was ahead of its time. Free-space light communication is now technically feasible if the application can tolerate occasional outages caused by fog, dust, atmospheric turbulence, and other path disruptions.

Two developments raised lightwave communication from the theoretical to the practical. The first was the laser in 1960. A laser produces an intense beam of highly collimated light, i.e., its rays travel in parallel paths. The pulses from a digital signal trigger the laser on and off at the speed of the modulating signal. The second development was the refinement of glass to the point that it was sufficiently transparent to carry an optical signal. The design target at the time was a maximum loss of 20 dB/km. The decibel is a logarithmic scale. Each 3 dB of loss cuts the signal power in half, so a reduction of 20 dB would mean that after a distance of 1 km only about 1 percent of the original signal would remain. In 1970, Corning Glass Works reached the 20-dB/km threshold and focused the attention of the world on fiber optics as the next communications technology. Today, the technology has advanced to the point that it is possible to purchase fiber with a loss as low as 0.2 dB/km.

With a laser source that is triggered on and off at high speed, the zeros and ones of a digital communication channel can be transmitted through the glass waveguide to a detector that converts the received signal from light pulses to electrical. Figure 17-1 shows the elements of a lightwave communication system. Amplifiers or regenerators are spaced at regular intervals, with the spacing dependent on the transmission loss of the fiber and the system gain at the transmission wavelength. System gain is discussed in a later section. Older systems bring the light signal down to the electrical level for regeneration, but newer systems amplify at the optical level.

A standby channel, which assumes the load when the regular channel fails, protects most lightwave systems. The two directions of transmission are normally

FIGURE 17-1

Lightwave Transmission System

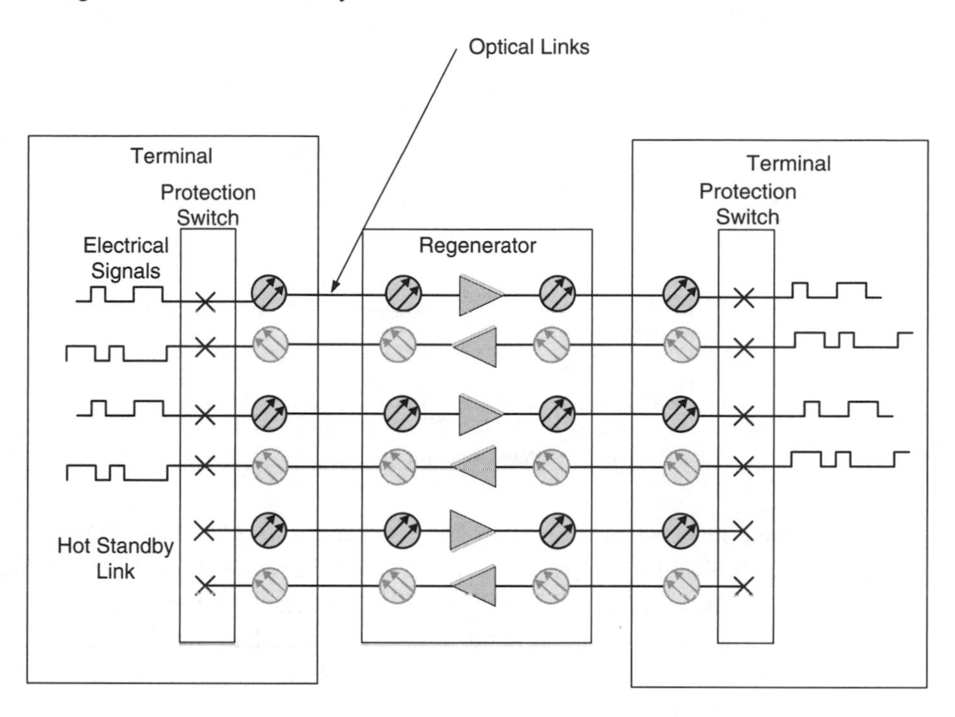

protected separately between the digital signal input and output points. If a failure occurs, the protection equipment switches the signal to a new combination of cable, terminal equipment, and repeaters. Fiber has two protection schemes, *unidirectional path-switched ring* (UPSR) or *bidirectional line-switched ring* (BLSR). In a UPSR configuration, the nodes send on two counter-rotating rings, but all equipment receives the signal on the same ring. If the working ring fails, the receiving equipment switches to the other fiber pair. This method provides full path redundancy, but bandwidth reuse between nodes is not possible because the spare fiber must be ready to carry the entire load. A BLSR ring allows traffic to travel between nodes in the shortest route, so bandwidth can be reused between pairs of nodes.

The advantages of lightwave accrue from the protected transmission medium of the glass fiber. The optical fiber attenuates the light signal, however, and as Figure 17-2 shows, the loss is not uniform across the spectrum. Lightwave communication can use three low-loss regions or *windows*. Loss disturbances labeled OH^- result from hydroxyl ion absorption.

The earliest fiber-optic systems used the 850-nm window because suitable lasers were first commercially available at that wavelength. Slower speed LANs also use this window because LEDs, to operate at that wavelength, are economical. As lasers became available at 1300 nm, applications have shifted to this

FIGURE 17-2

Spectral Loss for a Typical Optical Fiber: Loss Disturbances Labeled OH^- Result from Hydroxyl Ion Absorption

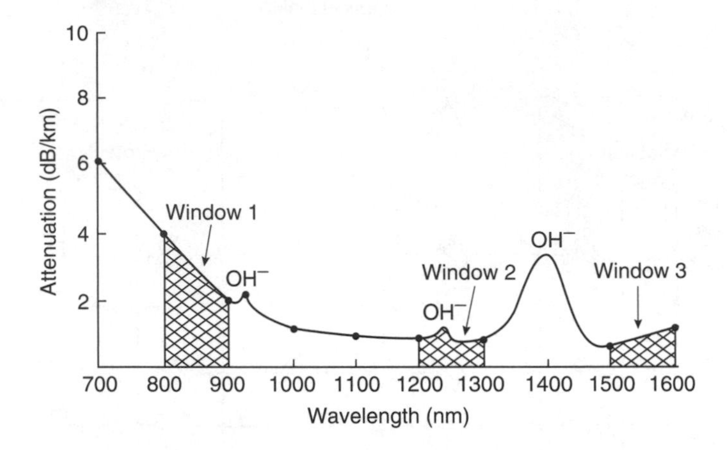

wavelength because of its lower loss. Single-mode fiber (SMF), discussed later, exhibits slightly lower losses in its third window at about 1550 nm. Table 17-1 lists the principal wavelength windows used in fiber transmission.

The first commercial fiber-optic system installed in 1977 operated at 45 Mbps (DS-3) with repeaters required at 4-mile (6.4-km) intervals. Current systems operate throughout the SONET/SDH range, up to and including OC-192, which has a line rate of 9.95 Gbps. If this much bandwidth were populated with 192 DS-3 signals, each of which carries 672 voice channels, an OC-192 system would carry 129,024 voice channels. As mentioned earlier, DWDM systems can multiply that capacity 40 times or more and the industry is beginning to consider the move up to OC-768.

LIGHTGUIDE CABLES

A digital signal is applied to a lightguide by pulsing the light source on and off at the bit rate of the modulating signal and the pulses propagate to the receiver at slightly less than the speed of light. The lightguide has three parts: the inner core, the outer *cladding*, and a protective coating around the cladding. Both the core and the cladding are of glass composition; the cladding has a greater refractive index so that most of the incoming light waves are contained within the core. Light entering an optical fiber propagates through the core in *modes*, which are defined as the different possible paths a lightwave can follow.

Two types of optical fiber are manufactured: single mode and multimode. In *single-mode fiber*, light can take only a single path through a core that measures about 9 microns in diameter, which is about the size of a bacterium. (A micron is one-millionth of a meter.) *Multimode fibers* (MMF) have cores of 50 to 200 microns in diameter. The original MMF standard had a 62.5-micron core diameter and it

TABLE 17-1

Single-Mode Fiber Wavelength Bands

Band	Description	Range (nm)
O	Original	1260 to 1360
E	Extended	1360 to 1460
S	Short Wavelength	1460 to 1530
C	Conventional	1530 to 1565
L	Long Wavelength	1565 to 1625
U	Ultralong Wavelength	1625 to 1675

is still specified for some applications. More recently, the industry has shifted to 50-micron core because the bandwidth is higher. MMF is used almost exclusively in customer premise applications. SMF is more efficient at long distances for reasons that we discuss below, but the small core diameter requires a high degree of precision in manufacturing, splicing, and terminating the fiber. Despite the greater precision needed, SMF is less expensive than multimode, primarily because of the vast quantities of single mode manufactured.

Lightwaves must enter the fiber at a critical angle known as the *angle of acceptance*. Any waves entering at a greater angle can escape through the cladding as Figure 17-3 shows. The reflected waves take a longer path to the detector than those that propagate directly. The multipath reflections arriving out of phase with the main signal attenuate the signal, round, and broaden the shoulders of the light pulses. This pulse rounding is known as *modal dispersion*. It can be corrected only by regenerating the signal. If a light pulse spreads so far that the trailing edge of one pulse merges with the leading edge of the next, bit errors result. The greater the core diameter, the greater the amount of modal dispersion. SMF propagates only one mode of light and therefore does not suffer from modal dispersion.

Both SMF and MMF are subject to another form of dispersion called *chromatic dispersion*. The term chromatic comes from the multiple light wavelengths that

FIGURE 17-3

Light Paths through a Step Index Optical Fiber

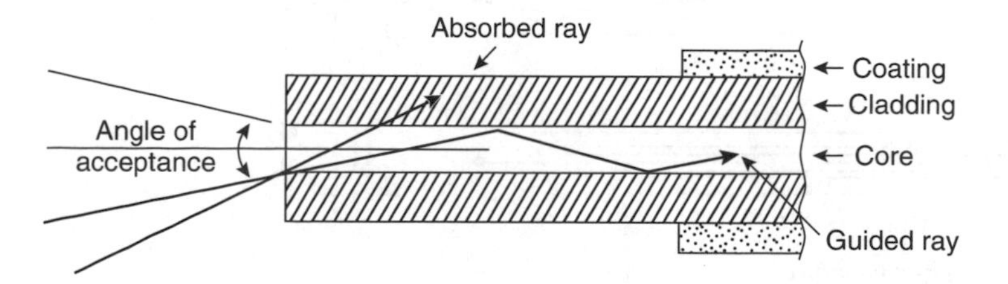

propagate through the core. The amount of dispersion is a function of the quality of the laser. High-quality lasers emit a narrower band of wavelengths, resulting in less dispersion.

A third type of dispersion is known as *polarization mode dispersion* (PMD). PMD is caused by small variations in the shape of the fiber core. When light travels down a fiber, some polarization modes are at angles to each other. If the core is not perfectly symmetrical, one mode travels faster than the other, resulting in pulse spreading. Both chromatic and polarization dispersion increase as the square of the bit rate.

MMF is classified by its *refractive index* into two general types, *step index* and *graded index*. With step index fiber the refractive index is uniform throughout the core diameter. In graded index MMF the refractive index is lower near the cladding than at the core so that lightwaves propagate at slightly lower speeds near the core than near the cladding, which result in lower dispersion. Figure 17-4 shows wave propagation through the three types of fiber.

Loss in fibers is caused by absorption and scattering. *Absorption* results from impurities in the glass core, imperfections in the core diameter, and the presence of hydroxyl ions or water in the core. The water losses occur most significantly at wavelengths of 1400, 1250, and 950 nm, as Figure 17-2 shows. *Scattering* results from variations in the density and composition of the glass material. These variations are an inherent by-product of the manufacturing process. SMF has its lowest chromatic dispersion at the 1300-nm wavelength, but minimum loss is at 1550 nm, which has led to the development of *dispersion-shifted fiber*. Dispersion-shifted fiber moves the minimum chromatic dispersion wavelength to the 1550-nm window to provide the lowest combination of loss and bandwidth at the same wavelength.

Nearly all carrier applications and some customer premise applications use SMF. LANs often use the 850-nm window because LEDs, to operate at that

FIGURE 17-4

Wave Propagation through Different Types of Optical Fiber

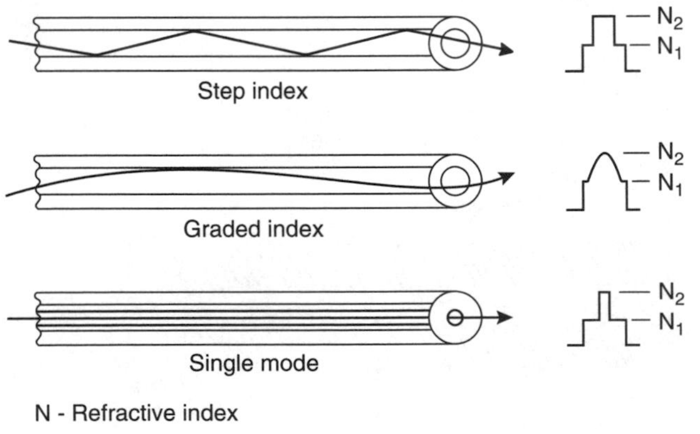

Step index N_2 N_1

Graded index N_2 N_1

Single mode N_2 N_1

N - Refractive index

wavelength, are economical. As lasers became available at 1300 nm, applications shifted to that wavelength because of its lower loss. SMF exhibits slightly lower losses in its third window at about 1550 nm. Although the lasers are more expensive in this window, it is used in DWDM systems because *erbium-doped fiber amplifiers* (EDFAs) operate in the 1550-nm window and the lower fiber loss helps offset the losses of the DWDM equipment.

Most metropolitan fiber-optic systems use 1300-nm wavelengths because the components are less expensive than in the 1550-nm window. In addition, the distance between termination points is shorter and amplification is generally not needed. In the long-haul network the extra cost of the components is offset by the lower loss of that wavelength. The 1550-nm window is divided into two bands. The most commonly used is C band, which has a wavelength of 1530 to 1565 nm. The L band, using longer wavelengths, is 1565 to 1625 nm. An additional window from 1350 to 1530 nm has not been commercially developed yet, but when components are developed for that band, the potential bandwidth of a fiber is in the order of 50 THz. Table 17-2 lists the ITU-T fiber types and their principal applications.

Manufacturing Processes

Glass fibers are made with a variety of processes: modified chemical vapor deposition (MCVD), outside vapor deposition (OVD), plasma-activated chemical vapor deposition (PCVD), and vapor-axial deposition (VAD). To illustrate, the MCVD process starts with a tube of ultra-pure silica about 6-ft long and 1.5-in diameter. The silica is doped with germanium to create the required refractive index profile. The tube is rotated over a flame of controlled temperature while a chemical vapor is introduced in one end as Figure 17-5 shows. The vapor is a carrier for chemicals that the heat deposits on the interior of the glass. The deposited chemicals

TABLE 17-2

ITU-T Fiber-Optic Cable Types and Applications

ITU-T Type	Description	Wavelength	Application
G.651	Multimode fiber with a 50-micron core	850 to 1300 nm	Short range
G.652	Nondispersion-shifted fiber	1310 to 1550 nm	General purpose telephony, CATV
G.652C	Low water peak nondispersion-shifted fiber	1285 to 1625 nm	Metropolitan and access networks
G.653	Dispersion-shifted fiber	1310 to 1550 nm	Superseded by G.655
G.654	1550-nm loss-minimized fiber	1500 to 1600 nm	Long haul undersea
G.655	Nonzero dispersion-shifted fiber	0.2 dB/km at 1550 nm	Long-haul backbone, DWDM

FIGURE 17-5

Modified Chemical Vapor Deposition

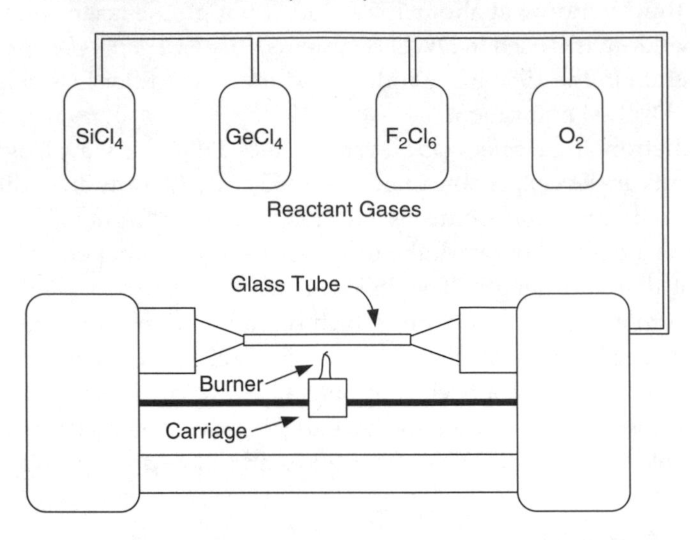

form a tube composed of many layers of glass inside the original tube. The OVD process deposits high-purity glass on the outside of a ceramic rod, which is removed after the process is complete. When the deposition process is complete the tube is collapsed under heat into a solid glass rod known as a *preform*. The preform is placed at the top of a drawing tower where the fiber is heated to the melting point and extruded through a die into a hair-thin glass strand, as Figure 17-6 shows. The fiber is then coated with a protective substance that may be colored for ease in identifying individual fibers in a cable.

The strands are then tested, segregated by quality, and multiple strands are wound together around a strength member and enclosed in a sheath. Like copper cable, fiber cable sheaths are made of polyethylene and can be enclosed in armor to protect against damage. Fiber-optic cable is suitable for direct burial, pulling through conduit, suspension from an aerial strand, or submersion in water. Two predominate methods are used for cabling: loose tube and tight-buffered. In *loose tube* construction, the fibers are placed in buffer tubes, which are usually gel-filled to keep moisture out. In *tight-buffered* design, the fibers are wrapped much like a copper cable. One advantage of loose tube is that it is easy to drop off fibers at branching points without affecting the other fibers in the sheath. It is also easier to identify and administer the fibers. The tight-buffered method takes less space in conduits and raceways. Loose tube is usually preferred for outdoor applications because the fibers are loosely coupled, which allows them freedom of movement during expansion and contraction and during installation.

Fiber comes from the factory with no splices. The size of the reel imposes a practical limit on the cable length, but the objective is to avoid splices by ordering the cable in reel lengths sufficient to span the normal connection points, such as central

FIGURE 17-6

Fiber Drawing Tower Diagram

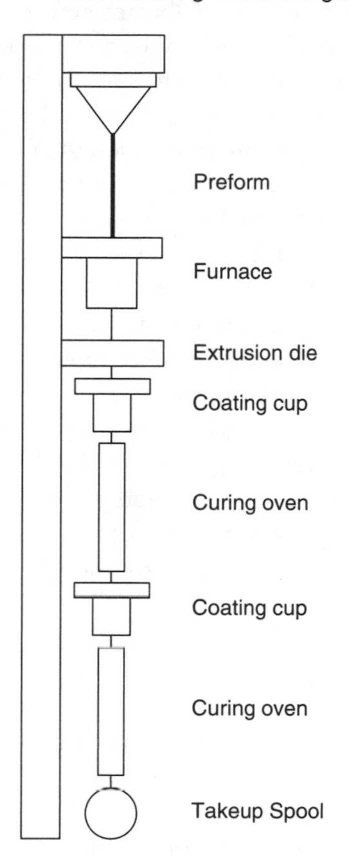

offices, manholes, and amplifier locations. Where splicing is necessary, it can be done by either adhesion or fusion. In the adhesion process, a technician places fibers in an alignment fixture and joins them with epoxy. The fusion method employs a splicing fixture that precisely aligns the two ends of the fiber under a microscope and fuses them with a short electric pulse. The transmission loss of a mechanical splice is usually higher than a fusion splice, but lower skill and less expensive equipment are needed. After splicing, the loss is measured to check the splice quality. Splices are made with enough slack in the cable that they can be respliced if necessary, until the objective loss is achieved. A good splice should have 0.1 to 0.3 dB of loss.

Cable Connectors

In midspan, fiber cables are joined by splicing, but at terminal locations they are connectorized for coupling to terminal devices and for ease in rearrangement. The physical structure of connectors is important because of the close tolerances that are required to match the cable to the transmission device. Connectors are made from thermally stable materials and have tightly locking keyed parts and polished mating surfaces.

Connector performance is judged by two criteria: the amount of insertion loss and the amount of reflection attenuation or return loss. A reflection is light that travels down the lightguide, strikes a discontinuity, and reflects toward the source, causing instability or errors. The acceptable amount of loss depends on the application. LAN applications, which usually use multimode cable, are more tolerant of connector performance. The loss of a LAN connector can usually be as much as 1.0 dB and still remain within the loss budget. A suitable connector for common-carrier applications should have an insertion loss of 0.5 dB or less and a return loss of at least 40 dB.

Couplers are made either free hanging or for bulkhead mounting. Connectors use an epoxy and polishing arrangement for termination. The fiber is stripped and inserted into the connector. The ends are then polished to an optical finish. The durability of the connector is important. The connector holds the ends of the fiber in alignment, and it must remain so even under the strain of pulling or sideways motion. Several different connector styles are available, but the most common is the ST type. The fiber is inserted into a precision pin that has a spring-loaded mechanism to press the pin against a mating connector. The connector has a strain-relieving device attached to the outer jacket of the fiber-optic cable. ST connectors are available for both MMFs and SMFs.

FIBER-OPTIC TERMINAL EQUIPMENT

A fiber link consists of a fiber pair terminated in a transmitter and receiver. The light transmitter employs either a LED or a solid-state injection laser diode as its output element. Lasers have a greater system gain than LEDs because their output is higher and because the light is more highly collimated, resulting in less dispersion. Lasers can be more tightly coupled into the fiber than LEDs, which illuminate the entire core. The transmitter may either vary its intensity between defined levels or trigger the light on and off in step with the electrical signal. The latter method is prevalent. The primary advantage of a LED transmitter is its lower cost. LANs, which do not need high system gain, usually use LED transmitters.

The light receiver is an avalanche photodiode (APD) or positive–intrinsic–negative (PIN) diode that detects the light pulses and converts them to electrical pulses, which the receiver reshapes into square wave pulses. A lightwave regenerator has back-to-back receiver and transmitter pairs that connect through a pulse reshaping circuit. Fiber-optic systems accept standard digital signals at the input, but each manufacturer develops its own output signal rate. Error checking and zero suppression bits are inserted to maintain synchronization, monitor the bit error rate, and determine when to switch to the protection channel. Because of differences in the line signals, lightwave systems are usually not end-to-end compatible between manufacturers unless they meet SONET/SDH standards, and even then OAM&P systems may not interoperate.

Optical Amplifiers

EDFAs amplify light in the C band, extending from 1530 to 1565 nm. EDFAs are independent of the modulation method, handling both analog and digital signals, whether baseband or DWDM signals. They can handle speeds well in excess of the 10-Gbps signals commonly used today and have been tested at bit rates in excess of 5 Tbps. A typical EDFA offers about 50 dB of gain over bandwidths of about 80 nm and have power outputs as high as +37 dBm or about 5 W. To increase the bandwidth of EDFAs, work is progressing on telluride-based EDFAs, which amplify in the 1565 to 1625-nm L band. L band amplifiers extend the total bandwidth of an EDFA to more than 110 nm. Extending the bandwidth of EDFAs offers significant benefits. In addition to providing more wavelength division multiplexing (WDM) channel capacity, it also enables manufacturers to use broader wavelength spacing, enabling them to use less-expensive components to separate wavelengths. To complicate the process, however, the wider the band the more important it is to provide gain-flattening filters to equalize the loss across the wavelength band.

Although EDFAs are the most common device in use today, they do not cover the 1300-nm window, which has led to the interest in amplifiers that do. One such device is the praseodymium-doped fluoride fiber amplifier (PDFFA). Another technology is the Raman fiber amplifier, which amplifies in both windows. When an intense light is injected into the fiber, the Raman Effect excites the photons to a higher energy state, making the fiber act as an amplifier. The result is to distribute amplification along the length of the fiber instead of at fixed locations. Raman amplifiers in conjunction with EDFAs will enable the use of a broader band of wavelengths. Manufacturers expect that fiber bandwidth will be approaching the 50-THz theoretical limit by 2010.

Wavelength Division Multiplexing

The capacity of a fiber pair can be multiplied by using WDM. WDM assigns services to different light wavelengths in much the same manner as frequency division multiplexing applies multiple carriers to an analog medium. Different wavelengths or "colors" of light are selected by using light-sensitive filters to combine light wavelengths at the sending end and separate them at the receiving end. Coarse WDM (CWDM) systems apply about 16 wavelengths to a fiber. DWDM systems today use 100-GHz spacing between wavelengths, generally providing 40 channels per fiber in the C band. In future, 50- or 25-GHz spacing will likely be possible with improved filters and interleavers. Coupled with dual band amplifiers, more than 100 wavelengths or *lambdas* can be transmitted. The vision of optical networking is to manipulate these wavelengths in the optical domain as TDM devices manipulate electrical bandwidth today.

Because the filter introduces loss, DWDM reduces the distance between regenerators and limits the path length by the wavelength with the highest loss.

When engineers design lightwave systems, they normally provide enough system gain to compensate for WDM even if it is not used initially.

A variety of filtering techniques are employed in optical networking. They are important not only for DWDM, but also for optical add–drop multiplexers as discussed in the next section. The most common filtering technique is known as *fiber gratings*. Variations in the refractive index of the fiber core cause discrete wavelengths to be either passed or reflected. The quality of the filters is one of the most important issues in advancing the optical network. Increasing fiber capacity will eventually mean spacing wavelengths more closely. The filter skirts must be steep to avoid interference between channels. The ultimate capacity of a fiber using a combination of high-speed multiplexing over DWDM fiber has hardly been glimpsed.

Wavelength Add–Drop Multiplexers (WADM)

The capacity of a single fiber pair is so great today that it may easily exceed the entire bandwidth needed between the points on a transcontinental network. Some means is needed to drop specific wavelengths in the same manner that a TDM add–drop multiplexer manipulates electrical bandwidth. Such a device allows some wavelengths to pass through while diverting others to an alternate path. The vacated lambda on the other side of the multiplexer can be replaced by an inserted signal.

Three types of WADM are possible. The simplest type is a fixed WADM in which certain channels are filtered out and recombined while the other channels are passed through the multiplexer. A reconfigurable WADM is the next stage of development. This device allows the provisioner to select the channels under manual or software control. The ultimate device is a flexible WADM that can respond to changes in traffic demand. This type of device could sense failures or congestion and reroute traffic accordingly. The application dividing line between a flexible WADM and an optical cross-connect is indistinct, but the technology used is different. WADMs achieve their wavelength manipulation by the use of filters. Optical cross-connects, as we will discuss next, redirect the lightwave by means of an optical switch.

Optical Cross-Connects

It should be clear from our earlier discussions that expanding the network indefinitely with SONET/SDH is not practical. For one thing, digital cross-connects have a finite capacity. When this is exceeded, carriers have no alternative but to deploy multiple DCSs. This is undesirable for several reasons, not the least of which is the cost of multiple optical–electrical–optical (OEO) conversions. As DCSs are interconnected, additional ports are used for the interconnection, reducing the effective capacity of the DCS.

DCS systems are effective when the signal is at the electrical level, but bringing the light signal down to an electrical signal at main branching points is expensive and requires floor space to house the regenerating equipment. Optical networking equipment has the objective of routing signals at the optical level wherever possible. An optical switch or cross-connect (OCX) operates like a DCS, but at the optical instead of the electrical level.

The next generation optical networks require a new architecture based on retaining SONET/SDH functions, while eliminating the equipment layer. The key to making this architecture practical is an economical and scaleable OCX. Switching in the optical domain retains the benefits of SONET/SDH while removing many of its limitations. Optical switches can provide such functions as multiplexing, provisioning, signaling, service restoration, performance monitoring, and fault management. The core of the network is optical, while the edge remains electrical with traditional interfaces as shown in Figure 17-7. The optical core of the network is OC-48 or OC-192 while the edge uses IP or ATM over SONET/SDH bandwidth building blocks.

FIGURE 17-7

Optical Cross-Connect System

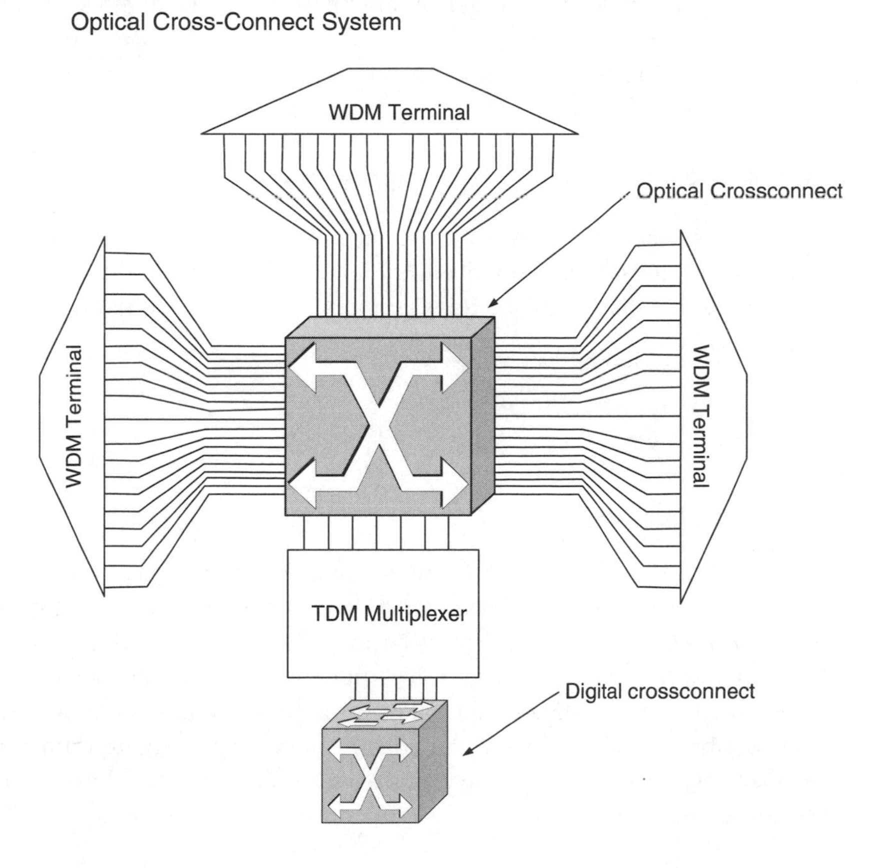

Optical cross-connects or switches, as they are sometimes known, switch lambdas to form point-to-point connections in the optical domain, transcending the limits of the electrical domain. Optical networking enables carriers to sell wavelengths without giving up the entire capacity of the fiber. This enables carriers to offer services to their customers without being limited to the digital hierarchy. Optical networking offers the ability to route and groom in the optical domain. Carriers want to be able to pick bandwidth out of the optical domain, add to it, drop it, or shift it to another wavelength to groom wavelengths into the total capacity of a fiber.

Ultra Long-Haul Equipment

Undersea cables generally have much longer regenerator spacing than the 400- to 500-km spacing that is typical of long spans. Partly, this is due to the high quality of cable that can be justified to increase spacing; also, it is because undersea fiber-optic systems can be more easily optimized since amplifiers and regenerator locations do not have to be selected to coincide with existing structures. The same is desirable in land-based systems. Ideally, carriers would be able to route lambdas for as much as 4000 km without regeneration. The technology to accomplish this is known as *ultra long haul*.

Several techniques are under development to achieve these kinds of distances. One is forward error correction to detect and correct bit errors through a complex process of adding overhead bits. Other techniques include new gain-flattening filters for EDFAs allied with Raman amplification.

LIGHTWAVE SYSTEM DESIGN CRITERIA

Fiber-optic system design is a balance between capacity requirements and costs, which include the cable, terminal equipment, regenerators and amplifiers, and construction and engineering. The three primary criteria for evaluating a system are:

- information transfer rate
- system attenuation and losses
- cutoff wavelength

Information Transfer Rate

The information transfer rate of a fiber-optic system depends on the bandwidth, which in turn depends on dispersion rate and the distance between terminal or repeater points. Manufacturers quote bandwidth in graded-index fiber as a product of length and frequency. For example, a fiber specification of 1500 MHz-km could be deployed as a 150-MHz system at 10 km or a 30-MHz system at 50 km. Special purpose fiber-optic systems intended for short-range private data transmission have low bit rates and typically use cables with considerably more bandwidth than the application requires.

System Attenuation and Losses

System gain in fiber optics is the algebraic difference between transmitter output power and receiver sensitivity. For example, a system with a transmitter output of –5 dBm and a receiver sensitivity of –40 dBm has a system gain of 35 dB. From the system gain, designers compute a *loss budget*, which is the amount of cable loss that can be tolerated within the available system gain. Besides cable loss, allowances must be made for other elements such as these:

- splice loss including an allowance for future maintenance splices
- connector losses
- temperature variations, which may cause variations in transmitter output power or receiver sensitivity
- measurement inaccuracies
- current or future WDM
- aging of electronic components
- safety margin

These additional losses typically subtract about 10 to 12 dB from the span between terminal points, which leaves a loss budget of about 25 dB for cable. Cable cost depends on loss, so system designers choose a cable grade to match the loss budget.

Sample Performance Margin Calculation

Table 17-3 shows how to calculate the performance margin of a fiber-optic system. The first step is to calculate the total attenuation, which is the sum of three elements: the loss of the cable, the loss of connectors, and the loss of any splices.

The next step is to calculate system gain, which is the algebraic sum of the transmitter output power minus the receiver sensitivity. For safety's sake, an operating margin is stated to allow for deterioration of the transmitter with component aging. If the manufacturer does not state a margin, 2.0 dB is typical for LEDs and 3.0 dB for lasers. If the manufacturer states a margin for the receiver, this should also be added. In addition, an allowance is made for future repair splices.

The performance margin is the system gain minus the cable loss, operating margin, and repair splice allowance. If the performance margin is not enough, the designer may have to specify lower loss cable or higher performance transmitter and receiver.

FIBER OPTICS IN THE LOCAL LOOP

Now that the interexchange network is converted to fiber optics, the LECs and CATV companies are turning to the next potential application, the local loop.

TABLE 17-3

Performance Margin Calculations

Cable length (km)	**60**
Losses	
Fiber loss (dB/km)	1.2
Total fiber loss (dB)	72
Individual connector loss (dB)	1
Number of connectors	4
Total connector loss (dB)	4
Individual splice loss (dB)	0.4
Number of splices	7
Total splice loss (dB)	2.8
WDM allowance (dB)	4
Total attenuation (dB)	82.8
System gain	
Transmitter output power (dBm)	35
Receiver sensitivity (dBm)	−65
System gain (dB)	100
Performance margin	
Operating margin	3
Repair splice allowance (dB)	0.8
Performance margin (dB)	13.4

The rationale behind fiber-in-the-loop (FITL) are broadband applications such as video-on-demand that cannot be served with conventional facilities. Most cable companies have converted their systems to HFC to support such applications as HDTV, Internet access, and IP telephony. If ILECs are to compete in this arena, they must provide fiber local loops.

Local loop fiber will likely assume one of several architectures. One is to replace feeder cable with fiber. Fiber extends from the central office to a serving area interface. The loop from the serving area interface to the customers' premises remains copper cable, which simplifies the interface problem and resolves the issue of feeding power to the customers' station. Cable companies use a similar architecture, extending fiber trunk cable from the headend to neighborhood nodes where the optical signal is converted to electrical and applied to coaxial distribution cable. This option is often called fiber-to-the-curb (FTTC).

The second local loop fiber optic option provides fiber direct to the customers' homes (FTTH). In some applications, two fibers are installed, one for voice and data communication and the other for video; in other plans, a single fiber is installed, using WDM or different transmission windows to separate the directions of transmission.

Passive Optical Network (PON)

A promising method of providing FTTH is the PON. This technique places all the active equipment in the central office. A passive signal is brought to the residence, either directly to the home or to the curbside. The same fiber could be connected to several residences, with the signals to and from the different premises multiplexed by TDMA.

A PON uses passive splitters to break the capacity of the backbone fiber into multiple wavelengths. Today, as many as 32 full-duplex channels can be carried on a single fiber to a terminating point, where the channels are split and routed to the served customers over a short length of fiber. A PON requires no power, which makes it practical for bringing fiber to the curb. Just as LECs bring cable to a centralized terminal and serve multiple subscribers from one terminal, a PON splitter can serve multiple subscribers over fiber without bringing individual fibers all the way back to the central office. The optical path is independent of bit rate, modulation, and protocol.

At the central office, the fiber connects to an optical line terminal (OLT), which is a device that converts the digital TDM signal from the central office into a multiwavelength lightwave signal. The light from the OLT is launched into a fiber, where it travels to a splitter at the distant end. The splitter separates the wavelengths, which are routed to the customer's premises over a single fiber. At the customer's end, the PON terminates in an optical network unit (ONU). The ONU separates the signal into its components, which may be voice, Internet access, video, Ethernet, or any other digital service. Conceptually, PON architecture is similar to DSL, except that the components are optical instead of electronic. More details on PON are included in Chapter 32, "Metropolitan Area Networks."

OPTICAL NETWORKING APPLICATION ISSUES

Lightwave communications systems have applications in both private and public communications systems. The primary LEC and IXC applications ride on SONET/SDH. For noncarrier applications SONET/SDH is also available, as are FDDI, fiber channel, and LAN protocols. The following lists representative fiber applications:

- long-haul transmission systems
- intercontinental and undersea transmission systems
- trunking between local central offices
- metropolitan area backbone systems
- digital loop carrier feeder systems
- local area networks
- cable television backbone transmission systems

- private campus backbones
- interconnection of PBXs with remote switch units
- short-haul data transmission systems through noisy environments
- Fibre Channel for high-speed computer communications
- intelligent transportation systems, such as smart highways with intelligent traffic lights, automated tollbooths, and changeable message signs
- process automation in factories and industrial plants
- diagnostic image transmission in telemedicine applications

The high cost of right-of-way often stands in the way of private fiber-optic systems, but the advantages of this medium make it attractive for private applications. A major impediment to many applications is the common carriers' refusal to offer dark fiber. Most common carriers offer to lease bandwidth in any increment, but where the application requires dark fiber, many decline to provide it except to other carriers. Nevertheless, as companies are able to obtain right-of-way, they can install fiber for countless applications. The following sections discuss the variety of ways companies can apply fiber optics.

Campus, Intra-, and Interbuilding Backbone

Fiber optics is an excellent medium for a campus or building backbone and most LANs now employ a fiber backbone. Fiber optics not only provides bandwidth, but also offers security and noise immunity that no other medium can match. Any campus or riser cable system should at least consider the potential future need for fiber optics. Either fiber pairs should be installed for future expansion or empty conduit should be installed to support future fiber. Most current applications use MMF, but as speeds increase, single mode is becoming more common, particularly as applications such as Fibre Channel and gigabit and 10G Ethernet gain acceptance. Companies installing fiber today should consider installing both varieties in separate cables. If today's applications call for MMF, subducts should be placed in conduit to provide a future pathway for SMF.

Fiber to the Desktop

The experts agree about using fiber as a backbone in a building or campus network, but the question of carrying it all the way to the desktop is still controversial. UTP is about the same price as fiber, has enough bandwidth for most applications, and is easier to install and apply. However, fiber has much greater bandwidth, and although the industry is working on higher category UTP, the connectors are not so mature as fiber connectors are. UTP must be installed to every desktop and the question is whether to install fiber as well. Fiber optics is an ideal transmission medium for LANs, but the terminating equipment is twice as costly for fiber as for UTP. The fiber premium will undoubtedly decrease as the volume increases. In today's environment, fiber to the desktop can be justified only if the application has a genuine need for high bandwidth, extended range, or if there is an overriding consideration, such as security or need for noise immunity that mandates the use of fiber.

If fiber is not feasible today, should it be installed today and left dark to support a future application? The answer to this question depends on economics. Many buildings are difficult to wire, and placing a composite fiber and UTP cable to desktops may make economic sense because so much of the cost is in installation labor. Other buildings are designed for modular furniture that may be rearranged before the fiber is even used. The question to evaluate here is whether the location of future applications can be foreseen reliably enough to justify the expense of fiber optics.

Evaluation Criteria

Fiber-optic equipment is purchased either as an integrated package of terminal equipment and cable for specialized private applications or as separate components assembled into a system for trunking between switching nodes. For the former applications, which include local area, point-to-point voice, data, and video networks, the evaluation criteria discussed below are not critical. In such systems the main question is whether the total system fits the application. In all fiber-optic systems the questions of reliability, technical support, cost, and compatibility are important. The following criteria should be considered in evaluating a system.

System Gain

In selecting lightwave-terminating equipment, the higher the system gain, the more the gain available to overcome cable and other losses. The cost of a lightwave system relates directly to the amount of system gain. High-output lasers and high-sensitivity diodes are more expensive than devices producing less system gain. The least expensive transmitters use LEDs for output and have less system gain than lasers. When the limits of lightwave range are being approached, obtaining equipment with maximum system gain is important.

Cable Characteristics

Cable is graded according to its loss and bandwidth. The cable grade should be selected to provide the loss and bandwidth needed to support the ultimate circuit requirement. For systems operating at 100 Mbps or more on MMF, bandwidth becomes the limiting factor. The cable composition should be selected with inner strength members sufficient to prevent damage when the cable is pulled through conduits or plowed in the ground. Armoring should be considered where sheath damage hazards exist.

In private applications the core size of multimode cable is an important consideration. EIA/TIA standards specify both 50/125- and 62.5/125-micron cable. (The 62/125 designation means the cable has a core of 62.5 microns and an external diameter of 125 microns.) If the application has not been selected and the cable is being placed for future applications, the safest choice is 50/125-micron cable.

Wavelength

With present technology the most feasible wavelength to choose in MMF is 1300 nm unless the equipment specifications state otherwise. FDDI specifications call for 1300-nm cable with a bandwidth of at least 500 MHz-km. Cable should be purchased with a 1550-nm window if circuit requirements will ultimately justify the use of WDM. For most applications 850 nm should be avoided because of its greater loss.

Light Source

The two choices for light source are laser and LED. A laser has much higher power output than an LED and can operate at higher bit rates. LEDs are lower cost and have a longer life, but they produce a wider beam of light and have a wider spectral width, which means that a broader range of light wavelengths are transmitted compared to a laser.

LEDs are typically used where the distance between the terminals is short; normally 10 km or less, and the bandwidth of the signal is lower than about 150 Mbps. LEDs are generally satisfactory for local networks. In long-haul networks where long repeater spacing and high bandwidth are important, lasers are always used.

Wavelength Division Multiplexing

The question of whether to plan a fiber-optics system with future WDM designed into the transmission plan is a balance between future capacity requirements and costs. WDM can multiply the capacity of a fiber pair many times for little additional cost or can convert a single optical fiber into a full-duplex mode of operation by transmitting in both directions on the same fiber. It accomplishes this by reduced regenerator spacing, however, which is important in long systems but unimportant on systems that do not require an intermediate regenerator. On very short systems the cost of the WDM equipment may be greater than the cost of extra fibers.

CHAPTER 18

Microwave Radio Systems

Before the development of fiber optics, point-to-point microwave was the workhorse of telecommunications transmission. Commercial microwave was an outgrowth of World War II radar technology. AT&T began constructing its transcontinental microwave network in 1947, 2 years after the war ended. The first transcontinental microwave call was placed in August 1951 and 3 months later, the first direct-dialed transcontinental call was placed. By the time fiber arrived on the scene, the North American continent was laced with networks of microwave routes. AT&T owned about 2000 microwave terminal or repeater stations, which the fiber-optic boom of the 1980s rendered obsolete.

Although fiber has displaced it in the long-haul network, microwave is far from dead. Companies compete vigorously for permission to use radio spectrum that was once considered useless. The definition of microwave is rather arbitrary, but it occupies the spectrum from about 2 or 3 GHz to the threshold of light. Satellites, which we discuss in the next chapter, are microwave repeater sites in the sky. Wireless LANs, radar, multipoint data services, radio astronomy, microwave ovens, and countless other services occupy the spectrum. Private organizations use microwave for extensions of their enterprise networks and to control devices at remote sites. Cellular and PCS providers use point-to-point microwave for backhaul and backbone links. Microwave frequencies are in high demand for wireless Internet access and for replacement for the wired local loop. Although the frequencies above 40 GHz are little used today, the FCC has adopted rules regarding their use and as the technology advances, new services will be brought into play.

One of the major benefits of fiber optics is that it freed a lot of microwave spectrum that was once used for point-to-point service, and can now be used to provide services for which its characteristics are ideal.

MICROWAVE TECHNOLOGY

The microwave bands constitute several times the bandwidth of the lower frequencies. Figure 18-1, which is a logarithmic scale, shows the span of the radio frequency (RF) spectrum. The range defined as microwave does not have precise limits, but it is generally considered as the super-high and extremely high frequency bands, which span the range of 3 to 300 GHz. Table 18-1 lists the RF ranges and their common designations. The infrared light spectrum borders the upper end of the EHF band and the shorter the radio wavelength, the more it assumes the characteristics of light. The usable portion of microwave keeps pushing upward, with so-called "wireless fiber" operating at 38 GHz. Considering that the frequency spectrum below microwave occupies at most a total of 3 GHz of bandwidth, you can see that microwave provides at least 100 times the bandwidth that is available below the microwave bands. This submicrowave bandwidth supports all the radio and television broadcast services, virtually all two-way radio, and countless other radio services.

FIGURE 18-1

The Radio Spectrum

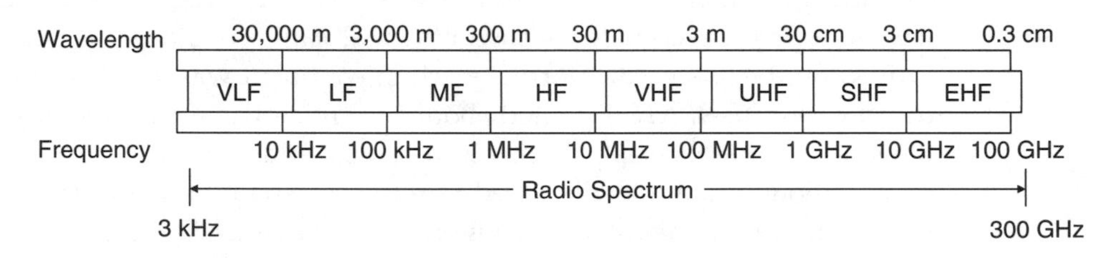

TABLE 18-1

Frequency Spectrum Designations

Frequency	Band
10 kHz to 30 kHz	Very low frequency (VLF)
30 kHz to 300 kHz	Low frequency (LF)
300 kHz to 3 MHz	Medium frequency (MF)
3 MHz to 30 MHz	High frequency (HF)
30 MHz to 300 MHz	Very high frequency (VHF)
300 MHz to 3 GHz	Ultra high frequency (UHF)
3 GHz to 30 GHz	Super high frequency (SHF)
30 GHz and above	Extremely high frequency (EHF)

Spectrum Licensing

Although the available microwave bandwidth is enormous, it is still a limited resource, and users must coordinate microwave paths to prevent interference. Because of congestion in metropolitan areas, it is often impossible to obtain frequency assignments in the lower end of the band. For years, the FCC regulated point-to-point microwave services under two distinct parts of its rules: Part 21 for common carriers and Part 94 for private operators. Since these services share many frequency bands and use the same type of equipment, the rules are now consolidated into a new Part 101.

Part of the microwave spectrum is licensed, which means FCC authorization and sometimes competitive bidding is required to gain the right to specific frequencies. Licensees are required to make frequency utilization studies to ensure that they will not interfere with existing services. Other portions of the spectrum are unlicensed. The ISM bands at 2.4–2.5 and 5.725–5.850 GHz are unlicensed, although the FCC imposes limits on radiated power. Licensing regulations and variables pertaining to power limits, modulation methods, and antenna gain vary from country to country.

Radio Signal Propagation

Different parts of the RF spectrum propagate through the atmosphere by different means as shown in Figure 18-2. Very low frequencies follow the Earth's surface, and are used for extended-range communications. For example, the U.S. Navy uses VLF to communicate with submarines at sea. Frequencies above VLF and below about 30 MHz are reflected by the ionosphere, and are therefore suitable for ranges beyond line of sight, a condition known as *skip*. Ionospheric reflections vary with atmospheric conditions, and therefore are not reliable for communication paths. Services that make use of skip must have access to several frequencies in different ranges so that an alternate frequency can be used if communication is ineffective at one frequency.

Frequencies higher than about 30 MHz can be focused to travel in straight lines, which makes them ideal for short communications. The higher the frequency, the more a radio signal takes on the characteristics of light and travels as a line-of-sight space wave. At microwave frequencies, the antenna focuses radio energy so a maximum amount of energy is radiated toward the receiving antenna. Provided a microwave system is properly engineered, it is a cost-effective and reliable method of communication.

When designing microwave systems, engineers take into account the following conditions that affect signals:

- *Free-space loss*, which is the attenuation the signal undergoes as it travels through the atmosphere.

FIGURE 18-2

Radio Wave Propagation

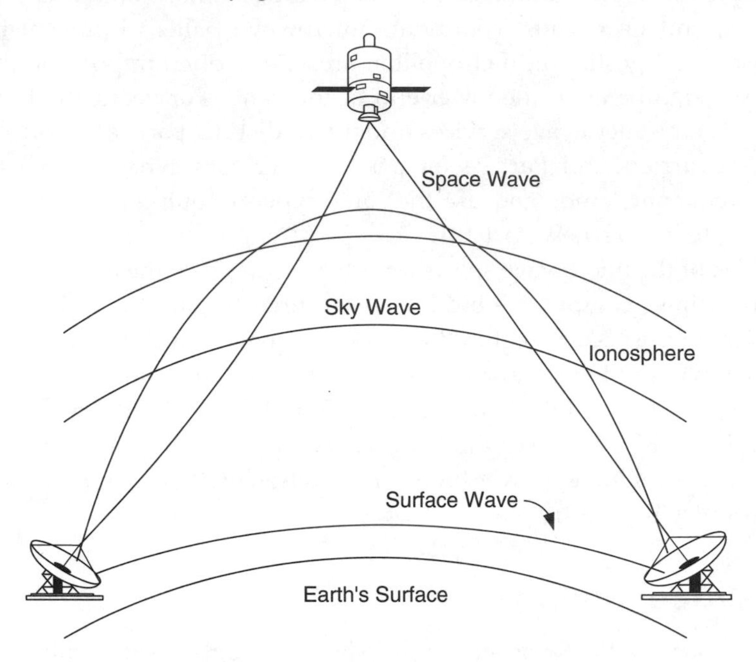

- *Atmospheric attenuation* is closely related to free-space attenuation. Changes in air density and absorption by atmospheric particles and water density attenuate the signal.
- *Reflections* can occur when the signal traverses a body of water or a fog bank. The signal takes multiple paths, which arrive at the receiving antenna out of phase and cause the signal to fade.
- *Diffraction* occurs as a result of the terrain the signal crosses.
- *Rain attenuation* occurs when raindrops absorb or scatter the microwave signal. The effect is greater at higher frequencies and varies with the size of the raindrops. Larger drops are more detrimental than fine mist.

Both analog and digital microwave systems are available, but as with other telecommunications technologies current products are largely digital using complex modulation schemes from 4 to 128 QAM to support bandwidths ranging from one DS-1 to OC-3/STM-1. A digital microwave system consists of three major components: the digital modem, the RF unit, and the antenna. The RF unit is typically connected directly to the antenna, and mounted on a rooftop, tower, building mast, or tripod. Systems with the RF unit separated from the antenna are also available. In such systems the antenna connects to the RF unit with waveguide, which is a rectangular or round section of low-loss pipe.

The antenna focuses and radiates the signal to the receiving location. To comply with zoning restrictions that regulate appearance, the antenna is sometimes mounted indoors behind glass. The glass must be chosen to avoid loss. Loss varies with the thickness of the glass, and lead compounds in some types of glass attenuate the signal. The angle of incidence or the angle at which the signal penetrates the glass affects the amount of loss. Reflections also attenuate the signal.

Microwave antennas are susceptible to snow and icing. A radome can be installed on the antenna to keep ice and snow from affecting the parabolic shape of the antenna. Heaters prevent interfering elements from building up on the radome.

In point-to-point microwave, the distance the signal travels is a function of system gain. As with fiber optics, system gain is the algebraic difference between the transmitter output and the receiver sensitivity. In long-haul microwave, repeaters, which are transmitter and receiver units connected back to back, extend the range. Passive repeaters, which resemble large billboards, can be used to reflect the signal when line of sight cannot be obtained.

MICROWAVE SYSTEMS

The general principles of microwave radio are the same as those of lower frequency radio. An RF signal is generated, modulated, amplified, and coupled to a transmitting antenna. It travels through free space to a receiving antenna where a receiver captures a portion of the radiated energy and amplifies and demodulates it. The primary differences between microwave and lower frequency radio are the wavelength and behavior of the radio waves. For example, VHF television channel 2 has a wavelength of about 6 m (20 ft). To gain the maximum efficiency, a half-wave antenna receiving element is about 3 m (10 ft) long. A 4 GHz microwave signal has a wavelength of about 7.5 cm (3 in.), so an effective antenna at microwave frequencies is small compared to those at lower frequencies.

Since microwave frequencies behave similarly to light waves, they can be focused with large parabolic or horn antennas similar to the antennas shown in Figure 18-3. Unlike lower frequencies where radio waves cannot be focused narrowly enough to prevent them from radiating in all directions, microwave stations can operate in physical proximity on the same frequency without interference.

On the minus side of the ledger, microwaves have some of the undesirable characteristics of light waves, particularly at the higher frequencies. The primary problem is fading, the causes of which are multipath reflections and rain attenuation. Multipath reflections occur when the main radio wave travels a straight path between antennas, but a portion of it reflects over a second path as shown in Figure 18-4. The reflected path is caused by some changing condition such as a temperature inversion, a heavy cloud layer, or reflection off a layer of ground fog. The reflected wave, taking a longer path, arrives at the receiving antenna slightly out of phase with the transmitted wave. The two waves added out of phase cause

FIGURE 18-3

Parabolic Horn Antennas (Photo by Author)

FIGURE 18-4

Direct and Reflected Paths between Antennas

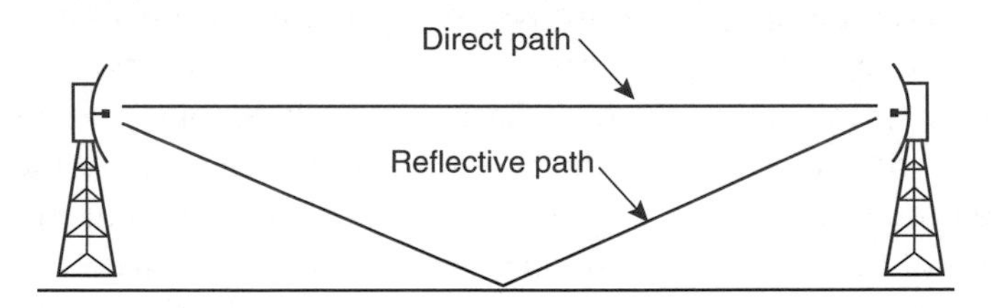

a drop in the received signal level. Several companies are working on a technique known as MIMO (multiple input–multiple output) to take advantage of multipath reflections. By segmenting the signal and transmitting and receiving on multiple antennas, MIMO uses reflections to obtain a reliable signal beyond line of sight. Chapter 21 discusses MIMO in more detail.

A second cause of fading is heavy rain, which absorbs part of the transmitted power. The effect is most detrimental at frequencies higher than about 10 GHz. Diversity, as described in a later section, alleviates the two primary causes of microwave path disruption, which are fading and equipment failures.

Microwave Repeaters

Point-to-point microwave routes are established by connecting a series of independent radio paths with repeater stations. Line of sight is required between the transmitting and receiving antennas for all microwave systems except MIMO and those that use forward scatter techniques to transmit beyond the horizon. Repeater spacing varies with frequency, transmitter output power, antenna gain, antenna height, receiver sensitivity, number of voice frequency channels carried, free space loss of the radio path, and depth of expected fading. At the high end of the band, repeaters are sometimes spaced as close as 2 km (1.2 miles). At the low end of the band and given sufficient antenna height, repeater spacing of up to 160 km (100 miles) is sometimes possible, but 40 to 50 km (25 to 30 miles) is more typical. The transmit frequency is usually shifted at the output of each repeater so it cannot feed past one repeater to the receiver of the next one in line.

Modulation Methods

Microwave systems are modulated with either digital or analog FM or AM signals. Most of the radio systems being installed today use digital microwave modulated with QAM. A point-to-point radio system regenerates the signal at each repeater point. If the incoming signal is sufficiently free of errors, digital radio provides a high-quality, low-noise channel. Unlike analog radio, which becomes progressively noisy during fades, digital radio remains quiet until it fades to a failure threshold, at which point the bit error rate (BER) becomes excessive and the radio is unusable.

Although each repeater regenerates the signal, errors are cumulative from station to station and cannot be corrected unless the radio employs forward error correction. Therefore, the errors that occur in one section repeat in the next section where additional errors may occur, until finally the signal becomes unsuitable for data transmission. For voice, however, the errors have little effect.

Spread spectrum modulation is one method of improving error performance. Spread spectrum is not technically a modulation method, but is a means of spreading the modulated signal over a broad band of frequencies. As an interesting historical note, the original patent for spread spectrum was awarded to Hollywood movie actress Hedy Lamarr during World War II. The method, known as *frequency hopping*, synchronized transmitter and receiver frequencies with perforated paper rolls similar to a player piano. The method became a top-secret method of preventing enemy interception of radio signals. A second method known as *direct sequence spread spectrum* (DSSS) spreads the signal across a broad band and the receiver picks its discreet signal from among a cacophony of signals. Cellular uses this method for CDMA and we discuss it further in Chapter 20.

Bit Error Rate

The most important measure of digital microwave performance is the BER. BER is expressed as the number of errored bits per transmitted bit, and usually is abbreviated as an exponential fraction. Microwave system-gain specifications are often quoted at a BER of 10^{-6}, which is one error per million transmitted bits, a BER that is generally accepted as the highest that can be tolerated. At a BER of 10^{-3} a radio is considered failed, although voice transmission is not noticeably impaired at this error rate.

Diversity

To guard against the effects of equipment and path failure, microwave systems use protection or diversity. Engineers often space receiving antennas a few meters apart on the same tower. This *space diversity* system protects against multipath fading because the wavelength of the signal is so short that the phase cancellation that occurs at one location will have little effect on an antenna located a few feet away.

Another system is *frequency diversity* in which a separate radio channel operating at a different frequency can assume the load of a failed channel. When fades occur they tend to affect only one frequency at a time, so frequency diversity provides a high degree of path reliability. The primary disadvantages of this method are the use of the extra frequency spectrum and the cost of the additional radio equipment.

FCC rules do not permit frequency diversity in most non-common-carrier frequency bands. Therefore, many microwave systems use *hot-standby* diversity. In a hot-standby system, two transmitter and receiver pairs are connected to the antenna, but only one system is working at a time. When the working system fails, the standby unit automatically assumes the load. Hot-standby protection is effective only against equipment failure. Hot standby cannot protect against fading and absorption, which affect the microwave path between stations.

The received noise level in an analog radio or BER in a digital radio initiates transfer to the backup. When the noise or BER on a protected channel reaches the threshold, the switch initiator sends an order to the transmitting end to switch the input signal to the protection channel. Technicians can initiate switches manually to clear a channel for maintenance. In any protection system, some loss of signal is experienced before the protection channel assumes the load. This signal loss is called a *hit*. Many systems can perform a hitless switch when a channel is manually transferred, but if equipment fails or fades, degradation will be experienced in the form of noise, excessive data errors, or both.

Protection systems protect working channels on a one-for-one or one-for-*N* basis with *N* being the number of working channels on the route. The FCC does not permit one-for-one protection where the application requires more than one radio channel because this method is wasteful of frequency spectrum. Figure 18-5 illustrates the three applications of protection—frequency diversity, space diversity, and hot standby.

Microwave Impairments

Microwave signals are subject to impairments from these sources:

- Equipment, antenna, and waveguide failures
- Fading and distortion from multipath reflections
- Absorption from rain, fog, and other atmospheric conditions
- Interference from other signals

FIGURE 18-5

Microwave Diversity Systems

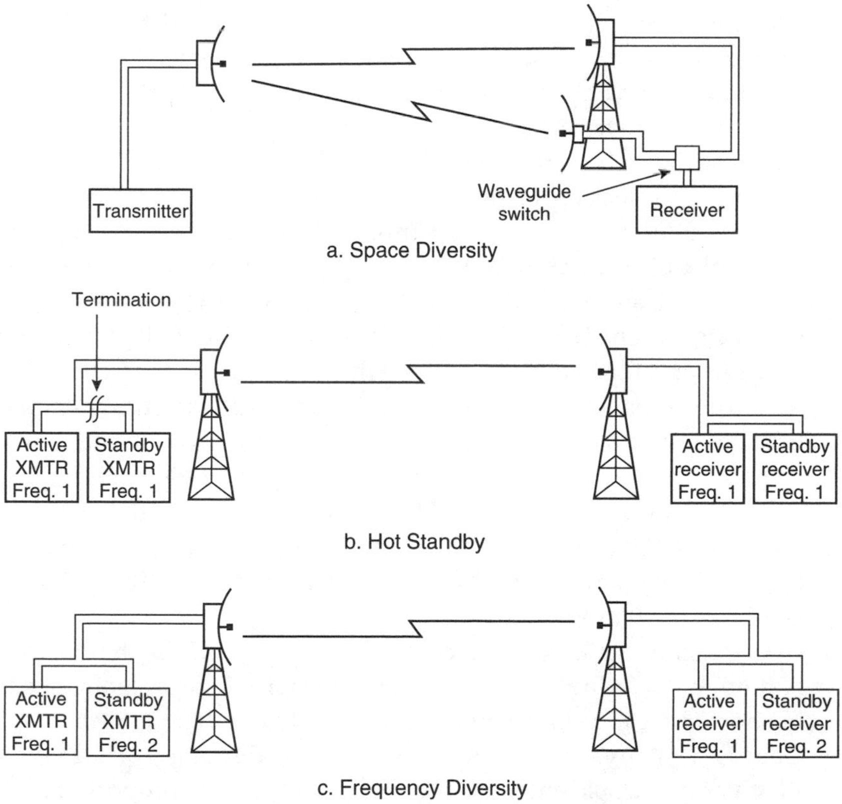

a. Space Diversity

b. Hot Standby

c. Frequency Diversity

Reliability is expressed as percent availability, or uptime, which is the percentage of the time communications circuits on a channel are usable. The starting point on a microwave path calculation is to determine the number of hours of path downtime that can be tolerated in a year. For example, 8 h/year of path outage would equate to 99.91 percent availability from the following formula:

$$\text{Percent Availability} = 1 - (\text{outage hours}/8760\ \text{h/year})$$

Because of path uncertainties, a satisfactory reliability level is attainable only with highly reliable equipment. Fortunately, equipment reliability has progressed to the point that equipment failures cause little downtime, and those failures that do occur can be protected by hot-standby diversity. Most private microwave systems require at least 99.99 percent availability. Bear in mind that microwave path failures do not usually last long. An hour per year of outage caused by rain fades is more likely to occur as 600 outages of 0.1 min each than as one failure of 60 min.

Microwave Path Analysis

Microwave path reliability is less predictable and controllable than equipment performance. The first factor to consider in laying out a microwave system is obtaining a properly engineered path. The path designer selects repeater sites for availability of real estate, lack of interference with existing services, accessibility for maintenance, and sufficient elevation to overcome obstacles in the path. Path loss and antenna height standards are defined in TIA/EIA RS 252-A.

The first step in path analysis is to prepare a balance sheet of gains and losses of the radio signal between transmitter and receiver. Gains and losses are measured in decibels or dB. Absolute measures of signal level are measured in decibels compared to 1 mW (0 dBm). One milliwatt is equal to 1×10^{-3} W of power. A signal of +30 dBm is equal to 1 W, a signal output that is typical of many microwave transmitters. The worksheet in Table 18-2 shows a sample microwave path calculation. The following explains the elements that comprise the calculation and provide information from which a similar worksheet can be constructed.

The transmitter output power is obtained from the manufacturer's specifications. For systems in the lower end of the microwave band, power is often 5 W or more; for systems in the higher end of the spectrum power is a fraction of a watt. Power outputs of +10 to +30 dBm are common at 18 and 23 GHz. Remember that 1 W is +30 dBm, so with the logarithmic nature of the decibel scale, a reduction of 10 dB is a reduction factor of 10. Therefore, +20 dBm is 0.1 W, and +10 dBm is 0.01 W or 10 mW.

Antenna gain can overcome the low powers in the 18 to 23 GHz band. Although the industry commonly uses the term antenna gain, it is somewhat of a misnomer because an antenna is a passive device that does not amplify the signal. Gain is a relative term compared to the performance of a free space mounted dipole, or isotropic antenna. The amount of gain is proportional to the physical

TABLE 18-2

Worksheet for Analyzing a Microwave Path

Path length (mile)	5.2
Frequency (GHz)	23
Gains	
Transmitter output power (dBm)	20.5
Antenna gain (dB) transmit	46
Antenna gain (dB) receive	38
Total gains	104.5
Losses	
Free space	138.2
Atmospheric	0.8
Antenna alignment	0.5
Safety factor	0.5
Total losses	140
Unfaded receive signal level	–35.5
Receiver sensitivity (dBm)	–74.5
Fade margin (dB)	39

characteristics of the antenna, primarily its diameter. For a given diameter of antenna, the gain increases as frequency increases and wavelength decreases. Because the wavelength is short at these frequencies (1 cm, or slightly less than 0.5 in., at 30 GHz), antennas with gains in the order of 40 dB or more are readily available without the high cost of antennas and tower structures that large antenna diameters require because of the stress imposed by high winds.

Antenna gain is obtained from manufacturers' specifications. The size of the antenna is one variable that can be changed to improve path reliability. Note that antenna gain operates in both transmitting and receiving directions. It is not necessary to use antennas with identical gains for both ends of a microwave path.

The factor that has the most influence on path loss is *free-space attenuation*, which can be calculated by the formula:

$$A = 96.6 + 20 \log P_L + 20 \log F$$

where

A is free-space attenuation in dB
P_L is path length in miles
F is frequency in GHz

In addition to free-space attenuation, *atmospheric losses* are caused by absorption of the signal by oxygen molecules and water vapor in the atmosphere.

This factor should not be confused with rain attenuation, which is covered later. Generally, atmospheric losses can be estimated at 0.12 dB/mile at 18 GHz and 0.16 dB/mile at 23 GHz.

The *antenna alignment factor*—usually about 0.5 dB—is a factor that designers include to reflect the imperfect alignment of antennas. When antennas are installed, they are aligned on major radiation lobes. Temperature changes and tower shifting because of wind stress may cause the alignment to drift slightly, so this factor is added to provide a safety margin. Besides the antenna alignment factor, other safety factors should be added to account for other imperfections. Usually another 0.5 dB of loss is added to be conservative.

Gains and losses are algebraically added to find the *unfaded received signal level*. This is the nominal signal level that should be received at the receiver input. The manufacturer supplies the receiver sensitivity figure as the minimum signal level that will provide a BER of 10^{-6} or better. If the unfaded received signal level is added algebraically to the receiver sensitivity, the result is the *fade margin*, which is the amount of fading the signal can tolerate.

As discussed earlier, fading results from signal absorption and multipath reflections. At frequencies above 10 GHz the raindrop size is a significant fraction of the signal wavelength (wavelength is 3 cm or 1.2 in. at 10 GHz). The rain rate that will attenuate the microwave signal by an amount equal to the fade margin is called the *critical rain rate*. The most important factor is not so much the amount of rain that falls, but the nature of the rain. The larger the raindrops and the more intense the rainfall, the greater the attenuation.

Rain absorption is most significant in areas of heavy rainfall with large drop size such as the Gulf Coast and southeastern United States. Conventional diversity is not effective against rain absorption because rain fading is not frequency selective. The most effective defenses are frequency diversity using a lower band such as 6 GHz, if permitted by the FCC, or improving the fade margin by using large antennas and closely spaced repeater stations. The easiest method of obtaining rainfall data is from the microwave manufacturers, who usually can estimate the number of minutes and frequency of outage that will be caused by rain in your part of the country.

Fresnel Zone Clearance

If insufficient clearance exists over buildings, terrain, or large bodies of water, the path will be unreliable because of reflection or path bending. The amount of clearance required is expressed in terms of *Fresnel zones*, which are imaginary elliptical zones surrounding the direct microwave beam. The first Fresnel zone is calculated by the formula:

$$\mathrm{FZ}_1 = 72.2\sqrt{\frac{D_1 \times D_2}{F \times D}}$$

where

FZ_1 is the radius of the first Fresnel zone in feet
D_1 is the distance from the transmitter to the reflection point in miles
D_2 is the distance from the reflection point to the receiver in miles
F is the frequency in GHz
D is the length of the signal path in miles.

To illustrate the principle of Fresnel zone calculations, refer to Figure 18-6, in which a signal is beamed between two buildings 5.2 mile apart with an obstruction 2.1 mile from one transmitter. The first Fresnel zone is calculated to be 16.8 ft. For best results, the clearance over the obstacle should be one Fresnel zone, but satisfactory results will usually be obtained if the clearance is at least 0.6 Fresnel zone, which in this case is 10 ft. If the clearance is insufficient, multipath fading will result.

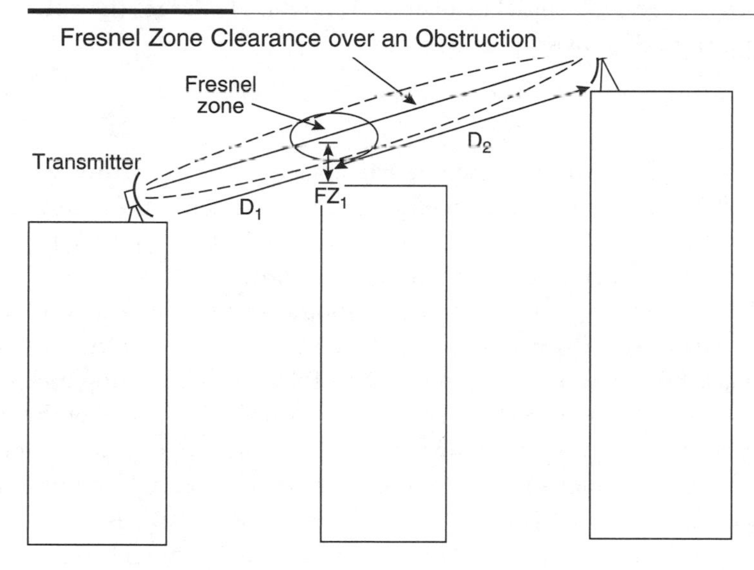

FIGURE 18-6

Fresnel Zone Clearance over an Obstruction

$$FZ_1 = 72.2\sqrt{\frac{D_1 \times D_2}{F \times D}}$$

$$= 72.2\sqrt{\frac{3.1 \times 2.1}{23 \times 5.2}}$$

$$= 72.2\sqrt{\frac{6.51}{119.6}}$$

$$= 72.2\sqrt{.0544}$$

$$= 16.8 \text{ feet}$$

D = 5.2 miles
D_1 = 2.1 miles
D_2 = 3.1 miles
F = 23 GHz

Multipath Fading

When the main wave and a reflected wave arrive at the receiving antenna slightly out of phase, the phase differences reduce the received signal level. Multipath reflections usually do not affect all frequencies within a band equally, which results in signal distortion within the received passband. Distortion is of particular concern with digital microwave, which is susceptible to a higher BER under multipath fading conditions. One way of minimizing the effects of distortion is to use an *adaptive equalizer*, which is a device inserted in the receiver to cancel the effects of distortion within the passband. Digital radio specifications usually include the *dispersive fade margin*, which states the tolerance of the radio for the frequency selective fades that cause received signal distortion. The dispersive fade margin is typically in the range of 50 to 60 dB.

Both frequency and space diversity are effective defenses against multipath fading. Other defenses include an effective path profile study with proper site selection and sufficient tower height to provide adequate clearance over obstacles. Also, the use of large antennas focuses the transmitted signal more narrowly and increases the received signal level at the receiver. The larger the antenna, however, the more rigid the tower must be.

Interference

Adjacent channel and overreach interference are other microwave impairments. *Overreach* is caused by a signal feeding past a repeater to the receiving antenna at the next station in the route. It is eliminated by selecting a zigzag path or by using alternative frequencies between adjacent stations.

Adjacent channel interference is another potential source of trouble in a microwave system. Digital radios, particularly those using QAM, are less susceptible to adjacent channel interference than PSK and FM analog radios because of the bandpass filtering used to keep the transmitter's emissions within narrow limits. Multichannel radio installations usually employ cross-polarization to prevent adjacent channel interference. In this technique, channel combining networks are used to cross-polarize the waves of adjacent channels. Cross-polarization discrimination adds 20 to 30 dB of selectivity to adjacent channels.

MICROWAVE ANTENNAS, WAVEGUIDES, AND TOWERS

Microwave antennas are manufactured as either parabolic dishes or horns, and range in diameter from less than 1 m for short, high-frequency hops to 30 m for Earth station satellite service. At lower frequencies, microwave antennas are fed with coaxial cable. Coaxial cable loss increases with frequency; therefore most microwave systems use *waveguide* for the transmission line to the antenna. Waveguide is circular or rectangular, with dimensions designed for the frequency range. At 18 to 23 GHz, and sometimes in lower frequency bands, the RF equipment mounts directly on the antenna, which eliminates the need for waveguide.

Multiple transmitters and receivers can be coupled to the same waveguide and antenna system by *branching filters*. *Directional couplers* are waveguide hybrids that allow coupling of a transmitter and receiver to the same antenna. This technique often is used in repeater stations to permit using one antenna for both directions of transmission.

Antennas are mounted on rooftops, if possible. If more elevation is needed, they can be mounted on towers. Antennas must be precisely aligned. They are first oriented by eye or calculated azimuth and then adjusted to maximum received signal level. In orienting antennas, it is important to know the calculated received signal level and to ensure that the received signal is within 1 or 2 dB of that level. Without this benchmark it is possible that the antenna will be oriented on a minor signal lobe instead of the main lobe.

Manufacturers supply microwave towers in guyed and self-supporting configurations. Self-supporting towers require less space and must be designed more rigidly to support the antenna against the effects of weather. If enough land is available, a less expensive guyed tower can be used. The larger the antenna diameter, the more rigid the tower must be to prevent flexing in the wind. Tower rigidity is important because excessive flexing can disorient the antennas.

MICROWAVE APPLICATION ISSUES

With the availability of low-cost, short-haul equipment, microwave technology has come within the reach of many companies and can quickly pay back the initial investment in savings of common-carrier facilities. Microwave finds its most important applications in the following:

- Trunking between central offices
- Bypass T1/E1 circuits from private companies to IXCs
- Studio to transmitter video links
- Access services (Chapter 8)
- Temporary or emergency restoration of cable or fiber optic facilities
- Connecting PBXs in a metropolitan network
- Interconnecting local area networks
- Crossing obstacles such as highways and rivers

Evaluation Considerations

In applying short-haul digital microwave, consider the following factors:

- What alternatives are available? Is microwave more cost effective than common carrier facilities such as fiber optics and leased T1/E1?

- How much bandwidth is required now and in the future? The greater the amount of bandwidth, the more expensive the systems, although it is generally less expensive to purchase spare capacity with a new system than it is to add capacity later.
- What level of availability is needed to make the system feasible? With the spacing between terminals and the rainfall factor, is it possible to obtain the required availability factor with short-haul microwave?
- Is there line of sight between the terminal locations with sufficient clearance over intervening obstacles? If not, repeaters may be required. If repeaters are required, is the necessary real estate available?
- Where will the equipment be located? The most desirable location is on rooftops. If necessary, small towers can be constructed on the rooftops. Separate, ground-mounted towers are expensive and should be avoided if possible.
- What kind of specialized technical assistance will be required? Most companies require assistance with path surveys, license applications, and frequency coordination. Often, the equipment vendor can supply these.

The factors of reliability, power consumption, availability, floor space, and the ability to operate under a variety of environmental conditions are important with microwave as with other telecommunications equipment. Besides these considerations, which are covered in previous chapters, the following factors also must be evaluated:

- *System gain*: When a microwave signal radiates into free space it is attenuated by losses that are a function of the frequency, elevation, distance between terminals, and atmospheric conditions such as rain, fog, and temperature inversions. System gain is the amount of free space loss that a system can overcome. Receiver sensitivity is a measure of how low the signal level into the receiver can be while still meeting noise objectives in an analog system or BER objectives in a digital system. For example, if a microwave transmitter has an output power of +30 dBm (1 W) and a receiver sensitivity of –70 dBm, the system gain is 100 dB. With other factors being equal, the greater the system gain, the more valuable the system because repeaters can be spaced farther apart. Given the same repeater spacing, a microwave radio with higher system gain has a greater fade margin than one with lower system gain. System gain can be improved in some microwave systems by the addition of optional higher power transmitters, low noise receiver amplifiers, or both.

- *Spectral efficiency*. Microwave radio can be evaluated based on its efficiency in using limited radio spectrum. The FCC prescribes minimum channel loadings for a microwave before it is type accepted. Within the frequency band, the license granted by the FCC limits the maximum bandwidth. Where growth in voice frequency channels is planned, the ability to increase the channel loading is of considerable interest to avoid adding more radio channels. Spectral efficiency in both analog and digital radios is a function of the modulation method. The controlling factor is noise in analog radio and BER in digital radios.

- *Fade margin*. The fade margin refers to the amount of fading of the received signal level that can be tolerated before the system crashes. A crash in an analog radio is defined as the maximum noise level that the application can tolerate. In a digital microwave, fade margin is the difference between the signal level that yields a maximum permissible BER (usually 10^{-6}) and the crash level (usually 10^{-3}). Analog radios fade more gracefully than digital radios. In an analog system as the received signal drops, the channel noise level increases, but communication is usable over a margin of about 20 dB. The margin between acceptable and unacceptable performance of a digital radio is narrow—about 3 dB.
- *Protection system*. The user's availability objective determines the need for protection in a microwave system. Equipment failures and fades affect availability. Equipment availability can be calculated from the formula:

$$\text{Percent availability} = \frac{\text{MTBF} - \text{MTTR (100)}}{\text{MTBF}}$$

Availability as affected by fades can be determined by a microwave path engineering study. It is possible to calculate percent availability within a reasonable degree of accuracy for both fades and failures, but it is impossible to predict when failures will occur. Therefore, diversity may be necessary to guard against the unpredictability of failures even though the computed availability is satisfactory.

Another factor weighing in the decision to provide diversity is the accessibility of equipment for maintenance. Some short-haul microwave is mounted in an office building where it can be accessed within a few minutes. On that basis it may be reasonable to provide spares and to forego diversity to save money. In a system with remote repeaters, diversity usually is needed because of the difficulty in reaching the site in time to meet availability objectives.

- *Standard interfaces*. Digital microwave systems should be designed to connect to a standard digital signal interface such as DSX-1, DSX-2, or DSX-3. Systems designed for the operational fixed band sometimes use nonstandard interfaces such as 12 or 14 DS-1 signals. Special multiplexers are required to implement these interfaces.
- *Frequency band*. Frequency availability often dictates the choice of microwave frequency band. Where choices are available, the primary

criteria are the number of voice channels required, the availability of repeater locations, and the required path reliability. As stated earlier, path reliability decreases with increasing RF because of rain absorption. Reliability can be improved by decreasing the repeater spacing or increasing the antenna size.

- *Path engineering*. A microwave path should not be attempted without an expert path survey. Several companies specialize in frequency coordination studies and path profile studies and should be consulted about a proposed route. Sites should be chosen for accessibility and availability of real estate and a reliable power source. Engineers choose tower heights to obtain the elevation dictated by the path survey. The antenna structure must support the size of antenna in a wind of predicted velocity. Wind velocities and icing exposure for various parts of the country are specified in EIA/TIA-222 Structural Standards for Steel Antenna Towers and Antenna Supporting Structures.
- *Environmental factors*. Frequency stability is a consideration in evaluating microwave equipment. FCC rules specify the stability required for a microwave system, but environmental treatment may be needed to keep the system within its specifications. Air conditioning usually is not required, but air circulation may be necessary. Heating may be required to keep the equipment above 0°C. Battery plants lose their capacity with decreasing temperature. Therefore in determining the need for heating, designers should remember that battery capacity is lowest during abnormal weather conditions when power failures are most apt to occur.

CHAPTER 19

Satellite Communications

The telecommunications industry experienced some spectacular failures around the end of the twentieth century and the start of the twenty-first, but few could compare to satellite companies Iridium and Teledesic. Iridium opened for business in late 1998 with a series of low Earth orbit (LEO) satellites, which delivered phone connectivity anywhere on the face of the Earth without the delay of a geosynchronous satellite. Led by Motorola, Iridium spent some $5 billion, but did not attract enough subscribers to make it pay. In 2000, Iridium filed for bankruptcy and was on the verge of destroying its satellites until a group of private investors acquired them for $25 million. With lower rates and a tolerable financial structure, the service remains alive through its network of 66 satellites.

Teledesic did not get quite that far. Its original plans called for a network of 840 satellites. This dropped to 288 satellites, then in February 2002 Teledesic announced that it was planning 20 medium Earth orbit (MEO) satellites. In July 2003 the company gave up its frequency slots and put the venture on hold. A merger with another MEO company, ICO, was considered and dropped. ICO had raised $3.1 billion before filing for Chapter 11 protection in 1999. A group of investors acquired control of ICO's assets for $1.2 billion. The network is not yet in service at this writing.

Satellite communications had its beginning in 1962 with AT&T's launch of its Telstar 1. Orbiting the Earth in about 2 h, Telstar was visible from the Earth station for less than half an hour, as the antennas followed its track across the sky. Although Telstar demonstrated that the technology was feasible, low-orbiting satellites were abandoned in favor of geosynchronous satellites (GEO). Orbiting the equator at an altitude of 22,239 miles (35,790 km) GEO satellites appear stationary to the Earth station.

The LEO was still a good idea, however, because sending a radio signal up to a repeater station that far from the Earth results in a round-trip delay of about

270 ms. Most data protocols can work around this kind of latency, but for voice many people find the delay disconcerting. The growth of fiber optics has captured the voice market except for those countries where nothing but satellite is available.

Although the delay makes satellites less than ideal for voice communications, they have plenty of applications that cannot readily be filled by any other alternative. Communications satellites are used for global positioning, communications with ships at sea, telemetering data from trucks in transit, and for many other applications where the user is either moving or in a remote area or both. Another application, direct broadcast television, was a long time in coming, but now high-quality TV can be received, even in remote locations. Very small aperture terminal (VSAT) enables users to mount small antennas on rooftops to run a multitude of applications such as point-of-sale, which need low bandwidth facilities distributed over a wide range.

SATELLITE TECHNOLOGY

Figure 19-1 shows the relative position of the three orbit classes—LEO, MEO, and GEO. At geosynchronous orbit, the satellite travels at the same speed as the Earth's rate of spin, so the orbiting vehicle remains at a fixed position with relation to the Earth station. From geosynchronous orbit, three satellites can theoretically cover the Earth's surface, except for the polar regions, with each satellite subtending a radio beam 17° wide. The portion of the Earth's surface that a satellite

FIGURE 19-1

Communication Satellite Orbits

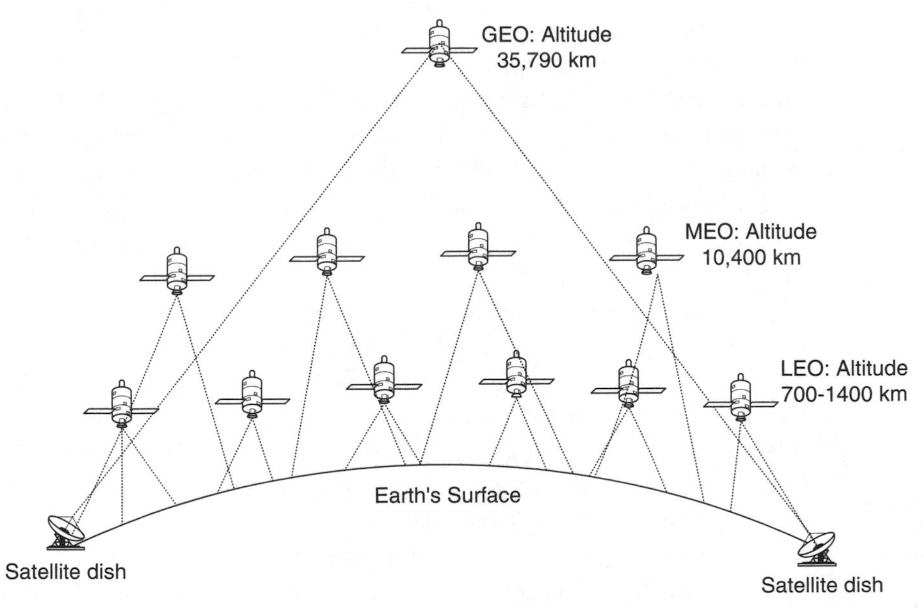

illuminates is called its *footprint*. MEO and LEO satellites have the advantage of low delay, but at the expense of needing more satellites and tracking antennas to provide the same coverage.

Satellites fall into three general categories—domestic, regional, and international. Domestic satellites carry traffic within one country. Regional satellites span a geographical area, such as Europe, and international satellites are intended for intercontinental traffic. Although undersea fiber-optic systems carry most of the international voice traffic, international television is still a large and growing market for satellites. International satellite communications are provided by Intelsat Ltd, which became a private company in 2001 following 37 years as an intergovernmental organization. Intelsat operates more than 20 satellites and serves about 150 countries at this writing.

As Table 19-1 shows, the frequencies available for communication satellites are limited. The lower frequency is always used from the satellite to the ground because Earth stations' transmitting power can overcome the greater path loss of the higher frequency, but solar battery capacity limits satellite output power. The 4- and 6-GHz C-band frequencies are the most desirable from a transmission standpoint because they are the least susceptible to rain absorption. Satellites share the C-band frequencies with common carrier terrestrial microwave, requiring close coordination of spacing and antenna positioning to prevent interference. Interference between satellites and between terrestrial microwave and satellites is prevented by using highly directional antennas. Currently, satellites are spaced about the equator at 2° intervals.

The Ku-band of frequencies has come into more general use as the C-band becomes congested. K-band frequencies are exclusive to satellites, allowing users to construct Earth stations almost anywhere, even in metropolitan areas where congestion often precludes placing C-band Earth stations. The primary disadvantage of the Ku-band is rain attenuation, which results in lower reliability. With identical 2° spacing for both C and Ku-bands, most satellites carry transponders for both bands.

Ka-band satellites are becoming more attractive as the lower frequencies become congested. Although the higher frequency of Ka-band means a higher probability of fading, it is possible to use smaller antennas and inexpensive Earth stations, which offset fading to some degree. Although considerable bandwidth

TABLE 19-1

Principal Communication Satellite Frequency Bands

Band	Uplink	Downlink
C	5925 to 6425 MHz	3700 to 4200 MHz
Ku	14.0 to 14.5 GHz	11.7 to 12.2 GHz
Ka	27.5 to 31.0 GHz	17.7 to 21.2 GHz

is available in Ka and higher frequencies, further development is needed before these come into general use.

Satellites have several advantages over terrestrial communications. These include:

- The receiving station can be mobile, which offers coverage from a single satellite.
- Within the coverage range of a single satellite, circuit cost is independent of distance.
- Impairments that accumulate on a per-hop basis on terrestrial microwave circuits are avoided with satellites because the Earth-station-to-Earth-station path is a single hop through a satellite repeater.
- Sparsely populated or inaccessible areas can be covered by a satellite signal, providing high-quality communications service to areas that are otherwise difficult or impossible to reach.
- Coverage is also independent of terrain and other obstacles that may block terrestrial communications.
- Earth stations can verify their own data transmission accuracy by listening to the return signal from the satellite.
- Because satellites broadcast a signal, they can cover wide areas.
- Large amounts of bandwidth are available over satellite circuits, making voice, video, and high-speed data circuits available.
- The satellite signal can be brought directly to the end user, bypassing the local telephone facilities that are expensive and limit bandwidth.
- Multipath reflections that impair terrestrial microwave communications have little effect on satellite radio paths.

Satellites are not without limitations, however. The greatest drawback is the lack of frequencies. If higher frequencies can be developed with reliable paths, plenty of spectrum is available, but atmospheric limitations may prevent their use for commercial-grade telecommunications service. Other limitations include:

- Multihop satellite connections impose delay that is detrimental to voice communications and is generally avoided.
- Path loss is high (about 200 dB) from Earth to satellite.
- Rain absorption affects path loss, particularly at higher microwave frequencies.
- Frequency crowding in the C-band is high with potential for interference between satellites and terrestrial microwave operating on the same frequency.

The rapid growth of fiber-optic systems has had an adverse effect on satellites' share of the telecommunications market, but the technology shows no signs of dying. Though the satellites' market share may be dropping, the traffic carried by communications satellites continues to increase and will do so in the future.

Satellite Systems

A satellite circuit has five elements as shown in Figure 19-2: two terrestrial links, an uplink, a downlink, and a satellite repeater. The satellite has six subsystems described below:

- physical structure
- transponder
- attitude control apparatus
- power supply
- telemetry equipment
- station-keeping equipment

Physical Structure

The size of communications satellites has been steadily increasing since the launch of the first commercial satellite in 1965. Size is limited by the capacity of launch vehicles and the need to carry enough solar batteries and fuel to keep the system alive for its design life of 5 to 10 years. Advances in space science such as launch vehicles that can carry greater payloads are making larger satellites technically feasible. A large physical size is desirable. Not only must the satellite contain the radio and support equipment, but it must also provide a platform for large antennas to obtain the high gain needed to overcome the path loss between the Earth station and the satellite.

Transponders

A transponder is a radio relay station on board the satellite. Transponders are technically complex, but their functions are identical with those of terrestrial microwave

FIGURE 19-2

Communication Satellite System

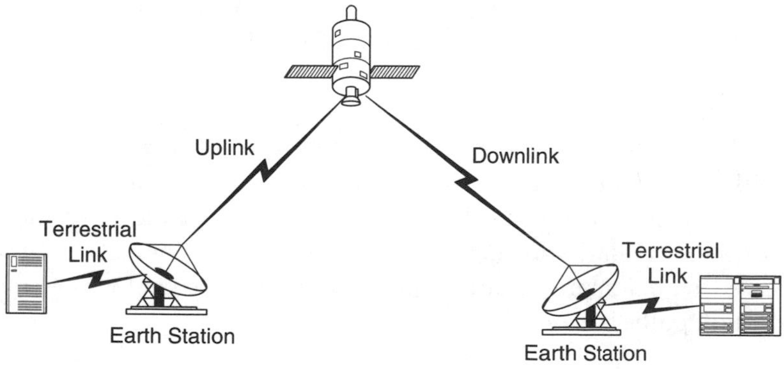

radio relay stations. A receiving antenna picks up the signal from the Earth station and amplifies it with a low noise amplifier (LNA), which boosts the received signal. The LNA output is amplified and converted to the downlink frequency. The downlink signal is applied to a high-power amplifier, using a traveling wave tube or solid-state amplifier as the output device. The output signal couples to the downlink or transmitting antenna. Solid-state amplifiers are popular because of their high reliability. Most satellites carry multiple transponders, each with a bandwidth of 36 to 72 MHz. For example, Echostar 9 (also known as Telstar 13 and Intelsat Americas 13), which was launched in August 2003, has 2 Ka-, 32 Ku-, and 24 C-band transponders. It covers the United States from 121° west longitude and operates with 120 W of power.

Attitude Control Apparatus

Satellites must be stabilized to prevent them from tumbling through space and to keep antennas precisely aligned toward Earth. Satellite stabilization is achieved by two methods. A spin-stabilized satellite rotates on its axis at about 100 rpm. The antenna is despun at the same speed to provide constant positioning and polarization toward Earth. The second method is three-axis stabilization, which consists of a gyroscopic stabilizer inside the vehicle. Accelerometers sense any change in position in all axes and fire positioning rockets to keep the satellite at a constant attitude.

Power Supply

Satellites are powered by solar batteries. Power is conserved by turning off unused equipment with signals from the Earth. On spin-stabilized satellites, the cells mount outside the unit so that one-third of the cells always face the sun. Three-axis stabilized satellites have cells mounted on solar panels that extend like wings from the satellite body. Solar cell life is a major factor that limits the working life of a satellite. Solar bombardment gradually weakens the cell output until the power supply can no longer power the on-board equipment.

A nickel–cadmium battery supply is also kept on board most GEO satellites to power the equipment during solar eclipses, which occur during two 45-day periods for about an hour per day. The eclipses also cause wide temperature changes that the on-board equipment must withstand.

Telemetry Equipment

A satellite contains telemetry equipment to monitor its position and attitude and to initiate correction of any deviation from its assigned station. Through telemetry equipment, the Earth control station initiates changes to keep the satellite at its assigned longitude and inclination toward Earth. Telemetry also monitors the received signal strength and adjusts the receiver gain to keep the uplink and downlink paths balanced.

Station-Keeping Equipment

Small rockets are installed on GEO vehicles to keep them on station. When the satellite drifts, rockets fire to return it to position. The tasks that keep the satellite on position are called station-keeping activities. The fuel required for station keeping is the factor, with solar cell life that limits the design life of the satellite. Moveable antennas track MEO and LEO satellites, so station-keeping equipment is not required.

EARTH STATION TECHNOLOGY

Earth stations vary from simple, inexpensive, receive-only stations that individual consumers can purchase to elaborate two-way communications stations such as the one in Figure 19-3 that offer commercial satellite service. An Earth station includes microwave radio relay equipment, terminating multiplex equipment, and a satellite communications controller. The Earth stations for MEO and LEO satellites link to the PSTN for dial-up telephone service.

Radio Relay Equipment

The radio relay equipment used in an Earth station is similar to terrestrial microwave equipment except that the transmitter output power is considerably

FIGURE 19-3

Satellite Earth Station on Majuro (Photo by Author)

higher. In addition, antennas up to 30 m in diameter in GEO Earth stations provide the narrow beam width required to concentrate power on the targeted satellite.

Because the Earth station's characteristics are more easily controllable than the satellite's and because power is not limited on Earth as it is in space, the Earth station plays the major role in overcoming the path loss between the satellite and Earth. Path loss for GEO satellites ranges from about 197 dB at 4 GHz to about 210 dB at 12 GHz. In addition, the higher the frequency, the greater the loss from rainfall absorption. Therefore the uplink always operates at the higher frequency where higher transmitter output power can overcome absorption, while the lower frequency is reserved for the downlink.

GEO antennas are adjustable to compensate for slight deviations in satellite positioning. Antennas at commercial stations are normally adjusted automatically by motor drives, while inexpensive antennas are adjusted manually as needed. Thirty-meter antennas provide an extremely narrow beam width, with half-power points 0.1° wide. LEO and MEO antennas are moveable to track the satellite in its orbit.

Satellite Communications Control

A satellite communications controller (SCC) apportions the satellite's bandwidth, processes signals for satellite transmission, and interconnects the Earth station microwave equipment to terrestrial circuits. The SCC formats the received signals into a single integrated bit stream in a digital satellite system or combines FDM signals into an analog FM signal. The multiplex interface of an Earth station is conventional. Satellite circuits use either analog or digital modulation, with interfaces to frequency division and time division terrestrial circuits.

Access Control

Satellites employ several techniques to increase the traffic-carrying capacity and provide access to that capacity. Some FDMA satellites divide the transponder capacity into multiple frequency segments between endpoints. One disadvantage of FDMA is that users are assigned a fixed amount of bandwidth that cannot be adjusted rapidly or easily assigned to other users when it is idle. In addition, the guard bands between channels use part of the capacity.

TDMA time-shares the total transponder capacity. Earth stations transmit only when permitted by the access protocol. When the Earth station receives permission to transmit, it is allotted the total bandwidth of the transponder for the duration of the station's assigned time slot. A master station controls the access or the Earth station listens to which station transmitted last and sends its burst in a preassigned sequence. Each Earth station receives all transmissions but decodes only those addressed to it. TDMA provides priority to stations with more traffic to transmit by assigning those stations more time slots than it assigns to low-priority stations. Therefore, a station with a growing amount of traffic can be allotted a greater share of total transmission time.

Demand-assigned multiple access (DAMA) is an alternative to preassigned multiple access. DAMA equipment keeps a record of idle radio channels or time slots. Channels are assigned on demand by one of the three methods—polling, random access with central control, and random access with distributed control. Control messages are sent over a separate terrestrial channel or contained in a control field in the transmitted frame.

Signal Processing

The SCC conditions the signals for transmission between the terrestrial and satellite links. The type of signal conditioning depends on the service provider and may include voice compression, echo cancellation, forward error correction, and digital speech interpolation to avoid transmitting the silent periods of a voice signal.

GEO SATELLITE TRANSMISSION

Much of the previous discussion is of only academic interest to those who use satellite services. However, satellite circuits and terrestrial circuits have different transmission characteristics.

Satellite Delay

The quarter-second delay between two Earth stations is noticeable in voice communications circuits, but most people become accustomed to it and accept it as normal if the circuit is confined to one satellite hop but delay affects many data protocols. TCP/IP, for example, sees a delay in packet acknowledgement as a sign of congestion or packet loss and may force retransmission of packets that have actually been correctly received. Furthermore, TCP has a slow-start characteristic that sets its packet acknowledgement window at a single packet and gradually opens it. Therefore, performance depends on the transfer rate and the round-trip delay.

Performance problems begin to occur when TCP/IP operates over what the industry calls a "long fat pipe." For example, a T1 satellite channel has a bandwidth delay product of 100 outstanding TCP segments of 1200 bytes each, which requires a large buffer to contain unacknowledged packets. RFC 1323 discusses recommended TCP modifications to improve performance on satellite circuits. One solution is "spoofing," in which the Earth stations acknowledge the packets to the endpoints and communicate between themselves in the satellite link with a protocol that is not delay susceptible. Satellite circuits are unsatisfactory for VoIP because the combined delays are well outside the recommended range.

Rain Absorption

Rain absorption has a dual detrimental effect on satellite communications. Heavy rains increase the path loss and may change the signal polarization enough to impair the cross-polarization discrimination ability of the receiving antennas. Rain absorption can be countered by these methods:

- choosing Earth station locations where heavy rain is least likely
- designing sufficient margin into the path to enable the circuits to tolerate the effects of rain
- locating a diversity Earth station at a sufficient distance from the main station

Technical considerations may limit the first two options. Transmit power and antenna gain from the satellite can be increased only within the limits dictated by the size of the satellite and the transmit power available. Locations with low precipitation cannot always deliver service where required. These considerations mandate the use of Earth station diversity at higher frequencies.

Sun Transit Outage

During the spring and fall equinoxes for periods of about 10 min/day for 10 days, the sun is positioned directly behind the satellite and focuses a considerable amount of high-energy radiation directly on the Earth station antenna. This solar radiation causes a high noise level that renders the circuits unusable during this time. If the outage cannot be tolerated, traffic must be rerouted through a backup satellite.

Carrier-to-Noise Ratio

Satellite transmission quality is based on carrier-to-noise ratio, which is analogous to signal-to-noise ratio on terrestrial circuits. The ratio is relatively easy to improve on the uplink portion of the satellite circuit because transmitter output power and antenna gain can be increased to offset noise. On the downlink portion of a circuit, the effective isotropic radiated power (EIRP), which is a measurement of the transmitter output power that is concentrated into the downlink footprint, can be increased only within the size and power limits of the satellite or by using spot beams to concentrate signal strength.

REPRESENTATIVE SATELLITE SERVICES

In this section, three different types of services are discussed to illustrate the versatility of communications satellites. LEO is a new service that avoids the delay inherent with other satellite services. Maritime radio service is an excellent example of a service that cannot be provided in any other feasible way: communication with aircraft and ships at sea. VSAT replaces conventional terrestrial communications and offers the advantage of bringing signals directly to the user without requiring the last link in a communications path—the local telephone loop—that is often expensive and bandwidth limiting.

Low Earth Orbiting Satellite

Iridium launched commercial voice, fax, paging, and narrow-band data service in 2001, using a network of 66 operational satellites plus six in-orbit backups. The system operates at an altitude of 780 km (485 miles). Its Earth station links and intersatellite links operate in the Ka-band. Voice is digitized and compressed to 2.4 Kbps. Dial-up data can be transmitted at a maximum speed of 2.4 Kbps, with Internet access at 10 Kbps. Short messages of up to 160 characters can be received on the telephone. In addition, Iridium offers short-burst data message service, which supports messages up to 1960 bytes.

Satellites are connected via intersatellite links to the four nearest neighbors and to an Earth station. Calls can be connected to landline telephones or to other satellite phones. The satellites use the L-band (1616 to 1626.5 MHz) for communication with the subscriber terminals. Subscribers can receive and place telephone calls and pages while roaming anywhere in the world that the service is authorized.

International Maritime Satellite Service

Inmarsat began life in 1979 as an intergovernmental organization. In 1999, it was converted to private ownership with headquarters in Britain. The company provides phone, fax, and data communications to more than 287,000 ship, vehicle, aircraft, and other mobile users through a network of nine satellites. Inmarsat provides service for all types of ocean-going vessels ranging from merchant ships to yachts. The system has a network of coastal Earth stations that can communicate with ships at sea. The ship's Earth station mounts above decks and automatically stays in position with satellite tracking equipment.

Inmarsat offers a variety of fleet services, providing both ISDN and mobile packet data service (MPDS). MPDS charges by packet volume and not for the time spent online, which makes it convenient for real-time access. The service can be used for video conferencing, store-and-forward video, remote monitoring, chart and weather updates, telemedicine, and distress and safety signals.

In the past, the principal methods of communication from ships were telex and high-frequency radio, which were unreliable and expensive. Now data circuits are replacing those modes of communication. Voice circuits replace the high-frequency ship-to-shore radio that often suffered from poor reliability. Ship locations can be monitored precisely through polling equipment. Distress calls can be received and rebroadcast to ships that are in the vicinity but out of radio range. Broadcasts such as storm warnings can be made to all ships in an area.

Through their I-4 satellites, Inmarsat provides Broadband Global Area Network (BGAN) service, which supports up to 432 Kbps for Internet access, mobile multimedia, and similar applications. BGAN is also compatible with third-generation cellular service.

Inmarsat is the wholesaler of satellite airtime, which is sold by hundreds of partners. In addition, Inmarsat provides satellite services to commercial airliners and corporate jets. Inmarsat's Swift64 offers Mobile ISDN- and IP-based MPDS connectivity at 64 Kbps. United Airlines has announced the use of Swift64 to provide video surveillance of the secured flight deck via satellite. Scandinavian Airlines System is offering high-speed Internet access on several of its long-range aircraft. The availability of satellite services for Internet access will become increasingly common in the future.

Very Small Aperture Terminal

VSATs are named for the size of the transmitting antennas, which are much smaller than those used in conventional Earth stations. VSAT antennas are normally 1.8 m (6 ft) or less in diameter, which makes them easy to conceal on rooftops and in areas with zoning restrictions. A VSAT network is star-connected with a hub at the center and dedicated lines running to the host computer as shown conceptually in Figure 19-4. The hub has a larger antenna, often 4 to 11 m in diameter, aimed at the satellite. Hubs are complex and expensive, so only the largest organizations can justify a privately owned hub. Usually, the VSAT vendor

FIGURE 19-4

A VSAT System

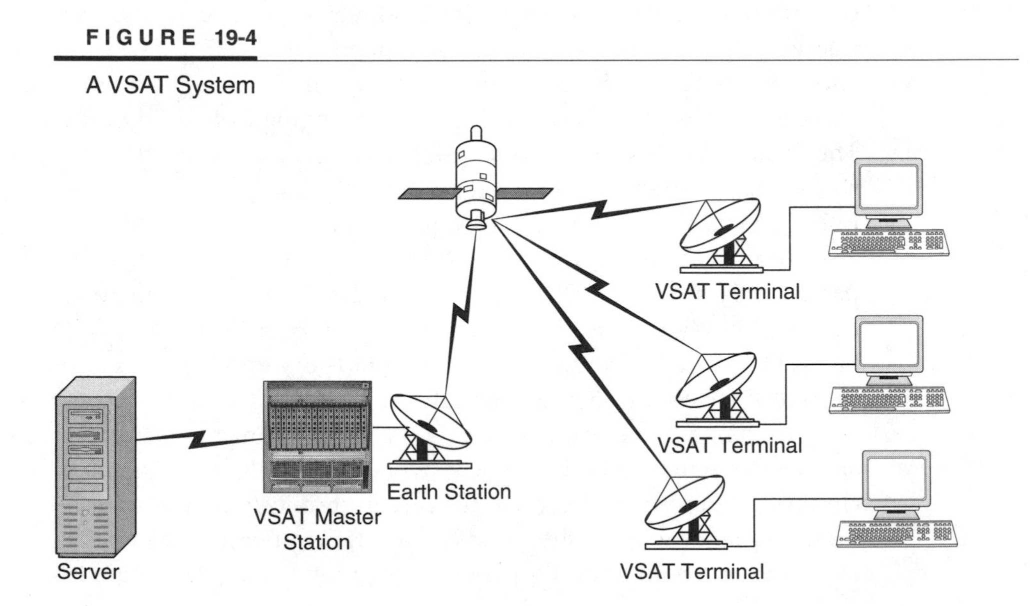

owns the hub, or one organization owns it and shares it with others. Not only is a shared hub more cost-effective for most companies, but it also relieves the company of the necessity of managing the hub, which may require one or two people per 100 nodes. Generally, a privately owned hub is feasible only when 200 or more remote stations share the service.

The hubs control demand assignment to the satellite and monitor and diagnose network performance. Demand is allocated in one of the four ways—pure aloha, slotted aloha, TDMA, or spread spectrum. The first three methods generally are used on Ku-band and the last on C-band. Pure aloha is an inefficient method of regulating access. Stations transmit at will, and when their transmissions collide they must retransmit. Slotted aloha is somewhat more efficient in that stations can transmit only during allotted time slots. TDMA and spread spectrum are the most effective ways of allocating access. VSAT provides bandwidth as high as T1/E1 and as low as the customer needs to go. It is used for voice, video, and data transmission.

The remote station has an antenna and a receiving unit, which is about the size of a personal computer base unit. The receiving unit contains a modulator/demodulator, a packet assembler/disassembler, and a communication controller. The remote transmitter operates with an output power of about 1 W. The receiver uses a low-noise amplifier.

The primary application for VSAT is data, although it can also carry voice and video. Typically, C-band VSATs carry 9.6 Kbps data, and Ku-band VSATs carry 56 or 64 Kbps data; some systems carry a full or fractional T1/E1. Most applications are two-way interactive. The primary advantage of VSAT is its ability to support multiple locations. For a few locations, the terrestrial link from the host computer to the hub plus the investment in remote stations may make VSAT prohibitively expensive. As the number of remote sites increases, however, VSAT becomes more attractive.

SATELLITE APPLICATION ISSUES

In one sense, satellite applications diminish as terrestrial and undersea fiber-optic circuits become more plentiful and economical, but satellite services are still uniquely suited for many applications. The heavy investments the major providers are making show that they expect demand to remain healthy.

Satellite Service Evaluation Considerations

Satellite space vehicle evaluation criteria are complex, technical, and of interest only to designers, owners, and manufacturers of satellites and on-board equipment. Therefore, this discussion omits these criteria. Likewise, common carrier Earth station equipment evaluations are omitted from this discussion. Evaluation criteria discussed in Chapter 18 on microwave equipment generally apply to satellite services except that multipath fading is not a significant problem in satellite services. In addition, alarm and control systems in terrestrial microwave are different from those used in satellite systems.

The following factors should be considered in evaluating satellite services and privately owned Earth station equipment.

Availability

Circuit availability is a function of path and equipment reliability. To the user of capacity over a carrier-owned Earth station, equipment reliability is a secondary consideration. The important issue is circuit reliability measured as percent error-free seconds in digital services and percent availability within specified noise limits for analog services.

These same availability criteria apply with privately owned Earth stations, but the carrier can quote availability based only on path reliability. Equipment availability depends on MTBF and MTTR and must be included in the reliability calculation. The frequency and duration of any expected outages because of solar radiation or solar eclipse should be evaluated. Availability figures of 99.5 to 99.9% are typical.

Data Bandwidths

Satellite carriers typically provide transponder bandwidth in 128 Kbps segments. The number of stations that can be supported in this bandwidth depends on the amount of activity. VSAT is generally cost-competitive with frame relay in large networks and offers a similar type of service, but with somewhat lower throughput for a given bandwidth of the access facility. The applications of VSAT are similar to those for frame relay with certain exceptions. First, the cost of a frame relay access channel varies with the distance from the carrier's point-of-presence. VSAT is not distance-sensitive except for the cost of a backhaul circuit to carry data from the Earth station to the customer's site. The backhaul circuit can be a point-to-point circuit, IP, or frame relay.

Access Method

Satellite carriers employ several techniques to increase the information-carrying capacity of the space vehicle. Techniques such as DAMA can result in congestion during peak load periods and the possibility that Earth station buffer capacity can be exceeded or access to the system blocked. Users should determine what methods the carrier uses to apportion access, whether blockage is possible, and whether transmission performance will meet objectives.

Transmission Performance

The carrier's BER, loss, noise, echo, envelope delay, and absolute delay objectives should be evaluated. To support TCP/IP, the carrier should provide spoofing. Except for absolute delay, which cannot be reduced except by using terrestrial facilities to limit the number of satellite hops, satellite transmission evaluation should be similar to terrestrial circuits.

Earth Station Equipment

Earth station equipment is evaluated against the following criteria:

- equipment reliability
- support for specific protocols such as TCP/IP
- technical criteria, such as antenna gain, transmitter power, and receiver sensitivity, that provide a sufficiently reliable path to meet availability objectives
- antenna positioning and tracking equipment that is automatically or manually adjustable to compensate for positional variation in the satellite
- physical structure that can withstand the wind velocity and ice-loading effects for the locale
- the availability of radome or deicing equipment to ensure operation during snow and icing conditions

Network Management Capability

Network management is important in VSAT networks where many Earth stations are under the control of a single hub. The service provider should be able to reconfigure the network rapidly from a central location. Monitoring and control equipment should be able to diagnose problems and detect degradations before hard faults occur. The network management package should collect statistics on network use and provide information for predicting when growth additions will be required. Determine whether the network provider can service all network components including routers.

CHAPTER 20

Mobile, Cellular, and PCS Radio Systems

In 1946, cartoonist Chester Gould tickled the fancies of readers of the comic strip Dick Tracy, with a two-way wrist radio. Such a device was then impossible outside the realm of fiction. Today, a two-way wristwatch radio is not only feasible; it hardly rates a shrug from anyone who has kept pace with technology. NTT DoCoMo produces a wristwatch phone that would have boggled Mr. Gould's imagination had he visualized the technology that is crammed into such a small space. The company's wrist 3G cell phone far outstrips Dick Tracy's phone, which was a mere two-way device that lacked the signaling and roaming capabilities of a cell phone, let alone Web browsing. The next version is even more futuristic. Although it is not on the market at the time this is written, the company claims that it will signal incoming calls with a vibration in the bones of the index finger. If that is not dazzling enough, the users poke their fingers in their ears and the vibrations turn to voice.

It is difficult to remember when owning a cell phone was a matter of prestige; a tool for the affluent, but that was a scant two decades ago. Wander any place on the face of the earth today, and you are apt to see people with tiny transceivers clutched to their ears. Satellites have removed humanity's last refuge from the reach of the telephone and users are embracing the service enthusiastically. Cellular and PCS (personal communication service) have demonstrated that the cost and lower quality of ubiquitous telephone service is more than outweighed by the convenience. The FCC gave the technology another boost, at least as far as users are concerned, by requiring cellular providers to support number portability.

Cellular started out as a separate technology from PCS, but many instruments today can hop from 850 MHz cellular bands to the 1.9 GHz PCS band and the users do not know the difference, nor do they care provided the quality is reasonable. Users willingly tolerate quality that is often inferior to the PSTN in favor of the convenience, proving that toll-quality connections are not always needed.

The history of mobile radio is short compared to other voice technologies. The first documented use of mobile radio was by the Detroit Police Department in 1921 and the technology thrived during World War II. Mobile telephone service began in 1946, but the transceivers then were primitive tube-type devices that filled half of a car's trunk. A year later Bell Laboratories developed the concept of cellular radio, but vacuum tube circuitry was too bulky to make it practical, so cellular had to wait for the transistor's reduced size and power drain. Also, the switching systems available at the time lacked the intelligence needed to control the cell sites and handoff calls to another cell as the vehicle moved. Cellular telephones became feasible only with solid-state transceivers and electronic switching systems.

Meanwhile, conventional two-way mobile radio advanced rapidly. Police, taxicabs, utilities, farmers, and construction workers all rely heavily on conventional mobile and handheld radios. Radio is one of the enabling technologies supporting vast industries that would otherwise be impractical. The airlines, public safety, and all companies that use radio dispatch are based on access to the airways.

Citizens band radio has long been available to the public at low cost, but mobile and handheld radio connected to the PSTN was too costly for general use. As we will discuss later in this chapter, the early versions of public mobile radio shared a limited frequency resource and either manual operation or an awkward signaling arrangement. As solid-state electronics advanced, radios shrank in size, power consumption, and cost, but cellular was the development that turned mobile radio into a household utility.

Technology has not only decreased the size of radios, it also has found ways to pack more channels into limited frequency spectrum. Digital radio, which once required more bandwidth than its analog counterpart, now provides as much as a 20-to-1 bandwidth advantage in the same spectrum. Spread-spectrum technology enables transceivers to use the same frequency without mutual interference, and the push to ever-higher frequencies results in more directionality, which enables channel reuse with less physical separation than the lower frequencies require.

CONVENTIONAL MOBILE TELEPHONE TECHNOLOGY

The term mobile radio often is used synonymously with mobile telephone. Although the two services use the same technology and equipment, they differ in these ways:

- Mobile telephones use separate transmit and receive frequencies, enabling full-duplex operation. Mobile two-way radios operate either on the same frequency in a simplex mode or on different frequencies in a half-duplex mode.

- Mobile telephones are connected directly to the telephone network and can be used to originate and terminate telephone calls with billing rendered directly to the mobile telephone number. Mobile radio, if connected to the telephone network, connects through a coupler to a telephone line. Billing, if any, is to the wireline telephone.
- Mobile telephones signal on a dialing plan that is compatible with the nation's plan. Mobile radios use loudspeaker paging or selective signaling that does not fit into the national dialing plan.

Specialized mobile radio (SMR) is a hybrid service that has many of cellular's characteristics, but a different network architecture and frequency assignment. SMR, originally set up as a local dispatch service, has roaming capability with either an interconnected or a dispatch mode. In the interconnected mode, sessions connect through the PSTN. Dispatch mode provides two-way over-the-air communications. This mode is sometimes called "push-to-talk" a service that some cellular carriers also offer. The service is available in a trunked mode, which enables the mobile device to search for an available channel.

As an aid to understanding cellular radio, it is instructive to review the operation of its predecessor, conventional mobile telephone service. In Part 22 rules, the FCC authorizes Public Mobile Service in the ranges of 152.03 to 152.81, 157.77 to 158.67, 157.77 to 158.67, 454.025 to 454.650, 459.025 to 459.650 MHz bands. Coverage in these frequency ranges is essentially line of sight with the lower frequencies providing the widest coverage. The FCC requires that service providers on these channels must also provide rural telephone service if a subscriber requests it.

Conventional mobile telephone service suffers from several drawbacks that led to its replacement. In its heyday, demand greatly outstripped the channels available. Also, a mobile telephone channel is a large party line with the disadvantages of limited access and lack of privacy. Handoff capability is not part of the service, so when a vehicle leaves a coverage area, quality deteriorates. When the signal drops off, the user must end the conversation and call on a different channel.

A metropolitan mobile telephone service area has transmitters centrally located and operating with 100 to 250 W of output. Because of the difference between mobile and base station transmitter output power, common carriers often install receivers in more than one site to improve coverage, as shown in Figure 20-1. These receivers are called *voting receivers* because a central unit measures the relative signal-to-noise ratio of each receiver and selects the one with the best signal. This improves the power balance between the mobile, which has relatively low output and a low-gain antenna, and the base station unit. Most coverage areas have several radio channels. Transceivers can shift between channels within the same band, but not between bands.

For the first 20-plus years of mobile telephone use, the LECs operated the service manually. Users placed calls by lifting the handset and keying the transmitter

FIGURE 20-1

Diagram of Conventional Mobile Telephone Service

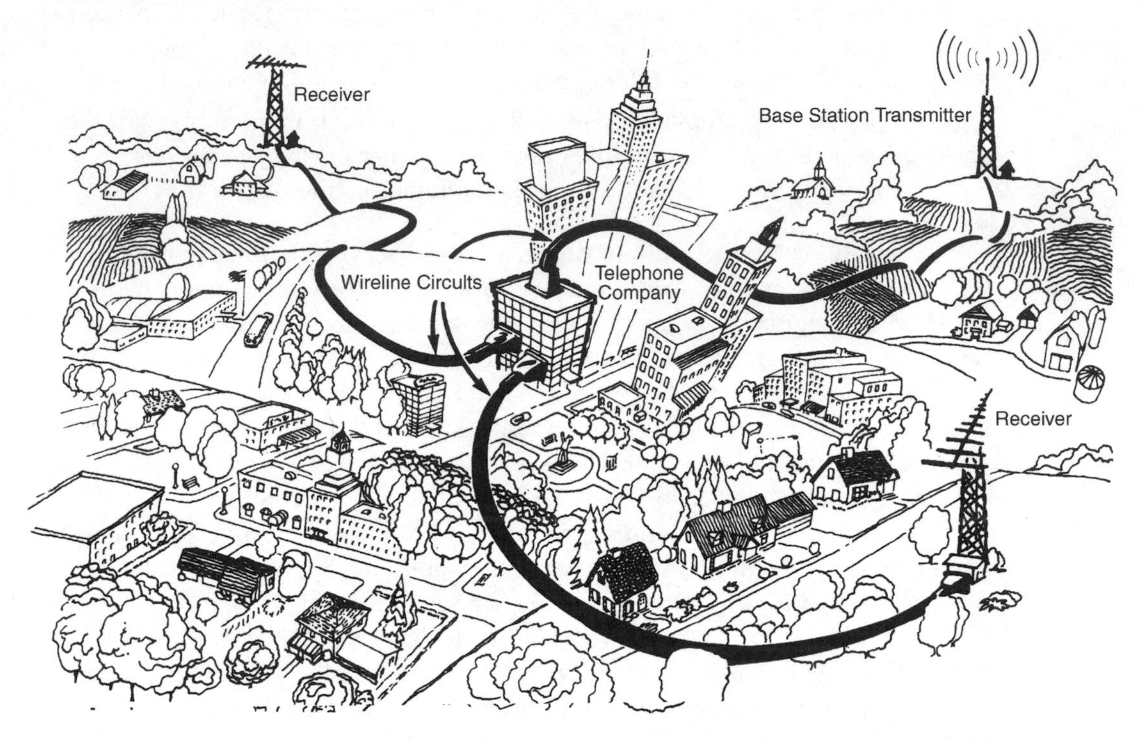

on momentarily to signal the operator. The operator connected and timed the call to a wireline telephone or other mobile unit. In 1964, AT&T introduced Improved Mobile Telephone Service (IMTS) to align mobile telephone service more closely with ordinary telephones. The IMTS mobile receiver automatically seeks an idle channel and tunes the transceiver to that channel. IMTS lacks the roaming characteristics of cellular, however. Mobile users have a designated home channel and can be called only while they tune that channel. When they leave their home areas, they must inform potential callers of what channel they are monitoring, or they cannot be called.

With both manual and IMTS systems, the base station configuration presents several disadvantages. The coverage area of a base station is more or less circular. The actual coverage area depends on the directionality of the antenna system and on the terrain. Obstructions are a problem with ordinary mobile telephone service. A hill some distance from the base station typically creates a radio signal shadow on the side away from the transmitter. When the user leaves the coverage area of the channel on which the call was established, the call must be terminated and re-established on another channel. Cellular and PCS are not perfect in their coverage patterns, but they are far superior to conventional mobile telephone service.

PRIVATE MOBILE RADIO SERVICE

Mobile radio operates in one of three modes—single-frequency simplex, two-frequency simplex, and duplex. Both of the simplex modes use push-to-talk operation. When the transmitter button is in the talk position, the receiver cuts off. In a single-frequency mode, the mobile units and base unit send and receive on the same frequency. In a two-frequency operation, the base transmits on one frequency and receives on another; the mobile units reverse the transmit and receive frequencies.

In a duplex mode, the RF carriers of both the mobile and the base are on for the duration of the session. The base station usually uses separate transmit and receive antennas, but most mobile units use the same antenna to transmit and receive. A filter separates the transmitter's RF energy from the receiving transmission line. The transmit and receive frequencies must be separated sufficiently to prevent the transmitter from desensitizing the receiver. Duplex operation provides mobile radio units with the equivalent of a wireline telephone conversation.

To improve mobile radio coverage, which is apt to be spotty in mountainous terrain, repeaters are often employed. If a session between a mobile and a base station is set up through a repeater, two sets of frequencies are used—one between the mobile and the repeater and the other between the repeater and the base station.

The base station often is mounted at a remote location to improve coverage and therefore must be remotely controlled. Control functions, including keying the transmitter, selectively calling the mobile unit, and linking the audio path between the base station and the control unit, may be carried over landline or point-to-point radio. Siting of the base station is critical. Because obstructions adversely affect mobile signals, it is advantageous to mount the antenna as high as possible. To overcome the effects of fading, it is desirable to use high transmitter power. Both of these, however, must be balanced with the objective of frequency reuse. On crowded frequencies, interference from distant stations becomes a problem, and a user may capture two base stations simultaneously.

Mobile-to-mobile communication is easy to administer in single-frequency simplex operation because the mobiles can hear each other. In a duplex operation, the mobiles can hear only the base station unless the base retransmits the signal from mobile units. The retransmission of the mobile signal, which is called talk through, permits mobiles to communicate with one another. It is necessary for the base station to monitor a mobile-to-mobile conversation and to disconnect the path when the session ends.

Mobile Unit Signaling

The simplest form of signaling a mobile unit is voice calling. The base station calls the mobile unit's identification, and the mobile unit responds if the operator is within earshot. On a crowded radio frequency, the constant squawking of the speaker can be annoying, which leads to the need for some form of selective calling.

The simplest form of selective calling relies on the receiver's squelch circuit. The squelch is the circuitry that deactivates the receiver's audio in the absence of a received carrier. Several receivers can operate on the same frequency by assigning subaudio tones to break the squelch of the desired receiver.

For a few stations, the tone-activated squelch is satisfactory, but it does not work well on a crowded channel. Busy channels can use a selective calling system similar to that used in mobile telephone can be employed. The receiver has a selector that responds to a series of audio tones or a digital code. The digital codes operate through a standard known as binary interchange of information and signaling (BIIS). When the user's unit is signaled, the selector rings a bell, which can also activate an external signal such as honking a horn or turning on a flashing light.

Trunking Radio

If many users contend for a single communication channel, the channel occupancy may be high, but service will be poor. Trunking radio employs multiple channels to improve service to a group of users. The best example of trunking radio is a cellular system, but many private and public safety radio systems also use trunking to improve service. SMR uses trunking to provide dispatch services for its customers. A trunking system designates one channel as the calling or control channel, and all idle receivers tune to that channel. The control channel can be the next idle channel in a sequence, or a talking channel may be designated as control. Some loss of efficiency occurs when units must switch signaling channels, so high-usage systems reach a point at which it is more efficient to have a dedicated signaling channel.

Trunking systems must resolve the occasional conflict of two stations signaling simultaneously. Two methods are commonly used—polling and contention. Both systems work the same way that channel-sharing methods in data communications systems work. In a polling system, the base station sends a continuous stream of polling messages to all mobiles on the channels. In a contention system, the mobile unit listens for an idle channel before transmitting. If two mobiles transmit simultaneously, the base station recognizes the collision and informs the mobiles of it. They then back off a random time before again attempting to transmit.

Mobile Radio Design Objectives

Private and public mobile radio and paging systems share a common set of design objectives. The design process is not precise because of the unpredictable nature of radio waves. Obstructions and fading cause the principal disturbances. Designers of point-to-point radio systems have tools such as high power, directional antennas, and diversity to compensate for the effects of disturbances. The mobile radio designer is at a disadvantage because omnidirectional antennas are needed to reach roaming users. The nature of the remote unit may make it difficult to increase power because of the resulting increase in battery drain. Also, the remote unit frequently operates in a high-noise environment (for example, ignition noise), and most important, the base station is attempting to communicate with a target that is constantly moving. As a result, no mobile radio system gives quality that is consistently as good as that provided by wired telephones. This section discusses some techniques that designers use to generate satisfactory mobile radio service.

Wide Area Coverage

The major objective in mobile system design is to provide coverage that allows a mobile unit to move through a defined area without loss of communication. Cellular is one way of accomplishing this. Another way is *quasi-synchronous* operation. FM receivers have a tendency to be captured by the strongest signal. Quasi-synchronous, which is also called *simulcast* employs adjacent transmitters operating on frequencies that are slightly offset. The offset is small enough that the resulting beat frequency is inaudible. The one with the strongest signal captures the mobile unit.

A second method of achieving wide-area coverage is a receiver voting system. When a mobile unit initiates or responds to a call, the receivers in the coverage area compare signal strength and determine which unit has the best signal. That unit and its associated transmitter establish communication with the mobile until it moves beyond the coverage area and captures another transmitter.

Adequate Signal Strength

Mobile units live in a hostile environment of high noise; most of which is man made, with the predominant source being auto ignition and charging systems. Portable units often are carried inside buildings where the signal may be attenuated, or other noise sources, such as elevators and industrial machinery, may interfere with the signal.

Another source of signal loss is fading. A stationary user may experience a slow fade due to gradual changes in atmospheric reflection. A moving user on the edge of the coverage area is likely to experience a fast fade as the signal bounces off buildings, trees, and other obstructions. As the vehicle moves, the frequency of the fade changes. If the fade is fast enough, it can be tolerated and sounds much like a noisy signal. A fade of about 10 cycles per second has the greatest adverse effect on the user and makes communication practically impossible. A user on the fringes of the coverage area often can improve communication by stopping the vehicle and edging forward to a high signal strength location. A move of only a few inches can make a great difference in received signal strength.

Fading is often frequency selective. If a band of frequencies is transmitted, one range of frequencies may be subject to heavy fading while the other suffers little or not at all. The range of frequencies that are subject to similar fading effects is called the *coherence bandwidth*. Because of the frequency-selective nature of fading, designers can use frequency diversity to establish a reliable communication path.

Since VHF and higher frequencies have a short wavelength, it is often possible to counter the effects of fading by using space diversity. In a space diversity system, antennas are mounted at a distance that reduces the probability of a similar fade striking both antennas simultaneously. In a vehicular radio, for example, one antenna might be mounted on the front fender and another on the rear.

Wave Propagation

Radio waves propagate by one of three methods—the ground wave, the tropospheric wave, and the ionospheric, or sky, wave. Each of these acts differently on different frequency ranges and has a significant effect on the propagation characteristics of the signal.

The ground wave guides radio frequencies below about 30 MHz. The signal follows the contour of the earth and diminishes with distance. Ground wave communication is effective for short distances and has little effect on VHF and higher frequencies. The tropospheric wave is effective at VHF and higher frequencies but has little effect below 3 MHz. At microwave frequencies, the tropospheric wave can be used to communicate beyond the line of sight.

The ionospheric wave is most effective below VHF frequencies. The ionosphere reflects high frequencies in a frequency-selective manner. With ionospheric reflection, skip signals can travel well beyond the range of the ground wave. Skip conditions can result in excellent communication capability most of the way around the world with low power, but the communication path is unreliable and difficult to predict. For mobile radio, skip is undesirable because it interferes with frequency reuse.

CELLULAR TELEPHONES

By the late 1960s all the technical elements for cellular were in place, but in the United States the technology languished through lengthy regulatory debates. A major issue was frequency spectrum. Plenty of unused spectrum existed in space reserved for UHF television. In 1974, the FCC designated part of the UHF spectrum between 800 and 900 MHz for cellular, but they delayed approval of the service pending hearings on the monopoly aspects of cellular. At the time, the LECs had a monopoly on landline services, and many interveners urged the FCC to open cellular to competition. In 1978 an AT&T subsidiary named Advanced Mobile Phone System (AMPS), installed the first cellular radio demonstration system in the United States in Chicago. The FCC addressed the monopoly issue by breaking the spectrum into two segments, allocating half to a non-wireline carrier and the other to the LEC.

Cellular caught on rapidly and exhausted the initial frequencies. The FCC expanded the band to 824 to 849 MHz in the mobile to base direction, and 869 to 894 MHz base to mobile. The lower half of each band, called the A band, is designated for wireline carriers, which are defined roughly as local exchange companies.

The upper half, or B band, is designated for non-wireline carriers, which are non-LEC common carriers. With the mergers and acquisitions that have occurred since the original allocations were made, this distinction has all but disappeared. The FCC grants licenses in both bands to serve a Cellular Geographic Serving Area (CGSA). A CGSA corresponds to a Metropolitan Statistical Area (MSA), which is a major metropolitan area defined by the Office of Management and Budget.

Cellular Technology

Cellular overcomes most of the disadvantages of conventional mobile telephone. A coverage area is divided into hexagonal cells as shown in Figure 20-2. Frequencies are not duplicated in adjacent cells, which reducess interference between base stations. It also allows the carrier to reuse frequencies within the coverage area with a buffer between cells that are operating on the same band of frequencies. This technique greatly increases the number of radio channels available compared to a conventional mobile telephone system, which uses a frequency only once in a coverage area. The cells are smaller in urban and larger in rural areas.

Figure 20-3 illustrates the general cellular plan. The carrier selects the number and size of cells to optimize coverage, cost, and total capacity within the serving area. FCC rules and regulations do not specify these design factors; they are up to the service provider. The mobile units are frequency agile; that is, they can shift to any of the voice channels. To operate on either digital or analog channels,

FIGURE 20-2

Cellular Serving Area Frequency Plan

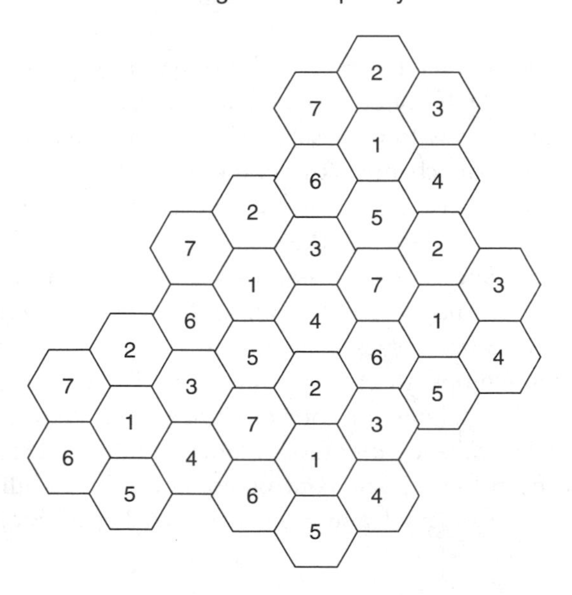

FIGURE 20-3

Cellular Serving Plan

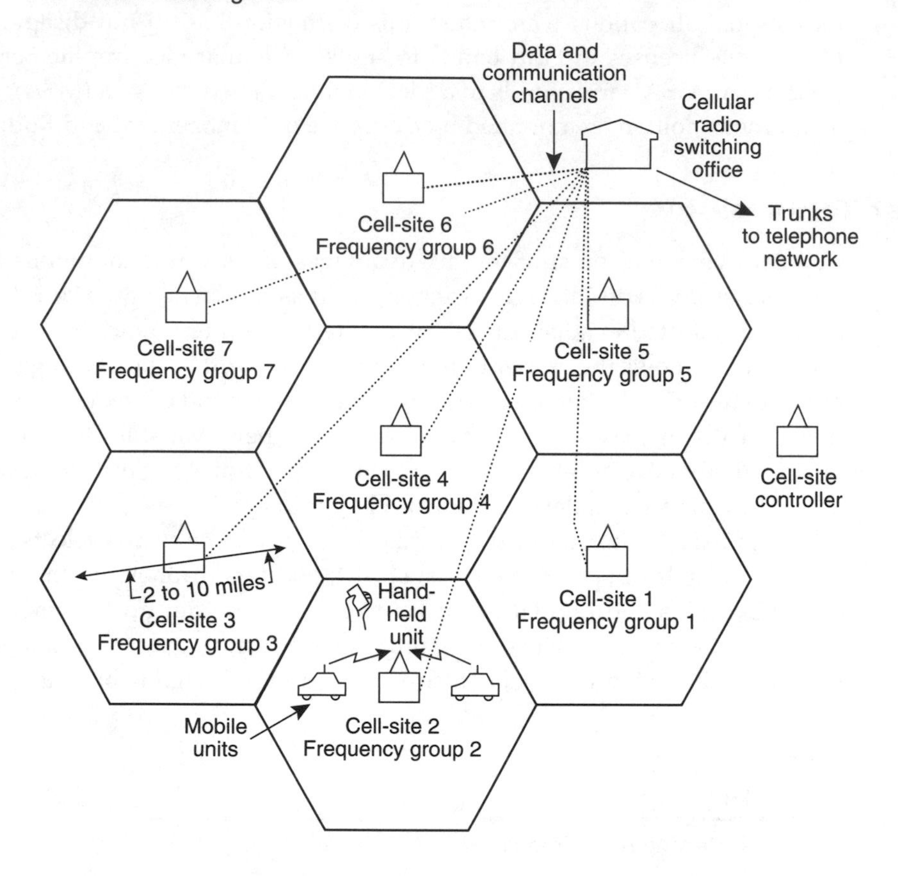

the mobile units also must be dual mode. A dual-mode unit responds to a digital channel first if one is available. If not, it falls back to analog.

Mobile units are equipped with processor-driven logic units that respond to incoming calls and shift to radio channels under control of the base station. Each cell site is equipped with transmitters, receivers, and control apparatus. One or more frequencies in each cell are designated for calling and control. For incoming and outgoing calls, the cell-site controller assigns the channel and directs a frequency synthesizer inside the mobile unit to shift to the appropriate frequency.

An electronic central office serves as a mobile telephone switching office (MTSO), and controls mobile operation within the cells. The cell-site controllers connect to the MTSO over datalinks for control signals, and voice channels for talking. The MTSO switches calls to other mobile units and to the local telephone system, processes data from the cell-site controllers, and records billing details. It also controls hand-off so a mobile leaving one cell switches automatically to a channel in the next cell.

Cell-Site Operation

A cell site has one radio transmitter and two receivers per channel, the cell-site controller, an antenna system, and voice and data links to the MTSO. The cell shape is roughly hexagonal because that shape provides a practical way of covering an area without the gaps and overlaps of circular cells. As a practical matter, cell boundaries are not precise. Directional antennas can approximate the shape, but the MTSO switches a user from one cell to another based on signal strength reports from the cell-site controllers. The hand-off between cells is nearly instantaneous, and users are generally unaware that it has occurred. The hand-off, which takes about 0.2 s, has little effect on voice transmission aside from an audible click, but data errors will result from the momentary interruption. As many as 168 channels per cell can be provided, with the number of channels based on demand.

Cell sites provide coverage with the relatively low power of the cell-site transmitters. FCC rules limit cellular transmitters to 100 W output, with higher power used only if necessary to cover large cells. At the UHF frequencies of cellular radio, transmission is line of sight, so careful planning is needed to define the coverage area of the individual cell while minimizing the need to realign cells in the future.

A minimum of one channel per cell is provided for control of the mobile units from the cell-site controller. The controller directs channel assignments, receives outgoing call data from the mobile unit, and pages mobile units over the control channel. When the load exceeds the capacity of one channel, separate paging and access channels are used.

The cell-site controller manages the radio channels within the cell. It receives instructions from the MTSO to turn transmitters and receivers on and off, and it supervises the calls, diagnoses trouble, and relays data messages to the MTSO and mobile units. The controller also monitors the mobile units' signal strength and reports it to the MTSO. It scans all active mobile units operating in adjacent cells and reports their signal strengths to the MTSO, which maps all working mobile units. This map determines which cell should serve a mobile unit when hand-off is required.

Supervisory audio tones (SAT tones) prevent a mobile unit from talking on the same frequency at separate cell sites. The base station sends one of three SAT tones, and the mobile loops it back. If the SAT tone returned by a mobile unit is different from the one sent by the cell site, the MTSO will not accept the call.

Mobile Telephone Switching Office

The MTSO is essentially a digital end office with a special purpose generic program for cellular radio operation. Not all MTSOs are local switching systems; some products are designed specifically for cellular radio. The objective of most service providers is to offer cellular radio features that are essentially identical to wireline telephone features. The MTSO's tasks can be grouped into three categories: connection management, mobility management, and radio resource management.

The MTSO links to the cell-site controller with data circuits for control purposes and with four-wire voice circuits for communication channels. When the cell-site controller receives a call from the mobile unit, the controller registers the dialed digits and passes them over the datalink. The MTSO registers the dialed digits and switches the call to the telephone network over an intermachine trunk or to another cellular mobile unit within the system. When mobile-to-mobile calls or calls from the local telephone system are placed, the MTSO pages the mobile unit by sending messages to all cell-site controllers.

The MTSO receives reports from the cell-site controller on the signal strength of each mobile unit transmitting within the coverage area. Data is relayed to the MTSO to enable it to decide which cell is the appropriate serving cell for each active unit. The MTSO also collects statistical information about traffic volumes for allowing the system administrator to determine when to add channels. In addition, the MTSO stores usage records for generating bills.

Mobile Units

The cellular transceiver is a sophisticated device that can tune all channels in an area. The major components of a typical mobile unit are shown in the block diagram in Figure 20-4. The transmitter and receiver are coupled to the antenna

FIGURE 20-4

Block Diagram of a Cellular Transceiver

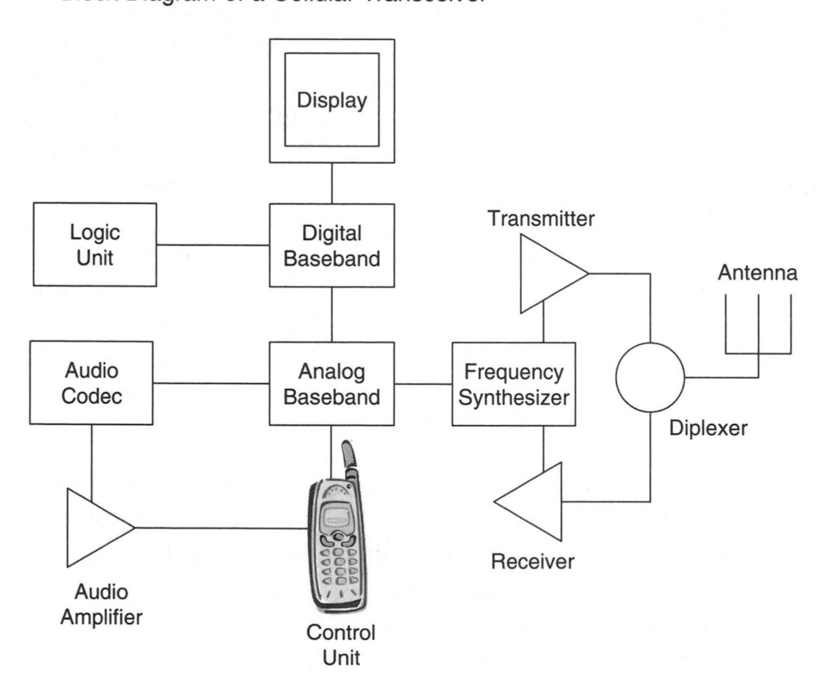

through a *diplexer*, which isolates the two directions of transmission so the transmitter does not feed power into its own receiver. A frequency synthesizer generates transmit and receive frequencies under control of the logic board. The synthesizer generates a reference frequency from a highly stable oscillator and divides, filters, and multiplies it to the required frequency.

The most complex part of the mobile unit is the logic equipment. This system communicates with the cell-site controller over the control channel and directs the other systems in the functions of receiving and initiating calls. These functions include recognizing and responding to incoming signals, shifting the RF equipment to the working channel for establishing a call initially and during hand-off, and interpreting users' service requests. The logic unit periodically scans the control channels and tunes the transceiver to the channel with the strongest signal. The user communicates with the logic unit through a control unit, which consists of the handset, dial, display unit, and other elements that emulate a conventional telephone set.

The unit also includes a power converter to supply the logic and RF equipment with the proper voltages from the battery source. FCC rules limit the transmitter to 7 W output, but many mobile units use 3 W or less, and handheld units are a fraction of a watt to reduce battery drain. Battery drain is particularly important in handheld or portable units. The FCC segregates mobile units into three power classes ranging from 0.6 W to the maximum power permitted.

Another module contains the unit's 32-bit binary electronic serial number (ESN), which theoretically prevents fraudulent or unauthorized use of a mobile unit. In practice, however, toll thieves intercept the ESN and duplicate it so they can masquerade as the registered owner. The ESN is used to program the mobile identification number (MIN), which is communicated to the cell-site controller with each call for comparison with the MTSO's database. The unit's 10-digit calling number and a station class mark are also built into memory and transmitted to the cell-site controller with each outgoing call. The class mark identifies the station type and power rating. The FCC assigns each carrier a system identification code (SID). The SID and MIN are programmed into the phone when it is activated.

When you turn on the cell phone, it listens for a SID on the control channel. If it cannot find a tone it turns on a no service display. If the phone receives a SID it compares it to the home SID and transmits a registration request to the MTSO. If the SID does not match, the phone knows it is roaming and registers that on its display. When the MTSO receives a call for a particular cell phone, it knows which cell site has the best signal, and assigns a frequency pair to the session. The MTSO communicates the frequency assignment to the cell phone over the control channel and rings the phone.

Signal Enhancements

Some buildings are constructed with so much steel and concrete that cellular signals cannot penetrate. The cellular industry provides two solutions: booster amplifiers and microcells. A booster amplifier amplifies the channel range of the nearest cell site, retransmitting it inside the building. The amplifier retransmits signals from handheld units to the cell site. A microcell is a low-power cell site that is dedicated to a building. An antenna system is extended throughout the building to transmit and receive low-power signals. The microcell has its own set of channels, and is connected to the MTSO with a T1/E1 connection like any other cell site.

Roaming

Every cellular operator provides a home area in which normal cellular rates apply. Roaming, which may carry a premium charge, enables a mobile unit to move outside its normal service area. The cellular operator must know about a roaming user to route calls to the proper service area. Roaming service is enabled by use of SS7 between the MTSOs. The MTSO maintains home location register and visitor location register databases. The visitor location register keeps track of visiting stations that are roaming within the MTSO's domain. The identity and location of roaming stations are validated with the home MTSO over the SS7 network. The carrier may extend roaming capability throughout a wide service area by linking its switches over SS7 networks with switches of other service providers, enabling users to roam without special procedures.

Digital Cellular

The initial AMPS system used FDM analog modulation with channels spaced at 30 kHz. This provides space for 832 channels, 790 for voice and 42 for signaling, registration, and calling. To make more efficient use of spectrum, the service providers are converting to digital. Cell sites are equipped with some analog and some digital transceivers. The subscriber units must be dual mode to communicate on either analog or digital channels. In addition to dual-mode mobile phones, the industry offers dual-band phones that can communicate on either the cellular or the PCS bands. PCS, as discussed in the next section, operates in the 1900 MHz band.

The technology is further complicated by the fact that three different modulation methods are in use throughout the world. TDMA was the first digital cellular method used in the United States. Three digital channels occupy the space of one analog channel, tripling the capacity of the spectrum. The FCC elected to let the market set the standards, so several carriers adopted CDMA, which uses spread spectrum. In most of the rest of the world, Global System for Mobile Communications (GSM) is the dominant digital cellular modulation scheme. PCS-1900 is the North American version of GSM. TIA has standardized TDMA as IS-54 and IS-136 and CDMA as IS-95.

A major advantage of FDMA was its universal compatibility, at least in North America. The TDMA/CDMA controversy means that either subscriber stations must be compatible with at least two modulation schemes or users will be unable to roam freely or to change carriers without changing the subscriber set. Under the analog FDMA system a user has exclusive use of a channel in a cell site for the duration of a session. The inherent half-duplex character of voice communications, plus the frequent pauses in most conversations wastes much airtime, which CDMA can utilize. Using voice compression, TDMA divides each 30 kHz channel into three 8 Kbps segments, tripling the capacity of the spectrum. Eventually, TDMA proponents expect to double the capacity of digital channels, providing a six-to-one improvement. The control channels remain analog, but eventually digital control channels will be used.

CDMA uses spread-spectrum techniques to multiply spectrum capacity. CDMA encodes 64 channels into 1.25 MHz of spectrum. Each channel uses an orthogonal code, with all channels transmitting across the same bandwidth. The CDMA receiver picks the appropriate code out of the signal and demodulates it. CDMA is not as susceptible to multipath fading as FDMA and TDMA. Another advantage of CDMA is its hand-off method, which is a make-before-break arrangement that eliminates the signal dropout that is common with TDMA and FDMA.

GSM uses a TDMA/FDMA method of dividing the radio spectrum not unlike D-AMPS, but it uses a signaling protocol that is based on OSI and ISDN standards and is therefore incompatible with TDMA. The call setup and teardown method is a modified Q.931 standard, which is the protocol used in ISDN. The layer-2 protocol used for transferring information across the radio channel in frames is a modified version of LAPD. GSM also encrypts the calls to ensure privacy. In North America GSM is used in PCS and by Nextel.

PERSONAL COMMUNICATIONS SYSTEM (PCS)

The frequency spectrum allocated to PCS in the United States is shown in Figure 20-5. Some C block licenses are split into multiple licenses. In Europe the service is called personal communication network (PCN). The frequency ranges are similar, but the coding methods are incompatible. Even if the North American carrier has chosen GSM as its coding method, the incompatibilities prevent the phones from working unless they have been designed for international use. PCS is similar in many ways to cellular, but with several notable exceptions including these:

- Its frequency spectrum is higher—in the 1900 MHz band.
- PCS frequency slots are 200 kHz wide and are divided into eight time slots.
- The cell size is smaller and power is lower.
- The frequency spectrum for PCS in the United States was auctioned off and divided into multiple segments to avoid some of the monopolistic characteristics of cellular.
- The system is intended to bundle other services such as caller ID, paging, and e-mail. Some service providers offer small-screen video.

FIGURE 20-5

PCS Frequency Block Plan

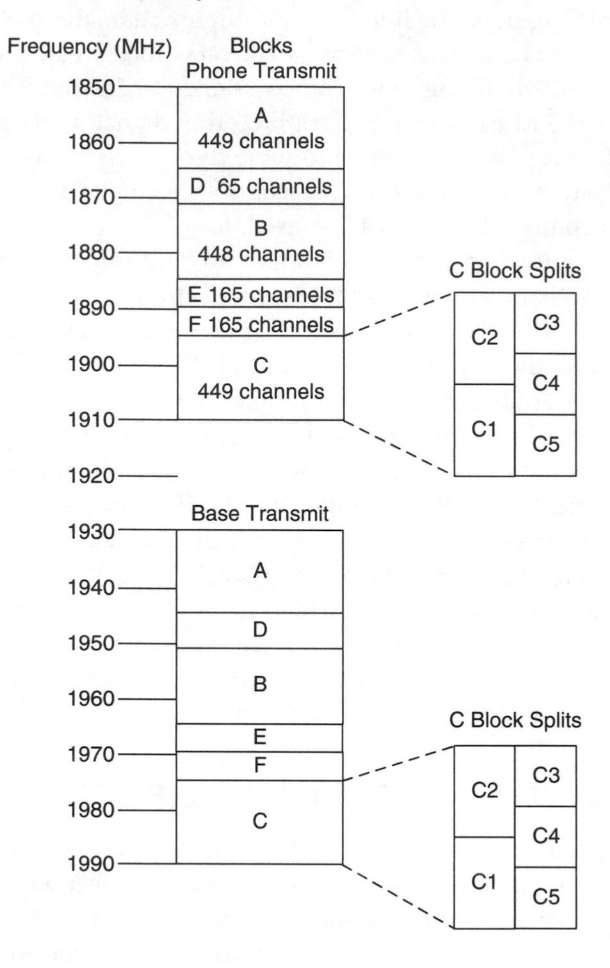

The PCS spectrum is divided into licensed and unlicensed portions, which enables private wireless users to shift to an adjacent PCS frequency when they roam outside the bounds of the unlicensed system. It is also divided into broadband and narrowband segments. Narrowband PCS operates in the frequency ranges of 901 to 902, 930 to 931, and 940 to 941 MHz bands to support such services as two-way paging and other text-based services. Handheld devices that double as a cell phone and personal digital assistant (PDA) can provide wireless e-mail over a narrowband link. The service is also used for wireless telemetry and automatic meter reading.

In the United States, the FCC issues licenses for two 30 MHz channels per MTA (major trading area) and one 20 MHz plus four 10 MHz channels in the basic trading area (BTA) band, with a minimum of three competitors in each market.

In addition, the C and D blocks in each band are reserved for protected groups: rural telephone companies, small business, and women and minority-owned businesses. In theory at least, the additional competition should result in better coverage, more innovative services, and competitive prices. PCS has no mandatory service standards. Competition drives the cost and features included in carriers' service offerings.

NEXT GENERATION CELLULAR

The cellular industry can mark at least three distinct generations of cell phones. The first was analog FDM, some of which are still in operation. The second is digital, but 2G cellular is unsuitable for high-speed data. The industry is now touting third-generation cellular (3G) as the answer to Web surfing from a handheld mobile device. The vision is of a single global system that permits worldwide roaming and broad data bandwidth for applications such as Internet browsing.

The third generation has been under development since 1992 when the ITU began work on its IMT-2000 standard. IMT stands for International mobile telecommunications and 2000 was supposed to be the year in which it became available. Some also said that it would be the data rate in Kbps to fulfill its promise of data transmission at as much as 2 Mbps. A third interpretation was the frequency band in which it was to run. Some products have the ability to carry VoIP sessions, enabling their use as mobile wireless terminals with connectivity to the company's internal network. Another trend is handsets that can operate in either WiFi or cellular, enabling users to communicate over private and public systems with the same handset. The small screens and keyboards of handheld devices put limits on the applications. This is causing manufacturers to scale back their plans to something called 2.5G cellular, which has data rates in the order of 384 Kbps.

The technology of IMT-2000 is CDMA spread over a 1.25 MHz channel, supporting as many as 64 sessions. The actual number of simultaneous sessions a channel can support depends on the traffic volume. The initial data rate objective is 144 Kbps from a moving vehicle and 384 Kbps at pedestrian speeds. To support 2 Mbps a different modulation method is needed. ITU is considering five standards for the radio interfaces for IMT-2000.

- *CDMA direct spread*: This interface, UMTS or WCDMA, is direct-sequence CDMA with information spread over a bandwidth of about 5 MHz.
- *CDMA multicarrier*: This interface, CDMA2000, uses channel bandwidths of 1.25 and 3.75 MHz, which joins three 1.25 MHz channels. The single-channel version is also known as CDMA2000 1X EV-DO (evolution data optimized).
- *CDMA TDD*: This version, time division duplex (TDD) has bandwidth of 5 MHz.

- *TDMA single-carrier*: This interface, also called universal wireless communication-136 (UWC-136) is intended for evolving TIA-136 (PCS) technology to 3G.
- *FDMA/TDMA*: This interface, also called digital enhanced cordless telecommunications (DECT) is a TDM duplex transmission method.

All of the major carriers are planning 3G, and some are rolling out products. The announced data rates, however, fall quite a bit short of the 2 Mbps promise. The early versions are called 3G, but there is little to distinguish them from 2.5G devices. The CDMA–GSM controversy will not be resolved with 3G. The CDMA carriers are favoring EV-DO and the GSM carriers favoring WCDMA. EV-DO is deployed on existing cell sites and provides typical data rates in the range of 600 Kbps downstream and 300 Kbps upstream. It is expected ultimately to open the way for migration from circuit-switched cellular to VoIP.

Several technical hurdles remain to be overcome before 3G can reach its potential. The first is frequency spectrum. The FCC is considering the use of 1710 to 1770 and 2110 to 2170 MHz bands, which other services now occupy. Another problem is that the circuitry needed to sustain high-data rates imposes high battery drain. One solution to battery drain is to reduce the data rate, perhaps by a factor of four or more. This will limit some of the expected applications while still providing a respectable speed for Web surfing. Although this is far from the promise of 3G, it is still considerably more than dial-up landline services.

Another issue is the protocol. Wireless Application Protocol (WAP) is effectively a de facto standard for using wireless technology for Internet access. Its developers intended to create a global wireless protocol that works across differing network types. WAP is optimized for information delivery to devices such as mobile phones and PDAs. It supports most wireless networks including cellular digital packet data (CDPD), CDMA, GSM, TDMA, and others. WAP devices run microbrowsers, which are browsers optimized for small file sizes. Such devices can run with the low memory and low-bandwidth constraints that are typical of handheld devices. WAP uses a language similar to HTML known as WAP markup language (WML). WML is designed for small screens and navigation without a keyboard.

The final threat to 3G comes from other wireless technologies, primarily 802.11a, b, and g (WiFi) and 802.16 (WiMAX). Today WiFi is widely available in "hotspots," often at little or no charge. The WiMAX vision is for ubiquitous wireless coverage over a broad range. The user's device for such networks is most likely a laptop computer, which does not have the keyboard and display limitations of the small handheld.

Fourth Generation

Even before the 3G controversies are resolved, the industry is looking forward to the fourth generation, which will offer speeds more in line with what users have come to expect on the office network. High-speed downlink packet access (HSDPA) is a packet-based data service with data rates of as high as 20 Mbps over a 5 MHz link. The network will have an all-IP core and QoS for four traffic classes:

- Conversational class (voice, video, video gaming)
- Streaming class (multimedia, video on demand, Webcast)
- Interactive class (Web browsing, database access)
- Background class (e-mail, downloading)

The objectives of 4G are to support new applications such as entertainment and video conferencing more effectively, provide global mobility, and scalability. The entire network will be packet switched as opposed to today's network, which is largely circuit switched. Speeds will eventually reach 100 Mbps and will support applications that today's networks cannot handle.

MOBILE RADIO APPLICATION ISSUES

Cellular radio and PCS are rapidly becoming as essential as ordinary telephone service and in many parts of the world, cellular development exceeds that of landline telephones. This section includes considerations that users of cellular radio service should evaluate, in terms of both the service and the mobile radio equipment.

Cellular mobile telephone equipment is evaluated on much the same basis as other telecommunications equipment. Reliability, coverage, battery life, and the ability to obtain fast and efficient service are of paramount concern. Cost is also an important consideration. With the cost of airtime cellular service more costly than landlines, most carriers offer attractive pricing options that effectively eliminate roaming and long-distance costs.

Security

Regular mobile radio is inherently an unsecure medium. Anyone with a receiver tuned to the mobile frequency can eavesdrop on any conversation, and there is little practical means of preventing it aside from scrambling. Cellular radio offers inherent security that may be sufficient but still is not interception proof. If a session takes place in one cell, it is easy for someone to intercept it. If the vehicle is moving, an eavesdropper may have some difficulty resuming reception of the conversation when it hands off to another cell. The smaller the cells, the more frequent the handoff, and the less likely it is that the entire session can be monitored. A vehicle following the vehicle under observation can, however, monitor the entire session.

TDMA digital cellular is somewhat more secure than analog because the equipment for monitoring is not as readily available. Anyone with the motivation and the proper equipment can still eavesdrop on a TDMA session, and since the control channels are analog, interception of ESNs is not difficult. CDMA is inherently more secure than TDMA. GSM uses a unit identification protocol that is more secure from fraud. The mobile unit receives an identification number from a pool just for the duration of the session. Its voice channels are TDMA/FDMA, and with the right equipment, are subject to interception.

Coverage

The coverage area is one of the two or three primary considerations most users review in evaluating a mobile radio system. Coverage can best be evaluated by taking a test drive and making calls from the areas you plan to drive through, paying particular attention to the fringes. The carriers' coverage maps are a good way to compare the home and roaming areas, but are not reliable indicators of coverage on the fringes. Note which areas are in the home area and which are in the carrier's service area, but for which a roaming charge is applied. The ability to roam internationally with a compatible handset will be an important factor for many users.

Calling Plan

Cellular radio charges are based on duration of both originating and terminating calls. Carriers charge for total airtime and add on other message charges such as roaming and long distance. Therefore, both originating and terminating calls are charged to the terminating mobile number. Some carriers do not charge for the first minute of incoming calls, which can make a big difference in the cost for some users. Most carriers offer plans inclusive of long-distance charges and many offer reduced cost or free service outside peak calling times of to other users on the same plan. Other calling features include call forwarding, caller ID, call waiting, three-way calling, and detailed billing.

Features and Accessories

The life of a cell phone is, according to the industry, about 2.5 years. Nothing in the circuitry affects the life, however. The instrument itself will last for many years, but greater miniaturization and feature changes seem to render the old phones obsolete. Operational features such as speed dialing, voice mail, text messaging capability, and call and usage logs are so universal that they hardly enter into the choice of phone. Now the selection decision often hinges on color faceplates, color screens, Bluetooth connectivity, built-in camera, headsets, and PDA functions with a full keyboard.

Cellular Emergency Calls

Calls from a cellular phone to a PSAP have long been a problem compared to wired phones because of the lack of relationship between the phone number and its physical location. Callers from unfamiliar territory often find it difficult to direct emergency vehicles to their locations. The FCC ordered a two-phase approach to the problem of handling E-911 calls. Similar techniques are being used in Europe for calls to the emergency mobile number, 112.

The first phase requires the service provider to identify the cell site from which the call was received. That narrows the geography somewhat, but it is far from pinpointing the actual location. Phase 2, mandated for 2005, requires service providers to pinpoint the location within 100 m. Two techniques are used. One relies on equipping the handset with a global positioning satellite (GPS) chip so the mobile unit is continually broadcasting its position. There are several objections to this process. One is that GPS signals are not reliable indoors. Also, older cell phones cannot be retrofitted and are rendered obsolete. The third objection is intrusion into privacy. For a variety of reasons, many people do not want their locations tracked.

The alternate solution is network based. The cellular provider detects the cell site with devices that pinpoint the location of the handset by measuring time differences between signals received in adjacent cells. For this method to be effective, more than one site must receive the signal from the handset placing the call, which is not always possible in rural areas. Neither method is foolproof, but it is a major step forward in resolving a long-standing problem.

Wireless Number Portability

FCC rules require wireless carriers to permit their customers to switch to another wireless carrier without a number change. At this writing the rules for changing from a cellular to landline phone and vice versa without a number change have not been issued. The change cannot always be made without a telephone instrument change, so it is essential to review this factor in addition to any term commitments before making the change. It may be necessary to obtain a new telephone and buy out the existing contract.

CHAPTER 21

Wireless Communications Systems

Advances in radio technology are the vanguard of the technical marvels of the past few years. The wireless vision is appealing: a device small enough to fit in your pocket with all the features wrapped into one package. It works as a telephone anywhere in the world. When you are at your desk it associates itself with the office wireless system and elsewhere it latches onto cellular or PCS if they are available and if not, it homes on a friendly satellite. You can send or receive calls without knowing anything more than how to dial the call and pay the bill. The same instrument doubles as a text pager and e-mail appliance with speeds and capacities approaching that of a desktop computer. Within the limits of its miniaturized keyboard and screen, it is a full-featured web browser and may support video. Its battery has plenty of energy to last at least a day or two, and eventually, when the technology manages to solve a few problems, its energy source will last for the life of the device. And all of this will cost a mere few hours' wages.

Not long ago this would have seemed to be a futile aspiration, but few people would doubt the potential today. Except for limited battery life, most of this is within reach now, although with proprietary software because the standards are not mature. The explosive growth of cellular is only one indication of the demand for breaking the tether to the wired network. In the wide area, mobile data terminals (MDTs) range from the PDA to a full-scale keyboard and monitor. Although wide-area radio networks operate down to the desktop, a separate in-building network avoids the usage fees of public networks.

Wireless is today among the hottest of the telecommunications technologies, thanks to a mélange of mature, emerging, and incipient protocols. In 1997, IEEE completed work on the first version of 802.11 wireless LAN (WLAN) protocols. These are intended primarily for in-building communications, but they are often applied in the metropolitan area. The 802.11 family of protocols is an excellent example of the vigor that standards bring to the industry. Before the standards were adopted a number of products using wireless and infrared were on the market, but none was widely accepted.

The original 802.11 protocols covered three different media: infrared and direct sequence and frequency-hopping spread spectrum. The FHSS option was limited to 2 Mbps and the infrared option suffered from the inability to penetrate walls. The standard that invigorated WLANs is 802.11b, also known as wireless fidelity or WiFi, which operates up to 11 Mbps, using DSSS. After approval of 802.11b, new products flooded the market. The expansion continues with 802.11a, which operates in the 5 GHz unlicensed band at speeds up to 54 Mbps, and 802.11g, which operates at 2.4 GHz with speeds comparable to 802.11a.

Wireless LANs can be found everywhere. Boeing jets are being equipped for WiFi through its Connexion unit. Airports, hotels, and coffee shops offer wireless connectivity for a fee. "War chalkers" mark the presence of unprotected WiFi to direct "war drivers," who roam the streets looking for free and unobstructed Internet access. Users can download hot-spot maps of open access points (APs), many of which are deliberately left open by enthusiasts who hope to create a grassroots public wireless infrastructure. WLANs are cheap to buy and easy to set up, creating headaches for corporate network managers who have little ability to control wireless purchases, and fear with complete justification that rogue wireless APs may inadvertently open security holes.

The frequency bands that the 802.11 products use are unlicensed. With high-gain antennas they are used in the metropolitan network for point-to-point service. Coming in the future is WiMAX, which claims to extend the reach of microwave frequencies to about 30 miles. With QoS protocols added, such a network can support voice as well as Internet access. As we will discuss in Chapter 32, WiMAX depends on the use of non-line-of-sight (NLOS) technologies, which permit wider range than traditional radio operating at microwave frequencies.

The use of WiFi and WiMAX is not confined to data. Technically, VoIP can be carried as an application on a wireless network as easily as it can on wired networks. The major PBX manufacturers that for years used TDM base stations for their wireless phones are switching to a VoIP product running over 802.11. Without standard QoS protocols, which the 802.11e group has under consideration, the products either use proprietary protocols or rely on sufficient bandwidth to avoid operational problems. VoIP over WiFi and WiMAX is a definite industry direction. Major issues remain to be resolved, however before they can be integrated into commercial networks. These include roaming, billing, compatibility, technical standards, and pricing, not to mention E-911 support. WiMAX is intended as a metropolitan area network, but its techniques can be used in the LAN, which is WiFi territory.

The motivation for this awakened interest in wireless technologies is twofold from the user's standpoint: improved mobility and reduced cost. Nurses and physicians, production personnel, managers, food service workers, technicians, maintenance personnel, and dozens of other such occupations need to be telephone accessible, yet free from cords and pagers. Once people leave the confines of the plant, their needs do not change dramatically, but the dividing line between the private and public communications arenas is distinct. Ultimately, a person will have a single number that is reachable from anywhere at any time. The ideal will not be realized immediately. For the immediate future, the handoff from wireless to PCS will not be seamless. The ideal of a universal, transportable number will be achieved in time, but not until some technical and practical problems are solved.

WIRELESS LAN TECHNOLOGY

The wired LAN is suitable for office workers who sit at the desk all day, but many workers are mobile. A wireless laptop or PDA enables people to receive e-mail while in a conference room, keep their calendars synchronized, or pop into the office file server to pick up an occasional file. Much of the communication that now requires a voice call can be eliminated or simplified with inexpensive and ubiquitous wireless data. Also driving the need for WLANs is the difficulty of wiring certain buildings. Historical buildings that cannot be altered, those built of masonry, and those constructed before computers were conceived impose wiring difficulties that the WLAN can potentially overcome.

WLAN applications can be divided into three categories: mobile/portable, building to building, and desktop. The applications for each of these are different, as are the alternatives that the market offers. The array of 802.11 products can be confusing as applications overlap to some degree. Table 21-1 lists the protocols in the 802.11 family. We do not attempt to distinguish between those that are adopted and the ones that are under development because the field is changing so rapidly. The sheer quantity of 802.11 protocols can be somewhat daunting, but only three of them, a, b, and g, are complete operational network protocols.

Unlicensed Radio Bands

WLAN and voice products use unlicensed spectrum in two bands: ISM and Unlicensed National Information Infrastructure (UNII). No license is required to operate FCC type-approved equipment in these bands. The spectrum is limited and each band eventually fills up, forcing new users to higher bands. The main drawback to the lower bands is the fact that they are shared with many other services. For example, microwave ovens, cordless telephones, and baby monitors operate in the 2.4 GHz band. Table 21-2 lists the frequencies in the unlicensed bands and their power limitations.

Besides potential interference, the 2.4 GHz ISM band has another major drawback. In the United States, the bandwidth is sufficient for only three nonoverlapping channels. In a large building it is difficult to devise a coverage plan that allows users to roam while avoiding interference. Interference may come from an unexpected quarter. Radio waves travel in a spherical pattern, so in a multitenant building interference may come from floors above and below with no way to

TABLE 21-1

802.11 LAN Protocols

802.11:	The original 1 and 2 Mbps standard in the 2.4 GHz band plus infrared
802.11a:	WLAN in the 5 GHz band, with a 54 Mbps data rate
802.11b:	WLAN in the 2.4 GHz band, 11 Mbps data rate
802.11c:	Provides information for bridge operation
802.11d:	Protocol to enable 802.11 hardware to satisfy regulatory requirements in various countries
802.11e:	Enhances the MAC layer for QoS features, such as prioritizing voice or video traffic
802.11f:	Recommends interaccess point communications to support functions such as roaming
802.11g:	WLAN in the 2.4 GHz band with 54 Mbps data rate
802.11h:	Provides dynamic channel selection and transmit power control for devices operating in the 5 GHz band
802.11i:	Enhances MAC layer to overcome weaknesses in Wired Equivalent Privacy
802.11j:	Enhancements to conform to Japanese rules on radiated power, spurious emissions and channel sense
802.11k:	Radio resource measurement of wireless LANs
802.11m:	WLAN maintenance
802.11n:	High throughput LAN protocol
802.11p:	Wireless access in the vehicular environment
802.11r:	Fast roaming fast handoff
802.11s:	Wireless mesh networking
802.11t:	Wireless performance prediction test methods and practices
802.11u:	Interworking with non-802 networks such as cellular
802.11v:	Wireless network management

TABLE 21-2

Unlicensed Frequency Bands in the United States

Band	Frequency Range	Max Effective Isotropic Radiated Power
ISM-900	902 to 928 MHz	4 W (+36 dBm)
ISM-2.4	2400 to 2483.5 MHz	4 W (+36 dBm) for multipoint, 200 W (+53 dBm) for point to point
ISM-5.8	5.725 to 5.850 GHz	200 W (+53 dBm)
UNII Indoor	5150 to 5250 MHz	200 mW (+23 dBm)
UNII Low Power	5250 to 5350 MHz	250 mW (+24 dBm)
UNII/ISM*	5725 to 5825 MHz	200 W (+53 dBm)

*Overlaps with ISM-5.8 but does not require spread spectrum.

control it. The 802.11a protocol operates in the 5 GHz band. Besides the greater data rate, 802.11a has as many as 24 nonoverlapping channels, which can provide greater coverage. The 5 GHz band is less subject to interference from other applications, but it has a narrower range.

802.11 Protocols

The original 802.11 protocols were not widely accepted because the speed was too low to satisfy users who were accustomed to at least 10 Mbps. IEEE approved the 802.11b protocol in late 1999 and products began appearing in 2000. It uses a DSSS physical layer in the 2.4 GHz unlicensed band and supports data rates as high as 11 Mbps. The second protocol to market was 802.11a, products for which began appearing in 2001. It supports data rates of 6 to 54 Mbps in the UNII band of 5.25 to 5.85 GHz. Because of the higher frequency, its range is not as great as 802.11b, but the faster speed comes closer to matching the speed of wired Ethernet. The third protocol, 802.11g, combines the high data rate of 802.11a with the better transmission characteristics at 2.4 GHz.

The data rates stated for these protocols may raise unrealistic expectations on the part of users who are accustomed to the speed of wired LANs. Under the best circumstances throughput probably will not exceed 60 percent of the data rate and it may be substantially less. When the signal is weak, the protocols automatically downshift to lower speeds to limit the amount of packet retransmission. The objective is to achieve a 100-m radius, which approximates the area covered by wired LANs. Under ideal circumstances this range may be achieved, but the signal strength drops when it penetrates walls. Metal studs inside the walls further contribute to loss. Multipath reflections from walls and furniture may also cause a drop in signal strength.

WiFi protocols are based on Alohanet, which the University of Hawaii devised to communicate between Oahu and the other islands. A radio-based network must cope with the hidden transmitter problem, which occurs when stations cannot always hear each other. They have no way of knowing when another station is transmitting, resulting in simultaneous transmissions and mutilated packets. Without some method of control, collisions grow as traffic increases and throughput deteriorates because of bandwidth consumed by retransmissions.

WiFi uses a collision-avoidance protocol. A station gains permission to transmit by broadcasting a short ready-to-send (RTS) packet that contains the address of the receiving station and the intended packet length. If the addressee is prepared to receive the packet, it transmits a clear-to-send (CTS) packet. All of the stations set backoff timers with the expected packet duration and do not attempt to acquire the network during that interval. The length of the backoff timer expands exponentially when a station senses that the medium is busy.

The layer-2 protocol handles packet acknowledgment instead of relying on higher layer protocols as standard Ethernet does. The addressee checks the CRC and sends the acknowledgment. If the transmitting station does not receive an acknowledgment in the expected interval, it assumes the packet was lost or mutilated and retransmits. The extra overhead of the RTS/CTS process is one reason for lower throughput in a WLAN. Another reason is automatic data rate reductions built into the protocol. Stations with a weak signal are forced to lower their transmission rate until it reaches a sustainable level. Also, a WLAN is a shared medium so performance is less than the switched connections users experience with their wired networks.

The protocol calls for a point coordination function, which is used on time-sensitive media such as voice. Although 802.11 is designed primarily for data, nothing precludes it from carrying voice provided time-sensitive frames receive higher priority. The protocol does not provide any standard method of quality of service. Nothing in the 802.11 protocols requires the network to recognize and handle 802.1p and q prioritization as a wired Ethernet may. Voice operation, even more than data, requires handoff between radio cells, which either requires APs from the same manufacturer, or compliance with 802.11f. IEEE is working on 802.11e, which is intended to provide MAC-layer QoS.

Wireless Network Architecture

A WLAN consists of one or more computers equipped with wireless NIC cards and at least one AP that forms a bridge between wireless and wired networks. The AP consists of the radio, an Ethernet interface, a WAN link, and bridging software conforming to the 802.1d bridging standard. The wireless NIC is built into many laptops. External NICs are available as a PCMCIA card, a PCI bus card, or a USB-connected transceiver.

Enterprise wireless networks are evolving toward a full-mesh architecture in which APs communicate with each other to manage the connectivity and roaming capability of the supported stations. Mesh network standards are being developed under IEEE 802.11s, which is incomplete at this writing.

Operation Modes

The 802.11 protocol defines two modes: ad hoc and infrastructure. Two or more wireless computers can communicate peer to peer without an AP using the *ad hoc mode*. In the *infrastructure mode*, which most corporate networks use, at least one AP is required. If the network is constructed with multiple cells, calls in progress can move from one base station to another. The *basic service set* (BSS) is an AP serving wireless end stations. An *extended service set* (ESS) is a network of two or more BSSs connected to form a subnet. Figure 21-1 shows three BSSs with overlapping coverage areas forming an ESS. The APs are typically connected to the wired network with Ethernet connections, although wireless can be used.

Access Points

The AP contains the RF and control circuitry for the wireless node. APs come in three varieties: bridge, router with NAT, and router plus bridge with NAT. A bridge AP can connect a wireless network to a wired network. The router type is used to share an Internet connection. The NAT feature isolates the internal network from the Internet. Router plus bridge devices are intended to share an Internet connection with both wired and wireless computers.

FIGURE 21-1

802.11 Extended Service Set

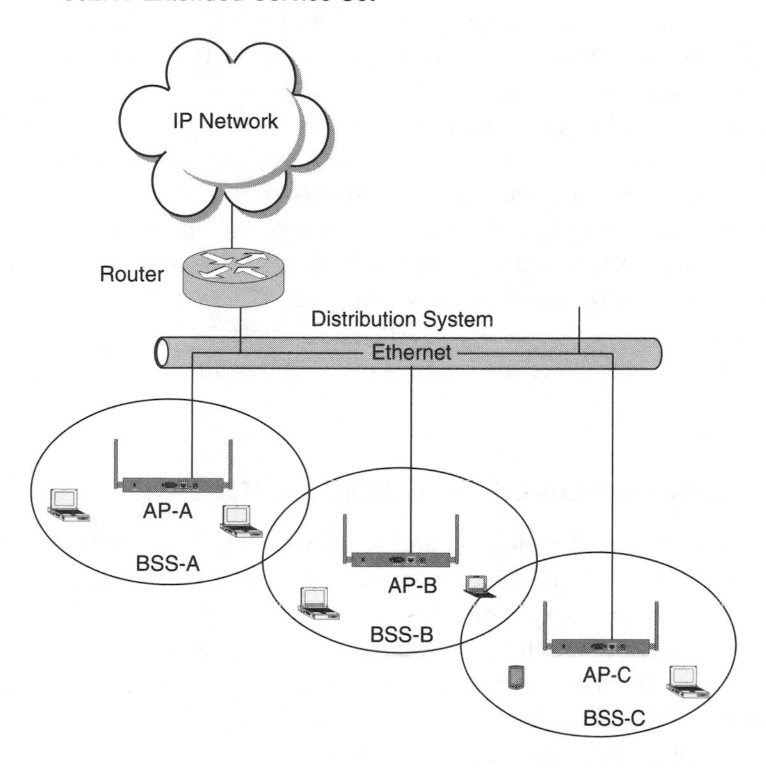

To communicate with an AP, a station must synchronize its clock with the AP's clock. The AP periodically transmits beacon frames that contain the clock value. When a station attempts to join a BSS, the AP must authenticate the station, synchronize its clock, and provide information that the station needs to communicate. The protocol provides two options for enabling a station to join a cell: passive scanning and active scanning. In the passive scanning method, a station waits until it receives a beacon frame. In the active scanning mode a station transmits a probe request frame and waits for the AP to respond.

When the station is recognized, it goes through an authentication process and then an association process. Authentication means that the station proves its right to join the network through a process that is discussed later in the "Security" section. After authentication, the association process locks the station onto the BSS. If a station moves to a different BSS, it retains its authentication but it must undergo reassociation. When a station detects multiple wireless signals, it joins a BSS based on received signal strength or user selection. The station tunes to the radio channel and periodically surveys other channels to determine if another has better performance.

It is important to consider the distinction between small office and home office (SoHo) and enterprise WLANs. The former are intended for plug and play, do not support roaming, and have little in the way of coverage guarantees. The user plugs in the AP and it either works or it does not. If it does not, the AP is moved around until coverage is acceptable. Enterprise WLANs do not remain static like wired LANs do, so it is difficult to meet service levels. Keeping the radio channels tuned can be time consuming. In some products a central controller automates channel assignment and adjusts power levels as necessary. Monitoring tools can verify service levels for each WLAN cell, determine AP loading, and balance the loads, making channel assignments to avoid conflicts. Some products provide site survey tools to help the designer place APs to optimize coverage.

Network designs with closely packed APs will undoubtedly require more than three channels, which means using 802.11a or a combination of 2.4 and 5 GHz products. Dual-technology APs accommodate both frequency bands, but setting these up can be tricky because of the different coverage ranges of the two frequencies.

Orthogonal Frequency Division Multiplex (OFDM)

OFDM, which is used in 802.11a and g, is the key to several technologies in both wired and wireless domains. OFDM uses a spread spectrum technique to distribute data over multiple carriers that are spaced at precise frequencies. The same technique is used in discreet multitone DSL, in which the signal is spread across 256 channels. An OFDM system spaces the frequencies as closely as possible while remaining orthogonal, that is mathematically perpendicular, while achieving minimum interference between subchannels.

In wireless applications, OFDM protects against the intersymbol interference (ISI) that comes from multipath delays by transmitting bits in parallel on separate ODFM channels. The oscillators at the transmitting and receiving ends must operate at exactly the same frequency to prevent the subcarriers from interfering with each other. The 802.11a protocol sends a training sequence at the start of each packet using four subcarriers. These are modulated with known data to produce pilot tones that enable the transmitter and receiver to adjust their oscillators. The 802.11a protocol uses binary phase shift keying (BPSK) for 6 to 9 Mbps, QPSK for 12 to 18 Mbps, and QAM for speeds from 24 to 54 Mbps.

Wireless Security

Security is the foremost issue in implementing WLANs. Most WLAN equipment comes from the factory with authentication configured as an open system, which means that any client may associate with the AP. The original 802.11 products use a protocol called wired equivalent privacy (WEP). WEP relies on a secret key to deter unauthorized stations from eavesdropping or joining the network. The WEP standard specifies the use of 40-bit keys, which are not difficult to crack. At the time WEP was developed, the U.S. government restricted the export of products with airtight encryption. Some manufacturers extend WEP to 128 bits, which makes it practically invulnerable to brute force attacks, but WEP is subject to other pitfalls. It uses a stream cipher called RC4. If a text string is available in encrypted and plain text form, the encryption key can be calculated with little difficulty. Shareware programs available on the Internet can be used to break WEP security. WEP also does not support key management, which requires users to change keys manually. Since this is a tedious process, keys are rarely changed. WEP is generally considered to be good enough for home use, but unacceptable for enterprise networks.

The WiFi Alliance resolved the security issue temporarily with a protocol known as WiFi Protected Access (WPA). WPA uses the Temporal Key Integrity Protocol (TKIP) to address the weaknesses in WEP. In addition to key management, WPA supports 802.1X Extensible Authentication Protocol (EAP) for station authentication. EAP relies on an authentication server such as RADIUS, (short for remote authentication dial-in user service), to subject the station to an authentication process. A wireless client sends an authentication request to an AP, which repackages the request and sends it to the RADIUS server. The server examines the request, and if authenticated, it informs the AP, which in turn informs the client that it is authorized. The process requires a user database, which could be a listing of users and passwords, digital certificates issued by a public-key infrastructure (PKI), or a Lightweight Directory Access Protocol (LDAP) directory.

To enhance security further, IEEE in 2004 approved 802.11i, which adds the Advanced Encryption Standard (AES) to the protocol family. These enhancements provide the platform for secure communications, but many off-the-shelf products, particularly in the SoHo category, support only WEP. Most products default to no privacy and the user must take action to turn privacy on, so many wireless networks are wide open. WiFi hotspots around most metropolitan areas provide free access and many introduce an insidious form of identity theft. Users can be lured into entering a credit card number over an open network where an identity thief can capture every keystroke. This fraud with countless variations is known in the industry as "phishing."

Even with WEP's weaknesses, most APs have security features built in. Most provide for MAC layer access control and encryption. Most also include a DHCP server which supplies stations with an internal IP address that cannot be reached from the Internet. APs broadcast the service set identifier (SSID), which client stations must know to associate with an AP. SSID broadcast can be turned off, but often it is left on. If hackers see the SSID being broadcast, they know the rest of the security settings probably have not been changed and the network is open.

High-Speed WLANs

Even though users sometimes overestimate their need for speed in LANs, low speed is one of the main factors that limit the growth of wireless LANs. As a result, proposals are in the works for LANs running up to 100 Mbps. One of the most interesting, which is under consideration for 802.11n, is the use of multiple input–multiple output (MIMO) antennas. Figure 21-2 shows conceptually how MIMO works. The input signal is divided into multiple streams—two are shown here. Each stream is transmitted on a different antenna at the same frequency. Phased-array antennas feed the signal into multiple elements and process the delay on each feed to reduce the beam width. Multipath reflections bounce off various obstructions and reach the receiving antennas out of phase. The receiver unscrambles the signal and combines the echoes to recover the original signal. MIMO exploits multipath, which is usually considered an impairment.

Wireless Bridges

A wireless bridge is a radio connection between two network segments. Wireless bridges generally have a range of a few hundred yards depending on the terrain. Most bridge products operate in the 2.4 and 5.7 GHz unlicensed bands, but some operate in unlicensed 24 GHz spectrum, which provides for greater throughput. The bridge gains its range by using high-gain antennas and power up to the maximum permitted. Three types of bridge are readily available:

- *Wired-to-wireless Ethernet*: This type links an Ethernet port on a device to an AP. These are useful when a device such as a printer has an Ethernet port, but cannot accept a NIC or USB transceiver.
- *Workgroup bridges*: This type of bridge connects wireless networks to wired Ethernet networks.
- *Combined AP and wireless bridge*: Some products can be configured as a bridge or an AP, but not simultaneously.

Infrared LANs

Although infrared wireless products are far less common than radio, they have a certain appeal because they have high bandwidth and do not require any form of

FIGURE 21-2

Multiple Input–Multiple Output (MIMO)

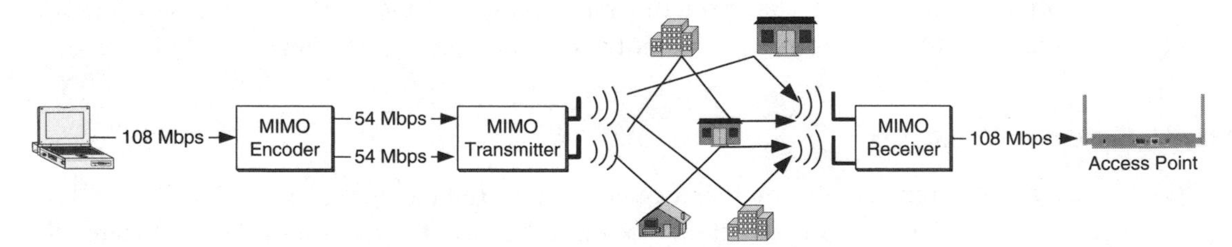

FCC type approval or licensing. Infrared products use one of three methods of distributing signals:

- *Line-of-sight* uses infrared transceiver pairs set up in a manner similar to point-to-point microwave. This method is used for building-to-building communications.
- *Reflective infrared* bounces the infrared signal off ceilings, walls, and floors to blanket an area.
- *Scattered infrared* uses a diffused signal that also bounces off walls and ceilings to cover an area.

Scattered and reflective infrared are primarily used for desktop wireless applications. The directionality of the signal and the inability to penetrate walls generally makes infrared impractical for mobile/portable operation. Line-of-sight infrared has a wider bandwidth than spread spectrum. Reflective and scattered infrared are low-speed systems.

WIRELESS VOICE

Most major PBX manufacturers offer a wireless adjunct to their systems, either as part of the product line or as a third-party add-on. Buildings can be equipped with an internal network that has most of the aspects of a cellular system. Miniature base stations are placed throughout the building and connected to TDM ports in the PBX. The base station controller directs base station assignments and handoff. The handset is a mobile version of proprietary desk phones and has most of the same features, although with a limited set of feature buttons.

When the FCC allocated space for the PCS band, it sandwiched 20 MHz of unlicensed spectrum between the two licensed portions of the band. The proximity is designed to allow frequency-agile units to hop from the unlicensed to the licensed spectrum so they can be used for PCS communications. Although this spectrum, like ISM, is unlicensed, it is an exclusive use segment, making it easy to find noninterfering frequencies.

A wireless installation consists of a central controller that interfaces the office telephone system, and controls a selection of remote base stations or cells. Figure 21-3 shows how the system is connected. The controller is either a standalone device that connects to analog ports on the PBX, or an integrated unit that provides station features over the radio link.

The base stations in a TDM network are self-contained low power, multi-channel radio transceivers. As with WLANs, locations are placed throughout the building to provide satisfactory coverage. A typical base station has a maximum range of about 700 ft; often less depending on the building structure. The base stations can be equipped with multiple radio channels, the quantity of which is chosen to support the expected number of simultaneous conversations. The base stations hand off to one another in a manner similar to cellular.

FIGURE 21-3

Wireless Voice Network

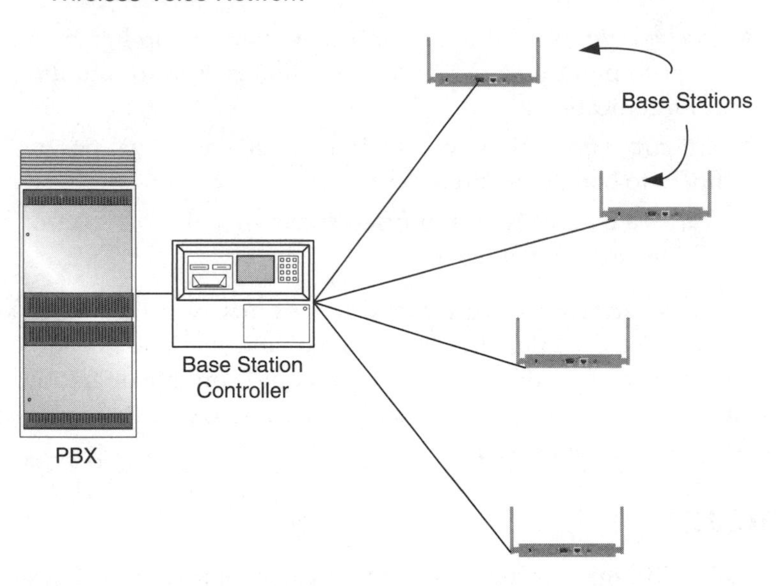

The portable instruments are either universal devices that emulate an analog telephone, or they are proprietary stations that work with the manufacturer's PBX. Analog stations are the least expensive, but they suffer the drawbacks of any analog telephone: features are activated by dialing codes. Proprietary instruments offer button access to features, but the physical dimensions of the instrument may limit the number of buttons. Some products are designed so the portable device is the handset of the regular desk instrument. The user can pick up the handset and walk away from the instrument, making a seamless wired-to-wireless transition. Another option is a docking station that effectively converts the wireless handset to a desk instrument. Docking stations typically include additional feature keys, an integrated speakerphone, and a battery charger.

The trend today is to use VoIP, running wireless phones over an 802.11 network. WiFi has several inherent characteristics that introduce complications that do not affect wired LANs. One is the collision-avoidance system that adds to delay. Another is packet loss, particularly when users roam between APs. QoS protocols are proprietary, at least until IEEE completes its work on 802.11e, which is not complete at this writing. Figure 21-4 shows a typical VoIP over WLAN network. The architecture is similar to TDM products, except that the interface between the APs and the PBX is a gateway and control is placed in a server that also regulates QoS. Voice packets are identified and prioritized in the server and scheduled ahead of data packets. If the voice load is heavy, it can have an adverse affect on data throughput.

PBX integration is not required in all cases. Wireless VoIP systems can communicate between stations internally, opening the way for miniaturized products

FIGURE 21-4

Wireless VoIP

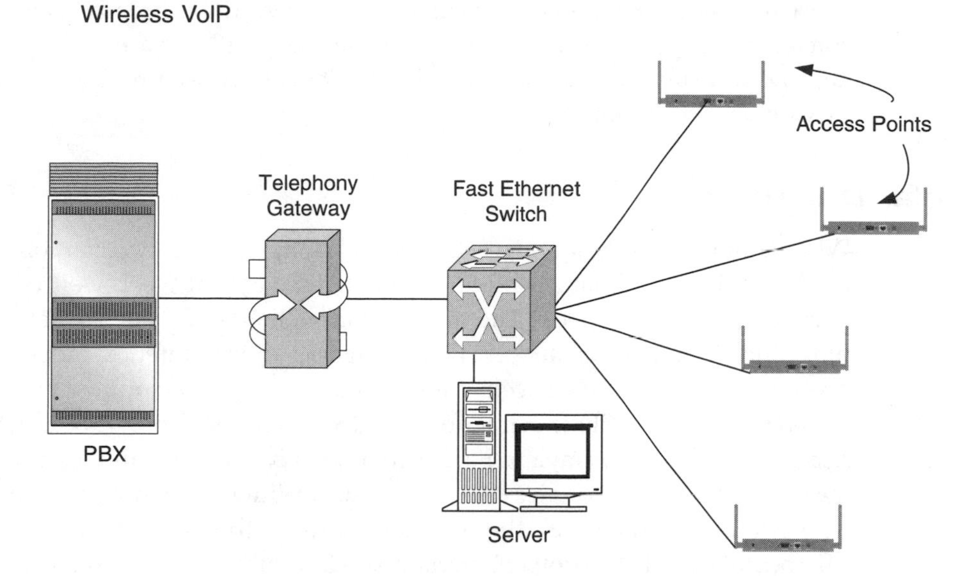

that can easily be stuffed in a shirt pocket or even clipped to the lapel. Such products are ideal for broadcasts such as a code blue in a hospital. The industry is working on developments that go by the name of ultra-wideband (UWB) to provide faster speeds and better coverage than 802.11. The objective of UWB, which should reach the market in 2007–2008 period, is to provide stable coverage and QoS for voice and short-range applications.

WIRELESS MOBILE DATA

Mobile and portable data applications require technology that allows the user to link to a remote computer or server over a wireless network. Parcel delivery companies now use wireless data extensively. Many police departments use MDTs, with which they can review arrest records and warrants without involving a dispatcher. Taxicab and service companies, utilities, and any company that dispatches personnel can save time, improve accuracy, and eliminate telephone calls with MDTs. Private radio systems use MDTs for such purposes as linkage between a law enforcement agency and a vehicular or law enforcement database.

The simplest and least expensive form of MDT is the alphanumeric pager. Provided the application can get by without an answer, pagers are an effective way of sending short messages using a service known as short message service (SMS). The length is limited, depending on the service, to about 160 characters—long enough for many e-mails, but truncating longer ones. At the other end of the scale are terminals with full QWERTY keyboards and laptop computers equipped with RF modems.

Within a narrow range of a building or campus, privately owned spread spectrum radio is a good alternative for wireless data. Broader ranges require some form of public network. The most common methods are dial-up over cellular, CDPD, general packet radio service (GPRS), or service provided by an SMR or private packet carrier.

Cellular Dial-up

While data transmission over cellular is simple, some precautions must be observed. First, ordinary commercial modems may not work well with cellular, particularly at high speed. The modulation methods of these modems are complex, and do not gracefully handle the vagaries of interference, fades, and signal dropout. Handoff between cells causes a momentary interruption, and may drop the connection. If a file is being downloaded when the dropout occurs, it may be necessary to start over again. Furthermore, many modems are designed to operate only after they recognize dial tone, which cellular does not provide.

Many products on the market address the problems. For example, Microcom's MNP-10 protocol starts a session with a handshake at 1200 bps to negotiate the connection. The protocol then gradually increases the transmission speed to reach the point of maximum throughput. Paradyne with its Enhanced Throughput Cellular protocol takes the opposite approach. It starts at a higher speed and reduces if necessary. This method of communication is shown in Figure 21-5 along with CDPD.

Most cellular providers deliver short e-mail messages directly to compatible cell phones. For occasional use and mobile file transfer, cellular is a good alternative, but it is not suitable for interactive applications. Most cellular operators levy

FIGURE 21-5

Mobile Data Through Two Options: CDPD and Dial-up Cellular

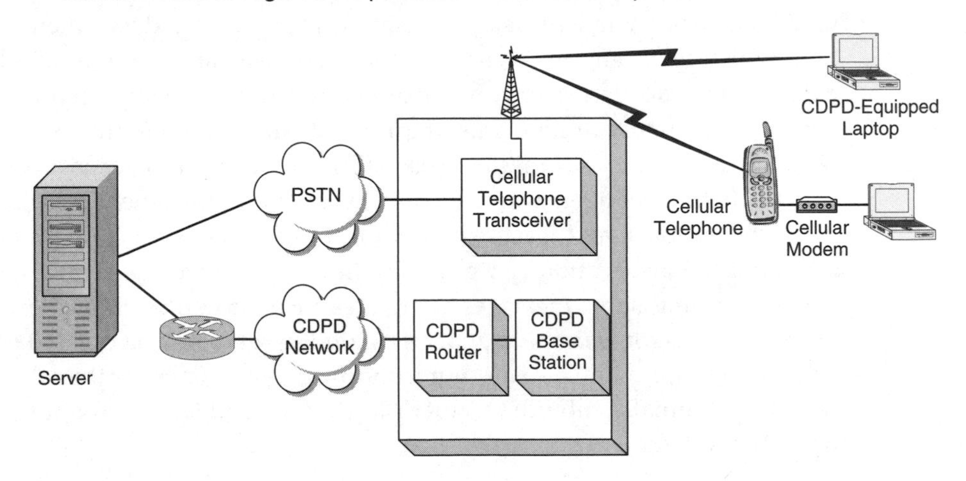

a 1-min per-call minimum charge, which means paying for a transaction that takes only a few seconds. Also, the setup time is long compared to other alternatives. Services that offer always-on capability are far more convenient, and furthermore, reduce the load on the cell sites. Cellular has the advantage of good coverage. In general, its applications are similar to those of the PSTN: it is good for occasional use, facsimile, and file transfers, but it is poor for short, bursty transactions.

Cellular Digital Packet Data (CDPD)

A major advantage of cellular is its coverage, which makes CDPD an attractive alternative. CDPD is a packet-switched data service that rides on top of cellular, and uses idle analog channel time. Because of the half-duplex nature of voice sessions and frequent pauses, phone conversation has many dead spots where the channel capacity is not being used. CDPD uses this idle time by hopping between channels, inserting packets as needed, and reassembling them at the other end. CDPD can be added to existing cell sites at a moderate cost. To avoid the problem of slow data speed during heavy voice usage, most service providers dedicate some channels to CDPD. Charging is by the kilobyte instead of by the minute, and the always-on capability eliminates the call setup time. This makes it good for short, bursty messages such as point of sale, dispatch, package tracking, telemetry, and e-mail. It also supports IP multicast service, which enables an organization to broadcast information to traveling users. CDPD is available in most metropolitan areas.

CDPD operates at 19.2 Kbps using ISO Connectionless Network Protocol (CLNP) or TCP/IP. Actual throughput is around 12 Kbps. Both ends of a session must use the same protocol since CDPD does not support a gateway function. While TCP/IP is an excellent protocol for wired services, it is not optimized for mobile use. The protocol is not equipped to handle a mobile subnet where the IP address is periodically changing. Also, TCP/IP returns frequent acknowledgment packets, and although these are short, they add to the cost of a session. The CDPD network provides a full-duplex connectionless network service (CLNS), a datagram service in which the network routes each packet based on the destination address and its knowledge of the network topology.

The link-layer protocol is called mobile datalink protocol (MDLP). It provides framing, sequencing, and flow control, and detects and recovers from frame loss. It defines the window size, which controls the number of frames that can be sent or received in a single block.

To implement CDPD, carriers install mobile database stations (MDBS) which retrieve packets from the wireless network, and a mobile data intermediate system (MDIS) which routes them based on its knowledge of end station locations. The subscriber device is an RF modem, which may be a PCMCIA card in a laptop. Frames are picked up by the MDBS and handed off to the MDIS.

Mobile stations use a protocol called digital sense multiple access with collision detection (DSMA/CD) for access to the network. The access method is similar to

Ethernet's CSMA/CD. A station wishing to transmit listens to the outbound channel to determine if a carrier is present. If not, it transmits a packet. If the network is busy, it backs off and attempts a short time later. The maximum time a station can hold the channel is about 1s. CDPD can support several active channel streams at one time.

A major advantage of CDPD is its coverage. The main population areas of the country have good cellular coverage, and if the carrier elects to overbuild the network with CDPD, data coverage can be equivalent to voice. Its mobility management service manages roaming and tracks the locations of end stations with a process similar to cellular roaming. Unless the carrier has equipped all cell sites with CDPD, which may not always be the case, the coverage may be narrower than regular cellular. Also, roaming to nonaffiliated areas can be a problem if the carriers have not negotiated an interconnection agreement.

General Packet Radio Service

GSM and its North American counterpart I-136 PCS do not work well for data transmission. Because the signal is already highly compressed, data is limited to 9.6 Kbps and PCS suffers from the same limitations that hamper dial-up cellular. The GSM counterpart to CDPD is GPRS, not to be confused with GPS. GPRS is a 2.5G always-on connection that is overlaid on GSM. The major use is for IP, but it is also capable of operating with X.25. The service carries a flat monthly fee that entitles the user to a block of capacity, measured in megabytes. It is a popular service to support combined PDA and voice appliances such as the popular Blackberry.

GPRS has a theoretical maximum speed of 171.2 Kbps, but in practice connections are much slower because the service provider limits the number of timeslots available to support data. The connection is somewhat faster than straight dial-up and makes much better use of the bandwidth because channels are shared among multiple users. Also, signaling and data traffic do not use the GSM network except for lookup in the location register database. The service operates in three modes:

- Class A: Simultaneous voice and data transmission
- Class B: Automatic switching between voice and data
- Class C: Manual switching between voice and data

The user appliance must be designed for GPRS; existing GSM phones do not work, but many wireless PDAs and pocket PCs can connect to the service. One goal of GPRS is to overcome the message length limitations of SMS. Regular GSM imposes a 160-character limit on SMS messages. The service supports any IP service such as Web browsing, chat, file transfer, and applications such as FTP and TELNET. The service is particularly effective for applications such as keeping appointment calendars synchronized and obtaining e-mail messages while on the move. Attachments can also be accommodated within a limited range of file formats. GPRS is not without drawbacks. Since the user device has an IP address, it is subject to all of the ravages of the Internet such as viruses, denial-of-service attacks, and spam. When users are paying for usage by the megabyte, unwanted traffic becomes particularly irksome.

To implement GPRS, the service provider equips the service area with a gateway GPRS service node (GGSN), which is the gateway between GPRS and the data network. The nodes are connected over a backbone IP network that routes the traffic to nodes to which the mobile unit is tuned. Figure 21-6 shows the architecture of a GPRS network.

Private Wireless Packet Carrier

Private packet radio carriers serve some metropolitan areas. These services can be a good way to handle mobile and portable data, except that the service is not generally available everywhere. Where it is available, the coverage range may not match cellular or CDPD. The same factors must be considered in evaluating these carriers as in evaluating CDPD.

RADIO PAGING

Paging is less sophisticated and less costly than cellular radio, but is growing at a fast pace. Developed under the centralized transmitter and receiver plan of conventional mobile telephone service, one variety of radio paging offers dial access from a wireline telephone to a pocket receiver. The readout is a beeping sound or vibration to alert the user to an incoming message. Digital pagers allow the caller to send a callback number or numeric message to a digital readout.

FIGURE 21-6

GPRS Architecture

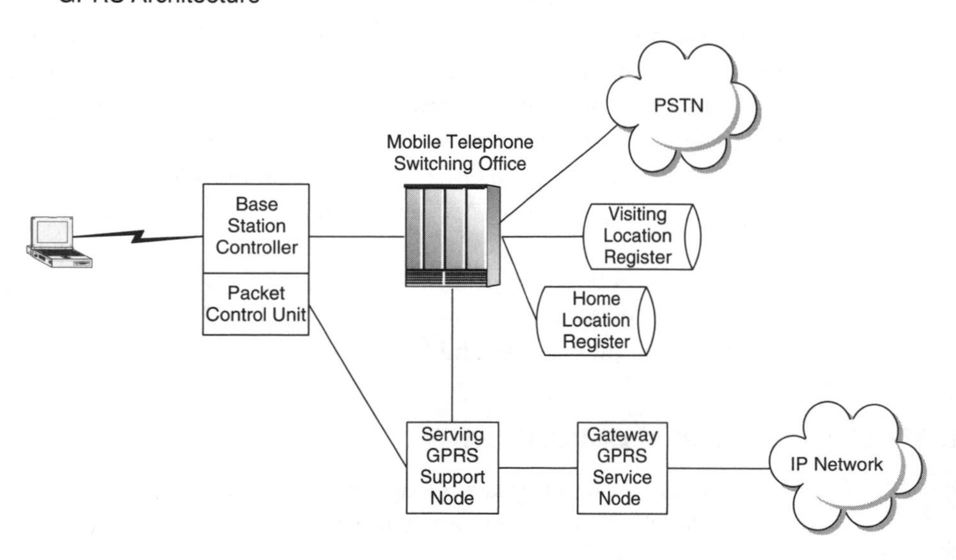

Several interesting advances have been made in radio paging in the past few years. One useful feature is alphanumeric paging, in which the caller can send a text message in place of the alternative, which is either beeping the user or sending a numeric message such as a callback number. To activate alphanumeric paging the caller keys in a message and it displays on the pager readout. Pagers are available with one or more readout lines. By pressing a button, the user scrolls the message in serial fashion across the readout if it is too long to fit in a single screen. Alphanumeric paging can reduce the number of cell phone messages at a significantly lower cost.

Alphanumeric paging is made easier with a paging server, which can be operated off a LAN. Any user on the network can access the server, send the message to it, and the server outdials to the paging system. Features available include sending messages to e-mail, screening calls, and outdialing messages from only selected stations.

Another innovation is answerback paging. When a message has been received, the recipient answers by pressing a button on the pager and the pager sends an acknowledgment to the sender. Some pagers enable the user to send messages to other one- and two-way pagers, fax machines, and Internet e-mail. Combined with devices such as the PDA, the two-way pager can be a compact device for keeping in touch.

Paging competes with SMS for short alphanumeric messages. The main advantage of paging is its lower cost of sending the same message to multiple recipients. Paging also delivers messages in a shorter time than SMS messages, which may take several minutes to arrive.

WIRELESS APPLICATION ISSUES

Wireless is not a universal replacement for wired voice or LAN service, but it fulfills a major requirement that many users have for mobility. One of the first requirements is to condition users' expectations. A wireless network is not the equivalent of wired Ethernet and client software does not always behave predictably. The nature of the application must be considered. Is coverage required for e-mail or other types of messaging, calendar synchronization, file transfer, point of sale, or other applications? If so, a WLAN may be an ideal solution, but expectations should not be too high, particularly if the network is intended for both voice and data.

The range of wireless products is so broad and the claims often so exaggerated that it is difficult to see how each fits. Figure 21-7 shows in broad terms the various wireless alternatives and the market they are expected to serve.

WLAN Design and Application

Two approaches can be taken to designing a WLAN. One is to purchase and connect a few APs and find out if users complain. If they do, add more APs.

FIGURE 21-7

Wireless Service Positioning

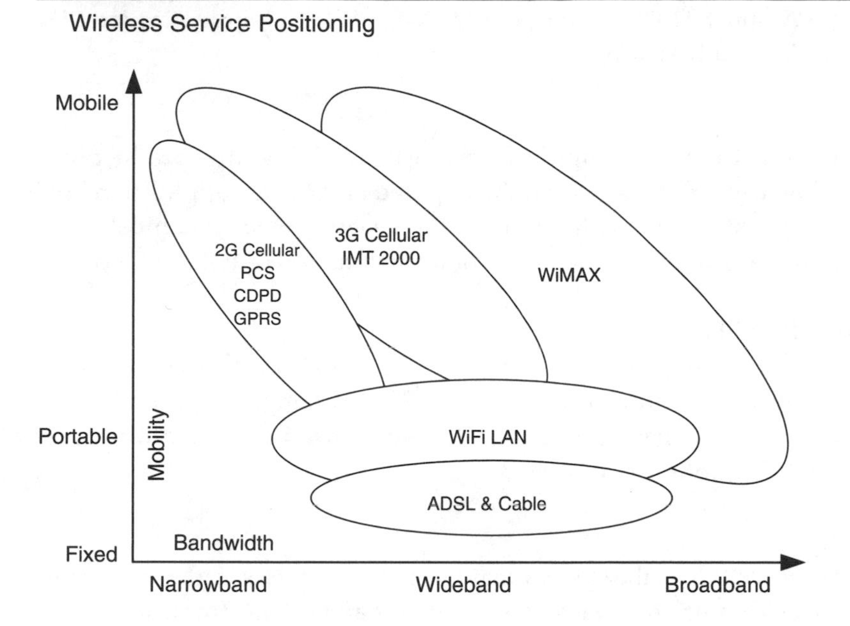

This approach works fine in smaller offices, particularly where the building is not shared with other companies. A more detailed engineering approach is recommended for enterprise networks. The first step is to map existing signals to detect potential interference. Then APs are chosen to provide coverage and the channels and radio power levels are set to minimize interference. Products are selected to meet the company's cost and administrative objectives. Enterprise wireless systems can support much of the administrative task of setting power levels, collecting usage information, and detecting dead spots that are revealed by signal loss. Theoretically, it should be possible to mix and match hardware, but if so, it pays to have a system integrator on the hook for resolving any irregularities. The industry trend is toward the use of mesh networks to avoid the dead spots that otherwise result.

Security

The number one concern with a WLAN is security. Wireless signals are not confined to the building so it is mandatory that strong encryption be enforced and that users are properly authenticated. Enterprise LANs must be secured by containing both wired and wireless segments within a firewall perimeter, minimizing exposure to the Internet. The RF link must be secured with WPA. WEP must be eliminated and all client software upgraded to WPA. Policies must be set to prohibit rogue APs that may cause interference or inadvertently open security holes. A process for distributing and changing encryption and authentication keys is essential. Review the security provisions built into the product. Does it support WPA and 802.11i as well as 802.1X for WLAN authentication? Does it support LDAP and RADIUS?

Range

What is the area the application must cover? Does the product have the range to allow users to roam within the required area? Is the application building to building, desktop, or mobile/portable? If the latter, what area must be covered? Which frequency range, 2.4 or 5 GHz or a combination of both will be used?

Service Availability

For mobile and portable applications, what alternatives are available locally? What carriers provide CDPD or GPRS? Is roaming required? If so, does GPRS or CDPD cover the required area? Do specialized carriers offer packet radio? What do the various services cost?

Robustness

How well does the system handle signal dropouts, fades, interference, and other such irregularities? How well does it handle long frames? How does it recover from interruptions? Can it recover if it is interrupted in the middle of a file transfer? What kind of speed is available? Does it automatically adjust to optimize throughput?

Administrative Support

How many system integrators are available to support the application in the local area? Do they have experience with wireless data? Be sure to check references. Can the system be globally configured or is it necessary to visit each AP individually? How is RF monitoring done—continuously or on schedule? Does the software support AP transmit power management and dynamic channel assignment? Does it detect interference and coverage holes? Does the system have dynamic spectrum management to change power and coverage in response to changing RF conditions? Can it change channel assignments automatically to avoid interference and optimize AP performance? Does the product measure traffic and automatically balance the load across multiple APs?

Wireless Voice Design and Application

This section deals with evaluation criteria for wireless voice. The first issue to deal with is whether to use TDM or VoIP products. VoIP appears to be the direction most manufacturers are going, but the products are immature and the protocols are proprietary. Consequently, it is risky to mix products from different manufacturers except with the help of an experienced system integrator. The following criteria should be considered in evaluating products.

Frequency Band

Determine which frequency band the system works in. Is hop-off to PCS available or desirable? Is the product likely to interfere with an existing or planned WLAN? What other building occupants use the frequencies?

Coverage

Determine the coverage area of wireless cells. How well does the signal penetrate walls? Can users roam from one cell to another? The narrower the coverage range, the more cells that will be required to cover a building or campus. Evaluate coverage between floors in buildings such as hospitals that may have a great deal of concrete and steel in their structure. Cells with less coverage may enable more consistent reuse of the available channels since the same frequencies can be reused in a different part of the building.

Station Sets

Are the telephone sets proprietary? Analog station sets are universal and connect to any analog port on any telephone system. Proprietary sets connect to proprietary ports, and offer the advantage of button access to features. Note that even wireless IP phones are likely to be proprietary until standards are more mature. Determine what features are available and whether they match the organization's needs. Determine the talk and standby time of the batteries. Determine the need for features such as docking stations.

Integration

How does the system integrate into your telephone system? What is the interface to the PBX: analog, TDM digital, or IP? Are proprietary phone sets available? Are multiline sets available? Does the display show the same information that a wired set does? Within an area in the building, how many calls can be carried simultaneously?

CHAPTER 22

Video Systems

Today's video systems bear little resemblance to cable television (CATV) systems of yore. Those delivered a dozen or so channels to households in areas that could not receive signals off the air. Today, cable systems deliver at least 70 channels with a combination of entertainment and information services, bringing broad bandwidths into homes and businesses and threatening to upset the traditional method of providing telephone service. Over the past few years, cable companies have rebuilt their networks into two-way broadband systems. The Internet was the first application, but once that was completed, some of the bandwidth earmarked for IP was available for video-on-demand. Cable has captured more than half the broadband access market in the United States, and now it is poised to begin competing with the LECs to deliver telephone service.

As the cable providers prepare to compete with the telephone companies, they are faced with competition of their own from satellite providers. Satellite video is digital and delivers better signal quality than the analog channels that the cable providers offer. Most cable providers also offer digital channels, and the broadcasters are converting their analog signals to digital. The FCC has mandated a transition to digital by the end of 2006 for broadcasters that use the airways, but that deadline will probably be extended.

While entertainment is the prevalent use of video today, videoconferencing is an important facility for businesses and desktop video will soon be routine. Distance learning for schools and telemedicine for healthcare organizations are emerging applications. As with most technologies, the key to expansion has been the development of standards. For years, the videoconferencing manufacturers produced equipment with proprietary standards, which meant they could not interoperate. Then ITU-T completed its work on H.320 standards, which enables equipment to interoperate over dial-up or dedicated digital facilities. For packet networks, H.323 standards enable video communication across IP and frame relay.

AT&T demonstrated its Picturephone at the 1964 New York World's Fair, but it turned out not to be practical. Like many other product developments, however, it was merely ahead of its time. Conferencing equipment built into desktop computers is becoming a regular feature in most companies and many homes. With the growth of broadband connections to most homes and businesses, video will become a common enhancement for telephone calls.

VIDEO TECHNOLOGY

Video signals in North America are generated under the National Television Systems Committee (NTSC) system. In Europe they are generated under two different and incompatible standards: PAL (phase alternate line) and SECAM (sequential couleur avec memoire). An analog video signal is formed by scanning an image with a video camera. As the camera scans the image, it creates a signal that varies in voltage with variations in the degree of blackness of the image. In the NTSC system, the television raster has 525 horizontal scans, forming a raster that is composed of two fields of 262.5 lines each. The two fields are interlaced to form a frame (Figure 22-1). The frame repeats 30 times per second; the persistence of the human eye eliminates flicker. Since the two fields are interlaced, the screen is refreshed 60 times per second.

On close inspection, a video screen is revealed to be a matrix of tiny dots. Each dot is called a picture element, abbreviated *pixel*. The resolution of a television picture is a function of the number of scan lines and pixels per frame, both of which affect the amount of bandwidth required to transmit a television signal. The NTSC system requires 4.2 MHz of bandwidth for satisfactory resolution. Because of the modulation system used and the need for guard bands between channels, the FCC assigns 6 MHz of bandwidth to analog television channels.

FIGURE 22-1

Interlaced Video Frame

Active Fields
Heavy lines represent first field.
Light lines represent second field.
Dotted lines represent retrace lines that are blanked out.

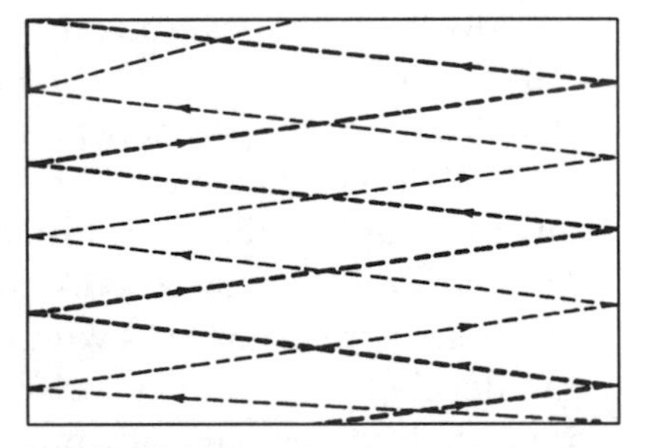

Inactive Fields
Heavy lines show retrace pattern from bottom to top of screen during vertical blanking interval. Heavy lines represent first field and light lines the second.

The signal resulting from each scan line varies between a black and a white voltage level (Figure 22-2*a*). A horizontal synchronizing pulse is inserted at the beginning of each line. Vertical pulses synchronize the frames as Figure 22-2*b* shows. Between frames the signal is blanked during a vertical synchronizing interval to allow the scanning trace to return to the upper left corner of the screen.

A color television signal has two parts: the *luminance* signal and the *chrominance* signal. A black and white picture has only the luminance signal, which controls the brightness of the screen in step with the sweep of the horizontal trace. The chrominance signal modulates subcarriers that are transmitted with the video signal. The color demodulator in the receiver is synchronized by a color burst consisting of eight cycles of a 3.58 MHz signal that is applied to the horizontal synchronizing pulse (Figure 22-2*a*).

When no picture is being transmitted, the scanning voltage rests at the black level, and the television receiver's screen is black. Because the signal is

FIGURE 22-2

Synchronizing and Blanking in a Television Signal

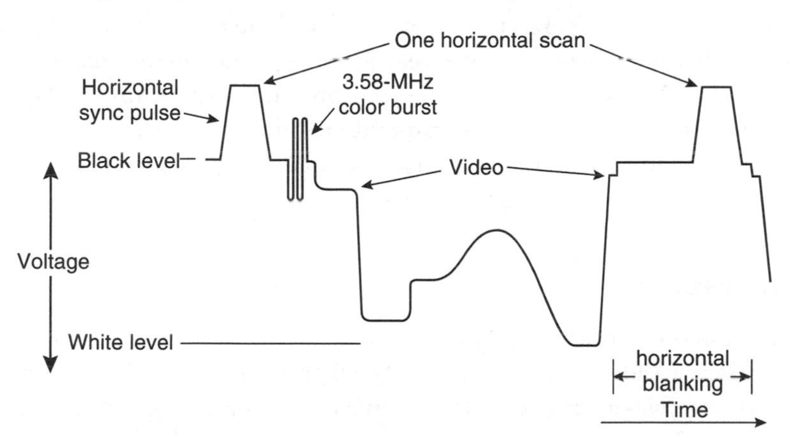

a. Voltage levels during one horizontal scan. As image is sanned, voltage varies from reference black to reference white level.

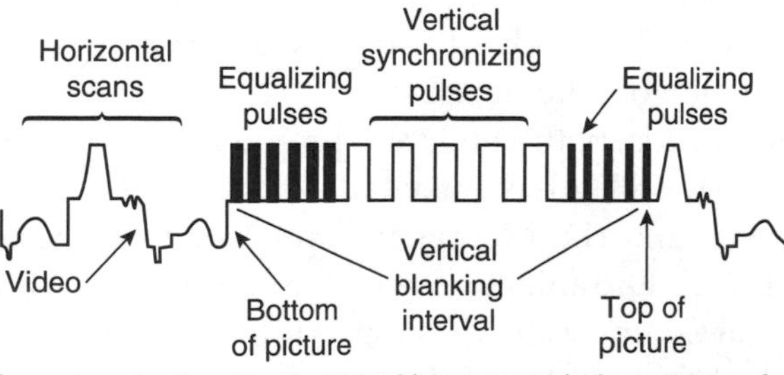

b. Vertical synchronization. Vertical blanking occurs during retrace of scanner to top of screen.

amplitude-modulated analog, any noise pulses that are higher in level than the black signal level appear on the screen as snow. A high-quality transmission medium keeps the signal above the noise to preserve satisfactory picture quality. The degree of resolution in a television picture depends on bandwidth. Signals sent through a narrow bandwidth are fuzzy with washed-out color. The channel also must be sufficiently linear. Lack of linearity results in high-level signals being amplified at a different rate than low-level signals, which affects picture contrast. Another critical requirement of the transmission medium is its envelope delay characteristic. *Envelope delay* is the difference in propagation speed of the various frequencies in the video passband. If envelope delay is excessive, the chrominance signal arrives at the receiver out of phase with the luminance signal, and color distortion results.

Digital Television

As with telephone transmission, analog impairments can be avoided by converting the signal to digital. An uncompressed studio-quality signal is about 270 Mbps, so efficient use of the bandwidth means the digital signal must be compressed. Using MPEG-2 compression, as many as 10 digital channels can be squeezed in the 6 MHz space formerly occupied by one analog channel. This greater channel efficiency is the reason the FCC has mandated the conversion to digital TV. MPEG-2 is the same compression algorithm as DVD disks use. Although some slight decrease in quality results from the compression, it is hardly noticeable and it overcomes many of the impairments such as interference and ghosting that affect analog transmission.

Video Compression

Compression algorithms rely on the fact that a video signal contains considerable redundancy. Often large portions of background do not change between frames, but analog systems transmit these anyway. By removing redundancy, digital signal processing can compress a video signal into a reasonably narrow bandwidth. The signal can be compressed to occupy as little as 64 Kbps, but the quality falls short, even for videoconferences, which need at least 384 Kbps, or six channels for reasonable quality. T1 bandwidth is needed for a signal that is approximately equal to the quality of a home VCR.

Many standards, both public and proprietary have been developed for video compression. ITU's H.261 is intended for videoconferencing, and is discussed later in more detail. The Motion Picture Experts Group, a working group of the International Organization for Standardization and the International Electrotechnical Commission, develops MPEG standards for encoding both video and audio. The principal standards of concern in telecommunications are MPEG-1, -2, and -4. Other standards describe multimedia content and delivery.

MPEG-1 compresses a video signal into bandwidths of up to 1.5 Mbps. The resolution is 288 lines per frame, which is home VCR quality. It is satisfactory for broadcast use if the scene does not have too much action. MPEG-2 codes studio quality video into bit streams of varying bandwidth. The most popular studio signal, known as D-1 or CCIR 601, is coded at 270 Mbps and can be compressed into about 3 Mbps. Scenes with high activity such as sports broadcasts require bit rates of about 5 or 6 Mbps. MPEG-3 was intended for HDTV, but it was discovered that MPEG-2 syntax was sufficient. MPEG-4 is an enhancement for multimedia transmission.

Video is compressed by predictive coding and eliminating redundancy. Flicking by at 30 frames per second, much in a moving picture does not change from frame to frame. Interframe encoding recognizes when portions of a frame remain constant, and transmits only the changed portions. Intraframe encoding provides another element of compression. The picture is broken into blocks of 16 × 16 pixels. A block is transmitted only when pixels within that block have changed. Otherwise, the decoding equipment retains the previous block and forwards it to the receiver with each frame. Predictive coding analyzes the elements that are changing, and predicts what the next frame will be. If the transmitting and receiving codecs both use the same prediction algorithm, only changes from the prediction must be transmitted, not the complete frame. Approximately every two seconds, the entire frame is refreshed, but in intervening frames, only the changed pixels are transmitted. The use of predictive coding requires a high-quality transmission facility. Lost packets in video-over-IP can have a detrimental effect beyond the information loss of one frame because of the predictive nature of the coding algorithm.

Encoding systems are classified as lossless, or lossy. Information in a lossless coding system can be restored bit-for-bit. A lossy encoding system transmits only enough to retain intelligibility, but not enough to ensure the integrity of the received signal. Lossy systems work well for sending still scenes, but motion can cause a tiling or smearing effect as the receiving codec attempts to catch up with the transmitter. The higher the compression and the more vigorous the motion, the greater is the smearing effect.

High Definition Television (HDTV)

The present NTSC television standard was defined in 1941 when 525-line resolution was considered excellent quality and when such technologies as large-scale integration were hardly imagined. The standard was advanced for its time, but it is far from the present state of the art. Larger cities are running out of broadcast channel capacity. Although not all channels are filled, co-channel interference prevents the FCC from assigning all available channels. Large television screens are becoming more the rule than the exception, and at close range the distance between scan lines is disconcerting. The 525 scan lines of broadcast television and the 4:3 width-to-height ratio of the screen, called its *aspect ratio*, limit picture quality. With wide-screen movies and the growing popularity of large-screen

TABLE 22-1

Comparison of NTSC Video to HDTV

	NTSC	HDTV (ATSC)
Total Lines	525	1125
Active Lines	486	1080
Sound	2 (Stereo)	5.1 (Surround)
Aspect Ratio	3:4	9:16
Max Resolution	720 × 486	1920 × 1080

television sets, the definition of the current scanning system is considerably less than that of the original image. All of this leads to a transition to HDTV.

HDTV has been introduced in most of the larger markets in the United States, but it has not been widely accepted yet because it requires replacement of existing television sets. Table 22-1 compares NTSC video with the new standard, which is known as Advanced Television Systems Committee (ATSC). The resolution of HDTV is far superior to NTSC. It has six times the number of pixels and supports wide screen television.

Video on Demand

A driving factor for digital television is video on demand (VoD). Today's CATV systems deliver all channels to every residence, and unwanted or unauthorized channels are trapped out or scrambled. VoD refers to a broad spectrum of services that are delivered over a broadband medium. Entertainment, information, and education services are examples of VoD services that the service provider transmits on order. Subscribers can choose movies or programs they want via an on-screen menu, and control the sessions with VCR-like functions such as stop, fast-forward, and rewind.

VoD may save a trip to the neighborhood video rental store, but the main question is delivery. The two principal delivery methods are via broadband cable or over DSL. Speed is the limiting factor with DSL. Reasonable quality can be obtained with the popular ADSL, but studio quality HDTV requires at least 3 Mbps, which requires VDSL. The limiting factor here is its range, which is about 4000 ft (1200 m). Delivery over cable is not too different from a range standpoint, however, to achieve the broadband speeds required without compromising quality, the streaming video signal must be brought to a neighborhood center over fiber. Cable providers offer VoD to digital service subscribers, which so far are a minority. Digital television sets are readily available, but for now most television sets are analog, which means a converter is required. Video servers can also be an obstacle. The amount of data that must be delivered to fill digital pipes to thousands of simultaneous users requires servers capable of storing many terabytes of data.

CABLE TELEVISION SYSTEMS

Cable television systems have three major components: headend equipment, trunk cable, and feeder and drop cable. Figure 22-3 is a block diagram of a conventional CATV system. All channels that originate at the headend are broadcast to all stations, which means the operator must scramble or block premium channels the user is not paying for. Headend equipment generates local video signals and receives signals from a variety of sources including off-the-air broadcasts, communication satellites, or microwave relay or fiber-optic connections from other providers.

The analog signals from the headend are modulated to a channel within the bandwidth of the cable, which may be as great as 1 GHz. Headend equipment applies the signal to a trunk cable to carry the signal to local distribution systems. The trunk cable is equipped with broadband amplifiers that are equalized to carry the entire bandwidth. Amplifiers have about 20 dB of gain and are placed at intervals of approximately 500 m. Amplifiers known as *bridgers* couple the signal to feeder cables.

FIGURE 22-3

Conventional CATV System

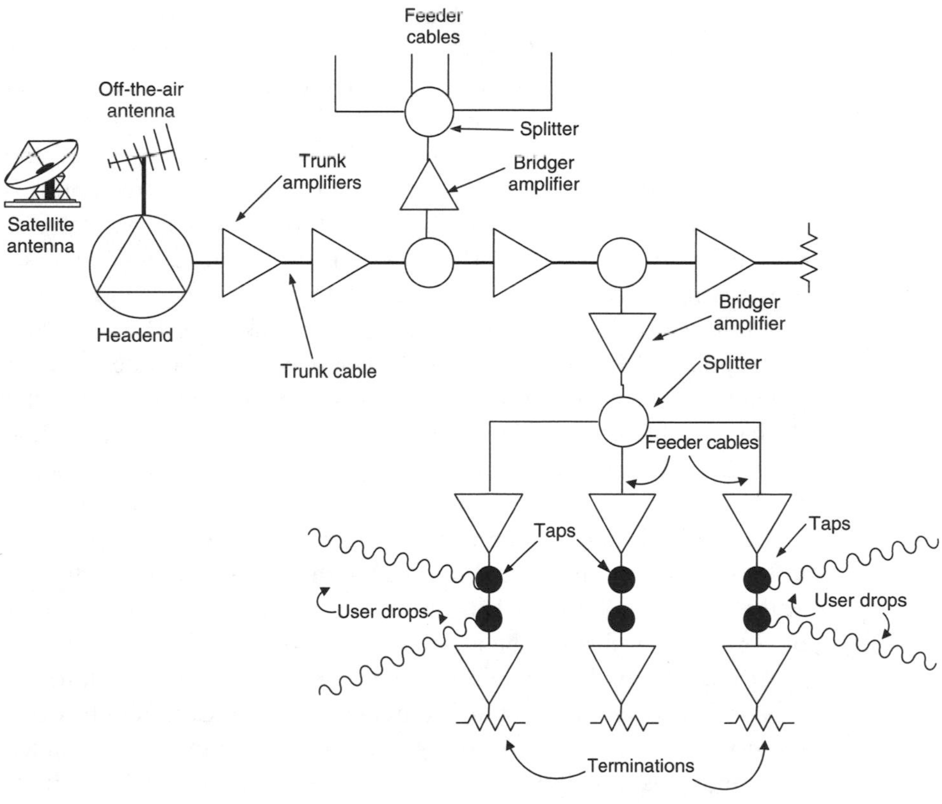

Amplifiers contain automatic gain control circuitry to compensate for variations in cable loss. Power is applied to amplifiers over the coaxial center conductor. To continue essential services during power outages and amplifier failures, the cable operator provides redundant amplifiers and backup battery supplies. Because the CATV signals operate on the same frequencies as many radio services, the cable and amplifiers must be free of signal leakage since a leaking cable can interfere with another service or vice versa. As with other analog transmission media, noise and distortion are cumulative through successive amplifier stages. Analog impairments limit the serving radius of 70-channel CATV to about 8 km from the headend.

Bridger amplifiers split feeder or distribution cable from the trunk cable. Multiple feeders are coupled with splitters or directional couplers, which match the impedance of the cables. The feeder cable is smaller and less expensive and has higher loss than trunk cable. Subscriber drops connect to the feeder cable through *taps*, which are passive devices that isolate the feeder cable from the drop. The tap must have enough isolation that shorts and opens at the television set do not affect other signals on the cable.

Hybrid Fiber Coax (HFC)

As cable technology improved, the bandwidth increased and more channels were added. Some operators added upstream channels to provide additional revenue-generating services such as alarm monitoring, but the real revolution came with Internet access. Although the bulk of Internet traffic flow is downstream, a substantial amount of upstream bandwidth is needed. The conventional CATV model has either limited or no upstream bandwidth, and routing every channel past every subscriber does not work for Internet access because of the shared nature of the medium. As Internet traffic increases, the response time increases, generally in proportion to the number of subscribers. The answer is to rebuild cable networks with an HFC model similar to the one in Figure 22-4. Both entertainment and access bandwidth are brought to neighborhood centers on a fiber backbone. The entertainment channels are applied to the distribution cable as always, but the Internet channels are combined to a lesser number of subscribers. The response time on the shared portion of the bandwidth is limited by controlling the number of subscribers that a node serves.

The HFC architecture gives cable companies control over shared bandwidth to make it suitable for other services. One logical candidate is telephone service. Once the cable infrastructure is in place, the incremental per-subscriber cost of VoIP is small. As shown in Figure 22-4, the IP bandwidth is delivered to a router at the headend. Voice packets are routed to a media gateway, which is controlled by a softswitch that can be located anywhere.

This method of delivering telephone service has certain advantages that exist today by regulatory fiat, but may not persist. Congress has elected to exempt IP services from the many taxes and fees that it imposes on telephone service. Although it is impossible to foresee what Congress will do in the future, nothing

FIGURE 22-4

Hybrid Fiber Coax Architecture

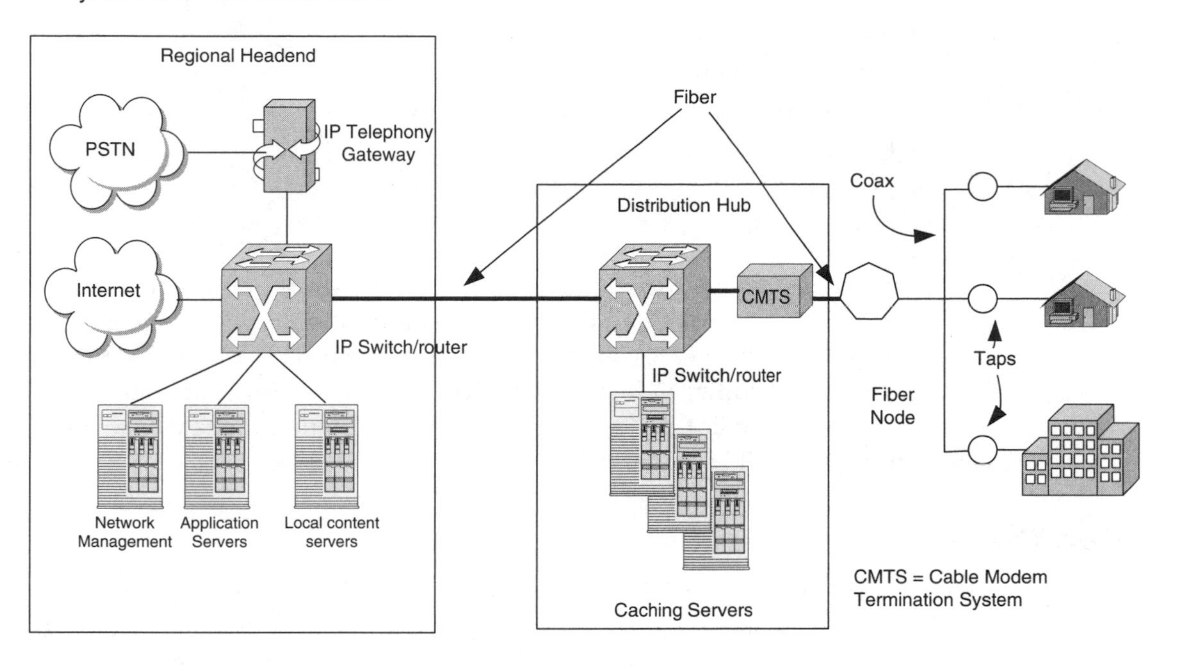

suggests that this exemption is permanent. Another issue is exemption from equal access. Today the cable companies are not required to permit equal access to their facilities, which means competing service providers cannot rely on cable as an access medium. Either of these would affect the profitability of local telephone service over cable.

VIDEOCONFERENCING EQUIPMENT

Videoconferencing has considerable appeal, primarily as a substitute for travel, but it has yet to become a mainstream application. The primary drawback has been cost. Not long ago, the cameras, monitors, codec, and formal conference room setup could easily exceed $100,000 and the payback was difficult to quantify. Recently, equipment costs have dropped to a fraction of their previous levels and a conference unit can now be placed almost anywhere. The transmission costs, however, remain high. A reasonable videoconference requires 384 Kbps, or six BRI channels. Add usage costs to that and conferences, particularly multipoint conferences, are still expensive.

A two-way videoconference over IP requires about 400 Kbps of full-duplex bandwidth. With a stable private network in place, the quality is as good as ISDN. The Internet is not a suitable medium where true conference-quality video is required, so most external conferences will need an ISDN gateway.

A videoconference facility has some or all of the following subsystems integrated into a unit:

- Video codec
- Audio equipment
- Video production and control equipment
- Graphics equipment
- Document hard copy equipment
- Communications

A full description of these systems is beyond the scope of this book, but we will discuss them here briefly to illustrate the composition of a full videoconferencing facility. Personal video communications equipment is readily available and if the network has sufficient bandwidth and QoS, it is far less elaborate than a formal conference facility, and will likely be the application that enables videoconferencing to fulfill its promise.

Video Codec

The codec converts the analog video signal to digital and compresses it for transmission. At the distant end the process is reversed. Codecs for dial-up conferences must support H.261. Many also support H.263 coding for IP conferences.

Audio Equipment

Most analysts believe that audio equipment is the most important part of a videoconference facility. In large videoconferences it is often impossible to show all participants, but it is important that everyone hear and be heard clearly. Audio equipment consists of microphones, speakers, and amplifiers placed strategically around the room. Sometimes speaker telephones are used, but with less satisfactory results than codec audio. The codec robs a portion of the bandwidth to transmit the audio, so some products allow the operator to reduce the amount of audio bandwidth as a way of improving the video. When IP conferencing is used the audio must be multiplexed on the bit stream because lip sync cannot be preserved with a separate audio channel.

Video and Control Equipment

Video equipment consists of two or three cameras and associated control equipment. The main camera usually is mounted at the front of the room and often automatically follows the voice of the speaker. Zoom, tilt, and azimuth controls are mounted on a console, where the conference participants can control them from a panel with a joystick. A second camera mounts overhead for graphic displays. The facility sometimes includes a third mobile camera that is operated independently. A switch at the console operator's position selects the camera. Usually, one monitor shows the picture from the distant end and another shows the picture at the near end. In single-monitor conferences the near end can be viewed in a window using the monitor's picture-in-picture feature.

Digitizing and encoding equipment compresses full motion video or creates freeze-frames. In addition, encryption equipment may be included for security. Other equipment can freeze a full motion display for a few seconds while the participants send a graphic image over the circuit. Sometimes digital storage equipment enables participants to transmit presentation material ahead of time so graphics transmission does not waste conference time.

Graphics Equipment

Videoconference facilities may include graphics-generating equipment to construct diagrams. Some systems provide desktop computer input so that tools in the computer can be used for generating graphics.

Communications

Digital communication facilities are required for videoconferences. ISDN is required for connections over the PSTN. Two 64 Kbps B channels and separate signaling channel provide enough bandwidth for an acceptable conference, but at least 384 Kbps is required for conference quality. Where an IP network with sufficient bandwidth and QoS is available, it is a good medium for conferences over the internal network. Frame relay is an excellent platform for videoconferencing if the access bandwidth is sufficient.

ITU-T H.320 Video Standards

Before the ITU-T H.320 standards were developed, manufacturers used proprietary standards, which meant that both ends of a videoconferencing session had to use the same equipment. Now, interoperability is assured by use of standards from the H.320 family. Table 22-2 lists the standards included under H.320.

The amount of bandwidth supplied in the transmission facilities must be in multiples of 64 Kbps, known as P × 64 (pronounced P times 64). P can be from 1 to 30 64 Kbps channels; i.e. a single DS-0 up to full E1. Two options are offered. The full common intermediate format (CIF) offers frames of 288 lines by 352 pixels. This is approximately half the resolution of commercial television, which is 525 lines by 480 pixels. The second alternative is quarter CIF (QCIF), which is 144 lines by 176 pixels. The modulation method is the discreet cosine transform (DCT) algorithm.

TABLE 22-2

H.320 Videoconferencing Standards

Standard	Function
H.320	Umbrella recommendation
H.221	Data formatting, framing, and demultiplexing. Defines how bits are identified as video, audio, and control, and how they are fitted into frames
H.230	Control and indication signals. Defines commands for control and diagnostics
H.231	Multipoint control unit using digital channels
H.242	Transmission and handshake protocols. Cell setup, transfer, and disconnect procedures
H.243	Multipoint handshake. Communications between the codec and a multipoint control unit
H.261	Syntax and semantics of the video bit stream. Defines the video coding algorithms, forward error correction, and picture format
H.263	Video coding for low bit rate communication
G.711	3.1 kHz PCM audio, 64 Kbps
G.722	7 kHz audio, 48, or 56 Kbps
G.728	16 Kbps low-delay CELP speech, 2300 Hz audio
T.120	Protocol suite for data conferencing

H.323 Video Standards

The public Internet is not a sufficiently stable medium for reliable high-quality videoconferences, but for some it is good enough. Even better is the enterprise IP network or frame relay. The network can be designed with sufficient bandwidth and equipped with QoS protocols that make videoconferencing over IP an excellent alternative and considerably less costly than dial-up.

We discussed H.323 signaling in Chapter 12. Those protocols work for either voice or video. The benefit of H.323 is its simplicity once it is set up and working. The standard RJ-45 Ethernet jack is ubiquitous, and can accept video without concern about how many channels are needed. A rollabout video unit can be hooked to any jack just like a laptop computer. One of the driving applications is desktop videoconferencing. The equipment is contained in a standard desktop computer that has a small video camera mounted on top of the monitor. The conferencing equipment is mounted on a board that plugs into a computer expansion slot, or it is an external box that plugs into a board in the computer.

If the network is designed for video, the result is an economical method of conducting a personal videoconference. Conferences are spontaneous, with no need to schedule a conference room. The addition of a picture to the call allows the parties to pick up the nonverbal content of a session by seeing expressions. The screen is generally too small and the camera angle too narrow for group conferences, but for small groups it is excellent. Most products permit the users to share and view computer files over the network. As desktop video gains acceptance, it will be an effective tool for enhancing the quality of telephone calls and for enabling users to collaborate on files.

VIDEO APPLICATION ISSUES

Entertainment is likely to remain the primary driving force behind video into the future, but business, healthcare, and educational uses of television will be increasingly important. As CATV provides a broadband information pipeline into a substantial portion of American households, the growth of nonentertainment services is expected.

Security

Television security applications take two forms—alarm systems and closed circuit television (CCTV) for monitoring unattended areas from a central location. Many businesses use CCTV for intrusion monitoring, and it is also widely used for intraorganizational information telecasts. Alarm services have principally relied on telephone lines to relay alarms to a center, which requires a separate line or automatic dialer. The expense of these devices can be saved by routing alarms over a CATV upstream channel, but to do so it requires a terminal to interface with the alarm unit. As described earlier, a computer in the headend scans the alarm terminals and forwards alarm information to a security agency as instructed by the user. H.323 video can also be used to deliver narrow-band IP video for security monitoring.

IP Services

Many CATV companies offer two-way data communications over their systems. As discussed in Chapter 8, the cable company allocates bandwidth for upstream and downstream using the DOCSIS protocol. VoD is delivered over IP bandwidth, using upstream channels to order services and control the delivery.

Control Systems

Two-way CATV systems offer the potential of controlling many functions in households and businesses. For example, utilities can use the system to poll remote gas, electric, and water meters to save the cost of manual meter reading. Power companies can use the system for load control. During periods of high demand, electric water heaters can be turned off and restored when reduced demand permits. A computer at the headend can remotely control a variety of household services such as appliances and environmental equipment. CATV companies themselves can use the system to register channels that viewers are watching and to bill for service consumed. They also can use the equipment to control addressable converters to unscramble a premium channel at the viewers' request.

Opinion Polling

Experiments with opinion polling over CATV have been conducted. For example, CATV has been used to enable viewers to evaluate the television program they have just finished watching. The potential of this system for allowing viewers to watch a political body in action and immediately express their opinion has great potential in a participatory democracy.

Streaming Video

The QoS requirements that may preclude using the Internet for the video transmission medium do not apply to streaming video. Enterprises can use the Internet for one-way broadcasts to employees and customers. Streaming video has endless applications in training, education, entertainment, and other such purposes.

Videoconferencing

Companies that had never considered videoconferencing are investigating it more closely as the economics become more compelling. The first line of justification is generally replacement for travel, but as organizations adopt video as a way of doing business, the need for economic justification disappears. Videoconferencing makes it economical for more people to participate directly because travel is eliminated. A major advantage of videoconferencing where formal conference rooms are provided is scheduling. When users must reserve a facility, meetings must begin and end on schedule, which is an added benefit.

Evaluation Considerations

As the cost of equipment drops, companies may find they have backed into the videoconferencing business without a plan. The result may be under-utilized equipment and network facilities, or the inverse, which is an overloaded network. Needs and expectations should be thoroughly assessed before embarking on videoconferencing. The issues discussed in this section should be considered and the necessary controls imposed to increase the chances of success.

Type of Transmission Facilities

The initial issue to resolve is the telecommunications medium. BRI ISDN service is the default method where conferences use the PSTN. For higher bandwidths, inverse multiplexing may be required. Some units have the inverse multiplexer built in, while others use an external i-mux. Most digital PBXs can support PRI on the PSTN side and BRI toward the videoconferencing endpoints. Where IP bandwidth is available, H.323 video is an excellent alternative. The typical company in the market for H.323 video will have an existing internal network with spare capacity, or one to which capacity can be added inexpensively. A major advantage of IP conferencing is the fact that equipment can be relocated easily by plugging into a live Ethernet jack. The terminal equipment may be interchangeable for H.320 conferences. If external conferencing is required, a H.323-to-H.320 gateway will be required. If the video and audio signals are separate across the gateway, lip sync problems are apt to result.

Single- or Multipoint Conferences

Videoconferences are classed as point-to-point or multipoint. With terrestrial facilities, the distance and number of points served have a significant effect on transmission costs. Large companies with a significant amount of multipoint conferencing can often justify the cost of an MCU. Companies that use multipoint conferencing only occasionally can use bridging services offered by the major IXCs. The IXCs offer "meet-me" conferencing in which conferees dial into a bridge. The control unit receives inputs from all locations, and sends each location the image that has seized the transmitting channel. The transmitting channel is allocated by one of three methods: under control of the conference leader, under time control in which each location gets a share of the time, or by switching automatically to the location that is currently talking. The latter method is the most common, but it requires a disciplined approach.

Videoconferencing Equipment

The following issues should be evaluated in selecting videoconferencing equipment:

- Will the equipment used be videoconferencing appliances or PC-based?
- Will the conference be set up in formal conference rooms or from desktops?
- What level of quality is needed? Is a highly compressed signal satisfactory? Does the system offer full 30 frames per second or some lower factor?
- Can the system support multiple video formats (NTSC, SECAM, PAL)?
- Does the system support still graphics?
- How easily can information be brought into the conference? Is information sharing fundamental to the equipment or is it an add-on?
- Will fixed or portable equipment be used? Do the applications require frequent equipment relocation?
- Does audio ride on the video facility? Is wideband audio available?
- Is single-point or multipoint communication required? If multipoint, will the user supply its own multipoint control unit or use facilities offered by the IXC?

System Integration

Videoconference equipment is sometimes an assembly of units made by different manufacturers. To ensure compatibility, it is advisable to obtain equipment from a vendor who can integrate it into a complete system.

Security

The type of information being transmitted over the channel must be considered. If proprietary information is discussed during conferences, encryption of both video and audio signals may be required, particularly if IP is the transmission protocol.

Public or Private Facilities

Private videoconference facilities have a significant advantage over public access systems. Public facilities are unavailable in many localities, which may preclude holding many videoconferences. The travel time to a public facility offsets some of the advantages of videoconference. Unless a private facility is used frequently, however, public facilities are usually the most cost-effective option.

PART FOUR

Customer Premise Systems

Customer premise switching systems are evolving under the influence of data communications and the personal computer. Until recently, PBXs and key systems had a closed structure. All of the elements of the CPE switch from the telephone down to the processor have, in most products, been proprietary. Such closed systems are contrary to the principles under which IP networks have evolved. Proprietary data protocols are surrendering to the impetus of the PC, the LAN, and the Internet and are yielding to open protocols and hardware interfaces. CPE switches, PBXs and key systems, are evolving in the same way into a new generation that runs on an open platform and communicates with other switches using IP.

The products contained in this part are the only telecommunication devices most users come into direct contact with or care about. As with everything else associated with telecommunications, the excitement is all about VoIP. The industry suffered a severe downturn in the early years of the twenty-first century and VoIP may revive it as organizations replace their aging phone systems. To some degree, the industry has been a victim of its own success. The key systems, PBXs, fax machines, and desktop devices that comprise customer premise equipment work well and have practically nothing to wear out. As a result, the only motivation for replacing them is to gain new features. TDM-based equipment has reached the point that all of the easy features are contained in practically every manufacturer's product line. That is a dangerous situation for the manufacturers because products that cannot be differentiated easily are effectively commodities and in a commodity market price becomes the differentiator.

VoIP raises both possibilities and problems that did not exist before. The industry is exploiting the possibilities to the hilt and downplaying the problems, which mostly resolve around security and quality issues. These could be resolved with a stable public IP network that has the ubiquity of the Internet but the stability and security of the PSTN. Such a network will probably arise, but it is not here today, and consequently VoIP is most effective on a private IP network. If the weakness of TDM systems lies in their closed structure, it is also one of their strengths. Call processing and feature support have evolved through decades of development, resulting in a platform that is practically bulletproof as long as power does not fail.

Customer premise switching systems occupy a significant portion of this part. We begin with a discussion of station equipment and features that are common to all types of voice switches regardless of whether the technology is TDM or IP. After those two chapters we have separate chapters on TDM and IP switching systems. Although the latter is getting all the attention, the vast majority of systems in service use TDM technology, and it will be many years before all are replaced.

Automatic call distribution is a key application of both types of switches, and is the subject of Chapter 27. Following that, we have chapters on voice processing, messaging systems, and facsimile.

CHAPTER 23

Station Equipment

The least exotic and complex element of the telecommunications system is the ordinary telephone, yet its importance is often unappreciated. The lowly POTS phone has driven the shape of the entire network. Because of the enormous numbers of instruments in service, the PSTN is designed to keep the telephone simple, rugged, and economical. Over the history of telephony, the telephone set has been much improved, but the fundamental principles that make it work have changed little since Alexander Graham Bell's original invention. The primary changes have been improvements in three areas: packaging to make telephones esthetically appealing and easy to use, signaling to improve the methods used to place and receive telephone calls, and transmission performance to improve the quality of the connection between users.

Until recently, the only universal telephone has been the analog POTS phone. Businesses survived for many years with analog phones behind key systems, PBXs, and Centrex. When these switches were electromechanical, their features were few. Such features as there were—primarily pickup and hold—were on keys that were connected to a central unit over bulky 25-pair cables. As switching systems improved, the fat cables slimmed down to one or two pairs, but users had to activate features by dialing codes that few people could remember. As a result, features were underused unless the enterprise invested in proprietary telephones that assigned the features to buttons.

Today, the most common enterprise telephone is a proprietary feature phone, but it is one devoid of intelligence; its brains reside in a central switch. In homes and many small businesses the POTS phone still prevails, but that is changing. IP telephones now entering the market can offer a degree of vendor-independence that has never been available in the past. The transition to intelligent telephone instruments is just beginning, but over the next several years, the traditional interaction between humans and their telephones is likely to change in ways that are difficult to predict. This chapter discusses different types of telephones, station protection equipment, and how they function in the telephone network.

TELEPHONE SET TECHNOLOGY

The telephone is inherently a four-wire device. Transmit and receive paths are separate in the handset, but they are electrically combined to interface the two-wire loops that serve most residences and many businesses. Figure 23-1 is a functional diagram of a telephone set.

Elements of a Telephone Set

A transmitter in the handset of a POTS phone converts the user's voice into fluctuating direct current. The most common type of transmitter has a housing containing tightly packed granules of carbon that are energized by DC voltage. The voice waves vary the degree of compactness of the granules, changing their resistance and the amount of current that flows. This fluctuating current travels over metallic circuits to a line circuit in the central office. The line circuit digitizes the signal and connects it through the office as discussed in Chapter 14. The line circuit at the other end of the connection converts the signal back to fluctuating AC to drive the receiver in the handset.

The receiver has coils of fine wire wound around a magnetic core. The current variations cause a diaphragm to move in step with changes in the line current. The diaphragm in the receiver and the carbon microphone in the transmitter are *transducers* that change fluctuations in sound pressure to fluctuations in electrical current, and vice versa. Many new telephone sets substitute electronic

FIGURE 23-1

Functional Diagram of Telephone Set

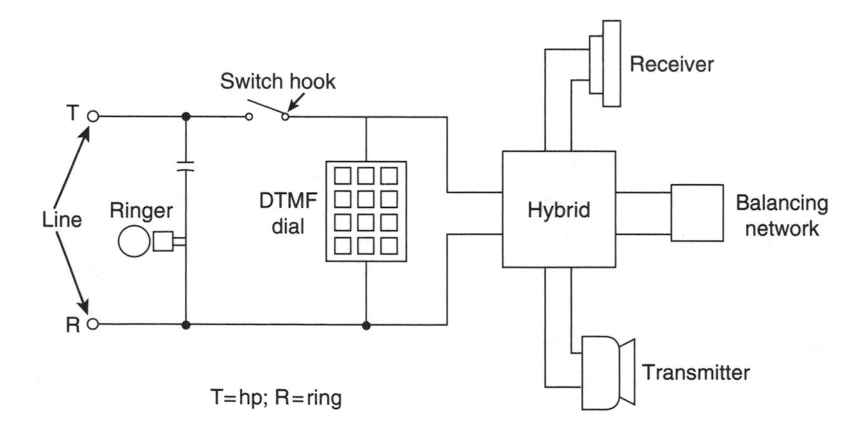

transmitters for the carbon units that have been used for more than a century. The carbon transmitter is still common, however, because it is inexpensive and rugged.

The telephone set contains a *hybrid coil* that converts the four-wire handset to the two-wire cable pair. The hybrid is a passive device that couples energy from the transmitter to the line and from the line to the receiver. By design, the isolation between the transmitter and receiver is less than perfect. It is desirable for a small amount of the user's voice to feed back into the receiver as *sidetone*, which is the feedback effect that makes the handset sound alive. Sidetone helps to regulate the user's voice level. With too little sidetone, users tend to speak too loudly; given too much, they do not speak loudly enough. The lack of sidetone, incidentally, is a major reason that so many people unconsciously speak loudly on a cell phone.

The switch hook disconnects everything but the ringer from the line when the telephone is idle, or on-hook. When the user lifts the receiver off-hook, the switch hook connects the line to the telephone set, and line current flows, which signals the central office to return dial tone. When the telephone is on-hook, the ringer is coupled to the line through a capacitor that blocks DC line current. The capacitor is large enough to pass the 20 Hz AC ringing signal. Many modern sets use an electronic tone ringer instead of the older gong-and-clapper bell.

The dial connects across the line behind the switch hook. Rotary dials, which are rare, but still in service, interrupt the flow of line current in step with the dial pulses. Tone dials, known as dual tone multifrequency (DTMF), use a 4 × 4-tone matrix (Figure 23-2). Each button generates a unique pair of frequencies, which

FIGURE 23-2

DTMF Tone Matrix

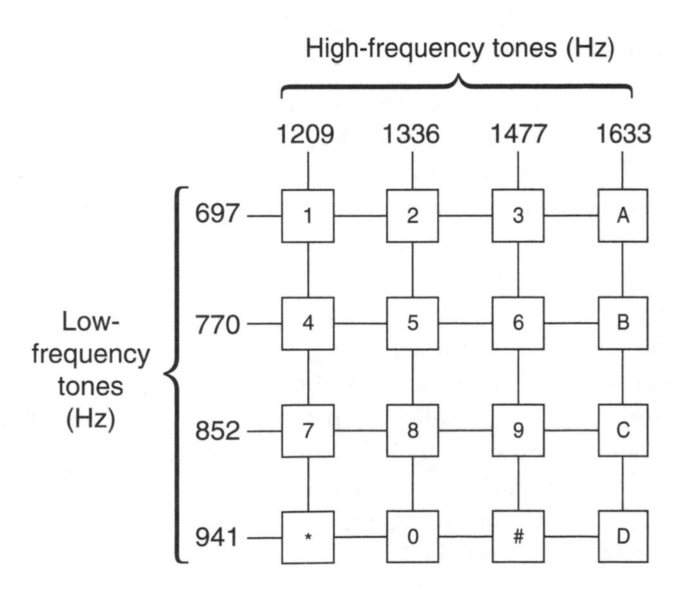

a DTMF receiver in the central office detects. Ordinary telephones send only three of the four columns of DTMF signals. The fourth column appears only on some telephones that need to transmit special codes such as preemption signals.

The two wires of a telephone circuit are designated as tip and ring, corresponding to the tip and ring of manual switchboard cord plugs. The central office feeds negative polarity battery over the ring side of the line. When the receiver is off-hook, current flows in the line. The resistance of the local loop, which is a function of the wire gauge and length, controls the amount of current flow. For adequate transmission, the telephone needs at least 23 mA of current. If the line current exceeds about 60 mA, the volume may be uncomfortably loud for some listeners. In addition to transmission considerations, DTMF dials need at least 20 mA of current for reliable operation. The telephone set, the wiring on the user's premises, the local loop, and the central office equipment interact to regulate the flow of current in the line and the quality of local transmission service.

TELEPHONE SET TYPES AND FEATURES

Telephone sets have evolved over the past several years so that even many analog sets emulate the features that proprietary business sets and cell phones offer. Such features as speed dial, speakerphone, last number redial, hold, conferencing, call logging, and intercom are widely available. As discussed in Chapter 12, most LECs offer calling party ID over analog lines using the ADSI protocol. The subscriber's telephone or an external caller ID box detects and displays the caller ID message, which the central office sends between the first and second rings. The set typically stores the details of the last 25 or so calls and makes it easy to return the call by scrolling through a list of callers.

Proprietary Digital Telephones

Analog telephone sets are adequate for residential use, but in a business environment they have undesirable limitations. The problem is feature access. Some types of users, hospital patients and hotel guests for example, do not need feature access, but others require many or most of the features discussed in this section. If users are equipped with analog phones they must dial feature access codes, which most people do not remember. The solution is to assign buttons so the feature is accessed directly.

Many telephones have features assigned to context-sensitive soft keys, which are labeled by a line on the display above a row of buttons. When the telephone is idle, for example, the call pickup and do-not-disturb buttons might be displayed. When the phone rings the call transfer button might appear, and when the user picks up the handset to place a call, the forward-calls button could display. Not only do soft keys reduce the size of the telephone set but they also eliminate the need to label buttons.

The following is a list of the most common features that developers provide in multiline telephone sets. Most of these same features are provided in IP telephones, which are discussed in the next section.

Call pickup. By pressing a button users can pick up a call that is ringing at another telephone in the group without the need for a line appearance.

Call hold. A call can be placed in a centralized holding circuit while the telephone is used for another call. In an analog phone the call is held at the telephone and the line is not available for a second call.

Supervisory signals. Lamps show when a line is ringing, in use, or on hold.

Automatic hold recall. After a call has been left on hold beyond a programmable period, the telephone emits a warning tone.

Conferencing. A station user can bridge two or more lines for a multiparty conversation.

Do not disturb. The station user can press a button that silences the bell and sends incoming calls to a coverage position such as voice mail.

Message waiting. A lamp on the telephone set shows that a message is waiting in voice mail.

Station display. Proprietary telephones may display date and time, last number dialed, elapsed time on the call, etc. When the phone is in use, a second incoming call displays.

Call park. A call can be placed in a parking orbit and retrieved from any phone by dialing the park code.

Intelligent Telephones

Telephones are becoming easier to use and contain several features that give them the appearance of being intelligent devices. Despite appearances, however, analog and proprietary digital telephones are by design nonintelligent endpoints. All of the real intelligence involved in call setup is contained in the central switch, whether it is a central office, PBX, or key system. The lack of endpoint intelligence limits the choices users have in handling calls. VoIP and the related protocols SIP and ENUM are increasing the flexibility users have in controlling calls, the intelligence for which is contained in a server.

Outwardly, an IP telephone looks much like a standard phone. Some vendors have attempted to diverge from standard designs, but the ergonomics of telephones have been proven through years of experience, and most people prefer the familiar. All major PBX vendors offer IP telephones, and in addition, many independent products are available.

Besides vendor independence, IP telephones have several advantages. For one thing, since they have versatile displays, sometimes with graphic capability, they can be used to retrieve and distribute information in new ways. For example, a short text message might appear daily on all displays to remind employees of an upcoming event. Employees would see the message every time they looked at the telephone and would have no ability to lose or delete the message. Telephone displays have limited information capability, so they are not a universal information distribution medium, but rather are suited to short text messages such as delivering call center statistics to managers.

Aside from the displays, which tend to be somewhat larger and more colorful on IP phones, most of the differences between IP and conventional telephones lie behind the scenes. The versatility that protocols such as SIP offer lies in placing some call control intelligence in the telephone itself.

The flexibility of an IP telephone is its strength and weakness, the latter because it requires setup. An analog telephone, by contrast, is strictly plug-and-play. If the line works, the telephone works even though some of its features may require configuring. A proprietary digital phone is configured centrally. Most manufacturers keep the setup out of the users' hands because this enforces uniformity and reduces assistance calls. IP phones are designed to promote user preference and programmability. They can be set up to a uniform standard and the configuration tools denied to the users, but by doing so, many inherent advantages of an IP phone are lost.

ISDN Telephones

From the outside, it is difficult to tell an ISDN telephone from a proprietary digital or IP telephone. Some of the important differences lie in the characteristics of a BRI line with its two 64 Kbps B channels and one 16 Kbps D channel. The external D channel enables the BRI phone to support most of the features of a proprietary digital phone. In most products the display is small because the functions are limited to those provided from the central office or PBX.

BRI phones are available with either a U interface, which has a built-in NT1, or an S/T interface, which requires an external NT1. The phones typically include dedicated function keys for features such as hold, conference, and transfer. Protocol support is an important issue with ISDN phones. Although the BRI protocol is a standard, central office switches employ multiple versions of the standard, so it is necessary to get assurance that a phone will work in the required environment. As with IP phones, setup is an issue with ISDN phones. Self-configuring phones can be obtained to detect the switch and automatically configure the SPID.

Conference Room Telephones

An ordinary speakerphone is unsatisfactory for use in larger conference rooms because it may lack sensitivity for voice pickup from all parts of the room, or it may clip parts of the conversation because it operates in half-duplex mode.

Half-duplex means the device switches from send to receive during conversations, with the loudest talker capturing the circuit.

The best conference room telephones operate in full-duplex mode. The manufacturer should certify speakerphones as IEEE 1329 Class 1 full-duplex systems. They process the voice in a manner somewhat similar to an echo canceller to enable parties to carry on a normal conversation with no clipping and few or no dead spots around the room. Such phones may clip during the first few seconds of conversation because they operate in half-duplex mode initially while evaluating the echo characteristics of the room and adjusting accordingly. The conference room environment may tax even the best telephones. Extraneous noises such as air conditioners, projector fans, and large reflective surfaces such as whiteboards and windows may set up an acoustic reverberation, all of which may confuse the phone. To improve pickup in large conference rooms, extension microphones are generally available. Conference room telephones typically operate behind analog telephone lines, either directly from the central office or from a PBX. Some products use proprietary digital ports or IP.

Cordless Telephones

Cordless telephones are not to be confused with wireless telephones. Cordless telephones have a base station that is connected to a central office line. A cordless phone is usable within the range of the base station, which is usually limited to a few hundred feet. Wireless telephones connect to a private or public wireless network of multiple base stations. The effective range of wireless is much greater than cordless because wireless hands off from one base station to another, much as cellular phones do. Wireless telephones are frequency agile so they seek a vacant channel out of several available frequencies. Cordless telephones are set to the frequency of the base station, although channel-scanning models are available.

Cordless phones operate in one of three frequency bands. Older phones operate in two channels in the 46 to 49 MHz range, with one frequency used for base-to-portable, and the other range used in the other direction. Privacy can be a problem with older cordless telephones because anyone using a scanner or a phone on the same frequency can eavesdrop on private calls.

The current generation of telephones operating in the 900 MHz and 2.4 GHz ISM bands contains safeguards against false rings and unauthorized calls. The base-to-portable link is authenticated with a code from the portable unit so the base responds only to a unit with the correct code. The code is randomly selected from about 65,000 codes, making accidental detection from another phone on the same frequency unlikely. Frequency-hopping spread spectrum phones are common and make unauthorized detection difficult.

The European Telecommunications Standards Institute (ETSI) developed a second-generation cordless standard that is known as the Digital Enhanced Cordless Telecommunications standard (DECT). DECT in Europe operates in the 1.88 to 1.90 GHz bands. In North American personal wireless telecommunications (PWT) standards from TIA are based on DECT. PWT operates in the unlicensed 1.91 to 1.92 GHz bands, which fit between the transmit and receive frequency bands of licensed PCS. DECT defines the radio connection between devices and does not specify standards needed for complete mobility such as switching, signaling, and management.

The purpose of DECT/PWT is to provide short-range wireless, usually inside buildings, using unlicensed spectrum. A major difference from conventional cordless is that is supports handoff between multiple base stations, which enables roaming. The principles are similar to voice over 802.11, but it uses frequency division instead of IP.

Answering Equipment

Answering equipment varies from ordinary telephone answering sets to elaborate voice mail-like equipment. Answering sets are available in a variety of quality levels, and are no more difficult to install than ordinary telephones. Most current models use digitized voice instead of the microcassette tapes that were common with older systems.

Features that many users find important include:

- Multiple outgoing messages
- Multiline capability
- Selective message save and delete
- Remote message retrieval
- Message time/date stamp

Because the central office does not relay answer supervision over the local loop, answering machines include timing circuitry to determine when the calling party hangs up. Most answering machines detect the caller's voice; when a silent period of more than a specified length is detected, the machine assumes the caller has hung up, and disconnects. If a caller hangs up when the answering message is first heard, the machine may hold the line busy for a time while it completes the announcement and times out.

STATION PROTECTION

Telephone circuits are occasionally subject to high voltages that could be injurious or fatal to the user without electrical protection. Lightning strikes and crosses with high voltage power lines are mitigated with a station protector, which is diagrammed in Figure 23-3. Protectors use either an air gap or a gas tube to conduct high voltage from either side of the line to ground if hazardous voltages occur. The telephone is insulated so that any voltage that gets past the protector will not

FIGURE 23-3

Diagram of a Station Protector

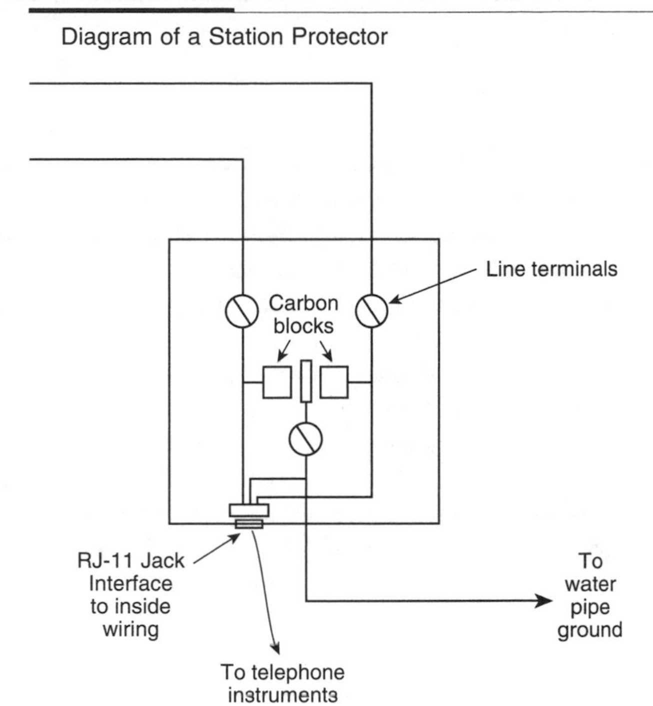

injure the user. The LEC places protectors, which also may form a demarcation point with customer-owned wiring. Protector grounds are connected to a ground rod, metallic water pipe, or other low resistance ground.

LINE TRANSFER DEVICES

Transfer devices fall into two categories: those that operate off distinctive ring from the central office, and those that answer the phone, recognize an incoming signal, and transfer the line to the appropriate terminal.

Distinctive Ring Devices

Many LECs offer distinctive ring, which provides multiple telephone numbers on a single line. Each number has a different ringing code combination. Many fax machines have the ability to respond to distinctive rings, or external devices are available to switch the call to the appropriate port based on the ringing signal. For example, a one-bell signal could ring the telephone, two-bell the fax, and a third combination could connect to a modem. The device recognizes an off-hook signal from any station and connects it to the line, disabling it toward the other devices to prevent barge-in. Distinctive ring is a cost-effective method of sharing lines that otherwise have low usage.

Line Switchers

Another type of line-sharing device is the fax/modem switch, which may have ports for telephone, fax, and modem. These devices answer the telephone and listen for a signal from the incoming line. If the device detects a CNG (calling) tone, which is an 1100-Hz tone transmitted by a fax machine, it sends the call to the fax machine. The CNG tone is repeated every 3.5s for approximately 45s. If a modem tone is heard a modem is switched to the line, and if no tone is heard the telephone is switched to the line. The device sends a ringing signal to the called equipment.

Some of these devices are not completely transparent to the calling party. For example, the receiving modem normally answers the line with a modem tone. This cannot work with a line switcher because it relies on the calling device to identify itself first. Also, the device returns answer supervision to the calling station regardless whether the line is ultimately answered or not. This may result in charges for long distance calls that are not completed.

STATION APPLICATION ISSUES

Since the FCC began to permit users to attach any registered device to the telephone network, the market has become flooded with special-purpose telephones. The price of general-purpose sets is often a clue to their quality. Many inexpensive instruments provide poor transmission quality and fail when dropped. At the high end of the scale, price usually is a function of features or decorator housings.

The main advantage of an analog telephone is its universal connectivity. Most of the world uses the same interface for a POTS line, and the RJ-11 jack is common so telephones and modems can be taken almost anywhere with a high expectation that they will work. The same may not be true for ISDN phones despite the fact that the standard is international, and it definitely is not true in the case of proprietary digital sets. Eventually, IP phones should be universally adaptable, but for now many of the features operate with proprietary protocols.

Evaluation Considerations

All apparatus connected to the network in the United States must be registered with the FCC to guard against harm to the network. Other countries have similar regulations, or in rare cases may prohibit connection.

Telephones

Analog telephones, cordless phones, answering machines, fax/modem switches, and the like are low-cost items that are purchased based on price and features.

The garden variety POTS set has practically disappeared from the market. Now, telephone sets can be obtained with dozens of optional features and with auxiliary equipment such as clock radios that have nothing to do with the telephone itself. Special features are available as either parts of the telephone or as add-on adapters.

Proprietary Telephone Sets

Proprietary sets are purchased as part of the key system or PBX, and with rare exceptions are usable only with the system they are designed for. The telephone set is the only contact most users have with the telephone system, and since most systems support the same features, the look, feel, and operation of the telephone are critical in selecting the switching system. One of the first issues is the number of lines the instrument supports. Single-line proprietary sets have a limited number of feature buttons, but more important, they cannot access some critical features (e.g., conferencing) that require multiple-line sets.

Unlike IP telephones, the features are centrally controlled and the user can change few functions from the telephone itself. Speed dial numbers stored in the telephone set are user programmable, but nearly all feature keys are fixed in the switch software and the user cannot change them. One exception is distinctive ringing tones, which users can program to make it easy to recognize their ring in a close environment. This lack of user programmability is not an inherent characteristic of TDM switches. The features can be made user programmable, and in some systems they are, but user control of features tends to increase support costs and may cause some features to be under-used. Lacking a feature button, nearly any programmable feature can still be activated from the telephone set with dial access codes.

The cost of the telephone set is a function of the number of buttons and the size of the display. The display is rated by the number of lines and characters it supports. Most products allow any feature to be associated with any button. The buttons can also be used for lamp displays. For example, secretaries may have their clients' line lamps on their sets so that they can see when the client is on the telephone or has it in do-not-disturb mode. Buttons may be either fixed or softkeys. The most flexible sets have fixed buttons for the most frequently used features such as hold, reserving softkeys for features that are rarely used or change with context. Expansion modules should be available for sets requiring more than the standard number of buttons.

Speakerphone is a standard feature of most proprietary telephones, but it may be desirable to disable it in software because it is disruptive in some office environments. Some phones purport to have full-duplex speakerphones, but this claim must be examined closely to verify that it is not a fast-acting half-duplex telephone.

Automatic call distributors (see Chapter 27) have telephones with special characteristics. The need for headset adapters is universal. A call center phone normally does not have a handset, which means some other method of switch hook control is needed. This is usually assigned to buttons. Normally, speakerphones are not used in a call center environment. The status of calls in queue may be displayed by lamp flashes on the instrument, with various flashing speeds indicating threshold levels.

IP and SIP Sets

Sets intended for VoIP may be either proprietary or universal. In either case, most of the characteristics of proprietary sets that are discussed above also apply to these. In addition, several other issues are unique to IP phones. One of the most important is how the telephone is configured. Most phones can be configured from a Web browser. Some systems support configuration from a TFTP server. When the phone powers up it gets its IP address through DHCP and its domain name from the server.

Next is the question of what protocols the instruments support. Many multiprotocol phones support proprietary protocols as well as H.323, SIP, and MGCP. This provides the capability of migrating from one protocol to another as the standards advance. Some phones offer the choice of an IP or a circuit-switched configuration. Others support Web protocols such as WAP and HTTP. The process for upgrading protocols should be examined. Is it telephone-by-telephone, or can a group of phones be updated from a server?

The physical characteristics of the set are important. Most sets should have an integrated dual-port Ethernet switch so a PC and telephone can share a single switch connection. The phone should support IEEE 802.3af Power-over-Ethernet to get centralized power from the LAN. Phones should also support 802.1p MAC-level QoS and IP type-of-service. Many systems have color screens, often touch screens, to access information sources such as stock market quotes and weather reports as well as an online help menu. Examine the display under daylight conditions with local power disconnected to determine if the display is visible during power outages.

The method of serving analog devices should be evaluated. Single-unit analog adapters should be available for stand-alone devices such as fax machines. A multiport gateway may be needed to support clusters of analog stations.

Most manufacturers offer *softphones,* which emulate a regular phone on a computer display. Since the softphone is usually used from an alternate location such as at home or on the road, a single number should serve both phones, with the calls flowing to either the softphone or the desk phone depending which one is logged on. The softphone should support a headset connection via USB port. Softphones may support conference call setup by drag-and-drop on the computer display. If the system supports LDAP, it should be possible to place calls by choosing from an online directory with a mouse click.

Conference Room Phones

Conference room phones are second only to fax machines as products organizations cannot do without. Ordinary speakerphones are suitable for small rooms and informal conferences, but few can compare to specialized conference room phones for sensitivity and clarity. Many such phones require analog ports, but some products can interface a proprietary digital port, and IP conference phones are readily available.

The best way to select a conference room phone is to try it in the intended environment, with people walking about the room and checking how well they can be heard on the other end. Before trying the phone, however, products should be screened. Is it a full-duplex device? What size conference room is it optimized for? Can it dynamically adjust to room conditions? Does it have a display? Does it support extension microphones? If it is an IP phone, can it be plugged into an Ethernet port and configure itself?

IP conference phones have many advantages. With a display, the phone should be able to identify who is speaking from the distant end, and it should facilitate setting up a multiparty conference, something that is difficult with an analog device. IP connections are subject to many impairments that have no impact from digital or analog phones on a TDM connection. The main impairment is echo, which can result from router configuration. If carefully selected, installed, and configured, a conference room phone is a vital business asset.

Answering Sets

Answering sets have many special features that should be considered before purchasing a unit. Among the most important features are:

- *Battery backup* for continued operation during power failures.
- *Call counter* to display the number of messages recorded.
- *Call monitoring* capability so incoming calls can be screened over a speaker.
- *Digital announcement and recording* to eliminate the need for tapes.
- *Remote-control recording* so the announcement can be changed from a remote location.
- *Ring control* so the number of rings before the line is answered can be adjusted. Some systems answer the telephone on the second ring if messages are present and on the fourth ring if they are not. This permits the owner to check for messages without paying for a long distance call.
- *Selective call erase* to allow selective erasing, saving, and repeating of incoming messages.
- *Multiple-line capability* so one device can answer more than one line.
- *Synthesized voice readout* so the answering machine can announce the time and date that each call was received.

CHAPTER 24

Customer-Premise Switching System Features

The main purpose of both TDM and IP customer-premise switching systems is identical. They connect voice users in the office to one another and to external users over a pool of shared trunks. The differences lie in the way the two types of switch accomplish their objectives. Telephone functions have evolved over more than a century and users are accustomed to how they work. Each new generation of switches has brought features that its predecessor could not support, but it also supported the older features as well. The result is a suite of features that most customer-premise systems support regardless of technology.

If trunk pooling was the only purpose of CPE switches, they could be quite simple devices, but PBXs, and to some degree key systems, have a more important role of unifying the office. CPE switches do for voice what the LAN does for data; they bind the office with shared applications. It is logical, therefore, to consider unifying voice and data into a single system that supports both. To some degree this is possible, but the requirements of voice and data are much different and most users care little or nothing about how the mission is accomplished. They do care about features, however. Users expect similar features whether the platform is TDM or IP. As we will discuss in Chapter 26, the strength of IP lies in its ability to provide features that TDM switches find difficult or impossible.

The office switching system is a strategic investment for most companies. A PBX ties your company with your customers, suppliers, the public, and other parts of the organization. The features it has and the way it is set up and administered have a significant effect on how those on the outside view your company. This chapter discusses the principal features of CPE switching systems.

KEY SYSTEM FEATURES

Most manufacturers offer separate key system and PBX product lines, although the functions may be nearly identical, at least in larger line sizes. Most TDM manufacturers also offer a cross between a PBX and a key system known as a *hybrid*. Table 24-1 lists the principal differences between these three types of system, but these distinctions are not absolute. The feature differences between product lines are many and varied, and some applications such as voice mail are universally available on all systems. This section discusses features that are common to key systems, which can be defined as CPE switches having the central office lines terminated on telephone set buttons. In such systems, users answer incoming calls by pressing a line button. If the call is for someone else, it is announced over an intercom. To place an outgoing call, the user selects an idle line by pressing a button. The physical size of the telephone set eventually places upper size limits on a key system. When the system terminates more than a dozen or so lines it is easier for users to dial 9 to access a line than to cope with all the buttons on the phone. Key systems and hybrids often use the same hardware and software platform. The distinction between the two is a matter of how the system is set up.

Key Telephone System Features

At one time key systems were electromechanical devices that had a limited set of features. New electronic systems have far surpassed these limitations, but the features of the last generation of electromechanical key systems defined the way

TABLE 24-1

Comparison of Key Systems and Hybrids with PBXs

Feature or Function	Key System	PBX	Hybrid
Lines terminate on	Telephone buttons	Pooled	Telephone buttons or pooled
Direct inward dialing	Not available	Available	Available
Listed directory number calls	Transfer by attendant or auto attendant	Transfer by attendant or auto attendant	Transfer by attendant or auto attendant
Paging through the telephone	Available	Not available	Available
Centrex compatibility	Available	Not available	Available
Remote switch unit	Not available	Available	Not available
Networking	Not available	Available	Not available
CO trunking	Analog or BRI	Analog, T-1, or PRI	Analog, T-1, PRI, or BRI
Automatic route selection	Not available	Available	Available, but limited
Centralized attendant	Not available	Available	Not available

users expect key telephones to work. The principal features are designated with illuminated buttons. A dark button indicates an idle line and a solid light denotes that the line is in use. A slow flash indicates an incoming call and a fast flash shows that the line is on hold. These buttons and lamps define the following features, which are common to all key systems:

- *Call pickup*: Any station can access a line by pressing a line button.
- *Call hold*: A hold button (usually red) can be pressed to hold the line in the central unit. By contrast, the hold button on a POTS phone holds the line in the telephone so the line cannot be used for another call.
- *Intercom*: A common path shared by all telephones is used to announce calls.
- *Supervisory signals*: Lamps show when a line is ringing, in use, or on hold.
- *Common bell*: A bell common to all lines signals an incoming call. A slow lamp flash shows which line is ringing.

The central control unit is known as a *key system unit* or KSU. Electronic KSUs can support many additional features that are characteristic of most key systems. The list below is in addition to the telephone set features such as last number redial, message-waiting lamps, speakerphone, call logging, speed dial, etc. In addition to the flashing lamp call indications, call status information may be displayed on the telephone.

- *Automatic line selection*: When the user picks up the phone, an outgoing line is selected automatically.
- *Bridged call appearance*: The same extension number can be terminated on multiple phone sets.
- *Call drop*: A call can be terminated without hanging up the receiver.
- *Call forwarding*: Users can forward their calls to another station in the system.
- *Call park*: This feature places a call in a parking orbit so it can be retrieved from any telephone in the system.
- *Call transfer*: An incoming or outgoing call can be transferred to another user.
- *Callback*: If someone transfers a call to an extension that does not answer after a set number of rings, the call returns to the original station.
- *Camp-on*: Users or the attendant can send an external call to another telephone even if it is busy. The callee hears a faint camp-on tone. When the user hangs up, the camped-on call rings at the station.
- *Conferencing*: Stations can bridge two or more lines together for a multiparty conversation.
- *Distinctive ringing*: Different ringing tones enable users to distinguish between internal and PSTN calls.

- *Do not disturb*: Users can press a button that silences the bell and prevents intercom calls from reaching the station.
- *Forward all calls*: Users can redirect all calls to another station or destination.
- *Forward on busy or no answer*: Users can redirect calls to another station or destination if the line is busy or does not answer.
- *Held-line reminder*: After a call has been left on hold for a specified period, the telephone emits a warning tone.
- *Missed-call indicator*: A list of unanswered calls is displayed on the telephone.
- *Music on hold*: While a call is on hold, music or a promotional announcement is played.
- *Mute*: A mute button on the telephone disables the microphone.
- *Paging*: Stations can page over the telephone speaker.
- *Privacy*: Prevents other stations from picking up a line that is in use. In some systems privacy is automatic unless the user presses a privacy release key.
- *Station restriction*: Stations can be assigned to different classes of service for restricting long distance calls.
- *Voice call*: A user can place a call directly to the speaker of another user's telephone.
- *Volume control*: The volume of the handset, speaker, and ringer can be adjusted.

Caller ID can be provided in some key systems using one of two methods. With analog lines, callers are identified with the ADSI protocol. Some key systems support BRI ISDN, which is also capable of caller ID. In either case, the call identification extends to the station display when the call is transferred.

Voice mail is readily available in key systems. Since DID is not a key system feature, calls are either transferred to the user's voice mail manually, or an auto attendant prompts the caller to dial by name or extension number.

Key systems are often used across Centrex lines to provide features that require dial-access codes on POTS telephones. Many LECs offer lines with Centrex-like features such as transfer, conferencing, and third-party add-on. These features are activated by sending a momentary on/off hook flash toward the central office to signal for second dial tone. Flashing the switch hook of a key system telephone signals the KSU to return second dial tone from the key system, not from the central office. Therefore, most key systems provide a feature button, typically called "flash" or "link" to flash the PSTN line. This feature is essential in any key system that is intended to work behind Centrex. PBXs rarely provide this feature because they are intended to interface trunk-side connections in the central office, and these do not respond to switch hook flashes.

PBX FEATURES

This section discusses the main features that most PBXs and many hybrids, both TDM and IP, support. These features are in addition to the key system features discussed in the previous section. Two features on the key system list, flash and paging, are generally available on hybrids but unavailable on PBXs. Although the features in this section are common to most PBXs, users will find operational differences among products. There are also differences in whether the features are standard or an extra-cost option.

Direct Inward Dialing (DID)

DID offers station users the ability to receive calls from outside the system without going through the attendant. The LEC's central office contains a software table with the location of the DID trunk group. When a call for a DID number arrives, the central office seizes an idle trunk and outpulses the extension number, usually with DTMF tones or over a channel. DID is effective in reducing the load on PBX attendants. It also enables users to receive calls when the switchboard is closed. DID is provided on both analog and digital trunks. On analog trunks a separate trunk group is required. Digital trunks may be provisioned as tie lines between the PBX and the central office to provide two-way DID. PRI trunks offer call-by-call service selection, enabling any trunk to be used for any purpose.

Automatic Route Selection (ARS)

Most PBXs terminate a combination of public switched and private trunks on the system. For example, in addition to local trunks, the PBX may terminate T1/E1 lines to the IXC, FEX lines, and tie trunks to another PBX. Educating users about which service to use is impractical, particularly as rates vary with time of day and terminating location, and the dialing plan varies with the type of service. It is a reasonably simple matter, however, to program route selection into the central processor of the PBX. With ARS, sometimes called least-cost routing (LCR), the user dials the number and the system determines the preferred route and dials the digits to complete the call over the appropriate trunk group.

The most sophisticated ARS systems can screen calls on the entire dialed number, but some simple systems, typical of hybrids, can screen on only the NPA and prefix. The ability to screen on the entire number is important for many companies. With it, for example, it is possible to allow users to dial some 900 numbers, but deny others. If a company has an IP gateway that enables it to call other company numbers in an overseas location, the ARS can route those calls to the gateway, and domestic calls to the PSTN. ARS can also select the trunk group based on class of service. One class could call internationally only over IP trunks, while international calls for another class always use the PSTN.

A related issue is digit insertion and deletion. Some services, such as FEX, may require the PBX to insert or delete an area code for correct routing. Telephone service is easiest for users if they always dial the same way regardless of the route the call takes. For example, if the PBX has FEX trunks to another area code, the user would dial the area code, but the PBX would strip it off before passing the digits forward to the FEX trunks if the central office does not require all 10 digits. Users cannot be expected to understand the logic of this arrangement.

Networking Options

Most PBXs offer networking options, which allow multiple PBXs to operate as a single system. Networking is available on most PBXs, but it is rare in hybrids. Call-processing messages pass between PBXs over a separate data channel using IP messages or some form of common channel signaling. With the networking option, call-processing information such as a station's identification and class of service travel across the network to permit features to operate in a distant PBX as they do in the local system. This feature is known as *traveling class mark.*

The objective of networking is to provide complete feature transparency, which is the ability of users to have the same calling features across the network as they have at the main PBX. For example, users want to be able to camp on a busy station, regardless of whether it is in their PBX or in a distant system, and they want to share a voice-mail system across the network. Some features do not work across a network in some products. Call pickup, for example, enables a user who hears a ringing telephone to press a button and bring the call to his or her telephone. The lack of this feature across a network is usually unimportant since users are normally in separate locations and cannot hear the bell. Some companies, however, start with separate systems in separate locations and later merge them. The PBXs are collocated in the same equipment room and remain networked together. If features such as call pickup do not work across the network, users in one work group must be assigned to the same switch, which often requires moving people from one PBX to another and possibly changing numbers.

TDM and IP PBXs have significant differences in the way they implement networking. Figure 24-1 illustrates some of the differences. In the top half of the figure three TDM switches are used. Each switch has a unique number range from a different central office and a local trunk group to that office. The database in each switch contains the details on each number range in its domain. Its ARS knows which trunk group to use to reach an extension in either of the other switches, but it does not contain the translations for the stations in the other switches.

The IP configuration, by contrast, has three servers, each with an identical database. If one of the servers fails, the other servers can support its stations, which are attached directly to the LAN behind the routers. Each server has direct access into the IP network, but the connectionless nature of IP enables any switch

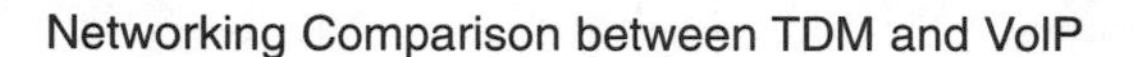

FIGURE 24-1

Networking Comparison between TDM and VoIP

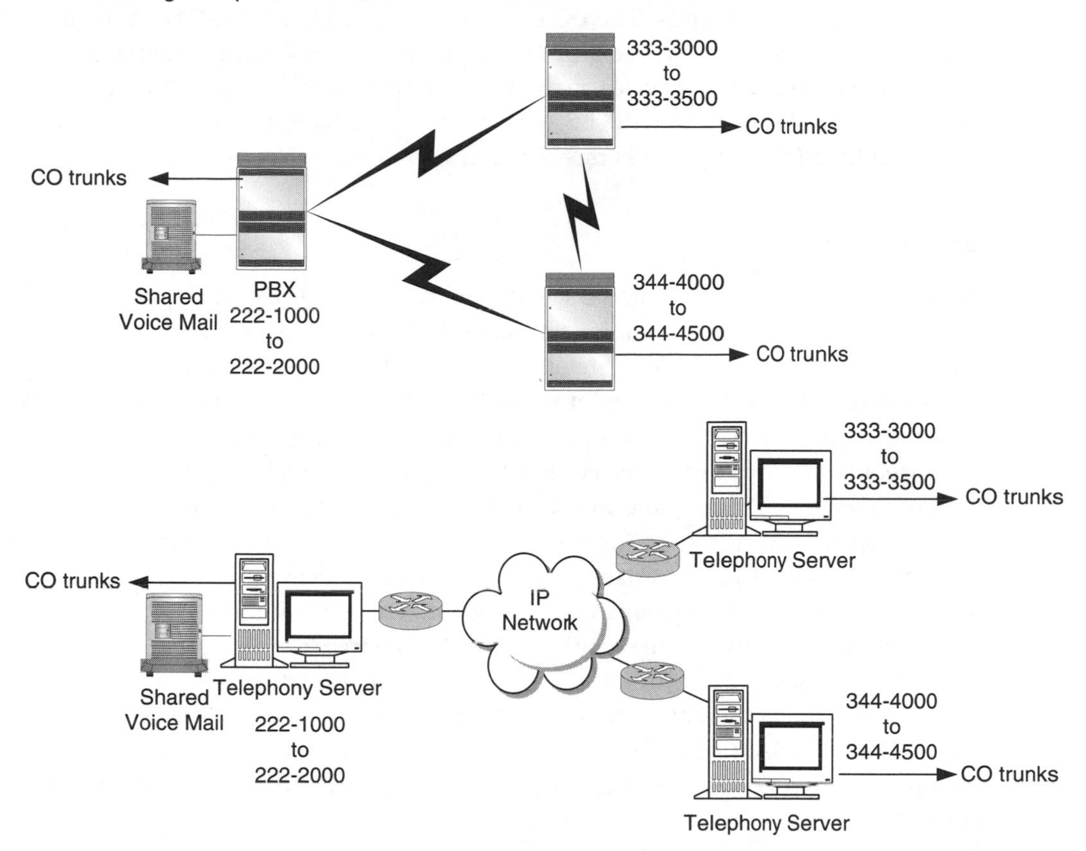

to set up a path to either of the other switches. This configuration provides survivability that the TDM model lacks. It also enables users to move between switches while retaining their telephone numbers.

QSIG

TDM PBXs use a proprietary protocol similar to ISDN for networking between their own products, but networking with switches of another manufacturer is impractical. The QSIG protocol, named after Q.931 ISDN signaling, is designed to support feature transparency and sharing of common resources such as voice mail between disparate products in a private integrated services network (PISN). The official ISO name for the protocol is private signaling system No. 1 (PSS1). QSIG separates the bearer channels from the signaling, which uses a separate packet-based signaling channel. QSIG can also be used with VoIP, where it offers the advantage of potentially reducing the number of hops needed for a call.

The first layer of the protocol is called QSIG basic call, which supports transparency between multivendor nodes. All products that claim QSIG compliance must support basic services. The BC feature set is intended for call control, but a higher layer known as QSIG generic function (GF) or QSIG supplementary services supports additional services such as calling line identification. QSIG capability is important for users with PBXs from different manufacturers to network them together. QSIG support is rarely found in hybrids.

Station Restrictions

An important feature of every PBX is its ability to limit the calling privileges of selected stations. Even companies that leave employees' extensions unrestricted normally require toll-restricted telephones in public locations such as waiting areas and lunchrooms. The type of restriction varies with manufacturer, but it is possible with most systems to restrict incoming, outgoing, and any type of long distance. One class of employee could be given international access, for example, while others are restricted to domestic calls. Some systems can restrict down to a specific telephone number. All restriction systems should be able to restrict selected area codes and prefixes. Area code restriction is necessary to prevent users from calling certain chargeable numbers, such as 900 numbers and to certain area codes that are known destinations for toll thieves.

Many systems provide an override feature that enables a user to dial an access code and identification number. This removes the restriction from a phone for the current session, and restores it when the call terminates. Another common feature is time-of-day restriction, which leaves phones open during working hours, but restricts them after hours.

Follow-Me Forwarding

With the increasing importance of telecommuting, several manufacturers are offering this feature, which allows the user to receive telephone calls at home, on a cell phone, or in a remote location such as a conference center. The user keeps the PBX informed of his or her location, and the PBX forwards calls accordingly. With caller ID and the appropriate programming, the system can screen calls as well, and forward calls from only selected users. Forwarding can be selective depending on time of day and day of week. At the user's option, the system can be programmed to ring to different destinations simultaneously or sequentially in patterns that can be changed for different days or times. This feature is available from both TDM and IP systems, usually as an extra-cost option. It requires a central server that typically is accessed with a browser over the Web or a private IP network. The protocol is usually proprietary on TDM systems and either proprietary or SIP in IP systems. The most effective products link the application to an electronic calendar.

When a user is available via e-mail, some applications can download voice mail as e-mail file attachments that the user can play back on a laptop computer. This is usually a feature of unified messaging (UM), which is discussed briefly later in this chapter and in more detail in Chapter 29.

Call Detail Recording (CDR)

This feature, sometimes known as station message detail recording (SMDR), in combination with a call accounting system provides the equivalent of a detailed toll statement for PBX users. Many businesses require call detail to control long distance usage and to spread costs among the user departments. The CDR port is a serial connection that outputs the raw call details in ASCII using a proprietary field format. A call accounting system, which is discussed later in this chapter, connected to the serial port parses the detail, rates the calls, and formats various management reports such as budgetary detail and individual toll statements.

Voice Mail

Voice mail (see Chapter 28) is available as an optional feature of all PBXs and hybrids and is one that is almost invariably applied. When a station is busy or unattended, the caller can leave a message, which is stored digitally on a hard disk. The station user can dial an access and identification code to retrieve the message. Most voice-mail systems include automated attendant, an option that enables callers with a DTMF dial to route their own calls within the system. Incoming calls are greeted with an announcement that invites them to dial the extension number if they know it or to stay on the line for an attendant. Most voice-mail systems also support dial by name for callers who reach the automated attendant and do not know the extension number.

Dialed Number Identification System (DNIS)

Offered by IXCs along with T1/E1-based toll-free services, DNIS provides the equivalent of DID for toll-free calls. If multiple toll-free numbers are terminated on the same switch, DNIS digits are sent with the call to identify the number dialed to the PBX so it can route the call to the appropriate station or group. DNIS enables an organization to have several toll-free numbers and to route each call to a different station, UCD or ACD hunt group, voice mail, or any other destination within the PBX. For example, if a company has different ACD groups for sales, service, and order inquiry, it can assign each of these groups a different toll-free number and use DNIS to route the calls appropriately. This alternative is often more effective than using an auto attendant to answer the call and offer a menu of choices.

Direct Inward System Access (DISA)

The DISA feature enables external callers to dial a telephone number and password to gain access to PBX features. The DISA port can be restricted to limit calls to internal extensions, tie lines, local calls, or any other restriction level used in the PBX. If the DISA port is unrestricted, callers can gain access to long distance services. DISA helps reduce credit card calls by enabling users outside the PBX to access low-cost long distance services.

Security is an obvious problem with DISA. It is one of the most prevalent targets for toll thieves, who use it to place calls at the company's expense. The best practice is to disable DISA. If it must be used, managers should change the password frequently and check the call accounting system for evidence of misuse.

$N \times 64$ Capability

With the growth of video conferencing, it is often desirable to dial more bandwidth than an ordinary BRI connection provides. Conference-quality video usually requires at least 384 Kbps, which is six 64 Kbps channels. A PBX with $N \times 64$ capability enables the user to dial as many channels of contiguous bandwidth as required.

Centralized Attendant Service (CAS)

CAS enables attendants at one location to handle attendant functions for remote PBXs over a network. Although each PBX has its own group of trunks, all calls routed to the attendant flow to the centralized location over the network. The attendant can terminate the call to any station or hunt group. A related feature is *release-link trunk*, which enables the PBX to release the attendant trunk after setting up the call. Without this feature the trunk is tied up for the duration of the call.

Power-Fail Transfer

Unless a PBX is configured to run from batteries or from an uninterruptible power supply, a commercial power failure will cause the system to fail. The power-fail-transfer feature connects central office trunks to standard DTMF telephones. Since most PBXs require ground-start trunks, provisions must be made to operate from loop-start telephones. This can be accomplished by two methods: use a separate loop-start-to-ground-start converter or equip the telephones with a ground start button. The former method is prevalent.

Power-fail transfer is an inexpensive and effective way to obtain minimum service during power failure conditions. Even users of systems with battery backup or UPS should consider power-fail transfer to retain some service if the PBX itself fails. Some manufacturers offer power-fail transfer for digital or ISDN trunks, which can enable the owner to avoid analog trunks.

Automatic Call Distribution (ACD)

ACD enables PBXs to route incoming calls to a group of service positions. Typical applications are sales and customer service positions. Incoming calls route to an agent position based on logic programmed into the switch. Calls can be routed based on the toll-free number that was dialed using DNIS. The caller's telephone number may be delivered by the network and used to route calls, or an automated attendant or call-prompting software in the switch can prompt the caller to select from a menu of routing options.

When agent positions are idle, the call routes to an agent immediately. If all positions are occupied, the ACD holds calls in queue and notifies the caller by recorded announcement that the call is being delayed. Calls can be overflowed to other agent groups, routed to voice mail so the caller can request a callback, or handled in a variety of different ways, which Chapter 27 discusses in more detail. ACD is one of the most important features in a PBX, and is included in more than three-fourths of the systems shipped.

Uniform Call Distribution (UCD)

UCD distributes calls evenly among a group of stations. When one or more active stations are idle, incoming calls are directed to the station that is next in line to receive a call. When all stations in the UCD group are busy, incoming calls are answered with a recording and held in queue. When a UCD station becomes idle, the call that has been in queue the longest is directed to the station. In many UCD systems, a station user can toggle between active and inactive status by dialing a code or pressing a feature button. Compared to ACD, UCD is unsophisticated, lacking the supervisory, management, and reporting features that an ACD offers. Chapter 27 discusses UCD further.

Unified Messaging

This feature, which is discussed in more detail in Chapter 29, integrates the PBX with voice mail, fax, and e-mail so that messages can be viewed and handled on a PC screen. The feature may also enable users to translate messages from one format to another. For example, e-mail messages may be read in synthesized voice if the user is calling from a telephone and wants them read out. Eventually, with improvement in speech-to-text software, voice-mail messages will be converted to e-mail or fax. Currently, speech to text enables users to speak limited commands to read, forward, and delete voice mail messages from a telephone.

Emergency Service Interface

Most of the developed world has adopted a special dialing code such as 999 or 911 for universal access to emergency services. The basic service enables the PSAP to hold up the line so it can be traced in case the caller is unable to report the address of the emergency. Enhanced emergency services contain a database that associates telephone numbers with street addresses. The street address is often not a fine enough distinction, however. Users who dial the emergency code from hotels, apartments with a shared-tenant PBX, campuses, and multibuilding developments and the like may be difficult or impossible to locate. Therefore, a trend is toward reporting the station identification to the PSAP so it can be associated with the room number or building name. The service is known as private system automatic line identification (PS/ALI). The PRI feature in most PBXs can relay the station identification to the LEC, which passes it to the PSAP. PBXs lacking PRI can use a CAMA trunk as an alternative. The need for this feature is particularly acute in PBXs with remote switch units that can be located in a different PSAP's jurisdiction from the host PBX. In this case, a separate trunk group to the local central office is usually provided. The ARS is programmed so it always seizes a local trunk when the emergency code is dialed from a station served by the remote.

A related feature that is important in any large PBX is the ability to route emergency calls simultaneously to the console attendant or a security position. The law requires that the call route directly to the PSAP, but management wants to be informed of any such call. It is also a useful feature in hotels, campuses, schools, hospitals, and other organizations where people may place emergency calls inadvertently or as a prank.

Multitenant Service

PBXs that provide service to users from different organizations can use multitenant software to give each organization the appearance of a private switch. Multitenant service is a software partition in the PBX. Separate attendant consoles can be provided, and each organization can have its own group of trunks and block of numbers.

Property Management Interface

Hotels, hospitals, dormitories, and other organizations that resell service often connect the PBX to a computer to provide features such as checking room status information, disabling the telephone set from the attendant console, and determining check-in or check-out status. The PBX provides information to the computer, and accepts orders from the front desk via computer terminal. The PMI is a specialized type of computer-telephony integration (see Chapter 27).

Uniform Dialing Plan (UDP)

UDP software in a multi-PBX network enables the caller to dial an extension number and have the call completed over a tie line network without the caller's being concerned about where the extension is located. The PBX selects the route and takes care of station number translations. UDP software is effective only among PBXs of the same manufacture although it can work with QSIG-compatible PBXs if they are so equipped.

Simplified Message Desk Interface (SMDI)

SMDI is a standard way of interfacing a switch to peripheral equipment such as voice mail. The voice mail connects to the PBX over analog ports or line-side T1 and to the SMDI with a serial connection. For a call going to a messaging unit such as voice mail, the SMDI link indicates the port the call is using, the type of call, information about the call such as the source and destination, and the reason the call is forwarded such as busy or no answer. The SMDI is an open protocol for interfacing voice mail to the switch as an alternative to the manufacturer's proprietary interface.

PBX Voice Features

As all PBXs are designed for voice switching service, they have features intended for the convenience and productivity of the users. Not all the features listed below are universally available, and many systems provide features not listed. This list, in addition to the key system features discussed earlier, briefly describes the most popular voice features found in PBXs.

- *Automatic call trace*: Harassing or nuisance call can be traced to the origin by dialing an access code.
- *Call blocking*: Users can selectively block calls such as specific extensions, numbers, or calls from particular trunk groups.
- *Call coverage*: Users can have one or more coverage paths to direct how calls route when the called station is busy, does not answer, or is in do-not-disturb status. External calls can take a different path than internal calls.
- *Executive override*: This feature allows a station to interrupt a busy line or preempt a long distance trunk.
- *Forced account code*: On long distance calls, this feature prompts callers to enter an identification code, which is registered on the CDR. It is often used in colleges and universities where roommates share the same extension number. Many professional organizations use account codes to allocate calls to clients.
- *Hoteling*: A station user can temporarily move to another location, log in, and have station features including the extension number follow to the new location. Intervention from the administrator is not required.

- *Paging access*: The PBX can be equipped with paging trunks that connect to an external paging system. The trunk is reached by dialing an extension number or trunk access code. Zone paging, which allows paging in specific locations rather than the entire building is available on most systems.
- *Personal call routing*: Users can define routing of incoming calls based on variables such as time of day, calling number, etc.
- *Portable directory number*: Allows a user on a networked PBX to move from one switch to another without changing the telephone number.
- *Priority ringing*: A distinctive ring is used for calls from specified numbers.
- *Recorded announcements*: This feature provides announcements for vacant and disconnected numbers.
- *Trunk answer any station*: This feature allows stations to answer incoming trunks when the attendant station is busy.
- *Whisper page*: A user can bridge into a call and speak to the local user without the other end hearing.

Attendant Features

Most PBXs have attendant consoles for incoming call answer and supervision. The attendant can also act as a central information source for directory and call assistance. The console is either a specialized telephone instrument or a PC running a console program. The latter is increasingly popular because it can be easily integrated with a directory. The following features are important for most consoles and represent only a fraction of the features available.

- *Attendant controlled conferencing*: Attendant can set up multiport conference calls.
- *Automatic timed reminders*: Alerts the attendant when a called line has not answered within a prescribed time.
- *Busy lamp field*: When the station is busy or in do-not-disturb mode, an LED associated with the station is lighted.
- *Direct station selection* (*DSS*): Allows the attendant to call stations by pressing an illuminated button associated with the line. The line button shows busy or idle status.
- *Directory features*: Attendants with PC-based consoles may be able to search by first and last name, department, and extension.
- *Night service*: Calls are automatically transferred to an alternate destination when the console is closed. In many systems this feature is sensitive to time of day and day of week.

System Administration Features

System administration is a costly element of every PBX, so features that ease the administrator's job are valuable. The following are some of the more popular features.

- *Automatic set relocation*: Allows users to move their telephones from one location to another without the need to retranslate. The administrator gives users a code and instructions to carry the set to the new location, plug it in, and dial the code. When this is complete the system moves the station translations to the new port. This feature is inherent with IP systems, which may enable a user to log in from any available Ethernet port.
- *LDAP synchronization*: Enables the system to update its PBX and voice mail database from customer's LDAP directory. Eliminates or reduces redundant database entries. The application may also permit the administrator to work translations in software in advance, and then upload them to the PBX.
- *Network move*: Similar to automatic set relocation, this feature works across a network, where automatic set relocation works only in the same PBX.

CALL ACCOUNTING SYSTEMS

All PBXs, most hybrids, and many key telephone systems include a CDR port that receives call details at the conclusion of each call. The call details can be printed or passed to a call accounting system for further processing. The CDR output of most systems is of little value by itself because calls are presented in order of completion and lack rates, identification of the called number, and other such details needed for control of long distance costs. Call accounting systems add details to create management reports, a complete long distance statement for each user, and departmental summaries. The primary purposes of a call accounting system are to discourage unauthorized use and to distribute costs to users. They also have other uses in some companies. For example, a supervisor may use the CDR record to check the effectiveness of an employee's outgoing sales calls.

Most call accounting systems on the market are software programs for PCs. CDR data either feeds directly into an on-line PC or it feeds into a buffer that stores call details until it is polled. A buffer makes it unnecessary to tie up a PC in collecting call details. If the power fails, the battery backup in the buffer retains the stored information.

In multi-PBX environments, a networked call accounting system may be required. These systems use buffers or computers to collect information at remote sites and upload it to a central processor at the end of the collection interval. If long distance calls can be placed from one PBX over trunks attached to another, a tie line reconciliation program may be needed. The tie line reconciliation program uses the completion time of calls to match calls that originate on one PBX and terminate on trunks connected to another. Networked PBXs send originating station identification over the signaling channel to a remote PBX. If the remote PBX is equipped to extract the calling station identification from the network and associate it with the CDR output, the need for tie line reconciliation is eliminated.

Most PBXs can output to the CDR port any combination of long distance, local, outgoing, and incoming calls. The amount of detail to collect is a matter of individual judgment, but sufficient buffer and disk storage space must be provided to hold all the information collected.

CALL ACCOUNTING APPLICATION ISSUES

Application information for PBXs and key systems is included in the next two chapters. This section covers application information for call accounting systems.

Call Accounting Evaluation

Most PBXs today are purchased with a call accounting system that is normally programmed and supported by a third-party manufacturer. The following are some criteria for selecting a call accounting system.

Reports

The main reason for buying a call accounting system is for its reports. Evaluate factors such as these:

- What kinds of special reports are provided? Do they meet the organization's requirements? Examples are unused extensions, long or short duration calls, unused trunks, and calls to emergency numbers.
- Can reports be distributed over the Internet or a company intranet?
- Can users access their reports with a browser?
- Are custom-designed reports possible?
- Is it possible to export report information to an external program, such as a spreadsheet or database management system, to produce custom reports?
- Are traffic reports produced? If so, are they accurate?
- Are management reports, such as inventories, provided?

- What kind of manual effort is needed to produce reports? Does it require a trained operator, or can clerical people perform the month-end operations with little or no formal training?
- Is tie line reconciliation required? If so, does the manufacturer support it?

Operational Issues

Most call accounting systems are not completely automatic. The functions required are downloading the call data from buffers (if they are used), rating calls, and producing end-of-period reports. The most effective systems provide drag-and-drop capabilities for setting up and scheduling reports and distributing them to users.

Features

Many call accounting systems provide features that are of extra value. Common features are toll fraud alerts, telephone directory, and equipment inventory. Some high-end systems offer telemanagement packages, which typically include service orders, repair, and inventory in addition to directory and call accounting.

Vendor Support

As with most software packages, vendor support is important for installing and maintaining the system. Evaluate the vendor's experience in supporting the package. Determine whether the vendor has people who have been specifically trained. Evaluate the amount of support the package developer has available and what it costs. Some vendors sell ongoing support packages, and where these are available, the cost-effectiveness should be evaluated.

Call Rating

Most call accounting packages have call-rating tables based on V&H (vertical and horizontal) tables. These divide the United States and Canada into a grid from which point-to-point mileage is calculated. Tables must be updated regularly as rates change. Also, consider that many companies do not need absolute rate accuracy. To distribute costs among organizational units, precision is usually not required. Many long distance rate plans use rates that are not distance sensitive, so V&H rating accuracy is not required. The rating tables identify the called location, so if rating tables are not used, the called city and state will not be printed on toll statements unless the vendor offers an abbreviated table. Determine facts such as these:

- What kind of rating tables does the manufacturer support?
- How frequently are tables updated?
- What do updates cost?

- What IXCs' rates does the package support?
- How are intrastate rates calculated?
- Do you need to bill back to user departments with high accuracy?

Capacity

Call storage equipment is intended to maintain information on a certain number of calls. When buffer storage is full, it must be unloaded and calls processed. Usually, the system must store at least 1 month's worth of calls. Evaluate questions such as these:

- How much storage space is required?
- What is the capacity in number of calls, both incoming and outgoing?
- How much growth capacity is provided?
- Is storage nonvolatile, so if power fails calls are not lost?

CHAPTER 25

TDM Customer-Premise Switching Systems

PBXs have evolved in a path parallel to central offices. The earliest versions were manual switchboards, followed by electromechanical dial, and then stored program systems that matured into the current TDM architecture. As with central offices, PBXs are on the cusp of a transition from TDM to VoIP and for much the same reasons. The legacy TDM technology is still adequate for most offices and in the absence of compelling reasons for replacing existing systems, the transition is likely to be gradual. Most manufacturers have modified their TDM systems to support VoIP, which further postpones the transition to an all-IP system because upgrades to most existing systems can provide equivalent services. Moreover, many of the justifications for replacing existing systems are difficult to quantify in tangible terms. We will discuss these considerations in the "Applications" section at the end of the next chapter.

All CPE equipment manufacturers are touting their IP systems to the point that the TDM alternatives are obscured in the hoopla, but TDM switches serve the majority of CPE lines and will continue to do so for the next several years. Feature development has reached its zenith with TDM systems, however, as manufacturers concentrate their development efforts on their IP products. The primary interest most companies have in TDM systems is preserving the existing investment, the life span of which has historically been in the 7-to-10-year range.

This chapter discusses the architectures and configuration of the three TDM product lines on the market: key systems, hybrids, and PBXs. Although the distinction between the product lines is somewhat blurred, if an organization requires more than about 100 central office line and station ports, a PBX is usually most effective because of its greater line, trunk, and intercom capacity. Hybrids can grow to about 250 ports, and are satisfactory for midsized offices at a lower cost than PBXs. Key systems are used in smaller line sizes. When the system supports more than a dozen central office lines it is set up with pooled trunk capabilities and defined as a hybrid.

TDM KEY AND HYBRID ARCHITECTURES

Key and hybrid systems are typically wall mounted in cabinets similar to the one in Figure 25-1. Many small systems have a fixed size and cannot be expanded. These are designated by their line and station size such as 3×8 or 6×16. When the system reaches capacity, it must be replaced if further growth is needed, often preserving only the telephone instruments. Larger hybrid systems may also be designated by line and station size, but the basic unit is expandable. These systems are also housed in wall-mounted cabinets, but they have expansion slots to accept additional line and station cards.

All key and hybrid systems support loop-start central office lines. Some high-end systems support ground start, T1/E1, BRI, and PRI trunks. Some also support tie lines between other switches, and may support networking either between themselves or with the manufacturer's PBX product line. All hybrids and key systems are processor controlled with either a proprietary or a commercial microprocessor. The generic program is usually stored in a chip, which is replaced to upgrade to a later software release. This is in contrast to PBXs, in which the program is upgraded in the field from tape or CD-ROM. Key and hybrid line and station translations are typically stored in flash memory, which enables them to restart quickly in case of power failure.

FIGURE 25-1

Nortel MICS Key System (Photo by Author)

KTSs usually include one or more intercom lines. These are used for station-to-station communication—in smaller systems the intercom is primarily for conversations between the attendant and the called party. In larger systems multiple intercom paths are provided so several internal conversations can be held simultaneously. Most systems provide a built-in speaker so the intercom line can be answered without using the telephone handset. Optionally, the user can lift the handset for privacy. The number of intercom lines provided is a feature that distinguishes a PBX from a hybrid. Many hybrid systems support a limited number of intercom paths, which may preclude their use in offices that require a large amount of internal calling. Most PBXs have enough time slots that virtually all stations can be talking simultaneously.

While the attendant can answer and transfer calls from an ordinary telephone, a special telephone is often provided. The attendant has all the features of regular stations and may also have a busy lamp field (BLF) to show which stations are occupied and a DSS field, which allows the attendant to transfer calls to stations by pushing a button instead of dialing the station number. To support the attendant, many systems include paging either to an overhead system or to telephone speakers. The paging system is accessed by pushing a button or dialing a code and can be divided into zones if the building is large enough to warrant it. Many systems provide for parking a call so a paged user can go to any telephone, dial a park number, and pick up an incoming call.

Several manufacturers produce multiline systems that do not require a KSU. Most KSU-less systems require one pair of wires per line, which limits the size of the system to four or fewer lines. The primary advantages of KSU-less systems are low cost and ease of installation. Anyone who knows how to install a single-line telephone can install KSU-less telephones because they do not require setup, which makes them ideal for small offices and residences. The primary drawbacks of KSU-less systems are limited expandability and lack of features. Since the systems have no KSU, the only features available are those contained in the telephone set itself. Some KSU-less systems also lack an intercom path, which means calls cannot be announced over an intercom as they are with most key systems.

Voice mail is provided almost as a matter of course in practically all key and hybrid installations. Other add-on peripherals such as ACD and call accounting are available for many product lines.

PBX ARCHITECTURES

PBX technology has progressed through three generations and is now starting the fourth generation of IP PBXs. First-generation systems were manual switchboards, which were common in the first half of the twentieth century. Second-generation systems used wired logic and analog step by step or crossbar switching fabric. Second-generation telephones were nonproprietary rotary dial or DTMF analog sets. If key features were needed with a second-generation PBX, a separate key telephone system was required. The third generation introduced stored program control processors. Some used analog switching fabric and others were TDM. The processor-controlled logic enabled PBXs to support proprietary telephones, which controlled a limited number of key telephone features. All systems supported POTS phones, which required users to dial feature access codes. Proprietary telephones either used an ISDN-like interface with a separate signaling channel, or they signaled over wires separate from the talking path.

All of today's third-generation switches employ TDM switching technology and support both analog and proprietary digital telephones. TDM PBXs are the de facto standard of the industry, having been operation for nearly three decades. Products have continually improved with feature and hardware enhancements and have reached the stage where the risk is slight in buying a PBX from any reputable manufacture and distributor. Proponents of IP PBXs contend that TDM PBXs are obsolete, and eventually they will be, but TDM systems meet the communications needs of most users and will continue to do so for several years.

All TDM PBXs are processor controlled. Some use proprietary processors and some use industry-standard microprocessors. PBXs are rated by busy hour call attempts (BHCA) and busy hour call completions (BHCC). A call attempt is any event that accesses the processor. This includes functions such as switch hook flashes and feature accesses to hold, transfer, conference, park, or any such event that occurs during the normal course of a call. The number of call attempts an office requires is difficult to estimate, but the BHCA and BHCC factors are convenient metrics for comparing products.

The switching matrix is evaluated by busy hour hundreds of calls seconds (BHCCS). Traffic engineers use CCS in evaluating trunking requirements. A discussion of traffic engineering is beyond the scope of this book, but it is covered in the companion volume, *The Irwin Handbook of Telecommunications Management*. A circuit occupied for an hour is busy for 3600 s or 36 CCS. If a switching matrix is nonblocking, it is capable of supporting all stations in an off-hook condition simultaneously on station-to-station calls. Trunk groups are never engineered to permit all stations to connect to trunk calls simultaneously, so the switching network would never be overloaded by trunk calls. Since each call involves two stations, a nonblocking switch must support half the quantity of stations times 36 CCS per station. To illustrate, assume a PBX has 100 stations. If it is nonblocking, the switching network should support 50×36 or 1800 BHCCS.

The architecture of a TDM PBX is similar to that of a TDM local switching system. As Figure 25-2 shows, PBXs have a line side and a trunk side. The interfaces are contained in hardware modules that fit into slots in a card cage. The line and trunk ports connect to the TDM switching matrix, which makes the connections under processor control. The features and configuration information are contained in memory. Smaller systems use flash memory for the program and configuration information. Larger systems usually employ volatile memory for the program and for station and trunk translations. Generic programs are upgraded

FIGURE 25-2

TDM PBX Architecture

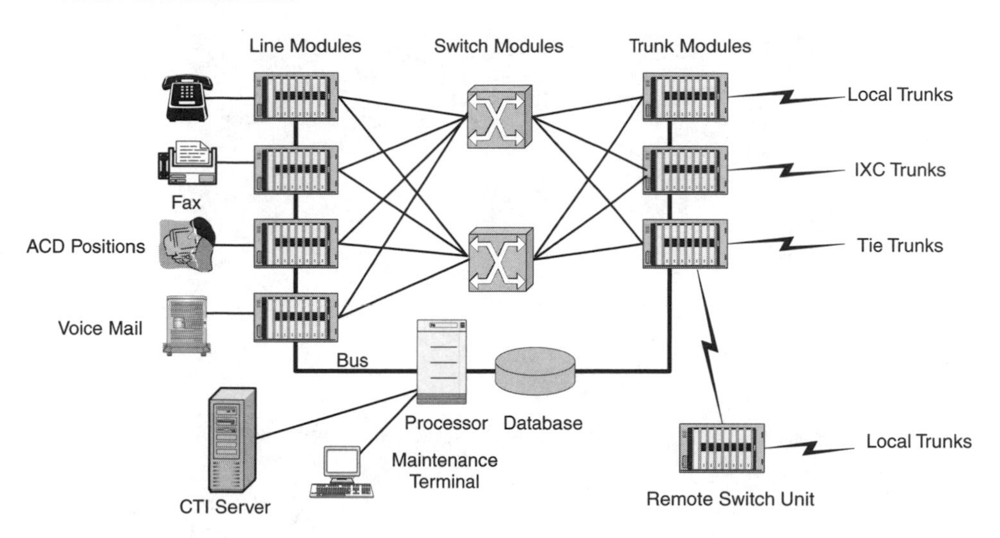

from tape or CD. Larger systems usually provide tape drives for backing up the program and configuration, which must reload if the power fails.

The expansion cards plug into the PBX's backplane, which ties the lines, trunks, and central control circuits to the switching fabric and busses over which the circuit elements communicate. Although the structure of PBXs is similar, there are significant differences in products on the market and the way features are packaged and sold. PBXs grow incrementally up to the total capacity of the system. The main module contains common equipment cards such as processor, memory, TDM switch, power supply, and tape and CD-ROM drives. These fit in dedicated slots that cannot be used for any other purpose. System recovery following power failure is much slower than key and hybrid systems, typically requiring 3 to 10 min. Figure 25-3 shows a cabinet stack from a Nortel SL-100 PBX.

Added to the main module are modules that support line and trunk cards. These plug into a backplane that connects to the TDM switch modules. Figure 25-2 shows separate line and trunk modules for clarity, but most systems have universal card slots—that is, either line or trunk cards can plug into any slot. Up to a given line size, any port can connect through the TDM switch to any other port. At some point, depending on the manufacturer's design, direct port-to-port connectivity becomes impractical and a center-stage switch is required to connect modules.

Station Connections

Stations connect to the line ports through a 64 Kbps circuit that runs on twisted-pair wire. The typical digital station range is in the order of 1500 to 2500 ft (460 to 760 m).

FIGURE 25-3

Nortel SL-100 PBX (Photo by Author)

Analog telephones connect to analog ports, or in some systems to digital ports through an analog adapter. PBXs have at least two different types of line interface cards, analog and digital. Most systems also support BRI cards, which are typically used for videoconferencing. Digital line cards support proprietary telephones that work only with that manufacturer's system. Analog and ISDN cards support telephone sets that are independent of the PBX manufacturer. Although ISDN telephones should work with any manufacturer's PBX, ISDN standards do not define all of the features that the system may be capable of. Therefore, ISDN sets from the PBX manufacturer will usually provide features that other manufacturers' telephones cannot support.

Line card density, which ranges from eight to 32 ports, is a distinguishing feature among products. High-density cards allow for smaller cabinet size, which is usually a plus. In smaller PBXs, however, high-density cards may result in more spare ports than the owner would normally purchase. For example, if the system has 32-port cards and 33 ports are required, 64 ports must be purchased, leaving nearly half of them unused.

Trunk Connections

PBXs, like central offices, interface the outside world through trunk circuits that exchange signals with other switching systems through a variety of signaling interfaces. TDM systems support both analog and digital trunks, but as the network evolves toward all digital, analog trunk interfaces for the PBX are gradually disappearing. Transmission quality is better on digital trunks and they take fewer card slots. Analog trunks are usually provided over three separate trunk groups: incoming to the main listed number, outgoing, and DID. A separate type of analog trunk card is required for DID trunks in most PBXs, although some manufacturers offer universal trunk cards that will support either DID or two-way CO. Incoming and outgoing analog trunks can be combined into a two-way trunk group. Calls to the main listed number route to the console, and any trunk can be seized for outgoing calls. Analog trunk cards contain from 4 to 16 trunks per card. Analog trunk cards support two-way central office trunks and foreign exchange lines. Trunks to the IXC are normally digital.

T1/E1 trunk cards support 24 or 32 circuits. Some PBXs use a single type of card for T1/E1 or PRI; others have separate card types. It is important to understand the difference between a line-side and trunk-side T1/E1. Line-side connections, which are always the case with analog trunks, allow the user to flash the central office and get second dial tone to activate features such as conferencing and transfer. Trunk-side connections do not offer this capability. Most key systems and hybrids allow the user to flash the central office, but PBXs do not. Therefore, even though the central office may provide line-side features toward a PBX, they cannot be accessed. For example, users may want incoming calls transferred to cell phones. With a line-side connection, an attendant can flash the line, receive second dial tone, and transfer the call. The connection is made in the central office and the line is released. A PBX with T1/E1 trunk-side connections can also transfer calls, but only if the switch has been configured for trunk-to-trunk transfer. Many PBXs have this feature deactivated for reasons that will be explained later, but if it is permitted, two trunks are tied up for the duration of the call. If the PBX has PRI trunks, and if both the PBX and central office are equipped for antitromboning, one of the two trunks will be released. Line-side T1 cards are used to connect the PBX to peripherals such as interactive voice response and voice mail.

Most LECs offer both PRI and digital trunks. The latter may be connected to the line side or the trunk side of the central office, but PRI is always a trunk-side connection. Depending on the LEC, non-PRI T1 trunks may be set up to provide two-way DID, offering a service similar to PRI, but with in-band signaling. The main impediment to the all-digital PBX trunking network is the premium prices that many LECs charge for digital or ISDN trunks, but even this is disappearing as competition lowers trunk prices. Most LECs now offer digital trunks, either as separate two-way and DID trunks or combination trunks that can be used for either DID or outgoing service. Digital trunks have two drawbacks. One is their lack of scalability. When additional trunks are required, they are added a full T1/E1 at a time. The other is their greater vulnerability to failure since a single failure can kill all the trunks to a PBX.

Signaling compatibility is an important issue in connecting a PBX to a central office. If analog trunks are used, PBXs use ground start trunks to prevent glare as discussed in Chapter 12. DID trunks are used for incoming service only, so they may be loop start from the central office, with the DID digits passed with DTMF signaling. Two-way DID trunks are normally connected to the central office as tie lines using E&M signaling. PRI trunks use the D channel for signaling.

As discussed in Chapter 15, PRI in North America provides 23 64-Kbps B channels plus one D channel. The rest of the world uses E-1, with 32 channels, of which two 64 Kbps channels are reserved for signaling and controlling. PRI is preferable to T1 because, among other advantages, it supports caller ID and call-by-call service selection, which permits the PBX and the central office to determine for each call what type of service is needed. For example, if a PBX supports video conferencing, multiple channels on the PRI will be needed to provide the desired degree of picture quality. The PBX and the central office set up the appropriate bandwidth by exchanging messages on the D channel. Many PBXs use BRI on the line side to support video conferencing.

PBXs require an access digit, usually 9, to connect station lines to central office trunks. When the user dials 9, the PBX seizes an idle central office trunk and connects the talking path through to the station if the station is permitted off-net dialing. The station hears central office dial tone as a signal to proceed with dialing.

Trunk-to-trunk transfer enables a user to link an incoming trunk to an outgoing trunk and hang up. The feature is available in most PBXs, but it is often disabled to prevent misuse. For example, unauthorized callers may request confederates on the inside to connect them to long distance trunks. The feature has many authorized uses, however, such as transferring a call to a cell phone, so many companies activate trunk-to-trunk transfer, but restrict it to local calls and tie trunks or to a particular class of service.

Tie Trunks and Networking

Organizations operating multiple PBXs have two alternatives for interconnecting them, tie trunks and networking. Tie trunks can be analog or digital, terminating on the trunk side of the PBX. Analog tie trunks are rare today because most organizations have both voice and data networks running over digital circuits. If VoIP is not used in the PBX, voice and data can share a T1/E1 by splitting the line through an add-drop multiplexer as shown in Figure 25-4. This configuration avoids the QoS issues of VoIP, but it does not allocate bandwidth dynamically. If the bandwidth balance is not optimum, the multiplexer must be reconfigured. Nevertheless, it is an inexpensive way of sharing a digital line with legacy equipment.

Tie trunks are generic, and can be set up between PBXs of different manufacture. Signaling is usually E&M and it supports no feature transparency beyond call origination and termination. A call into one PBX can terminate to a station in the other PBX, and users can transfer calls across the tie lines, but sharing a voice mail requires QSIG networking. If tie trunks terminate in a single location they are often accessed by dialing a digit, such as 8, which connects them to the distant PBX. Many multi-PBX organizations have a separate dialing plan for each system plus a single organization-wide dialing plan. The PBX then is then programmed to provide the translations necessary to reach the distant number over the tie trunk network. This feature is called *uniform dial plan*. To avoid the need for users to understand the dialing plan, many organizations use the PBX's ARS to dial the necessary codes. Users dial the number, and the PBX selects the route and dials any additional digits.

Most companies network their PBXs to obtain feature transparency. Proprietary networking protocols may be used between PBXs of the same manufacture, or, as discussed in Chapter 24, QSIG may be used between otherwise incompatible products. Networked PBXs provide service equivalent to a central

FIGURE 25-4

Voice-Data Line Sharing with an Add-Drop Multiplexer

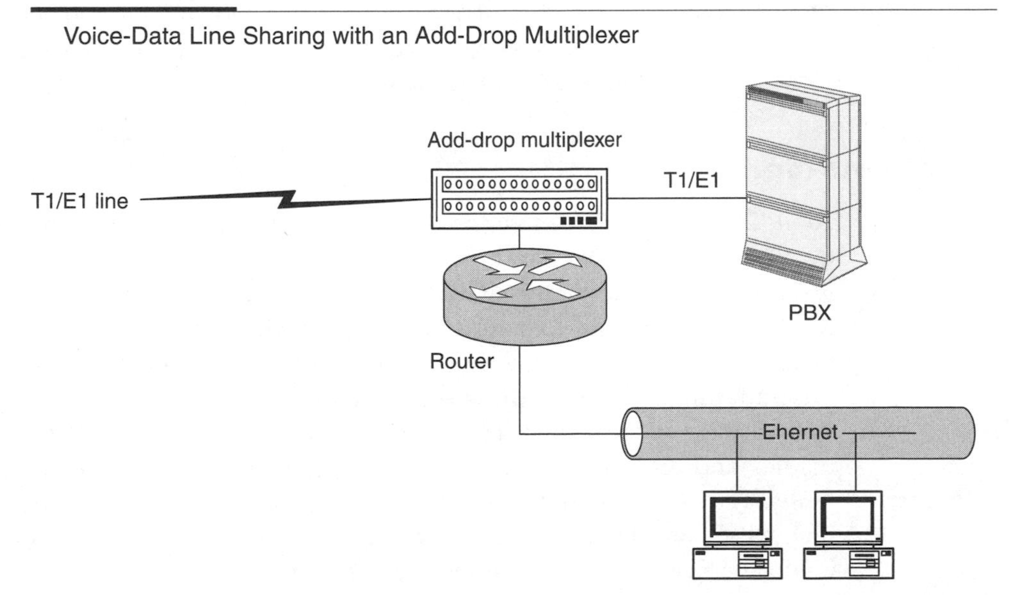

PBX and RSU except that each PBX contains a separate database and is inherently survivable. If the link to the main PBX is lost, features are lost, but local switching is unaffected.

Distributed Switching

Many PBXs offer remote switching units. An RSU extends line and sometimes trunk interfaces over T1/E1 lines to secondary locations. The main advantage of an RSU as compared to networked PBXs is that the remote stations are in every respect equivalent to those in the host. All configuration and management are centrally controlled, in contrast to networked PBXs, each of which is separately administered. All processing is in the central unit and connects to the remote over an umbilical. If the remote is in a separate calling area it may be equipped with local trunks.

On the down side, if the umbilical fails, the remote is dead. Some remotes have survivability features that permit limited stand-alone operation. A survivable remote typically has a copy of the line and trunk database so the system operates with reduced capabilities. If incoming calls terminate on the central unit during normal operation, they will be lost during emergency conditions.

Remote units have several advantages compared to networked PBXs:

- Only one processor and software set are needed. This is usually less expensive than maintaining separate systems.
- Administration is from the host. All database changes are made on the host switch.
- Wiring costs are reduced on a campus. It is often less costly to install a remote than to cable from the central site.
- Total feature transparency is achieved. Users in the remote location share the same voice mail, numbering plan, and trunks as the central site, and have access to exactly the same features.

Administration

All PBXs are administered from a maintenance and administration terminal (MAT). In some products a dumb terminal connects directly to the PBX through a serial port, but many current PBXs provide Ethernet connections to a PC. The MAT terminal is used to enter station and trunk translations in a command language that is unique to the product. Most products offer an optional PC-based program that allows the administrator to make changes on a graphic screen and upload them to the switch. The same terminal and associated printer are used to display and diagnose maintenance messages that indicate hardware or trunking faults. The same terminal is used to collect statistics such as processor activity, and line, trunk, and feature usage. Most PBXs support remote administration either over Ethernet or dialup to a remote maintenance port.

The maintenance and administration process is designed to be closed and proprietary. It is kept out of the hands of users as one of the ways the configuration is controlled to support management's objectives. The process is complex enough that training in factory-authorized schools is required to become certified in administering the system.

Redundancy

Modern TDM PBXs are inherently reliable. Total failures are rare, but not unheard of. Reliability can be increased through redundancy. Some products offer redundant processors while others offer redundant switching matrices, power supplies, and common equipment items. With full redundancy and uninterruptible power supplies, TDM PBXs can achieve reliability in the order of central offices.

Emergency Communications

Because PBX stations are tied to a location, it is easy to convey location information to an emergency center. Fixed station locations are changed only through order activity, which gives the administrator an opportunity to update the emergency database. This is in contrast to IP stations, which may change location without management's involvement. Systems employing remotes that do not have separate trunks, must take precautions to ensure that the true location of the reporting station is transmitted to the PSAP.

Computer–Telephony Interface

Most PBXs provide an open interface for limited call control from an external processor. The physical interface and the command language are proprietary to the manufacturer, and allow application programmers to connect the PBX to a standard application program interface (API). The most common are telephony API (TAPI), which is for connecting a PC running MS Windows to telephone services and telephony server API (TSAPI), which is a server-based interface. CTI is discussed in Chapter 27.

Wireless Capability

Many organizations have classes of users who must roam the building. Wireless systems allow use of the telephone anywhere in a building or within a restricted range on a campus. Two types of wireless systems are available. One type plugs into analog ports on the PBX, and gives the user capabilities of analog telephones. Proprietary wireless systems provide the features of digital telephones including multi-line capability and button access to features. Wireless phones are discussed in Chapter 21.

TDM CPE Application Issues

Nearly every business that has more than a handful of stations is in the market for a key system, hybrid, PBX, or its central office counterpart Centrex. PBXs are economical for some small businesses that need features such as restriction, networking, and ARS that are beyond the capabilities that most key systems provide. Very large businesses may use central office switching systems of a size that rivals many metropolitan public networks. Between these two extremes lie hundreds of thousands of PBXs.

PBX Standards

PBXs, almost by definition, are proprietary in their internal switching and features, but trunk interfaces are standard T1/E1 or analog trunks that are covered in ANSI/TIA/EIA-464-C. This document is a detailed set of standards for transmission, signaling, framing, and private network synchronization. It describes digital and analog interfaces to a local central office in detail. The interfaces are so well documented that incompatibility is practically non-existent.

QSIG, as discussed in Chapter 24, is an ISO standard for networking PBXs of disparate manufacture. SMDI, also discussed in Chapter 24, is a Telcordia Technical Reference TSR-TSY-000283, *Interface Between Customer Premises Equipment; Simplified Message Desk and Switching System*. Both of these are open interfaces for connecting peripherals to proprietary PBXs. The TAPI and TSAPI standards for connecting CTI peripherals are open standards, but the protocols are unique to each switch, which publishes its own APIs.

Evaluation Considerations

In choosing a key system or PBX it is important that you understand exactly what you want it to do. The variety of key and hybrid systems on the market is so vast that managers must carefully evaluate their requirements before selecting a system. The differences between systems are often subtle, and differences in function and support are not apparent until you have lived with the system for several months. This makes it important to check references carefully. Considerations that are important in some applications will have no importance in others and the buyer should weigh them accordingly.

The first consideration in selecting any office phone system is to determine how many station and line or trunk ports are needed initially and to accommodate growth. The following are some general rules, but be aware that exceptions are many, and product lines are changing constantly, which may invalidate some of these distinctions:

- If more than 24 central office trunks are required, favor a PBX or a hybrid.
- If fewer than eight central office trunks are required, favor a key system unless the system will grow significantly.

- If the system will never grow beyond three or four lines and about eight stations and if PBX features are not needed, consider a KSU-less system.
- If ACD or voice processing is required, favor a PBX or a hybrid, with a PBX providing superior features.
- If half the total system traffic is intercom, favor a PBX or, depending on size, a hybrid.

Line and Trunk Interfaces

Every system must conform to the standard EIA-464 interface to a local telephone central office and must be registered with the FCC for network connection. In addition, interfaces such as these should be considered:

- PRI and BRI interfaces
- Computer–telephony integration interface
- QSIG interface
- T1/E1 interface to external trunk groups or to internal devices such as remote access servers
- IP line and trunk interfaces (Chapter 26)

A key consideration in evaluating any system is the type of terminals it supports. All systems have, at a minimum, a two-wire station interface to a standard analog DTMF telephone, either directly or through an analog adapter. Ordinary telephones are the least expensive terminals and because of the quantities of stations involved in a large PBX, inability to use standard telephones can add significantly to the cost. The standard analog telephone falls short as a user device in most offices, but it is usually preferable in residential facilities such as hospitals, hotels, and dormitories.

Proprietary telephones are the most practical way of accessing integrated key telephone features. A proprietary terminal makes some features, such as call pickup and transfer, easier to use by assigning features to buttons to avoid the switch hook flashes and special codes required with analog telephones.

These features, among others discussed in Chapter 24, should be considered in evaluating a PBX or key system terminal interface:

- Proprietary or nonproprietary telephone interface
- Number of lines and characters on the telephone set display
- Station conductor loop range (in a campus environment)
- Integrated key telephone system features
- Message waiting or nonmessage waiting analog line card
- Availability of BRI interface

Voice Mail

Voice mail evaluation is discussed in Chapter 28. Once available only in PBXs and some hybrids, voice mail is now one of the most desired features among users purchasing new key systems.

Wireless Capability

Many organizations need wireless to enable some users to roam all or part of a building. Wireless systems that support analog telephones can be used with any PBX or key system, but the ability to use proprietary telephones may be important if button access to features is required.

ISDN Compatibility

PRI capability is an essential feature for all PBXs and many hybrids. Key systems and hybrids may support BRI toward the central office. Determine whether ISDN compatibility is likely to be required within the life of the system, and if so, whether the system can be purchased or retrofitted to interface either BRI or PRI lines. If ISDN is not used, ADSI may furnish equivalent service on key systems and hybrids. BRI line interfaces are needed in many PBXs to support video conferencing.

Administrative Interface

All PBXs have some form of administrative interface through a terminal or attached PC. The ease of use of this interface differs significantly among products. The most difficult products to use have a command-driven terminal interface. At the other end of the spectrum are systems with graphical user interfaces that allow users to make point-and-click changes. Key systems and hybrids are administered by plugging a laptop computer into a serial interface or in some systems a telephone can be used for system setup.

The ease of changing classes of service and telephone numbers is an important evaluation consideration. If an easy-to-use maintenance terminal can control these, it is possible to add, remove, and move stations and change features such as restrictions without using a trained technician.

The degree to which a PBX can diagnose its own trouble and direct a technician to the source of trouble is important in controlling maintenance expense. It is also important that a system has remote diagnostic capability so the manufacturer's technical assistance center can access the system over a dial-up port.

Cost

The initial purchase price of a PBX or key system is only part of the total lifetime cost of the system. All systems require maintenance and administration, and the method of accomplishing them can be significantly different among products. As with all types of telecommunications apparatus, the failure rate and the cost of restoring failed equipment are critical and difficult to evaluate. The most effective way to evaluate them is by reviewing the experience of other users.

Installation cost is another important factor. One factor is the method of programming the station options in the processor. Some systems provide such options as toll call restriction, system speed calling, and other such features in an external database that can be uploaded to the switch.

Maintenance costs may be significant over the life of the system. The best way to evaluate maintenance cost is to request a quotation on a maintenance contract, which most vendors offer. Cost savings are possible with systems that provide internal diagnostic capability. Virtually all PBXs provide remote diagnostic capability so the vendor can diagnose the system over an ordinary telephone line. These features can offer cost savings in hybrids, but are less important in key systems.

Power Failure Protection

During power outages, a PBX or key system is inoperative unless battery backup or a UPS is provided. Some systems include emergency battery supplies, while others are inoperative until power is restored. Lacking battery backup, the system should at least maintain its system memory during power failures.

The system should include a power-failure transfer system that connects incoming lines to ordinary telephone sets so calls can be handled during power outages. The method of restarting the system after a power failure is also important because of the time required to get the system restarted. Some systems use nonvolatile memory that does not lose data when power fails. Other systems reload the database from a backup tape or disk, which results in a delay before the system can be used following restoral.

Key System Considerations

Capacity

Key telephone systems should be purchased with a view toward long-term growth in central office lines and stations. This specified size figure is the capacity of the cabinet, and further expansion may be expensive or impossible. Some systems can grow by adding another cabinet, but it also may be necessary to replace the power supply and main control module. With some systems it is possible to move major components, such as line and station cards, to a larger cabinet to increase capacity. Most key systems use plug-in circuit cards. These are less costly than wired systems, which must be purchased at their ultimate size. The number of internal or intercom call paths also should be considered.

Station Equipment Interfaces

Many key telephone systems support only a proprietary station interface so analog telephones cannot be used. The lack of a single-line interface is not a disadvantage in many applications, but some companies need to connect modems or facsimile machines to key system ports. An important feature for many users is upward compatibility of telephone sets and line and trunk cards across the manufacturer's entire product line. This capability reduces the cost of converting from a key system to a PBX and enables users to keep their instruments, which not only reduces cost, but also minimizes retraining.

Key Service Unit versus KSU-Less Systems

Some systems support from two to four lines without a KSU. For small systems these can be effective, providing many of the capabilities of a key telephone system without the need for a central unit. KSU-less systems have disadvantages, however, which make them inappropriate for many installations. First, they have limited capacity, so they are usable only for small locations and cannot grow. Second, they usually lack intercom paths, on-hook voice announcing, and other features that are essential in a multiroom office.

Centrex Compatibility

Key systems are often used behind Centrex. Many Centrex features cannot be activated unless the key system is Centrex compatible. For example, call transfer requires a switch hook flash to get second dial tone. To make a key system Centrex compatible, it must have a special button to flash the central office line. Many key systems are provided with buttons to make them directly compatible with Centrex features.

Number of Intercom Paths

A nonblocking switching network is one that provides as many links through the network as there are input and output ports. For example, one popular key system has capacity for 24 central office trunks, 61 stations, and eight intercom lines. The system provides 32 transmission paths, which support calls to and from all 24 central office trunks. The eight intercom paths limit intrasystem conversations to eight pairs of stations. A nonblocking network provides enough paths for all line and trunk ports to be connected simultaneously. In this system, if all central office trunks are connected, of the remaining 37 stations, only eight pairs can be in conversation over the intercom paths. Although this system is not nonblocking, it meets an important test of having sufficient paths to handle all central office trunks and intercom lines.

PBX Considerations

Universal-Shelf Architecture

Universal-shelf architecture permits various types of line and trunk cards to be installed in any slot. Lacking this feature, slots are dedicated to a particular type of card. It is, therefore, possible to have spare slot capacity in the PBX but have no room for cards of the desired type. Check to determine if all port cards are universal or whether some specialized types require dedicated slots.

Switch Network

A key evaluation consideration is whether the switch network is blocking or nonblocking. Blocking networks are acceptable, but may require additional administrative effort to keep them in balance. Also, consider the number of BHCAs the PBX is capable of supporting to determine if the processor limits the capacity of the system. Some manufacturers use the term "virtually nonblocking" to indicate that there are nearly as many time slots as stations.

Redundancy

Organizations that cannot tolerate PBX outages can improve reliability by purchasing redundant systems. Several levels of redundancy are available. The lowest level provides redundant processors. Higher reliability can be achieved with redundant power supplies and switching networks. Even with redundancy failures will still occur, but reliability should be much higher than with a nonredundant system.

Application Programming Interface

An open architecture interface is important for future computer–telephony applications that will be appearing in the next few years. Determine if the PBX has such an interface, and if the standards are readily available to developers. Consider that outside developers will apply the greatest amount of development effort to the most popular PBXs.

CHAPTER 26

IP Customer Premise Equipment

TDM PBXs are proprietary in all their elements from telephone instruments to operating systems. This strategy does not play well in the data communications industry where standards are set in an open environment, and proprietary protocols usually do not survive. PBXs are bowing to IP, perhaps the most open of all protocols. PBX manufacturers have been talking about the fourth-generation PBX for several years, but not until the IP PBX did the architecture depart significantly from the previous one. Despite pronouncements of its demise, however, the TDM PBX is still alive and will remain so for many years. IP PBXs now account for more than half the new line shipments and are expected to grow to half the systems in service by about 2010. TDM PBXs are on the downward slope of their life cycle, but the embedded base assures that they will survive for at least another decade and perhaps longer.

The longevity of TDM PBXs is extended by their support of IP line and trunk interfaces. This architecture, known as an IP-enabled PBX, provides an evolutionary step for its owner. Besides the IP-enabled TDM PBX, the industry produces two other architectures: IP PBX and converged IP/TDM PBX. Both PBXs and key systems are available in these product lines and their features are similar. A major departure between IP PBXs and conventional systems lies in the operating system. Most systems use a commercial microprocessor, but TDM operating systems are proprietary, whereas IP systems run on a commercial operating system—Microsoft Windows and Linux are the most popular. Also, IP systems use a server-based architecture in contrast to the cabinetized processor in a conventional PBX.

Another major point of departure is in the telephone sets. TDM systems all support multibutton display sets, but they are dumb telephones with the intelligence residing in the central processor. IP telephones are intelligent endpoints, at least to some degree. The fact that IP is an open protocol, however, does not mean that any IP phone can be used on any IP PBX with the expectation that all functions will work seamlessly. Each manufacturer chooses the features it will implement and the method of implementing them. The software resides in the telephone and the central processor and the telephone must interoperate with the processor. As with any programmable device, interoperability is possible, but it is assured only after extensive testing. This situation will probably persist until SIP is more widely deployed.

We begin this chapter by exploring the differences between the three configurations. Following that we will discuss features that are unique to IP PBXs and the reasons one model might be preferred to another. Next we will look at issues in applying them such as network requirements and security. The chapter concludes with a discussion of considerations in evaluating and applying IP PBXs.

IP PBX ARCHITECTURES

IP PBXs have three distinct architectures: IP-enabled TDM PBX, IP PBX, and converged IP/TDM PBX. Each has advantages and drawbacks and the choice depends on the application and the enterprise that is applying it. This section discusses the principal architectural differences.

IP-Enabled TDM PBX

An IP-enabled PBX is a TDM system that is modified to accept IP line and trunk cards. Each card fits into a universal slot and connects to an Ethernet port on the station or trunk side. Figure 26-1 shows this architecture. An IP-enabled PBX differs from the other two architectures in that it contains a conventional processor

FIGURE 26-1

IP-Enabled PBX

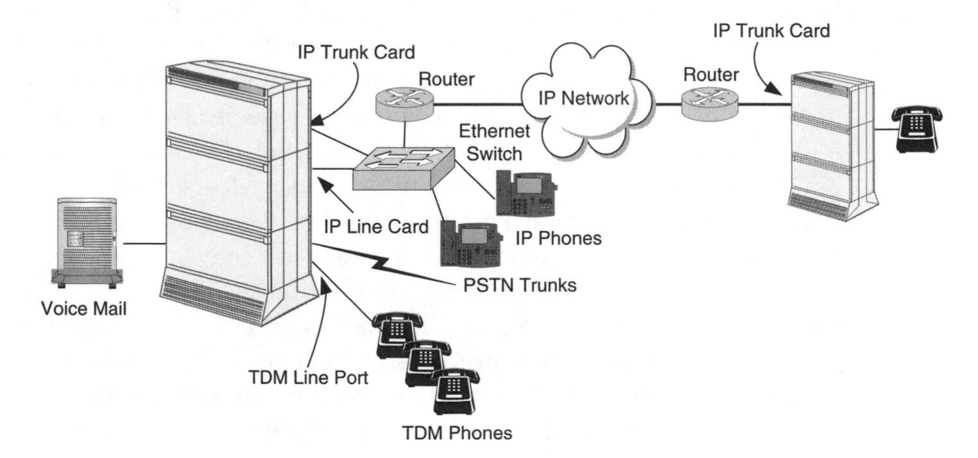

as opposed to a telephony server. This alternative is an appropriate way to get VoIP benefits while preserving the investment in existing equipment.

IP Lines

IP line cards plug into a TDM bus and connect to an Ethernet port on the station side. IP line cards include a gateway/gatekeeper function that contains the MAC and IP addresses of all IP stations. The IP addresses can be static or obtained from a DHCP server. The number of simultaneous sessions a line card can support depends on the TDM bus. Sixteen or 32 ports are typical, but most products can cascade multiple cards together to expand the group size. Each IP telephone is assigned an extension number off the PBX, and except for the IP connection, is indistinguishable from any other extension. The PBX side of the connection is channelized. An IP station-to-IP station call connects peer-to-peer through the data network without using any of the TDM capacity. An IP station-to-TDM station or trunk call is converted to TDM and switches through the TDM matrix.

When a station is on a call that requires a TDM port, the processor assigns one for the duration of the session, and on termination, returns it to the pool of available ports. This pooled operation of ports is an improvement over TDM cards, where each station is served by a single port that resides on a single card. Although the IP stations are assigned fixed extension numbers, the relationship between the extension and the physical port is severed. This effectively expands the number of ports in the switch since stations are rarely all active simultaneously. The administrator can use traffic analysis to determine port requirements based on the number of simultaneous sessions. A typical application of IP lines is for off-site workers.

IP Trunks

Figure 26-1 shows two TDM PBXs connected via IP trunks. The trunk card fits in a universal slot, and as with the TDM line card, the quantity of trunk ports depends on the bus structure of the PBX. A typical system supports eight ports per card. The gateway function is built into the card and multiple cards can feed into the same IP network. The number of simultaneous sessions is limited by the number of TDM ports available plus the IP bandwidth, which may be shared with data.

The inherent multiplexing of IP makes better use of the bandwidth than channelizing it. Not only are the packets interleaved to use available capacity but voice also can be compressed. IP trunks typically use G.729a protocol, which compresses the bandwidth to about 8 Kbps with some loss of quality. Silence suppression further reduces the bandwidth because nothing flows during silent periods, which are typical of voice since it is half-duplex by nature. This configuration is particularly advantageous in wide-area networks where the capacity of a point-to-point T1/E1 line can be at least quadrupled, and the bandwidth can support a combination of voice and data.

VoIP Remote

Some products support a VoIP remote switch unit. In this option the stations are part of the main switch and share facilities such as voice mail over an IP connection that links the remote to the host switch. Except for the IP umbilical, a VoIP remote is equivalent to a TDM remote. Survivability is available with some products. Depending on the manufacturer's design, the remote may use IP telephones or the telephones may be proprietary TDM models identical to those used on the host PBX. The primary advantage of this architecture in contrast to a TDM remote is the ability to multiplex the umbilical to support both voice and data. Users are unaware of any difference.

IP PBX

The second alternative is an all-IP PBX, which has architecture similar to a central office softswitch. Most systems support the same call control protocols as a softswitch. H.323 is the most common and complete at this writing, but most manufacturers have SIP under development. An IP PBX is a client–server system running over a robust LAN and, in some cases, extended to remote locations over a WAN. The client software runs on the client telephones and the central intelligence resides on one or more servers. The servers can be located anywhere, but logically they co-reside in the location that has the majority of stations to take advantage of centralized uninterruptible power. Figure 26-2 is representative of many IP PBX architectures. The call processing software resides on a telephony server that goes by a variety of trade names. The server hardware platform uses a standard PC bus architecture, hardened to provide reliability. The network architecture is identical to a robust LAN, which makes it ideal for large buildings and campuses because the inherently distributed nature of the architecture eliminates the need for riser cables.

In contrast to TDM PBXs, which are centralized and self-contained, IP PBXs are distributed. Features may be identical to those in a TDM PBX, but the software runs on servers and it uses a commercial operating system such as Microsoft Windows or Linux. Ethernet switches replace the central switching matrix, typically sharing the LAN between voice and data. This moves the switching closer to the endpoints and if enough backbone capacity to remote locations is available, it avoids the station range limitations that are typical of a TDM PBX. An IP PBX switches all calls as IP except when they traverse the gateway to a TDM trunk group. In place of the per-station riser cable pairs that TDM systems require, IP PBXs use a backbone, often gigabit Ethernet running over fiber, and designed to support the combined voice and data traffic load. Most products are configured to support both voice and data on a single UTP wire run. An Ethernet switch is built into the telephone set to support both voice and data. A single wire run can be used to each workstation, which can save wiring costs in a new building or one that is being rewired. As discussed later, the voice and data segments are physically combined, but logically separated with VLANs. Each workstation

FIGURE 26-2

IP PBX

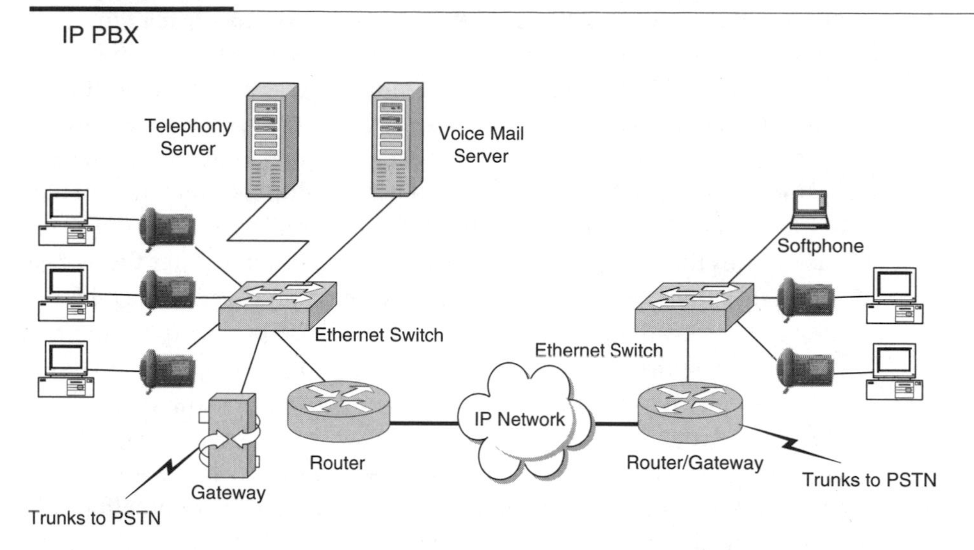

requires a dedicated switch port. The use of hubs is not recommended because of the potential of collisions and consequent delays.

Figure 26-2 shows only one of many possible configurations. In this example, PSTN trunks terminate on the gateway, but they could terminate on a switch, router, or server. Most manufacturers support T1/E1, PRI, and analog trunks. The system also can interface with IP trunks to another switch. Separate gateways may be employed as shown in the left side of the figure, or the PSTN gateway can be included as part of the router software. The telephony server, which handles call admission and control, typically includes the gatekeeper functions. In addition, other servers may include voice mail, a SIP-to-H.323 gateway, a signaling server, and an emergency call server. Omitted from the figure are the firewalls that are required if the system connects to the Internet. Many products support redundancy by running duplicate servers. The servers are typically high-quality devices running RAID disk arrays.

The IP PBX has two alternatives for remote offices. The least complex, which Figure 26-2 shows, is a router with limited call-processing capability. Local trunks are usually terminated on the router, particularly if the remote is outside the local calling area. Under normal conditions, the main server handles all calls, both station-to-station, and trunk. During network failure conditions, the router serves as a limited-function call-processing server. The second alternative is to equip the remote location with a full-function telephony server. If the two locations are equipped for synchronized processor backup, the servers can share the load, and in case of failure either server can serve all stations. Each endpoint on the network can register with multiple gatekeepers to protect against failure.

IP telephones resemble conventional TDM telephones and the features may be similar as well. Although IP is an open protocol, telephones from different manufacturers are likely to be incompatible because of the method of implementing features. SIP telephones are intended to resolve this lack of compatibility, but manufacturers may still choose to implement proprietary features to differentiate their products from their competitors. As with other LAN devices, the IP telephone contains a MAC address and can be plugged into any port in the system. It obtains its IP address through DHCP and registers with the gatekeeper, which associates the IP address with the extension number. After the system is initialized with its extension number, the gatekeeper can recognize the telephone even if it is logged in at a distant office as long as they are part of the same network and not isolated by a firewall. This portability is one of the major advantages of the IP PBX. The telephone can work from home or a location on the road, although it is not a simple matter of plug-and-play. The PBX must be protected by a firewall, and opening the firewall to VoIP may expose the network to hackers. If the bandwidth is sufficient, a remote telephone can operate over a VPN, although these are not simple to set up, and the voice quality may vary. As with cell phones, though, the convenience may offset the occasional poor quality connection.

This portability of extensions is one of the major advantages of an IP PBX and also one of its weaknesses because the telephone number is not associated with a fixed location. This can result in sending emergency vehicles to the wrong location if a user dials an emergency number on a phone that has been physically relocated to another jurisdiction. The solution is either a separate trunk group to which emergency calls are directed, or a server to identify the station and ensure that an emergency call is placed over a trunk group that identifies its physical address to the PSAP.

Analog Device Support

IP PBXs cannot support analog ports directly. Moreover, faxes and modems are intolerant of delay and since the signal is already compressed, further compression in the PBX may be intolerable. Therefore, IP PBXs need a way of supporting fax, modems, and analog telephones. Converged and IP-enabled PBXs do not have this issue because they simply assign such devices to appropriate hardware ports. For analog devices, most IP PBX manufacturers offer both an analog port server that serves multiple ports in a single location and an IP terminal adapter that supports a single device. The analog port server may require riser cable.

For fax transmission, two methods can be employed: fax relay and fax pass-through. A fax relay demodulates the T.30 fax signal at the sending gateway, packetizes it for transmission over the network, and reassembles it at the receiving end. Fax pass-through sends the fax through the codec, but only if G.711 is used with no voice activity detection (VAD) and no echo cancellation. Fax transmission does not work over low-bit-rate codecs. ITU recommends the T.38 protocol for fax transmission over IP. This protocol is discussed in more detail in Chapter 30.

Converged PBX

A converged PBX has both TDM and IP switching capability and has the characteristics of both of the other types. The logic behind this architecture lies in the fact that many organizations have classes of employees whose needs are satisfied by TDM or analog phones, so there is little point in providing IP phones. For example, analog phones are sufficient for most hospital patient rooms, student dormitories, and hotel rooms. Some such organizations may prefer IP phones for staff, but not for room phones. Moreover, IP PBXs do not handle analog phones, fax, and modems as effectively as TDM PBXs do. IP local and IXC trunks are rarely available, so TDM trunking is logically assigned to TDM ports.

Outwardly, a converged PBX looks like an IP-enabled PBX, but it has several important differences. The processor is server-based and likely runs a commercial operating system. It controls both IP and TDM stations, which are assigned according to users' needs. Unlike the IP-enabled PBX, a converged PBX does not have IP line and trunk cards. Effectively, it is separate IP and TDM PBXs connected through a gateway and controlled by a common processor. Architecturally, it is similar to central office switches that support both IP and TDM stations.

VALIDATING EXPECTATIONS

As with most new technologies, IP PBXs have been promoted with a barrage of hype. Many people are attracted to IP on the assumption that since it is newer it is better, and that an investment in the older TDM technology will have a truncated service life. Others favor the IP PBX because they understand IP, and believe that its open characteristics are superior to the closed and proprietary nature of the TDM PBX. These claims and others we will discuss in this section all contain a thread of truth, but the differences between TDM and VoIP should be examined and understood and matched with the user requirements.

First is the question of TDM obsolescence. The assumption is that manufacturers will eventually discontinue new software releases for TDM PBXs and focus their efforts on IP. That will undoubtedly happen, but probably not within the expected service life of a TDM PBX. Offsetting this is the fact that IP PBXs are immature and changing rapidly in response to problems that are discovered in practice. TDM PBXs have historically had a life of 7 to 15 years. Contrast this to servers, routers, and Ethernet switches, which have had service lives of less than half that. It is risky to conclude with any assurance that IP PBXs will have a longer service life. IP has the advantage, however, of supporting hardware expansion in smaller increments than TDM PBXs. For example, servers can be replaced without the need to replace or upgrade Ethernet switches and telephone instruments.

The open versus closed architecture issue is also subject to question. As we discuss later in the security section, IP PBXs are subject to the same security hazards of a TDM PBX, and if connected to the public Internet, add numerous risks to which TDM PBXs are immune. Moreover, the fact that the transport mechanism runs on IP does not mean the IP PBX is not proprietary. Most manufacturers use some proprietary protocols, particularly as new standards are still in process. If the feature delivery protocol used in the system is proprietary, then the telephone sets are proprietary. The hardware consists of open server platforms and Ethernet switches, which can reduce the cost of ownership, but most organizations will require the vendor to assume full responsibility for system integration, and this will most likely mean using components from the system manufacturer.

Cost Reductions

An IP PBX can be less costly than TDM, but only if the right conditions exist. The LAN infrastructure must be capable of voice support. This means the switches must be fast Ethernet, layer 2/3 capable with support for VLAN and QoS protocols such as type of service (ToS) or DiffServ, and have sufficient upstream bandwidth to support the load. The wiring infrastructure must be sufficient to support the switches, which means at least category 5 cable properly terminated. If the LAN is replaced, any expected savings from an IP PBX probably disappear, but if the LAN must be upgraded anyway, the incremental cost of supporting an IP PBX may not be significant. To make the system survivable, however, requires that every active element have backup power. Many organizations do not back up their wiring closets on the theory that the desktop computers are not backed up, so it is not worth backing up the LAN. With VoIP, these assumptions must be reexamined.

Wiring costs can be reduced for VoIP because of the need for only one wire run per workstation. Most existing buildings are already wired with at least two wire runs per station, so this advantage occurs only in new workspace. Here, the organization needs to consider whether it wants to install only one cable per station. A second run may be needed to support analog devices or another data peripheral. Since the labor cost of dual wire runs is little more than a single run, this factor may disappear on closer inspection.

Another advantage touted for IP PBXs is a reduction in long distance costs. This cost reduction depends on several assumptions: that the existing IP network has excess bandwidth, that the hardware is all capable of running compatible QoS protocols, and, perhaps, that the public Internet can be used. As we have discussed earlier, the Internet is an unreliable transport medium. Sometimes it may be good enough, but it may at times be unusable. Furthermore, until a protocol such as ENUM is approved, mapping phone numbers to IP addresses will require either the enterprise or the service provider to maintain a separate database. Moreover, as long distance costs have dropped, most companies large enough to see a significant cost reduction from using spare network capacity have already seen their long distance costs fall dramatically. The most probable place to reduce long distance costs is on international calls, but that assumes that the data network has spare capacity.

Another advantage that IP vendors claim is reduced costs from having only one network to administer and allocating maintenance costs to both voice and data. Reduced costs can definitely be achieved, but they are not automatic, and offsetting to some degree is the fact that a centralized TDM network is easier to plan and control than a decentralized VoIP network. Expertise in data network administration does not translate readily into voice, which is more than just another application. As applications are added to a network, unanticipated results often occur. Users generally have higher reliability expectations for voice than for data, which they know will probably experience unexplained anomalies from time to time. Although users tolerate data delays, they are apt to be unforgiving of voice quality degradation. Troubleshooting quality complaints is one of the most difficult issues in managing VoIP because of the transitory nature of the cause. WAN congestion is a likely cause of quality degradation, but if this is resolved by provisioning additional bandwidth, the economic justification may disappear.

Quality Monitoring and Testing

The method of testing the data network may not be sufficient for voice. Traditional WAN test equipment measures such factors as delay, bit errors, and packet loss, but good call quality does not necessarily follow, particularly with small sample size. Packet loss generally arises from network congestion and this typically happens in bursts. Codecs attempt to mask packet loss by adding noise or interpolating missing packets, which reduces voice quality. The ideal location to monitor impairments is from inside the network such as a gateway and it may be necessary to add probes to obtain better quality samples. The IP telephone also may accumulate statistics that are indicative of quality problems. The point is, maintenance and administrative costs for voice do not disappear when voice is put on the data network. Conservative analysts will assume that costs will remain the same, and hope for pleasant surprises.

One method of evaluating performance is Real-Time Control Protocol XR (RTCP XR), which provides VoIP performance metrics. RTCP XR uses probes and analyzers to measure performance. Endpoints exchange RTCP XR messages, calculating such metrics as packet loss, signal level, noise, and echo return loss. To make the protocol effective for quality measurement, it must be incorporated as an integral part of the IP phones, gateways, and other VoIP elements.

The vendor's design must be examined for potential quality problems, and after installation, the network must be tested to simulate the way users will communicate. For example, in Figure 26-1 assume a call arrives from the PSTN on the incoming trunk group and is intended for a TDM station in the other PBX. The call is converted to IP and probably compressed to traverse the IP trunk group to the other switch. When it arrives at the second switch it is converted to TDM to connect to the station, but the called station is busy, so it wraps to another TDM port, converts to IP again, and returns to the original switch, where it routes to voice mail. The voice mail connects to the main switch over TDM ports, so the call is converted again, and then is again compressed for recording on the voice mail. The caller, hearing the callee's greeting, presses 0 to escape to the coverage position, which is at the second switch. By the time the call undergoes two more conversions, the distortion, jitter, and latency will probably make it difficult to communicate; yet this arrangement works without noticeable impairment with TDM switches. IP products can be designed to avoid this circuit hairpinning with its multiple conversions. For example, the gateway at the called PBX could determine that the called station will not answer, and return a message to the host PBX to connect the call directly to voice mail, but does it? The vendor should be questioned about such potential quality problems before you approve the design.

Look for bottlenecks and other traps. Gateways are always a potential source of blockage. Traffic analysis should assist in sizing the gateway to avoid blocked calls. A potential trap lies in use of the spanning tree protocol. If it is left enabled in switches, topology changes could cause outages of about a minute while the switch databases are updated. Calls in progress would not be cut off, but endpoints would be unable to communicate with the server. Issues such as these can be avoided by developing operational procedures that consider real-time traffic.

Feature Enhancements

When the claims for VoIP have been examined and the hype appropriately filtered, significant advantages remain. When vendors discuss feature enhancements, unified messaging—which is discussed further in Chapter 29—is one of the first mentioned. This feature, which blends voice mail, e-mail, fax, and sometimes instant messaging into a package of attractive functions, has been available for several years on conventional PBXs. It is, however, easier to implement on VoIP.

For multisite organizations, the main advantage of the IP architecture is its support for distributed communications. This has the potential of combining both voice and data into a single architecture with a unified dialing plan and access to a common set of features and functions. The balkanization of today's telephone systems often separates users into islands of feature functionality. The ability to operate features such as transferring and forwarding calls and voice-mail messages across the organization is impeded by the lack of functionality or the need to understand obscure button pushes and switch hook flashes. Unification can be accomplished with TDM equipment, but without integrating voice and data, it is difficult to justify the cost in small remote offices. Other features that are rare or difficult with TDM are capable of improving productivity. This section discusses them briefly.

Enhanced message display. The improved display of VoIP sets can provide useful information. For example, all voice mail headers can be displayed so users can scroll through them and listen to messages out of sequence. Users can obtain a visual indication of message length and fast forward or replay with controls on the telephone set. Also, the telephone can display text messages so a user who is on another call can receive a message or press a button to return a preprogrammed text message to the caller.

Simplified station moves. An IP endpoint has no fixed physical relationship between the desktop device and the telephony server. Therefore, an IP telephone can be recognized anywhere in the network and system administrators do not need to be involved in relocating IP telephones and other peripherals. A user can plug an IP telephone into any jack in the same VLAN and the telephone number and all features will be automatically reassigned. This feature can save money, but many organizations will not perceive the benefit because they choose to keep control of station moves.

Remote access. VoIP supports working from remote or mobile locations. Users who have broadband connections can have their IP telephones at home or a remote office and have access to the same features and functions they have at the desk.

Presence notification. Users can broadcast their status to others in the community, notifying them of availability and their preferred means of being contacted. The system can update this dynamically from the state of the telephone and electronic calendar or it can follow static rules. Find me/follow me functionality can be implemented.

Meet-me conferencing. Users can schedule audio conferences on demand and send invite messages to participants. The conferees can advertise to the network whether they have video capability.

Access to information. Information from a database can be displayed on the telephone using an integrated thin-client browser.

Other applications include logging of personal call statistics on the telephone as well as ease of activating station features such as conferencing and forwarding. Such features are easy to implement on a desktop computer that is integrated with the phone system, but many employees such as those in health, manufacturing, retail, and education do not have a full-function desktop PC. The IP telephone, some of which even have an optional mouse and keyboard interface and a UDP port for a headset, is often an ideal way to provide information for such employees.

VoIP NETWORK REQUIREMENTS

Before installing VoIP on an existing network, it must be thoroughly evaluated for capacity and to ensure that the hardware supports the necessary protocols. To assess the load that voice adds to the network, you need to determine the number of simultaneous sessions during peak periods. As we discussed in Chapter 25, a nonblocking network assumes that all of the stations are connected simultaneously. The worst-case assumption is to use G.711 protocol, which encodes voice packets to 64 Kbps. With a payload of 20 ms of voice per packet this translates to about 80 Kbps per voice session. At this rate, fast Ethernet can handle as many as

1160 simultaneous calls on a full duplex link; gigabit Internet has ten times that capacity. If voice will overload the LAN, capacity can usually be added for minimal cost, but adding capacity to the WAN is a different matter.

To assess the WAN capacity for adding voice, you will need minimum, maximum, and average statistics for bandwidth utilization, latency, jitter, and packet loss. The network provider typically quotes average statistics, which are useless because the network must be designed to accommodate peaks. If it is not, excessive packet loss and jitter will reduce quality during congested periods and users will complain. The most effective VoIP products can assess these variables dynamically and roll over to the PSTN when the results fall outside the acceptable range. In the WAN voice can be compressed to about 8 Kbps, which, with packet overheads, translates to a network load of about 18 Kbps per session. If the routers support RTP header compression, the headers can be reduced from 40 bytes to 2 bytes. Also, determine whether the endpoints support VAD, which also goes by the name of silence suppression. VAD reduces the load on the WAN almost by half.

Table 26-1 compares the bandwidth requirements of various compression algorithms assuming no header compression. To illustrate the calculation, here is the formula using G.711:

64 Kbps × 10 ms/8 bits/octet = 80 octets of packet payload + 40 octets of header = 120 octets per packet × 100 packets per second = 12,000 octets per second/1000 × 8 = 96 Kbps per call.

The second line in the table shows how sending two 10 ms voice samples per packet, which is typically used with VoIP, reduces the load on the network. Header compression reduces the packet length from 120 octets to 82, which reduces the bandwidth requirements accordingly.

If the organization has an existing PBX, its statistics can be used to estimate the additional load that the voice will impose on the network. Most PBXs collect

TABLE 26-1

Bandwidth Requirements for VoIP

			Header Bytes					
Compression Algorithm	Bandwidth Kbps	Packet Payload (Octets)	IP	UDP	RTP	Total Packet (Octets)	PPS	Kbps/Call No. Header Compression
G.711	64	80	20	12	8	120	100	96
G.711	64	160	20	12	8	200	50	80
G.726	32	80	20	12	8	120	51	49
G.729	8	20	20	12	8	60	50	24

trunk statistics and they may provide station usage information that can be interpolated into call volume. Many PBXs can register internal calls on the CDR and these may be used to assess the load on the LAN. Fortunately, voice traffic falls into predictable patterns. Peak volumes fall outside expected boundaries only during storms and other unusual events.

Traffic Classification and Prioritization

Balancing traffic on the network can be a tricky proposition. Since real-time applications must be prioritized, if too much traffic is imposed on the network, data response time increases. The effect is particularly critical on the WAN, where the peak voice load, if not properly balanced, could impair data communications.

MAC Layer Classification

The network must have a method of classifying real-time traffic to ensure that it remains unimpaired. This classification is ideally handled by a policy management system, but policy management is largely proprietary and will remain so until standards are developed. The other alternative is to have the application itself dictate the appropriate class at the origin. A variety of methods are available to classify real-time traffic, a combination of which is appropriate. Classification starts at the endpoint, using 802.1p at the MAC level to identify voice frames. The 802.1p protocol provides a total of eight traffic classes and eight user priorities. The highest class is network control, which takes precedence because of its role in controlling the network. Voice and video are given controlled delay, which is the next highest. From there the classes descend to the lowest, background class, which is suitable for such applications as games. The standard is implemented in network interface cards in response to priority requested by the application. In order to get end-to-end QoS, all the network elements have to support the standard. The MAC layer does not provide guaranteed delivery of data units, but the nature of Ethernet in a LAN is such that the probability of data loss is minimal. The advantage of using 802.1p to classify traffic is simplicity. Layer 2 switches can read the information and forward frames with the prioritization intact, but when frames pass through a router, the Ethernet header information is stripped off.

At the network layer, IP phones must set the ToS or DiffServ priority bits in the IP headers. Layer 3 switches and routers can identify RTP packets by the protocol and by the UTP port range, which is 16383 through 32767.

Router Queue Scheduling Protocols

In the absence of a prioritization scheme, routers process packets first-come-first-served, which is not suitable for time-sensitive packets. A VoIP network must support one or more of the router queue-scheduling protocols discussed in Chapter 13: priority queuing, custom queuing, weighted fair queuing, or weighted random early detection.

Fault Tolerance

The methods of making an IP PBX fault tolerant are little different than in a TDM system: redundancy and path diversity. IP PBXs have more single-points-of-failure than TDM systems, and they are vulnerable to external attacks that cannot penetrate conventional PBXs. Therefore, as we discuss in the next section on security, a fault-tolerant PBX must have an intelligently conceived and flawlessly executed plan for defeating intruders that enter through the Internet.

Servers, gateways, routers, firewalls, and switches must be case hardened to provide reliable service. They should have redundant elements wherever feasible and cards should be hot swappable, which means the device does not have to be turned off to replace a card. Path diversity should be provided wherever economical. Figure 26-3 shows a LAN that is configured with path diversity. Each workgroup switch can reach the server farm through at least two backbone switches so that failure of any of the upstream links will have minimal effect on traffic flow. For path diversity to be effective, the links should not run more than 50 percent utilization so they have capacity to carry the entire load if necessary.

FIGURE 26-3

Path Diversity in a VoIP LAN

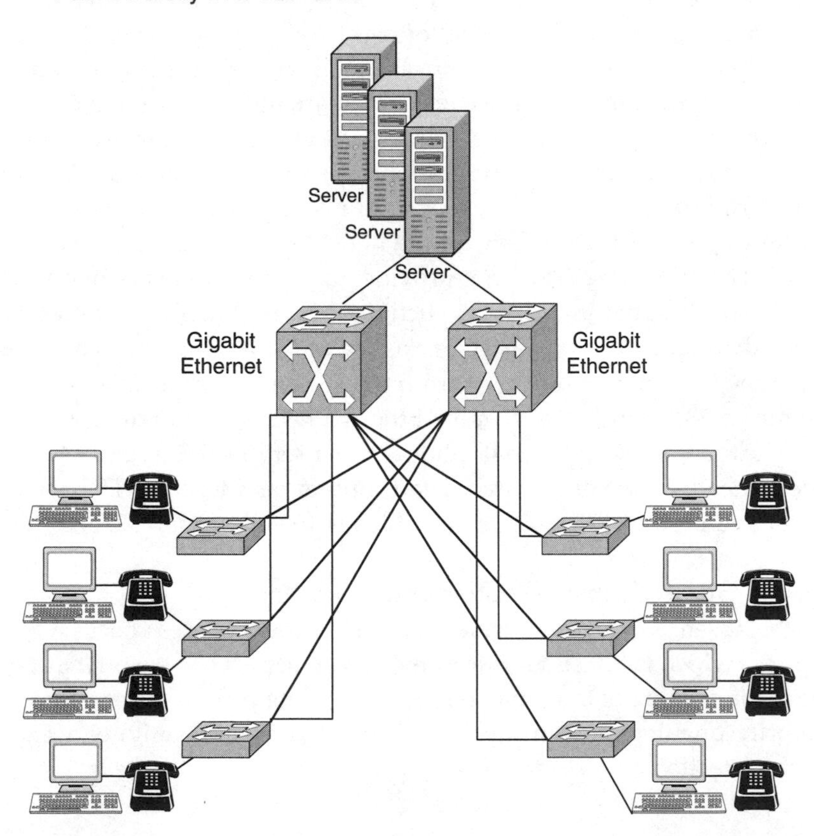

The fault-tolerant PBX must have failover resiliency. This means that an IP station can register with more than one call server. If an IP phone cannot link with its primary server, it can register with a secondary server, the address of which is programmed into the telephone.

VoIP PBX SECURITY

Telecommunications managers have long been concerned with the security of TDM PBXs. Toll fraud was rampant in the 1990s, but most managers learned how to lock down the PBX to keep intruders out. The IP PBX brings with it all of the vulnerabilities of a TDM PBX plus those that plague the Internet. The threat is easier to control if the PBX is not connected in any way to the public Internet, but once the connection is made the telephone system is exposed to the viruses, worms, Trojan horses, and denial-of-service attacks that plague the Internet. The risk is aggravated by the fact that so many individuals know how to attack IP networks. Furthermore, an IP PBX has many individual components: servers, gateways, switches, routers, and telephones that are all subject to attack. These systems run standard software such as H.323, MGCP, and SIP that have a well-known structure.

IP switches are vulnerable to the following kinds of attacks:

- *Denial of service*. The PBX is overwhelmed by a flood of requests.
- *Unauthorized access*. Unauthorized personnel may physically access servers and other devices such as firewalls, routers, and switches. Toll fraud may be committed from rogue telephones that are automatically registered.
- *IP spoofing*. Incoming traffic may masquerade as a known benign entity.
- *Operating system*. Commercial operating systems such as Windows and Linux are subject to attack.
- *Application*. The intruder exploits a weakness in the way the application is configured; for example, weak authentication.
- *Protocols*. Open protocols may have vulnerability in the way they are configured.

Physical security is the first level of defense in any network. Following that, remote-access modems and vulnerable features such as DISA, trunk-to-trunk transfer, and off-system call forwarding are either turned off or restricted to a limited class of users. Outsiders are prevented from dialing through the PBX by locking down such access doors as dialing through voice mail or transferring to a trunk access code. The so-called "social engineering" is an issue with all telephone systems. A persuasive toll thief calls the console attendant, masquerades as a company employee, and persuades him or her to place a long distance call. Low long distance rates have reduced the profitability of toll theft, but the theft will persist and constant vigilance is required to mitigate the threat.

Keeping the switch connected only to trusted networks is the best way of mitigating the risk, but that limits the effectiveness of the IP PBX for most organizations. To allow off-site users to connect to the switch, Internet connectivity is required. The solution is careful design with security paramount. A properly configured stateful firewall is essential, but to permit off-site access, holes must be opened. A stateful firewall is a packet-filtering device that maintains call-state tables. Traffic can cross the firewall only if it complies with the access control filters or if it is part of an established session in the state table. The firewall problem is complicated by NAT and the fact that VoIP has separate media and signaling paths. NAT must allow the external signaling path to connect to the signaling server, but after the call is set up, the media flows to an endpoint with a different address. Firewalls must recognize two open ports as part of the same session.

Calls between IP telephony devices use UDP port numbers higher than 16384. Opening all of the ports that endpoints might use weakens the firewall. A hacker with a port scanner could quickly find the open ports and launch a DoS attack. A voice-aware firewall dynamically opens and closes ports on a connection-by-connection basis. For maximum security, VPNs should be used for off-site locations to tunnel through the firewall. The VPN adds latency as the packets go through the encryption process, but it impedes hackers.

The voice and data portions of the network should reside on separate VLANs with the telephones behind a voice-aware firewall. Segmentation is necessary, not only for security but also for separating the packet flows for QoS. Maintaining segmentation is difficult with softphones, which by their nature run in the data segment, but must have access to both the voice and data segments through a single Ethernet port. A softphone is more vulnerable than an IP phone because it runs over an operating system that has well-known vulnerabilities to worms and viruses. Other applications such as unified messaging require access to both voice and data segments.

Not all the threats are external. The IP PBX is more vulnerable to internal eavesdropping than a TDM system. With a TDM PBX, digital lines are difficult to tap because bridging on the cable pair will usually kill the circuit. If the PBX is kept secured, the risk of anyone intercepting calls from a digital phone is low. With an IP PBX, an intruder with a packet sniffer can get access to conversations at switch points. Not only can test equipment permit eavesdropping, it can even store and replay conversations. An internal intruder who wishes to harm the company could easily do so by capturing telephone conversations of key executives and releasing them to news media or competitors. The most effective cure for eavesdropping is encryption, although many products do not provide encryption of RTP packets.

Security is a critical portion of the IP PBX design and is fully as important as the design of the network. A VoIP system vendor should be expected to take the necessary steps to make the system as invulnerable as possible. It is not reasonable, however, to expect the PBX vendor to cure security problems that already exist in the data network. If the existing network is to be used for both voice and data, a thorough security audit should be performed. In addition, the VoIP network should include the following elements to reduce the vulnerability:

- Install a voice-optimized stateful firewall that is carefully configured.
- Use host-based intrusion detection to alert management to potential attacks.
- Segregate voice and data traffic with VLANs.
- Restrict access to all networking components including servers, switches, routers, and firewalls.
- Consider encryption, at least for key personnel.
- Inspect the vendor's network design and review all steps taken to prevent attacks. Consider using one of the many tools available on the Internet to probe the network for open access.

IP PBX APPLICATION ISSUES

In this chapter we have attempted to show that VoIP has attractive aspects that can be translated into savings and productivity gains for the company that applies it intelligently. The gains that VoIP brings are not automatic. Cost reductions and personnel reductions may result from the technology, and enterprises that have inadequate systems today will doubtless be impressed with the improvement. Enterprises with well-functioning PBXs will probably see little gain until they introduce new applications.

The highest payoff from VoIP will be for companies that have a substantial number of home workers. If workers have a broadband connection and the equipment to set up a VPN tunnel to headquarters, VoIP can provide an excellent remote access vehicle. Call centers are a good application. Part-time home workers can help relieve the call peaks that affect service in many call centers, and they have the additional benefit of helping with local regulations that mandate travel reductions during peak hours.

IP PBXs will also be advantageous to distributed organizations such as banks, school districts, and government at all levels. Such organizations typically have data networks in place, but find it costly to integrate the telephone system, particularly to locations that need less than a full T1/E1 of bandwidth.

We have not attempted to discuss which of the three architectures is best because this depends completely on the using organization. A company with an adequate TDM system can gain VoIP benefits by adding IP station and trunk cards. VoIP technology can be added incrementally without a wholesale upgrade of the LAN. An IP-enabled PBX would probably not be the best choice in a new or "greenfield" application. There, the decision is whether to extend VoIP to all stations or leave a portion of them on TDM or analog. In the latter case, the converged PBX is probably the best solution, particularly if it replaces a PBX of the same manufacture and some of the existing TDM telephones can be retained.

Another major consideration is the character of the organization itself. Some organizations embrace new technology and others prefer to remain with the proven. How quickly the promised benefits can be realized depends on the organization and its implementation plan.

Evaluation Criteria

In selecting VoIP technology, the following is a sample of questions and issues that should be explored.

Technical

As the network grows, will the VoIP application scale with it? In what increments does the system grow? What is the maximum number of voice stations it supports?

Does the system have redundant call controllers? Do they load balance?

If the main controller fails, can phones rehome to the remote controller?

How long does a cold restart require?

What call control software does the system use: SIP, MGCP, H.323, proprietary?

How many telephony servers can be combined in a cluster?

Does the system employ any proprietary standards? Has interoperability been tested with other manufacturers' equipment? If so, which products?

Can cards in the media gateway be hot-swapped? Is it necessary to take the system down to upgrade it?

How does the system handle facsimile?

Does the existing LAN/WAN infrastructure support VoIP? Will it be necessary to upgrade switches, add bandwidth to access circuits, etc.?

How does the system support audio conferencing?

How many simultaneous connections does the product support to the PSTN? What reroute provisions are made?

Does the system have wide-band audio codec support?

Can the system recognize fax and modem signals and provide them dedicated bandwidth?

Does the application support emergency dialing such as E-911/999? Is the calling station ID sent to the PSAP? If a station moves to another location that is served by a different PSAP, how is the database updated?

Do the phones have a full alphanumeric keypad?

Can the telephone screens display images? Do they strip graphics content to narrow the display down to text-based information?

Do the telephones support silence suppression?

Do the telephones include a call log and missed-call indicator?

Quality

How are time-sensitive packets recognized and prioritized?

What QoS protocols does the product support?

Can IP calls rollover to PSTN trunks if the IP link is down or congested?

Is it necessary to reestablish the calls or do they reconnect automatically?

Does rollover to the PSTN occur if quality drops below limits for latency and packet loss?

Security

Does the system support RTP media encryption? What effect does encryption have on latency?

Does the system have any open TCP or UDP ports?

Is the system susceptible to DoS attacks? How are calls in progress affected? How are the phones affected—do they reset?

Administrative

What kind of statistics does the system support—jitter, latency, codec use, CPU and memory utilization, call volumes?

Can phones be reconfigured or otherwise administered via SNMP?

Does the vendor or manufacturer actively encourage application providers for VoIP applications?

What kind of diagnostic tools are required and provided for troubleshooting? What kind of statistical information does the system provide for evaluating service and calculating trunking needs?

Can all the systems in the cluster be managed simultaneously from a single management application?

Does the system support bulk transactions such as upgrading or adding or deleting a large number of phones and users?

Can it all be managed from a graphical user interface or do some of the functions have to be performed with a command line interface?

How much training is required to learn the management system?

Can the user profile be transferred to a phone by logging into that telephone?

To what degree does the system provide event and trap logging?

CHAPTER 27

Automatic Media Distribution Systems

An effective call center is essential to companies that need to maintain telephone contact with their customers and sometimes with employees and suppliers as well. The first ACDs were specialized standalone devices that served the single purpose of handling incoming calls, sending them to the first available agent, and holding callers in queue if no agents were immediately available. If the agents needed an office telephone, they often had two instruments on their desks: one for the call center and one for the office. The PBXs of that era provided uniform call distribution (UCD), which sent calls in round-robin fashion to a group of agents, but lacked functions most call centers need such as a comprehensive body of statistical reports and routing strategies that vary with factors such as agent skill, call load, and time of day.

PBX manufacturers, recognizing the need for ACD in smaller offices, began adding the capability to their systems, with each enhancement adding features and functions that emulated standalone products. As voice processing capabilities improved, call centers were augmented with voice mail, interactive voice response (IVR), and voice recognition. As these products evolved from leading edge to mainstream, PBX-based ACDs improved their functionality and surpassed standalone ACDs in market share. To a large degree this was because standalone ACDs cost about three times as much per agent seat as PBX systems, particularly where the office needed a PBX anyway. Now, the product lines of both standalone and PBX ACDs are continuing to evolve under the influence of VoIP.

The main factor driving the current evolution is the nature of the contact center. No longer are customers confined to the telephone for obtaining product information and support. Many contacts are shifting to the Internet in the form of e-mail and chat. The most effective companies do not try to channel their customer contacts; they respond to the customer's preference. As this transition occurs, the call center evolves into a multimedia contact center where the agents' expertise is available to customers communicating via voice, fax, e-mail, and interactive chat session.

Customers can easily abandon calls and shift to a competitor so managers must offer a response time that callers perceive as reasonable. When customers must wait for one medium, the most effective systems give them an alternative for getting service, or at least induce them to wait. This is where the traditional ACD falls short. ACDs are capable of handling telephone calls, but they are unaware of calls arriving by other media. The consequence of this is that the contact-handling process must be divorced from the PBX or standalone ACD and integrated into a multimedia call server. This architecture, illustrated in Figure 27-1, bears more than a coincidental resemblance to an IP PBX.

The opportunities for creativity in the contact center are unlimited, and each new wave of equipment brings innovations that make it easier to do business through application of technology. This chapter discusses media distribution equipment and features. We look first at the traditional call center that specializes in voice calls, and then at the ways it can transition to a multimedia contact center.

CALL DISTRIBUTION TECHNOLOGY

An incoming call center can be set up with ordinary key telephone equipment, which is the way smaller companies handle calls to a defined work group. Calls arrive on trunk hunting lines, and an agent pushes a button to answer the call. If incoming calls exceed the available agents, someone must interrupt a call in progress to answer it and put it on hold. If several calls are on hold, there is no easy way to know which call arrived first. Also, workload distribution depends on the actions of the agents and how effectively they are supervised. Managers need information to assess service quality and workload, and this method does not provide it. While many companies still use key equipment for distributing calls, any organization with more than a few answering positions finds that the cost of some form of machine-controlled call distribution pays for itself. Call distribution can be effective for even a single answering position if it eliminates the need for the callee to interrupt a call in progress to answer another call and put it on hold. The prevalent way of handling these calls is ACD, but the industry also offers UCD and call sequencers, which are discussed later.

An ACD has the following major components:

- Trunks
- Switching unit
- Agent positions
- Voice processing equipment
- Announcement equipment
- Supervisory and monitoring equipment
- Management software

FIGURE 27-1

Multimedia Contact Center

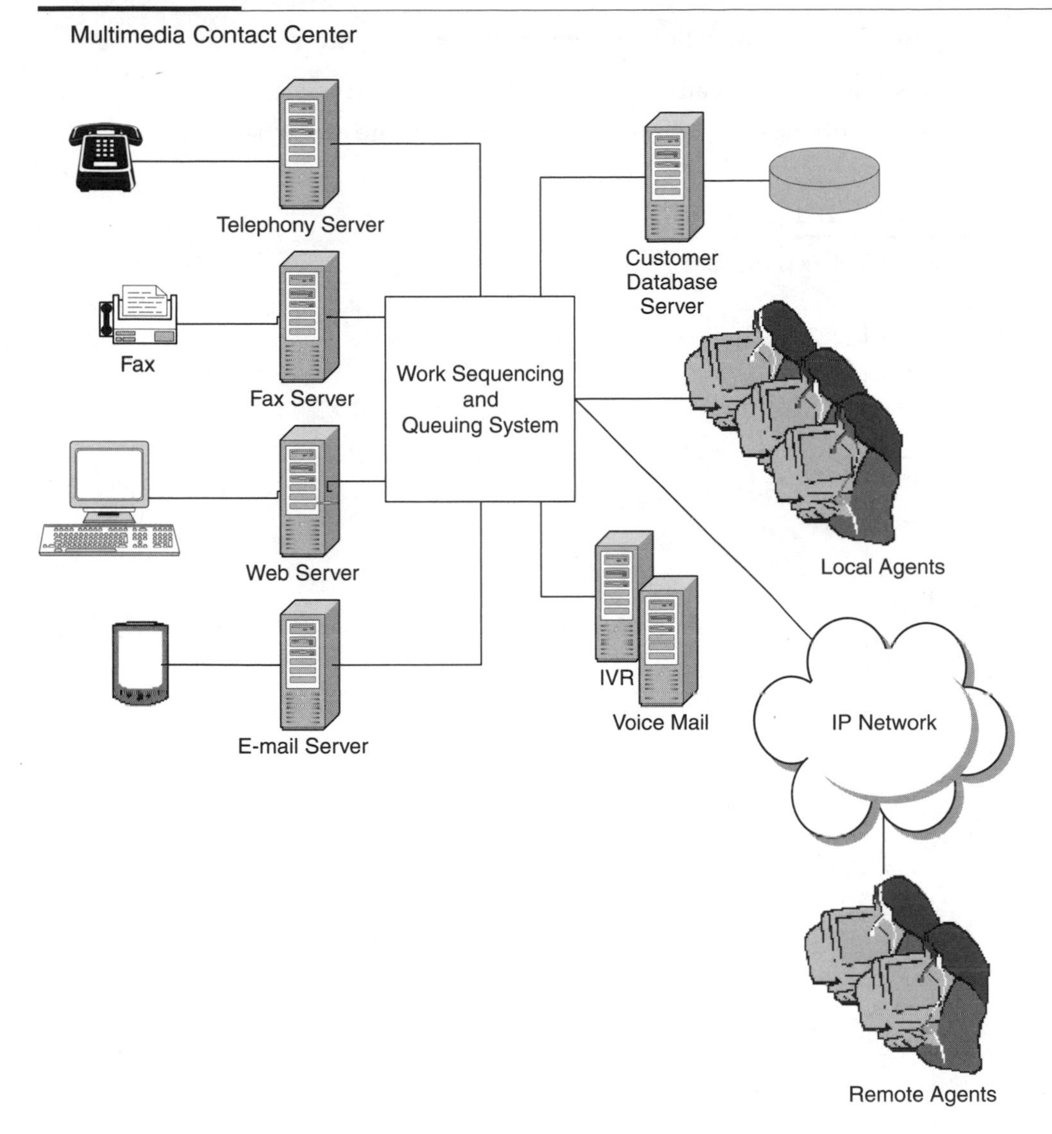

The functions of an ACD are to answer calls, and if possible to identify callers and find out what they want. If agents are available, the ACD routes the call to the appropriate position. If no agents are available, the ACD queues the calls, provides music and announcements while callers are in queue, and overflows queued calls to another queue or voice mail after a programmable interval. It collects and process calls statistics, although detailed processing and reporting is usually assigned to an outboard processor.

Call-Handling Elements

Every ACD call has the following generic process, as Figure 27-2 shows:

- Answering the call
- Identifying or determining who the callers are and what they want

FIGURE 27-2

Generic Call Flow Process

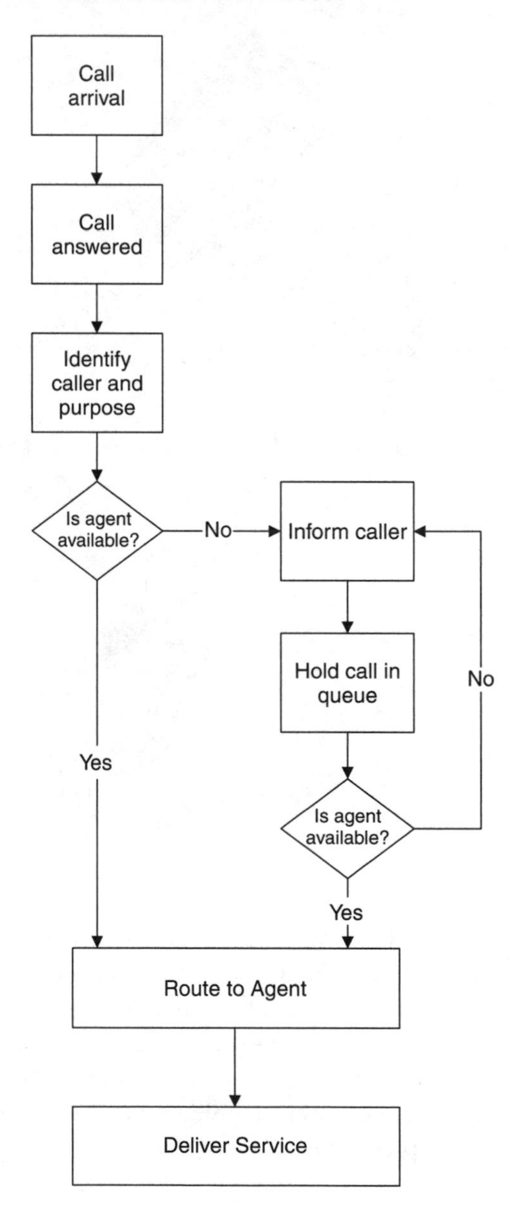

- Queuing or holding the call for an agent if none is immediately available
- Informing callers, which means providing announcements about their status and providing other information while they wait in queue
- Routing the call to the appropriate service agent when one is available
- Delivering service
- Terminating the call

The key to successful call center operation is the sequence of handling these functions. If the system identifies the purpose of the call early in the process it can handle the call more intelligently than if it is identified later. The network and the call center equipment offer several identification strategies, which *The Irwin Handbook of Telecommunications Management* covers in more detail. Briefly, technology can assist in several ways. The caller can be identified by CLID or ANI. The former applies to local trunks, and the latter to toll-free lines. Callers can also identify themselves by dialing digits, such as an account number. An IVR or a call prompting application captures the digits and uses them to route the call. Certain classes of callers can be given a special DID or toll-free number to dial, with the latter captured by DNIS.

Callers can be prompted to select the service they want through automated attendant or IVR. Many companies provide different telephone or toll-free numbers for different services in order to route calls without using automated attendant. For example, one number is sales, another customer service, another is order inquiry, and the ACD routes the call to the appropriate queue. With early caller identification, calls can be routed much more effectively. The least effective strategy is to route the call directly to an agent without prescreening. If the agent receiving the call is incapable of providing the service the call must be transferred, which contributes to added delay, customer annoyance, and extra cost. A major objective of effective call center design is to avoid call transfer.

Service Delivery

Although service delivery is the end of the call center process, considering it first can lead to some interesting alternatives. Only a human can deliver some services, and in call centers where this is the case, the objective of the process is to route the call to a qualified agent with a minimum of delay. Not incidentally, another objective is to keep the agents at a comfortable level of occupancy. (Occupancy in a call center is the percentage of time that agents are in work time, which consists of talking on calls and wrapping up after completion.)

If callers can satisfy their requests without involving an agent, they may be willing to do some of the work themselves. The IVR, which nearly every financial institution uses, is the best example of caller-directed service delivery. IVRs enable callers to read account balances and transfer funds 24 h/day. Customer service is improved and labor costs are reduced. By thinking closely about alternatives, many forms of mechanized service delivery are possible. Here are some examples:

- IVR or audiotex can deliver work schedules.
- IVRs can deliver order status or backorder confirmation.
- Outdialing equipment can remind patients of scheduled appointments.
- An audiotex bulletin board can deliver information about job openings.
- Technical service callers can be offered a trouble diagnostic tree while they wait in queue.

Callers should be offered a way of transferring to an agent after they have listened to the information available. For example, a job applicant who wants more detail should be able to press a key to transfer to a human agent.

A multimedia contact center departs from a traditional call center, which assumes that callers arrive via telephone. In a contact center the same inquiries can arrive by e-mail, fax, or a chat session over the Web. In a traditional call center, such contacts can be routed to agents based on skill level and availability. If the ACD is not integrated with the alternate form, however, agents have to remember to busy out their telephone sets so they are not interrupted with a voice call. Furthermore, a multimedia contact center can provide statistical information on all types of arriving contacts, which an ACD cannot do.

Informing

Technology can improve customer satisfaction by keeping callers informed while they wait. The most basic form is through queue announcements, which can be delivered by digital announcement systems, IVR, or even by humans who can enter the queue to provide information or entertainment. Music-on-hold is one form of information—it informs callers that they are still connected in queue. Some companies provide product information through a recorded source to callers while they wait.

Some effective call centers provide intelligent queue announcements, which inform callers of their queue position and expected wait time. An external device, usually an IVR, obtains information about queue conditions from the ACD's MIS channel, performs the necessary calculations, and reports it to the callers in conjunction with the regular queue announcement. The routing script may give the caller the opportunity to leave a callback message in voice mail. While expected-wait-time announcements are appealing to callers, they can also be confusing if the algorithm for predicting length of wait is deficient, or if the center prioritizes certain types of calls. If other calls receive higher priority, callers may hear a predicted wait time that becomes progressively longer and abandon the call.

Routing

The objective of an ACD is to connect a caller to an available agent that has the skills to fulfill the request. To do an effective matching job, it is evident that the system needs to know something about both the callers and the agents. Agent call status is readily available from the switch. Caller information is collected from a variety of sources such as caller ID, DNIS number, caller-dialed digits, speech recognition, and automated attendant menu selections.

Conditional Routing

The most flexible systems permit the administrator to write a *routing script*, or *vector*, as some manufacturers call it, to vary call routing under different conditions. Typical conditions that routing scripts can support include length of time in queue, number of agents logged on in a queue, number of waiting calls, time of day or day of week, and other such variables. Writing a vector is similar to writing a simple computer program using if/and/then logic. For example, a routing script might read:

If calls in queue >10,
And agents logged on <8,
And oldest waiting call >2 min,
Then overflow to backup queue

Skill-Based Routing

ACDs often are equipped with an automated attendant on the front end to assist in call routing and voice mail on the back end to handle overflows and give callers the option of leaving a message instead of waiting. Usually a call center is divided into several *splits* or groups. When a call arrives at the head of a queue, if agents are available the ACD routes the call directly to the most idle agent unless the caller is connected to an announcement before being routed. The greatest call-handling efficiency is gained through using larger groups.

The most effective ACDs maintain a skill database on each agent, match the caller's needs to agents' skills and availability, and route to the most idle agent with matching skills. With skill-based routing, it may be possible to keep all agents in a single large group, which can improve efficiency. The skills could be anything from language proficiency to technical knowledge. Skill-routing systems provide several levels of ability, and give the administrator tools to set up a routing algorithm. Skill-based routing software may be included in the ACD's generic program, in an IVR, or in an outboard server.

Overflow

One feature distinguishing an ACD from a UCD or a call sequencer is its ability to overflow calls to other queues. The simplest option, timed overflow, routes calls based on the amount of time the user has waited in queue. For example, a queue might be programmed to play an opening announcement, hold the call in queue for 20 s, play a second announcement, and hold the call again. The system might loop through this set of instructions until a specified amount of time or number of loops has elapsed, and then route the call to another queue.

Timed overflow is not an effective routing strategy because you may overflow the call into a queue with worse congestion than the one in which the caller is waiting. To avoid this, some ACDs provide *look-ahead routing*, in which the ACD looks at conditions in the next queue before overflowing. An ACD with *look-back routing* checks to see if congestion has improved in the original queue, and if so, returns the call. The function can be built into the routing script if it has that capability. The script above might be modified to read:

If calls in queue >10,
And agents logged on <8,
And oldest waiting call >2 min,
And if service level in backup queue ≥97,
Then overflow to backup queue
Else play announcement 3
If digits = 1, then voice mail

Service level is a percentage figure that expresses management's objectives as a percentage figure. If the management sets an objective of answering calls within 20 s, a service level of 97 means that 97 percent of the calls were answered in that time. The script above would overflow to the backup if the service level in that queue was equal to or greater than 97. Otherwise, it would route the call to voice mail and invite the caller to press 1 to leave a callback message. If the caller presses any other digit, the call returns to queue. Some ACDs queue the caller on both the original and the overflow queue and route the call to the first available agent in either.

An ACD with conditional routing can overflow based on one or more of several variables besides time in queue:

- *Number of calls waiting*: For example, a routing script might check the alternate queue and overflow if less than a specified number of calls are waiting.
- *Time of day/day of week*: Calls might overflow from a center in the east to one in the west after 5:00 p.m. and on weekends, but not during the day unless the load exceeds a threshold.
- *Caller priority*: Arrival on a priority trunk group or dialing a special number such as a frequent flyer number might entitle the caller to different treatment than callers on non-priority groups.
- *Length of oldest waiting call*: Calls might overflow only after the oldest call in queue has been waiting more than a specified number of seconds.

This retains calls in the primary queue for service, and overflows them only when the wait exceeds a threshold.

ACD FEATURES AND OPERATION

To be recognized by the system, agents must log on to a particular split or skill. In some systems agents can log onto more than one split. The act of logging on does not automatically mean the agent is available to take a call. In most ACDs the agent also must press a status key on the telephone set. The system identifies agents as being in one of several states. Typically, the software recognizes the following states and tallies statistics accordingly:

- *Available*: When an agent is available, the system can deliver the next call.
- *Busy*: The busy state indicates that the agent is currently handling a call.
- *After-call work or wrap-up*: Most ACDs permit the agent to spend a variable or fixed amount of time after each contact completing paper work before the next call arrives.
- *Unavailable*: This state is used when the agent is temporarily away from the position, typically during breaks.

The call-center supervisor usually has a terminal to monitor the status of service and load. The screen lists agents by name, shows their current state, and displays a summary of their production statistics. In some systems the display also shows how long agents have been in their current state, which is useful for determining when they may be misusing wrap-up or unavailable time. The terminal also displays information about service levels, typically registering the percentage of calls answered within the objective time. If an ACD is being used as one element of a multimedia call center, these states and the resulting statistics may be meaningless because agents may have to use the unavailable state to prevent a voice call from arriving while they work on a chat or e-mail session. Multimedia systems detect which medium the agent is working on and log all statistics appropriately.

Most ACDs also provide for service observing. Monitoring may be silent or accompanied by a tone that is audible to the agent but not the caller. Call monitoring is a sensitive issue, and company policy or state law may prohibit monitoring unless the agent is notified.

Telephone Instruments

The agent's telephone set is important to call center effectiveness. Most PBX-integrated ACDs use a specialized telephone set for the agent terminal. The set is equipped with a headset jack, and is programmed to deliver calls without the need to operate the switch hook. Multibutton digital or IP sets provide for functions such as logging in and out and changing state. Displays provide the agent with such information as caller identity, the trunk group on which the call arrived, length of time the caller waited, and number of calls currently in queue.

Outbound Calling

Many ACDs (also standalone predictive dialers) are equipped for outbound calling. A system with outbound capability dials numbers from a database and connects the called party to an agent only after dialing is complete. Some systems have answer detection capability and connect to an agent only after the called party answers. Although answering machines and voice mail sometimes fool such systems, most systems detect busy and unanswered calls and store the numbers for later retry.

If dialers wait for an agent to be idle before initiating the call, the productivity of the agents will suffer because of call setup time and because many lines are busy or do not answer. *Predictive dialers* place calls in advance of agent availability, using statistical predictions of when the next agent will become available. If an agent is not available within seconds of the time the called party answers, the callee perceives the session as a nuisance call and hangs up.

ACD Statistical Reporting

A major part of an ACD's value lies in its reporting capabilities. ACD reports are either part of the generic program of the switch, or they are produced in an outboard computer that attaches to the ACD's MIS channel. Reports produced in the ACD are not user programmable, and are therefore less flexible than those produced in an outboard computer. Reports produced by most systems fall into the following categories:

- *Agent reports*: These reports provide statistics for individual agents. Included may be amount of time logged on, number of calls handled, average talking and wrap-up time per call, number and duration of outgoing calls, hold time on calls, and amount of time in an unavailable state. Supervisors can use agent reports to determine how agents perform with respect to each other, group averages, or objectives.
- *Group reports*: Groups, splits, or queues as they are variously known, are identifiable workgroups that are assigned a particular function. Group reports show group totals and averages for the variables discussed under agent reports. They also provide information that does not apply to individual agents such as average speed of answer, numbers of abandoned calls, average and maximum length of time to abandon, and length of the longest waiting call. Group and agent statistics together give administrators insight into service levels and information to calculate staff requirements.

- *Skillset reports*: Systems equipped with skill-based routing normally report results for the skillset, which may include agents that are members of multiple groups. These reports could enable a manager to determine service levels for a specific customer group that has been defined as a skill such as service to callers who speak a particular language.
- *Trunk reports*: Most large call centers have several trunk groups on which the ACD collects statistics. Reports such as call volume and length, number of abandons, trunk-group usage, and so on permit administrators to determine whether service or cost trends are patterned to a specific trunk group.
- *Routing script reports*: These reports show information such as the stage in the routing script at which callers abandon. These statistics can be used to evaluate the scripts' effectiveness. A high rate of abandonment at a particular point may show that the script is confusing.
- *Real-time reporting*: ACDs provide MIS channels that enable supervisors to see queue status in real time on a desktop computer. Group statistics show such variables as the number of calls in queue, the length of average and longest wait, number of abandoned calls, and other such variables during the current reporting period (usually 15 min to 1 h). Agent reports show for each agent the current status (talking, available, unavailable, wrap-up), how long the agent has been in that status, number of calls handled, and number of outgoing calls during the reporting period.

Most supervisors' displays show alerts in different colors. For example, thresholds could be set for length of time in an unavailable mode, which is normally used for relief periods. The name of an agent that was within the prescribed interval would display in green, one that was a few minutes over would be yellow, and one that had overstayed the upper threshold would be red. Color displays aid the supervisor in monitoring staff and costs with a quick glance.

Networked Call Centers

In a company with multiple call centers, efficiency can be achieved with networking—often enough to repay the cost of facilities between the centers. If ACDs are connected with IP, PRI, or a proprietary networking interface, they can exchange load and workforce information across the D channel, switching calls across the network to available agents. An ACD experiencing an overload can check other centers, reserve an idle agent, and pass the call across the network.

The major IXCs offer a call allocation service for distributing calls to ACDs that are not fully networked. In the static call allocation model, the administrator informs the IXC's SCP of the percentage of calls to allocate to each center. With dynamic call allocation, the ACDs are networked together with a signaling channel, or they use the PSTN to obtain status information and communicate it over a D channel to the IXC. With call-by-call allocation, the ACD informs the SCP on a real-time basis of which center should receive the next call.

VoIP IN THE CONTACT CENTER

As the call center evolves into a multimedia contact center, the ACD loses effectiveness because it is intended for routing telephone calls and does not lend itself well to other media. A contact center needs the same call handling and control capabilities for other media, but the characteristics of other media are different. Telephone calls are handled in real time, one agent per call. Chat sessions are also handled in real time, but a proficient agent can handle two or three sessions simultaneously because of the half-duplex nature of the medium and the time it takes for the customer to respond. Customers have response time expectations for fax and e-mail contacts also, but these are not real-time media so delay expectations are different than for voice and chat. It is also much easier to route fax and e-mails to specialists. Most companies auto-respond to e-mail to assure the sender that it was received, but good customer service requires handling the request expeditiously.

Contacts can be handled through a multimedia server that recognizes the type of medium and routes it to the available agents with the appropriate skills. A TDM switch may be able to handle this through its computer-telephony integration (CTI) interface. Whether this can provide full multimedia capability depends on the degree of control CTI permits. The server needs to know the same information for each type of media as the ACD does for phone calls. This is where ACD falls short by itself. To handle media other than phone calls, agents must put themselves into unavailable status, but unavailable time is not counted in work-time statistics. It is used for relief periods and when time thresholds are crossed, alerts pop up on supervisors' displays.

Most ACDs provide for sub-coding unavailable time, but this is an unnecessary extra step if the server is keeping track of statistics. For example, if the server sends a chat session to an agent it shows that agent in work status and unavailable for any telephone calls until the session is complete. It may, however, monitor the nature of the session and send another chat to the same agent. The periodic and end-of-period statistics display the same information by type of media. Furthermore, the mathematical models that are used for work-force management are optimized for voice calls, but do not support mixed media.

Even if the CTI interface does provide call control suitable for a multimedia center, VoIP brings other advantages that are the result of its distributed nature. Media center agents do not have to be collocated with a central switch. Neither do they have to be located in a networked environment. They can be located in a neighborhood center, at home, or wherever the IP bandwidth is sufficient to handle the contact. In fact, sufficient bandwidth is not necessarily a firm requirement because the server can distribute non-real-time contacts such as chat to those agents who are known to have substandard connections, while sending voice contacts to local agents.

OTHER CALL DISTRIBUTION ALTERNATIVES

Although the ACD is an appropriate vehicle for most call centers, other call distribution alternatives are available. VoIP enables service providers to offer hosted ACD, one version of which is central office ACD. Most ILECs and many CLECs offer CO-ACD in conjunction with their Centrex service offering. This section also discusses UCD and call sequencers, which were mentioned in the introduction to this chapter. The differences between ACDs, UCDs, and sequencers are significant but not always apparent. Table 27-1 lists the most important differences.

TABLE 27-1

Comparison of Call Distribution Products

	ACD	UCD	Call Sequencer
Basis for distributing calls to agents	Based on time. Next call is sent to the least busy agent. Can route based on skills	Based on sequence. Usually top-down or circular hunt	Does not distribute calls. Alerts agents to incoming calls by colored lamps
Statistical information	Provides real-time information on supervisor's terminal plus standard and custom printed reports	Limited to line utilization and other basic reports	Provides limited reporting
Call overflow between groups	Programmable overflow based on variables such as time in queue, priority, workload, etc.	Does not overflow	Overflow is not applicable
Features	Features are virtually unlimited. Can integrate with computer for customized call treatment	Feature set is limited. Offers few enhancements	Feature set is limited, but different products provide enhancements not found in others
Telephone system integration	Furnished as standalone or part of a PBX, hybrid, or key system	Furnished as part of a PBX, hybrid, or key system	A standalone device that operates with a key system

Central Office ACD

With CO-ACD agent sets and supervisory consoles are mounted on the user's premises and extended from the central office on local loops or over IP. The user's capital investment is reduced, and the LEC takes care of maintenance. If the CO-ACD is trunk rated, the LEC applies a software restriction to limit the number of incoming trunks. If the service is not trunk rated, a CO-ACD can offer the advantage of reducing trunk costs for the user. This can, however, be a disadvantage because the user may lose some control. For example, one way of handling temporary overloads is to let incoming calls ring several times before answering and queuing them. This way, toll-free costs are reduced and the caller may be less critical of service because the length of time in queue is reduced. Another way of limiting calls is to choke down the number of trunks to return busy signal to the callers. Unless the LEC offers complete customer control of the system software, some of this flexibility will be lost.

The features of a CO-ACD are similar to the features of a standalone unit. The advantages and disadvantages compared to customer-owned equipment are essentially the same as those of Centrex versus a PBX.

Hosted VoIP ACD

Companies that want the benefits of ACD without the capital investment have another alternative that VoIP enables. In a hosted ACD, the agents reside wherever they have sufficient bandwidth to the Internet. This could be at home or on company premises. The agents connect to the host, which provides the facilities of a call center. This may include both inbound and outbound call control, IVR, automated attendant, and other voice processing features. Incoming lines terminate at the host, which checks agent availability and routes the call accordingly. To keep the calls off the public Internet, they can be sent over frame relay or a private IP network.

Uniform Call Distributor (UCD)

UCD, which is a standard feature of most PBXs and hybrids, can improve call handling. A proprietary PBX station set serves as the agent telephone. The UCD routes incoming calls to the first available position. If no agent is available, the UCD routes calls to an announcement, holds them in queue, and sends the oldest waiting call to the first available agent. In most UCD systems, the caller hears only the initial announcement and listens to silence or music-on-hold after that. UCDs relieve agents of the need to interrupt a call in progress to answer another call and put it on hold, but they lack the ability to balance workload among agents, nor do they provide more than basic management reports.

The UCD supervisory terminal is less sophisticated than with ACDs. Most UCDs use a display telephone for the supervisory terminal, so the real-time reports and agent status that an ACD provides are not available. The routing and overflowing capabilities are also less flexible with a UCD. The flexible response features discussed earlier are generally not available with a UCD. Reports are limited to what is programmed into the system and are considerably less useful than those provided by a full ACD.

In most PBXs the UCD software is a standard feature, and ordinary telephone sets can be used for the agent terminals. If the company does not have an extensive call center and does not require extensive reports, UCD is an inexpensive way to improve customer service. Also, the UCD can take advantage of features that the PBX provides such as voice mail.

Call Sequencer

Call sequencers are the least sophisticated call distributors. Unlike ACDs and UCDs, which deliver calls to idle agents, sequencers operate in conjunction with a key telephone system. They answer the call after a programmed number of rings, play an announcement, and play music while the caller waits. A display shows which call has been waiting the longest. Although customer service and productivity usually are improved by answering and holding calls automatically, there is no way to ensure that the workload is evenly distributed. Sequencers may be programmed to go into night mode automatically and follow a strategy such as answer the line, play a message, and hang up.

Sequencers are most effective when the work group has other duties besides call answering. For example, a dispatch center might receive service requests from customers, make remote tests, and dispatch service people. The only portion of the job that involves answering calls is the incoming requests, which may be only a small fraction of the total process. The sequencer is valuable because it can answer calls when all agents are occupied and indicate which call has been waiting the longest.

MEDIA DISTRIBUTION APPLICATION ISSUES

Selecting the correct call-distribution system and applying it intelligently is critical to the success of the contact center. In most applications, the arriving calls are customers or potential customers, and the organization's objective is to treat them professionally and promptly. This section discusses some aspects of selecting and evaluating a call-distribution system.

Evaluation Criteria

Most of the criteria listed for evaluating conventional and IP PBXs apply to call distribution equipment. For example, the questions of redundant processors, backup battery supply, universal port structure, cabinet capacity, and other such criteria should be evaluated along with the criteria discussed here.

Type of Contact Distributor

The first issue to resolve is whether the application involves more than just telephone calls. If the contact center is now or is evolving into a multimedia center, consider carefully whether the contact control should be outside the PBX. IP PBXs by their nature distribute call control to a server, but this is not characteristic of the PBX or standalone ACD. The following is a list of generalizations, most of which have exceptions in some product lines.

If the center handles multimedia contacts, a server-based architecture is likely to be most effective.

If distributed agents are required, VoIP will likely be the most effective way of connecting them.

If the office is served by a key telephone system, a call sequencer may be sufficient. Most hybrid key systems can be equipped with an ACD add-on.

If the incoming call load is handled by a pool of people for whom answering incoming telephone calls is only one of a list of duties, choose a call sequencer.

If the objective is handling incoming telephone calls, the office is already served by a PBX, detailed reports are not important, low cost is important, and workload equalization is unimportant, choose a UCD.

If the center is limited to handling telephone calls, detailed reports are important, and the call center must be closely integrated with non-ACD operations in the office, choose a PBX with integrated ACD.

If the center is large and needs few PBX functions, choose a standalone ACD. The decision whether to use server-based call control depends on the multimedia scope of the center.

If controlling capital investment is important, consider a hosted or CO-ACD. The latter is effective if the organization uses Centrex.

In selecting a product, it is important to match the product to the application bearing in mind that much of the improvement achieved with a call distributor may come from changing existing procedures.

System Architecture

The architecture of most call-distribution systems is similar within a particular category, but there are important differences among systems.

- *Open architecture*: CTI may be required if the call center specializes in voice. Otherwise, an open IP system should be considered.
- *Traffic capacity*: ACD agent stations are designed for a much higher level of call arrivals and port occupancy than regular PBXs. As with other switching systems, the principal evaluation criteria are expressed in terms of peak hour call completions per hour and CCS per line.

- *Type of display*: The display for both agents and supervisors can affect productivity. Supervisory displays can show critical criteria such as low service levels in red. Supervisory graphs can be displayed with the evaluation criteria highlighted in different colors. Split-screen displays permit the operator to view more than one call at a time and are particularly useful to a supervisor who monitors multiple queues. Agents can receive important information about calls waiting and other such criteria on their telephone displays or PCs. Applications are available to scroll results in a ribbon across workstation screens.
- *Networking capability*: Large organizations with multiple centers may be interested in networking, which is the ability to connect the centers with a TDM or IP network. Networked systems should pass call-specific information such as caller identification from one system to another.
- *Queuing and routing algorithms*: The most flexible systems provide granular tools for programming call flow. It is important to know how many steps can be programmed in the call routing script. It is also important to know what criteria, such as time of day, day of week, number of calls in queue, and length of oldest waiting call can be used as the test points in the script. Determine whether the system has conditional routing and if skill-based routing is available.
- *Failover resiliency*: If the system supports this feature, determine how long the lag time is before the telephone can set up a session with an alternate server. Also, determine whether the feature is automatic or requires administrator intervention to activate it.

External Interfaces

Most products require external system interfaces. For example, consider whether the system must interface with a database on a mainframe or server and how they will be interconnected. Callers are given several choices for completing their calls if the system has interfaces to voice mail, automated attendant, or an IVR. The system may require an interface to an outbound telemarketing unit, or the outbound unit may be included as a portion of the ACD.

Features

The PBX features discussed in Chapter 24 are equally important for call distributors. The following additional features should be considered in evaluating systems.

- *Overflow capability*: Consider the degree to which calls should overflow to an alternate queue. If overflowing is required, determine whether it can be based on length of time in queue, position in queue, trunk group over which the call arrived, caller identification, or other criteria.

- *Outbound calling capability*: Whether a feature of an ACD or of a standalone unit, outbound calling capability is important to companies that have telemarketing or other functions such as collections that require a volume of outgoing calls. The system should be able to interface with a user database. Other important features include computer-controlled pacing of agents, predictive dialing, some type of reliable detection of callee answer, and rescheduling of unanswered calls.
- *Web-enabled interface*: With the increasing amount of business being done across the Internet, this feature enables the caller to communicate with an agent without leaving the Web browser. Generally, three options are available:
  - Click a button on the Web page to leave a callback message. An agent calls immediately.
  - Hold a voice-only conversation with an agent using VoIP. The agent can guide the customer through a price list, take an order, etc.
  - Hold a chat session with an agent.
- *Wrap-up strategy*: Many applications require time for the agent to wrap up the call after completing the transaction. For example, it may be necessary to complete an order before taking the next call. ACDs should offer one or more ways to determine the duration of wrap-up:
  - *Manual*. The agent presses the available key to accept another call.
  - *Forced*. The system provides a set amount of time for wrap-up, and then forces the agent to available status.
  - *Programmable*. The ACD supervisor can create a program to vary the amount of wrap-up time.

The strategy for handling wrap-up can have a significant effect on productivity. With the manual method, an agent can extend wrap-up time and reduce productivity. With the forced system the machine fails to recognize the variability in the amount of wrap-up time needed for different types of transactions. If the system is programmable it can vary the amount of wrap-up time allotted for each type of transaction.

Reporting Capabilities

Most companies receive as much value from the reports a call distributor creates as they do from the call handling process. The following factors should be considered:

- *Record versus status orientation*: A record-oriented system provides call status after completion. A status-oriented system provides calling information in real time. With a record-oriented system it is possible to determine how long a caller waited in queue after the transaction is finished, but not while it is in process.

- *Data accumulation*: Systems have different periods of time during which they can accumulate statistical information. Hourly and sub-hourly information is valuable for making immediate force adjustments or changing switch parameters such as wrap-up time. Daily and monthly information is useful for load forecasting and making long-term force adjustments. Consider how the information is accumulated and displayed.

  Can information be extracted in some standard file format to analyze in a spreadsheet or database management system?
  Is information stored in a volatile medium such as RAM, and if so, is it lost in case of power failure?
  When calls overflow does the system collect information on how the call is handled in the second queue?
  If the call is transferred or routed to give the caller an opportunity to leave a voice-mail message, are the statistics continuous, or is the call treated as a fresh call and statistics counted as new?

- *Management software*: Many systems provide software for management analysis. These programs fall into three categories:
- *Scheduler*: This software, which may be combined with forecasting software, helps the supervisor prepare schedules. The program analyzes force requirements by hour of the day and day of the week and optimizes available staff to meet the requirements.
- *Tracker*: Tracking software dynamically reviews service levels based on the number of staff logged on and the volume of calls arriving. Force requirements are calculated based on service objectives. The software keeps track of absences and number of agents required per hour and calculates predicted service levels based on workload and the available force.
- *Forecaster*: Forecasting software reviews historical data and predicts call volume, workload, and force requirements.
- *Transaction audit trail*: A transaction audit trail timestamps each event in a transaction from start to finish. The system identifies the agent and position handling the call and records the times the call was transferred or put on hold. The amount of wrap-up time is registered, and any telephone numbers, such as supervisor or wrap-up position, dialed during the transaction are recorded. If the call fails for such reasons as inadequate facilities, excessively long wait, or poor queue management, this feature helps diagnose the cause.
- *Report types*: The following reports are typically produced by the more sophisticated systems:

  Percent abandoned calls
  Percent all trunks busy

Percent calls answered in X seconds
Percent all positions busy
Percent position occupancy
Longest waiting time in queue
Average number of calls in queue

- *Report production*: Consider whether the reports are produced in real time or only after the fact. Real-time reports should display on the supervisor's console while calls are in progress, although they may be refreshed at intervals of 10 s or so. Most call distributors include basic reports as part of the system's generic program. These are usually not user programmable and seldom satisfy all users' requirements. Many call distributors provide a port to an outboard processor, which is usually a personal computer. The following questions should be considered:

  Are the fixed reports provided with the system sufficient? If not, does the ACD provide for download to an outboard processor?
  Is statistical information stored for some period, or is it lost each time the report is printed?
  Can the user select the reporting period—for example, hourly, half-hourly, or quarter-hourly update?
  If agents shift between queues, do their production statistics follow them as individual agents, or do the statistics remain as part of the queue? Is it possible to track both?

Call Center Supervisor's Information

A major reason for acquiring a call distributor is to provide management with real-time information about workload and service. Some systems use a personal computer or terminal for the supervisor's console. Low-end systems may use a display telephone or audible or visual signals or provide no information at all to inform the supervisor when certain thresholds are exceeded. The following criteria should be evaluated:

What type of display is provided?

Does the display alert the supervisor to critical situations with different colors, intensity, inverse video, or other means?

If the display is variable, can the user change it to suit individual preference?

Are thresholds programmable?

How frequently is the information on the display refreshed? Can the user change the refresh interval?

Is the information display programmable or fixed?

Does the display show the supervisor how long each individual agent has been in the current state?

Can the supervisor remotely force the agent from one state to another—for example, from wrap-up to available?

Can the supervisor monitor calls without being detected?

If the system provides an audible monitoring tone to the agent, does the caller hear it also?

Queue Management

The degree of flexibility the system offers in managing the way calls are routed to particular queues and the way agents are assigned to queues is fundamental to call center management. Low-end systems may have only one fixed queue and little or no ability for the administrator to reprogram the system. High-end systems have complete flexibility in changing routing to meet changes in workload conditions. Consider the following:

Can call routing be changed? If so, is the programming language easy to understand?

Is a method provided for off-line testing of contact routing?

Can the routing be changed in real time?

Can the system queue chat sessions along with voice sessions?

If the system provides priority treatment for certain classes of callers, how is the priority recognized? How many priority classes are there?

When a call is given priority treatment, how is the priority administered? For example, is the caller moved to the head of the queue, placed at the end of the first third, and so on?

Can the system look ahead to evaluate congestion before overflowing to an alternate queue? If so, does the system predict waiting time in the next queue based on a dynamic evaluation as opposed to merely counting calls?

After the system looks ahead to an alternate queue, can it look back to the primary queue?

If a call returns to a previous queue, is it given priority treatment based on total waiting time?

Can the supervisor remotely assign agents to another queue to relieve congestion?

Can an agent log on to more than one queue?

Can the queue routing be varied by time of day, day of week, and other such variables?

Incoming Call Handling

The least-effective call distributors provide little flexibility in handling incoming calls. Calls are answered and placed in queue, and the caller may not hear anything until an agent is available. The most flexible systems offer callers a choice. These are some questions that should be evaluated:

What does the caller hear after the initial announcement: music, promotional announcements, silence?

Is a second announcement available? How many different announcements can be programmed in a routing script?

Does the system offer skill-based routing? How are skill levels identified in the ACD?

Can the system vary the announcement based on time in queue or other such variable?

Does the system inform callers of their position in queue?

Can the caller choose to exit and leave a message in voice mail? If so, does caller retain the original queue position?

Can announcements on hold be varied by the queue, or are they common across the entire system?

CHAPTER 28

Voice-Processing Systems

Many people have a love–hate relationship with voice processing. Despised by many, voice mail and the auto attendant are mainstream products, even in the smallest of offices. We love the convenience of retrieving voice messages when we are away from the office but rail at the frustration of landing in voice mail when we are in a hurry to reach someone. We endure the endless prompts of badly designed automated attendants, but enjoy the convenience of checking our account balance 24 h a day. Voice-processing systems are here to stay; yet they are the most misused of telecommunications technologies. Properly designed and applied, voice-processing systems can save a great deal of time and relieve phone tag by making their users accessible while they are away from the phone, but improperly used systems generate phone tag and frustrate customers.

Whatever antipathy people may feel toward voice processing has little to do with the devices, but with the way their designers and users apply them. The fundamental principle, frequently violated in practice, is to use voice technologies to make it easier for others to communicate with you, not as a way to avoid communication. Regular updates to personal greetings can convey information about how you can be reached in contrast to "I'm either on the phone or away from the desk," which explains nothing. SIP proponents plan to enhance accessibility by using servers that can be updated with Web browsers to enable people to advertise their presence. It will be interesting to see whether SIP makes a beneficial change in those who use technology to avoid disclosing their presence.

In the context in which we use it in this chapter, voice processing includes the following technologies:

- voice mail,
- automated attendant,
- interactive voice response (IVR),

- digital recording,
- speech recognition.

These technologies are listed in order of their maturity. Voice mail has become such a mainstream application that few offices do not have it. Automated attendant, a close cousin of voice mail, enables callers to route themselves by selecting from a menu.

IVR carries the process a step further by enabling callers to converse with a computer by using the DTMF pad on their telephones. As anyone who has designed or used such a system knows, however, the triple-letter key combination of a DTMF pad is limiting. One answer is speech recognition, which allows the users to speak commands instead of using the telephone keypad. Speech recognition is immature, but it is steadily improving. Digital recording is required in many organizations that take orders of different kinds over the telephone. Products in this category compress, record, and index the voice stream, so a session can be retrieved for verification of what the parties said.

Even given the occasional misapplication, voice-processing technologies have enormous potential for improving customer service and cutting costs. Voice mail makes it possible to leave messages across time zones, keeping voice inflections intact, and avoiding the misunderstandings that so often result from passing messages through a third party. IVR makes it possible for people to retrieve information outside normal working hours and without enduring the delays that often occur during peaks. The organization saves labor costs, employees are relieved of the mind-numbing task of repeatedly delivering the same information, and customers do not have to wait.

The automated attendant enables callers to route their own calls quickly to the appropriate department without lengthy oral exchanges and enables the company to eliminate the cost of handling the call manually. Everyone benefits from voice processing if the applications are chosen carefully and administered intelligently.

This chapter discusses the elements of the five principal voice-processing technologies and describes the features available. The applications sections discuss typical uses and precautions to observe in applying them. The chapter also covers digital announcers. Although these are not technically voice-processing devices, they have many of the same elements and are used behind PBXs and ACDs to deliver queue announcements to callers.

VOICE MAIL

Voice mail caught on slowly at first, but is now almost universally accepted and the market offers so many alternatives that it is practically a commodity. The feature packages are so similar that it is difficult for manufacturers to differentiate their products. The switch manufacturers generally do a better job of product integration than third-party developers, but as voice mail migrates from proprietary systems to servers, the proprietary aspects diminish.

Voice mail has the potential for improving internal communications in most organizations. A high percentage of telephone calls are uncompleted because the called party is away from the desk or on the telephone, and the frustrating game of telephone tag begins. Properly used, voice mail enables people to communicate asynchronously by exchanging voice-mail messages. Calls can be exchanged across time zones, messages can be left and retrieved quickly when only a few minutes are available between meetings or flights, and the group broadcast feature enables a manager to convey information to everyone in the workgroup with a single call.

A voice-mail system is a specialized computer application that compresses and stores messages and personal greetings on a hard disk that also holds the prompts. A processor controls the storing, retrieving, forwarding, and purging of files. Voice-mail systems connect directly to a PBX bus or through station ports. T1/E1 connections are common for large systems. Increasingly, the voice-mail application is migrating to servers, which may communicate with either a VoIP host or an IP-enabled computer by exchanging control messages across an IP network.

Voice-mail service is available from service bureaus or behind a PBX or key system. The products and features are similar except for the scale and the method of identifying the called number. Service bureau systems are either stand-alone or integrated with a central office switch. Private voice-mail systems are integrated with the PBX or key system. Figure 28-1 shows a Nortel Call Pilot system that is integrated with an SL-100 PBX.

FIGURE 28-1

Nortel Call Pilot Voice-Mail System (Photo by Author)

Voice-Mail Integration

PBX-integrated voice-mail systems provide features that enhance their value. To achieve integration, the PBX and voice-mail system must be designed to interoperate. A fully integrated voice-mail system offers several features that require direct communication between the voice mail and switching system:

- *Return to attendant*. A caller, on reaching voice mail, can escape to a message center or switchboard operator by dialing 0.
- *Multiple greetings*. Voice-mail users may have different greetings for internal versus external calls or a different greeting when the phone is busy than when it is not answered. With time-of-day control the called party can vary call coverage and personal assistance options.
- *Called-party recognition*. If the called party does not answer and the call forwards to an extension that is covered to voice mail, the original called-party's greeting is heard.
- *Message-waiting indication*. When a message arrives in a user's mailbox the system activates a message-waiting light or applies stutter dial tone to remind users to retrieve messages.
- *Alphanumeric prompts*. The voice-mail system may send alphanumeric prompts to display phones.
- *Security*. The telephone system and voice mail may interact to prevent would-be hackers from dialing invalid extension numbers to place fraudulent calls.
- *Caller identification*. The name of the person calling from within the system or the number of an external caller to the system equipped for caller ID is inserted into the message header.
- *Message forwarding*. The user can forward a voice-mail message to another station with or without comment.

The first of these features allows the caller to escape from the voice-mail system and leave a message with a personal attendant, such as a departmental secretary. Lacking this feature, a caller can reach only the PBX attendant. The second feature, multiple greetings, conveys valuable information to the caller and allows the callee to treat callers more personally. If a caller hears a greeting that states the called party is on the telephone, the caller's action may be much different than if he or she hears a message stating that the callee is out of the office. Furthermore, many companies prefer as a matter of policy that an attendant answers external calls, restricting automatic voice-mail answering to internal calls.

The third feature, called-station recognition, is a major flaw of poorly integrated systems. With full integration, if calls cover to voice mail, callers always hear the greeting of the party they originally called, even if the call forwards to another station that is also forwarded to voice mail. With some systems, the caller hears the greeting of someone other than the callee, which is confusing.

Message-waiting light illumination greatly affects voice mail's utility to the station user. With an electronic telephone that displays the light on a feature button, the user may need to do nothing more than press the button to retrieve messages. Without the feature, the user must periodically call voice mail to check for waiting messages, which results in delays in returning calls. For users who are rarely in the office, message-waiting light illumination is of little or no value, but for users who are frequently in the office, it is an essential feature. Some systems substitute a short burst of stutter dial tone as an alternative, which is useful for off-premise stations or telephones that lack a message-waiting lamp.

Alphanumeric prompts on display telephones are easier to use than the audible prompts that many systems provide. Users can press buttons to replay, forward, delete, or save messages. Security is a matter of great concern to all companies that have a voice mail. If the proper restrictions have not been installed, toll thieves transfer through voice mail to an outside line at the expense of the company. Integrated voice-mail systems exchange information with the PBX processor to prevent such transfers. Internal caller identification inserts the calling party's name into the message header. The user can listen to headers and choose the most important ones to handle first.

Developers use three principal methods of achieving integration. The first is through a direct datalink from the voice-mail processor to the PBX processor. The PBX manufacturer may have a proprietary interface that it uses for its voice-mail product and may license it to others. Many PBXs also support SMDI, which exchanges messages with the voice-mail system indicating the callee's identity, the calling number, and the message desk terminal to which the incoming call was delivered.

A second alternative is through emulation of a display telephone. The PBX sends the calling number over the data channel from the switch to the telephone. The third method is integration through a CTI interface. The first method is faster and more efficient than the second, but it may be proprietary to the PBX manufacturer. Third parties can reverse-engineer a display set to integrate voice mail with another system, but the method is slower than the datalink method. Furthermore, some features are available only through processor integration. CTI integration offers all the advantages of processor integration. Its main drawback is the expense of the PBX's CTI interface. The extra cost may be considerable unless the interface is needed for other purposes. With this configuration, the voice-mail application resides in a server. The same server may also include IVR functions and connect to the PBX over a LAN.

Voice-Mail Features

The following is a list of the most important features in voice-mail systems in addition to those listed above as integration features.

- *Alerting*. The message header is transmitted to a pager. The service is intended for use with short message service (SMS).
- *Audiotex*. A "voice bulletin board" feature that permits the caller to choose from a menu of announcements.
- *Automatic purge*. System automatically purges over-age messages to recover storage space.
- *Broadcast*. User with this feature enabled in the class-of-service can send a group message to all box holders in the system.
- *Call recording*. The user can press a button to record a conversation on voice mail. A short announcement informs users that the call is being recorded.
- *Class-of-service*. Defines how a user may operate or interface with the system including such variables as the amount of storage allocated and the length of time messages can be stored before they are automatically deleted.
- *Disk usage report*. System informs administrator of the percentage usage of the fixed disk storage unit.
- *Distribution lists*. Permit users to establish lists of mailbox holders who receive messages when the appropriate code is dialed.
- *External device activation*. The voice mail can be programmed to activate external relays, e.g., turning on the heat in the office.
- *Forms*. Callers can be prompted to fill out a form such as registrant for an event in voice mail.
- *Future delivery*. Allows you to postdate the delivery of a voice-mail message. Useful for sending reminders of things to do in the future.
- *Guest mailbox*. Permits the system administrator to establish mailboxes for customers and temporary users.
- *Individual user profile*. Permits individual users or classes of service to establish variables such as length of greeting, coverage for return to operator feature, and length of message retention.
- *Mailbox full warning*. Informs the user when the mailbox is close to reaching capacity.
- *Message priority*. Permits callers to designate a message as high priority. System reads out high-priority messages first.
- *Off-premise message notification*. The system can be programmed to try a list of telephone numbers to deliver a voice-mail message. If the message is not delivered, it notifies the users of the attempts and the nondelivery of the message.
- *Out-calling*. System can call subscribers at a predetermined telephone number, cellular telephone, or pager to notify them that they have received a message.
- *Undelete*. Deleted voice-mail messages can be restored to the voice mailbox.

Unified Messaging

One deficiency with voice mail is the fact that messages are stored and retrieved in the same sequence in which they arrived. Some systems allow the user to listen to the headers and play the most important messages first, but the process is cumbersome compared to e-mail in which the computer displays all messages. This makes it easy to skip through messages and save, delete, and forward them in any sequence. Unified messaging brings these benefits, which are inherent to e-mail, voice mail, and fax. A unified-messaging client runs on the desktop computer and links to the voice-mail server over the office network. It enables the user to handle voice mail, e-mail, and fax messages on the computer screen with no regard to the device that created them or will retrieve them. Using keyboard and mouse the user can replay, save, delete, forward, and otherwise handle and dispose off all types of messages.

Manufacturers have been touting this technology for several years, but it has not quite caught on. The main problem has been return on investment. Unified-messaging products are purchased with hard dollars and, for the most part, are justified with soft dollars. Managers find it difficult to demonstrate that unified messaging improves productivity for most workers. For workers who stay at their desks most of the day, voice mail without unified messaging is quite adequate. The commands for replaying, saving, deleting, responding to, and forwarding messages may not be intuitive, but once they are learned, the benefits of handling them on a PC screen instead of the telephone are, at best, marginal.

Traveling and mobile people, on the other hand, can make good use of unified messaging. Voice messages and faxes can be attached to e-mail and retrieved from a laptop computer. A voice synthesizer can read out file attachments, including fax-to-text conversion, if necessary. Eventually, speech-to-text technology will reach a point that voice mail can be converted to e-mail, but for now it is easy enough to forward voice mails across a network as a file attachment. Speech recognition also plays a part in unified messaging by enabling users to speak commands instead of using the telephone keypad. This relieves users of the need to remember which keys to press, but it may not work perfectly from a cell phone in a marginal coverage area or over a compressed VoIP circuit.

Eventually, unified messaging will become a standard feature of voice mail. The main inhibitor today is its cost. It is a proprietary application and, although it is mostly software, it has not reached the point that mass production has brought the cost within easy reach.

Networked Voice Mail

A large multisite operation can often bring its employees closer together with networked voice mail. If PBXs are of the same manufacture and are networked together, one PBX can host the voice mail and others can share it. Shared voice mail requires networking software in the PBXs. If the products are of different manufacture and if both are compatible with QSIG, this may offer a viable alternative for shared voice mail.

A second method is networking the voice-mail systems themselves as shown in Figure 28-2. This method is effective when each site has many users. Messages are stored on the local voice-mail system, which reduces the network traffic compared to a shared voice-mail system. Networked voice mail permits users to have a mailbox in a distant PBX and have the messages delivered to their own voice mailboxes. For example, a sales manager could provide local telephone numbers for key clients in several different cities. Managers with staffs in more than one location can maintain distribution lists and forward calls across the network.

Networked voice-mail systems use point-to-point lines, an IP or frame relay network, or dial-up to transport messages between locations. Messages can also travel across the Internet. Since the message is being sent as a file, the delay inherent in the Internet is not an important issue, but packet loss will degrade usability. Systems from different manufacturers can be linked through a standard known as audio message interchange service (AMIS). AMIS standards exist for both analog and digital interchanges. The latter is the most effective because the digitized messages can be forwarded across the network without being converted to analog from the disk and then converted back to digital again for storage at the user's location. Note that the AMIS interface must be provided separately for each pair of PBXs, so universal networking requires both products to support the standard.

FIGURE 28-2

Networked Voice-Mail Systems

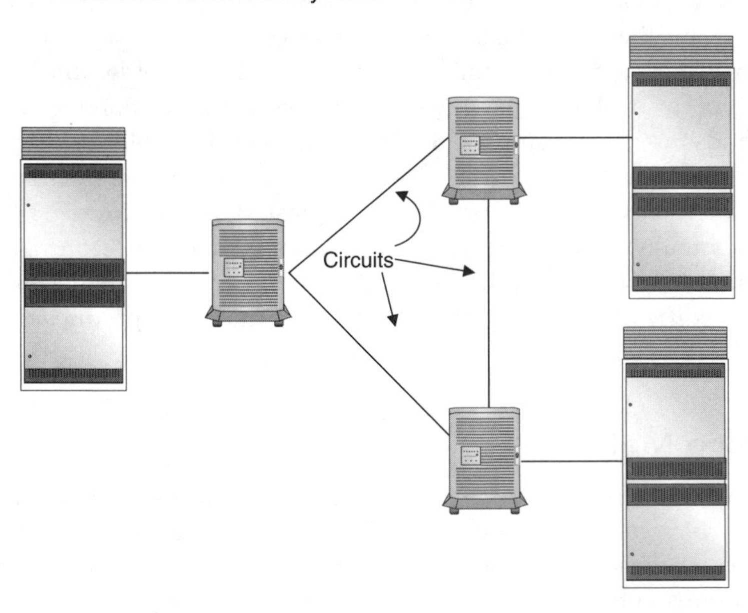

Stand-Alone Voice Mail

In a stand-alone configuration, users dial into the voice mail over the PSTN. Many service bureaus and almost all LECs and cellular providers offer voice mail to outside callers. Service bureau voice-mail systems normally have large capacity—often sized in tens of thousands of mailboxes. If the voice mail is integrated with the central office switch it can cover a busy or unanswered line. Many service bureaus do not associate the voice mailbox with a specific telephone line. The user is given a telephone number that connects the caller into the box directly. In this respect it is little different than an answering machine.

Service bureau voice-mail systems are effective for users who need a few mailboxes and cannot justify the purchase of a private system. They are also effective for users who are out of the office a great deal and have no need for message-waiting lights to indicate call arrival. The lack of message-waiting lights is one of the chief drawbacks of stand-alone systems for people who are regularly in the office, but most systems can send a message to a pager or send an e-mail to signal the arrival of a message. A second drawback compared to integrated voice mail is the dead-end nature of the medium. When a call enters the mailbox, the caller's only options are to leave a message, hang up, or possibly send a message to a pager.

The earliest service bureaus were exclusive to voice mail, but many now offer unified-messaging features such as reading e-mail over the telephone with text-to-speech and forwarding voice-mail messages to an e-mail address as a file attachment.

AUTOMATED ATTENDANT

The automated attendant is the most misused and maligned of the voice-processing technologies. The system answers the telephone and offers the caller a menu of choices such as "dial 1 for sales, 2 for service, 3 for engineering," and so on. A properly designed system can save time for callers, but systems with lengthy prompts and endless menus frustrate callers. The automated attendant is usually a feature of a voice-mail system, but the two are separable. IVRs can also be programmed to support automated attendant functions.

Many companies prefer not to have their customer calls answered by an automated attendant. Different numbers or separate trunk groups can be configured so that the automated attendant answers employee calls, while those to the main number go to the operator. In most systems, pressing 0 returns the caller to the attendant. Most systems allow callers to bypass the prompts and route directly through to the desired function, which works well if the caller knows the prompts.

Automated attendant can substitute for DID. The system prompts the caller to enter the extension number if known, dial by name, or wait for an operator. This is an effective tool if the caller knows the extension number or the name of the person they are calling. If not, an operator must be available to transfer the call.

Name dialing is available on most automated attendants and IVRs. The caller dials enough digits of the last name to satisfy any ambiguity. The system responds with the called-party's name, often in the person's own voice. Large companies are likely to have several people with the same last name. The system may prompt for the first name or offer a menu of choices.

The second use of automated attendant is to enable callers to route their own calls to an answering position by dialing a code from a menu of choices. The system manager must take care in creating the menu, because it is here that callers experience frustration. If too many choices at a menu level or too many menu levels are offered, callers will be unable to remember them and have to start over or return to the attendant. A good policy is to permit no more than four or five choices at each menu level and no more than two or three menu levels. The system should be set up to allow callers who know the menu to dial the code immediately without waiting for the prompt.

INTERACTIVE VOICE RESPONSE

IVR prompts callers to enter information such as an account number via DTMF dial and passes it to a server. The server processes the request and passes it back to the IVR, which reads the information to the caller. Banks and credit unions use IVR to enable customers to obtain their account balances without waiting for an agent. In addition to financial institutions, IVR can support many applications where telephone service representatives receive simple requests from callers, key them into a terminal, and read the results from the screen. Examples are informing customers of backorders and prompting them to confirm or cancel their order, registering people for events, and delivering information such as status of a pending public assistance request. Properly designed, the application can save time for both caller and the company.

A typical session involves customer identification by account number and password as the first step, after which the IVR presents a menu of choices. Within a range of choices callers can send instructions to the computer or query the database. The IVR responds with audio announcements, confirming the transaction or delivering the information to the caller. One of the earliest applications was LECs' directory assistance. After receiving a request from the customer and looking up the number, the operator transfers the call to an announcement unit, which completes the transaction. LECs' intercept services are another example. When the caller reaches an intercepted number the system responds by announcing that the old number is disconnected and is reading the new.

IVR voice announcement technology falls into two categories. The most versatile method is voice synthesis, in which the characters coming from the computer are formed into words by a device that emulates the human larynx. The second form is a series of words stored in the memory or on the disk, which are triggered into the voice stream as needed. The synthesizer method is understandable, but the voice may sound accented. The stored voice method sounds natural, but the vocabulary is limited to what is stored, whereas a voice synthesizer can pronounce nearly every word.

IVR should be considered as more than just a voice-response unit. If Web access is available, that is usually a more effective way to deliver information, but for callers without Web access, an IVR is an alternative. Here are examples of IVR applications:

- *Automated order entry and status inquiry.* Callers can access databases to place an order from a catalog, check availability and price, and check the status of pending orders.
- *Status verification.* Callers can verify various types of information such as credit cards, employment, insurance, and other such information that can be delivered after authenticating the caller. Frequent fliers, e.g., can enter their account number to determine their current award status.
- *Time reporting and employee information.* Field personnel can report time to a system that tracks attendance, provides information on accumulated leave, and other such employee information.
- *Dealer locator service.* The IVR can prompt callers to enter their ZIP code for the name of the nearest dealer.
- *Fax-on-demand.* The voice-response unit can lead callers through a menu of document choices. When the document is selected, the system prompts for a fax number and sends the fax immediately.
- *Student registration.* Colleges and universities use IVR to enable students to register for classes from a DTMF telephone. Callers can use IVR to register for seminars and other events.
- *Appointment rescheduling.* Clinic patients can reschedule their own appointments by calling into an IVR and identifying themselves by patient number. Clinics can also use the outdialing capability to remind patients of appointments.
- *Frequently asked questions.* Support organizations can forward callers to an IVR to listen to the answer to questions. For example, a bar association lists attorney specialties and a hospital can offer self-help information to callers.

IVRs can be stand-alone or they can connect to a PBX over analog or digital ports. For more versatility they can connect through a CTI port. The CTI connection is important when callers transfer from the IVR into a call center to complete additional transactions. The IVR should forward enough information that the agent does not need to ask the caller for his or her account number because it is already known. In addition, if the agent sees the nature of the caller's activity in the IVR, it offers clues about the reason for the call. With an integrated IVR, callers can even be routed in ways that they least expect. For example, a caller with a delinquent account might be routed to the collections department when the account is accessed through the IVR.

DIGITAL RECORDING

Many companies must record calls as a way of confirming later what the parties said during the session. Stockbrokers, hospitals that offer advice over the telephone, catalog companies, and other such organizations sometimes need to retrieve the details of a call to resolve disputes. Digital recorders provide an array of voice channels that are digitized, compressed, and stored on a disk. Usually, the disk has removable media for archiving the calls. The products also time stamp the calls and provide output devices that can be used to retrieve the call. The system may provide for time synchronization from an external source such as GPS or radio station WWV.

Recorders are available to connect to either analog or digital lines and to trunks of all types. They may monitor extensions or trunks in real time through a passive tap. Calls can be retrieved based on variables such as channel, start time, call length, dialed number, caller ID, or ANI. Some products convert recordings to .wav files for e-mailing, burning to CD, or playing on an audio player. Products also typically report statistics on numbers and durations of calls.

SPEECH RECOGNITION

The dream of a product that does near-perfect speech recognition has existed for a long time and eventually the product will arrive. Speech recognition has improved steadily as a result of better algorithms and faster computers, but it is not quite there yet. Dialects, accents, vocabulary, and inflection all complicate the problem of teaching a computer to recognize voices. Then there are homonyms such as two, too, and to, which can be distinguished only through context. Speech recognition takes computer power and plenty of it, but the potential rewards are enormous.

Telephony has many applications for speech recognition. The telephone instrument suffers from its keypad limitations. Dozens of products are available today to work around the shortcomings, but they have difficulties of their own. The PDA with a screen-operated stylus keyboard is one approach, but it is slow and awkward. Miniaturized keyboards are plentiful, but hardly suitable for anything but clumsy typing. The real answer will come through speech recognition.

At today's state of development a computer can be trained to recognize a wide vocabulary from a single speaker with a reasonable degree of accuracy. Equipment can understand simple spoken commands from a broader range of callers and can recognize numbers and some simple words. This often makes it possible to replace an automated attendant. For example, a dialog could read "for sales press or say 1, for customer service press or say 2," and so on. This method can solve the problem of directing calls for users who do not have DTMF telephones. High-end products can do an even better job of recognizing words spoken by random speakers with a high degree of accuracy. Managers should watch speech recognition closely and be prepared to adopt it when it can help them reduce costs or improve customer service.

Speech recognition does a reasonable job of replacing the DTMF pad for making selections from a menu where the range of responses is narrow. In a library, e.g., the system could ask the caller to speak the department he or she wants. If the caller says "science" the system could prompt "do you want social science or physical science?" The answer is in a predictable range, allowing the system to carry on a dialog until it is clear where the call should go or until an attendant intervenes. When it comes to communication with the telephoning public, the computer's vocabulary narrows considerably. Computers can recognize numbers with a reasonable degree of accuracy regardless of dialect, and it is easy to read them back for confirmation. Speech recognition can support directory access by asking callers to say the last name of the person they want to speak to and then reading back possible alternatives. The result is fallible, but the application is improving and will get better.

The primary uses for speech recognition today are account number recognition and call routing. As the technology improves, the applications will replace some of the mind-numbing tasks of today. Directory assistance is high on the list. Customer service centers can greatly multiply their effectiveness by doing a better job of screening and routing calls. For example, a speech recognition system could ask several questions that enable it to route calls to an expert. By contrast, today's expert agent applications in an ACD have limited screening capability. Verizon uses speech recognition to handle customer repair requests. The application is impressively natural sounding and undoubtedly saves the company money, but callers find it much less effective and more time consuming than talking to a human.

Speech recognition has already had some positive impact on telephony, but compared to its potential, the surface has merely been scratched.

DIGITAL ANNOUNCERS

Digital announcers provide information announcements in call centers, announcements-on-hold, intercepted number announcements, and a variety of such tasks. Most large PBXs need to route callers to announcements—frequently if they support call centers. Voice mail can provide occasional or infrequent announcements, but digital announcers permit multiple callers to listen to the same announcement simultaneously, which many voice-mail systems do not support.

PBX manufacturers offer digital announcers as cards that plug into universal slots. Systems are also available as stand-alone units that connect to trunk or analog station ports.

VOICE-PROCESSING APPLICATION ISSUES

Users sometimes perceive voice mail as just a convenient answering machine, which has a tendency to limit its usefulness. A properly chosen and applied system can improve the effectiveness of the company's staff by making them more accessible to customers and other employees. Voice mail has become so effective that it is indispensable to most organizations. Its applications are too numerous to list and, in most cases, are self-evident. Some creative uses that are not so obvious are:

- A hospital offers voice mail for its patient rooms, allowing patients to receive messages while they are asleep, in surgery, etc., and play them back at their leisure.
- A public telephone service provider allows a caller who reaches an unanswered or busy telephone the option of leaving a voice-mail message for later delivery. This is a valuable service for callers who cannot afford to wait.
- A hospital posts its job openings in voice mail, segregated by licensed, nonlicensed, and office occupations. Callers can listen to descriptions of the jobs available and route themselves to the employment department to get more information.
- A district school provides its teachers with a nonintegrated voice mailbox so that parents can leave messages for callback without interrupting either staff or teachers during class hours. The system forwards the messages as .wav file attachments across the data network.
- A food processing plant provides guest mailboxes for its growers to inform them of recommended planting schedules and the availability of seeds and fertilizers. During the harvest season, daily tonnage quotas are left in voice mail. Not only does this system eliminate calls to farmers who are away from the telephone, but also makes it possible for the processing plant to verify that the message has been retrieved.

To gain maximum benefit from voice mail, an organization must ensure that its employees do not abuse the service. The most frequent abuse is the tendency of some users to hide behind voice mail, using it to answer all calls, which they can return at their convenience. When used this way, voice mail aggravates, instead of alleviating the problem of telephone tag.

Improperly designed and applied voice mail and automated attendant systems have led to the creation of the term "voice-mail jail," which refers to the situation in which a caller is locked in the system and cannot escape to get personalized assistance. A well-designed system should be brief and clear in its instructions and should give callers a choice.

Many companies are reluctant to use voice mail with their customers because of the impersonality of the service. The first contact a company has with a new customer usually should not be through voice mail, but when the relationship is firmly cemented, voice mail can be as advantageous to a customer as it is to employees. The key is to leave the caller in control. Many companies answer outside calls with an attendant, but offer the caller the option of being transferred to the user's voice mail. Even if messages are taken manually, the receptionist can read them into voice mail or send them by e-mail, which is usually a more effective means of distributing messages than writing them on message slips.

Evaluating Voice Mail

Evaluations of voice mail, automated attendant, and audiotex systems should be based on features and on the criteria listed below, which discusses the most important features.

Port and Storage Capacity

Voice-mail systems are sized by the number of ports and hours of storage. The number of ports limits the number of callers that can be connected simultaneously. Ports are required for both leaving and retrieving messages and are occupied while callers listen to prompts or users update their personal greetings. The number of ports required can be calculated from the number of accesses and the holding time per access during the busy hour. (See *The Irwin Handbook of Telecommunications Management* for an explanation of how to calculate voice-mail port and storage requirements.) Designing the system with the correct number of ports is important. Having too many ports increases the cost of the voice-mail system. Having too few restricts the ability of callers to reach voice mail and blocks users when they attempt to retrieve messages. Ports normally are added in increments of two to four depending on the manufacturer. Large voice-mail systems may connect to the PBX over T1/E1 ports.

The type of integration often affects the efficiency with which the voice-mail system uses ports. A popular feature on many systems is outcalling, which is the ability of the voice-mail system to dial the user's pager or cellular phone. This feature requires ports to handle both incoming and outgoing calls. A tightly integrated system can assign any port to either type of call, but some systems may require separate incoming and outgoing groups of ports, which increases the total ports needed.

Voice-mail systems that are associated with traditional PBXs use physical ports for the connection. Each time a message crosses a port, it is converted between voice and whatever compression algorithm is used. Too many compressions and decompressions can affect intelligibility. For example, when a caller leaves a voice-mail message it is digitized, compressed, and stored. When the user retrieves it, it is decompressed and converted to analog. If it is forwarded to another user across a network and covers to voice mail on the distant system, it is digitized again. If the network is IP, it may be converted to a different compression algorithm. It is important to understand exactly how the manufacturer and vendor set up the system to avoid quality degradation.

The amount of storage required is highly variable and depends on the number of users and the number of stored messages per user. It also depends on whether disk capacity varies with the announcement messages and how efficiently the manufacturer packs the disk. Efficient packing algorithms compress the silent intervals so that the disk space is not wasted. As a rule of thumb in sizing voice-mail systems, the average user takes about 3 to 5 min of storage for greetings and messages. A system serving 100 users, therefore, would require 5 to 9 h of storage. Heavy users or those who are not careful to purge their old messages may require more than 5 min of storage on the average.

The voice-mail system's purging algorithm also affects storage efficiency. The more effective systems inform the administrator when the disk is reaching capacity. Over-age messages are manually or automatically purged after a specified time. In addition, most systems provide for different classes of service and can vary the amount of storage a user can have before callers receive a full-box message. Hard disks are so inexpensive that message storage may be of little or no concern. Some manufacturers charge for storage and, even though the hard disk has plenty of capacity, the owner must pay to have it unlocked.

Integration with a PBX

The most important consideration in evaluating and using a voice-mail system is the degree to which it integrates with the PBX. This is not to say that integration is always necessary; often the application needs a stand-alone voice-mail system. For most office workers, however, an integrated system provides essential features.

Compression Algorithm

All digital voice-mail systems use some form of compressed voice technology to digitize and store messages and announcements. Part of the compression is gained by pause compression and expansion, in which the duration of a pause is coded instead of the pause itself. Voice-mail systems use pause compression to speed or retard playback. If a listener wants a faster playback, the system shortens pauses; for a slower playback, it lengthens them. In addition, the specific technology used for digitizing the voice affects the efficiency of the system. The more the voice is compressed, the less natural it sounds and the less storage space it occupies. The best way to evaluate this feature is to listen to the voice quality of several different systems and compare voice quality. Pay particular attention to conditions that might result in excessive analog-to-digital conversions.

Port Utilization

The number of ports determines how many users can leave and retrieve messages simultaneously. Not all ports are necessarily available for full voice-mail use. Some systems require dedicated ports for automated attendant. Some systems require separate ports for outgoing messages such as calls that the voice-mail system places to a paging or cellular radio system.

System Reports

System reports can be used to determine how efficiently the system is being utilized. Reports should provide information such as the following:

- number of messages sent and received by a user;
- average length of messages by user or group;
- percentage of disk space used;
- busy-hour traffic for various ports;
- number of times all ports are busy;
- message aging by individual mailbox;
- number of messages not deleted after specified time.

Networking Capability

The cost of voice mail can be reduced by networking multiple voice-mail systems to act as an integrated unit. From the system administrator's standpoint, an important issue is how to maintain distribution lists across the network. Some systems exchange messages that automatically update the other systems on the network. Another important function of networked voice mail is how systems exchange messages. In some systems, the voice-mail system can transmit messages as IP packets.

Security

The voice-mail system must prevent callers from dialing invalid extension numbers and thereby connecting through to trunks. The system administrator should be able to force password changes and to require nontrivial passwords of a minimum length.

Evaluating Unified Messaging

An important issue in unified messaging is how and where the messages are stored. This is not uniform among products. Voice mail, e-mail, and fax usually arrive on different servers, although many voice-mail products also support fax reception. An important issue in evaluating products is whether the message remains stored on the server it arrived on or whether it is transferred or copied to a different server under certain conditions. Some products combine e-mail and voice mail into a single server that also retains any faxes that arrive via voice mail. If faxes arrive on a separate server, which is often a third-party product, integration with unified messaging will probably require custom software development.

Another issue to consider is how unified messaging is priced. Regular voice mail carries a cost for the platform plus the hours of storage and number of ports. Some vendors also charge a fee per user, but many do not. Unified-messaging pricing is usually based on the number of users. This enables the customer to start with a few users to determine if the product pays off and expand it if it does.

In addition to the per-seat license, the customer must provide a server. The administrator's main issue is how to control storage. Many users are not diligent about deleting messages and the storage capacity can be exceeded unless the system has some kind of automatic purge.

Evaluating Automated Attendant

In applying an automated attendant design, administrators should consider its effect on callers. Frequent callers usually have no problem navigating the menus, but a first-time caller may be baffled by the variety of choices. Any company considering an automated attendant will be well advised to study the application carefully, design it intelligently, test it thoroughly, and listen to the comments of callers.

Automated attendant substitutes for DID, particularly outside business hours. It reduces labor for console attendants and, in many cases, it is advantageous to the customer. A good menu is a fast way of reaching the right destination in a call center. Employees and frequent callers can be told that when the company's main number is answered they can immediately dial an extension number, which is faster than talking to an attendant.

Evaluating Audiotex

The audiotex feature of voice mail is a useful means of distributing information. Callers receive a menu of choices. They can select two or three levels of menu before reaching the desired information. For example, a university might disseminate class information via audiotex by listing the major courses of study—science, liberal arts, engineering—on the first menu, the field of study—biology, botany, chemistry—on the second menu level, and class—freshman, sophomore—on a third. If the menu choices are no more than the caller can easily remember, this can be an effective way of delivering information. The system should be designed to enable callers to interrupt the menu by pressing a special key such as "#" to repeat. The most important criteria in evaluating audiotex are the types of host interfaces, application development tools, and local databases supported.

Evaluating IVR

Many of the same criteria as voice mail can be used to evaluate IVR. A major exception is the fact that the interface between the host computer and the IVR must be programmed. Shrink-wrapped software is available for some applications, but most are custom programmed—often by a system integrator that provides the IVR hardware and the company's software application.

Voice quality, ease of programming, and quality of vendor support are important considerations in IVR. Many systems provide application-development tools that provide a high-level language the developer can use. In addition, the following criteria should be considered with IVR:

- *Hardware platform.* Determine how many ports are available, what the maximum capacity is, and how much it costs to expand. What kind of platform does the system run on? What operating system does it use?
- *Software.* Check the features that the system supports. Does the manufacturer have a good record of introducing new features to keep pace with competition? What is the cost of new software releases? Has the application been programmed by a third party who makes it available for a fee or is it necessary to train or hire an application developer? If the latter, how much does the development kit cost and how easy is it to learn? What kind of debugging tools are available?
- *Integration.* Determine how the system interfaces to the PBX or telephone system. Is it a proprietary interface that integrates with the PBX's processor or is it a standard analog telephone interface? The former is more versatile, but the latter can be transported to other applications more easily and may integrate with any PBX. Determine how easily a caller can transfer to a live attendant. The system ideally should not require callers to re-identify themselves when transferring from IVR to an attendant.
- *Speech generation method.* Does the system synthesize speech or use stored speech fragments? How natural does the speech sound? Can words or phrases unique to your operation be added easily?
- *Growth capability.* How many ports does the system support? Does it have the processor power to handle as many simultaneous callers as there are ports? Are you given the tools to measure and verify performance?
- *Reporting capability.* Does the system provide usage statistics? What form are they in? Can they easily be translated into service-related reports? Can the statistics be extracted while the system is active?

Evaluating Speech Recognition Equipment

The primary concern with speech recognition is its accuracy, which can be tested by observing the result with a variety of callers. Any system purchased today will undoubtedly be improved in the future, so determine the developer's commitment to continued research and development and find out how the updates are introduced and at what cost.

Pay particular attention to the application. The best results will be achieved with simple applications that require recognizing a limited number of short words such as numbers, alphabet, department names, etc.

Evaluating Digital Recorders

The primary factors in evaluating a digital voice recorder are:

- *Security*. Can voice sessions be kept secure from unauthorized access while still making them available to supervisors who need to review the session? Are recordings tamper-proof so they can be admitted to court sessions if necessary?
- *Accessibility*. How easy is it to find the session from the client workstation?
- *Clarity*. Is the voice compression algorithm good enough that voice can be retrieved with a high degree of intelligibility?
- *Archiving*. How easy is it to archive the recordings onto long-term storage media such as digital audiotape or DVD?
- *Maintainability*. Is system adequately alarmed? Does the system provide disk mirroring or other methods of fault tolerance? Is it secure against data loss during full disk conditions?

Evaluating Digital Announcers

Digital announcers are available as stand-alone devices that can interface to any PBX or public access line and as integrated devices that fit into a card slot in a PBX. It is important to evaluate the application carefully before buying an announcement system. The primary criteria in evaluating digital announcement systems are:

- Storage capacity.
- Number of ports.
- Voice quality.
- Method of integration with the PBX—through the bus or through a port.
- Method of storing the announcement. Is it on volatile memory? How is it protected against power outage?
- Method of updating the announcement—locally, remotely, or through a professional service.

As with voice mail, the primary criteria in selecting a system are the amount of message storage and the number of channels required. It is also important to decide how the message will be changed. A permanent message such as an informational message about a company's product line probably would be recorded professionally and rarely changed. The message might even be encoded on a chip before it leaves the factory. Conversely, a school providing announcements about closures due to weather will want to record messages locally and, perhaps, from a remote location.

The control circuitry should determine when to connect the caller to the announcement. Some messages play only from the beginning; in other cases the listener can barge in on the message wherever it happens to be in a cycle. For example, if a promotional or informational message is being played, it would be appropriate to break in at only certain spots and the message would play only once. Messages on hold might play repeatedly while the caller is on hold or the system might vary the message depending on how long the caller waits.

Voice quality is another important consideration. As with voice mail, the higher the degree of compression, the lower the storage cost and sometimes the lower the voice quality. The best way to evaluate quality is to listen to a sample.

CHAPTER 29

Electronic Messaging Systems

Electronic messaging has transformed commerce and our personal lives to such a degree that it is difficult to remember when we did not have instantaneous message and file transfer at our fingertips. E-mail is the most frequently used application on the Internet, which is rather astounding considering the popularity of the Web, which occupies second place. E-mail is revolutionizing the way people communicate. No longer is it necessary to fax a document, wait for the postal service, or even walk down the hall to another office. Information can be transferred instantaneously between any computers that have access to an e-mail server, either a public server or a private server attached to the office LAN.

The world has had e-mail for many years, but it needed the Internet to bring the ease of use and near-universal connectivity that we now have. The first e-mail systems were isolated islands. Some were host based, residing on a mainframe or minicomputer, and useful only for internal communications. For example, IBM provided PROFS or SNADS for users operating on an IBM mainframe and the UNIX operating system provided e-mail as a standard feature. Communicating outside the office network, however, was neither easy nor inexpensive. Value-added carriers such as AT&T and MCI provided e-mail as did database companies such as AOL and CompuServe. If you knew the addressing protocol, you could usually transmit messages to someone on another network, but it was difficult to be sure the message would arrive.

E-mail was one of the earliest applications for the Internet. Simple Mail Transfer Protocol was devised to permit two UNIX computers connected with IP to interact with each other's internal mail application. SMTP uses a text-based dialogue between computers to transfer mail.

The ITU-T's answer to e-mail is the X.400 Message Handling System (MHS) protocol. Public e-mail systems such as those operated by the major IXCs and most international carriers use X.400 and it is the common language that many proprietary systems speak. In contrast to SMTP, X.400 is not simple. It is, however, an international standard that can operate between companies and across borders. Before the Department of Defense turned over the Internet to private operation, X.400 was the main method of interconnecting otherwise incompatible systems. Both X.400 and Internet e-mail have similar features. The earliest versions did not support file transfer, but today you can attach any type of information such as voice and video clips, documents, and spreadsheets to an e-mail message with little more effort than a mouse click.

One of the major attractions of Internet e-mail is the fact that practically anyone can get an address that is easy to use and remember, and if you are willing to put up with advertising, it may even be free. The other major value of e-mail is its inherently asynchronous nature. Messages are stored in a server until the recipient is ready to pick them up. Internet e-mail is a client-server application. The client software resides on desktop computers and the servers are either internal to a company or operated by an ISP. Internal e-mail servers receive mail for the organization and distribute it internally, usually after checking it for viruses.

To some degree, e-mail and voice mail have similar characteristics. Both permit information transfer without the simultaneous availability of the participants. Both work across time zones. They can broadcast messages to defined user groups, and with unified messaging the two are linked to enable users to choose the device that they want to use to retrieve messages. Voice mail is more transitory than e-mail, which retains a written record of the communication. This is a major advantage of e-mail, except for those who later need deniability.

E-mail has transcended organizational boundaries. Individuals throughout the world exchange e-mail messages at a cost that is practically zero. Instant messaging (IM), once the province of teenagers, is moving into the enterprise, where short messages can pop up on computer screens, wireless phones, or PDAs. This chapter discusses some of the applications and the protocols under which e-mail operates. We review the issues involved in implementing such applications as electronic mail, IM, electronic data interchange (EDI), and data interchange between similar applications such as word processors.

MESSAGING SYSTEM OVERVIEW

The terms messaging system and e-mail are often used interchangeably, but to be strictly accurate, e-mail is only one of the types of information that a messaging system can transport. For example, scheduling, calendaring, forms, workflow, inventory, and business documents that are transported over EDI are examples of applications that can be built on a messaging system. Some form of e-mail exists for nearly every computer operating system in existence, and nearly every desktop computer today runs an e-mail client.

Figure 29-1 is a generic diagram of an electronic messaging system. *Message transfer agents* (MTAs) are the servers that move e-mail across the backbone network. The backbone can be a public backbone such as the Internet or a private

FIGURE 29-1

A Generic Messaging System

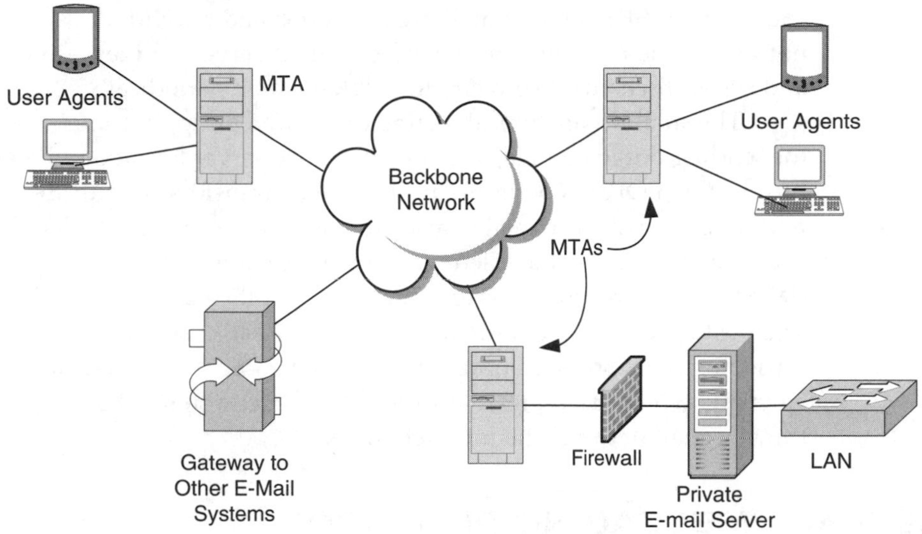

network running IP or some other protocol. The *user agent* (UA), also known as the client, is used at both ends of a mail system to create, send, receive, and manage messages. Microsoft Outlook, for example, is a popular client. Many others are available; some can be downloaded without cost. The client software links with one or more MTAs to transfer messages. The client has four main functions:

- Lists headers of all of the messages in your mailbox. The header identifies the subject, sender, time, date, and message size.
- Provides a method of opening and reading messages and file attachments.
- Provides a structure for creating and sending messages.
- Provides a structure for attaching files.

A client wishing to send a message first links to a server that recognizes it as an authorized user. The server identifies the client by user name and password, and then gives the client permission to transfer the message. If the addressee resides on the same server, the message is moved to the addressee's mailbox. If the addressee is not local, the server queries DNS to find the address of the addressee's SMTP server. Then the message is forwarded across the backbone. All of the MTAs in the message transfer process must be capable of supporting all of the features the client requested or the transfer will fail. For example, if the messaging system is capable of carrying voice clips, but one server in the chain lacks that capability, the feature will be unavailable. If the addressee is unidentified or unreachable, the server sends an e-mail message to the sender.

In the Internet mail system, the MTA resides in ISPs' mail servers. When client software is set up, it contains the name and address of the mail server, which could be an internal server running software such as Microsoft Exchange, or it could be an ISP's server. Small organizations and individual users normally cannot justify the cost and complexity of a mail server, so they use the ISP's server and set up their client software to check for mail periodically.

The mail server typically runs two different applications: the SMTP server for sending outgoing mail, and the post office server for inbound mail. Post Office Protocol 3 (POP3) is a simple protocol that transfers mail to the client without retaining a copy. Once mail is read from a POP3 server, it is deleted, although the user can save it on the client computer. An Internet Message Access Protocol (IMAP) server retains messages on the post office server until they are deleted. The post office server maintains a list of all clients' user names and passwords for retrieving mail. Message transfer between incompatible MTAs must go through a gateway, which translates the protocols of the systems. For example, gateways transfer mail between the Internet and X.400.

THE X.400 MESSAGING PROTOCOL

ITU-T has adopted X.400 as its standard message-handling protocol. X.400 is more popular in Europe than in the United States, where Internet mail over SMTP became the accepted method of e-mail communications. Some international carriers implement all or part of X.400. The protocol provides for both a UA and an MTA, but many carriers choose to provide a proprietary user interface.

The power of X.400 is procured at a sacrifice of simplicity. It is a complex system to set up and administer, which is the reason that many host-based and proprietary mail systems use their own protocol and use X.400 for message exchange between systems. X.400 is an open architecture, but it is impractical for LAN application because of its overhead and complex addressing scheme. Each X.400 MHS is an independent domain, and these must be connected by agreement between the carriers. If you subscribe to mail from one carrier, you will be able to send mail to customers of another carrier only if they have established an interconnection agreement and a gateway to pass messages. Many X.400-to-Internet mail gateways are in use because Internet mail is so much more prevalent.

X.400 Addressing

X.400 uses a multilevel hierarchical address as illustrated in Figure 29-2. The character string can be daunting until one understands its structure. At the top level is the root, followed by country, which is a mandatory part of the address. The next level can be A for an administrative domain or P for a private domain. An administrative domain is usually a service provider such as AT&T Mail. Large companies may use a private domain in lieu of a service provider. At the next level is

FIGURE 29-2

X.400 Addressing Structure

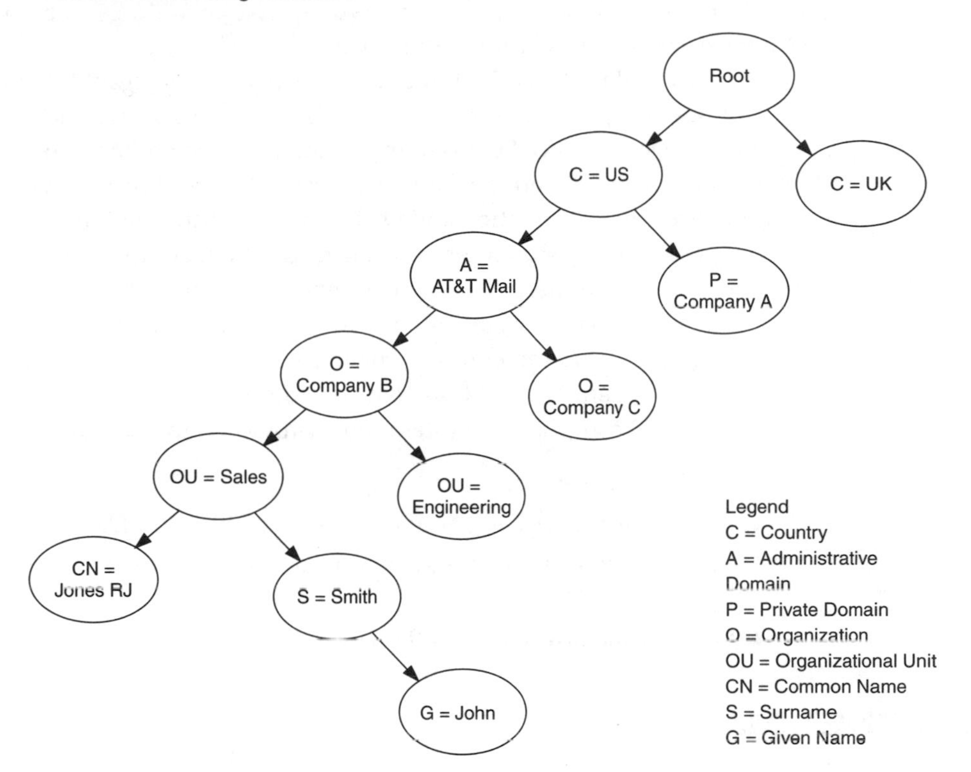

organization unit (OU), which is a unique identifier referring to a company's private messaging system. Below the organization companies may identify localities and organizational units of several levels. The lowest level is the CN (common name), which can optionally be stated as given name (G) and surname (S).

The X.400 addressing system is flexible, but verbose. The protocol allows for address aliasing, which can provide a shortcut address across directory levels. Users are unlikely to understand the addressing well enough to create their own aliases, so expert assistance may be needed. Most effective UAs provide shortcut addresses.

X.500 Directory Service for X.400

ITU's X.500 protocol contains information about X.400 addresses. The directory is hierarchically organized along X.400 lines by country, organization, organizational unit, and person. An entry at each level has certain attributes, some of which are optional. Any organization can implement a directory provided it adheres to the structure. The X.500 standard does not cover the contents of the directory, so companies can maintain any type of information desired in the database, filtering and screening to keep parts of it private. A company could, for example, publish certain e-mail addresses, but keep everything else including telephone numbers and mail stops private within the company.

Directory information is stored in a *directory systems agent* (DSA), which can represent one or a group of organizations. The interface to the DSA is known as a *directory user agent* (DUA). DSAs communicate with each other using the Directory Access Protocol (DAP) and the Directory Service Protocol (DSP). DSAs use DSP to exchange directory information and DAP to retrieve information. Users not knowing the e-mail address of another user can access the DUA much as they might call directory assistance to get a telephone number. After they obtain the address, they can store it in the e-mail application or in the central directory in an enterprise e-mail system. DAP is too complex for most small intracompany mail systems. These systems generally use LDAP, which is discussed later.

To be X.500 compliant, a system must provide four elements:

- A DUA for user access
- A directory service agent that speaks the DAP and DSP protocols
- Basic directory functions such as information storage, retrieval, and updating
- Security including user authentication

INTERNET MAIL

As e-mail becomes mission critical in many companies, the message volume the typical worker must contend with grows, and handling it becomes a problem. Hackers and other miscreants regularly use e-mail to propagate viruses, worms, and Trojan horses through firewalls and on to their victims' computers. (A Trojan horse is a program that performs some detrimental action while pretending to do something else. For example, fake login programs prompt users for logins and passwords.) Unwanted e-mail, popularly known as "spam" permeates the network, clogging servers and wasting bandwidth with transmissions that are useless or worse. To cope with this growing problem, e-mail systems are becoming more intelligent, classifying mail based on information in the header, known noxious sites, key words, and other such methods of prioritizing and discarding messages. These techniques, called rules-based messaging, enable users and administrators to filter messages that might otherwise inundate them. Unfortunately, some benign messages are filtered out in the process. Many e-mail systems temporarily quarantine messages, leaving it up to the users to decide whether to discard them.

Many application programs such as word processing and spreadsheets are messaging system enabled. A mail-aware program knows that a messaging system exists, and how to send data over it. It does not know how to process the data. A mail-enabled application generates e-mail messages and uses the message transfer system to transfer information through the network. A mail-enabled application can do everything that a mail-aware program can, plus it can route, receive, and authenticate messages.

Simple Mail Transfer Protocol

Although comparative statistics are not available, there is little question that the quantity of messages sent on Internet e-mail far exceed those on X.400 networks, even in European countries. SMTP is an easy and unsophisticated mail system that was first set up between UNIX computers. Figure 29-3 shows how UNIX computers transfer mail between each other. To send a message, the sender types the message into the UA, which passes the message to the MTA. Each computer can reach a DNS, which contains the IP address for every user on the network. The MTA transfers the message to the other computer via a TCP/IP link, locally over Ethernet or to a distant network over the Internet. The two MTAs carry on a simple dialog to establish communication. The message transfers to the receiving MTA, which stores it in the user's mailbox. If the addressee is unknown, the receiving MTA returns a message to that effect.

By itself, SMTP can handle only seven-bit ASCII text, which means that additional software is needed to enable a message to carry a binary file. The IETF developed a protocol known as Multipurpose Internet Mail Extensions (MIME), which converts files into a format that SMTP can handle. All mail servers that enable users to attach files to messages implement MIME.

FIGURE 29-3

Simple Mail Transfer Protocol

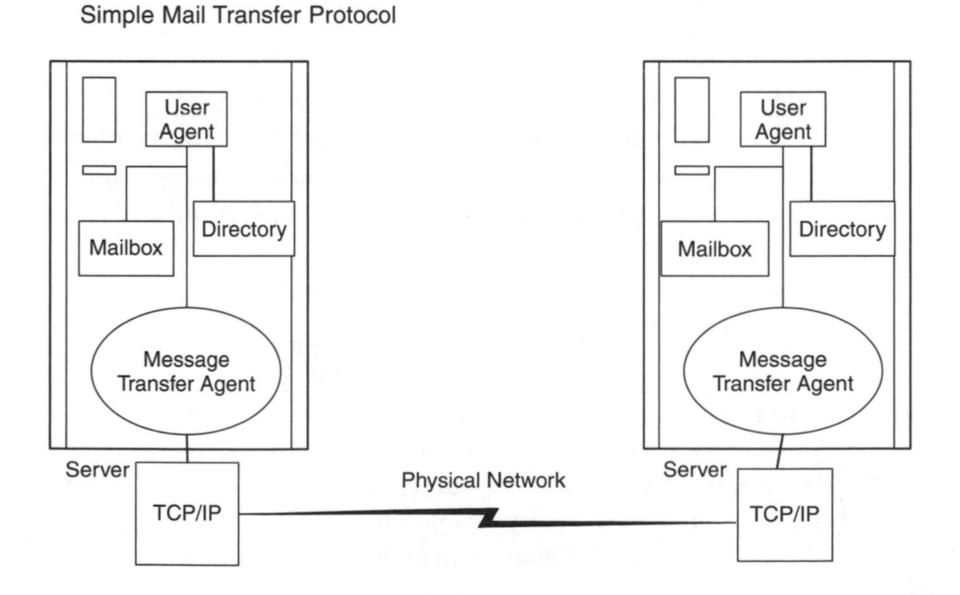

A major difference between SMTP and X.400 is in the language used. X.400 communicates using ASN.1, which is a formal language for describing messages to be exchanged between network devices. It communicates in binary, in contrast to SMTP, which communicates in text form. For example, a communication between an e-mail client and a POP3 server uses simple commands such as USER, PASS, and LIST to log in and retrieve messages. With a little practice, anyone can look at the details of SMTP message exchanges and see what is going on.

Lightweight Directory Access Protocol (LDAP)

The X.500 protocol is too cumbersome for many organizations to use, but some method of maintaining directories is required. Every organization has its internal directory, often printed on paper, and usually obsolete before it can be distributed. On-line directories are the preferred method because they can be updated daily if necessary. An on-line directory is easy for people to use, but it needs structure to be used in machine communications. Applications must be able to find someone regardless of location to send an e-mail, instant message, or voice call. This usually depends on multiple interacting directories. Most enterprises have multiple directories to manage the various applications that define their employees' telecommunications resources. Even products from the same manufacturer such as a PBX and its voice mail have separate databases that must be separately updated. By the time a new employee is equipped with logins, passwords, authentication and authorization for devices such as telephone, voice mail, e-mail, an ID badge, a user account on one or more servers, a door access code, pager, cell phone, and a company credit card, the number of accounts that must be established is overwhelming. The answer, increasingly, is LDAP. With a properly configured directory, an HR person could add a new employee, update the directory, and automatically populate other directories.

LDAP defines:

- A protocol for accessing directory information
- A model defining the format and character of the information
- A namespace defining how information is referenced and organized
- A model defining the operations one can perform such a search, add, delete, etc.
- A model of defining how data may be distributed and referenced

LDAP is now in Version 3, which supports non-ASCII and non-English characters for international operation. LDAP enables users to look up information from an online directory or e-mail contact list by entering the name or other request. From a single request, LDAP can search multiple directories and return various ways that the same person or information could be located. LDAP also plays a key role in directory-enabled networks (DEN), which is a multivendor initiative to simplify the management of complex networks through directories.

LDAP directories are arranged in a hierarchy starting with a global name and working downward in more detail such as department, division, employee name, and so on. It can also link a person's name to other information such as preference for being located. The LDAP hierarchy starts with the distinguished name (DN). The DN is divided into domain components (DC), organizational unit (OU), and common name (CN), ending with an attribute type and value. Here is an example of how an LDAP directory might appear:

DN = McGraw-Hill Corporation
DC = Book Division
OU = Professional Trade
CN = Mary Smith
Address: 1333 Burr Ridge Parkway Burr Ridge, IL 60527
Phone: 630-555-1234
Fax: 630-555-4321
e-mail: marysmith@mcgraw-hill.com

The directory could be further enlarged to include such information as cell phone number, supervisor, emergency contact information, preferred method of being contacted, normal work schedule, deviations from the schedule, etc. Some information can be kept private, while opening the rest to public view. The directory information resides on a local server, which can be configured to pass information to other servers such as one at corporate headquarters. LDAP can search multiple directories or use URLs for directory look up. Accessible information might reside in a Web page, a catalog, a human resource directory, or other such servers, and LDAP can check them all. LDAP can reside on some endpoints. SIP telephones, for example, can have LDAP capability. SIP for Instant Messaging and Presence Leveraging Extensions (SIMPLE), a protocol for notifying IM users of status changes, can also be implemented from SIP devices. Calendar and scheduling information can be kept in the directory or accessed from other applications such as an electronic calendar. Companies can keep records on customer purchases and buying habits and associate them with entries in their LDAP directory.

Every e-mail program has a personal address book, but the question is how you reach someone whom you have never communicated with before. LDAP servers index data in trees and you can use a variety of filters to select a person or group you want. Users can set permissions to allow only certain people to access the database and to keep some information private. Servers also have an authentication service to identify the person requesting information from a directory.

Instant Messaging

IM, once the province of teenagers, friends, and family members, is finding a place in the enterprise as products gain business features. IM is an IP-based application that can be used with mobile devices such as digital cell phones and PDAs as well as desktop computers. In some ways it is reminiscent of the old bulletin board systems that people often set up as a sort of private Web site. The value of IM to most users is its immediacy. With e-mail you have no idea whether the other person is online and available to exchange messages. With IM, users can set up buddy lists and determine instantly whether others in their circle are online and available to chat.

IM has become an important collaboration tool for many enterprises, providing such features as chat rooms, conferencing, and whiteboards. IM can be used to schedule meetings, bounce questions off colleagues without undue interruption; maybe even while they are on the phone or in a meeting. It provides a low-cost method of enabling workers spread over a wide area to keep in touch and collaborate on projects. It can also be an effective and low-cost way of distributing information to multiple recipients. Sending a corporate announcement by e-mail causes it to land in multiple mailboxes and servers, and consumes disk space. IM can also be used for distributing information that people need quickly, such as notice of a delayed meeting or a server that must be taken down in a few minutes. Benefits such as these are motivating enterprises to invest in IM products.

In many ways, management is motivated to consider IM by more than just the potential for improving productivity. Even more crucial to many companies is the fact that IM is almost impossible to prevent. Offered free by Web portals as a way to pump up advertising revenue, IM enables anyone with Web access to set up a buddy list and open perilous security holes. Public messaging services are wide open and well documented so that hackers can easily use them to harvest addresses or scan for confidential information that should not be released outside the company. There is also danger that instant messengers can bypass firewalls to transfer worms and viruses and provide an access point for Trojan horses. Public IM is simply not secure enough for enterprise use. The first line of defense, therefore, is a firm policy regarding its use, coupled with firewall filters to ensure compliance.

Plenty of IM products are available to enable the enterprise to set up a system with the appropriate internal controls. Most products are priced on a per-seat basis, so they can be provided to those who need them and unavailable to others. IM can be either network based or device based. In a device-based system the user information is stored on the desktop device. In a network-based system it is stored in a server, which is most appropriate for enterprise IM. Network-based systems give all users access to the same services and information such as the contact list, addresses, and other personal information regardless whether they are logging in from home, office, or mobile phones. Remote users can communicate through a VPN tunnel, which may avoid firewall issues.

User authentication is essential. It is best accomplished by close integration with the corporate directory, ideally using LDAP. With an effective directory, users' presence can be indicated by status messages such as stating that they are out of the office or on vacation. The directory provides a way for users to communicate their reachability and communication profile as a means of controlling how others communicate with them. Users can receive messages even while off line/pending messages are delivered when they log in. Most messaging systems have a presence-awareness feature. The basic level indicates whether the user is online and willing to accept messages. Users can elaborate on the basic level with additional status messages. This brings up a privacy issue since users must be assured that their personal information is available only to trusted users. The fact that a person is on vacation or out of town, for example, would be provided to business colleagues and the family, but not to strangers.

Besides the ability to pop messages on other users' screens, chat rooms are the most valuable feature of IM products. With these, users can set up online meetings either informal or with a moderator. Chat rooms can be restricted to a limited group of users that are screened and approved, or they can be set up as public so anyone can browse. Users can click a button to "raise their hand" to be recognized. Sessions can be browsed and searched and replayed in real time or users can fast forward and rewind through the meeting. If it is a public meeting, all users can see it, but the moderator can password protect the chat room.

In addition to chat rooms, enterprise IM products have a variety of features such as the following that are among the most valuable:

- *News board*: This feature lets users post announcements for general consumption.
- *Content filtering*: Inappropriate language can be filtered out of chat sessions and online news boards.
- *File and screen sharing*: IM provides an easy way of sending spreadsheets and other documents as file attachments.
- *Whiteboard and screen sharing*: Participants in a meeting can all share a whiteboard or information on other users' screens.
- *Management and reporting tools*: Tools show information such as who is logged in, how many active meetings are in progress, and who is using tools such as whiteboarding and videoconferencing.
- *Polling*: IM can distribute polling screens requesting users to respond to multiple-choice polls. For example, several alternatives could be displayed for the company picnic and the system could collect and summarize the results.
- *LDAP integration*: The user directory can run under LDAP, which brings numerous benefits over and above IM. Some products have proprietary directories or run under a directory such as Netware/NDS.
- *Message encryption*: Most products support encryption to keep confidential messages out of unauthorized hands.

- *IM platform integration*: Some products can integrate with other IM platforms or servers that may be applied by business partners or other outside organizations. Some products can interface with public IM networks.
- *Scheduling calendar*: Some products provide an online scheduling calendar that can be used to reserve chat rooms and show who is using the space.

Unlike the IP protocol on which it runs, IM is proprietary. Products can be designed for interoperability, but in most cases some features will not operate between platforms. The IETF is working on Instant Messaging and Presence Protocol (IMPP), which is an architecture for simple presence and awareness notification. It specifies how authentication, messaging integrity, encryption, and access control are integrated. Message formats may also be proprietary. Extensible Messaging and Presence Protocol (XMPP) is an XML-based protocol for near real-time extensible messaging and presence that the IETF is working on to standardize message formats.

ELECTRONIC DATA INTERCHANGE

Global markets are changing the way we do business. To remain competitive, businesses must know their customers and respond to their needs. In international trade it takes an average of 46 documents to move products across boundaries. Traditional mail is too slow, and facsimile is impractical to authenticate for some applications, which raises the need for electronic document interchange. EDI is the intercompany exchange of legally binding trade documents. EDI enables companies to exchange a variety of business documents such as invoices, requests for proposals, and shipping and purchase orders over an electronic network.

EDI offers both cost savings and strategic benefits to the trading partners. Cost savings come from reduced labor costs, reduced stock levels that result from shorter document turnaround, and savings in telephone and postage costs. Improved response time is a major advantage. Companies can respond quickly to purchase orders, requests for quotes, and other such documents. In the past, trading partners developed document formats they could use between themselves, but it was impractical to make these arrangements with multiple trading partners. An international standard set of forms was needed. EDI was conceived in the 1970s, but its acceptance has been slow. Part of the reason has been that expensive, often mainframe-based software was required. Now EDI can be implemented on PCs at a much lower cost than mainframe or minicomputer systems. PC-based EDI can be integrated with e-mail and documents can be transferred across the Internet by taking appropriate security precautions.

EDI software performs two main functions. It maps the fields on EDI documents with fields in the application software, and translates the data to and from the EDI documents. EDI applications receive and prepare documents on the screen, allowing the user to fill in the fields from the keyboard. Mapping data between fields is one of the most difficult parts of implementing EDI. It may be possible to use a translation program to remap the fields from one document to another. In the worst case it may be necessary to print out the EDI document and rekey the information into the application, which defeats much of the purpose.

Documents can be transferred by a variety of methods. Many large organizations maintain EDI sites where documents can be transferred to a secure site by FTP/s or HTTP/s. EDI can be implemented through store-and-forward messaging over value-added networks or sent over the Internet as e-mail attachments. Portable document format (PDF) files are a convenient way of locking the content of file attachments to ensure that they cannot be altered. Value-added networks also offer document validation, security functions, and special reports. VANs may also convert protocols between the trading partners if they are incompatible.

EDI Standards

Four types of EDI standards are used:

- Proprietary
- Noncompliant and industry specific
- National, such as ANSI X.12
- International, such as EDIFACT (EDI for Administration, Commerce, and Transport)

EDIFACT is a complex set of EDI standards and rules developed under the auspices of the United Nations. Between ANSI X.12 and EDIFACT, standards have been developed for most documents used in electronic commerce.

ELECTRONIC MESSAGING APPLICATION ISSUES

The Internet has revolutionized e-mail, evolving into a wide-open worldwide backbone for fast and easy document delivery. It is also vulnerable as evidenced by the attacks that have infected millions of computers with worms and viruses and cost billions of dollars. Nevertheless, there is no turning back to the days when creating and sending e-mail was hardly worth the effort.

E-mail Evaluation

In selecting an e-mail system, the following are some issues that should be evaluated:

- *Adherence to standards*: All e-mail systems support SMTP, MIME, POP-3, and IMAP-4. Many systems also support LDAP. A messaging system may need to connect to an X.400 system for some types of communication.
- *Ease of setup*: E-mail server setup is not a trivial task with most systems. In large organizations with thousands of users, some method of generating

the accounts automatically should be provided. Some systems provide for importing accounts from other e-mail systems, databases, or ASCII files.

- *Security*: Determine whether the system supports authentication and encryption methods such as Secure Sockets Layer (SSL), which enables Web clients and servers to pass confidential information securely by encrypting it. The system should provide logs that the administrator can view to detect such matters as repeated ineffective attempts to log in. The system should provide an effective way of rejecting viruses and filtering spam.
- *Choice of client*: The using organization may standardize on client software because it is part of a groupware application. The client and the server may therefore be from different manufacturers. Determine whether they support the necessary protocols to interoperate.
- *Disk management*: Users are often careless about purging messages, which may consume disk space rapidly, particularly when large files are attached. The e-mail system should have tools for limiting message storage space and for notifying users to purge old messages.
- *Additional applications*: Many mail systems provide for other applications besides mail receipt and delivery. For example, a corporate bulletin board is one popular feature. Other common features of the client are document control, calendar, and contact manager.
- *Usage information*: Determine whether an audit trail through the system is needed. Is message receipting necessary? Should the system provide usage statistics for charging back to users or determining which departments are the high users?
- *User interface*: The user interface must be easy to use for composing, sending, receiving, and managing messages. Users should be able to cut and paste to and from word processors, attach files, easily, and move them seamlessly across the network.

Instant Messaging Evaluation

Although IM is similar to e-mail, it has advantages that warrant its application in many organizations. Although e-mail is supposed to be practically instantaneous, sometimes it is not. IM is immediate and works in a conversational mode. It can provide presence detection, and IM messages jump to the top of the priority list, which can be a disadvantage. The following are considerations in selecting and applying IM:

- *Security*: Enterprise-grade systems should provide for end-to-end encryption, authentication of participants, virus checking, and content filtering. Products should be capable of rejecting unwanted IM messages.

- *Peer-to-peer, public, or server based*: Server-based IM is more appropriate for larger organizations. Public IM raises security and privacy issues.
- *Interoperability*: The proprietary nature of IM may become an impediment. If trading partners use IM, compatibility of different systems may become an issue. The system should be capable of implementing SIP and SIMPLE as they are added to the network.
- *Directory integration*: The IM system must be able to link to the corporate directory, preferably on LDAP. It should be capable of detecting user presence and communications preferences.
- *Auditing and logging*: IM sessions should be logged and archived at the server for later retrieval and playback. This is particularly critical in organizations that are under a legal obligation to retain document communications.
- *Feature implementation*: The distinguishing factor among IM alternatives may be the features the product provides. Such features as whiteboard, voice and video conferencing, and sharing of computer applications such as documents and spreadsheets are important in most enterprises.

EDI Evaluation

To implement EDI, a company needs four principal elements:

- *User application*: This is the software that generates and receives EDI documents. The major partner in an EDI exchange may furnish the software. If not, review packages on the market, looking primarily for ease of use. Programs that operate under a graphical user interface are the easiest to use.
- *EDI software*: This software translates the user application to EDI message formats, provides the communications protocol for accessing the network, and translates incoming EDI messages to the user application. This software may be bundled with the user application software.
- *EDI hardware platform*: This is a computer for running the EDI software. In large companies it may be a mainframe or minicomputer. Most companies starting EDI implementation will run it on a PC.
- *Network*: If permissible, Internet is the easiest way to transfer messages. If a VAN is used, either a dedicated or dial-up access is used.

CHAPTER 30

Facsimile Systems

Most people would be surprised to learn that facsimile is one of the oldest forms of electrical information transmission, following close behind the telegraph. The first facsimile was a crude device consisting of two pens connected to two pendulums, joined by a wire. The apparatus was able to reproduce writing on an electrically conductive surface. The year was 1843, the same year that Congress appropriated $30,000 to build an experimental telegraph line from Washington to Baltimore. The telegraph expanded quickly, but facsimile languished for more than a century before it became a mainstream product. In a prophetic statement in 1872, the U.S. Postmaster General, Jonathan Creswell said, "The facsimile systems and the countless other applications of electricity to the transmission of intelligence yet to be made must eventually interfere with the transportation of letters by the slower means of post."

News media used facsimile to send wire photos and the military used it widely in World War II. These devices scanned the image and converted it to an analog tone—one tone for white and another for black. Some used photosensitive receiving media, which required developing the image like a photo. Other devices converted the analog tone to an electrical current that varied the heat of a stylus, which literally burned the image into paper. The receiving paper mounted on a rotating drum similar to a lathe, and as the image burned, a thin trickle of smoke rose from the machine.

Facsimile is one of the first and best illustrations of the potency of international standards. Fax technology improved steadily as solid-state electronics advanced through the second half of the twentieth century, but the standards were proprietary. Since fax transmission requires matched pairs of machines, this limited its application to communication between organizations that had deliberately chosen compatible machines. The first fax standards, Group 1, were published in 1968. These machines, now obsolete, required about 6 min to transmit a page. Long-distance telephone at the time cost about 40 cents per minute, so fax was neither fast nor cheap. Group 2 fax, introduced in 1976, cut the transmission time in half. Group 3 standards, published in 1980, were the turning point that caused fax to become indispensable. ITU published the final standards, Group 4, in 1984. Group 4 is intended for transmitting high-quality documents over ISDN lines. Today every organization and many individuals have G3 fax capability, either through a machine or over the Internet, but G4 has never caught on.

Group 3 facsimile is digital, making it a natural adjunct to the computer. *Fax-on-demand* (FoD) marries fax and IVR technology to enable callers to request technical documents and product information and have them sent to a fax machine. The Web is a more effective way of providing such information today, but many companies still offer FoD service. Most desktop computer modems can be used interchangeably for either data or fax, bringing fax capability down to the desktop. Many companies have internal servers to make facsimile available to everyone on the company network. Users can send and receive faxes without leaving their desks, and without the extra cost of a dedicated business line or an analog station port on the PBX. Fax documents can be converted to text with optical character recognition (OCR) software. It is not quite as convenient as receiving a document attached to an e-mail message, but it is fast, particularly when it comes to sending documents that originate on paper instead of a computer file. Fulfilling Mr. Creswell's prophecy, a facsimile machine can send a letter in a fraction of a minute for less than the cost of postage.

The Internet has changed the face of facsimile and will continue to do so. Besides eliminating much of the need for FoD, users find it easier to send documents as an e-mail attachment than to fax them. One of the main drawbacks of fax servers is the difficulty of delivering incoming fax to the desktop, which is not a problem with e-mail. Even if the original document is not stored in a computer file, scanners have become so inexpensive that it is often easier to scan and e-mail a document than to carry it to the office fax machine. Furthermore, for fax to be effective, it requires access to a phone line, or at least a switching device that recognizes the fax tone. If a document can be sent over the Internet, its incremental cost is essentially zero.

Even with the Internet and e-mail as alternatives, fax still has several advantages. You never have to think about compatibility. The documents cannot be edited, which is important for contracts, invoices, and such documents. It is easy to send paper documents, while retaining the original format and any graphics. In this chapter we explain how facsimile works and discuss the alternatives to a regular fax machine including computer fax boards, fax over IP (FoIP), and service bureaus that provide you with a fax number and send and receive faxes worldwide on your behalf.

FACSIMILE TECHNOLOGY

A facsimile machine has four major elements as diagrammed in Figure 30-1: scanner, printer, controller, and communications facilities. The scanner sweeps across

FIGURE 30-1

Block Diagram of a Facsimile Machine

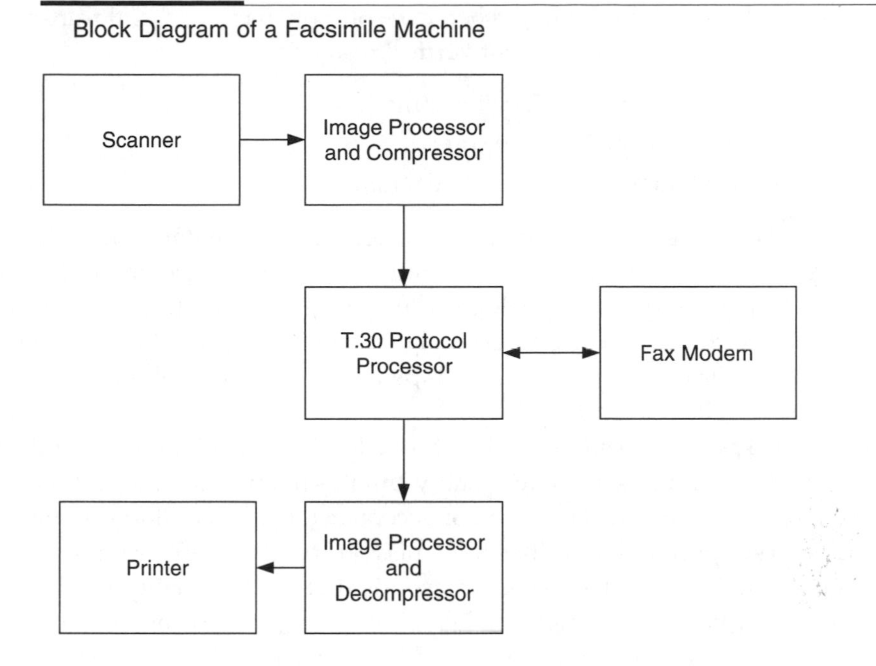

a page, segmenting it into multiple lines in much the same way that a television camera scans an image. The scanner output converts the image to a binary code. The digital signal is compressed and applied directly to a digital circuit in the case of G4, or through a modem to an analog circuit. Control circuitry directs the scanning rate and compresses solid expanses of black or white to reduce transmission time. At the receiving end the incoming signal is demodulated and drives a print mechanism that reproduces the incoming image on paper.

Facsimile Machine Characteristics

G3 equipment is categorized by modulation method, speed, resolution, and transmission rate. Digital machines produce a binary signal that is applied to the circuit through a modem. The digital transmission speed varies from as low as 2400 bps to 64 Kbps. The time required to send a page depends on circuit quality and the amount of information encoded. Typically, a single-spaced page of text takes a minute or less depending on factors that we will discuss later. Resolution is measured in lines per inch (lpi) and varies from slightly less than 100 lpi to nearly 400 lpi.

G3 transmits at a maximum data rate of 14.4 Kbps, and falls back in steps to a sustainable rate. On a noisy line, the rate may be as low as 2.4 Kbps. The original standard had a maximum speed of 9.6 Kbps, but the V.17 protocol increased the speed to 14.4 Kbps. Some products can transmit as high as 33.6 Kbps under the right conditions. The latter are known as "super group 3." Group 3 faxes use T.30

protocol to negotiate the connection and manage the details of the session. The T.4 protocol controls the coding scheme, resolution, page size, and transmission time. Group 3 fax has three different vertical resolutions:

- Standard: 98 lpi (3.85 lines/mm)
- Fine: 196 lpi (7.7 lines/mm)
- Superfine: 391 lpi (15.4 lines/mm)

The superfine resolution is not an official standard, but many machines support it. The horizontal resolution is 203 picture elements (PELs) per inch (8 PELs/mm). A PEL is analogous to a pixel in video transmission. Digital facsimile machines produce a binary signal that is either on (black) or off (white) for each PEL. The number of PELs and lpi determines the resolution and transmission time of digital facsimile.

G4 fax has a resolution of up to 400 lpi. It is intended for transmission of documents needing laser printer quality, but it is not popular except for special applications. One reason for its lack of acceptance is the fact that it requires an ISDN line to support its 56/64 Kbps transmission speed, and ISDN lines are rare, at least in North America. The second reason is that G3 fax quality is good enough for most applications. If high quality is required, the document can be converted at the receiving end with OCR, and if an image such as an original signature is required, the document can be scanned at the source, converted to a PDF file, and sent as an e-mail attachment.

The ITU-T standards specify the protocol, scanning rate, phasing, scans per millimeter, synchronization, and modulation method. Phasing is the process of starting the printer and the scanner at the same position on the page at the beginning of a transmission. Synchronization keeps the scanner and printer aligned for the duration of the transmission.

Scanners

Most machines use a flatbed scanner that converts an image to binary form. As the document moves through the feeder, the scanner detects light from the source and emits a 1 or 0 pulse, depending on whether the reflected light is above or below the threshold. A single line can be represented in 1728 bits or PELs. In standard resolution there are 1145 lines in the document, which equates to about 2 Mb of information. This amount of data would take nearly 2.5 min to transmit at 14.4 Kbps, so it must be compressed to reduce transmission time.

Transmission times vary with the line quality and the density of the information on a page. Many documents have expanses of white or black that are compressed by a process called *run-length encoding*. Instead of transmitting a string of zeros or ones corresponding to a long stretch of white or black, the length of the run is encoded into a short data message. By using run-length encoding, a digital facsimile machine compresses data into approximately one-eighth of the number of nonencoded bits, but the amount of compression depends on the character of the document. For example, a document with a border around it cannot be compressed vertically. The average typewritten document is compressed by a factor of at least 10:1.

The speed of a fax machine is often quoted in pages per minute. ITU-T publishes a reference page known as Test Chart # 1 that is standard for quoting transmission speed. Four percent of the standard page is black, and the rest is white. A page with more black content will take longer to transmit.

Facsimile Handshake

When fax machines connect, they go through a handshake process to resolve such variables as the speed of transmission, compression method, and special features the machines may have. The initial handshake takes about 16 s for fax machines up to 14,400 bps. They also pass messages at the end of each page to verify transmission, a process known as *retraining*. This function takes about four seconds per page. The G3 standard uses a compression method known as Modified Huffman (MH). Two other methods provide faster image transmission through additional compression. These are modified read (MR) and modified modified read (MMR). MMR is an advanced data encoding and compression scheme similar to that used in G4. Both of these still require 16 s for the initial handshake at lower speeds, but with V.34 modems the time is cut in half. The retraining interval is cut from 4 to 0.25 s. The handshaking time has a significant effect on facsimile transmission speed. As the modem speed increases, the transmission time does not decrease in direct proportion because the handshake remains fixed. Only when the faster MMR protocol is employed does improved transmission speed become significant.

When the receiving fax machine answers the line, the transmitting machine sends a CNG (calling) tone to signal the receiving machine that a fax machine is calling. Fax line-sharing devices recognize this tone and switch the line to the fax machine. The receiving fax sends a CED (called) tone to signal the transmitter that it is connected to a fax machine. At this point, the two machines send a series of signals to communicate such variables as paper size, modem speed, compression method, and resolution. The sending fax transmits its transmitting subscriber identification, without which the receiving machine refuses the session. These signals up to this point are sent at 300 bps.

Next, the machines exchange training signals, starting at 14.4 Kbps and gradually reducing speed if necessary. At 7.2 and 9.6 Kbps, the data is modulated on an analog carrier with QAM. Trellis modulation is used at 9.6 and 14.4 Kbps. When the transmitting machine receives confirmation that the session has been set up, it begins sending the image. The receiver checks each line of data for errors with an error correction mode. At the end of the page it slows down to 300 bps and either sends an end of message or a multipage signal to indicate that more is to come. At the end of the session, the transmitter and receiver hang up the line.

Printing

Fax machines either use plain paper or coated electro-sensitive paper. Coated fax paper comes in a roll, and is less desirable than plain paper fax. Its major disadvantages are that it curls up, and it discolors, but it is durable and the machine is cheap to manufacture. Coated paper deteriorates over time, so it must be run through a copier to make it permanent.

Plain-paper fax creates documents that have the same degree of permanence as printed output. Plain-paper fax machines are more expensive, but supplies are less costly, so for high-volume applications they are the preferable method. The principal plain-paper fax printing technologies are thermal film, laser, and inkjet. A thermal-film machine uses a page-width ribbon that melts onto the paper. These machines are also inexpensive, but the ribbon is good only for a single use. Laser and inkjet printers are identical to those used in regular office copiers and printers. Many products combine fax, copying, scanning, and printing into a single device.

FACSIMILE FEATURES

Some facsimile machines have features that make them complete document-communications centers, designed for attended or unattended operation. For example, stations can be equipped with polling features so a master can interrogate slave stations and retrieve messages from queue. In some machines the master polls the remote, after which the remote redials the master to send the document. Other machines have a feature called reverse polling, which enables the receiving machine to transmit a document on the initial poll.

Most fax machines have automatic digital terminal identification capability and apply a time and date stamp to transmitted and received documents. Machines can be equipped with document feeders and stackers to enable them to send and receive documents while unattended. Machines using coated paper often include cutters to separate the document at each page. Some digital facsimile machines contain memory to store digitized messages and route them to designated addressees on either a selective or a broadcast basis. Memory also enables fax machines to scan and store documents while they are dialing out, and receive documents before the printer is ready to start or when they are out of printing supplies. Most facsimile machines handle only standard letter or legal size paper, but some machines have a larger bed for handling oversized paper.

Besides the above, the following features are available as options on fax machines and servers:

- *Automatic dial directory*. This feature is the same as speed dialing on a telephone. Direct speed-dialing buttons are available in some machines. In others the directory is stored in memory and is recalled from a list.

- *Automatic redial*. When a busy signal is encountered the document is retained in memory or held in the document feeder. The machine continues to dial until the transmission goes through.
- *Confidential transmission*. The document is sent into memory in the receiving machine, and can be printed only by entering a code.
- *Custom cover sheets*. This feature is common with PC fax boards and with fax servers. The system creates a cover sheet with a few keystrokes by the operator and automatically fills in details such as sender's name, number of pages, return fax and telephone number, and custom graphics.
- *Distinctive ring detection*. The fax machine can be set up to respond to different ringing codes sent from the central office.
- *Document verification*. The system stamps each page transmitted to enable the operator to detect misfeeds.
- *Duplex transmission*. The system can send and receive simultaneously over the same connection.
- *Dynamic port allocation*. This feature limits fax broadcasting to certain ports so incoming ports will not be tied up with outgoing calls.
- *E-mail gateway*. This feature is normally found only in fax servers. The server links to an electronic mail system and uses it to receive and deliver fax messages.
- *Forwarding*. The system can receive a document and forward to a different machine without the loss of resolution that results from forwarding a fax document that has been printed.
- *Group fax*. This feature enables the user to send the same fax to an address list.
- *Halftone capability*. Enables the system to transmit photographs.
- *Message logging*. The system logs both failed and correctly received messages. The date, time of day, received telephone number, and other significant information are logged. The log can be interrogated or printed automatically.
- *Priority queuing*. In a fax server certain stations can be designated as high priority. Their transmissions take precedence over those of lower priority.
- *Receive alerts*. When a fax message is received, an alert pops up on the computer screen or telephone display to notify the user.
- *Transmission scheduling*. The system can schedule faxes for delivery at night when rates are lower. This feature is particularly important when large documents are sent to multiple locations. The system also can reschedule when delivery fails for some reason such as busy or no answer.

FACSIMILE ALTERNATIVES

The Group 3 fax machine is the most common way of faxing, but other alternatives are available. Nearly every modem available for or built into a laptop or desktop computer has fax capability. Accompanying software displays the fax on the screen, and it can be sent to a printer if required. Fax servers, FoD systems, and voice mail are all convenient ways of handling faxes. One of the most intriguing is Internet fax, largely because it costs little or nothing.

Facsimile over IP

As the price of fax machines has continued to drop, one drawback remains: the need to either dedicate a phone line to the fax or share it through distinctive ringing or a fax switcher. Analog phone lines can be eliminated completely by sending the faxes over an IP network, which can either be the Internet or private. ITU-T and the IETF worked together to devise two different standards for FoIP. Simple pass-through FoIP is covered by T.37 standards. Real-time FoIP, also known as fax relay, is covered in T.38.

The major difficulty with inserting an IP network between two fax machines is timing. Machines must exchange the various handshake signals within controlled time intervals or else the session fails. The fax pass-through mode is also known in ITU terms as voice-band data (VBD), which is the transmission of fax or modem signals over a packet network. The gateways must detect that the call is from a fax machine, but they treat it like a voice call. Low bit-rate codecs cannot be used for fax transmission. The fax gateways must ensure that they use the G.711 (full 64 Kbps) algorithm with no voice activity detection or echo cancellation. This method is simple, but it may not work if delay or jitter disrupts the timing.

A more reliable method is the fax relay, illustrated in Figure 30-2. In this method, the gateway carries on a T.30 dialog with the sending fax machine over the PSTN. The message is then packetized and sent over a VoIP network using RTP for the transport to the other end. There the process is reversed with the receiving gateway communicating with the receiving fax machine using T.30.

FIGURE 30-2

Fax over IP

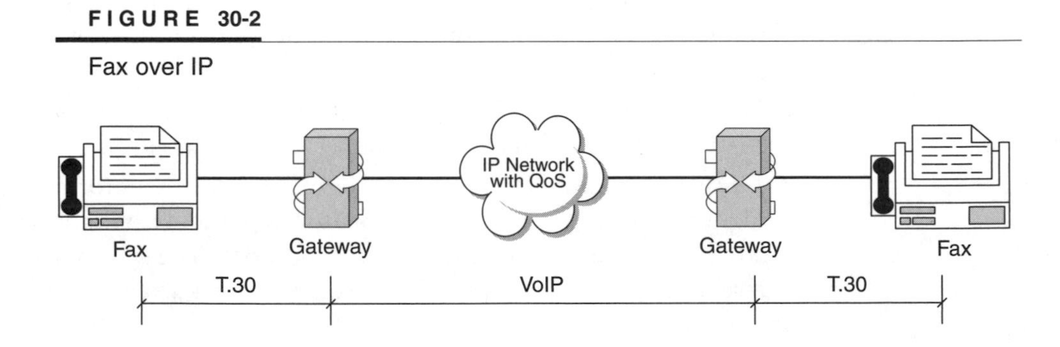

The gateways spoof the two machines into thinking they have a PSTN connection. As with the pass-through mode, the call must be identified as a fax call and QoS must be enabled to ensure sufficient bandwidth.

Facsimile via Internet E-mail

Fee-based fax-by-e-mail service is an alternative to IP-based fax, which is usually associated with VoIP. Several services use a Web portal to send and deliver faxes. The service provides you with a personal PSTN fax number that can be reached from anywhere in the world. In the receiving direction, the fax is converted to a graphic file and delivered as an e-mail attachment or the recipient can view it on the Web.

To send a fax you can send an e-mail to the service bureau, which converts it to a document and transmits it. You can also compose the fax on the service bureau's Web site. Electronic mail services can transmit documents with embedded graphics that are stored in their library. For example, a sender can store a logo and signature block and insert them in the transmitted document. Many of these services also offer fax broadcasting, in which documents are sent to a directory of users.

Fax Boards and Modems

A board plugged into an expansion slot in a desktop computer or a fax feature in a modem allows a computer to emulate a facsimile machine. Fax-enabled application software can transmit a document via fax by treating the fax board or modem as a printer. The user does not have to go through any special steps to convert the document for transmission. The application software takes care of the conversion and dials the recipient without any special action on the user's part except selecting the addressee.

Fax software offers features that are available only on high-end facsimile machines. Since the file to be transmitted is retrieved from a disk, a facsimile board can transmit a document or a list of documents to a list of different telephone numbers with minimal effort. The fax board can store an almost unlimited number of pages. Signatures, logos, and other graphics may be integrated from a separate file, but authentication requires additional steps. PC facsimile is also useful as a relay device. With conventional facsimile, each time a received document is rescanned and transmitted, it loses clarity. Since a PC facsimile is received and sent from a file, it can be relayed indefinitely without loss of detail.

Although a PC board is a satisfactory means of transmitting fax, receiving is another matter. Not all fax boards operate in background mode, and those that do may slow down the computer while a document is being received. Received documents can be displayed on either the screen or printer. Since smaller monitors are incapable of displaying an entire page in readable form, it is necessary to scroll the document, and the resolution may be too low to read the document without expanding it.

Although the cost of a fax board is less than a full-featured fax machine, it is less convenient for users to share, more time consuming for transmitting graphics, and less than adequate for receiving documents. It also requires access to an analog phone line, which is becoming less prevalent as more organizations move such applications to the LAN.

Fax Servers

In most offices everyone needs access to fax, which may result in queues at the fax machine. Furthermore, when many outgoing faxes are being sent the number rings busy to incoming callers. The solution to the busy fax machine is more lines and machines, or some form of fax store-and-forward capability. The latter is provided by LECs as a service, and it is available in fax servers and fax-enabled voice-mail systems.

Fax servers offer the following advantages over standalone fax machines:

- It is unnecessary to print documents before faxing them, which saves cost and time.
- The quality of received documents is higher because it is unnecessary to scan them if they are machine-readable.
- It eliminates human queues at the fax machine, and therefore increases productivity.
- Automatic delivery of received documents may be possible with one of the methods discussed later.
- The system automatically redials busy numbers and those that do not answer.
- Fax lines can be left free to answer incoming calls.
- Fax documents can be sent over the company network to file in another server, eliminating the need to go outside the system, and saving transmission costs.
- Fax can originate from within fax-enabled applications. Word processors, spreadsheets, and other applications treat fax servers and modems as printers.

Fax servers have some drawbacks that limit their application. Some of the principal disadvantages of a server compared to a standalone fax machine are:

- If the document is not already in a computer file, it must be converted with a scanner.
- Signatures are difficult to control and authenticate. Scanned signatures can be added to documents, but it is easy for someone other than the signer to add it.

- Bit-mapped fax files are slower to print than text. They can be converted to text with OCR, but the conversion process is time consuming.
- Automatic routing of incoming faxes is problematic in many applications.

The choice of modems to use with fax servers is important. Inexpensive fax modems are readily available, but they lack the features of more expensive modems. TIA/EIA has developed standards for fax modems known as Class 1 and Class 2 standards. The common fax modems that are available everywhere are Class 1 modems (TIA/EIA 578). They assume that most of the transmission features are built into the application software. Class 2 modems (TIA/EIA 592) build communication features into hardware. Class 2 modems have flow control capability to enable the receiving modem to control the transmission rate. The initial handshake between the two modems, which complies with ITU T.30 standards, establishes their ability to regulate transmission rates.

Class 1 modems assume that fax-timing specifications will be implemented in software. Class 2 modems handle timing from an on-board processor. The Class 2 standards provide copy quality checking, which is absent from the Class 1 specification. These and other features make Class 2 modems preferable for heavy fax server applications, but some devices the server communicates with may be unable to comply with Class 2 standards.

Inbound Fax Delivery

The delivery of incoming faxes from a server is a problem with few ideal solutions. With standalone machines paper output must be delivered manually, which is time-consuming and often results in delays. Theoretically, fax servers can deliver faxes directly to the desktop by one of three methods. The most reliable is direct inward dialing, in which each user has his or her own DID fax number. This method is common with voice-mail systems. It is the most reliable method, but it adds expense for DID numbers plus the additional cost of the hardware and software in the voice mail.

A second method is to scan the cover page with OCR software and pick the recipient's name out of the resulting file. This method is fine in theory, but reliability is low because of different ways the recipient's name may be written. Users report about 75 percent reliability with this method. The third method is using additional DTMF digits for routing the document. This method is reliable, but it requires the sender to know the receiver's personal code.

Group 4 Facsimile

ITU Group 4 facsimile standards support a high-speed, high-resolution digital system operating at speeds up to 64 Kbps using either switched 56 or BRI service. Resolution is from 200 to 400 lpi using high compression. G3 fax machines code each line individually, but lines are coded in G4 fax based on the coding of the previous line. Each line becomes a reference line for the succeeding line and only the changes are transmitted. The G4 coding standard is T.6. It achieves compression more than double that of G3 machines. G4 assumes a digital medium so it omits error checking. As discussed earlier, G4 fax is not a popular application because other alternatives are available, but where high-quality images are important, it provides better results than G3. For example, law enforcement agencies use it for picture and fingerprint transmission.

The G4 standard separates fax machines into three classes. All three classes support 100 PEL per inch horizontal resolution. Classes 2 and 3 have 300 PEL per inch resolution with options of 240 and 400 lpi. Table 30-1 lists the characteristics of the three classes of Group 4 machines. The quality of Group 4 fax is about as good as an office laser printer, which makes it suitable for transmitting letter-quality documents.

In addition to offering the advantages of high speed and improved resolution, G4 supports communications between word processors. Communicating word processors transfer data between their memories using a feature known as *teletex* (not to be confused with teletext, which is the transmission of information during the vertical blanking interval of a video signal). Group 4 facsimile standards make it possible to integrate facsimile and communicating word processors. To overcome the inherent disadvantages of each type of system, Class 2 and 3 machines send textual information in alphanumeric form and graphic information in facsimile form. Terminals capable of this form of communication are called mixed mode. Page memory is required for memory-to-memory transmission.

Facsimile Line-Sharing Devices

It is awkward to receive fax messages if the machine is not connected to a dedicated telephone line. Several methods of sharing the line manually are used, but

TABLE 30-1

ITU-T Group 4 Facsimile Characteristics

	Class		
Service	1	2	3
Facsimile	Transmit/receive	Transmit/receive	Transmit/receive
Teletex		Receive	Create/transmit/receive
Mixed Mode		Receive	Create/transmit/receive
		Resolution (lines/inch)	
Standard	200	200/300	200/300
Optional	240/300/400	240/400	240/400

none are foolproof. One way is to let the phone ring and answer it if the fax does not pick it up. Another way is to listen for the CNG tone and connect the fax if one is heard. Some devices answer the telephone with synthesized voice that instructs the caller to press a DTMF key to direct the call to a person. Otherwise, the device connects to the fax machine to initiate a facsimile session. More reliable are two different types of device that enable line sharing among fax, voice, and modem.

Some line-sharing devices monitor the line for a CNG tone and switch to the fax port if one is detected. Otherwise, the phone rings until answered. A more effective method is distinctive ringing, which most LECs offer, and which many fax machines support. Distinctive ringing assigns more than one telephone number to the same line, with each number assigned a different ringing code. A decoding box recognizes the ring and switches it to telephone, fax, or modem as appropriate. Most fax machines can recognize distinctive rings.

FACSIMILE APPLICATION ISSUES

Fax is undoubtedly the largest single data transmission application on the PSTN. The criteria discussed below apply primarily to the telecommunication aspects of facsimile. A discussion of the technical requirements and features of facsimile machines is beyond the scope of this book.

Document Characteristics

The primary factors in determining which group of facsimile equipment to select are the document volume and resolution required. Plain-paper fax offers superior quality compared to coated paper. Of the plain-paper devices, laser printers offer somewhat better quality than ink jet machines, although the difference is unimportant in most companies.

Many machines offer a high-resolution mode, which produces quality close to that of a letter-quality printer. If copy quality is important, a high-resolution machine should be selected. The need for halftones in documents should also be considered. Facsimile machines vary widely in their ability to handle halftones, with the level of gray scale varying from 0 to 64 shades of gray.

Labor-saving features such as automatic document feed, paper cutter, automatic dial and answer, document storage, and document routing should be considered. If confidential information is being transmitted, encryption should be considered.

Compatibility

When special features such as halftone transmission, networking, and polling are required the systems should be from the same manufacturer or fully tested with systems of other manufacturer. Compatibility between machines for basic sending and receiving is generally not a problem.

Document Memory

Most high-end fax machines offer memory for document storage. This feature is useful for unattended operation. If the machine runs out of paper, received documents can be stored in memory. Multipage documents can be sent without fear of paper jams. Documents can be sent to address lists that are recalled from memory. Transmission time is often reduced because the machine receives into memory so that transmission is not paced by the speed of the printer.

Fax Server or Fax Machine

The question of whether to use a fax server or fax machines may arise in some organizations. As a practical matter, a fax server will not replace all fax machines, but a server has several advantages that weigh in its favor:

- Faxes can be sent with a click of a button instead of printing the document and taking it to a fax machine.
- Faxes can be sent to groups quickly and with less manual effort.
- Lost time waiting for busy fax numbers is eliminated.
- Fax numbers are stored in one central directory, reducing personal fax directories.
- Faxes are of higher quality because they are sent directly from a high-resolution computer file.
- Faxes can easily be archived.
- If incoming faxes are distributed automatically, they can be kept private.
- Costs can be distributed using a standard call-accounting system.
- Faxes can be sent at off-peak times.

A fax server has disadvantages in addition to the cost that may outweigh the advantages in some organizations.

- A fax server can fax only computer files so paper documents must be scanned.
- Routing automatically to recipients is difficult and may be expensive.
- Some standalone fax machines will still be needed to print faxes.

PART FIVE
Telecommunications Networks

The final portion of this book is devoted to a discussion of practical networks that are the building blocks of the enterprise network. Once the province of organizations large enough to justify a mainframe computer, the Internet has brought network connections down to small organizations that use it to connect to business partners and customers and carry on electronic commerce. In this part we will discuss the services that a multitude of providers offer.

The industry has developed a vast collection of protocols that have been integrated into a group of services, some of which have been successful and others that flared briefly and fizzled, although not from the lack of hype. Some, like ATM, were expected to become almost the exclusive communications medium. ATM is widely used in the carrier backbone, but it failed to gain popular acceptance at the enterprise network. Much of the reason can be traced down to the LAN. Years ago the discussion centered over whether Ethernet and its contention protocol were robust enough to support industrial strength data communications. Numerous alternatives were advanced, but Ethernet has outstripped them all to the point that fast Ethernet is included in nearly every computer built today.

Ironically, 100 Mbps was considered so fast that FDDI was developed as a backbone protocol. Outside legacy networks, FDDI is rarely applied today because gigabit Ethernet is cheaper, faster, and simpler. Despite the fact that the original 10-Mbps Ethernet bandwidth is still more than most users need, arguments are being advanced for gigabit Ethernet to the desktop, while 10G Ethernet supports the backbone. Both protocols are moving outward from the LAN into the metropolitan network and discussions of a 100G protocol are beginning to surface.

One of the major factors driving network change is convergence. The concept has been promoted for more than a decade, but it appears that the economics and technical considerations are beginning to merge and it will shape future networks. Fully converged networks are still in the early-adopter stages, but the trends are clear. The obvious motivation for enterprise networks is saving money, but cost saving is often insignificant compared to other benefits that can be achieved by designing and managing the network to support multiple media.

This part of the book is about applications and how these services are selected and managed to bind the organization with its employees, customers, and business partners by means of the communications media it employs. The final chapter concludes with a glimpse at the future and how the networks and services will continue to evolve.

CHAPTER 31

Enterprise Networks

The telecommunications industry has no universal definition of an enterprise network. A network with no connections outside the enterprise could meet the definition because it is critical to the functioning of the organization. Usually, however, an enterprise network implies external connections to remote offices, employees, customers, business partners, patients, constituents, or anyone who has a business relationship that they conduct over the network. When such a network malfunctions, business relationships are impaired or crippled, so design and management are more than incidental functions.

Thinking about enterprise networks has changed over the past few decades. In the heyday of mainframe computers, the enterprise network was a way of linking selected employees to corporate databases. The scope and size of the mainframe limited networks to those organizations with major data-processing operations. Smaller organizations transmitted data over teletypewriters and the nearest thing to data storage was a piece of paper or a punched paper tape. The shift began when companies such as Digital Equipment Corporation and Hewlett Packard scaled the mainframe down to minicomputer size, paving the way for more organizations to mechanize selected applications.

Both mainframes and minicomputers had common characteristics that shaped the enterprise networks of the time. The intelligence resided in the central computer, which meant the only applications that could be used from the desktop were those that ran on the host. Some, such as word processing, were available from the central computer but these were provided only to the select few that had terminals. The closest the industry came to desktop computing was a timeshare machine on which users could write their own applications in a language such as Basic. The other property of centralized computer was the fact that their networks were proprietary and closed. IBM's SNA was at that time a complex collection of hardware and protocols that scaled to prodigious dimensions, governing a flock of dumb terminals and printers.

Ironically, IBM set in motion the events that made SNA obsolescent by introducing the PC and opening its architecture to outside developers. Virtually no one in 1982 could foresee the enormous influence the IBM PC would have on technology. It was a perfect partner for the LAN, the standards for which were developed in the same timeframe. It took several years for LAN technology to catch hold, but once it did, the shape of the network changed by quantum leaps. Freed from the limits of host-based software and faced with technology they could purchase under departmental budgets, users purchased desktop devices by the millions and enterprise network managers had no choice but to adapt.

Figure 31-1 shows the main evolutionary steps that brought networking to today's state. Enterprise networks of the past were single-purpose. To access multiple applications, users often needed separate terminals for each. With its low cost and versatility, the PC soon replaced dumb terminals. At first, it was equipped with special cards and terminal-emulation software. When in terminal-emulation mode, a PC was no different to the user than the dumb terminal it replaced, but it was possible to switch to desktop software, which was easier to use and had more features than anything available on the mainframe. The architecture at this stage was awkward. The terminal-emulation mode required hard cable: coax back to the controller or the mainframe. Twin-ax, coax, or RS-232 were common. The benefits of sharing files and peripherals mandated a separate LAN connection. This meant multiple cables to each desktop, which clogged conduits and crowded overhead ceiling space.

FIGURE 31-1

Evolution of the Enterprise Network

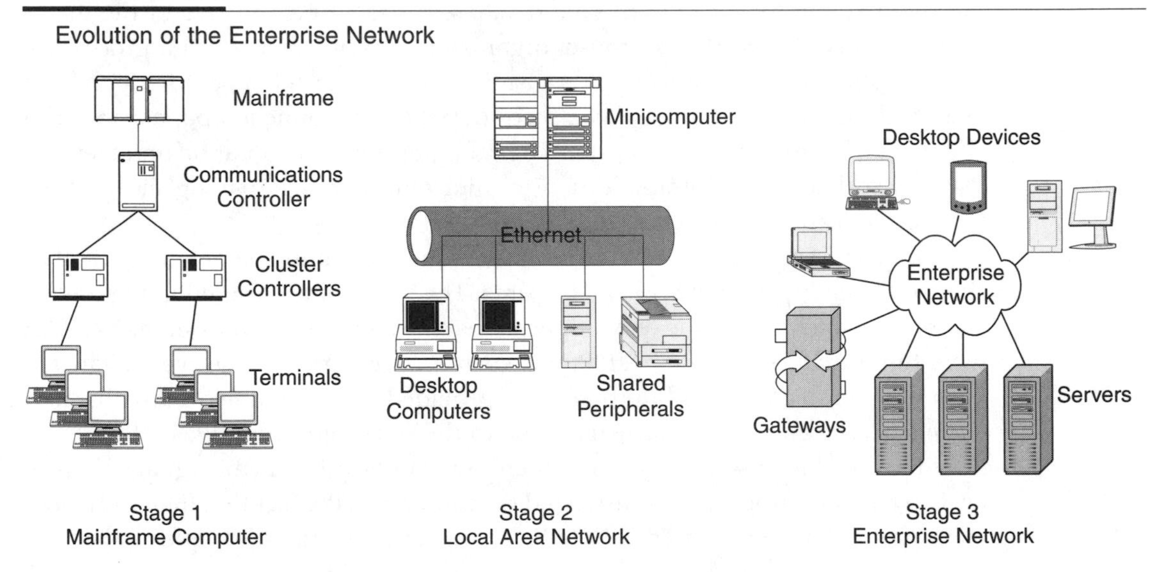

The next step in the evolution in the late 1980s and early 1990s was to connect the mainframe or minicomputer to the LAN so users could run its applications over the LAN, still in terminal-emulation mode. Sometimes this was as simple as installing a card in a minicomputer and loading software to link the desktop devices to the central computer. For the IBM mainframe, this represented a major point of departure because SNA was designed on the principle that all communications went through the host. IBM enhanced SNA with Advanced Peer-to-Peer Networking (APPN), a group of protocols that allows sessions to be established between peer nodes. At this stage of evolution, LANs were shared-media systems, but traffic was relatively light. The Internet was still the private enclave of academics, government, and a few businesses, and the bandwidth demands of the World Wide Web had not yet materialized.

The next stage of development brought enterprise networks to where they are today. The enterprise network provides a communications infrastructure that sustains the strategic objectives of the enterprise. Ideally, the network is a converged multimedia infrastructure that is flexible, easy to manage, and based on open standards. That ideal is reached gradually in most organizations because of the embedded base of existing equipment and the need to show a return on investment by migrating to a new model. No longer the exclusive terrain of large organizations, networks serve even the smallest enterprises and they are almost certainly connected to the public Internet. The network islands and silos of the past are giving way to an infrastructure that enables workers to communicate with and share information in real time with customers, business partners, and employees without regard to their location or preferred communications medium.

Both technology and user demand are driving this phase of network development. Organizations of a few years ago could survive comfortably without a technology plan because their data communications were internal and their voice communications used the PSTN or a private network. That degree of isolation is no longer feasible. The company that does not display its presence on the Web is invisible to a substantial portion of its clientele. Delivering information and services efficiently is the cornerstone of the enterprise network. The enterprise opens itself to its constituency by delivering an infrastructure that is available, secure, and easy to access. This means it is based on open standards and handles multiple traffic types with consistent quality and reliability.

Part of this evolution is driven by the concept of convergence, a subject we will discuss in more detail in Chapter 37. The networks of the past shared backbone facilities by dividing bandwidth, but the media remained separate. In the converged network, different types of traffic coexist seamlessly. The enterprise network must adapt to handle these multiple traffic types and support a broad range of applications. This requires classifying and prioritizing traffic so that one class does not the cripple the service level of another.

ELEMENTS OF THE ENTERPRISE NETWORK

The enterprise network designer chooses a fundamental standards-based architecture with close linkage between the architecture and the enterprise's strategies. The network is almost certainly based on the use of IP, but that does not mean it is converged at the outset or that it uses an IP backbone. Many networks run over point-to-point circuits, frame relay, and ATM. Enterprise networks are evolutionary. They must consider existing equipment and replace and upgrade it as economics dictate. The objective is a single network that is flexible, secure, easy to manage, and supports all communication needs. It adjusts to current and future applications, functions reliably, and delivers service at an affordable price.

The enterprise network focuses on its constituents: internal users, customers, and suppliers, while defeating hostile forces from the outside that have the objective of bringing it down. The network takes into account trends in the workplace. The current shibboleth is collaboration. Recognizing that teamwork occurs among people, network managers can do little to cause collaboration to happen, but they can ensure that the network supports it at the appropriate time. Today's work force is increasingly mobile and the enterprise network is called on to enhance mobility by supporting wireless applications and remote access.

The main attributes of the enterprise network are these:

- *Quality of service* (*QoS*). The network is designed, configured, and managed with QoS as its keystone. This means that traffic is classified at the source and prioritized end-to-end through an infrastructure that requires nothing of its users outside their normal method of operation. The ideal network is an extension of its applications and remains invisible to the users except for the jack in the wall.
- *Standards-based*. Recognizing that no single vendor can provide all the elements of the enterprise network, the design and structure are based on open standards. This does not preclude the use of equipment with proprietary elements, but cross-platform and cross-product interoperability are required. Furthermore, purchases, including those made with departmental budgets, must conform to the company's equipment and application standards.
- *Security*. The network is designed, configured, and managed with security and server protection paramount. It integrates both wired and wireless components and is designed to detect and prevent intrusion at the edge and to survive attacks without service interruption. Policies and procedures such as sanitizing files and applications are conveyed to users to protect the network core.
- *Reliability*. Fault tolerance is designed into the network. Components are manufactured to high standards of performance and are hot swappable. Management announces and adheres to service-level agreements (SLAs)

that give the users assurance that mission-critical applications will not experience unscheduled downtime.

- *Asset and investment protection*. Existing equipment is applied for maximum benefit. Duplicate infrastructures are avoided by adapting to new applications and new technologies without major changes or upgrades.

A major advantage of the mainframe computer of the past was the ability of the network manager to enforce uniformity. Today, equipment is so inexpensive and readily available that uniformity can be achieved only through policies and persuasion. Rogue equipment can be brought in unnoticed and open security holes that are difficult to detect and impossible to defend. Without control of the applications, it is difficult to deliver agreed-upon SLAs, so this means the enterprise needs to have plans and strategies for adopting new technology and applications. The enterprise must develop policies in two senses of the word. On one hand are written directives and procedures that ensure that the necessary controls are in place. The other is technical policy that provides the bandwidth and QoS that the applications demand.

Policy-based networking means permitting access to resources to those with the need to use them while denying it to those that do not. It means managing traffic flow, prioritizing applications, and delivering bandwidth-differentiated services according to the needs of the application. It implies that servers and applications are classified based on their position and behavior in the network. Policies are established to ensure that users have appropriate privileges. All of this requires more information than the network manager can possibly have at his or her fingertips, which implies mechanized support. We will discuss policy-based networking in a later section.

A technical structure must be in place to manage the network with internal resources, by outsourcing, or with a combination of both. Testing and managing networks, as we discuss in Chapters 38 and 39, is costly, but the cost of interruption can be devastating. Effective network management requires that performance be monitored in real time and that problems are detected and corrected before they have an adverse impact on service. The network operations center must be staffed with people with the appropriate skills and the necessary tools. Knowledge of traffic flows is essential, e.g., what applications communicate with other applications, at what times of the day, and with what bandwidth requirements. Besides performance monitoring, security violations must be detected and thwarted to prevent damage. Policy-based networking is in its infancy, but networks of the future will not survive without it.

The enterprise network is multiprotocol, multivendor, and multiapplication. The increasing globalization of business means the network is multilocation and multinational as well. The strategic implications of the network also mean that it is likely to be multicompany. Such a network is difficult to manage and control. As companies do business electronically with their strategic partners, conflicts arise in protocols, governmental regulations, and standards. The enterprise network ties the parts together into a unified design and architecture.

BUILDING BLOCKS OF THE ENTERPRISE NETWORK

This section discusses the elements of an enterprise network. These have been discussed in earlier chapters and are presented here with applications to show how they fit as an integrated whole.

Terminal Equipment

Figure 31-1 shows the evolution of terminal equipment interfaces to a host computer. Many legacy networks resemble the left side of the figure: a host computer controls a network of dumb terminals. The terminals connect through a cluster controller or multiplexer to the host, which contains the database and manages the network. The dumb terminal has practically disappeared, replaced by the personal computer—at first emulating the terminal and then operating as a peer as shown in the central part of the figure. Local area networks, usually using twisted pair wire, replace the coaxial and RS-232 wiring to controllers and multiplexers.

Local Area Networks

The foundation of virtually every network today is the Ethernet LAN. The basic building blocks—the NIC, high-quality UTP wire, switches, servers, network operating system, and routers—are becoming so common as to be practically commodities. The 100-Mbps LAN with layer 2 and 3 switches, VLAN and QoS capability, a backbone with sufficient bandwidth, and PoE provide capacity for nearly any application the enterprise adopts.

Circuits

The building block of the network of the past was the voice grade analog circuit. With conditioning it supports data and with signaling it supports voice. The limited bandwidth of the voice-grade circuit gave rise to digital data service (DDS), a digital service operating at speeds up to 56 Kbps. Based on the bit-robbed signaling of a T1 backbone, DDS is expensive and incapable of providing 64-Kbps clear channel circuits. As the demand for greater bandwidths increased, the major IXCs provided T1/E1 and fractional T1/E1 facilities and, increasingly, are delivering T3/E3 and fractional T3/E3. The trend clearly is toward providing point-to-point circuits in bulk. A T1 can be leased for the price of six or seven voice grade circuits; a T3 can be obtained for the price of about the same number of T1s. Anyone needing point-to-point bandwidth should evaluate fractional T1/E1 if a full T1 is not needed. Fractional T1/E1 is a point-to-point service between major metropolitan areas. Some LECs offer fractional T1/E1, but others offer digital service only in DS-0, DS-1, and DS-3 increments.

With the demise of the voice-grade analog circuit, bandwidth is provided in the standard SONET/SDH levels. This bandwidth may be divided and voice and data kept separate. Although this foregoes the advantage of efficient multiplexing, it is an inexpensive way of sharing facilities between voice and data. IXCs can split the bandwidth at the central office, using part of it for access to the long-distance network and the remainder for access to point-to-point, frame relay data services, or even local service. An IAD at the customer's premises splits the circuit into its component parts as shown in Figure 31-2. The information travels over an access circuit, which is usually T1/E1, to the IXC's central office, where it is separated into its component parts, typically in a DCS.

Common-Carrier Switched Facilities

Long-distance costs have dropped to the point that it is difficult to justify a private voice network except where it can be shared with data and ride for minimal cost or where the volume is high enough to justify a voice VPN, which we discuss in a later section. One of the theories behind divestiture was that competition would

FIGURE 31-2

Shared Access Facilities

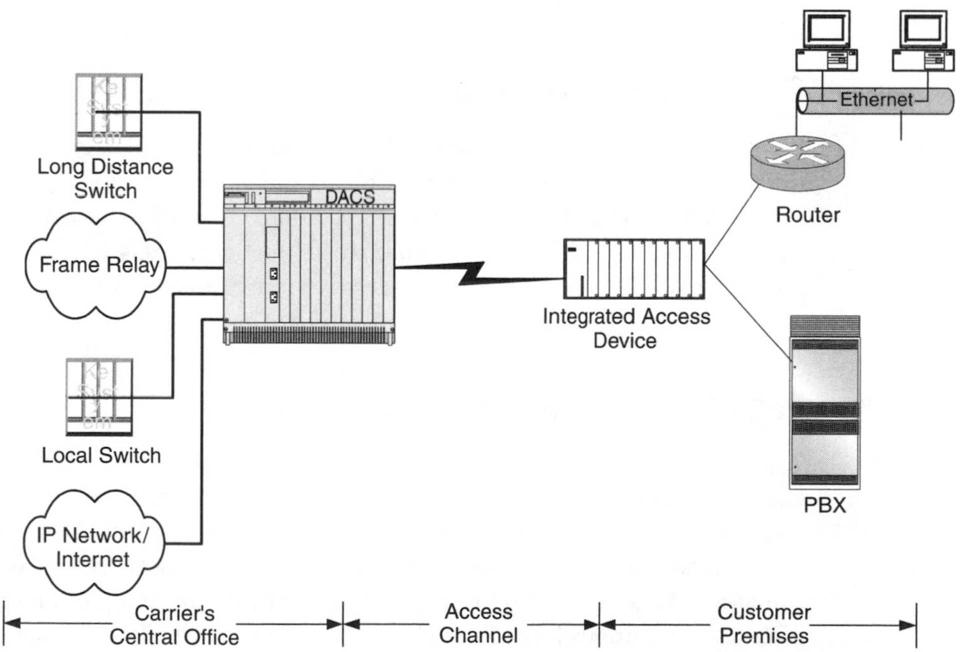

drive down long-distance costs, and that has proved to be true. In the past, switched long-distance networks offered two alternatives: wide area telephone service (WATS) for the heavy users and direct distance dialing for everyone else. Now, WATS has disappeared. Large users connect directly to the long-distance carrier's switch with T1/E1 facilities and smaller users negotiate reduced-cost long distance over switched access.

The long-distance market can be segregated into three groups of users. At the low end are small users who have no alternative but to use switched access. Mid-sized users can save money by using T1/E1 access for large sites and switched access for smaller locations. If a private data network serves the remote sites, long distance from those sites can be transported to the larger location and switched over dedicated access facilities. The largest organizations can usually justify a voice VPN, which offers the equivalent of a private voice network, but over common-carrier facilities.

Premises Switching Systems

The PBX is the most familiar kind of circuit switch in private networks. Larger networks use class 5 or tandem switches with a private system generic program, but most private networks use a PBX for the purpose. As voice and data converge, it becomes more effective to use one of the varieties of IP switching systems. Common-carrier switching services also are available. The most familiar type is the Centrex services that most LECs offer. Centrex systems can support the same T1/E1 long-distance services and special trunks that PBXs support. Centrex, also, is migrating toward IP.

Facility Termination Equipment

Private facilities are terminated in equipment that provides testing access, conditions the signal to meet the line protocol, and divides the bandwidth among the users. Digital facilities terminate in a CSU that converts the bipolar line signal to the T1/E1 format as well as other maintenance functions. CSUs with add-drop capability are available to divide the line. Access to individual channels is also obtained by terminating the line in a channel bank, an IAD, or a digital crossconnect.

PRIVATE VOICE NETWORKS

A private voice network tends to integrate the enterprise more tightly. Human endeavor revolves around communication, and the more personalized the communication, the more effective the organization. Multilocation companies can be more closely knit if they can dial each other with an abbreviated dialing code, transfer calls across the network, and all locations are part of the same directory. Small organizations have little choice but to use the PSTN, at least until they can justify a converged network. Large organizations can develop private voice networks by one of the two methods. The traditional method is to use an ETN. This arrangement connects switches together with a network of T1/E1 lines. The T1/E1 lines carry a fixed monthly rate that is independent of the amount of traffic. For companies with heavy usage, an ETN can be cost-effective. Long-distance costs have dropped, however, to the point that it is difficult for an ETN to compete with common-carrier pricing.

Of particular interest to many companies is the voice VPN. A voice VPN operates as if it is composed of private voice circuits, but is actually part of the IXC's switched network. AT&T's virtual network is Software Defined Network (SDN), MCI's is V-Net, and Sprint's is Voice VPN. A voice VPN depends on SS7 to link the various IXC switching nodes and to direct them to behave as if they were a private network. The service definitions are retained in the IXC's SCP database, which is queried over the data network that links the switches at the SSPs. (Refer to Figure 12-2 for this architecture.)

Stations in a VPN are defined as off net if they access the IXC through the LEC and on net if they bypass the LEC with a T1/E1 link to the IXC. Unlike conventional T1/E1 long-distance access, a VPN can both terminate and originate calls over the access line. This enables the IXC to bypass the usage charges the LECs impose on the originating and terminating ends of a call. Calls placed over the network are rated in three categories: on-net-to-on-net, on-net-to-off-net, or off-net-to-off-net. The on-net portion of calls does not incur access charges, which reduces the cost of the call.

When the IXC's switch receives a call setup request, it sends a message to its SCP database requesting instructions. The SCP checks for restrictions and service classifications, such as forced account code dialing, and returns a message to the originating switch, sends routing information to the switches in the connection, and directs the switches to connect the path.

Virtual networks are economical for large companies that have a considerable amount of on-net calling. Most of the features of a dedicated private network can be provided. For example, locations can call each other with an abbreviated dialing plan, calls can be restricted from selected area or country codes, and other dialing privileges can be applied based on trunk group or location. Special billing arrangements are provided. Call detail furnished online or on a CD/ROM enables the company to analyze long-distance costs in a variety of ways.

PRIVATE DATA NETWORKS

The services for implementing data networks are abundant. For convenience, they are classified as local area, metropolitan, and wide area or global services. That distinction is diminishing as Ethernet transcends its original distance limitations and ventures into the metropolitan network, which we discuss in Chapter 32. The WAN–metropolitan distinction is also fuzzy because the same protocols may be used for both. A private data network uses fixed point-to-point bandwidth or a common-carrier service such as frame relay (Chapter 34), ATM (Chapter 35), or a private IP network (Chapter 36). The Internet is also available for private network use, with security precautions discussed in the next section.

Data VPNs

The main attraction of connection-oriented services such as frame relay as a platform for private networks is the lack of concern about security. The Internet has spawned a type of data network known in the trade as a VPN, referred to here as a data VPN to distinguish it from a voice VPN. A VPN is a set of sites that communicate with one another over a public IP network while maintaining the security and management capabilities of a dedicated circuit or frame relay network. The basic functions of a VPN are membership discovery—who belongs to what VPN and the establishment of a secure tunnel through the network. VPN subscribers have the following objectives for the network:

- *Security*. The VPN is secure from unauthorized access to the same degree as a network implemented over frame relay or dedicated circuits.
- *Connectivity*. Any authorized site can use IP's connectionless capability to connect to other sites. New sites can be added quickly. Mobile users can access the network from remote locations. Also, the network can span multiple service providers where necessary.
- *Simplicity*. The network is easy to set up and manage.
- *Resiliency*. The network can respond rapidly to changing traffic patterns.
- *Scalability*. The network can scale to meet changing needs as the subscriber adds locations or connects to external users such as customers and business partners.
- *Quality*. The VPN can support multiple media including voice, video, and multicast with sufficient QoS. The service provider offers and adheres to SLAs based on worst-case scenarios as opposed to averages.

There is no single standard configuration for a data VPN, but the illustration in Figure 31-3 is typical. A VPN consists of a combination of authentication, tunneling, access control, and encryption that is designed to carry data securely over a public network. A network *tunnel* is a metaphor for the process of encapsulating the data of one protocol inside the data field of another protocol. Encapsulation permits data originated on one protocol to transit otherwise incompatible networks including a hostile network such as the Internet. It is not normally necessary to encrypt data that is carried exclusively on common-carrier networks because these are normally considered as secure. Nevertheless, many organizations encrypt everything including voice as a matter of course.

FIGURE 31-3

Data Virtual Private Network

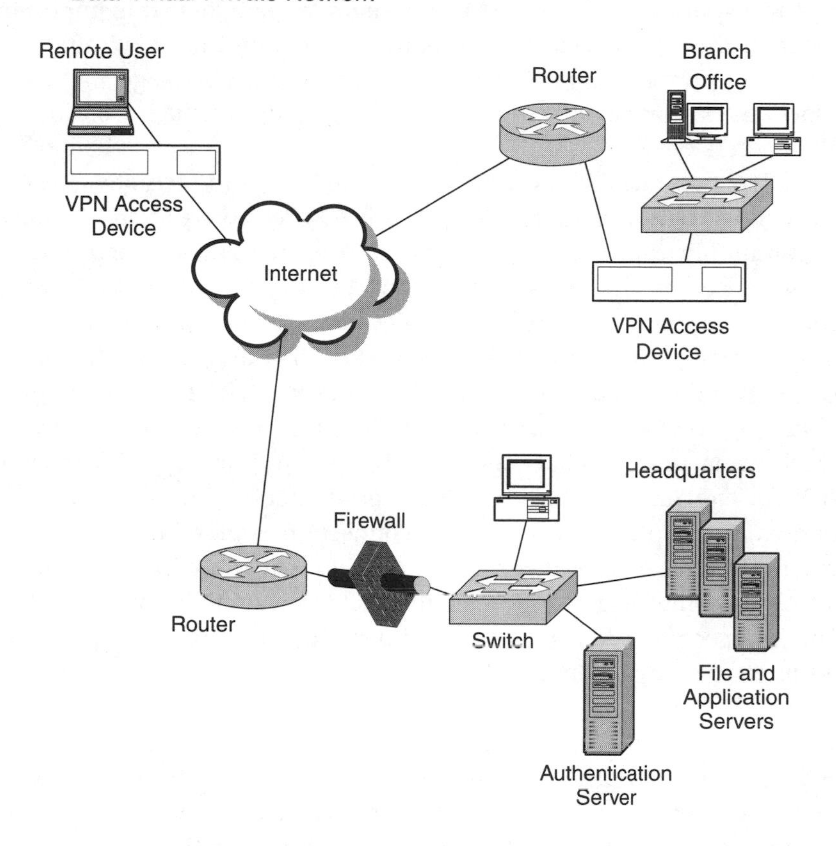

VPNs operate at either layer 2 or layer 3, with the latter comprising the majority of VPNs in service. In an L3VPN the service provider routes packets across the network based on customers' IP addresses. In an L2VPN the packets are forwarded based on L2 information or port information, and the customer is responsible for routing. An L3VPN is deployed over an IP network and emulates a multisite routed network. A layer 2 VPN emulates a bridged LAN connection over a VPN and is often called virtual private LAN service (VPLS). The industry uses the term *pseudowire* to describe the emulation of a native point-to-point service over a packet network.

Organizations can develop their own VPN or purchase the service from a carrier. A VPN over a private IP network consists of routers and access devices connected with leased circuits, frame relay, ATM, or other secure facilities, providing the private equivalent of the Internet. Such a network provides security that is equivalent to frame relay, but it is tricky to set up because in almost every case it will be called on to serve nodes that have direct Internet connections. One solution that may require less administrative expertise is to outsource the network to a common carrier that manages everything. Although the user sends IP traffic to the network, the carrier may wrap the traffic in cells or frames and transport it across an ATM or frame relay network. It is important to know whether the carrier uses any part of the Internet for the VPN traffic. If so, encryption is required. Several equipment manufacturers package the necessary VPN functions of routing, encryption, firewalling, and tunneling protocols into a single VPN access device.

The two main purposes of a VPN are to provide remote access and site-to-site connectivity. If a site-to-site VPN is used exclusively as an internal network, it is commonly called an *intranet*. If it is used for outsider access it is called an *extranet*. Remote access users can connect to the network with either a dedicated connection such as DSL or cable or they can use dial-up. The savings are particularly attractive for overseas users. Since the VPN may transit the Internet, it must have several methods for keeping the data secure. The most common method of obtaining the necessary security is to build encrypted tunnels through the network using protocols such as IPSec (IP Security) or Layer 2 Tunneling Protocol (L2TP). This method, known as the overlay model, achieves security at the price of greater complexity because of its connection-oriented nature. For sites to communicate with one another either a tunnel must be defined between sites or the site must communicate through a third site. The first alternative adds complexity and the second adds unnecessary traffic to the third site's access circuit plus adding the delay inherent in the extra hop.

IP Security

IPSec is a framework of standards that ensures private communications at the network layer. It introduces enhanced encryption and authentication through two modes: tunnel and transport. The tunnel mode encrypts the entire packet including the header. The transport mode encrypts only the payload. The encrypted packets in transport mode look like ordinary packets, so they can be routed transparently. In the tunnel mode, the encrypted packet is enclosed in a new packet, adding to the overhead, but further increasing security by keeping addresses private.

For IPSec to work, all of the devices on the network must share a common public encryption key. The authentication process must also be encrypted, so some method of bootstrapping the process is needed. This is generally done by the use of an independent digital certificate authority or by the two ends sending each other keys through some other secure media. After the initial authentication, the Internet Key Exchange (IKE) protocol allows the devices to negotiate a secure tunnel between them.

Authentication, Authorization, and Accounting (AAA)

AAA authentication is frequently used in a remote-access VPN with dial-up clients. When a call arrives, the request is forwarded to the AAA server, which verifies the caller's identity, authorization, and tracks the call for auditing or billing purposes. A common method of authentication is the industry standard remote authentication dial-in user service (RADIUS) server, which provides several methods of user authentication.

Tunneling Protocols

Tunnels are of two types, end-to-end and node-to-node. An end-to-end tunnel extends from the remote PC through the network to the server. In this configuration the encryption, decryption, and session setup happen at the ends of the connection. In node-to-node tunneling the tunnel terminates at the edge of the network. The traffic connects through edge devices such as a router or the VPN access devices shown in Figure 31-3, where it is encrypted and tunneled to a matching device at the edge of the distant network.

The primary tunneling protocols in use besides IPSec are L2TP and Point-to-Point Tunneling Protocol (PPTP). PPTP and L2TP support Password Authentication Protocol (PAP) and Challenge Handshake Authentication Protocol (CHAP).

PRIVATE NETWORK DEVELOPMENT ISSUES

The overview of the building blocks of telecommunications networks shows that there is no shortage of choices. This section discusses some of the principal design issues that managers must consider in implementing networks.

Security

Connections to the Internet raise security concerns for every network manager. External threats from viruses, Trojan horses, worms, and denial-of-service attacks are an infuriating reality that requires constant vigilance. A more subtle threat to security comes from internal users who can compromise security inadvertently or launch internal attacks that destroy files or access them without authorization. Rogue wireless nodes, laptops, PDAs, and cell phones that carry infections around the firewall and open discussions of confidential information on public IM are just a few ways users can jeopardize the best security precautions.

Network security involves the following issues:

- *Physical security*. Network and computer equipment and the circuits that connect them are physically secure. Equipment rooms and wiring closets are kept locked and the keys under control. The facilities are clean and free of debris and fire hazards.
- *Terminal security*. Access to the network is controlled. Dial-up circuits use a system such as dial-back, a hardware security device, or an authorization server to prevent unauthorized access. Passwords are controlled and revised periodically. Terminals and computers are kept physically

secure and the keyboards locked if possible. Laptops are secured so that if they are lost, they cannot be used to access the network.

- *Disaster recovery*. The network has a plan for restoring service in case common-carrier services fail or major equipment is lost because of fire, earthquake, sabotage, or other disasters.
- *Data security*. Firewalls are configured to close all unused ports. Information that is carried over wireless or public networks is encrypted and keys are guarded and transported only over secure media. Access is controlled with smart cards, biometrics, or RADIUS server. Intrusion detection devices at the edge alert management to attempts to penetrate or deny service to the network.
- *Policies*. The enterprise's policies for safeguarding information are clearly written, published, and regularly reviewed. Employees are sensitized to the nature of security threats and instructed on appropriate preventive measures.

Policy-Based Networking

A major issue in deploying a multimedia network is determining which users or applications receive preferential access to network resources. The industry uses the term *policy* to encompass the practices and systems needed to regulate access to network resources. Certain applications need priority access and all need predictable service levels. All network equipment including routers, switches, firewalls, and hosts must participate in a plan for discriminating between packets that can tolerate best-effort service and those that cannot and for identifying which applications users are permitted to employ.

Policies include the conditions under which the policy is activated and the action that is to be taken. The policy is enforced by the apparatus that applies the action under the appropriate conditions. The enforcer could be static, e.g., toll restrictions on particular stations, router access control lists, and authentication servers are examples of static enforcers. The allocation model may change with time of day and day of week and with the class of service assigned to users. For example, if the board of directors is having a video conference, its packets have a higher priority than a desktop video conference between two engineers.

Effective policy adapts to the needs of the moment. A company might choose to throttle Web access during peak hours, but allow it at other hours if the load permits. This requires some method of monitoring the traffic flow and taking remedial action if it falls outside objective bounds. For example, time-sensitive UDP packets must take precedence over TCP packets. They can be identified easily enough and prioritized with DiffServ and the path resources can be reserved with RSVP, but whether they should be is a different question. If the network is capable of rolling voice traffic over to the PSTN at peak usage times, this may have a less detrimental impact on the organization than, say, delaying HTTP traffic. The difficulty is how the network manager makes this determination, especially in an environment that changes dynamically.

Policy is unique to each organization and presupposes the availability of information and capability of control that are often beyond human capacity. A rough policy of sorts can be built with a firewall and router, but a policy-based network requires a policy server, which is a specialized device that monitors data flow and administers management's policies. The server is probably linked to an LDAP directory to assist in identifying users who may require priority treatment.

Policy standards are primitive today and largely proprietary. Consequently, the multivendor network may be impossible to manage end-to-end. The solution for many companies is to keep increasing bandwidth to the point that policy management is not needed. Moreover, effective policy administration presupposes that someone has a thorough understanding of everyone's needs and knows how to translate them into computer code. The policy server gives troubleshooters yet another place to look when things go wrong. For example, a complaint about slow response time could result from a policy that reserves a substantial part of the network's bandwidth for a scheduled video conference. These issues will be resolved as designers and managers gain more experience, but in the meantime, policy-based networking is more vision than actuality.

Future Compatibility

Network products and services are changing so rapidly that it is difficult to be sure the current design will be compatible with the future shape of the network. Voice and video over IP will be an important part of many networks in the future. Services that demand high bandwidth are here now and will grow. Some will have a strong impact and others may fizzle. The key to designing a network is to remain flexible—not locked into a single technology that will limit the organization's ability to follow the shifting telecommunications environment.

Vendor Independence

In the past, network managers relied on a single vendor for the mainframe computer and the network components. Having a single vendor to hold responsible for network performance is comforting, but not practical for most networks now that servers have replaced central computers. Network designers must determine whether standards are open enough to remain free of proprietary equipment and protocols. Many network components such as NICs, servers, and Ethernet switches are effectively commodities that can be purchased from almost anyone, but the greater the degree of vendor independence, the more the reliance of the enterprise on internal resources. The need for policy management and a network management system often precludes vendor independence. On the other hand, no vendor has a monopoly on technology, and niche players are more apt to have better prices or superior performance in one family of products because their developmental efforts are more concentrated.

Network Management Issues

Chapter 39 discusses principles of testing and managing networks. The hierarchical networks of the past had a significant advantage over the peer-to-peer networks of the present and future: they were easier to manage. Vendor-specific network management products make it possible to look into components and diagnose and sometimes clear trouble. As the network becomes multivendor, these management capabilities diminish.

Network Planning

Many managers today face a dilemma. Control of computing budgets is moving from the MIS department to the end users. Users purchase computers and wireless equipment today as they purchased office machines in the past: they are justified on an individual basis and the network manager may lack control over the applications. Organizations that have control over equipment standards are in the best position to plan the network. If the information in desktop devices is of any value, someone must also plan for such factors as security, regular data backup, and network capacity.

ENTERPRISE NETWORK APPLICATION ISSUES

With the trend toward obtaining digital circuits in bulk, as either full or fractional T1/E1, the most cost-effective networks are those that integrate different applications at the circuit level. Effective network convergence requires a systematic planning approach to choose which of the many alternatives are to be employed. Developing a network generally involves the following:

- *Identify the applications.* Present and future applications including voice, data, video, facsimile, imaging, and all other foreseeable communications services should be identified. It is not enough to consider only present applications. Knowledge of future plans and expected growth is essential.
- *Identify locations to serve.* The geographic location of all points on the network must be identified along with the makeup of the users at that location and their potential demands on the network.
- *Determine traffic volume.* The amount of traffic, both terminating and originating, should be identified at each location. Determine the volume, type, and length of data transactions. Determine the quantity of voice

traffic from sources such as common-carrier bills, traffic usage recorders, and call-accounting systems. Identify both on-net and off-net traffic. It is useful to create a matrix of traffic flows between locations.

- *Determine network type*. Each application will have an optimum network type to support it. Short-range, high-speed data applications are served by LANs. Routers and remote bridges can link geographically dispersed LANs with a common interest. Companies with large amounts of intra-company voice traffic should consider a voice VPN. Companies with multiple remote offices can consider a data VPN.
- *Develop network topology*. The topology of the network is based on the applications. Costs of alternative transmission methods are calculated, and where the volume and cost of traffic are enough to justify private circuits or a public network such as a virtual network, these are added to the design. Optimize the design by trying different combinations of circuits and by selecting alternative concentration points.
- *Develop security measures*. The network must be secure from unauthorized access. Develop plans to secure it physically and to prevent unwanted access. Where the public Internet is involved for private intranet and extranet communications, encryption and authentication are necessary.
- *Determine how the network will be managed*. Most networks use SNMP and a proprietary network management system to oversee the network. See Chapter 39 for further information on selecting and applying a network management system.

CHAPTER 32

Metropolitan Area Networks

The telecommunications industry categorizes networks as local, metropolitan, and wide area or global. The distinction between WAN and MAN is vague because many of the same protocols and services are used without regard to the span of the network. A MAN is a public network that bridges LAN and WAN, typically covering a span of 5 to 50 km. The area is roughly the size of a city, and in larger metropolitan areas multiple providers offer MAN services. A MAN has many of the same characteristics as a WAN, and shares some of the same applications. It could be considered simply as a scaled-down version of the WAN, but many of its requirements are distinctly different.

MAN technologies are borrowed from both the LAN and WAN and work primarily at the datalink layer. MAN services are delivered over fiber optics if possible, but copper and wireless alternatives also exist. Legacy technologies used for MANs include frame relay, ATM, Fiber Distributed Data Interface (FDDI), Fibre Channel, Distributed Queue Dual Bus (DQDB), and Switched Multimegabit Data Service (SMDS). Newer networks work over gigabit and 10G Ethernet applied directly to fiber and some of the wireless technologies we discussed in Chapter 21. Besides point-to-point circuits riding on various levels of TDM, The ILECs offer frame relay with bandwidths of up to DS-3. The ILECs' frame relay networks generally span a LATA which, depending on the territory, may encompass a metropolitan area or an entire state. The following are typical applications for a MAN:

- LAN-to-LAN connectivity—sometimes called transparent LAN service (TLS)
- Storage area networks (SANs)
- Telemedicine
- Delay-sensitive data such as VoIP and video conferencing

- Low bit-rate synchronous and asynchronous data such as telemetry and alarm services
- WAN access services
- High bit-rate data such as video-on-demand
- Traditional data and voice leased-line services

None of the legacy protocols is a good match for all of these applications, several of which are bursty. Conventional networks, usually provisioned over SONET/SDH are composed of fixed-bandwidth circuits that are optimized for voice, but do not scale well for data. Fixed-bandwidth circuits are acceptable for many applications, but emerging services have a high ratio of peak-to-average data flow. For example, a physician examining an x-ray may need to move 10 megabits or more of data while the image downloads to the display, but then demand drops to zero.

Dedicated facilities are satisfactory for data applications that have reasonably constant bandwidth requirements. If the load is predictable, fixed-bandwidth circuits are satisfactory, provided they terminate at a single destination. If the bandwidth requirements are variable, the user pays a penalty, either in slow response during high-demand periods, or in wasted circuit time during low-demand periods. Furthermore, the provisioning interval for changing or rearranging circuits is lengthy and costly and cannot respond quickly to changing applications.

MANs must meet a set of conditions, many of which are unique to the metropolitan area:

- Provide high reliability equivalent to that provided by SONET/SDH rings with fast failover in case of disruption.
- Deliver bandwidth on demand in simple increments that are readily configurable.
- Offer simple and fast provisioning.
- Offer tiered services with traffic classified and segmented on a per-user or per-flow basis.
- Provide SLAs with guaranteed latency, jitter, and packet loss.
- Enforce policy. If a customer signs up for a particular level of service, police traffic to ensure that the network provides that level and no more.
- Provide end-to-end security. Users' packets must be isolated from other users.
- Provide usage-based billing.
- Make efficient use of the fiber-optic infrastructure, while holding costs to an affordable level.

The telecommunications industry offers many solutions to meet metropolitan area needs. Several legacy services that we discuss in this chapter are available, but most fall short in meeting one or more of the objectives. The current trend is to fall back on the known and proven. Ethernet has long been the default LAN protocol but its range limitations have prevented it from spanning the metropolitan area. That is changing with modifications to the Ethernet MAC.

LEGACY MAN PROTOCOLS AND SERVICES

This section discusses the primary alternatives that are traditionally used for MAN service. We discussed SONET/SDH in Chapter 17 and with that background, will review its applications in the metropolitan area, and discuss the reasons it is less than ideal for multimedia communications. Fibre Channel is a broadband protocol that is optimized for SANs. FDDI is the most fully developed of the alternatives, but it is on the downward slope of its life cycle. SMDS is discussed because it is the LECs' broadband vehicle and has reasonable availability in some areas. Two other protocols that are often used in the metropolitan area, frame relay and ATM, are mentioned here, and discussed at length in Chapters 34 and 35.

SONET/SDH

The LECs deploy SONET/SDH across their fiber-optic backbone, which makes it the default candidate for point-to-point service. Its main strength lies in the fact that it is universally available in LEC central offices, and is deployed in a dual-ring topology that makes it almost failure-proof. SONET/SDH nodes are designed to wrap around a fiber failure in 50 ms or less. Figure 32-1 shows a typical configuration.

FIGURE 32-1

Metropolitan SONET/SDH Network

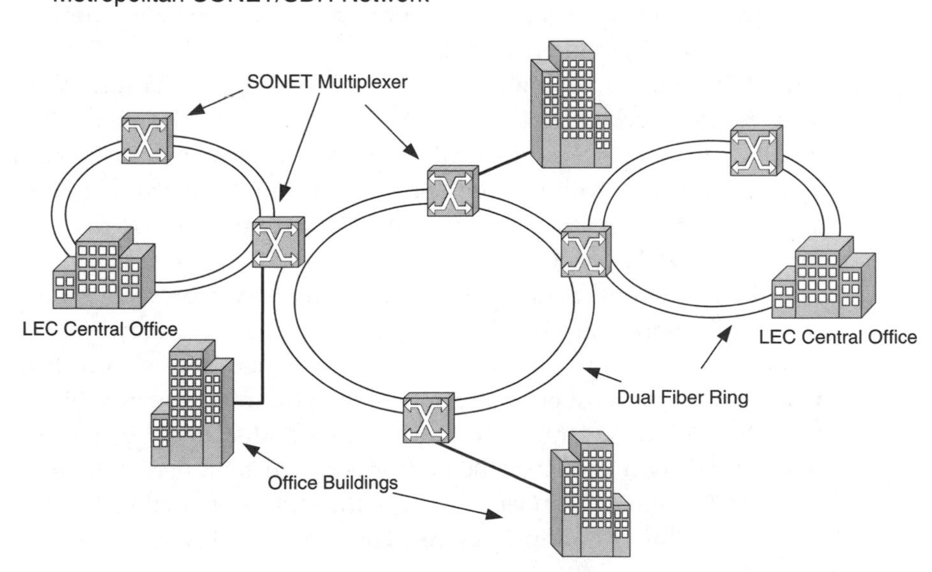

In a bidirectional line-switched ring (BLSR) traffic flows from node to node so bandwidth can be reused between any two nodes. In a unidirectional path-switched ring (UPSR) traffic circulates around the entire ring so bandwidth cannot be reused between nodes.

ATM can run over SONET/SDH to support multimedia traffic, but at a cost of greater complexity. Two less complex protocols are Packet over SONET (PoS) and Ethernet over SONET (EoS). PoS encapsulates IP packets in HDLC frames. This method has less overhead than ATM, but all traffic must be of the same general type. EoS strips away the Ethernet frame and encapsulates the payload in a Point-to-Point Protocol (PPP) frame. This method preserves the Ethernet frame structure, which eliminates the need for converting from Ethernet to another protocol and back to Ethernet again. The fixed bandwidth of SONET/SDH results in wasted capacity with either method. An entire OC-24 (1244 Gbps), for example, is required to carry gigabit Ethernet.

SONET/SDH has two major disadvantages for MAN service. The first is the fact that fiber is not available everywhere a subscriber may need it. It is usually available in large buildings, but a distributed organization such as a bank is unlikely to find that fiber is available to all of its branches. The alternative is to use T1/E1 over copper to those branches, requiring a protocol conversion. The second drawback is its lack of scalability. The slowest SONET/SDH channel runs at 1.7 Mbps and the next higher increment jumps to 51.84 Mbps. Besides being expensive and rigid for the metropolitan area, with SONET/SDH either the network may be over-provisioned and under-utilized or it is a bottleneck.

Fibre Channel

Fibre Channel almost falls outside the definition of a telecommunications technology. Its primary objective is to provide high-speed communications between computers and peripherals, primarily mass storage. It was developed to replace Small Computer System Interface (SCSI) as a high-speed protocol for applications such as data backup, recovery, and mirroring. Fibre Channel is the predominate datalink technology used in SANs. The standard interfaces are 1GFC, 2GFC, 4GFC, and 8GFC, with the number representing the speed in gigabits per second. The 8GFC standard was approved in 2004, with products expected on the market in the 2007–2008 period.

Fibre Channel is designed around a five-layer protocol model that permits communication over shielded copper wire, coax, or fiber optics. The industry association coined the term "fibre" to encompass both optical fiber and copper wire. Single-mode fiber can span 10 km. Multimode fiber with 62.5 μm core is limited to 1 km; 50 μm fiber can go twice that distance. Video coax and shielded twisted-pair wire can span 100 m. Copper wire is limited to 30 m.

Fibre Channel connects devices through a switching fabric that supports interconnection of multiple devices. This makes it ideal for a campus environment where several computers need to share access to multiple storage devices, printers, and other peripherals without being permanently and directly attached. The protocol is not limited to switched connections, however. It also supports point-to-point, loop, and switched topologies as Figure 32-2 shows. In the arbitrated loop configuration up to 126 devices can be daisy-chained in a closed loop.

The five-layer protocol is loosely based on the OSI model. The bottom layer, FC-0 is the physical layer. FC-1 is the transmission protocol, which covers byte synchronization, serial encoding and decoding, and error control. Fibre Channel shares with gigabit Ethernet an 8B/10B coding scheme, which converts 8-bit octets into 10-bit code groups. The 10-bit code groups must contain five ones and five zeros, four ones and six zeros, or six ones and four zeros. The purpose is to

FIGURE 32-2

Fibre Channel

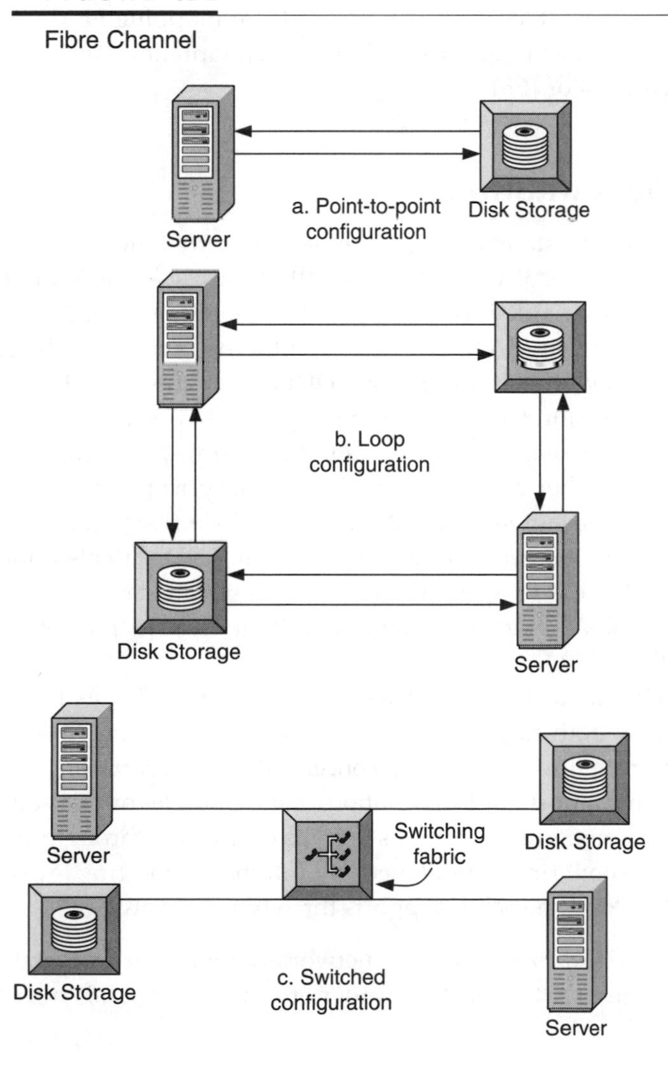

maintain clock synchronization by ensuring high transition density and no long strings of 0s and 1s.

FC-2 defines the frame format and specifies how data is transferred between nodes. The protocol defines three classes of service. Class 1 is a connection-oriented switched circuit providing the equivalent of a dedicated connection. Class 2 is a connectionless service without guaranteed delivery, but with frame acknowledgement. Class 3 is a connectionless nonacknowledged datagram service. Layer 4, FC-3, handles some of the functions of the OSI transport and session layers, and several functions that are not defined in OSI. One such function is striping, which enables multiple ports to operate in parallel to obtain additional bandwidth. The multicast function allows the same data to be sent to multiple destinations. The protocol also provides hunt groups, which enable incoming frames to hunt for an idle port.

FC-4 is the application layer, which provides buffering, synchronization, and prioritization of data. It also specifies methods for mapping Fibre Channel data to other protocols such as IP, ATM's AAL-5 (ATM adaptation layer 5), SCSI, and the logical link protocol of IEEE 802.

Fiber Distributed Data Interface

FDDI is a 100 Mbps standard that operates on either fiber or copper. In 1990 when the FDDI protocol was approved, 100 Mbps was considered high speed. The protocol was intended as a LAN backbone, although it has enough speed and range to serve as a MAN. With gigabit Ethernet and 10G Ethernet providing much greater bandwidth at lower cost, FDDI is now considered obsolescent, with applications largely limited to extending legacy networks.

FDDI operates over counter-rotating token rings; one ring is designated as primary, and the other as secondary. The primary ring carries data, while the secondary ring is in hot standby. With single-mode fiber, stations can be as far apart as 60 km and the network can span 200 km. FDDI can also ride on category 5 or higher UTP. The UTP application is sometimes called copper distributed data interface (CDDI), although the standard calls it twisted pair-physical medium dependent (TP-PMD).

The FDDI standard specifies two classes of stations. Dual-attachment stations (DAS) connect to both rings, while single-attachment stations (SAS) connect only to the primary ring through a wiring concentrator. Concentrators, which are FDDI devices that can support nonFDDI stations, can be single- or dual-attachment. The dual-ring appearance of DAS stations permits double the throughput since the station can send on both rings simultaneously, but the second ring is usually reserved for backup. The FDDI network supports three types of networks (see Figure 32-3):

- *Front-end networks.* These are networks connecting workstations through a concentrator to a host computer or other peripherals over the FDDI network.

FIGURE 32-3

FDDI Topology

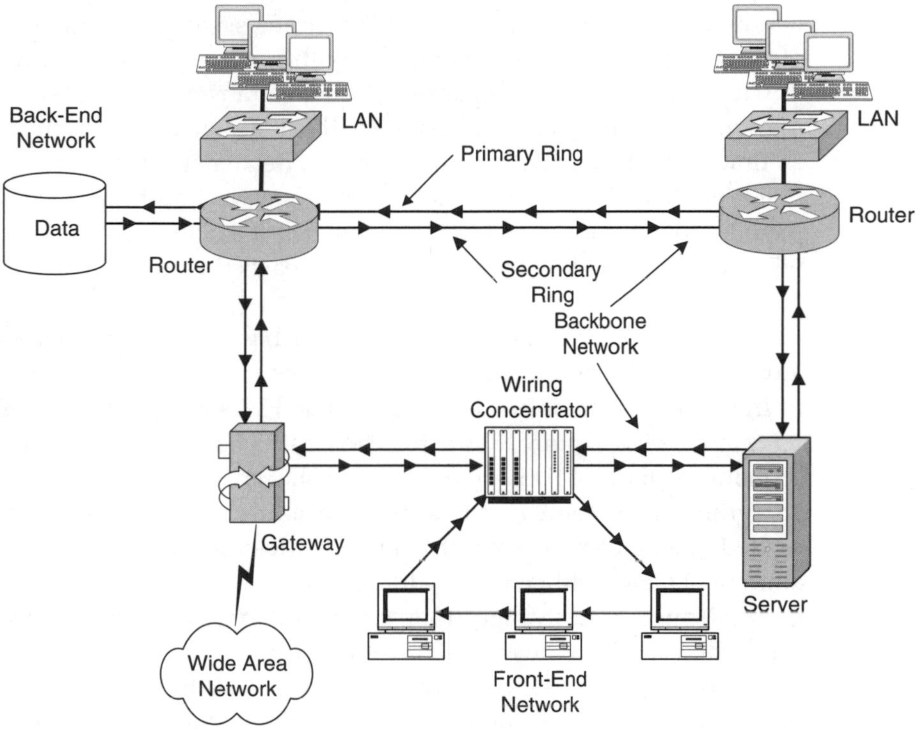

- *Back-end networks.* These allow connections from a host computer to peripherals such as high-speed disks and printers to be connected over FDDI to replace parallel bus connections.
- *Backbone networks.* These are connections between the main nodes on the network.

The dual-ring architecture provides an effective measure of protection. If a link or a station fails, the stations go into an automatic bypass mode to route around the failure. Stations on either side of the break loop the primary ring to the secondary ring to form a single ring of twice the diameter. FDDI operates much like 802.5 token ring except that multiple frames can circulate simultaneously on an FDDI network. Other differences are the line-coding format and the dual-ring configuration of FDDI, while token ring supports only a single ring.

Switched Multimegabit Data Service

SMDS is a high-speed data service that LECs and others offer for linking applications within a metropolitan area. The service can ride on ATM, although the initial designs used the IEEE 802.6 DQDB MAN protocol. DQDB is a ring protocol that uses a master control station. If the master station fails, any station can assume the role. It operates on a counter-rotating ring and uses a unique method of reserving bandwidth. The master generates 125 μs frames, each containing a fixed number of time slots. Each node is equipped with a request counter and packet countdown. When a node has a packet to send, it transmits a request upstream to the master. As the requests flow upstream, the nodes increment request counters and as time slots flow downstream, the nodes decrement their counters. Each node, therefore, knows how many slots were requested and how many were filled by passing slots intended for downstream nodes. If a node's request counter is set at *n*, when the *n*th slot goes by its counter has reached zero so it can send on the next empty slot.

SMDS is a service, the objective of which is to transport data with the any-to-any connectivity of telephone service. It is connectionless service, so unlike ATM or frame relay, there is no need to provision PVCs across the network. Both frame relay and ATM require PVCs for each pair of nodes, which makes them less economical than SMDS for full mesh connectivity. Packets are variable-length datagrams with source and destination addresses, which enables them to be carried independently across the network. Most LECs charge a flat rate for a given amount of bandwidth, with no additional cost for distance or usage. SMDS speeds in the United States range from 1.5 to 34 Mbps, the latter being the effective throughput of SMDS over a T-3 access circuit.

SMDS nodes use globally unique E.164 addresses. The service provider maintains a database of addresses that are validated to communicate with receiving stations. Nonvalid attempts are blocked. Because of its connectionless protocol, the station can connect to any device that will accept the transmission.

The principal application for SMDS is LAN interconnection, although it can be used for any data transport. It is attractive for companies with multiple sites that are interconnected with T1/E1 or T3/E3. SMDS has had little impact on the metropolitan market, mainly because it has not been widely enough deployed. It is expensive for carriers to implement and user demand is fulfilled in other ways, principally frame relay.

METROPOLITAN AREA ETHERNET

Protocols to extend Ethernet into the metropolitan area have been developed and most observers believe Ethernet will become the dominant MAN protocol. The reasons for Ethernet's popularity and its attractiveness in the MAN are easy to understand:

- Interface cards are installed on millions of computers worldwide.
- Many developers understand the protocol and it is robust and easy to implement.

- The end-to-end use of Ethernet eliminates the need to convert the LAN to another protocol such as ATM, SONET/SDH, or frame relay and back again.
- Service providers can build simple Ethernet networks over fiber without SONET or ATM and provide low-cost service in the WAN.
- OAM&P is simple with Ethernet. Multiple services can be accessed through a single user-to-network interface (UNI).

For these reasons plus the fact that legacy MAN protocols have multiple drawbacks, Ethernet in the first mile (EFM) is the dominant direction in the metropolitan network today. The protocol has great promise because it extends technologies that many manufacturers support and users trust. It is not suitable for a MAN protocol without modification, however, because it lacks provisions for QoS and service providers cannot offer the guaranteed SLAs that customers demand. Industry associations have developed extensions to the Ethernet protocol that permit it to operate outside the LAN. For Ethernet to be a viable carrier service it must include service attributes including a UNI, plus the capability of establishing virtual circuits and measuring service performance.

Gigabit Ethernet

From the datalink layer up, gigabit Ethernet uses the same 802.2 LLC as 10/100 Ethernet. The MAC is a scaled-up version of the CSMA/CD protocol used in 802.3 and operates in either half-duplex or full-duplex mode. Collisions are generally not at issue because the protocol is used in a switched environment. For the physical layer gigabit Ethernet borrows the 8B/10B encoding of Fibre Channel. It supports four different media as shown in Table 32-1. The multimode standard supports both long- and short-wave lasers. Only long-wave lasers, which span greater distances, are supported for single-mode. Multimode fiber is supported in both the 62.5 and 50 (m sizes, with the latter providing better performance. The shielded twisted-pair wire option is designed for short-haul

TABLE 32-1

Gigabit Ethernet Characteristics

Medium	Range	Abbreviation
Shielded copper wire	25 m	1000-Base-CX
Category 5 UTP	100 m	1000-Base-T
50 μm or 62.5 μm Multimode fiber optics	550 m	1000-Base-SX
Single-mode fiber optics	5 km*	1000-Base-LX

SX, short-wave laser;
LX, long-wave laser.
*Can extend to 10 km with long haul (LH) extension that is not part of the 802.3z standard.

data center applications. The UTP standard uses all four pairs of category 5e cabling, and uses a different encoding scheme than fiber optics.

Although gigabit Ethernet uses the same 8B/10B coding scheme as the FC-1 in Fibre Channel, it operates at a slightly different baud rate. The FC-1 layer receives octets from the FC-2 layer and maps them to a 10-bit pattern. This method improves the ability of receiving devices to recover clocking from the transmitted signal. Above the physical layer, gigabit Ethernet uses the standard 802.3 frame format, which enables slower speed devices to communicate over gigabit Ethernet without translation. The protocol provides a flow-control mechanism whereby a receiving station can send a frame to throttle back a sending station. Flow control operates only in a point-to-point mode, not through a switch.

10G Ethernet

The 10G Ethernet protocol is a scaled-up version of the 10/100/1000 protocols. It can be used in the backbone for aggregating slower links and it can be extended into the wide area as well as the MAN. It uses the 802.3 MAC, frame format and frame size, so it is an extension of the slower versions with two exceptions. First, it operates only in full duplex so no collision detection and recovery is required. Second, except for limited applications, 10G Ethernet operates only over fiber. The fiber protocol is designed to interoperate with SONET/SDH. The objective is to provide an end-to-end layer 2 transport without any reframing or protocol conversion.

The exception to the fiber-only medium is a copper protocol 802.3ak that was approved in 2004. That version operates over dual twinaxial cables and is intended primarily for data center use. Another standard proposed at the time of this writing is 802.3an. It is intended to support 10G Ethernet over UTP. It can reach 55 m on existing wiring. New wiring standards are needed to reach the 100-m range of category 5/6 wiring. Another task force is planning a standard 802.3aq for running 10G Ethernet over multimode fiber.

10G Ethernet defines two physical (PHY) layers: LAN and WAN. The WAN PHY is an extension of the LAN PHY, adding a SONET/SDH framer. The OC-192 interface is close to 10 gigabits, but they do not match exactly. The WAN interface provides a payload of approximately 9.29 Gbps. The specification defines seven port types: 10GBase-LR, 10GBase-ER, 10GBase-SR, 10GBase-LW, 10GBase-ZR, 10GBase-SW, and 10GBase-LX4. The copper specification adds an eighth interface, 10GBase-CX4, and the proposed UTP standard would be 10GBase-T. Figure 32-4 shows the defined configurations.

The supported ranges are:

- 1310 nm serial PMD for single-mode fiber up to 10 km
- 1550 nm serial PMD for single-mode fiber up to 40 km
- 850 nm serial PMD for multimode fiber up to 300 m
- 1310 nm serial wide wave division multiplexing (WWDM) PMD for single-mode fiber up to 10 km or multimode up to 300 m

FIGURE 32-4

Media for 10-Gigabit Ethernet

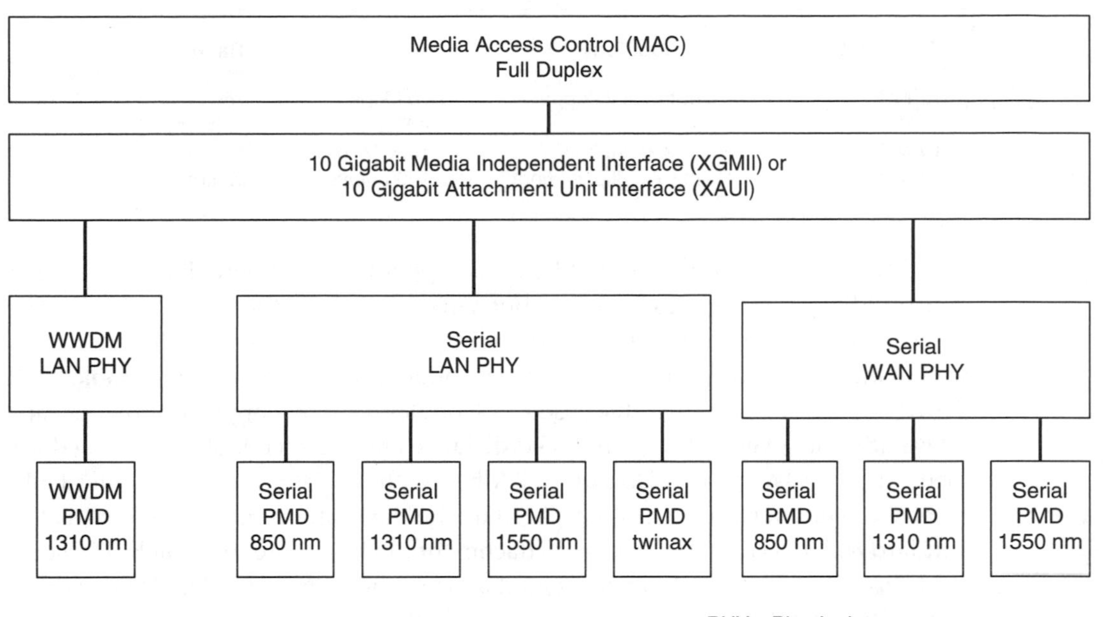

Ethernet in the First Mile

Some call it the first mile, some call it the last mile, and it is not even a mile at all, but just a metaphor for the link between users' premises and the central office that has always been a bottleneck. Ethernet in the metropolitan area is not new. Several Ethernet-based CLECs entered the market shortly after passage of the Telecommunications Act of 1996, but most failed in the technology downturn a few years later. The idea of extending Ethernet into the MAN is far from dead, however. In 2004, IEEE approved the EFM standard, 802.3ah.

The EFM vision is broadband access with a universal end-to-end technology for both residential and business subscribers. The standard defines two classes of service. Ethernet line service (E-Line) is a point-to-point virtual connection between two UNIs. Ethernet LAN (E-LAN) service offers multipoint connectivity. An Ethernet virtual connection (EVC) is an association of two or more UNIs. The market for E-Line is to provide site-to-site connectivity. The service is used to create private line services, Internet access, and point-to-point VPNs over the metropolitan area. The purpose of E-LAN is to create multipoint VPNs and TLS, which makes the entire metropolitan network look like a single LAN without the need for protocol conversion. With E-LAN service, new VLAN members can be added without involving the service provider. In both versions, multiple services can be multiplexed on the same UNI.

TABLE 32-2

Ethernet in the First Mile Technologies

Abbreviation	Topology	Speed	Range
EFMC	Point-to-point copper	10 Mbps	750 m
		2 Mbps	2700 m
EFMF	Point-to-point fiber	1000 Mbps	10 km
EFMP	Point-to-multipoint fiber	1000 Mbps	20 km

The standard supports both fiber and copper media. Where fiber is available to a building, it is the best answer, but where it is not, the standard provides Ethernet over copper. The standard defines an operations, maintenance, and administration (OAM) process plus three technologies that are shown in Table 32-2. EFMP is a PON, which we discussed in Chapter 17. The copper options encapsulate Ethernet over DSL with reduced distance and/or speed. SHDSL is used for speeds up to 10 Mbps or VDSL up to 2 Mbps within 2700 m. Unlike regular DSL, Ethernet over copper does not use ATM as the datalink protocol. The UNI is the standard 802.3 PHY, so the customer attachment can be a router or switch. The standard supports a variety of transport media including SONET/SDH, WDM, MPLS, and resilient packet ring (RPR) plus two alternative encapsulation techniques that are designed to reduce the number of MAC addresses that must be retained in switch tables: MAC-in-MAC and Q-in-Q. The need for these is discussed later.

The IEEE standards include provisions for prioritization, VLAN tagging, traffic shaping, bandwidth management, and resource reservation. The performance parameters include delay, jitter, and frame loss. The EFM standards do not include provisions for fault tolerance. That is included in the RPR standards, which we discuss in the next section, or Ethernet can be applied to SONET/SDH to take advantage of its fault tolerance. The multipoint PON option, depicted in Figure 32-5, is particularly appealing for residential networks because the bandwidth of a fiber is so far in excess of residential needs. The fiber runs through passive splitters with a single strand delivered to subscribers, which share the trunk fiber back to the central office.

VLAN Issues

When the LAN is extended to the metropolitan area, it takes on all of the characteristics of a LAN including some of the problems. One of these is the tendency of large networks to use a lot of bandwidth for broadcast and multicast traffic. The IEEE 802.1q VLAN protocol breaks large networks into smaller ones as a means of isolating broadcast traffic to VLAN members, which are recognized by port, MAC address, or IP address. The VLAN members are identified by means of a 12-bit tag known as a Q-tag that is inserted in the Ethernet frame. When VLANs are extended to the metropolitan area, the numbers of addresses they must retain

FIGURE 32-5

Passive Optical Network

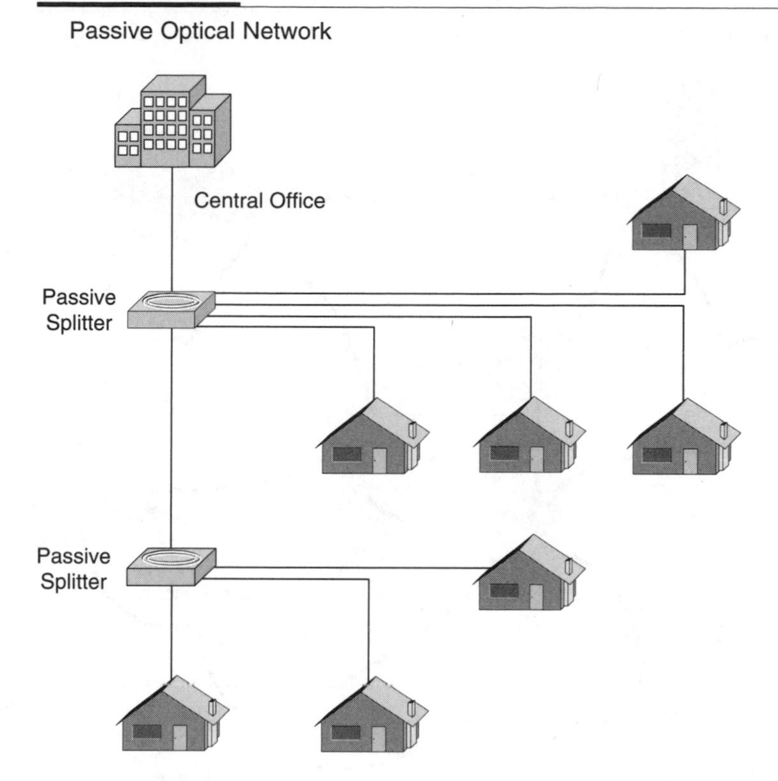

in their tables can overwhelm the core switches. The two headquarters shown in Figure 32-6 could each have several thousand MAC addresses and the branch offices increase the total. The switches in the core need to learn the MAC addresses of all the devices, resulting in huge tables. Eventually, the service can outstrip the switches' table capacity.

Several methods have been proposed to resolve the issue by adding tags to frames as they cross the ingress edge node and stripping them off at the egress node. The tag frames are forwarded only to the end host of the VLAN the originator belongs to. The Q-in-Q encapsulation method inserts an additional Q-tag into the Ethernet frame at the ingress node. In the MAC-in-MAC method the ingress nodes insert two additional MAC address fields to identify the source and destination MAC addresses. With these methods, the core switch has to learn only the MAC addresses of the edge switches.

Resilient Packet Ring

The RPR protocol, standardized in 2004 as IEEE 802.17, provides the service protection and resilience of SONET/SDH, while retaining the simplicity of Ethernet. RPR rings are dual counter-rotating rings that are optimized for data and designed

FIGURE 32-6

Metropolitan Area Ethernet

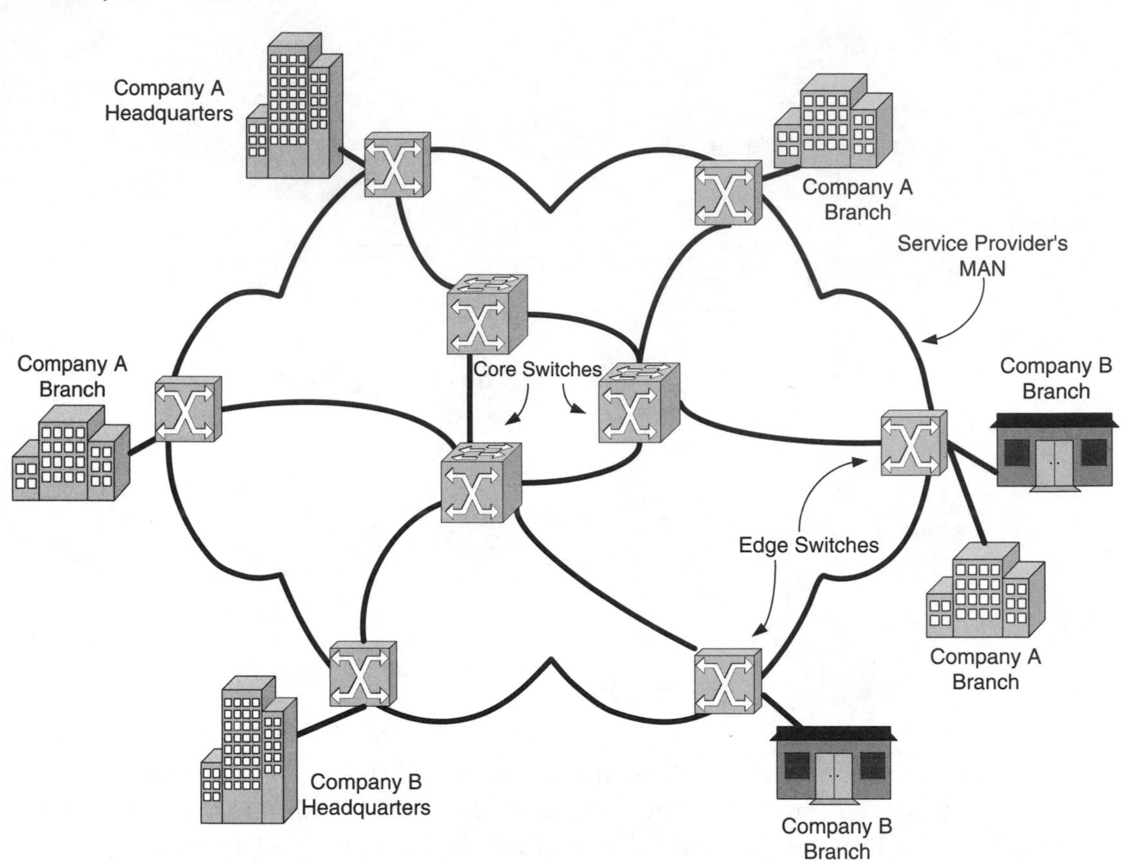

to wrap around a ring failure in 50 ms or less. Unlike DQDB, the ring has no master control node. The nodes negotiate for bandwidth based on a fairness algorithm that is programmed into each node. Each node is assigned a MAC address and can handle end-to-end or multicast traffic. The MAC protocol regulates access to the media and defines how the transmitting stations use the bandwidth. The RPR MAC has three service classes:

- Class A is a high rate CIR service. It supports guaranteed bandwidth, low latency, and circuit emulation. It is suitable for time-sensitive traffic such as voice and video.
- Class B is also a CIR service. Usage above the CIR, defined as excess information rate (EIR) is subject to the fairness algorithm. It is intended for business data applications.
- Class C is best-effort traffic; for example, consumer Internet access. The nodes negotiate for a share of ring capacity for this class of traffic.

FIGURE 32-7

Resilient Packet Ring Architecture

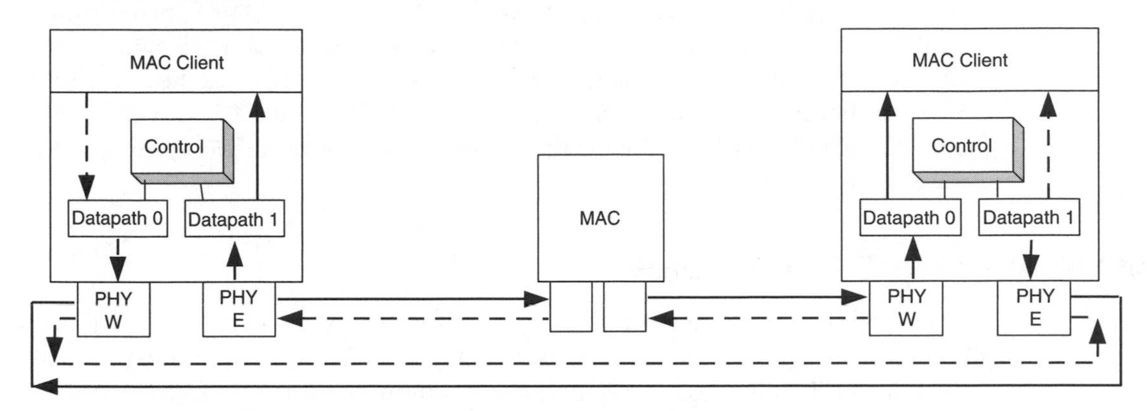

PHY = Physical Layer

Figure 32-7 shows the architecture of RPR. All nodes have appearances on dual counter-rotating rings. Each node has a topology map listing the location, status, and capabilities of other nodes. The node uses the topology map to determine the appropriate ring to use to reach the destination. When a node has permission to send, it inserts a packet on the ring. The MAC in the next downstream node determines whether to deliver it to its client or to forward it to the next station. Succeeding nodes forward the packet downstream until it either reaches the destination or exceeds the hop count limit. The destination node removes the packet from the ring.

The control units in the nodes participate in a fairness doctrine that ensures smooth delivery of traffic below Class A, which is not affected by fairness control. Class A traffic is subject to traffic shapers, which regulate large bursts. Stations calculate the amount of bandwidth they are allowed to consume (defined as the fair rate) based on the traffic they receive from downstream stations, the bandwidth that their client is consuming, and the capacity of the ring. The node calculates its fair rate by using the received fair rate and comparing it to its current usage. If a node determines that it is exceeding its fair rate, it reduces its fair rate value. The MAC transmits traffic in class order. When class B and C traffic is delayed, it triggers the fairness algorithm to request more bandwidth. The node generates fairness control messages indicating bandwidth requirements and ring congestion. The node considers itself congested when it cannot add the required amount of traffic to the ring.

Each MAC broadcasts topology messages to other stations on the ring. From these messages, each station creates its topology map. The stations operate independently without a master station and have the capability of dynamically adding and removing stations. When a new station joins the ring, it broadcasts its topology message, which triggers the other nodes to send messages of their own. The station learns about its neighbors by noting which received messages have traveled only one hop.

RPR can work over multiple physical layers including SONET/SDH. Traffic is sent in both directions on the ring to get maximum bandwidth utilization. This is in contrast to other ring topologies that require traffic to traverse the entire ring even though the destination is a ring neighbor. RPR sends the traffic directly to the destination and keeps the rest of the ring open for use by other stations, a technique known as *spatial reuse*.

MAN APPLICATION ISSUES

The metropolitan area network is in a state of transition as a result of customer shifts toward a fully converged network. When ATM was developed, it was expected to fill the backbone role and eventually migrate to the desktop, but its use is retarded by its complexity. Gigabit Ethernet, as a natural evolution of fast Ethernet, which managers already understand and trust, suffers from no such impediment. It can be shared, switched, and routed using the same topology as the slower speed products, which makes it the logical candidate for tomorrow's MAN. That change will be evolutionary, with metropolitan Ethernet gradually replacing other services.

The Ethernet MAN standards have not been in place long enough to see the outcome, but if history is any indication, prices will drop and demand will thrive as the protocol breaks its previous bounds. Most metropolitan areas have huge amounts of undeveloped fiber because of the building frenzy that followed passage of the 1996 legislation, but most of it serves the same high-density business district, leaving the rest served only with copper cable. Ethernet is a good alternative for data applications in those buildings that are served by fiber optics.

Demand for fixed-bandwidth circuits will continue well into the future. This need is adequately filled by SONET/SDH and the ILEC networks have ample capacity in place. The EFM standards coupled with RPR service protection can supplement SONET/SDH for a time, and will probably replace it at lower cost wherever fiber is available. The lack of fiber optics to homes and smaller commercial buildings means that T1/E1 will be around for many years, which in turn means that the all-Ethernet network for most companies will be years in the future.

The key to serving the metropolitan area in the future appears to be development of the multipoint alternative of the EFM standards. The copper alternatives are a stopgap. Although end-to-end Ethernet can be implemented over copper, that is the only real advantage of EFMC over DSL. To gain the minimum Ethernet wire speed of 10 Mbps means placing copper nodes within 750 m of the served subscribers and that will require major rebuilding of the copper cable network. Of the technologies discussed in this chapter, Ethernet has the brightest outlook for serving the MAN, but it is likely to come on gradually. The ILECs will serve the easy places first: that is the locations served with fiber already. They will also probably expand the fiber optics network by implementing PONs. Fiber provides more capacity than gigabit Ethernet can use, which will probably mean that 10G Ethernet will dominate in the MAN backbone.

Fibre Channel

Fibre Channel is designed for linking high-speed devices in a campus environment and that appears to be the market it will serve. Its switching capability enables it to create exclusive nonshared channels between devices for the duration of a session, which has always been the main benefit of the PSTN for low-speed applications. While Fibre Channel could fulfill some of the same functions as SMDS, it is not designed with the data security provisions that would make it suitable as a public protocol. Furthermore, to apply it in the metropolitan environment as a private network would mean obtaining dark fiber, which is not readily available except to common carriers.

Fiber Distributed Data Interface

FDDI was widely used for linking LAN segments through routers in the past, but its heyday, like its token ring cousin, is past. As Ethernet standards developed, the price dropped, and FDDI was no longer economical. It has higher throughput than fast Ethernet, but at higher cost. In the backbone it cannot compete with gigabit Ethernet, the price of which has fallen. FDDI will still be used in legacy network backbones for several more years, but it is not a good candidate for new network development.

Switched Multimegabit Data Service

As a LAN interconnection service, which is its primary use, SMDS has attractive aspects, but it is likely to disappear as the ILECs implement metropolitan Ethernet and RPR. In many regions the ILECs have not pushed SMDS because it serves much of the same market as frame relay, which is more widely accepted. The main advantage SMDS has over frame relay is its any-to-any connectivity without defining PVCs. In most enterprise networks, however, remote sites communicate mainly with headquarters, so site-to-site PVCs are unnecessary. That will change as multimedia applications gravitate to IP, which results in the central site becoming a choke point. SMDS is a more static configuration than Ethernet and its E.164 addressing scheme is less effective than Ethernet for such traffic.

CHAPTER 33

Wide Area Networks

This is the first of four chapters on WANs and their facilities. These facilities, point-to-point circuits, frame relay, ATM, and IP, are common to both MAN and WAN, but with differences. A WAN is a data, voice, or converged network that operates over carrier facilities to link the enterprise beyond the metropolitan area. Typically, a single carrier provides the facilities for the MAN portion of an enterprise network, but as the network scope moves into the wide area, multiple carriers are usually the rule even though one carrier accepts end-to-end responsibility. A WAN is global in scope and only rarely can one service provider deliver end-to-end service over its own facilities. IXCs can sometimes provide the local loop over facilities they own and control, but more often they rely on the LEC to provide the loops. Carriers offer service agreements that make this multiple-carrier environment transparent or invisible to the customers. Depending on the endpoints, some carriers will be able to assume responsibility for the entire wide-area portion of the network, but sometimes they will parcel out portions to other providers. This and the next few chapters focus on carrier services and how they are assembled into WANs.

Wide area voice networks were once common in large enterprises, but the economics of these are rapidly disappearing as long distance costs drop. The point-to-point services we discuss in this chapter are equally available for voice and data, and the circuits may be shared. The trend, however, is away from private voice circuits and toward either converged voice–data networks or virtual voice networks. Most of the discussion in this and subsequent chapters revolves around data networks and examines the considerations in running voice and video as data applications.

This chapter discusses requirements common to all WANs and reviews legacy network facilities that have applications in the enterprise network. These include point-to-point circuits, dial-up circuits, VSAT, and PDN. In addition, we briefly cover IBM's SNA, which is a practical architecture that served as the foundation for the OSI model and has matured to support mainframe applications over frame relay and IP. This chapter serves as the foundation for more detailed discussions of the most popular network services as the WAN continues to evolve to support converged voice and data.

WIDE AREA NETWORK REQUIREMENTS

Regardless of the application and irrespective of the WAN's purpose, the enterprise expects certain deliverables of the service provider:

- Reliability and availability are provided within defined limits.
- Costs are reasonable and predictable.
- Implementation and ongoing support from competent design and maintenance staff are provided to make the best use of vendor's services.
- Information is available to assist the customer in evaluating service levels, usage, and other criteria needed for adjusting the configuration.
- Billing is understandable, reliable, and not burdened with excessive fees and surcharges.

Multiple vendors may be involved in the service, but a common objective of most network managers is to place responsibility on the smallest number of vendors possible. Many carriers respond by offering single-point-of-contact service, which may involve managing not only the portions of the network they provide, but also the customer's terminating equipment. With managed network services the type of equipment is critical because the service providers are unable to manage every product on the market. Routers and switches may use proprietary protocols, but their operating systems and user interfaces are proprietary. A network management plan that relies on service-provider management may preclude certain choices of network equipment.

DATA NETWORK FACILITIES

The data communications designer has several decisions to make in selecting network facilities. The following are the primary issues in facility selection:

- *Bandwidth requirements*: If the network has a global reach, it is difficult to justify the cost of point-to-point facilities unless the utilization is kept to a high level. Low-bandwidth requirements are served more economically on usage-sensitive services.
- *Nature of the application*: The facility should be matched closely to the flow characteristics of the application. Bursty services with high peak-to-average bandwidth requirements are best suited to services that are optimized to deliver bandwidth on demand.

- *Service availability*: This factor often drives the network design because the desired service is not always provided in the terminating locations so the designer is forced to fall back to what is available. This factor is particularly relevant in international circuits.
- *Digital or analog*: Some vestige of analog circuits remains in the enterprise network, but tariff offerings such as fractional T1/E1 and the superior quality of digital circuits are driving the network from analog to digital.
- *Common carrier or value-added carrier*: A common carrier transports bandwidth to an interface point, but does not process the data. A value-added carrier not only transports data but also may process it or add other services such as store and forward, error correction, and authentication. The carrier also may provide other message-processing services such as e-mail, EDI, and message logging and receipting. The value-added carrier furnishes the user a dial-up or dedicated interface and provides the equivalent of a private network over shared facilities.
- *Switched or dedicated facilities*: The nature of the application determines whether switched or dedicated service is best. Dedicated facilities provide the convenience of always-on access, but they are costly and may raise security issues that switched facilities avoid. Circuit-switched data is slower and involves additional setup time, but it is appropriate for backup to dedicated service, occasional use, or for sessions to multiple termination points. Switched service is usage sensitive and international rates are much higher than domestic.
- *Terrestrial or satellite circuits*: Communications satellites offer a cost-effective alternative to terrestrial circuits, particularly where the endpoints are widely dispersed and located away from metropolitan areas. Because the tariff rates of terrestrial circuits are distance based, terrestrial circuits are less expensive over shorter routes unless the network has many endpoints. The break-even point depends on the topology of the network because cost is independent of distance within the coverage field of a single satellite. Satellite circuits are most effective with a large number of widely distributed endpoints. Satellite circuits are also effective for mobile applications such as ships at sea and long-haul transportation.
- *Circuit sharing*: Some applications do not use the full bandwidth of a circuit, so multiple applications can share the transmission medium.

Point-to-Point Network Facilities

Point-to-point circuits, sometimes called leased lines, are analog or digital, with digital circuits replacing analog except for growth of legacy networks. Circuits are

FIGURE 33-1

Billing Elements of a Dedicated Circuit

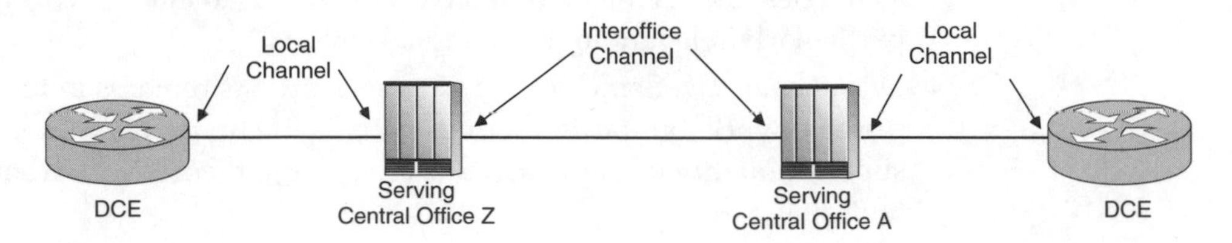

billed in three elements as shown in Figure 33-1. The IXC normally obtains the local access channels from the LEC or a CAP, and provisions the IOC over its own facilities. The IXC typically accepts end-to-end responsibility for provisioning, implementing, and clearing trouble. Unless the customer opts to obtain the local loops separately, the circuit is billed end-to-end. The cost of the IOC portion of the circuit is distance sensitive. The cost of the local access channel is not distance sensitive within a LEC wire center, but between wire centers a mileage charge applies.

The carrier designates one end of the circuit as the control end. The control office is responsible for coordinating implementation and trouble clearance. The control end, normally the end serving the customer's principal location, is known as the "A" end of the circuit and the termination is the "Z" end. The customer is insulated from the various circuit designations that may result when channels are connected through DCS systems over several different media to form an end-to-end facility.

Dedicated circuits have several advantages over carrier-provided shared facilities. First, the entire bandwidth is assigned to a single customer. Security is as tight as the physical security the carrier provides, which is usually high. The circuit is most vulnerable in the local channel where it may travel a cable pair to the customer's premises. Local channels served over fiber optics are difficult to tap, but an experienced technician can tap into channels provisioned over copper cable without much difficulty. Once the channel leaves the IXC's central office, it is multiplexed together with a huge bundle of services. Even if the fiber backbone were tapped, knowing which channel to select is difficult without access to circuit records. Point-to-point services are considered as highly secure.

Dedicated facilities are available in any bandwidth the customer is willing to pay for. At the low end of the scale are analog circuits that have the bandwidth of a single voice channel. From there, bandwidth can be obtained in increments up to the limits of SONET/SDH. One important characteristic in considering point-to-point bandwidth is its scalability. With the exception of fractional bandwidths, which are sometimes available, digital circuits do not scale well—that is, a DS-3 comprises 28 DS-1s, and is usually not economically feasible until at least seven or eight DS-1s are required. An OC-3 is three DS-3s, which scales reasonably well, but from there the hierarchy scales in multiples of four DS3s. This can either be an advantage or a disadvantage, depending on whether the enterprise can make good use of the extra bandwidth.

Dedicated facilities have lengthy provisioning intervals. A circuit within the continental U.S. often involves the IXC and two LECs. The IXC sets its provisioning interval to meet the worst-case LEC intervals plus some additional slack. When the circuit is international, the interval is much longer. The IXC can set up the IOC quickly within its facilities by sending orders to its DCS systems, but the need for intercompany coordination lengthens the interval. Carriers often quote "half-circuit" costs in international facilities, meaning that the cost is quoted to the meet point with the other carrier, and the subscriber gets two bills.

Point-to-Point Analog Facilities

In the past analog facilities and modems were the default private circuit, but they are disappearing in favor of digital circuits. Analog private lines have a nominal bandwidth of 4 KHz, with a usable bandwidth of approximately 300 to 3300 Hz. They are available in point-to-point or multipoint configuration and as two-wire or four-wire. The error rate should be in the order of 10^{-5} or better.

Analog facilities are available with channel performance enhancements, which may include signaling and improved transmission performance. In the latter case, carriers can condition circuits to control noise and to limit the amount of amplitude and envelope delay distortion within the voice pass band. Analog service is capable of supporting up to 19.2 Kbps full-duplex data on a four-wire line using private line modems.

Point-to-Point Digital Facilities

IXCs offer digital facilities in a variety of configurations. The carrier's service guide, which the FCC requires to be available on the Web, discusses service objectives and specifies the terms and conditions under which the service is delivered. These are normally quoted by percent availability and percent error-free seconds (EFS). Figures in the range of 99.99 percent are typical for both metrics, but the objective quoted may vary depending on the circuit length. These services, which sometimes go by the name of digital data service (DDS) are available in speeds of 2.4, 4.8, 9.6, 19.2, and 56/64 Kbps. Speeds below 56 Kbps were important at one time because they matched speeds of modems that were used on analog circuits. Today the use of anything less than 56 Kbps is rare.

The carriers transport digital circuits to their serving central office over their fiber-optic backbone. From there the customer has a choice of access. Most IXCs permit subscribers to provide their own access, in which case the IXC's responsibility ends at the point of interface (POI). This is typical of a customer who has obtained broadband access such as a DS-3 or an OC-3 from the LEC and uses it to carry multiple services. In this case, the IXC would bring the service to the LEC's end office. If the IXC has fiber optics to the customer's premise on one or both ends, it delivers the service over its facilities. If not, the next choice would likely be CAP or LEC fiber optics. If neither of those is available, the LEC would provide access over copper cable. Note that the only way 64 Kbps is delivered to the user's premises in North America is over T1 or greater bandwidth. If a single digital channel is required and T1 is not feasible, DDS will be delivered as a 56 Kbps service over copper. The carrier's service guide will provide interface requirements.

Sub-64 Kbps DDS uses a bipolar signaling format that requires the user's data signal to be converted from the usual unipolar interface of terminal equipment in a DSU. If the user's equipment can accept a bipolar signal and provide timing recovery, the data signal is coupled to a CSU in a CSU/DSU. Both units provide loopback facilities, so the local cable can be tested by looping the transmit and receive pairs together. The signal is fully synchronized from end to end. A DDS hub office concentrates data signals from multiple users and connects them to the long-haul network. The hub is also a testing point.

The major IXCs and some LECs offer fractional T1/E1 services. Fractional T1/E1 is economical up to some crossover point with full T1/E1, after which the full bandwidth is more cost effective. If you need more than six or eight circuits, then the full bandwidth costs about the same as fractional. LECs and CAPs can deliver the local access portion of T1/E1 over fiber-optic or copper facilities. If the local access circuit permits, T1 and fractional T1 can be offered as clear channel 64 Kbps service. E-1 is always clear channel. Some carriers provide central office multiplexing, and service protection measures such as diversity routing and routing over exclusively fiber-optic facilities.

T3/E3 and Fractional T3/E3

Where multiple T1/E1 circuits are needed between points, T3/E3 or fractional T3/E3 may be economical. T-3 service in North America is a full 28 DS-1s operating at 45 Mbps. Fractional T3/E3 consists of multiples of T1/E1. The fractional T-3 IOC must be routed over the same DS-3, or the DS-1s may arrive at slightly different times. Local channels may be routed over microwave or fiber-optic facilities provided by the customer or a CAP. Figure 33-2 shows how a DS-3 access circuit might be combined in either the LEC's or the IXC's central office to carry a variety of services. The variations on this arrangement are almost unlimited. For example, the figure shows separate bandwidth being routed to the PBX for local and long-distance access, but in a converged network the bandwidth could all terminate in a router.

Dial-up Circuits

The superiority of data networks notwithstanding, the PSTN still carries a significant amount of data. Circuit switching is not an ideal way of handling data, but it has one major advantage: it is available nearly everywhere in the world with a universal numbering plan. Wherever telephones are found, data can be

FIGURE 33-2

Shared Broadband Access

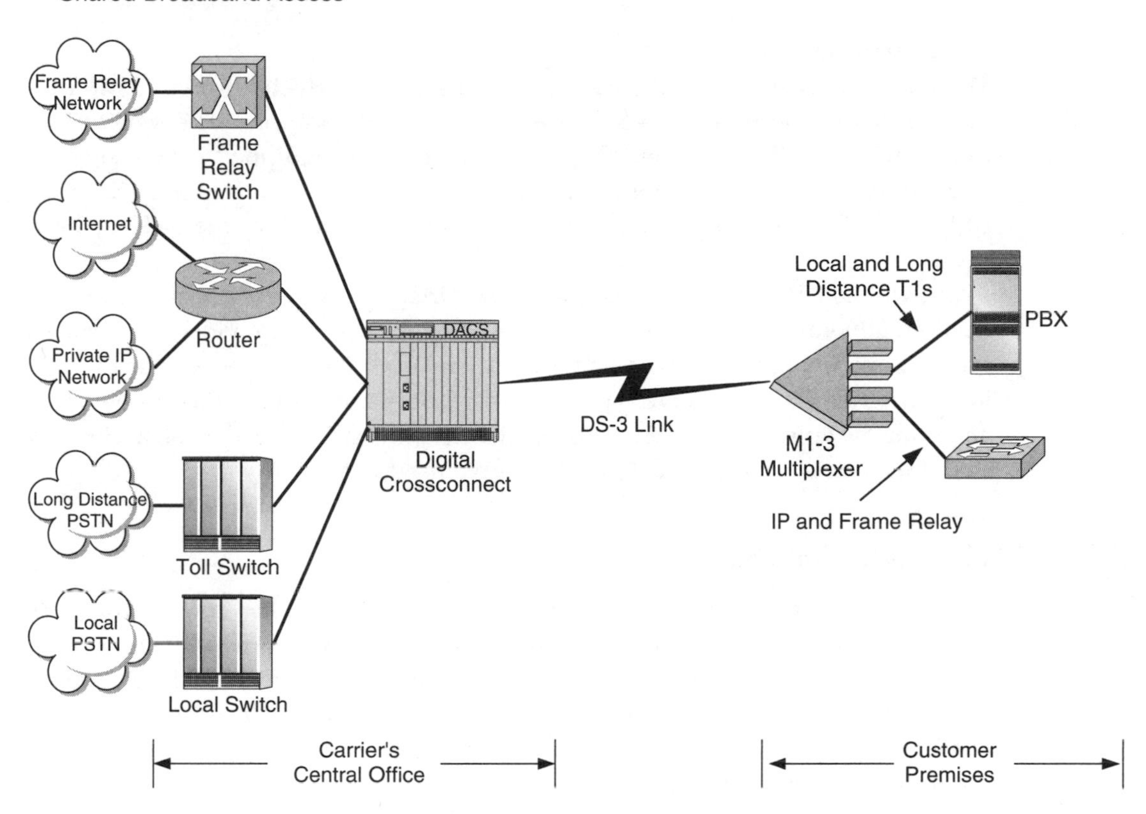

transported with respectable speed, reasonable cost, and easy setup. Dial-up data is effective under one or more of the following conditions:

- Data is exchanged among multiple terminating points.
- The application does not require an on-line connection.
- The use of data is occasional for limited periods each day.
- The connection setup time is not a limiting factor.

Facsimile is a data application that fits the above conditions perfectly. As a result, most facsimile is sent over the PSTN because of its universal availability. The same is true of the countless dial-up sessions into the Internet. Circuit switching is not ideal for these types of sessions, but it is convenient, the incremental cost is low, and security is tight. Although the PSTN may be less than ideal, many users have no alternative because broadband is unavailable or too expensive.

Dial-up data has become more effective in the last few years with the introduction of fast modems with data compression and error correction. V.92 modems are a commodity and are contained in most computers. For Internet access, V.92

modems may provide downstream data as high as 56 Kbps for Internet access, but for end-to-end transmissions it is limited to 33.6 Kbps.

Switched 56 Service

The major IXCs and LECs offer switched 56 services, which allow users to dial up 56 Kbps connections between properly equipped locations. The service is a four-wire connection into the LEC central office if it is equipped for the service. The subscriber's end of the connection requires a specialized switched 56 DSU/CSU. Signaling is carried over the least-significant bit of the 64 Kbps base signal.

Switched 56 is intended for applications that need an occasional digital connection, generally three or fewer hours per day. Examples are videoconferences, group 4 fax, file transfer, and data network backup. As with other switched services, switched 56 charges are based on usage time and carry a rate higher than an ordinary voice connection. This service was initially intended as a stopgap until ISDN became available. Now it is rarely used.

MULTIDROP NETWORKS

Point-to-point circuits are cost effective when the network has enough traffic to justify the cost of a dedicated circuit between two devices, but where the traffic flows in bursts or short transmissions, some method of sharing the circuit among multiple devices is needed. Polling is a common method of circuit sharing. Devices on the network buffer their transmissions until they receive a polling message that opens the gate to send. If a device has no traffic to send at the polling moment, it returns a negative response. The process is disciplined and the network performance is predictable when the load is known.

A multidrop network, illustrated in Figure 4-1, has a polling processor at the head end. A backbone circuit extends to the carrier's serving central office, where circuit legs to the terminating points are bridged together. The bridges are impedance-matching devices that funnel the traffic throughout the network, with all devices on the backbone hearing all of the traffic and sharing appropriate portions of the circuit. For example, a bank that has multiple automatic teller machines in an area can obtain local loops to the machines and bridge them in the central office to share a single circuit to the front-end processor. The front-end processor, or communications controller, is a computer equipped to handle SNA protocols. The host computer processes the data. The front-end processor's role is to relieve the host of teleprocessing chores.

Each station is assigned an address and connected to the network through a local controller. The host polls the stations by sending short messages to the controller. If the controller has no traffic from any of its attached terminals, it responds with a negative acknowledgment message. If it has traffic, it responds by sending a block of data, which the host acknowledges. Before the host sends data to the distant devices, it transmits a short message to determine whether the device is ready to receive. If it is ready, the host then sends a block of data.

Multidrop networks are designed as full- or half-duplex. In half-duplex networks, the modem reverses after each one-way transmission. In full-duplex networks, the devices can send data in both directions simultaneously. Polling is an efficient way of sharing a common data circuit, but it has high overhead compared to other alternatives. The overhead of sending polling messages, returning negative acknowledgments, and reversing the modems consumes a substantial portion of the circuit time. Throughput can be improved by using hub polling. In hub polling, when a station receives a poll, it sends its traffic to the host and passes the polling message to the next station in line. Hub polling is more complex than roll call polling and is not as widely used.

Although multidrop networks still have limited use, they have diminished considerably as applications evolved to the client-server model. When data transmission was in its infancy, networks tended to serve a single purpose. A bank branch, for example, might have separate networks for its tellers and its automatic teller machines. Now most networks are multiapplication and multiprotocol. Single-purpose networks still exist where the economics are attractive. For example, some state lotteries run multidrop networks to the end devices.

IBM Systems Network Architecture (SNA)

The most prominent example of multidrop architecture is IBM's SNA, which is a tree-structured hierarchical architecture, with a mainframe computer at the head. The architecture was first published in 1974 when processing was controlled by an expensive mainframe computer such as an IBM 360 that communicated with remote 3270 terminals over a network composed of voice-grade circuits. The architecture provides error-free communications over error-prone narrow-bandwidth analog circuits, while still providing the data integrity needed for corporate transactions such as bank transfers, reservations, and payroll.

SNA Physical Components

The physical components in an SNA network are hosts, communications controllers (also known as front-end processors), cluster controllers, and terminals. The host provides database access, computations, directory services, network management, and program execution. Front-end processors control the network, and cluster controllers manage the terminals and printers.

In the original network a mainframe computer runs a control program known as Advanced Communications Function/Virtual Telecommunications Access Method (ACF/VTAM), usually abbreviated VTAM. VTAM maintains a table of every device and circuit in its domain and controls information flow through the network. It establishes sessions and activates and deactivates

FIGURE 33-3

Physical and Logical View of IBM Systems Network Architecture

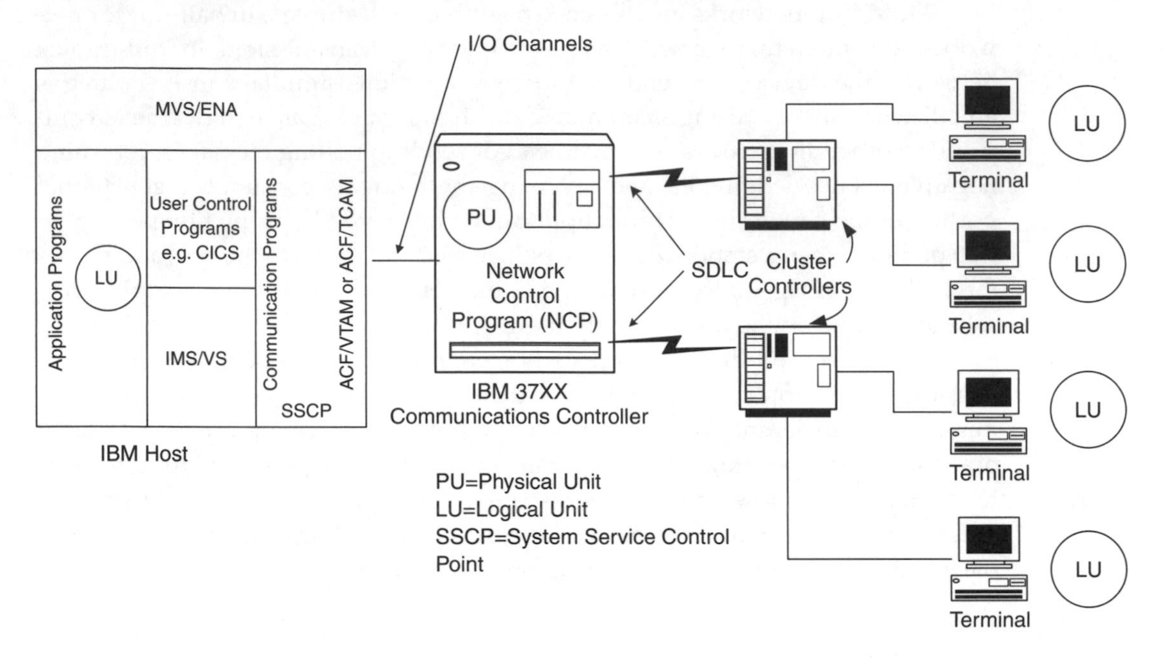

resources. Communications with these resources are handled through the minicomputer front-end processors, which in turn communicate with cluster controllers and terminals. Figure 33-3 is a diagram of the logical and physical elements in SNA. SNA establishes a logical path between network nodes and routes each message with addressing information contained in the protocol. The datalink protocol is SDLC, which bears close resemblance to HDLC.

Network Addressable Units (NAU)

SNA defines three types of NAU: logical units (LUs), physical units (PUs), and control points (CPs). LUs provide user access into the network and manage transmission of information among users. PUs monitor and control links and network resources. Although its name implies that it is a piece of hardware, a PU is a control program executed in software or firmware. PUs are implemented on hosts through access programs such as VTAM and on front-end processors by network control programs (NCPs).

The host NAU is categorized as a system service control point (SSCP). Each network contains at least one SSCP, which resides in an IBM mainframe computer. The SSCP exercises network control, establishing routes and interconnections between logical units. The NCP runs in a front-end processor such as an IBM 3745. Its purpose is to control information flow in the network. It polls the attached controllers, handles error detection and correction, and establishes circuits under control of the SSCP.

SNA's Layered Architecture

SNA is defined in layers that are roughly analogous to the layers in ISO's OSI model. SNA was the basis for much of the OSI model, but the layers do not line up exactly and it differs from OSI in several other respects.

Level 1, Physical, is not part of the SNA architecture. The physical interface for analog voice-grade circuits is ITU V.24 and V.31. The digital interface is X.21.

Level 2, Datalink Control, is SDLC. Later versions added support for X.25, token ring, Ethernet, frame relay, and FDDI. The SDLC frame, which is identical in structure to the HDLC frame in Figure 4-6, has six octets of overhead. The first octet is a flag to establish the start of the frame. This is followed by a one-octet address and a one-octet control field. Next is a variable length data field followed by a two-octet CRC field and an ending flag, which becomes the starting flag of the next frame. The control field contains the number of packets received to allow SDLC to acknowledge multiple packets simultaneously. SDLC permits up to 128 unacknowledged packets, which enables it to function with satellite circuits. This layer corresponds closely to ISO's datalink layer and the LAPB protocol used in X.25 networks.

Level 3, Path Control, establishes data paths through the network. It carries addressing, mapping, and message sequencing information. At the start of a session, the path control layer establishes a virtual route, which is the sequence of nodes forming a path between the endpoints. The circuits between the nodes are formed into transmission groups, which are circuits having identical characteristics such as speed, delay, and error rate. The path control layer is also responsible for address translation. Through this layer, LUs can address other LUs without being concerned with the entire detailed address of the other terminal. This layer is also responsible for flow control, protecting the network's resources by delaying traffic that would cause congestion. The path control layer also segments and blocks messages. Segmenting is the process of breaking long messages into manageable size. Blocking is the reverse—combining short messages so the network's resources are not consumed by small messages of uneconomical size.

Level 4, Transmission Control, is responsible for pacing. At the beginning of a session the LUs exchange information about variables, such as transmission speed and buffer size, that affect their ability to receive information. The pacing function prevents a LU from sending more data than the receiving LU can accept. Through this layer, SNA also provides other functions such as encryption, message sequencing, and flow control.

Level 5, Data Flow Control, conditions messages for transmission by chaining and bracketing. Chaining is the process of grouping messages with one-way transmission requirements, and bracketing is grouping messages for two-way transmission.

Level 6, Presentation Services, has three primary purposes. Configuration service activates and deactivates internodal links. Network operator service is the interface through which the network operator sends commands and receives responses. The management services function is used in testing and troubleshooting the network.

Level 7, Transaction Services, is responsible for formatting data between display devices such as printers and CRTs. It performs some functions of the ISO presentation layer, including data compression and compaction. It also synchronizes transmissions.

SNA lacks an applications layer as such, but IBM has defined standards that allow for document interchange and display between SNA devices. Document interchange architecture can be thought of as the envelope in which documents travel. DIA standards cover editing, printing, and displaying documents. The document itself is defined by document content architecture, which is analogous to the letter within the envelope. The purpose of the DIA/DCA combination is to make it possible for business machines to transmit documents with formatting commands such as tabs, indents, margins, and other format information intact. Documents containing graphic information are defined by graphic codepoint definition, which defines the placement of graphic symbols on printers and screens.

Advanced Peer-to-Peer Network (APPN)

The original SNA protocol was built on the assumption that devices such as terminals and printers lacked processing capability so all traffic flowed through the host. This condition was reasonable when terminals lacked intelligence, but as user devices evolved from dumb terminals to PCs, direct peer-to-peer communication became necessary. Cluster controllers could support only one SDLC link and could not communicate among themselves, so IBM developed a cluster controller modification known as PU 2.1. This enabled two controllers to be linked across an SDLC or dial-up connection without requiring a path through the front-end processor. PU 2.1 supports the physical connection, but it does not provide all the logical functions necessary for peer-to-peer communications.

To enable device-to-device communications, IBM introduced the LU 6.2 Advanced Program-to-Program Communications (APPC) protocol. LU 6.2 severs the SNA master/slave relationship between devices, permitting communication between peers. Either device can manage the session, establish and terminate communications, and initiate session error recovery procedures without involving an SSCP. APPC permits direct PC-to-mainframe communications, which enables the PC to transfer files without consuming excessive mainframe processing power.

The new SNA is called APPN. APPN defines two kinds of nodes: network nodes and end nodes. An end node, as the name suggests, can send and receive traffic, but data is not routed through it. A network node handles through traffic, and acts as the concentration point for end nodes. Devices on the network are named. To communicate with another device, an end node sends a bind message to its network node, which in turn broadcasts a query that passes through the network.

The node answering to that name responds to the query with a message that establishes the session and route.

PUBLIC PACKET DATA NETWORKS (PDNs)

PDNs using X.25 protocol were the predecessors of frame relay and IP networks. Packet-based networks use a mesh topology. Intelligent nodes are interconnected by circuits with the circuit routes chosen to optimize cost and reliability. Because each node has multiple connections, it has alternate methods of sending packets to the destination. A major difference between the protocols lies in the changes that have occurred in the backbone circuits over the years. PDNs were developed well before the fiber-optic backbone was in place. The X.25 protocol assumes narrow bandwidth, short, bursty messages, and vulnerability to errors. As packets flow across the network, each pair of nodes checks for errors before forwarding the packet to the next node. Therefore, latency is high. As fiber optics replaced long-haul microwave, the error rate was so low that link-by-link error correction became superfluous and frame relay replaced X.25.

Frame relay and X.25 are connection-oriented protocols, which means the carrier provisions a virtual circuit between nodes. At each node a table shows the route that was set up for the connection. Packets or frames have short headers that contain a virtual circuit identifier (VCI) with no need to indicate the destination address. IP, by contrast, is connectionless, so TCP/IP datagrams flow over paths that can change with conditions such as node failure or congestion.

In an X.25 network the nodes can be accessed by one of three methods:

- A dedicated X.25 link between the user's host computer and the data network
- A dedicated link between the data network and a packet assembler—disassembler (PAD) on the user's premises
- Dedicated or dial-up access into a PAD provided by the network vendor

In the first option, the user's host computer performs the PAD functions. In the second option the customer furnishes the PAD. The third option, dial-up access, is the least complex and is the only method economically feasible for small users.

PDNs were popular in North America until frame relay entered the market. In 2003 Tymnet, which was one of the two large PDNs, shut down. The other, Telenet, is active in other countries, but the services provided in North America have moved to frame relay or IP.

X.25 Protocol

X.25, built on the first three layers of the OSI model, specifies the interface between DTE and a packet-switched network. The physical layer interface is X.21. The link layer uses LAPB to control errors, transfer packets, and to establish the datalink. The network layer establishes logical channels and virtual circuits between the PAD and the network. X.75 protocol prescribes the interface for gateways between packet networks.

Packet networks offer both PVCs and SVCs. With a PVC a path between users is provisioned, and all packets take the same route through the network. With an SVC, the session is managed through control packets, which are analogous to signaling in a circuit-switched network. For example, a call setup packet would be used to establish the initial connection to the terminating device, which would return answer packets. The network uses control packets to interrupt calls in progress, disconnect, show acceptance of reversed charges, and other such functions.

VERY SMALL APERTURE TERMINAL (VSAT)

VSAT is a Ku band satellite service that is an excellent medium for a widely dispersed operation. Its pricing is not distance sensitive, so it is particularly effective in remote locations. Typical applications include LAN/WAN networking, Internet access, and videoconferencing. It is economical in rural areas and small towns where frame relay is not cost effective because of the access circuit cost. Another application for which VSAT is well suited is telemetering from mobile devices such as trucks, ships, and trains.

Figure 33-4 shows a typical VSAT arrangement. As discussed in Chapter 19, the VSAT terminal is a small device with a 90 to 120 cm dish antenna. Transponder space is leased from the satellite provider or obtained from the hub provider.

FIGURE 33-4

VSAT Network

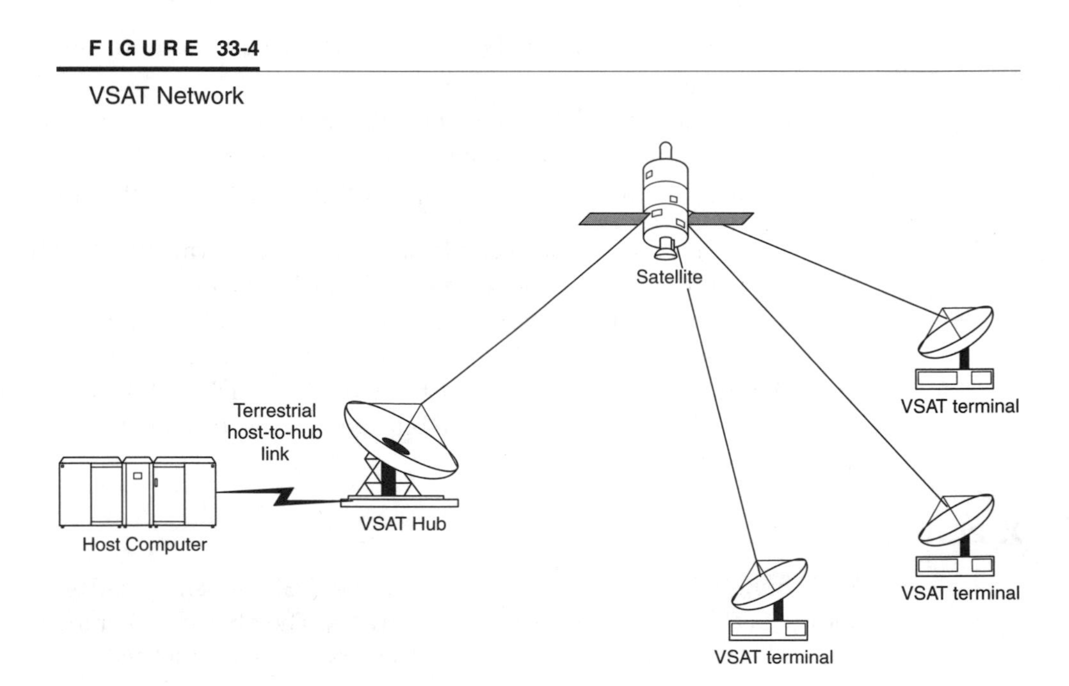

Large companies may be able to justify the cost of the earth station and hub, but smaller companies lease capacity from a service provider. Added to the cost of the transponder is a terrestrial extension from the hub to the host computer, called the backhaul circuit. VSAT bandwidth can range from as low as 1200 bps to T1/E1. For Internet access asymmetric service is offered with upstream bandwidth about one-fourth of downstream.

With all else equal, terrestrial circuits are more effective for data than satellite circuits because of lower delay and freedom from degradation during rainstorms. At longer distances, however, the greater cost of terrestrial circuits often offsets the disadvantages of a satellite. The primary disadvantage of a satellite circuit is the round trip propagation delay, which is approximately 0.5 s. Throughput is also reduced by the access method, which is typically slotted Aloha or TDMA. With TDMA, multiple terminals share the same frequency spectrum, but they cannot hear one another's signal. Therefore, within a time slot, stations with traffic to send may transmit simultaneously. The more heavily loaded the network, the higher the probability of collision.

WAN APPLICATION ISSUES

Data network applications can be separated into the following general types:

- *Inquiry/response:* This is typical of information services where a short inquiry generates a lengthy response from the host. Because the data flow is asymmetric, half-duplex facilities generally offer the greatest throughput on a dedicated private line. On digital facilities, the connection is inherently full duplex. Typical applications are airline reservations, streaming video, and on-line database sessions such as the World Wide Web. The operator keys a few characters into the terminal, and the host computer responds with a lengthy message that might be confirmation of a reservation, a printed ticket, or an information dump.
- *Conversational:* This mode, typical of terminal-to-terminal communication, is characterized by short messages that are of approximately equal length in both directions. Throughput is improved by using full-duplex operation. Conversational mode is typical of voice and videoconference over IP.
- *Bulk data transfer:* This is typical of applications such as storage area networks where large files are passed, often at high speed, in only one direction. This method is often used when a local processor collects information during the day and makes daily updates of a master file such as an inventory on the host.
- *Remote job entry:* In RJE remote terminals send information to a host. The bulk of the transmission is from the terminal with a short acknowledgment from the host. Half-duplex circuits may be the most effective form of transmission because the bulk of the information flows

from remote to host. Dedicated facilities, either leased lines or frame relay, are almost invariably needed for this kind of application. Many remote terminals, each of which is used only occasionally, may share a higher speed line to the host.

Evaluation Considerations

The criteria used to evaluate data networks differ significantly from those used to evaluate voice networks. For example, the short length of many data messages makes setup time, which is of little concern in voice networks, a factor that may rule out dial-up. Also, error considerations are important in data networks, but unimportant in voice. Circuit noise that is annoying in a voice network may render the channel unusable for data. For most data networks, the primary factors are reliability, standards compliance, manageability, and cost.

Circuit reliability is the frequency of circuit failure expressed as mean time between failures (MTBF). A related factor is circuit availability, which is the percentage of time the circuit is available to the user. Availability depends on how frequently the circuit fails and how long it takes to repair it. The average length of time to repair a circuit is expressed as mean time to repair (MTTR). The formula for determining availability is

$$\text{Availability} = \frac{(\text{MTBF} - \text{MTTR}) \times 100}{\text{MTBF}}$$

For example, a circuit with an MTBF of 1000 h and an MTTR of 2 h would have an availability of

$$\frac{(1000 - 2) \times 100}{1000} = 99.8\%$$

It is important to reach an understanding with the supplier of the conditions under which a circuit is considered failed. When the circuit is totally inoperative, a failure condition clearly exists, but when a high error rate impairs the circuit, it is less clear whether the service is usable. The error rate in a data circuit is usually expressed as a ratio of error bits to transmitted bits. For example, a data circuit with a bit-error rate of 1×10^{-6} will have one bit in error for every one million bits transmitted. Reliability and error rate have a significant effect on throughput. Most error correction systems initiate retransmission of a block and retransmissions reduce throughput.

Very Small Aperture Terminal

The most important factors in evaluating VSAT services are network availability and throughput. Availability varies by geographical region. Areas with heavy rainfall will be subjected to more outage than drier areas, and will have lower availability. MTTR is calculated on the same basis as discussed above, and is affected by the distance of the terminal from the nearest repair center. Dial backup can be used to improve availability. Throughput is affected by how heavily loaded the channel is. Since stations are unable to hear signals of other stations, collisions mutilate packets and require retransmission. Other considerations include the following:

- Capital investment may be a consideration. Some VSAT carriers carry the entire investment, charging only for the service at a rate that is independent of distance.
- Network management capability should cover not only the VSAT terminal equipment, but may also extend to the customer's equipment. To compete with frame relay carriers VSAT vendors may provide managed services.
- The method of quoting transponder capacity varies with the VSAT carrier. A typical method of quoting rates is based on a dedicated in route (remote to hub) and a portion of an out route (hub to remote). The amount of capacity allocated to the customer affects response time.
- Interface type may range from LAN to serial interfaces. The remote transceiver should be configurable over the network.
- Hub location is important because it affects the cost of the backhaul circuit. VSAT vendors can connect to nondistance sensitive services such as frame relay or the satellite network itself for the backhaul circuit. The customer should also determine the degree of redundancy in the hub.
- The network should be capable of supporting multiple protocols. For time-sensitive protocols such as IP, the network should support spoofing.

CHAPTER 34

Frame Relay

When frame relay reached the market in the early 1990s it was the right service at the right time and it immediately grabbed the attention of network managers. The data network choices at the time were X.25 and leased-line networks, which were fine for text-oriented applications, but they were unsuitable for client–server and LAN interconnection. The older protocols were based on the assumption that data transmission lives in a hostile environment composed of voice-grade circuits with a high error rate. X.25 checks errors on each link so a packet cannot advance until the receiving node pronounces it pure. The price of this error checking is latency and a need for processing power at each node.

In the late 1980s, several factors arose that rendered these older protocols and services inadequate for the new applications. The first was the decline of analog coaxial cable and microwave in favor of digital transmission over fiber optics. Error rates dropped by three or four orders of magnitude, broadband digital circuits suddenly became available, and carriers built fiber-optic links to large users' premises. The error rate on fiber is so low that link-by-link error control is unnecessary. The end devices handle flow control, error checking, and retransmission.

One result of this transition was the loss of predictability. The bandwidth requirements for terminal-based applications are easy to calculate, but as LANs, remote bridges, and routers became common, data flowed in bursts. A circuit linking a pair of LANs may sit almost unused until someone decides to move a massive file, and then we reach the classic tradeoff between bandwidth and speed of download. Even docile applications like e-mail suddenly become bandwidth hogs when it is easy for users to attach files and voice and video clips.

Frame relay was born from this set of conditions. When ITU-T began work on the protocol, they assumed it would be used on the ISDN D channel. Sixteen Kbps was a respectable amount of bandwidth in an era when 9.6 Kbps was considered fast, but the market was evolving away from narrow bandwidths and circuit switching and ISDN lost whatever attraction it had for data transmission. The new network paradigm of bandwidth on demand between peer nodes over high-quality circuits demanded a new service model. X.25 was designed for the old analog network, but frame relay, a better match for the new, has rendered X.25 obsolete. Frame relay evolved into a carrier service that was aimed at the objectives that it still fulfills: simplicity, economy, and bandwidth on demand within the limits of the access channel. Even though frame relay makes no guarantee of data integrity, its payload usually consists of TCP/IP packets that provide their own flow control and error correction mechanisms.

FRAME RELAY TECHNOLOGY

Frame relay was a natural evolution of X.25. By eliminating layer 3 functions and leaving error correction to the edge devices, the topology is similar to X.25, but latency is lower. A frame relay network accepts data up to the limits of the port speed, which the customer selects to meet its service requirements. The frame relay network is a complex of high-speed circuits and backbone processors, typically linked with ATM. As with X.25, the internal structure of the network is proprietary to the service provider. Each service provider specifies its own SLAs and establishes its policies for carrying or discarding traffic and providing usage information to the customer.

From a customer's standpoint, frame relay is a simple concept. Each location connects to the network over an access circuit that the carrier obtains from a local access provider. The access circuit is a dedicated channel that varies from a low of 56/64 Kbps up to as high as DS-3. Often, if the customer has dedicated access to the IXC's voice network, a portion of the bandwidth is split off with an ADM and shared with frame relay. At each customer location a router or a frame relay access device (FRAD) connects the customer's LAN to the access circuit. The customer selects a port speed per location, which establishes a ceiling on the transmission rate. The customer also selects PVCs between its locations. Some carriers offer asymmetric PVCs in which the bandwidth is greater in one direction than the other.

A typical frame relay network resembles the diagram in Figure 34-1. PVCs are shown as dotted lines in the figure. In this configuration, all locations feed into Location A, which is typical of many networks with a centralized computer. Note that Locations D and E also have PVCs to Location B. The subscriber can select PVCs to the point of creating a fully meshed network, the cost of which approximates the cost of leased lines. The customer determines the need for node-to-node traffic based on the amount and type of traffic. An alternative to connecting PVCs between sites is to feed all traffic to the central location and route site-to-site traffic back across the network.

Increasingly, companies are adding voice and video capabilities to their frame relay networks. The access circuit and port speed are adjusted to meet the higher demand and need for prioritization of time-sensitive traffic. Some

FIGURE 34-1

A Frame Relay Network (Access circuits and committed information rates are shown in brackets)

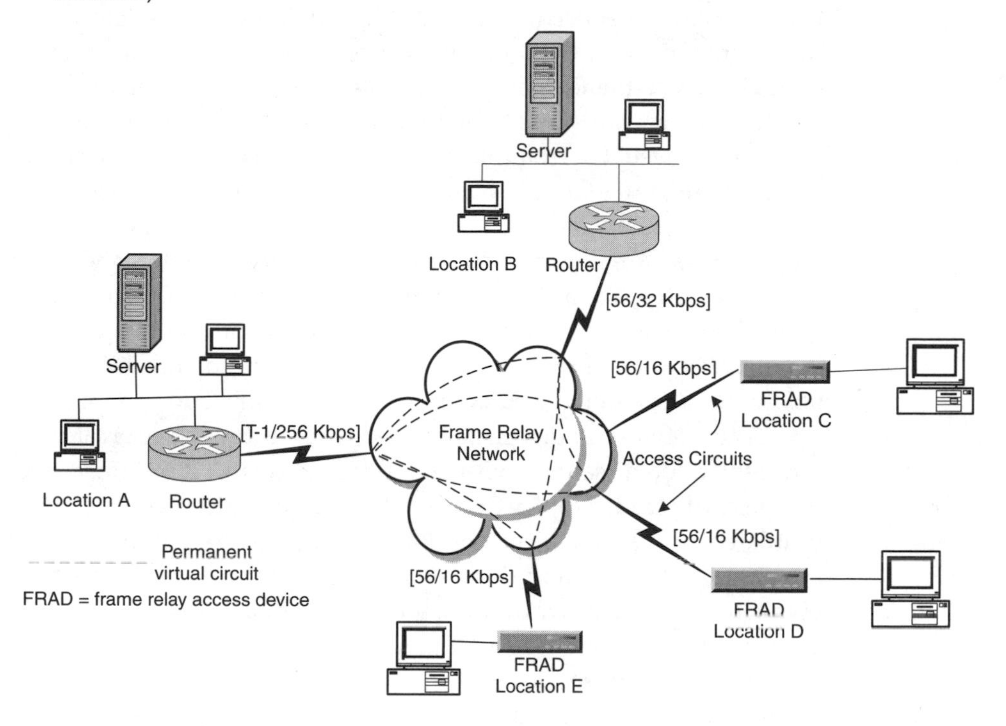

carriers offer different classes of service for networks that carry voice and video. The highest quality of service, real-time variable frame rate, is intended for time-sensitive traffic such as voice and SNA. The second class, nonreal-time variable frame rate, is intended for applications such as LAN-to-LAN and intranet traffic. The lowest class, available or unspecified frame rate, is for traffic such as e-mail and file transfer.

Another difference between frame relay and dedicated circuits or X.25 is the method of charging. Private line circuits are distance and bandwidth sensitive, so the longer the connection and the higher the bandwidth, the more they cost. X.25 is not distance sensitive, but it is usage sensitive. VANs typically charge by the kilopacket, so the quantities of data that flow across an Internet would make X.25 prohibitively expensive for LAN interconnection. Frame relay is not distance sensitive for domestic points, and instead of metering traffic, the pricing is based on a combination of three elements:

- Committed information rate (CIR).
- Port speed, which is the speed of the access port into the carrier's network.
- Access circuit, which ranges from 56 Kbps up to T3/E3.

The customer selects a CIR and port speed from the carrier's service offerings. The CIR defines the guaranteed transmission rate for each PVC. The access circuit is a dedicated private line from the customer's premises to the carrier's POP. As with other private lines, the access circuit length determines its cost. The carrier may have a POP for rating purposes that is different from its switch location. The key to frame relay pricing lies in the difference between the CIR and port speed. If the carrier's backbone capacity permits, the network transports data up to the port speed. For example, the customer can choose a 64 Kbps port speed and access circuit, but a CIR of, say, 16 Kbps. This makes the service effective for applications with bursty data because the network can carry the average data the customer needs. If the carrier has enough capacity, it allows the customer to send bursts as high as the port speed, with frames above the CIR marked as discard eligible (DE). When the network is congested, the protocol can discard DE frames. Some carriers employ a "leaky bucket" algorithm in which periods of traffic below the CIR become credits for bursts above the CIR.

The network nodes calculate the CRC to check for errors. If a frame is in error, the node discards it with no attempt to correct it or inform the endpoints of the discard. Nodes can also discard traffic to relieve congestion. If frames are mutilated or discarded, the endpoints recognize the failure and arrange retransmission. Customers can prioritize their traffic in some networks by marking their own DE frames. This way the customer can be sure that any frames exceeding the CIR are of low priority.

The network service provider manages its internal network to avoid congestion and to statistically multiplex the traffic of its customers throughout the busy periods. Therefore, even though the customer is paying only for the CIR, the network usually handles bursts up to the port speed. In fact, some carriers do not even quote a CIR unless the customer requests it. Customers can take advantage of the nature of frame relay by scheduling certain operations for off-peak hours. For example, large file transfers can be scheduled at night when the carrier's network has plenty of spare capacity, and even though much of the traffic may be well above the CIR, the probability of discard is slight.

Typically, a frame relay network costs from 25 to 50 percent less than an equivalent dedicated digital network. The cost difference depends on tariffs, distance of the nodes from the carrier's frame relay POP, and span of the network. The wider the network's span, the more attractive frame relay becomes. It is particularly effective for international branch offices.

Frame relay is a robust service. The most likely point of failure is in the access circuits. Once the traffic reaches the carrier's backbone network, disruption is unlikely. If the carrier loses a circuit or even a node, the traffic can be rerouted. To get the same degree of protection in a private line network would require full mesh architecture, which is too costly for many companies to justify.

The major LECs offer intraLATA frame relay. As long as MFJ restrictions on interLATA traffic exist, LEC frame relay must connect to an IXC to bridge LATA boundaries. That IXC might be the parent company's IXC subsidiary. LECs' service areas may be further defined by exchange boundaries with independent telephone companies. Some carriers have worked out network-to-network interface (NNI) arrangements to make the service area boundaries transparent to the customer. In other cases the subscriber must provide its own NNI between carriers.

The Frame Relay Protocol

Frame relay is a connection-oriented protocol with a variable length packet that can range as high as 4000 octets. Figure 34-2 shows the frame structure. The frame is created by encapsulating the layer 2 frame of the base protocol, excluding the flags and CRC, into a frame with a flag delimiter, two-octet header, and frame check sequence. The protocol provides a three-octet header, shown in the figure, but the third octet is rarely used. Each frame contains a 10-bit field known as the datalink connection identifier (DLCI), which identifies the connection. Since the protocol does not care whether the frame is error free, it can switch the frame to an output port as soon as it reads the address. The 10-bit DLCI number theoretically yields 1024 connections, but the numbers 1 to 15 and 1008 to 1022 are reserved, and 1023

FIGURE 34-2

Frame Relay Frame Format

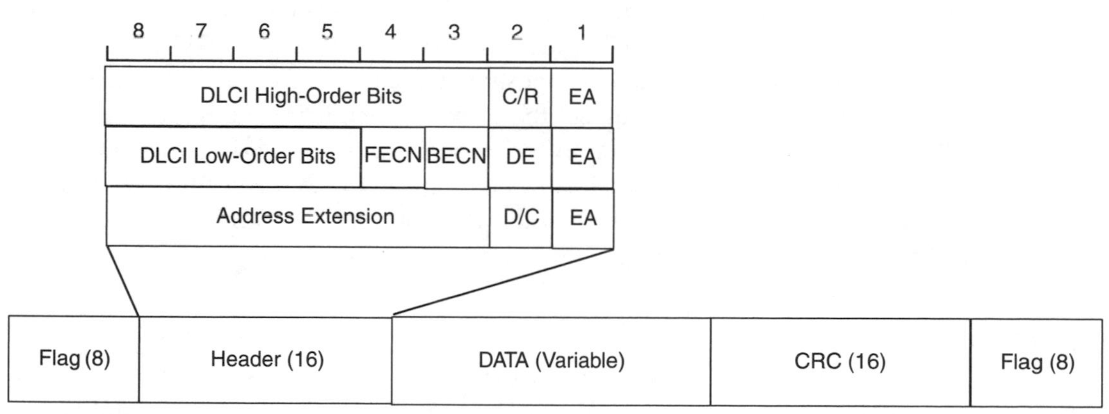

DLCI = Datalink connection identifier
EA = Address extension bit
C/R = Command/response bit
FECN = Forward explicit congestion notification
BECN = Backward explicit congestion notification
CRC = Cyclical redundancy check

is used for local management interface (LMI) as discussed later. The total number of available DLCIs is 992. The DLCI number is unique on its own link, and may be repeated elsewhere in the network. The carrier has the option of using global addressing, which keeps the DLCI unique; in which case the third address octet is needed after the network exceeds the number of addresses in the basic header. The combination of DLCIs forms the VCI. The subscriber can have several DLCIs configured at one location to enable it to communicate with multiple sites.

The C/R bit is not defined in the specification, so users are free to define it. A "0" in the EA bit indicates that the DLCI continues in the next octet. A "1" in the DE bit indicates that the frame can be discarded. Either the customer or the network can set this indicator. The data block is variable in length. The subscriber would normally use short blocks for discard-eligible traffic to avoid lengthy retransmissions, which add to congestion. If time-sensitive traffic such as voice and SNA are mixed with other traffic, it may be necessary to fragment data frames to prevent them from imposing excessive delay. Each frame relay switch node checks the frame check sequence and if it finds an error, it discards the frame. Likewise, it checks the DLCI and if it is not valid it discards the frame. If the frame passes these two checks, the node passes it to the next node.

Congestion Control

Frame relay does not support true flow control, but it has features that minimize frame loss. If the network becomes congested, the node buffers frames up to its capacity. After the buffers are full, it discards frames. A partial solution to buffer overflow is congestion notification, which both the frame relay switch and the router must support. Forward explicit congestion notification (FECN) bits are set when frames begin to encounter congestion in the forward direction. Downstream nodes detect the FECN bit and set the backward explicit congestion notification (BECN) bits in frames traveling in the opposite direction. When the router sees frames with the BECN bits set, it slows the rate at which it is feeding frames into the network. Not all carriers and routers support FECN/BECN, however, so it is not a reliable way of reducing congestion.

Status Polling

The customer's ingress router polls the carrier's access switch periodically to check the status of the network and DLCI connections. The switch and router exchange link integrity verification (LIV) packets at 10-s intervals to verify that the connection is still good. At 1-min intervals the devices exchange full status packets to verify which DLCIs are active. The oldest form of verification is LMI, which is a dedicated DLCI (1023) that is used for reporting connectivity status to the customer. If a DLCI is removed or lost for some reason, the LMI message reports the loss to the customer. LMI also provides a one-bit optional field for the network to report congestion to the subscriber if it is implemented. ANSI T1.616 and ITU-T

Q.933 are alternatives to LMI. Q.933 protocol provides status signaling for PVCs plus signaling to set up, maintain, and release SVCs.

Network-to-Network Interface

Unlike the public Internet where practically all carriers connect at NAPs and Internet exchanges, frame relay networks are isolated. Some, but not all carriers provide NNIs between their networks. This means that a customer has little choice but to use the same carrier for its entire frame relay service. Most LECs support NNIs with selected carriers; if not with the major IXCs, at least to their own networks where the customer needs service to multiple LATAs. As long as the RBOCs are restricted from transporting interLATA traffic, they are prohibited from expanding the reach of their LATA frame relay network, so the subscriber must provide the NNI. The RBOC may offer the NNI through its separated interexchange subsidiary.

Prospective customers should review what provisions may be lost with NNI connections. Many of the features such as LMI, usage reporting, congestion control, network management information, and the like may be blocked at the NNI and provide less than a complete picture of network operations.

Switched Virtual Circuits

SVCs are temporary connections between endpoints that are set up for the duration of a session. Although many carriers offer only PVCs, SVCs are part of the frame relay specification. SVCs are a good way of making any-to-any connections in organizations with sites too small to justify a PVC. SVCs overcome some of the drawbacks of carriers' provisioning processes, which typically cannot respond quickly enough to meet customers' needs. Occasional connections between locations may not justify a PVC, but an SVC connection can be set up easily. An SVC session consists of four operational states, call setup, data transfer, idle, and call termination. In the idle state the connection is active but no data is transferred. Although SVC service has some application, it is not widely deployed.

VOICE OVER FRAME RELAY (VoFR)

Frame relay is a stable and predictable medium that is suitable for voice provided the carrier holds end-to-end delay and jitter within acceptable bounds. Some carriers provide separate service classes for delay-sensitive traffic. Users have several alternatives for connecting VoFR. If the subscriber has an existing PBX or key system, routers or FRADS can be equipped for E&M tie lines, foreign exchange channels, or loop-start lines. The VoFR router or FRAD compresses the voice signal and packetizes it for transmission. VoFR can be cost effective for companies that have enough traffic to justify the additional cost of enabling the routers for voice plus the equipment to terminate the voice channels.

A second alternative is a direct connection from an IP PBX over a PVC to the other endpoint. It is possible to define a second PVC and designate it for voice or video, but that is unnecessary if the routers can prioritize time-sensitive packets. Of course, voice packets must not be marked DE and with a router protocol such as weighted fair queuing, data should receive a proportionate share of the bandwidth. If insufficient bandwidth is provided, data frames may experience excessive discard. Therefore, it is essential to provide a CIR that is high enough to accommodate all voice traffic at peaks plus enough bandwidth to handle all data traffic that cannot be delayed.

The principal issue with VoFR lies in the nature of the protocol. Since it is connection oriented, the traffic flows point-to-point. Separate PVCs can be defined to each pair of endpoints, but this increases cost if they are not otherwise required for data. The other alternative is to switch traffic in the router at the main node, but this doubles the latency.

Support for fax and modem traffic is also an issue for some customers. This traffic cannot be compressed because the DTE has already compressed it to the maximum. Therefore, the equipment should be capable of recognizing fax and modem traffic and forwarding it at an appropriate rate without attempting to compress it further.

SNA OVER FRAME RELAY

Many companies run mission-critical applications on SNA networks, but equip remote sites with LANs. The network connects the FEP to a LAN, using PCs for the terminal device. A multidrop network does not fit into the LAN-to-LAN environment of bursty traffic and multiple applications. This leaves two choices: connect the enterprise with separate networks for business and host computer applications, or combine SNA and business traffic on a single network.

The obvious solution is to deploy a multipurpose network, and frame relay is the popular choice. Frame relay is highly reliable, but SNA relies on a synchronized operation. Excessive delay can cause the session to time out or stations failing to receive acknowledgments within the expected interval presume the frame is lost and retransmit. The SDLC protocol also cannot cope with duplicate frames.

The industry has developed two general approaches for enabling SNA to run over frame relay, because the protocol does not care what the user data block contains. In the first approach, known as the FRAD mode, the router or FRAD communicates directly with the front-end processor as shown in the upper diagram of Figure 34-3. In the centralized router mode, shown in the lower diagram, a router at the processing center connects to the front-end processor through a LAN.

The FRAD mode bridges the remote LAN to the communications controller, replacing SDLC at the link layer with the LAN LLC2. LLC2 runs on cluster controllers and on NCP, where it is called Boundary Network Node (BNN). FRADs were developed to bridge 3270-type terminals or LANs to the communications

FIGURE 34-3

Alternatives for SNA over Frame Relay

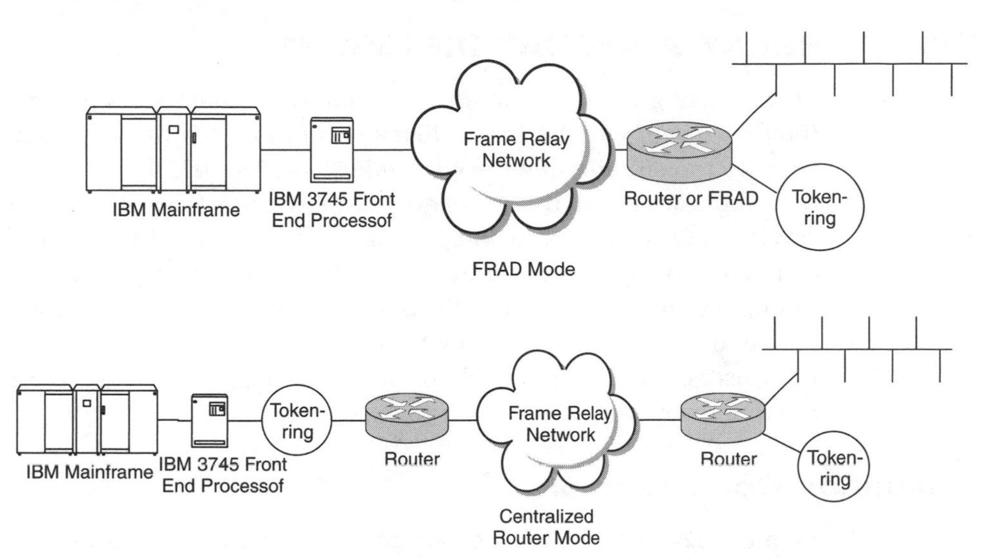

controller over a frame relay network by encapsulating LLC2 frames in the frame relay data block. A similar protocol known as Boundary Access Node (BAN) uses the RFC 1490 specification for bridged token ring. BAN enables Ethernet, token ring, or FDDI networks to bridge to the front-end processor.

The other method, known as *datalink switching* (DLSw), uses TCP/IP as the transport mechanism for SNA traffic. Instead of bridging LLC2 frames across the network, DLSw encapsulates the information and unnumbered information frames in IP packets and sends them to a centralized DLSw router for delivery to the addressee. The local router sends acknowledgments to the transmitting station, spoofing it into thinking the acknowledgment came from the destination and preventing it from transmitting duplicate frames. In this mode no changes to NCP are required because the entire protocol is implemented in routers. The DLSw protocol adds 52 octets of overhead to each frame, which is significant in applications with short data payloads. DLSw supports data compression, and all protocols on the network can share the same PVC. The central router handles the traffic at the central site, so that only frames addressed to the front-end processor reach it, which reduces the processing load.

Several issues must be addressed in handling SNA over frame relay. One is the need for prioritization of SNA data so it is not subjected to delay and discard. SNA can survive an occasional frame discard, but if the data consistently exceeds the CIR, the SNA network may bog down under a condition of excess packet discard. The solution is to ensure that the routers recognize and prioritize packets containing SNA data. Another issue is end-to-end delay. SNA is designed for point-to-point networks that have minimum delay. If LLC2 frames encounter excessive delay, the session can be dropped. Frame relay providers may offer a class of service specifically for SNA networks.

FRAME RELAY APPLICATION ISSUES

Frame relay grew more rapidly than almost any data service in existence. From the first meeting of the Frame Relay Forum in 1991, the service has developed from a curiosity that few people understood into the default protocol for most data applications. All router vendors support frame relay, the major IXCs offer the service both domestically and internationally, and the RBOCs and independent companies offer service within their territories. Carriers are currently attempting to migrate the service to MPLS-based IP networks. That transition will probably happen in the end, but for now the change is difficult to justify for many frame relay users, particularly in light of the fact that frame relay is as secure as a dedicated network, and IP networks raise security issues.

Evaluation Considerations

Frame relay services are not all equal. Each carrier publishes its own SLAs and provides its combinations of CIR, access, and port speed. This section discusses some of the issues that should be evaluated when applying frame relay and selecting a carrier.

- *Access circuit and port speed.* Determine the required speed of the access circuit and port. Determine at each location if 56/64 Kbps service offers enough bandwidth or if T1/E1 or fractional T1/E1 is required. Determine if access can be shared with some other service routed to the same carrier. For very high-bandwidth applications, determine whether the carrier offers T3/E3 access speed.
- *Network-to-network connections.* Some IXCs and LECs have signed NNI agreements that make local access to IXC frame relay networks more economical. For example, a company with both inter- and intraLATA connections on the same network may be able to obtain local access circuits from the LEC's frame relay network at a lower cost than using circuits to the IXC. If the carriers have not provided an NNI, the customer has the option of providing its own. Some features will undoubtedly be lost across the NNI. Inquire closely about service level reports and features between two carriers.
- *Point-of-presence.* The location of the carrier's POP is of concern to designers. In locations without a POP in the same city, the cost of the local access circuit can be high enough that frame relay is not cost-effective. If international service is required, determine what countries the carrier covers.

- *Committed Information Rate.* Selection of the CIR is one of the most important factors in determining the success of frame relay. If the wrong choice is made initially, the carrier can change CIR in a short interval. Review the expected maximum data throughput requirement of your network and the times of day that bursts of data are likely to occur. Review how the carrier handles bursts over the CIR. Determine the degree to which they buffer or discard packets when congestion occurs.
- *Network information.* A major advantage of frame relay over fixed networks is the amount of network information the carrier provides. Determine what reports are available, how often they are produced, and how they are obtained. For example, does the carrier provide data online over the frame relay network or the Web? Do they display it in graphical form?
- *Access method.* Determine whether frame relay access can be shared over existing T1/E1 lines, or if access can be obtained from LECs or CAPs. Is DSL access available? Is dial-up access needed in some locations, and if so, is it available from the carrier? Determine whether a router or FRAD will be used as the access device.
- *Congestion control.* Does the carrier implement FECN and BECN to handle congestion? What method does the carrier use to set the DE bit on frames transmitted at a rate above the CIR? Does the carrier implement any discard prioritization? Discuss the conditions under which frames are discarded.
- *Service level agreement.* The carrier should quote end-to-end availability and data delivery rate or throughput. What availability guarantees does the carrier make? What is their guaranteed percentage of frame delivery within the CIR? What is the average end-to-end latency? How long does it take to reconfigure the PVC and CIR? If the carrier fails to meet its SLAs, what credits does it provide? What exclusions from carrier liability are contained in the SLA?
- *Local management interface.* Some carriers have implemented LMI to provide network status information. Determine whether the carrier has implemented LMI, and if so, what functions are available such as alarm reporting.
- *Managed network service.* Major carriers offer this service, in which they accept end-to-end responsibility for setup and maintenance including access devices. Determine costs, what services are included, and what products they support.
- *Backup PVC.* This service provides the subscriber with the ability to recover from failure or isolation of its primary location. In case of failure the service is redirected to the backup location.

- *Automatic dial backup initiation.* The network should provide a reliable method to signal the customer's router of a loss of connectivity, including both loss of the link and end-to-end network connectivity through drop of LMI. Determine whether LMI is dropped if PVC connectivity is lost.
- *Disaster recovery plans.* The carrier should have a clearly documented plan discussing what disaster conditions are covered and how the carrier minimizes the probability of lengthy service failure. Some carriers offer access to an alternate node.
- *Service alternatives.* Does the carrier provide both SVC and PVC or only PVC? Is SVC service billed on connect time or quantity of data transferred? Does the carrier offer separate service classes for time-sensitive applications such as voice and SNA?
- *Access coordination.* The carrier should coordinate access with the LECs in all locations. Most carriers provide total access coordination in which the carrier bills for the access circuit in addition to handling provisioning and trouble reports. Does the carrier permit sharing the access circuit with voice?
- *Excess information rate.* Some carriers provide credits for information sent under the CIR. When frames are sent over the CIR, the carrier checks to determine whether credits are available, and if so, these excess information frames are sent without being marked as DE.

Quality of Service Measurement

Frame relay has several measurements of service quality that may not be specified by all service providers. The principal measurements and their abbreviations are:

- *Committed rate measurement interval (Tc).* The time increment used to measure information flow.
- *Committed burst size (Bc).* The maximum number of consecutive bits the network will guarantee to deliver without discarding data, measured over the interval Tc.
- *Excess burst size (Be).* The maximum number of bits the network will attempt to deliver for the user measured over the interval Tc.
- *Data delivery rate (throughput).* The percentage of frames that are delivered per unit of time.
- *Availability.* The percentage of time the network is available. Note that this figure should be measured end-to-end, which includes the access circuits on both ends of the connection.
- *Latency.* The round-trip time in milliseconds.

Subscribers need to examine the service definitions carefully to be sure they support the required grade of service. Service variables are often quoted as averages, whereas the subscriber is usually more interested in peak or worst-case conditions. Note also whether the SLA is quoted in terms of a true guarantee or merely as a performance objective.

Frame Relay Access

One of the first questions the manager encounters is whether to use a router or a FRAD for frame relay access. A router is a more versatile device. A FRAD sends data over only one route. The device must have large enough buffers to accommodate incoming data and to eliminate jitter if voice is used. The device should also have the ability to buffer low-priority traffic without discarding it while passing through high-priority traffic. If you plan to run voice over the frame network, the FRAD or router must be able to support voice. The following are some issues to consider in evaluating the access device for frame relay service, with an emphasis on voice and video over frame:

- Determine what protocols the device supports.
- Compare your requirements for LAN and WAN interfaces to the number of interfaces that the device supports. Many routers can interface multiple LAN connections. They also support multiple WAN interfaces that automatically synchronize to the network speed.
- Determine what network management protocols the device supports.
- Determine whether dial backup can be initiated through LMI and what variable such as loss of PVC initiates the backup.
- How many voice ports and what type of interface (e.g., FX, E&M, loop and ground start) does the device provide? How many and what type of data ports does it support? Does it support serial ports?
- Can the device mark packets as DE? What does it use to segregate DE packets?
- How does the device prioritize voice and other time-sensitive traffic such as SNA? Does it support WRED or WFQ or do voice packets move to the head of the queue and delay data packets? Does it support 802.1p and q, RSVP, and/or DiffServ?
- How many PVCs does the device support? Can data and voice operate over the same PVC?
- Can the device support different classes of frame relay service?
- Does the device support file fragmentation so time-sensitive traffic is not delayed waiting for long data frames?
- Does the device have a jitter buffer of appropriate size?
- Does the device support voice compression and silence suppression? What algorithms does it support?
- Does the device support echo cancellation?

- Does the device support Group 2 fax?
- Does the device support LMI? Q.933 status reporting?
- Does the device support SDLC local port spoofing and RFC 1490 encapsulation for IP and LLC2?
- What local statistics, such as port and PVC utilization, throughput, network availability, and excess bursting over the CIR, does the device collect and retain?

CHAPTER 35

Asynchronous Transfer Mode

ATM is a monument to the hazards of speculating about the future direction of telecommunications networks. Less than a decade ago most industry pundits were confident that ATM was the service of the future. It would span from desktop to desktop, handling all kinds of media as streams of short cells. Data would flow seamlessly between endpoints without a media conversion, delivering whatever bandwidth and service quality the application and the network negotiated. What happened to this vision? Several things. For one, ATM has proven to be unexpectedly complex and costly for a general-purpose protocol, especially considering the other alternatives. While developers were working on 155 Mbps ATM and trying to scale it down economically to 25 Mbps to the desktop, Ethernet did an end run. Fast Ethernet switching had three times the throughput at a fraction of the cost and on its heels came gigabit Ethernet. They are cheap, fast, and easy to implement; so much so that ATM is left with no place in the local network.

Another factor was the ascent of IP. A major attraction of ATM is its ability to handle time-sensitive media with service quality equivalent to circuit switching. IP networks lacked QoS guarantees, and were not seamless between the local and the wide area. The IETF, with representation from users and manufacturers directed their resources toward developing protocols that can provide QoS approaching the promise of ATM. Now, the major IXCs offer IP VPNs as their migration path from frame relay. Much of this backbone runs over ATM, but as we discuss in the next chapter, MPLS is gaining a foothold in the network core.

So what role does that leave for ATM? The answer is in the backbone for carriers and large enterprises, where ATM is alive and well. ATM is the service platform for frame relay, DSL, private line, IP, and carrier TDM switch interconnection. IP enhancements are threatening ATM in the backbone, but the standards for delivering quality over IP are still evolving, while QoS is inherent with ATM, a considerable amount of which is already operational. Although MPLS is coming on line to harden IP networks, the transition will take time. With the backbone network in mind as ATM's turf, let us turn to a high-level understanding of ATM, its method of operation, and its classes of service.

ATM TECHNOLOGY

ATM is a multiplexing and switching technology that is also known as broadband ISDN (B-ISDN). The ISDN term may give a wrong impression in this context because it has connotations of circuit switching. While ATM can behave like circuit switching in that it is connection oriented and provides guaranteed capacity and constant latency, it has the topology of a packet network. ATM carries the information payload in short PDUs known as cells. The reason it is called asynchronous can be understood by contrasting it to TDM, in which information streams are assigned to fixed time slots. If an application has data to send, it must wait for its time slot, even though other slots are unused. TDM wastes media capacity, but it gains simplicity in the process because the time slot is identified by its bit position and multiplexing is simple and inexpensive. ATM can make use of this empty capacity by multiplexing data asynchronously into time slots and attaching headers to identify the data flow as a stream of cells.

ATM cells are 53 octets long. Each cell has a five-octet header, and a 48-octet payload. Note that the header, which is shown in Figure 35-1, contains a virtual

FIGURE 35-1

ATM Header

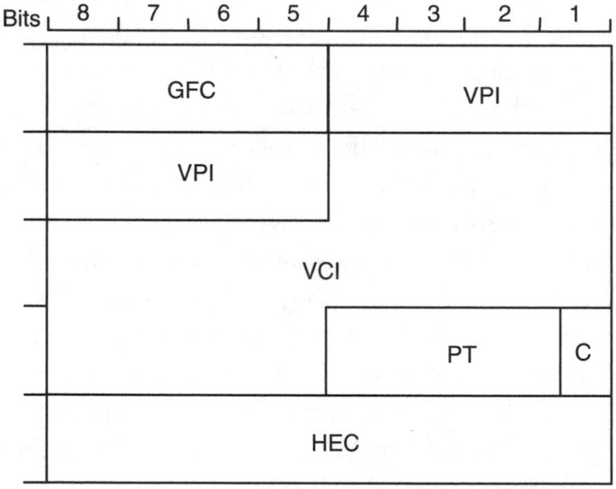

Bits 8	7	6	5	4	3	2	1
GFC				VPI			
VPI				VCI			
VCI							
VCI				PT			C
HEC							

GFC = Generic Flow Control
VPI = Virtual Path Identifier
VCI = Virtual Channel Identifier
PT = Payload Type
C = Cell Loss Priority
HEC = Header Error Control

FIGURE 35-2

ATM Virtual Paths and Virtual Channels

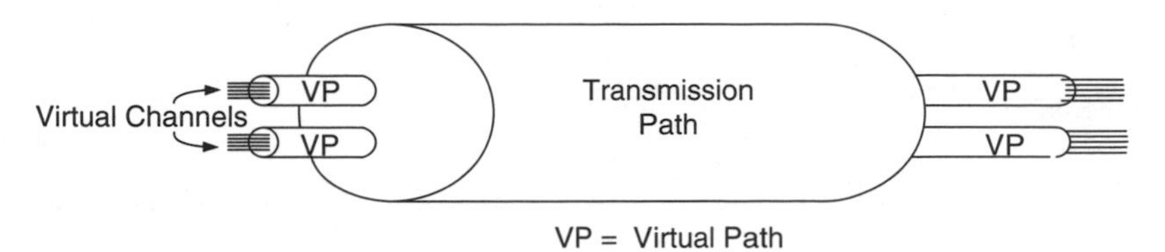

path indicator (VPI) and a virtual channel indicator (VCI). These correspond to ATM's two types of circuits: virtual paths and virtual channels. *Virtual channels* are analogous to virtual circuits. They are defined between endpoints, and share the bandwidth with other channels. *Virtual paths* are bundles of virtual channels as depicted in Figure 35-2. If two switches have many different virtual channels between them, they can bundle them into a virtual path connection.

ATM is optimized for multimedia traffic because of these unique characteristics:

- It provides multiple classes of service so the user can match the application to the required grade of service.
- It is scalable in link speeds from T1/E1 to OC-192 (10 Gbps).
- Switching is done in hardware, which results in low latency and minimal jitter.
- It supports virtual channels that are equivalent to circuit switching for time-sensitive traffic.
- It supports bandwidth on demand for bursty traffic.
- It is an international standard that is supported by a wide variety of equipment.

Connections between endpoints are either provisioned as PVCs or set up per session as SVCs. SVCs are set up with a signaling protocol and remain active for the duration of a session. In case of failure, SVCs can be dynamically rerouted. They are advantageous for direct connection between sites where the traffic volume is not sufficient to justify the cost of PVCs. With PVCs, each switch in the path must be individually provisioned for the PVC. Also, the path is static, so it lacks the resiliency of a connectionless service, but ATM also includes a connectionless service similar to SMDS.

Virtual channel connections (VCCs) are concatenations of virtual channels that carry a stream of cells in sequence over an end-to-end connection. When the virtual circuit is defined, the VCC control assigns the circuit to a VCI and a VPI. As the connection is set up through the switch serving a particular node, the switch must connect a VPI and VCI from an input port to a VPI and VCI on an output port. Figure 35-3 should help clarify this concept using two-octet VPIs

FIGURE 35-3

Virtual Path and Virtual Circuit Indicators

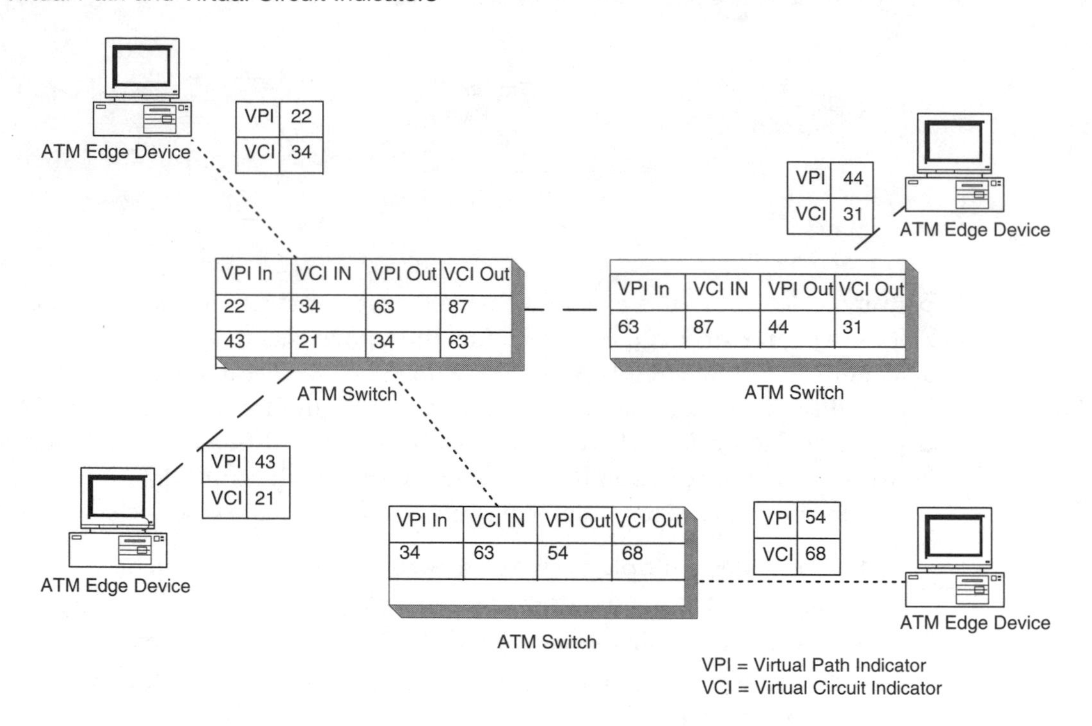

and VCIs. The ATM VPI is actually eight octets and the VCI is 16 octets long. The VPI and VCI are selected at the switch to keep track of the connections, and have no end-to-end significance.

Rationale for Fixed-Length Cells

ATM's short cell is the key to its ability to handle time-sensitive traffic without excessive delay or jitter. Since the cell length is fixed, an ATM switch needs to look only at the VPI and VCI in the header to switch the cell to an output port. Because of the fixed length cells, the performance of the network is more predictable than one based on variable-length frames, and buffers at the switch nodes are easier to manage. Furthermore, as the load increases long packets cannot delay time-sensitive packets. The circuitry can be programmed into an ASIC with minimal processing compared to routing.

The short cell works well for voice and video, but for data a five-octet header represents almost a 10 percent overhead. Data networks make more efficient use of bandwidth by using long packets where the header length is insignificant compared to the payload. The 53-octet cell length is not magic. It arose through compromise, not engineering analysis. When ATM standards were being designed, the data faction wanted a payload size of 64 octets, while voice advocates held out for 32 octets. They split the difference.

Complexity and cost aside, data network engineers prefer TCP/IP with its variable length packet. Long data packets increase throughput, but if multiple media share the network, the protocols must prevent long data packets from delaying short voice packets. If voice packets are forced to wait in queue while the router transmits long data packets, jitter increases to the point that it cannot be buffered without exceeding latency objectives. This means time-sensitive packets must be tagged and prioritized, steps that are unnecessary in ATM. Data engineers refer to the ratio between the ATM header and payload as the "cell tax," and the emphasis from the IETF is to use IP for all media. From an overhead efficiency standpoint, when an IP network carries voice, the header comprises a much higher portion of the total packet length than the cell structure. An uncompressed VoIP signal has 44 octets of header for 160 octets of voice, or 21.5 percent overhead. The overall efficiency of the network depends on the type of traffic it is carrying. If time-sensitive traffic predominates, ATM is more efficient than IP, but TCP/IP supports higher throughput for data.

ATM Network Interfaces

An ATM network is composed of switches connected by high-speed links. The user connects to the network through the UNI. Two types of UNI are defined. A public UNI defines the interface between a public ATM network and a private ATM switch. A private UNI defines the interface between the user and a private or public ATM switch. ATM switches within the same network interconnect through NNIs. A public NNI defines the interface between public network nodes. A B-ISDN Inter-Carrier Interface (B-ICI) supports user services across multiple public carriers. Figure 35-4 illustrates these interfaces.

Private network-to-network interface (PNNI) is a routing protocol that enables different manufacturers' ATM switches to be integrated into the same network. It is capable of setting up point-to-point and point-to-multipoint connections. PNNI automates routing table generation, which enables any ATM switch to automatically discover the network topology and determine a path to another switch. In determining the route, it uses such metrics as cost, capacity, delay, jitter, and active data such as peak and average load.

The headers are slightly different for NNIs and UNIs. Figure 35-1 shows the UNI header, which is identical to the NNI header except that the latter extends the VPI, and omits the generic flow control, which is sometimes used to identify multiple stations that share a single ATM interface. The payload type indicates in the first bit whether the cell contains user or control data. If it is set to one, the field contains control data. The cell loss priority bit indicates whether the cell can be discarded. The header error correction block checks the first four octets of the header for errors.

FIGURE 35-4

Public and Private ATM Network Interfaces

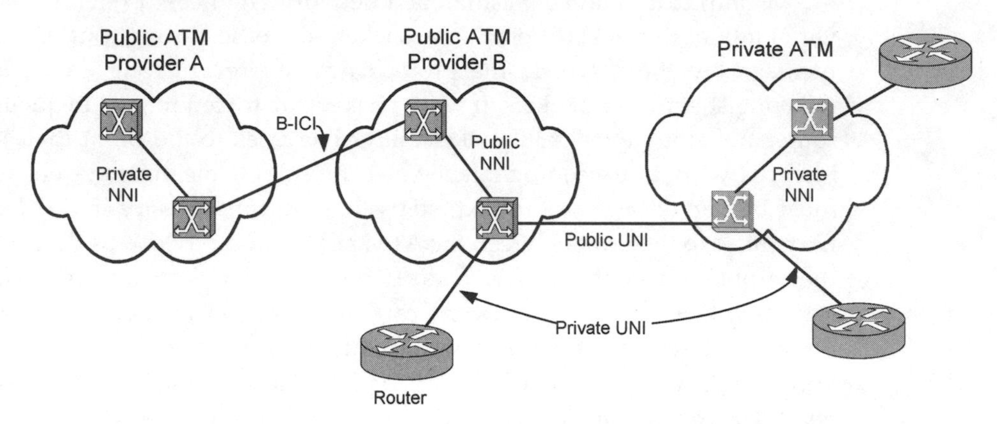

ATM Protocol Layers

Like all modern protocols, ATM is a layered protocol, but greatly simplified compared to others. Figure 35-5 shows the logical layers of ATM, all of which fit in the first two layers of the OSI model. Three planes: control, user, and management span all layers. The control plane generates and manages signaling messages, the user plane manages data transfer, and the management plane handles overall operation of the protocol.

FIGURE 35-5

The ATM Layers

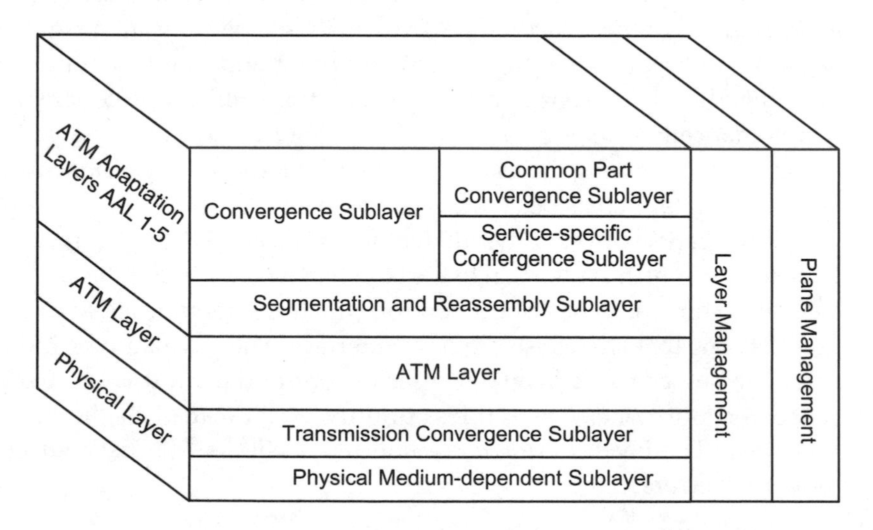

The physical layer is designed to operate over a variety of services such as DS-1, DS-3, SONET/SDH, fiber, twisted pair, or even radio. This layer packages cells according to the requirements of the physical medium. The media independent ATM layer multiplexes and demultiplexes cell streams onto the physical layer. The application fits on top of the ATM adaptation layer (AAL). The AAL allows ATM to statistically multiplex various traffic types. AAL is divided into two sublayers, the segmentation and reassembly (SAR) and the convergence. The SAR segments the user's data stream on outbound traffic and reassembles it inbound. The convergence sublayer protocols are different for the various types of information such as voice, video, and data. The AAL supports five classes of traffic as illustrated in Figure 35-6:

- *Constant bit rate* (*CBR*) for connection-oriented traffic such as uncompressed voice and video. This class provides low latency, jitter, and cell loss.
- *Real-time variable bit rate* (*RT-VBR*) for connection-oriented bursty traffic that requires close synchronization between the source and destination. Examples are packet video and compressed voice.
- *Non-real-time variable bit rate* (*NRT-VBR*) for connection-oriented traffic that is not sensitive to latency and packet loss. Examples are bursty data such as frame relay and LAN-to-LAN traffic where the application can recover from irregularities.
- *Unspecified bit rate* (*UBR*), which is a best-effort service where the application, such as e-mail and file transfer, does not require QoS.
- *Available bit rate* (*ABR*) is whatever is left over. It is also a best-effort low-cost service that is used for services similar to ABR.

The solid block at the bottom of Figure 35-6 represents traffic the network is committed to handle, and which requires a constant amount of bandwidth. Above it is VBR traffic that the network is also committed to carry, but which varies in bandwidth because of the nature of the application. The bandwidth left over is available, and can be provided at a lower cost because it does not require a firm

FIGURE 35-6

Utilization of Network Capacity

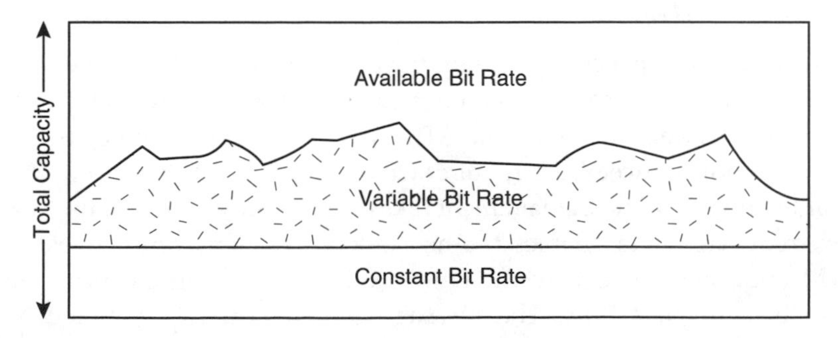

commitment from the carrier that it will be delivered within specified QoS parameters. UBR is a laissez-faire class of service providing best-effort delivery. It is less complex to set up than the other service classes, and is therefore often what the carrier provides by default. It does not have the guaranteed QoS of CBR and VBR service, and cannot be depended on for these applications. Service providers may interpret the traffic classes differently, so it pays to understand the carrier's definitions.

Corresponding to traffic classifications, the AAL is divided into four categories.

- AAL-1 is a connection-oriented service designed to meet CBR service requirements. It is intended for video, voice, and other CBR traffic. ATM transports CBR in a circuit-emulation mode. To preserve timing synchronization between endpoints, it must operate over a medium such as SONET/SDH that supports clocking.
- AAL-2 is for VBR applications such as compressed voice that depend on synchronization between endpoints, but do not have a constant data transmission speed. AAL-2 supports silence suppression, where AAL-1 transmits packets containing silence. AAL-2 supports both real-time and non-real-time traffic.
- AAL-3/4 supports both connection-oriented and connectionless data service. It is used by carriers to provide services such as SMDS.
- AAL-5 is for connection-oriented and connectionless data communications that do not require CBR or VBR stability, including services such as frame relay, LAN emulation (LANE), and multiprotocol over ATM (MPOA).

The AAL layer allows the network to provide different classes of service to meet the requirements of different types of traffic. AAL service differentiation is used only between the end systems and QoS is not based on the AAL designation of the cells. ATM switches can handle multiple sessions and classes of service simultaneously.

ATM Call Processing

When an ATM connection is set up the calling station asks for a connection to the called station. The calling station and the network negotiate bandwidth and QoS classifications, whereby the ATM network provides the QoS and the station promises not to exceed the requirements that were set up during the connection establishment. Traffic management takes care of providing the users with the QoS they requested, and enables the network to recover from congestion. When an ATM endpoint connects to the network, it sends a traffic contract message that describes the data flow. The message contains such parameters as peak and average bandwidth requirements and burst size. The network uses traffic shaping to ensure that traffic fits within the bounds of the contract. ATM switches can enforce the contract by setting the cell-loss priority bit in the header for excess traffic. The switches can discard such traffic during congestion periods.

ATM endpoints set up calls with a signaling-request message to their serving switch. This setup message contains a call reference number, addresses of called and calling parties, traffic characteristics, and a QoS indicator. Signaling messages are sent over the signaling AAL, which assures their delivery. The signaling messages are based on the Q.931 format, which consists of a header and a variable number of message elements. The destination returns a call proceeding message that contains the same call reference number plus a VPI/VCI identifier. A series of setup and call proceeding messages are exchanged while the network determines such matters as whether the called party is willing to accept the call. Switches use the PNNI protocol to discover the topology and link characteristics of the network. When a change such as a link failure occurs, PNNI communicates the event to all switches.

Public networks use an addressing system of up to 15 digits following E.164 standards. Private networks use a 20-octet address modeled after OSI network service access point (NSAP) addressing. The ATM layer maps network addresses to ATM addresses.

VOICE OVER ATM (VoATM)

The CBR and VBR traffic classes are designed for voice, video, and other time-sensitive traffic over ATM, collectively known as VoATM. Circuit emulation service (CES) enables circuits to be connected across an ATM network using CBR PVCs. It is intended for use by non-ATM devices such as PBXs or video codecs that need controlled bandwidth, end-to-end delay, and jitter just as if the devices were connected by a private line. CES, usually used with AAL1, allows these variables to be specified at the time of call setup.

The AAL1 specification provides two modes of operation, structured and unstructured. Unstructured CES extends all channels of a T1/E1 across the network in a single VC. The network does not look into the underlying channels of the T1/E1, but reproduces the data stream across the network without modification. This provides the same degree of stability as it would if sent over SONET/SDH, but this method is not bandwidth efficient because if some of the channels are idle, they are sent anyway. Structured CES, intended to emulate fractional T1/E1, splits the T1/E1 into multiple DS-0s, and transmits each one with a different VC. If the block size is greater than one octet, AAL1 uses an internal pointer to delineate the block size. This enables the ATM network to minimize bandwidth by using only the timeslots that are actually needed and allows the endpoints to be different, which improves utilization. CES services can be set up as either synchronous, which assumes that each end is individually clocked from a reference clock, or asynchronous, in which clocking information is transported in ATM cells.

VoATM Signaling and Call Setup

ATM uses two types of signaling for CES, in-band channel-associated signaling (CAS), and out-of-band common-channel signaling (CCS). In the CCS model, the network transports the signaling transparently. A PVC carries the signaling from end to end and the stations select a PVC to carry the voice channel. The network itself is not involved in the signaling—it merely transports the signals from the endpoints over the PVC. This method, sometimes called the transport model, would be used when the network provides a dedicated connection between two devices such as PBXs that signal each other using internal protocols. This alternative requires structured CES.

In the CAS method, also known as the translate model, the network interprets the signal. When a station requests service, the ATM network sets up an SVC with the requested QoS to the terminating endpoint. When a call setup request is received, the source ATM switch determines the route through the network based on the QoS requested. It uses the PNNI protocol to send a setup request through the network to determine whether each switch has the resources to support the connection. Once the connection is set up, it has the same degree of stability as a circuit-switched connection.

LAN EMULATION

Since ATM is a connection-oriented protocol, it is far from a seamless service for linking LANs with their variable-length frames and connectionless format. To serve the market for LAN-to-LAN interconnection, ATM uses LANE to make the ATM channel look like a bridge to the LAN protocols. LANE sets up SVCs across ATM networks to serve LAN clients that exist in each host. Data transmitted over ATM is encapsulated into cells, but ATM does not inspect cell contents. Therefore, it must have a method of mapping the underlying addresses to ATM addresses. LANE's function is to map MAC addresses to ATM addresses, encapsulate IP datagrams into ATM cells, and deliver them across the network. The LANE protocol defines the operation of an ELAN (emulated LAN), which is effectively a VLAN implemented across an ATM network. Multiple ELANs can be defined across the same network, but LANE operates at layer 2, so it is confined to creating bridged connections. If multiple ELANS need to communicate, external routers are required.

A LANE network, as illustrated in Figure 35-7, consists of a collection of servers that enable ATM to support functions that it lacks, such as broadcast capability. The figure shows separate servers, but the functions can be integrated into a single box. A LAN emulation client (LEC) in each host is the interface between the LAN and the ATM network. It can run in any device, such as an ATM edge switch, that has an ATM interface on one side and Ethernet on the other. Each ELAN acts like a broadcast domain.

FIGURE 35-7

LAN Emulation

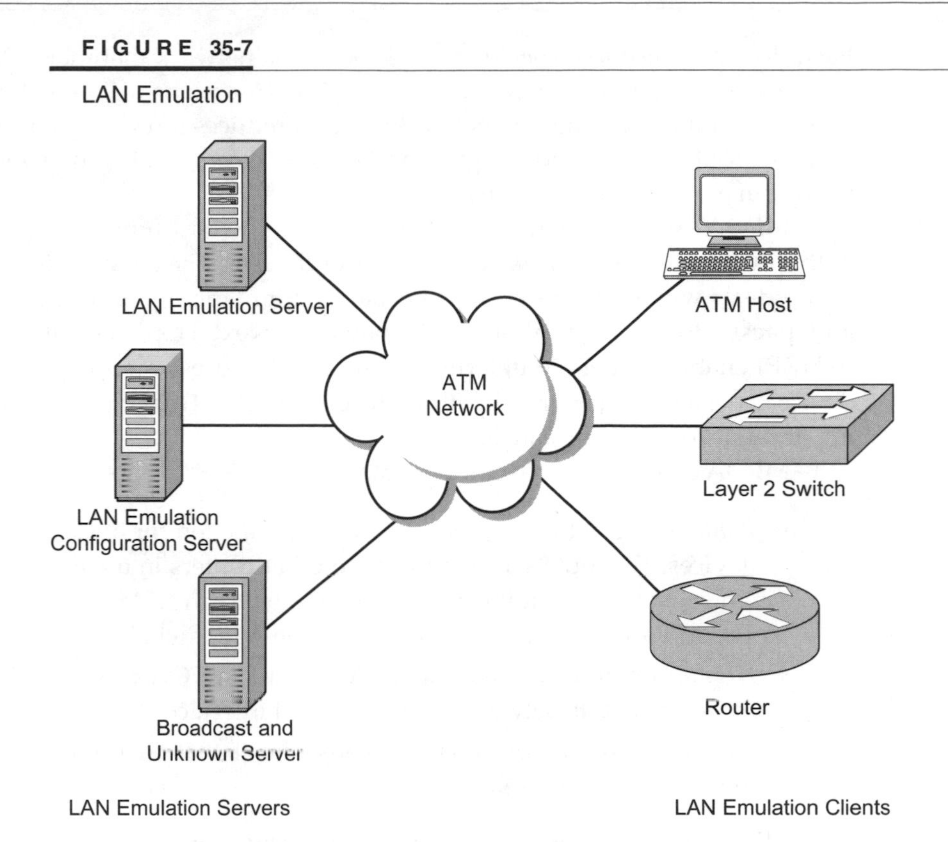

When a LEC joins an ELAN, it learns which ELAN it belongs to by communicating with a LAN emulation configuration server (LECS) to identify its LAN emulation server (LES). The LECS accepts requests from clients and informs them of the type of LAN being emulated and which LES to use. The LES is a central database that correlates all MAC addresses on the ELAN with their ATM addresses. Its function is similar to ARP on an IP network. If the LES cannot resolve the address, it uses the broadcast and unknown server (BUS) to broadcast frames to all of the LECs in a broadcast group. When the address is resolved, LEC sends the ATM address to the client.

Multiprotocol Over ATM

LANE operates at the datalink layer, and therefore enables ATM to operate as a bridge between LANs. The ATM Forum's MPOA operates at the network layer to enable routed networks to take advantage of ATM's low latency and scalability without the need for external routers. The objective of MPOA is to identify a flow between two network endpoints and affix a label to it that can be directly tied to an ATM virtual circuit. The network nodes then can forward packets based on labels rather than on IP address. The labels can be VCIs or frame relay DLCIs.

Each flow is related to a specific path through the network that makes IP behave as if it were connection oriented. In effect, MPOA assigns a flow between endpoints to a tunnel through the network. Since it reduces router hops and processing, the routers can handle significantly more throughput and networks can behave in a more predictable manner.

MPOA assigns two functions to the router. The host functional group deals with direct communication with the end user devices. The edge device functional group deals with functions such as virtual circuit mapping, route determination, and packet forwarding. A protocol known as Next Hop Resolution Protocol (NHRP) enables routers to determine IP-to-ATM address mappings so the edge device can establish a shortcut path to the destination. This limits router processing and improves performance.

MPOA consists of three components:

- *Route servers*, which perform the routing function for hosts and edge devices. The route server appears to other routers in the network to be a normal router, but it connects the session to an ATM virtual circuit. The route server can be embedded in the ATM switch.
- *Edge devices* connect traditional LANs to the MPOA network. They can forward packets between LAN and ATM interfaces.
- *ATM hosts* are MPOA-enhanced LANE hosts that are directly connected to the MPOA network.

MPOA is in many ways similar to LANE and requires LANE for its operation. At startup time, devices contact a configuration server that knows which devices are assigned to virtual networks. As devices are turned on to connect with the network, they register themselves with their servers so they can acquire address information and begin communicating.

ATM APPLICATION ISSUES

Although the predominate use of ATM is for carrier backbones, enterprise networks use it as a backbone. Its major advantage aside from its raw capacity is carrying multiple types of traffic. Frame relay and IP can carry multimedia traffic, but they lack the traffic management and signaling capability of dedicated circuits, which are ATM's strong points. Routers can be configured with an ATM backbone, a configuration that will become more prominent as voice and video are fed over high-speed routers. This section discusses some considerations in implementing ATM in an enterprise network.

The network design begins with collecting detailed information about current applications and traffic flows and how they will likely change in the future. If voice and video traffic will be added to the network, data from existing switches should be collected and organized into a matrix of originating and terminating points and traffic volumes between them. Analyze each application according to its sensitivity to delay, jitter, and packet loss as a way of determining ATM service class requirements.

Evaluating ATM Services

In choosing ATM services, here are some considerations:

- Will the network use private switches and circuits, a public ATM carrier, or a combination of both?
- Does the carrier offer both PVC and SVC? Is the charging based on usage, bandwidth, connect time, or what?
- Does the carrier support all of the traffic classes the application demands?
- What is the carrier's network topology? Does it have enough switches to ensure reliability? Where are its POPs and how do they meet with your network requirements?
- Does the carrier offer managed services for subscriber access devices and routers?
- Do the published SLAs with respect to end-to-end delay, delay variation, cell loss ratio, and other such QoS measurements meet the requirements of the application? How does the carrier measure the service, and what kinds of quality reports does the user receive?
- What are the carrier's traffic shaping and policing policies?
- Is routing in the network core static or dynamic?
- Can the carrier provide the necessary bandwidth? Can it handle peaks in compliance with the service agreement?
- What are the bandwidth increments? Does the carrier offer inverse multiplexing to increase the bandwidth?
- Can you internetwork between ATM and frame relay?
- What kinds of network management reports does the carrier provide?
- Does the carrier support MPOA? PNNI?

Addressing procedures and policies are also important to consider when selecting a carrier. Unlike the voice network, number portability is not implemented with ATM. If the network is completely private and will never be connected to other networks, then a private addressing plan can be used. Otherwise, it is necessary to obtain addresses that fit within the public numbering plan. Some considerations, which are not unlike the kind of problems you have with DID numbers, include these:

- Will the carrier give you an address base large enough to support anticipated growth?

- Can you get a contiguous range of numbers?
- If you change carriers can you retain the numbers?
- What happens if you have an address conflict with an adjoining network?

Evaluating ATM Equipment

Queue management is one of the primary factors distinguishing products on the market. Some products may support a limited number of queues, which will not be adequate for end-to-end QoS. Determine whether a switch is blocking or nonblocking—nonblocking meaning that the switching fabric can handle the capacity of all the input ports without cell loss. Look also at scalability and upgradeability. Is the switch capable of keeping pace with changes in ATM technology? What kind of switching architecture does it use—shared memory or self-routing? How does it handle output link contention?

- Look at bandwidth scalability. For example, can you upgrade a DS-3 link to an OC-3 link, and does this require rebooting the switch? In other words, can you add bandwidth without disrupting other services?
- Ease of setup is important and one of the distinguishing factors among products. Determine whether the equipment can be set up for all classes of operation from UBR to CBR.
- Interoperability will be important if the equipment is not all furnished by the same manufacturer. Insist on demonstrated interoperability of all required features from the network interface cards through the network of switches.
- Determine the degree of fault tolerance the equipment has. Are critical components such as processors and power supplies redundant? Are modules hot swappable?
- Does the product support integrated link management interface (ILMI)? ILMI enables equipment to monitor link health and configure addresses.
- How do the switches handle congestion? If congestion occurs, can the switch distinguish between high-priority traffic such as video and give it priority over traffic such as data that can be discarded with less effect on service?
- What is the process for routing around trouble? Does it implement PNNI to route around failures or congestion?
- What management tools does the carrier provide? Is the equipment SNMP compatible? Does it support ATM-RMON MIBs?
- Does the equipment use ATM Forum standards in all cases or are some standards proprietary?

CHAPTER 36

IP Networks

From the standpoint of carriers and user organizations alike, frame relay has been an enormously successful service. It is simple to set up, easy to understand, and provides statistics that network managers can use to assess performance and optimize CIRs and access circuit bandwidth. Frame relay has been so successful, in fact, that carriers are having a difficult time persuading their customers to migrate to the next step in evolution, which is an all-IP network. Some reticence is over security concerns. Frame relay traffic is internal to the service provider's network so packets do not flow over insecure paths. Most customers trust frame relay carrier's networks to the point of sending unencrypted data between their endpoints, but not so with IP.

Despite security concerns, IP networks have two major selling points that give them an edge over frame relay. A single access circuit can be used for both Internet and the enterprise's private network, and sites can be meshed without the need to define PVCs. Intersite connectivity is a major advantage for enterprises that have considerable internal traffic and intend to deploy VoIP or IP video. With frame relay the alternative to configuring multiple PVCs is to hub all traffic on the central site and route it back out to the destination. This increases traffic in the access circuit and increases latency and the complexity of traffic engineering.

On the surface, an IP network appears to be ideal for carrying voice and video traffic. Carriers provide Web portals that display their performance metrics, which are similar to frame relay and well within the bounds needed for isochronous traffic. As we discuss in this chapter, however, these metrics are usually based on monthly averages and do not include access circuits. IP metrics are edge to edge, not end to end, and the access circuit is where the choke points arise. Putting all traffic in one access pipe lets the customer use the bandwidth flexibly, but it is difficult to achieve a balance between end-to-end service quality and access circuit utilization.

If an IP network is to be effective for isochronous traffic, the originating network must classify the traffic and the carrier must prioritize and route it accordingly. The larger carriers use MPLS in their backbone to support differentiated traffic classes and to keep traffic from different users separated. An important difference between an IP network and frame relay or ATM lies in the carriers' difficulty of providing a guaranteed end-to-end QoS on IP. Any ISP can accept and carry IP traffic, and within its domain it can configure its routers to provide QoS for its core, but when traffic crosses into other domains assurance of service quality is lost.

MPLS standards as currently defined operate within a domain, which means QoS cannot be provided across domains unless the service provider has negotiated agreements with other carriers or has a footprint large enough to serve all the endpoints on the enterprise network. The second difficulty is in the access circuit. Few enterprise networks implement MPLS and the protocol does not define a UNI. Standards activity is under way to correct this problem, but it will be some years before it is complete.

QoS provisions do not pertain to most data traffic, so for data the main difference between IP and frame relay lies in security. IP networks can be used for transporting information that is not crucial to the organization; in fact, enormous amounts of unsecured information daily flows from corporate Web pages or attached to e-mail messages. IP networks cannot be trusted with sensitive information, however, so sessions must be secured by tunneling through the network with VPNs as discussed in Chapter 31. In this chapter, we enlarge on these with a discussion of MPLS VPNs, which have some interesting characteristics compared to VPNs running over native IP.

This chapter begins with a discussion of routing. In Chapter 6, we looked briefly at how routers and routing tables function. In this chapter, we will inspect routing protocols in more depth as a guide to understanding how IP networks function, with particular emphasis on how they achieve QoS. This is followed by a more detailed discussion of MPLS than the overview in Chapter 13. The chapter closes with an "Applications" section that provides information for selecting IP services and negotiating SLAs.

IP NETWORK ROUTING

Routers have two independent functions: control and forwarding. The control function inspects incoming packets, examines their headers, and consults routing tables to determine which output link to forward the packet over. Routers do not have a complete view of the path packets take from source to destination; they only know how to reach the next router. If the path is congested and the router's buffers are full, it discards packets and the customer arranges for retransmission.

Best-effort packet delivery is adequate and efficient for data, but it is unsatisfactory for isochronous applications. Since routing algorithms are based on packets taking the shortest path, UDP/IP cannot guarantee standards of reliability, packet loss, and delay. Voice can tolerate these irregularities within limits, and if the bandwidth is sufficient, the grade of service may be acceptable, but video quality drops noticeably with more than about 0.1 percent packet loss. To run these applications over a routed network, IP must take on some of the characteristics of a circuit-switched network with its reserved bandwidth and controlled delay.

Routers maintain tables to choose the optimum route for packet forwarding. Routing tables in simple networks are static and manually written into the router. Dynamic routing tables are updated periodically from messages transmitted across the network. This process of updating routing tables imposes a load on the network, but it has the advantage of enabling the routers to respond automatically to changing conditions such as a link failure, whereas static tables remain unchanged until they are manually updated.

Complex networks require dynamic routing protocols. If the destination is reachable from a directly attached active link, the router will use that route. Otherwise, it hands the packet off to a neighbor to boost it to the next hop. If a link fails or becomes congested, the routers communicate among themselves to update their tables. Routing protocols are classified as interior if they are capable of reaching destinations within a domain or autonomous system (AS). If they can span ASs, they are classified as external protocols. We will look briefly at the operation of the most common algorithms in these two categories.

Interior Gateway Protocols

IGPs, also known as intradomain routing protocols, are optimized for operation within a single AS. The most common interior protocols found on enterprise networks are RIP (Routing Information Protocol), OSPF (Open Shortest Path First), and IS–IS. RIP is the oldest routing protocol and the simplest, but it has limitations that make it unsuitable for complex networks. Routing protocols typically support only one layer 3 protocol at a time. As a result, if a network is running both IP and IPX, for example, multiple routers are required. Cisco supports a proprietary protocol, Enhanced Interior Gateway Routing Protocol (EIGRP) to allow the network to run multiple protocols at the same time.

Routing Information Protocol

RIP uses a distance-vector algorithm that selects a route based on the number of hops to a destination network or host. If the link serving the destination is directly attached to a router, the hop count is 0 and that would be the preferred route. If the destination is attached to a neighboring router, the hop count is 1, and so on. RIP propagates route information by broadcasting table updates to its neighbors at 1-min intervals. A route table for a new router can be generated by merely entering the IP addresses of its directly connected networks. The routers on the network exchange routing tables until they learn the topology and number of hops to the other routers. RIP has a limit of 15 hops. If the hop count over a particular route exceeds that, the destination is unreachable over that route. The protocol uses the hop count factor to advertize directly attached routes as unreachable. For example, if a route is silent for 180 s, the routers assume it has failed and set their hop counts to 16, which marks the route as invalid.

Refer to Figure 36-1, which is an enterprise network model consisting of three sites connected by frame relay, with the PSTN as a dial-backup alternative. Router 1 would advertize its four routes, the two Ethernets and ISPs A and B, to the other routers. For either of the two remote sites served by routers 3 and 4 to reach the server, the route would be across frame relay, through router 2, and across the Ethernet link to router 1. The 15-hop limit is fine for small networks of 10 or so subnets, but insufficient for more complex ones. Furthermore, hop count is insufficient to optimize complex networks. For example, router 1 has two routes to reach ISP B: direct and through router 2. Since the hop count is lowest over the direct link that would be the primary route despite the possibility that it might be more congested or have bandwidth limitations that make the two-hop route through router 2 more favorable. For that and other reasons, more complex networks use the OSPF protocol.

Open Shortest Path First

A major shortcoming of RIP, besides its limited hop count, is its inability to route based on any variable other than the number of hops. Consequently, it sends traffic to the preferred route without knowing whether another route is less congested. OSPF is a link-state protocol that uses a cost-vector algorithm to determine the optimum route. OSPF is capable of operating in a hierarchical network with the lowest

FIGURE 36-1

Enterprise Network Model with Multiple Routes

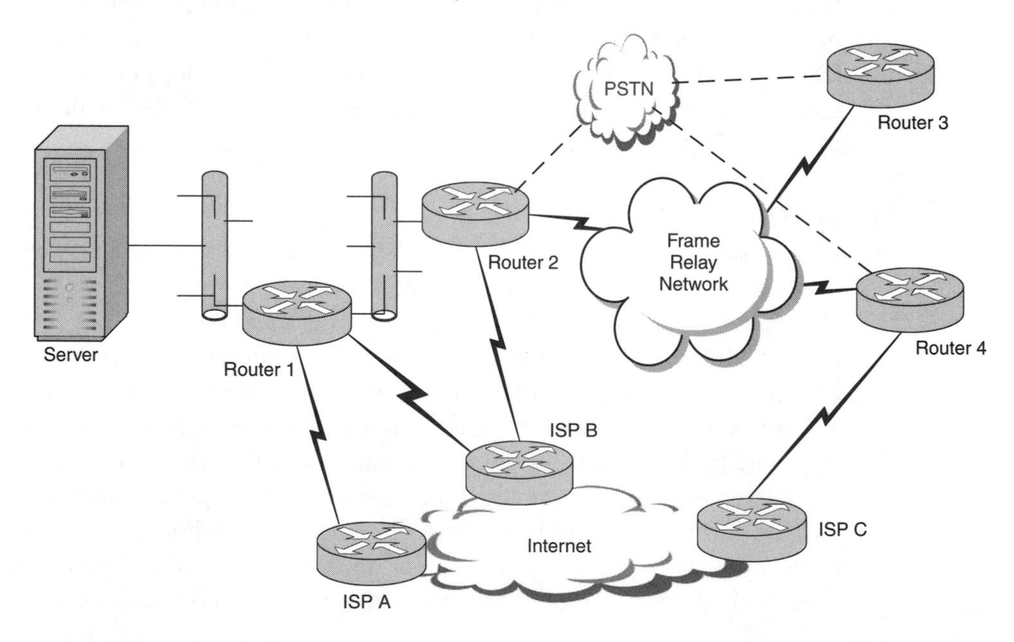

level, known as an area, which links to other areas through a backbone. The topology within an area is invisible to other areas. Each area has at least one border router to reach the backbone. An area that has only one border router is known as a *stub area*.

This partitioning of the domain reduces table complexity and reduces the amount of traffic between areas. Routers maintain databases of the entire area, including other routers on the networks and the cost of each router's connections. Instead of flooding the network with table updates as RIP does, routers exchange information about the domain only with their neighbors. During normal operations the routers exchange short hello messages to indicate that they are alive. When an event such as a link failure occurs, they exchange a series of link-state-advertisement messages to inform one another what happened and the routers adjust their tables accordingly. Table reconvergence may take several minutes.

OSPF's cost algorithm takes into account the total cost, including speed and quality variables, for all outbound interfaces. In Figure 36-1, the three main routers have dial-backup interfaces, which would be rated at higher cost than the frame relay interfaces and therefore used only in case of frame relay failure. If the DLCI between router 2 and router 3 is lost, but the link to router 4 is still available, router 1 would send its router 3 traffic through router 4 since that link would be less costly than the PSTN.

OSPF propagates reachability information to other routers, which enables them to simplify their routing tables. OSPF routers can also be configured as both internal and external. In addition, they can consider the ToS field in the IP header in making routing decisions. The three bits corresponding to delay, throughput, and reliability allow eight priority combinations. If the router has priority routes, it can base routing calculations on ToS and forward packets accordingly.

Integrated IS–IS

IS–IS is an ISO link-state protocol that was originally developed to support Connectionless Network Layer Protocol (CNLP), which is ISO's layer 3 datagram protocol. Integrated IS–IS has been adapted to support IP routing in addition to CNLP. Routers can be configured to run either protocol or both simultaneously. The integrated IS–IS algorithm is similar in operation to OSPF in topology and operation, but with a different addressing method. CNLP uses a 20-octet addressing method known as network service access point (NSAP). NSAP addresses provide the flexibility for the worldwide address space of IPv6.

ISO terminology is somewhat different from TCP/IP, but network devices have the same function. User devices, usually called hosts in Internet terms, are known as end systems (ESs) and a router is called an intermediate system (IS) in ISO terminology. IS–IS has two hierarchical levels. Routers are organized into level 1 groups called areas and areas are grouped into domains, which are linked with a backbone. L1 routers have information only about their own area plus a default route to the nearest L1–L2 router. L1 routers forward all traffic outside their area to an L2 router. L2 routers know the L2 topology and retain information about routes to L1 destinations, but do not know the L1 topology. A router can operate in either or both levels, i.e. it can perform inter-area or intra-area routing. OSI routing is divided into three parts:

- Routing exchanges between ESs and routers (ES-IS)
- Routing exchanges between routers in the same domain (intradomain IS–IS)
- Routing exchanges among domains (interdomain IS–IS)

Border Gateway Protocol (BGP)

Exterior routing protocols enable border routers located at the edge of the domains they serve to exchange routing information. Much Internet traffic is confined within an AS. This traffic is referred to as *local traffic*, with traffic between ASs called *transit traffic*. BGP's purpose is to manage the flow of transit traffic. BGP can carry local traffic within an AS, but between ASs, such as between ISPs and between domains in large enterprise networks, BGP is required. Routers that communicate with BGP are known as *BGP speakers*. An AS can contain as many BGP speakers as necessary for connectivity to neighboring ASs. BGP comes in two versions. Internal BGP is used for routing exchanges between routers that are part of the same AS. External BGP is used between ASs as shown in Figure 36-2.

In a BGP network, ASs are assigned group numbers. Tables in each router contain its neighbors' IP addresses and group numbers. Route selection is based on the shortest path within the AS. If neighboring routers have the same group

FIGURE 36-2

Border Gateway Protocol

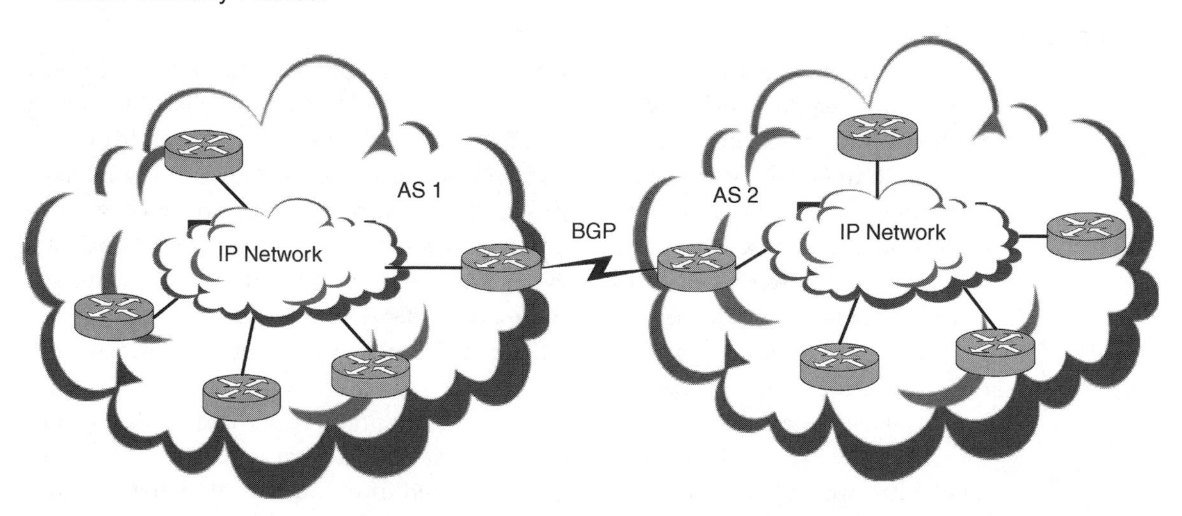

number, they are part of the same AS, so subnets within that system can be reached through either path. This multiple path capability gives BGP networks their robustness. BGP speakers do not load balance; they are configured to select the best route. Load balancing can be achieved by influencing the path selection based on available bandwidth and knowledge of the link metrics such as bandwidth and delay.

BGP routing tables can contain more than 90,000 routes, which is a daunting management job. To simplify the process, large ASs often use a *route reflector*, which is a router that is set up to send table updates to all routers within the AS. BGP uses *classless interdomain routing* (CIDR) to reduce routing table size. CIDR adds a masking block to the four-octet IP address to indicate the number of bits used for the network portion of the address. This enables routers to combine routes and simplify the table. When a router is first connected to the network, it exchanges the entire routing table with its neighbors over a TCP connection. Broadcasting massive table updates would consume an excessive amount of network capacity. Instead, when table changes occur, BGP routers send only the changed information to their neighbors. Through this exchange, each router maintains the current routing tables of its peers.

A BGP speaker is configured to evaluate different paths, select the best one, apply policy constraints, and advertise the routes to its neighbors. The administrator uses routing parameters called *attributes* to define routing policies. Policies are not contained in the protocol, but are controlled by the router configuration. For example, if one AS declines to carry transit traffic for another AS, it can block such traffic by advertising only routes that are internal to the AS. BGP can be configured to choose routes based on quality variables such as bandwidth, capacity, link dynamics, cost, and other such metrics.

Routing in an Internet

Internet complexity varies from a singe-domain AS in a corporate network up to and including the public Internet. Within the Internet, which consists of a large number of ASs, BGP is used between domains. Large enterprise networks also use BGP, but its complexity is overkill for smaller networks. The simplest networks use static routing tables or sometimes RIP to eliminate the need of manually changing tables if something happens to a link. The point is reached, however, where RIP will not support the applications. A prime example of this is when VoIP is employed and the network has link alternatives that must be selected on the basis of ToS bits.

The routing protocols we have just discussed are incapable of guaranteeing QoS connections through the network unless they are under unified control, and then only with limitations. For example, when routers reconverge, they may drop packets or delay them beyond the limits jitter buffers can contain. With a network such as the public Internet, transit traffic is handled by the routing decisions of multiple independent administrators. Those administrators are under no compulsion to consider QoS in making their routing decisions, and may not have equipment capable of supporting QoS. Therefore, connection quality within the Internet as it is presently constructed cannot be guaranteed.

IP service providers can support QoS within the bounds of their private networks. If the carrier is large enough, one provider may be capable of serving all of the requirements of an enterprise network; if not, the network manager can choose to use dedicated backhaul access links into that carrier's IP network. Carriers can also negotiate private agreements with other IP providers to fill in their coverage gaps. Even so, UDP/IP is incapable of supporting guaranteed SLAs. As packets traverse the network, each router inspects the address header and selects the next hop based on its routing algorithm. In Chapter 35, we discussed how LANE and MPOA can simplify the routing and forwarding process by analyzing packet flows and choosing a shortcut through an ATM network. These protocols still leave gaps in the quality equation. Most carriers are relying on MPLS to enable them to provide guaranteed SLAs in their IP networks.

MULTIPROTOCOL LABEL SWITCHING (MPLS)

MPLS, an encapsulation protocol for tunneling through an IP network, is a service-enabling technology that is becoming the cornerstone of VPN strategy for the major carriers. MPLS maps IP addresses to simple fixed-length labels, which are numbers that identify a data flow. A sequence of labels and links is called a *label-switched path* (LSP), also known as an *MPLS tunnel*. The LSP setup is unidirectional. Return traffic for the same session takes a different LSP. MPLS supports a variety of QoS functions including bandwidth reservation, prioritization, traffic engineering, traffic shaping, and traffic policing on almost any type of interface. These characteristics make MPLS superior to LANE and MPOA for use in a hardened VPN.

MPLS routing is greatly simplified because it happens only once at the edge of the network in the *label edge router* (LER). The LER attaches a label to each packet at the originating end. Instead of reading the entire header, routers read the label and route the call along a path that the label defines. When a packet enters the network it is assigned to a particular *forwarding equivalence class* (FEC), which is a group of packets to the same address that share the same QoS requirements. More than one FEC can be mapped to a single LSP. The routers internal to the network, known as *label switched routers* (LSRs), do not need to analyze the network layer of the packet. Once an edge router labels the packet, the rest of its journey is based on label switching. Each LSR examines the incoming label, uses it as an index in its *label information base* (LIB) to determine the next hop, and sends it to an outgoing interface with a new label. Label switching is done in hardware, which makes it much faster than routing. At the destination end of the network, an egress LER strips the label and sends the packet to the destination endpoint.

Figure 36-3 shows how label switching works. Ingress router 1 receives packets from Host A, determines the FEC for each packet, and from that determines the LSP. The LER adds the label to the packet and forwards it on the interface for that

FIGURE 36-3

MPLS Packet Network

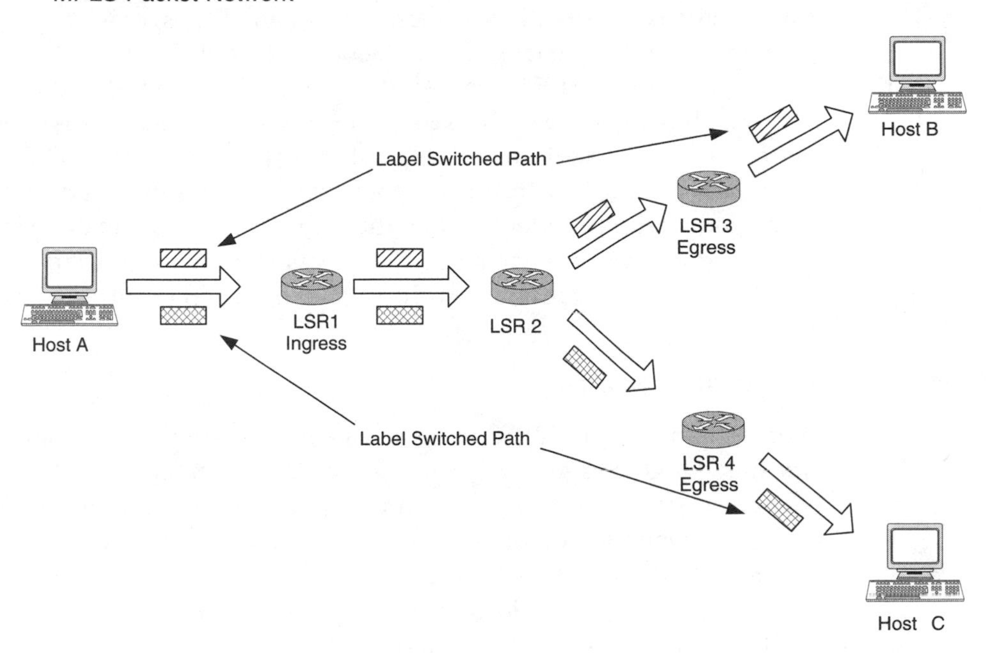

LSP, which happens to be the same for both. LSR 2 checks the incoming interface and label value, looks up the outgoing interface and label in the LIB, and forwards the packet to the next hop. The egress LERs strip the labels from the packets and forward them using the appropriate routing protocol, usually IP.

The labels themselves are flexible in content and depend on the layer 2 technology. In the case of ATM, the label could be a VPI/VCI or in frame relay the DLCI. The label could also correspond to a TDM time slot, a DWDM lambda, or the designation of a particular fiber if MPLS is applied directly to those media. For Ethernet and PPP, the label is added to the frame as a "shim" header that is placed between the layer 2 and layer 3 headers. When an LSR receives a label, its bindings create entries in the LIB. The contents of the LIB contain the mapping between the label and an FEC—that is the mapping between the input port and input label to the output port and label. These entries remain fixed until the label bindings are renegotiated. If LSPs are parallel, they can be routed together through a higher level LSP tunnel. The process of putting multiple labels on a packet is called *label stacking*.

Label Distribution

Labels are distributed through the network by several methods. Label Distribution Protocol (LDP) is an IETF recommendation that uses TCP/IP to distribute the labels by sending control messages over the link that will carry the data. The label can also be embedded in the contents of RSVP and in BGP messages. The labels are either set up before transmission begins, or they are established when a certain flow of data is detected. Labels are bound to an FEC as a result of a policy or an event that indicates the label should belong to that FEC.

A collection of MPLS-enabled devices is known as an *MPLS domain*. Within a domain, LSPs are set up for packets to travel based on FEC. This is either done hop by hop, in which case each LSR selects the next hop for a given FEC, or it is done by explicit routing in which the ingress LSR specifies the nodes. Explicit routes enable the service provider to set up the LSPs over a facility that is engineered and monitored to support the required QoS. Explicit routing allows the carrier to provide service within constraints such as bandwidth and delay.

MPLS Traffic Engineering

Traffic engineering in MPLS terms means the ability to vector traffic over a specific link. Large ISPs route data across the network based on a top-down view of network conditions and current and projected traffic flow. Traffic flow may be manually controlled or driven by an automated process similar to the policy process we discussed in Chapter 31. Traffic engineering enables the service provider to distribute the load over the appropriate network links while reserving bandwidth for particular classes of traffic.

MPLS has many advantages over ordinary routing, which often leaves some links congested and others underutilized, particularly while the routers update their tables. MPLS uses constraint-based routing, which takes into account such variables as bandwidth, delay, hop count, and QoS requirements. One alternative, which we discussed in the last chapter, is to run IP over ATM networks and let ATM provide the QoS. Although this works well, it is complex to set up and manage compared to using MPLS directly on the underlying physical facility such as a lambda. When an LSP is set up under traffic engineering constraints it is known as a *traffic-engineered LSP*.

Constraint-based Routed Label Distribution Protocol (CR-LDP) is a modification of LDP that allows the label to be distributed over an explicit route that has the required capabilities. Explicit routes can also be set up by using an extension to RSVP. CR-LDP distributes the label requests over TCP; RSVP uses UDP. The setup process is similar for both. A setup request is sent end to end from the ingress to the egress nodes and the response confirms the route and reservation.

Cell-Switched MPLS

In RFC 3035 the IETF defines procedures for enabling ATM switches to operate as LSRs. In this mode, ATM switches use IP addressing and run layer 3 protocols such as OSPF or IS–IS to determine routes and distribute labels over the cell-based infrastructure. No ATM-specific addressing or routing is required with this mode.

When ATM switches receive labeled packets, which are identified by the contents of the VCI and VPI fields, they segment them into cells and forward them across the appropriate interface. Effectively, the LERs are using ATM as the path between them, and the ATM switches function as routers.

The MPLS protocol operates on only a single domain. The protocol does not provide for an NNI and does not support interoperation between carriers to provide functions such as billing, failure notification, and network management. Such carrier interoperation is handled with private agreements.

MPLS VPNS

MPLS's VPN features enable multiple customer sites to interconnect across the service provider's IP network with the characteristics of a private network. Three major MPLS-based VPN architectures are in use:

- *Layer 3 VPNs*, which offer multipoint any-to-any service. These are also known as BGP/MPLS VPNs or RFC 2547bis.
- *Layer 2 point-to-point VPNs.*
- *Virtual private LAN service*, which offers multipoint any-to-any service for Ethernet endpoints.

MPLS VPNs use connectionless architecture, enabling the service provider to offer a range of value-added services such as Internet access, Web hosting, and Inter- and Intranet operations. Each customer site connects to an LER that maintains tables of VPN routes and services for authorized subscribers. The subscriber's routers do not need to support MPLS; they communicate with the LER using IP. This makes for simplified setup because the customer only needs to connect from the office router to the provider's edge router. The LER receives packets from the customer on a known interface that is identified by its VPN label. Inside the carrier's network, packets from multiple customers are mixed, but they can be delivered only to egress points that are defined as part of the VPN. Subscribers can use unregistered private addresses. NAT is required only if two VPNs with overlapping addresses need to communicate. Figure 36-4 shows two VPNs implemented across the carrier's IP backbone. Note that VPN 1 has both a private and public network connection between sites 1 and 2.

VPN membership is defined by a table in the LER that is known as a *VPN routing/forwarding table* (VPF). The VPF contains the IP routing table, the interfaces that use the table, and the rules for routing within the VPN. A site can belong to multiple VPNs. All of the information pertaining to those VPNs is contained in the VRF.

Most MPLS-based VPNs are implemented within a single service provider. Currently, multiprovider provisioning, billing, and fault detection are based on private agreements. Although services are end to end, MPLS functions that are available in the core do not extend to the access network.

FIGURE 36-4

An MPLS Virtual Private Network

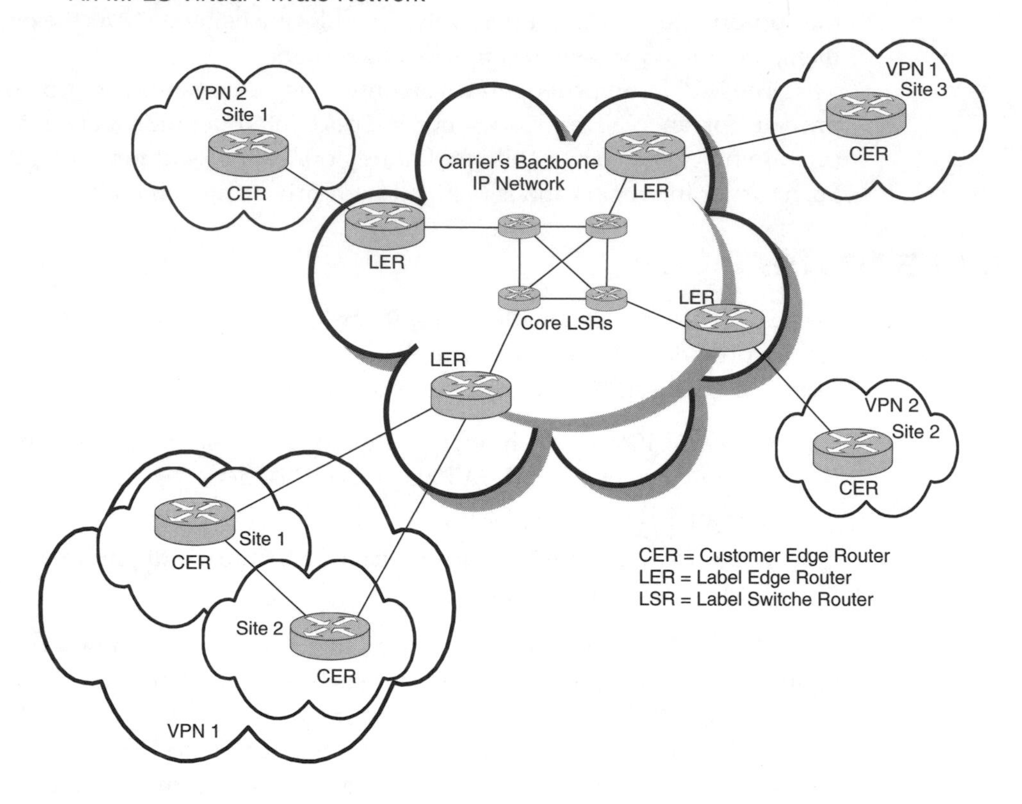

IP NETWORK APPLICATION ISSUES

The statistics that most carriers quote for their IP SLAs are specified within the network cores. Carriers typically quote percent availability, latency, and packet loss for city pairs. The quoted figures are generally within the requirements for VoIP, but they must be used with caution. First, they do not include the access circuits and second, they are based on averages.

Averages for most carriers are quoted 24 × 7 for a billing month. These metrics are adequate for data, but not for voice and video. Considering the fact that packets are most likely to be lost during congestion and the network is most apt to be congested during working hours, averages are insufficient for assessing quality for voice and video. If an IP network is to be used for real-time traffic, the application or customer test equipment should monitor key variables such as packet loss and jitter and fail over to the PSTN if necessary.

In the access circuit, the greatest efficiency is obtained by multiplexing various types of applications on the same circuit. Voice and video, with their short packet lengths and predictable characteristics, are not difficult to evaluate. When their contribution to the total load is known, the bandwidth available for data is easy to calculate. Circuit utilization must also be considered carefully. If utilization is too high, data packets, which have much larger packet sizes than voice and video, can delay time-sensitive packets outside the range of jitter buffers. Even though QoS variables may be within limits in the carrier backbone, degradation may occur in the access circuit.

Carrier Performance Metrics

Most carriers quote their performance metrics based on the criteria discussed in this section. The carrier typically uses active probes to measure network performance. The average is updated with new measurements at carrier-determined intervals. Measurements are typically made from edge to edge of the backbone. Not all carriers quote all of the metrics listed below. For example, some of the larger carriers do not include jitter.

- *Latency*: Latency or delay is the round-trip transmission time measured in milliseconds for a data packet to travel between two endpoints. It is affected by propagation delay, circuit bandwidth, and the forwarding rate of routers in the network. The carrier's SLA may quote one-way latency.
- *Packet loss*: This metric is defined as the percentage of packets lost in a transmission. It occurs primarily because router buffers overflow and packets are dropped because of lack of buffer space.
- *Backbone availability*: Availability is the percentage of time that the backbone is available to route packets. It is quoted as percentage of time the backbone is available on a 24 × 7 basis.
- *Mean time to repair*: This is the average time from the subscriber's report of failure until the service is restored.
- *Jitter:* This is the variation in packet arrival time expressed in milliseconds.

Evaluation Issues

The following questions should be explored in evaluating the carrier's quoted SLAs.

- Does the carrier provide real-time performance monitoring?
- Does the carrier provide Web-based real-time access to network metrics?
- Does the service run on a network separate from the carrier's Internet backbone?
- Are metrics calculated on a per-customer basis or for the network as a whole?
- What credits does the carrier offer for failure to meet its SLAs?

- Does the carrier interconnect its MPLS backbone with another carrier's network? Is the connection totally seamless or are some functions lost across the connection?
- Does the carrier support running multiple applications with different priorities and different QoS parameters across the same MPLS connection?

Ancillary Services

Most of the major carriers offer managed network services in which they handle router and firewall configuration for the customer's sites. The carrier may assume end-to-end responsibility for VPN configuration, either network based or CPE based. The advantage of network-based service is that the customer site can be a member of the VPN without any CPE. The following are some issues to consider:

- Does the carrier offer end-to-end managed services? For all sites?
- Does the carrier offer managed firewalling and intrusion detection?
- Does the carrier offer IP address management?
- Does the carrier offer secure remote access?
- Is IP multicast service available?
- Is dial access into the network available for remote users? What is the cost and what sites are available?
- Can the customer perform configuration, SLA monitoring, and report and review trouble tickets on the carrier's Web page?
- Does the carrier support dial-backup service?

Classes of Service

A major reason for using MPLS as opposed to IPSec VPN is the ability of MPLS to support different classes of service. The following issues should be explored:

- What classes of service does the carrier offer?
- Does the carrier charge different rates for different service classes?
- Does the carrier offer class of service across the network core or rely on overprovisioning the core to meet its SLA metrics?
- How is class-of-service pricing administered? Does the carrier set a ceiling on the percentage of time-sensitive packets? What happens when the ceiling is exceeded?
- Can the customer run multiple applications at different priorities with different QoS associated with each priority across the same MPLS connection?

CHAPTER 37

Network Testing

As networks evolve, maintenance strategies must follow, and in most networks that involves some degree of testing. An effective network design takes into account the method of isolating and clearing problems. A close relationship with key vendors is important. End-to-end maintenance responsibility may be turned over to a vendor, but costs can be reduced and problems cleared with less user impact if at least some service is handled in-house. Testing was simpler when the entire network was analog because a technician could measure loss, noise, envelope delay, and other such variables with a transmission impairment measuring set. Some analog testing is still appropriate, but as the network evolves to digital and broadband, network managers need different testing strategies. Complete failures are rare in today's network, but impairments are common and the source of the problem is seldom obvious. Many problems can be masked by adding bandwidth, but a few tests can often reveal a faster and cheaper solution.

To do an effective troubleshooting job on an IP internet, a technician needs to know which devices originate the most traffic, what route packets are taking, what protocols are being used, and what time of day the load peaks occur. A network management system, which we discuss in the next chapter, can provide some of the answers, but an NMS is not always enough. An NMS may display network maps, collect statistics, identify defective devices, and monitor network segments for errors and other irregularities, but almost every network needs to supplement the NMS with portable or centralized test equipment that can capture and inspect packets, displaying their contents in English.

Companies with complex networks find that testing capability is essential, not only to control costs, but also to restore failed service quickly. An investment in test equipment is wasted, however, unless technicians know how to use it to advantage. An effective management strategy requires trained personnel who know the topology, how the protocols function, and how the network looks and behaves under normal conditions. Even simple tests such as checking the lights on a CSU to determine whether the trouble is in company or common carrier facilities can reduce clearing time and maintenance costs.

Network testing has these objectives:

- to establish benchmarks that serve as references or confirm that design objectives have been achieved;
- to identify bottlenecks and defective or improperly configured equipment;
- to obtain assurance that trouble is reported to the appropriate vendor and to provide information to the vendor to reduce clearing time;
- to provide a high degree of network availability to users by preventing impairments and failures and restoring service rapidly when they occur;
- to verify carrier and vendor SLAs independently of their internal reports.

This chapter discusses test equipment that the enterprise network is most likely to need. Carriers and vendors are the primary users of bit-error rate testers (BERTs) and cable testers, but they find application in enterprise networks as well.

CIRCUIT TESTS

The foundation of the network is the physical layer. Contractors and carriers install most of the infrastructure, which the customer accepts only after thorough tests have demonstrated that it meets the specifications. You can verify the installation quality by reviewing the service provider's documentation or conducting some tests with internal staff. In either case, knowledge of the kinds of tests is essential.

Structured Cable Testing

In a structured cabling system, physical inspection of the wiring job is important, but the only way to check its electrical performance is by testing. At the most basic level, cable test equipment detects physical problems such as splits, reversals, and shorted pairs. Beyond that, cable testers measure all of the parameters of the corresponding EIA/TIA category. They normally report the numerical results of the test along with a pass or fail indication.

When Ethernet ran at 10 Mbps, installers could be somewhat sloppy in their twists and terminations and the network would still run, but at 100 Mbps problems appear, and gigabit Ethernet has even less margin for error. It is a well-known fact in the industry that a marginal installation degrades the performance of the highest quality materials. Cable testers are typically tools for the installation company, but they may be a good investment for the customer as well because things have a tendency to change over time. The crosstalk performance of a cable depends on maintaining the proper twist within and between pairs, and careless treatment of wire in a remodeling job, poor quality patch cords, wet jacks caused by spills, and other such disturbances can upset the balance and result in trouble.

Any cable installer should be prepared to provide complete test results showing conformance to the variables required by the specifications. The results should show attenuation (insertion loss), return loss, NEXT, ACR, and length. For multiconductor cables, power sum tests should be included. Propagation delay and delay skew may be included, although these are more likely to be a function of the cable than the installation job. Key to the testing job is a properly calibrated test set that is designed and selected to measure the category of cable under test.

Category 6 testers are known as level III, with earlier testers for category 5 and 5e known as level II and level IIe, respectively. It is essential that category 6 cable be tested with a level III tester because tests with a lesser test set will not measure all of the parameters of the standard. The category 6 standard includes the concept of a permanent link, which includes not only cable and jacks, but also the patch cords on each end. Many wiring systems purporting to be category 6 were installed before TIA/EIA approved the standard. It is important to realize that before approval of the standard, test sets to meet the standard were based on expectations that may or may not have been realized. Retesting of these older installations with current test equipment can identify nonconforming cable.

A key factor in selecting the test set is its storage method and capability. The best test sets can store the results in flash card memory, which can later be dumped into a PC and printed to form a permanent benchmark. Another factor is the availability of interfaces to test other media such as coax and fiber optics.

Fiber-Optic Testing

Organizations with an extensive amount of fiber may invest in test equipment for locating trouble and testing performance of the fiber. Testing in the optical domain is similar in many ways to structured wire testing. In fact, many cable testers offer adapters for fiber optics. Crosstalk between fibers is nonexistent, however, which eliminates the need for those kinds of tests. Testers typically measure insertion loss with a light source at one end of the fiber and an optical power meter at the other. The light source is usually LED-based for multimode and laser-based for single mode.

An optical time domain reflectometer (OTDR) completes the picture. This device operates on a principle similar to radar. It sends a light pulse down the fiber and displays reflected signals on a screen with a time scale calibrated in distance. Any impedance irregularities show as a blip. The OTDR may include a power meter for measuring loss with a far-end source. Some products contain a microscope for viewing the cleaved ends of the fiber.

Bit-Error Rate Test

Circuit impairments often manifest themselves as a high error rate, which a BERT can measure. The tester generates a pseudorandom signal that injects a known bit pattern into the network. Either the signal is looped back at the far end or a matched pair of devices is used end to end. BERT units typically measure bit errors, block errors, error-free seconds, bipolar violations, and framing or CRC errors. BERT tests require the line to be out of service.

Carriers usually run a BERT test before turning a line over to the customer, but the ability to perform such test independently of the carrier is often advantageous and may not require expensive test equipment. Most CSUs operating on T1 lines using ESF can provide equivalent information over the Fe channel, which monitors six-bit CRC, framing errors, and bipolar violations. CSUs that monitor the Fe channel can send diagnostic information to a printer or it can be retrieved from a storage buffer. Some routers can also be equipped with a BERT module.

REMOTE MONITORING (RMON)

RMON and the updated version RMON2 enable a protocol analyzer or network management workstation to collect statistical information from remote devices. A management workstation or protocol analyzer communicates with an RMON agent over TCP/IP to tabulate statistics. RMON probes or agents are either stand-alone hardware devices or software embedded in managed devices. RMON collects as many as nine types of information:

- *Statistics* accumulates such variables as numbers of packets, octets, broadcasts, collisions, errors, runts, jabbers, and distribution of packet sizes.
- *History* collects statistics based on user-defined sampling intervals.
- *Host* discovers what hosts are active on the network and tracks statistics such as packets received and transmitted for each of them.
- *Host Top N* selects the hosts that had the largest traffic or error counts based on statistics from the host group.
- *Alarm* allows managers to set alarm thresholds based on absolute or delta values. Alarms trigger other actions through the events group.
- *Events* operate with alarms to define an action that will be taken when the condition occurs. The event may write a log entry or send a trap message.
- *Matrix* logs statistics by pairs of nodes.
- *Filter* allows the manager to define the criteria that trigger an event.
- *Capture* stores captured packet data for protocol analysis.

RMON is limited to looking at layer-2 protocols. RMON2 can monitor all seven layers of the OSI model, which enables agents to track activity such as quantities of packets of different types.

PROTOCOL ANALYZERS

A protocol analyzer is the network administrator's bird dog, flushing out impairments that otherwise would require hours of cut-and-try. Protocol analyzers are essential tools for most networks. The analyzer looks deep into packets and exposes reasons the network is bogging down. They help technicians locate the source of logical malfunctions by providing statistics, setting traps, emulating devices, and offering expert assistance. A major value of a protocol analyzer is to serve as a training tool. Technicians can inspect live packets on real networks and understand the relationships between them.

The analyzer may run on a proprietary platform or on a laptop computer. Some analyzers are self-contained, while others separate the packet capture from the analysis function. The latter architecture enables the manager to distribute probes throughout the network and collect information centrally. Tests are fast, repeatable, and easy to customize. Protocol analyzers are available for a variety of purposes. Some such as Ethernet, frame relay, or IP analyzers are useful for enterprise networks, while others such as VoIP quality analyzers, SS7, and SONET testers are intended for designers and service providers. As could be expected, the price varies with the complexity, but some software is available in the public domain as freeware or shareware.

Protocol analyzers operate in a monitor or simulation mode. In the monitor mode, the test set is a passive observer of the bit stream. In the simulation mode, the circuit is opened so the test set simulates either the network or the CPE. Programmable units allow the operator to set triggers or test sequences to trap error conditions. Some devices operate in conjunction with RMON probes. Distributed probes connect to a network segment, capture statistics, and report the results back to the central device, which can be either a protocol analyzer or a network management system. Some probes can communicate with each other to exercise remote portions of the network.

The analyzer compiles statistical information to provide a view of what is going on in the network. For example, if a network carrying a combination of IP, IPX, and AppleTalk traffic shows signs of overload, the technician needs to know which stations are originating the bulk of the traffic. By displaying a line graph or pie chart of traffic distribution by the protocol, the technician knows where to start looking. Protocol analyzers not only capture data, but may also have artificial intelligence either built-in or in an outboard processor. Besides capturing data, such devices analyze it and suggest potential causes.

A good protocol analyzer enables the manager to determine how the network is performing exclusive of user and vendor opinions and reports, while gathering data to pinpoint bottlenecks and performance problems. It can simulate traffic loads and predict whether the network has the capability of supporting applications such as VoIP without incurring the expense of a field trial. A protocol analyzer facilitates functions such as these examples:

- developing benchmarks that can be used for analyzing performance changes or programming policy servers;
- troubleshooting user reports such as slow response time, printing problems, etc.;

- eavesdropping on traffic to detect deviations such as unauthorized Web browsing or instant messaging;
- detecting outside attempts to penetrate intrusion detection systems;
- generating traffic to test route tables, prioritization, and network robustness;
- discovering devices on the network and information about the devices such as IP address, subnet mask, and RMON capability;
- observing per-port utilization and error rates;
- evaluating vendor SLAs for QoS variables such as latency, jitter, and packet loss;
- obtaining statistics regarding the amount of bandwidth each protocol on the network consumes.

Protocol Analyzer Features

Most protocol analyzers have similar features and functions, although the methods of displaying and analyzing the results vary considerably. Few analyzers include all of the features listed below. Some units are specialized for particular functions such as LAN, WAN, or wireless testing, while others are adapted with software or add-on units.

Packet Capture

The analyzer connects to the network at a point where it can observe all the traffic. If the switch supports port mirroring, the analyzer can capture the traffic flowing through the monitored port. If the switch does not support port mirroring, a hub can be connected between the switch and the attached device, and the analyzer connected to the hub. Whatever the access method, the device captures packets, decodes, analyzes, and displays them. Figure 37-1 is a sample of captured data from Ethereal, which is an open-source protocol analyzer. The upper part of the screen is a running list of captured MPLS packets. By putting the cursor (the heavy black line in the center of the screen) on a packet, the second window shows decoded details such as MAC and IP addresses. The lower window shows the selected packet in hexadecimal.

Analyzers include filters, so you can select only packets that meet particular criteria such as traffic to a particular device or a conversation between devices. The ease with which this filtering is accomplished is one of the differentiating features among analyzers. Filters are particularly important on high-speed networks such as gigabit Ethernet because the capture buffer can fill so rapidly at full wire speed. Some analyzers have the capability of slicing long packets. Most of the information needed for troubleshooting is in the header, so the data block can be sliced off without affecting the analysis.

FIGURE 37-1

Packet Capture from Ethereal Protocol Analyzer

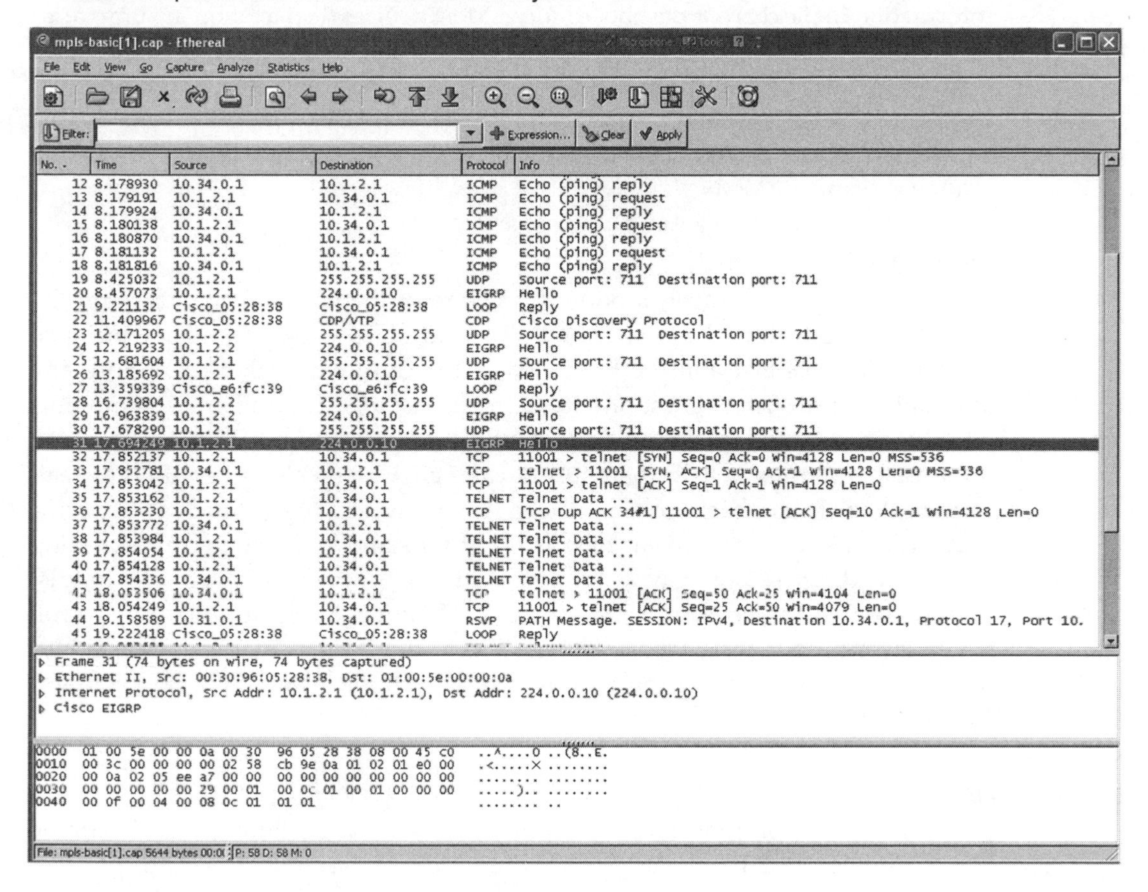

Packet Decode

All protocol analyzers decode the most common packet types such as TCP, UDP, ICMP, HTTP, FTP, and the like. Nearly any analyzer will decode any packet that runs over default ports, but they may have problems recognizing defined packets that use other than the well-known ports. Most analyzers support 200 or so of the most common protocols. If the network is running specialized protocols, the analyzer will capture the packets, but will not decode it. Most analyzers display the packet in hexadecimal, so the user can manually decode unrecognized packets. Some analyzers convert IP and MAC addresses to DNS names to make it easier to identify the device. Analyzers may also look into the packet to identify what type of application created it. This feature enables the analyzer to distinguish between applications such as VoIP, Microsoft Exchange, SQL Server, LDAP, etc.

Statistical Displays

Analyzers can summarize the captured data in several forms. Some display the information in pie chart or dashboard form. Most devices display tabular summaries of protocols and sessions between devices, often called "top talkers." Many also display statistics in real time including network utilization over a time period, protocol statistics, number of packets, frame size distribution, number of errors, runts and jabbers, and a matrix of traffic between nodes. Some devices are able to read and display RMON statistics.

Packet Analysis

Packet analysis is the most important function of the protocol analyzer and the distinguishing feature among products. The most effective devices have artificial intelligence tools that analyze the fields in the captured packets, diagnose any anomalies, and offer suggestions for remedial action. They identify symptoms such as slow servers, excessive ARP, multiple retransmissions, ICMP errors, and broadcast and multicast storms. Most devices analyze on either captured or real-time packets. In the former case, the technician can set a trigger such as a specific type of packet with a particular to or from address and look in detail at preceding and succeeding packets. On a real-time basis the protocol analyzer can be displaying results such as circuit and protocol utilization. Some products have a full seven-layer decode, which looks into the application. Some devices have standard and expert versions, with the latter offering additional analysis functions at a higher cost.

Traffic Generation

Many analyzers generate traffic to stress the network to its limits or to determine how the device treats particular protocols. In a multimedia network, e.g., the protocol analyzer may generate a mixture of different traffic types to enable the technician to verify that the real-time traffic is receiving a better QoS than other types. An analyzer should be able to stress the network with various types of traffic and packet sizes and show how response time and QoS vary as utilization increases. Some analyzers are able to simulate hacking attacks, so technicians can see if firewalls and intrusion detectors are functioning properly, particularly under load.

Node Discovery

A useful feature is the ability to automatically detect and map all devices on the network, showing their speeds, configuration, slot and port, etc. Some analyzers can create network diagrams in a graphics program such as Microsoft Visio, displaying the range, mask, and broadcast address of subnets. The analyzer may identify VLANs and distinguish between subnets carrying different protocols such as IPX, NetBIOS, and AppleTalk. If the analyzer supports SNMP, it should be able to display other information about a selected device such as its RMON capabilities.

Problem Discovery

Many analyzers can detect and display problems on the network such as duplicate IP address, incorrect subnet mask, devices that fail to respond to a ping, and other common problems. The analyzer may have the ability to ping all devices on the network and report on any device that is not responding.

VLAN Analysis

An analyzer with VLAN capability can determine which switch ports are assigned to which VLANs and which devices are connected to which switch ports. Switch ports that are used as trunks are displayed with the trunk protocol in use. They display such information as the VLAN number, description, subnets, etc.

Cable Analysis

Some devices can test patch cords, measure cable length, and discover problems such as opens, crosses, and splits. Some devices show the impedance and distance to an anomaly such as a sharp bend that affects impedance. Some devices also have a fiber interface.

Baselining

Baselining is an excellent technique for supporting network troubleshooting. When the network is operating normally, baseline information is captured and retained for later comparison. For example, most analyzers show network utilization. If utilization snapshots are taken at regular intervals they can be recorded to document normal performance. If users complain of slow response, subsequent snapshots can show the time at which overloads are occurring. By looking for the high-usage devices, the analyzer can help pinpoint those that are causing the problems.

WAN Analysis

Some devices have WAN interfaces that enable them to extend beyond the LAN and monitor traffic flow and errors. The analyzer may use remote probes to collect information. They can check access link congestion and monitor the network to validate the carrier's SLAs and plan for capacity changes. Some analyzers can auto-discover CIR and display information such as data flow and errors by DLCI or PVC.

Wireless Network Testing

Some LAN analyzers can operate on either wired or wireless networks. In addition to regular analyzers, numerous specialized wireless testers are available. Wireless LANs require the protocol analysis functions listed above, plus more that apply only to wireless. These include:

- detect and locate rogue nodes;
- display AP and device configuration information including ad hoc modes;

- identify what device is using each RF channel;
- check security settings on APs and mobile devices;
- check AP signal strength;
- optimize AP placement;
- check handoff delay between APs;
- display signal constellation and interfering signals across the bandwidth of the network on a CRT-type device or oscilloscope;
- check bandwidth utilization, transmit rates, number of devices using the network, etc., and locate bottlenecks such as overloaded access points.

In addition to the above, wireless testers may include coverage mapping and direction finding, so stations can be located based on their signal strength and azimuth.

DISTRIBUTED PROTOCOL ANALYSIS SYSTEMS

Complex enterprise networks can often profit from a sophisticated form of protocol analyzer that can perform the above functions and more on the WAN and attached LANs. The line between a network test system and a network management system, which we cover in the next chapter, is often indistinct, particularly in that both may use RMON probes. In general, a network management system uses an open protocol such as SNMP to detect alarms, show alerts, and maintain a record of attached devices.

A protocol analysis system performs the same varieties of tests that a protocol analyzer does, except on a larger scale. Often it is not practical to make manual tests with an ordinary protocol analyzer. For example, you may want to see how much traffic a device such as a switch or router can handle before it begins dropping packets. To simulate the effects of actual traffic requires variability of packet sizes and burst rates that would be difficult or impossible for a human to generate with protocol analyzers. You could test a network's capability of handling VoIP by mixing in with the normal data a varying load of VoIP packets over sessions of variable length. Vendors can use a test system to demonstrate the network's capabilities and managers can use it to verify that the vendor has not oversold. Developers also use test systems. For example, a developer could develop a script to test an entire RFC and report results.

Probes are distributed throughout the network to collect data and deliver it across the network in response to queries or when triggered by an event. The system enables the network manager to set up automated test scripts to execute tests repetitively in a predefined sequence. The system can be set up as either an active or a passive monitor. In the active state the system generates traffic; in the passive state it monitors traffic and reports link status and the health of routers and servers.

Perceptual Evaluation of Speech Quality (PESQ)

As we discussed in the last chapter, carriers' quoted performance statistics are based on averages and the customer has no assurance that speech quality will not deteriorate during peak load periods. Protocol analyzers can provide accurate end-to-end statistics over a period, but evaluation of speech quality in accordance with ITU-T P.862 is more accurate. Some test systems have the capability of generating and monitoring RTP traffic to qualify networks for VoIP. The system simulates various codecs and measures variables such as maximum, minimum, and average delay, jitter, and packet loss. In addition, it compares the monitored signal to a reference signal to deliver the mean opinion score (MOS), which closely approximates actual user perceptions.

TESTING APPLICATION ISSUES

The degree of testing and analysis that will be done in-house is unique to every network. An investment in test equipment, therefore, must match management's testing strategy and the support staff's ability to use it. Test equipment can be rented for occasional use or to evaluate a unit before purchase, but most companies that depend on the performance of their network will want to own the appropriate types of equipment.

Protocol Analyzer Evaluation Criteria

Protocol analyzers are packed with features and functions, the need for which depends on what the instrument will be used for. The first question is its purpose. Will it be used on LAN, WAN, or both? To be effective, the analyzer must have the processing power to read the addresses of the packets going by on the network segment, determine the packet type, capture the ones needed for analysis, and filter out unwanted packets. Nearly any device can do this on 10/100 Ethernet, but the packet throughput on gigabit Ethernet exceeds the capabilities of many devices. Standard analyzer functions include measuring network utilization, determining protocol distribution, decoding packets, and displaying the statistics in a readable manner. The sections below discuss evaluation criteria and suggest questions to ask about the analyzer.

Protocol Support

Be certain that the analyzer can recognize and analyze all the protocols the network carries. For each protocol supported, the system should be capable of decoding and displaying header information in human readable form. If some protocols must be decoded manually, the speed and accuracy of testing will be impaired. For example, unless you are good at converting hex to decimal addresses, it is not readily apparent that 856FBB04 translates to 133.111.187.4. Any analyzer will support the most common protocols, but there are hundreds of protocols in existence, more are being added regularly, and changes are being made to others:

- What protocols does the analyzer need to decode?
- Does your network have unusual protocols that the typical analyzer does not support?
- Can the analyzer recognize the traffic regardless of the port it uses?
- Does it need to support WAN protocols such as X.25, HDLC, SDLC, SNA, asynchronous, and PPP?
- Does the manufacturer regularly update the protocols the analyzer can decode? What is the process for keeping the analyzer updated?
- Is wireless support needed? If so, can it handle 802.11a, b, and g?

Hardware Interfaces

Will the analyzer be used exclusively for 10/100 Ethernet? If so, laptop software can probably do the job. For gigabit speeds and faster, a special device with a hardware interface will probably be needed to keep pace with the passing data in full-duplex mode. In addition, the analyzer must have buffer space sufficient to hold captured packets:

- What network interfaces does it need to support—Ethernet, frame relay, RS-232, gigabit Ethernet, 10G, T1/E1, wireless, ISDN BRI and PRI, DS-1/E1, DS-3/E3, etc.?
- Is the analyzer required to support gigabit speeds? If so, does it have the processing power to keep up with wire speed?
- Does the analyzer have the ability to filter and slice packets to reduce buffer requirements? Does it filter and slice packets before they are stored? Does the manufacturer support regular filter updates?
- Does it support RMON probes?
- Does it support VLAN?
- Does it have a fiber-optic interface?
- Can the device display errored packets that many NIC drivers discard?

User Interface

The user interface is one aspect that is unique to each analyzer and is the easiest to observe among the available products. The best displays give the user the ability to display statistics in a variety of different ways. One key factor is whether the analyzer runs on a laptop or in a hand-held device. A laptop gives the developer more latitude in creating a pleasing display, but it is not so convenient to carry around:

- How intuitive is the user interface? Is it easy to enable basic features?
- Does the device convert numerical addresses to device names?

- How complete is the help menu?
- Is it easy to export data to another application such as a spreadsheet?
- Can displays readily be changed such as from pie chart to graph?
- How effective are reports and statistics?

Traffic Generation

- Can it generate traffic at line rate to stress-test interfaces?
- Can it generate mixed media for evaluating traffic prioritization?
- Can it loop a test set at the far end for end-to-end link verification?
- Can it run in both active and passive modes?
- Can it be set to run various frame sizes, interface data rate, and utilization percentage of each type of packet?
- Does it have the ability to capture and replay packets?

Diagnostic Support

The best analyzers include some diagnostic capability to aid the user in interpreting data and detecting problems. Some systems are capable of interacting with Cisco's Service Assurance (SA) agents, which are router software applications that record a variety of measurements:

- What application-specific expert diagnostics are available? Does it detect the most common problems such as duplicate IP addresses and TCP retransmission errors?
- Can the system perform WAN analysis? Can it determine where bottlenecks exist?
- Can it measure server response time?
- Can it initiate SNMP trap messages and respond to get messages from the management workstation?
- Can it perform deep packet inspection at the application layer to detect viruses and DoS attacks?
- Can the device perform as an RMON agent?
- Does the device have standard and expert versions?
- Can the system interact with Cisco SA agents? To what degree do they interpret the information obtained from the router?

QoS Measurements

Most analyzers can monitor real-time traffic and evaluate some parameters by looping back the far end, but high-end analyzers will be needed to do a thorough job of measuring jitter and latency:

- Can the device measure VoIP quality?
- Can it simulate voice traffic?

- Does it provide perceptual quality measurements?
- How are devices synchronized for making jitter measurements?

Ancillary Functions

- Does it have a wireless application? Does it decode 802.11a, b, and g?
- Can it evaluate characteristics of the physical medium—wire and fiber optics both or only wire?
- Can the system determine which stations are producing the greatest amount of traffic and of what type?
- Can the system identify all nodes on the network and list their hardware and software addresses? Can it display these in graphic form?
- Can the system detect attempts to hack the network?
- Does the system support SNMP traps and triggers? What MIBs does it support?
- Can the system detect and display alarms such as log-on failures, retransmissions, etc., and notify the manager with a page or e-mail?

CHAPTER 38

Network Management Systems

As mission-critical operations migrate to LANs and distributed servers, the need for a comprehensive network management system (NMS) becomes compelling. In the days of the mainframe, network management was considerably less complex than it is today because only a narrow range of users had data devices and the load and performance were predictable. In today's network, virtually all office workers have desktop devices that they use for a variety of applications, many of which are outside the control of the IT staff. This raises some interesting issues. Does management provide bandwidth to meet all requirements—even personal Web surfing, downloading MP3 files, and playing games? Or should some applications be denied access to the network, and if so, what is the process for deciding which applications and users are admitted and which are rejected? These issues and the mechanics for administering them fall under the category of policy-based networking. Computerized tools for traffic shaping, policing, or enforcement, control the share of the network that certain user groups or applications receive. These devices, known collectively as *bandwidth managers*, can enable network managers to provide the necessary bandwidth for priority and delay-sensitive traffic, while avoiding the alternative of over-provisioning.

As networks become more complex and skilled labor to maintain them climbs in cost and shrinks in supply, both enterprise network managers and carriers find it necessary to mechanize network management where possible. The objective is to make networks self-managing to the extent feasible. Network equipment is increasingly designed to support network management. Software agents built into such devices as routers, switches, and multiplexers can report status and accept orders to reconfigure themselves under computer control. Humans are needed to make decisions that cannot easily be formulated into simple choices, but computers can react faster and weigh and discard alternatives at a speed that humans cannot match.

LANs and WANs present a challenge because they are so closely integrated with users' operations. The requirements for managing an enterprise network are not a great deal different from the management of a data center. The following functions, which are typical of data centers, must be provided:

- Capacity planning
- Access control and policing
- Change management
- Disaster planning and recovery
- Fault management
- Power conditioning
- Security management
- Service management
- Storage medium backup and recovery

Simple Network Management Protocol (SNMP) and Common Management Information Protocol (CMIP) are the principal open NMSs in existence today. SNMP is based on TCP/IP and CMIP on OSI. All major manufacturers support SNMP in most of their products. CMIP has been slower to gain acceptance because of its complexity. Its primary use to date has been in common-carrier networks where the additional complexity is accepted in exchange for its versatility.

NETWORK MANAGEMENT AND CONTROL

Network management is a specialized client–server application that filters and correlates alarms, alerts, and statistics to either make decisions or assist humans in maximizing network performance. The term *technical control* is often applied to centralized network management and control systems that monitor status and manage capacity in large networks. Besides providing testing capability, technical control centers include alarm reporting, trouble history, and, usually, a mechanized inventory of circuit equipment. ISO suggests dividing network control operations into five classifications:

- *Configuration management*. Retains records, and where feasible configures network equipment remotely. Retains a complete record of users, assignments, equipment, and other records needed to administer telecommunications apparatus. The function discovers nodes on the network and detects operational status and changes in configuration.
- *Accounting management*. Tracks vendor bills and distributes costs to organizational units based on resource usage.
- *Fault management*. Receives reports from users, diagnoses trouble, corrects trouble, and restores service to users. A major objective is to detect incipient trouble before it affects service.

- *Performance management.* Monitors service levels and measures response time, throughput, error rate, availability, and other measures of user satisfaction. Also collects network statistics such as packet quantities, runts, jabbers, and other trouble indications.
- *Security management.* Ensures that network and files are accessible only to authorized personnel. Assigns passwords and user numbers and detects unauthorized attempts to penetrate the network. Controls access through authorization and authentication.

Generically, an NMS consists of three elements: managed devices, agents, and management workstations. *Agents* are software modules that reside in the managed devices, which are sometimes called network elements. The managed devices connect to the NMS over the managed network.

The heart of the system is network management software that runs on management workstations. The workstation can be a PC running management software under Microsoft Windows, or a UNIX workstation running an application such as Hewlett-Packard's OpenView. The agent running in the managed device collects information, responds to commands, and provides information, either when polled or in response to an event. Managed devices can be programmed with *traps*, which asynchronously inform the NMS of the event that triggered the alert.

Management workstations are capable of drawing network maps and forming the interface between the managed devices and the operator. The workstation software stores and presents network information for the operator's use, correlates alarms, and aids the operator in isolating trouble. In most cases the workstation communicates with agents over the managed network, but when the network is down, it may be necessary to access certain devices over the dial network. The network management protocol includes a set of rules known as the *structure of management information* (SMI). The SMI defines relationships between management elements, organizes the network management data, and assigns identifiers to the variables.

The agent collects statistical information that can be used to monitor the configuration, health, and activity in the network. For example, a device may contain static information about itself such as its defined type, operational status such as up, down, or testing, and it may count statistics such as frames sent, received, and discarded. This information is stored in tables, counters, or as switch settings. This logical base of information is called a *management information base* (MIB). The MIB determines what the agent collects and stores; for example, packets sent and received, errored packets, etc. Each MIB consists of a set of *managed objects*. For example, system description, packet counts, and IP addresses are examples of managed objects. A managed device may contain multiple managed objects. Figure 38-1 shows a functional view of an NMS.

The above structure is common to most types of NMSs. SNMP is the preferred network management method with LANs and IP networks, but some WANs still retain a proprietary NMS, and many common carriers use CMIP.

FIGURE 38-1

Elements of a Network Management System

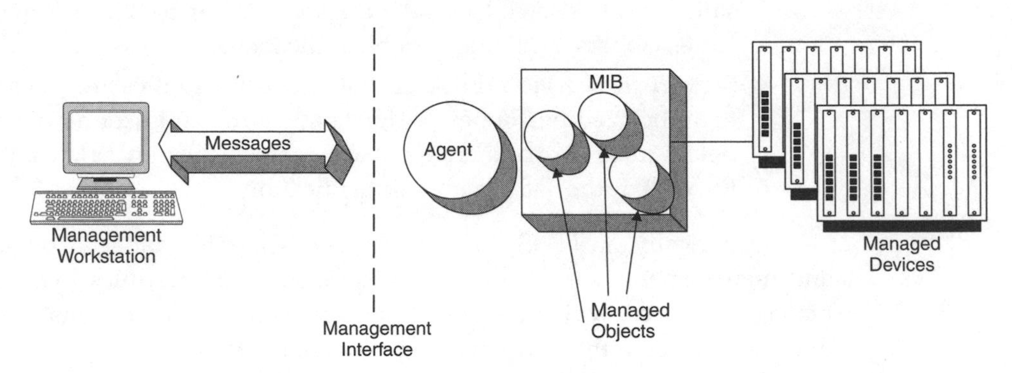

SIMPLE NETWORK MANAGEMENT PROTOCOL

SNMP is a protocol for the exchange of information over a TCP/IP network. The IETF developed the protocol to be a fast and easy way of collecting information about an IP network. It is not a complete network management specification because it lacks any kind of presentation interface. That is left to the developers that use SNMP as a foundation. SNMP runs under UDP as its transport mechanism to perform the following operations:

- Identify devices
- Aggregate traffic measurements
- Control devices
- Detect unreachable devices

SNMP uses two techniques for obtaining information from the managed devices: polling and interrupts. In the first method, a management workstation polls agents for information such as performance statistics and status. Management agents wait to be polled in all cases except for trap messages, which they send when triggered by thresholds and events. For example, a trap might be triggered when an alarm occurs or the number of error packets during a specified interval exceeds a threshold. Trap messages contain a minimum of information; the workstation polls to get the additional data needed to interpret the trap. Polling is one drawback of SNMP. As the size of the network grows the amount of capacity devoted to polling overhead increases. The overhead can be reduced by increasing the polling interval, but this limits the timeliness of information.

Fault correlation is a key issue in network management. When multiple alarms are received, the manager must determine which is the major fault, and which are sympathetic alarms that other devices initiate as a result of the major fault. Most NMSs provide a database that correlates alarms to help the operator determine the most probable cause of a failure.

MIB-I from the first implementation of SNMP, includes information such as the following:

- the number and type of interfaces the device has;
- the IP address of each interface;
- packet and error counts;
- system description of the managed device.

MIBs are defined in an English-like text using Abstract System Notation 1 (ASN.1) syntax. The MIB represents or describes the managed device, containing static information such as its vintage or dynamic information such as the number of packets handled. MIBs are defined to provide raw information, leaving computations to the management workstation. MIBs consist of basic information defined by the standard, plus proprietary extensions that most manufacturers add. These proprietary additions limit the management system's ability to collect and digest information in a multivendor network.
Version 1 of the protocol provides for six message types:

- *Get request*. Retrieves specific MIB objects.
- *Get response*. Sends the value of specific MIB objects in response to a get request.
- *Get next request*. Used to traverse the MIB tree or to get the next sequential MIB object.
- *Set*. Allows the management workstation to modify a MIB object to a new value or perform an action such as a restart.
- *Traps*. Sent from the managed device to the management workstation in response to some triggering event or threshold.
- *Inform*. Sent between management workstations.

Figure 38-2 shows how the SNMP protocol operates. The managed device responds to commands from the network management station with a get response. When it needs to report an event in response to a trigger, it initiates a trap message to the workstation. Trap types include the following:

- *Cold start*. The device has done a complete reboot, which may result in reconfiguration.
- *Warm start*. The device has done a warm reboot, which retains the existing configuration.
- *Link down*. A link connected to the device is inoperative.
- *Link up*. A previously failed link has been restored.

FIGURE 38-2

SNMP Command Protocol

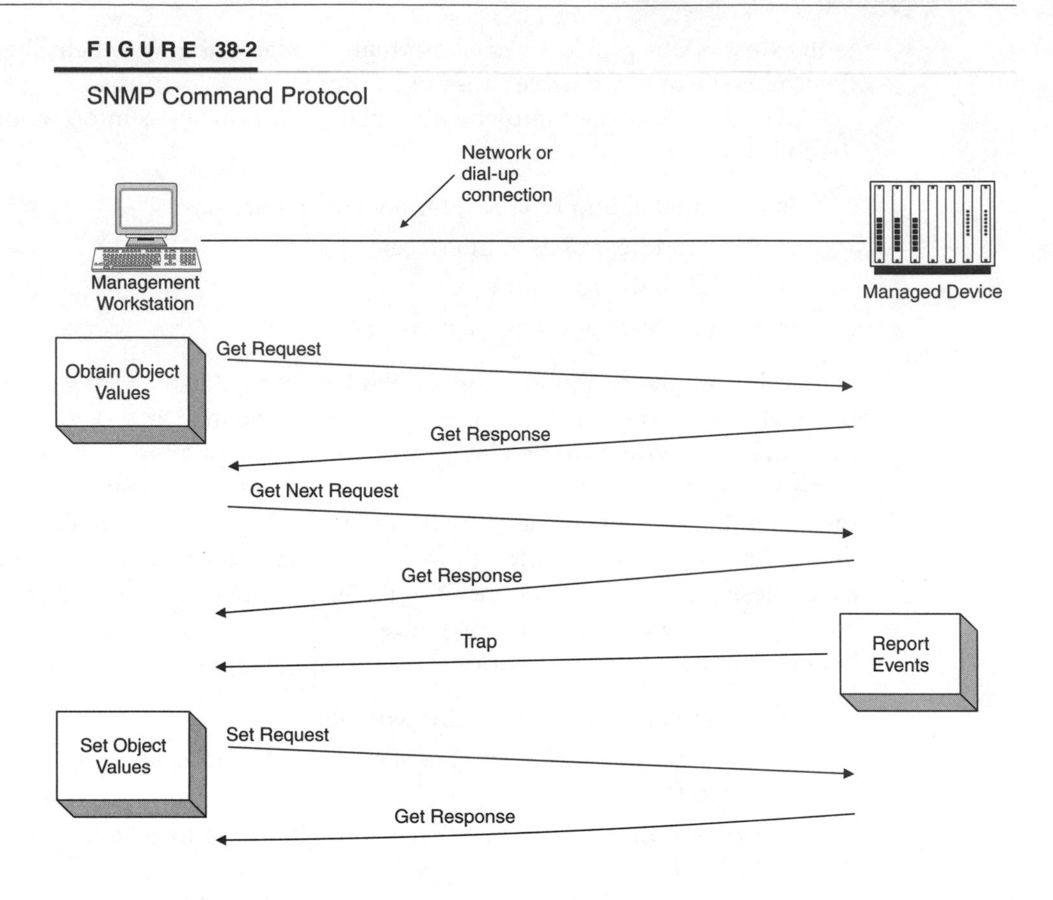

- *Authentication failure.* A workstation attempting to communicate with the device has failed to authenticate itself properly.
- *EGP neighbor loss.* A device linked with the exterior gateway protocol has been lost.

The protocol uses an identifier called a *community string* in combination with IP addresses to authenticate set and get requests. The community string is unencrypted, and is easy to duplicate, which leaves SNMP vulnerable to security violations. Any management workstation could intercept messages, duplicate them, and use them to send its own messages to a managed device. The security weakness in the system is the chief drawback of SNMP Version 1.

Each managed device contains agent software. The agent receives commands from the NMS, executes the request, and returns a response. It may also initiate a trap message when an alarm or event occurs. A *proxy agent* is one that does not reside in the same system as that which holds the MIB data. A proxy agent acts as a protocol translator to talk to the managed device and convert it to SNMP. Network managers have an objective of managing any device on the network from the same workstation, which may mean converting the incompatible devices to SNMP. This can be done with proxy agents, but proxy agents often do not implement all the information in the proprietary protocol. Furthermore, proxy agents may not implement all the SNMP functions.

MIB-II

MIB-II was devised to augment and provide information that was omitted from MIB-I. It is organized into the following groups:

- The *system* group is required for every device. It includes basic system information that enables nodes to identify and describe themselves. A management station can poll a device to extract its system information.
- The *interfaces* group describes the configuration and status of any type of interface to the managed device. Interfaces could be provided for a network such as frame relay, ISDN, ATM, or any other network type. The MIB provides information about the status of the interface and pertinent data such as traffic statistics and error counts.
- The *address translation* group maps the IP address to the physical address of the managed device. A table may be created manually or by using a protocol such as ARP.
- The *IP* group provides configuration and management information for hosts and routers. It contains such information as the default time-to-live counter by which misrouted IP datagrams are deleted. It includes IP routing, address translation, and forwarding tables. It also collects statistical information on IP traffic and errors.
- The *ICMP* group provides information back to the source about unreachable destinations, redirected messages, time-out information, and invalid information in IP headers. ICMP provides special service requests and responses to pings, which allows the workstation to determine the operational status of a device.
- The *TCP* group enables the workstation to detect TCP variables such as traffic and error statistics and TCP connection statistics. The group allows the network management station to view active TCP connections to see which applications are being accessed.
- The *UDP* group enables the management station to view UDP traffic statistics and to determine which UDP services are active.
- The *EGP* group provides information such as message counts and the identity of EGP neighbors.
- The *SNMP* group keeps count of SNMP variables such as the number of traps and get and set responses that occur.

SNMP Version 2

SNMP has numerous weaknesses in its initial version that led to proposals for SNMPv2. Chief among these is its lack of authentication and privacy provisions. The community string serves as a password in each PDU. A hacker monitoring the network with a protocol analyzer would be able to intercept SNMP messages, determine the community string, and communicate with agents to inflict harm. For example, a hacker might send a message to a mission-critical device such as a router telling it to disable its ports or shut itself down. As a result, many network managers disable the set command to prevent malicious interference.

SNMPv2 introduced several changes to improve security, but they were never widely accepted. Several subversions were published, but they eventually culminated in a new working group that was chartered in 1997 to develop consensus on Version 3. SNMPv3 is not a stand-alone management system. It adds essential security features that were missing from Version 1 and never widely implemented in Version 2.

The SNMPv2 inner PDU supports the five original commands from SNMP Version 1, plus two new operations: inform-request, and get-bulk-request. Inform-request enables management workstations to pass information between themselves. Inform-request is similar to a trap message in that the network management workstation sends the message without being polled. Unlike trap messages, however, the receiving station responds with a confirmation so the sender can be certain the message was received. Inform-request messages can be more complex than the trap messages ordinary devices send. Since the management workstation has intelligence, it can receive, digest, and report complex events that may be a summarization of many events sent from the managed devices.

A further deficiency with Version 1 is its inability to retrieve large blocks of data, such as a large router table, without multiple requests. If the agent is unable to respond to the entire get request, it responds with nothing. The get-bulk request enables the management workstation to retrieve a collection of variables; perhaps an entire table.

SNMPv2 adds several other enhancements. Here are some of the differences between Versions 1 and 2:

- The word "get" is dropped from responses. Instead of responding to a get request with "get response" the station simply returns "response."
- A manager-to-manager MIB is added to enable management workstations to exchange information.
- SNMP can be mapped to a variety of different transport protocols in addition to UDP.
- The data types in MIBs are expanded.
- MIB definitions allow the vendor to specify which of the MIB variables it supports.

A major problem with SNMPv2 is that to implement it all devices must be updated to the new version, which can be expensive if the upgrade is available for all devices, which often is not. Second, the management workstations themselves must be updated to SNMPv2. Proxy agents can be used to translate between versions if they are available. Another alternative is to use a bilingual network management workstation that supports both SNMP 1 and 2.

SNMP Version 3

SNMPv3 is intended to provide the security that SNMPv1 and SNMPv2 lack, but it is not a complete management system itself. Either a Version 1 or a Version 2 PDU is encapsulated in a Version 3 message. SNMPv3 introduces a security model that is designed to control access to managed devices and to provide message-level security through authentication, encryption, and timeliness checking. The access control system operates on SNMP PDUs, while the security system operates on SNMP messages. An introduction to Version 3 can be found in RFC 3410.

Version 3 uses a modular architecture to allow the standard to evolve easily. The major elements are an SNMP engine containing subsystems for message - processing, security, and access control. The engine accepts outgoing PDUs from the SNMP applications, encrypts them, inserts authentication codes, and sends them out on the network. The process is reversed on incoming messages. RFC 2271 discusses the security threats the model is designed to prevent:

- *Modification of information* prevents an unauthorized entity from altering in-transit SNMP messages including falsifying the value of an object.
- *Masquerade* prevents an unauthorized entity from assuming the identity of an authorized entity.
- *Message stream modification* prevents an unauthorized entity from reordering, delaying, or replaying a message.
- *Disclosure* prevents unauthorized eavesdropping on exchanges between SNMP engines.

Since security functions require authentication and encryption, the engine requires authentication and encryption keys for both local and remote users. As an additional security precaution, the protocol checks for the timeliness of messages. The device responding to a message inserts clock time so the receiving device can determine elapsed time between message transmission and reception. A delayed message could indicate that an unauthorized station has intercepted a message and delayed it while it was being modified.

COMMON MANAGEMENT INFORMATION PROTOCOL

CMIP is the other contender for the title of a universal standard network management protocol, but like the contention between OSI and TCP/IP for the network architecture standard, it is losing out to SNMP. The principal reason is that while nearly every manufacturer in North America supports SNMP, CMIP is more complex, at least as far as noncommon-carrier networks are concerned. It was designed to avoid the security holes that SNMP experienced in its early versions. It has several advantages over SNMP.

- It is a more secure system with built-in security that supports access control and authorization.
- Polling is reduced or eliminated with CMIP.
- Improved event filtering reduces the number of event messages (corresponding to SNMP's traps) that are sent to the management workstation.
- Unlike SNMP, which manages devices, CMIP manages relationships between devices.
- CMIP uses a connection-oriented transmission medium. With SNMP's datagram transmission method, data can be lost without the sender receiving notification.
- CMIP has more comprehensive automatic event notification. It has a larger number of alarm indications.
- More functions can be accomplished with a single request.
- It provides better reporting of unusual network conditions.

CMIP is closely related to the OSI model, and maps a MIB to each layer as Figure 38-3 shows.

The functions defined so far for CMIP outstrip SNMP's limited functions. They include:

- Access control
- Accounting meter
- Alarm reporting
- Event reporting
- Log control
- Object management
- Relationship management
- Security audit trail
- Security-alarm reporting
- State management
- Summarization
- Test management
- Workload monitoring

FIGURE 38-3

OSI Network Management

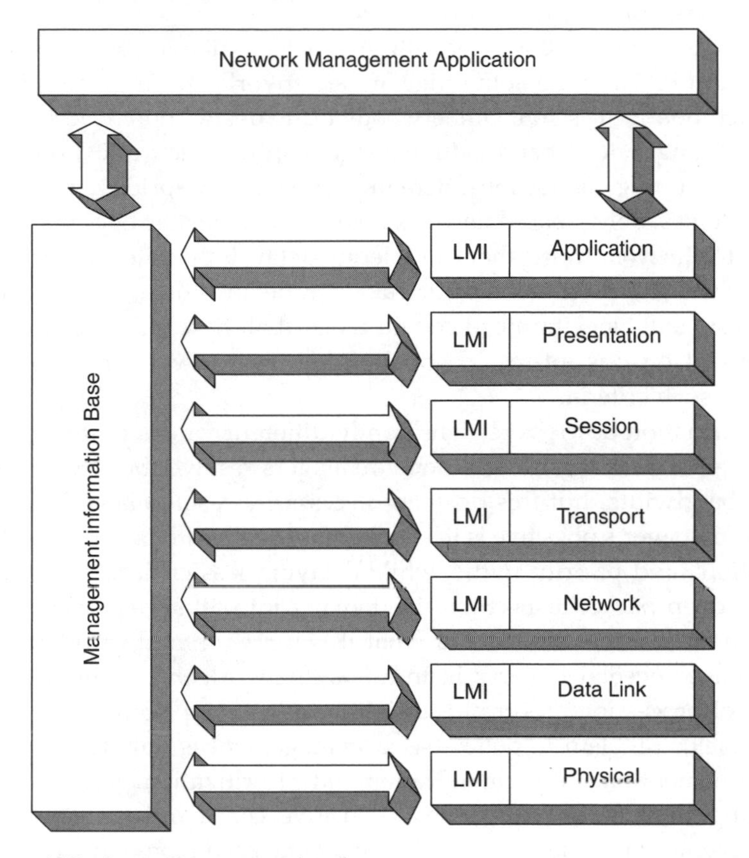

CMIP specifications called CMIP over TCP have been developed for running it over TCP/IP network, but these have not been applied to any great extent because most devices in an IP network support only SNMP.

BANDWIDTH MANAGEMENT

The increasing complexity of common carrier and enterprise networks makes it essential that the network run on autopilot to the maximum degree feasible. Trained personnel are expensive and cannot react to changing network conditions as quickly as a computer can. Furthermore, both private and carrier networks are losing their predictability for a variety of reasons. Users can move massive files at moment's notice, and they have no way of knowing how this affects other applications. The network is also increasingly required to support jitter-sensitive and delay-sensitive applications such as voice and video that previously rode on separate networks. These are more predictable than most data applications, but they must take priority or quality suffers.

To accommodate this increasing complexity, networks are applying traffic shaping and rate limiting at the edge of the network. Traffic is classified by a variety of methods such as ToS/DiffServ, source and destination IP address, IP subnet, UDP port range, etc. A bandwidth manager controls the traffic flow by dropping packets, dropping connections, denying access to low-priority traffic, and other such strategies. Policy-based network management has the objective of separating priority traffic from traffic that can tolerate delay. It operates under rules regarding priorities and permissions. The rules can be absolute as in the case of access control lists that block traffic of certain types at all times. They can also be conditional based on class-of-service, time-of-day, day-of-week, network utilization, and other such criteria.

Information flows have significantly different values to the enterprise, and should be handled accordingly. Some managers resolve the issue by over-provisioning bandwidth, but this does not necessarily resolve latency and jitter. The network manager's objective is to ensure that bandwidth is made available to key applications and priority traffic, while delaying less critical traffic. This can be done by deep packet inspection—i.e. looking into all seven layers to determine whom the packet belongs to and what it is carrying, and calculating the bandwidth that is needed to carry it. In a busy network this requires an enormous amount of processing power plus insight into relative priority.

A major problem in policy-based management is how to get organization-wide agreement on traffic classification and prioritization. This may be done by classifying the traffic as jitter- or delay-sensitive, confidential, mission-critical, and such designations. The classification may be based on who the sender and receiver are and how much bandwidth they need for their applications. The controls may be relaxed during low-traffic periods. For example, an organization might permit personal Web-browsing during lunch hours, but during peak periods only a select class of users might be permitted to access the Web. Access to some applications might be restricted based on organizational roles. For example, personnel files could be restricted to HR people and top executives. Traffic shaping supplements, but does not replace AAA servers, passwords, VPNs, access control lists, and other such methods of controlling access to sensitive information.

Policy-based network management is not a trivial task. It requires expensive equipment, constant attention, and organizational acceptance of the way the policies are administered.

CUSTOMER NETWORK MANAGEMENT (CNM)

The major IXCs offer CNM to their customers. CNM gives users the ability to access information on usage, accounting, fault, and performance on a near-real time basis. Through a CNM interface, customers can order new services and reconfigure existing ones in less time than it takes to order them through regular channels. Network managers can use the service to determine if their facilities are being used efficiently, and also to determine what services they subscribe to. Without CNM, phone bills are about the only source of information on what services a company has. Extracting information on private lines, frame relay circuits, access circuits, and the like from bills is often a daunting task, and getting the information from the carrier is difficult. CNM enables the user to extract the information from the carrier's database without going through conventional channels.

CNM is a MIB that enables the service to work through NMSs. Some NMSs are available with a CNM interface, but the lack of standards may make it difficult to obtain the desired result immediately. The initial implementations of the service for most users will likely be through vendor-provided application software.

NETWORK MANAGEMENT APPLICATION ISSUES

Single-segment networks and networks that reside within a single building can be managed effectively without an NMS, but in complex WANs, network management is essential. Ideally, managers would have a single system that manages the entire network from a single group of consoles, but that is rarely possible today except when all equipment is from a single manufacturer. Managers are faced with conflicting alternatives. Proprietary NMSs are available for proprietary networks, but they cannot be expanded to encompass LANs. Simple management systems are available for LANs, but they lack security, may overload the network with overhead traffic, and they are not easily extensible to proprietary WAN architectures. More secure and robust systems are becoming available, but these are not backward compatible with legacy equipment, and they increase the complexity and resources needed to manage the network.

Faced with these alternatives, most managers are forced to live with NMSs that are less than ideal. To minimize management difficulties today, the best approach is to obtain all equipment from a single vendor, or at least to require interoperability among the equipment you purchase. Few networks can realize that ideal in practice. As a result, network management will remain one of the major challenges that managers face.

Centralized vs. Decentralized Network Management

The benefits of centralization depend on the character of the network, the types of services being carried, the penalty for outages, the availability of trained people to administer the system at remote locations, and the nature of the organization that owns the network. If centralization can be justified, it is usually the most effective method of managing a network because diagnosing network trouble requires information that often can be analyzed more meaningfully from a central site. The people on the spot, however, are usually the best equipped to take corrective action when troubles occur. The issue of centralization is one that must be dealt with as part of the fundamental network design and reexamined as new services and equipment make it feasible to change the basic plan.

A parallel issue to that of centralization is whether testing will be done manually or automatically. As networks gain intelligence, automatic testing becomes more the rule than the exception. The network must be large enough, however, to justify the cost of automatic testing before it can be considered. Equipment to do automatic tests is advancing rapidly, and network management strategies should be reexamined periodically in the light of new developments.

Multiple Vendor Issues

Enterprises usually obtain their equipment from multiple vendors. Although this offers opportunities for cost saving, it thrusts a greater responsibility on the network manager to monitor service and develop techniques for dealing with multiple vendors. Compatibility problems become the manager's responsibility—a responsibility that can be exercised best by preparing precise procurement requirements and specifications when acquiring equipment and services.

An allied issue is the degree of internal network management expertise that an organization develops. Most major carriers offer managed networks but only within limits. Usually they support a narrow range of products, and the cost is high. A managed network may be preferable to developing internal resources, but this is a decision that should be made only after an analysis of the alternative of developing internal staff.

Organizational Considerations

Most companies have aligned their internal organizations with the realities of the new telecommunications networks so that the IT department administers both the data and voice network, but these are separate in some companies. In future, the integration of voice and data networks is inevitable for most large organizations. Though the two may be separated at the source, bulk circuit procurement is more economical and flexible than obtaining individual circuits. When failures occur, a single organizational unit will be the most effective in dealing with the problems of restoring services and rerouting high priority circuits over alternate facilities. In most organizations, the most effective structure will separate the information-generating entities from the information-transporting entities. As this is contrary to the way some companies are organized, this issue should be confronted as the character of the network becomes more integrated.

Another issue that must be dealt with is flexibility of network planning. Changing tariffs and rates of long haul carriers, access charges, and competitive local service make it essential that managers continually reexamine network plans. Plans should be flexible enough to enable the organization to react quickly as changes in tariffs and grades of service occur.

Network Management System Evaluation Criteria

As networks grow increasingly complex, NMSs likewise grow in complexity. Also, as networks are continually changing, NMSs must have the flexibility to accommodate the growth and rearrangement of the network without becoming so complicated that only experts can administer them. Network management can be simplified by using computers, and as with any mechanized system the user interface is essential to keeping it understandable. The primary issues in evaluating an NMS are:

- *Interoperability*. Ideally, the system should operate with equipment from several different manufacturers. In practice, manufacturers include functions in their MIB implementations that may confine interoperability with NMSs from other manufacturers to only basic functions.
- *Multiprotocol support*. The most effective systems are able to operate over the common transport protocols.
- *Security*. The system should provide adequate security to prevent unauthorized persons from deactivating equipment, intercepting confidential information, duplicating commands, or other such potentially harmful functions.
- *Ease of configuration*. The administrator should be able to configure the key elements in the network from a management workstation. The workstation should be capable of performing all the desirable functions without requiring a visit to the managed device to set switches, check light status, and other items of status and configuration.
- *Scalability*. As the network grows the management system should grow without overloading the network with additional management overhead traffic.
- *MIB definitions*. The management workstation should provide MIB definitions on demand.
- *Out-of-band management*. In addition to in-band management, the system should be accessible over the dial network to aid in diagnosing and restoring troubles during circuit failures.
- *Alarm correlation*. The system should aid the manager in determining which of multiple alarms is the primary fault, and which are sympathetic alarms. For example, a hardware alarm should be diagnosed as the cause of a group of circuit failures. The system should aid the operators in determining appropriate action for alleviating symptoms.
- *Manager-to-manager capability*. The system should provide for management workstations to communicate with one another. Subordinate workstations

should be able to summarize and condense information, and report it to workstations higher in the hierarchy.

- *Platform performance.* The platform on which the management workstation runs must be powerful enough to handle the volume of traffic and overhead.
- *Mapping capability.* The management workstation should be able to interrogate all devices on the network and compile a network map showing information such as IP addresses, system status, and alarms in color. The network layout records should be readily accessible and easily understood.
- *Output data.* The system should provide output data summarized and sorted for analysis or it should provide a port to an external system to accomplish the same objective. A complete trouble and event log should be provided. The system should be able to correlate events relating to a single case of trouble. The system should be able to export data to spreadsheets or other applications.
- *Web implementation.* The system should be capable of extending management functions over the Internet to devices located on intranets and extranets.

CHAPTER 39

Telecommunications Convergence

If the telecommunication experts are unanimous on anything, it is the conviction that all forms of communication are gravitating toward a single unified IP network. There is less unanimity about when this transformation will be complete, but few expect it to happen overnight. Both customers and carriers have too much invested in the PSTN to abandon it, and while the industry has solutions for most of the technical obstacles, the technologies are immature and changing steadily. Furthermore, the economics of convergence are insufficient to justify the capital expenditures of anything but an evolutionary transition.

Historically, enhanced features have started in private networks and gravitated to the public. While the world lingered over ISDN standards, for example, PBX manufacturers brought its features to the office. The telephones and protocols are proprietary, but the features are proven and available in every product line. The same is happening with the transition to IP. International Data Corporation (IDC) projects that about 1.4 million IP PBXs will be in service by 2008, but circuit-switched PBX lines will still outnumber IP lines by a factor of 3 to 1. Convergence faces fewer obstacles in the private arena because enterprise systems have shorter service lives than public and the customer has better control of the infrastructure.

Convergence in the public network, where switches typically have service lives of 20 years or more, is a different matter. The PSTN core is moving toward IP, but class 5 switches remain circuit-switched. ILECs are beginning to offer VoIP service over softswitches, but the motivation is more to meet competition than because of any technical advantage of IP. For what it does, the PSTN is good enough. It meets the needs of most residence and small business subscribers, and for voice alone it is superior to IP. Several factors are standing in the way of moving public voice and video to IP. The most important of these is the lack of a suitable infrastructure. The IP equivalent of the PSTN is the Internet, but it is not suitable for commercial-grade telephone service. The Internet is designed to be cheap and ubiquitous, but it is a chaotic model, not because of the behavior of the network, but because of the nature of the service. You can have low cost or QoS, but you cannot have both.

Some observers suggest that the QoS issue is overblown, pointing to cell phones as evidence that people are willing to forego service quality in exchange for convenience. This observation is accurate as far as it goes, but it ignores two issues. One is that business-class telephone service demands consistent quality that the Internet is incapable of providing. VoIP over the Internet can supplement the PSTN, but it is not a suitable business model to replace it. The second issue is the fact that every telephone session has two parties. Some users may decide to forego quality in favor of low-cost Internet connections, which may be suitable for family and friends, but enterprises cannot afford to impose poor quality on their customers. The telecommunications industry has elevated transmission quality to its present state through a long series of technological advances. To diminish it for the sake of expediency would be a serious mistake.

That said, the industry is evolving toward the converged network. It will be a gradual transition; one that will occur only after some significant barriers have been overcome. The obstacles will be resolved in time with more new protocols and a lot of industry work, particularly on the infrastructure. The Internet could be adapted for commercial grade telephone service, but several things would have to change. ISPs, which are competitive and independent, would have to agree to adhere to service and quality standards. Open interconnection with their competitors would need to occur and today's pricing model would have to change. The Internet is designed and constructed to offer best-effort service at low cost and that is inconsistent with the needs of time-sensitive applications.

Service quality notwithstanding, a growing amount of voice traffic is moving to the Internet as service providers such as Skype, Vonage, and AT&T's CallVantage offer subscribers service at prices so low that millions are signing up. So what if the service is not toll quality. If people can call halfway around the world for nothing over a broadband connection that they have already paid for, some sacrifice in reliability is a small price to pay. Cable companies with their broadband access into most residences and many small businesses are in a particularly advantageous position to offer VoIP service because they have control of the access channel.

VoIP in the public network is a disruptive technology. Although its impact has been slight so far, it forces the LECs to review their business models carefully. When a company can come from nowhere and furnish telephone service over a broadband connection with minimal investment, it causes concern for telephone companies and regulators alike. Regulators' traditional modes of taxation and control are impossible with Internet telephony, at least at this point. In 2004 the State of Minnesota attempted to impose telephone regulation on Vonage despite the fact that the PUC's control does not extend to interstate calls and it is impractical to segregate VoIP calls. This prompted the FCC to issue an order preempting the Minnesota order, but the issue has not been put to rest.

In this chapter we look first at the forces that are driving toward a converged network. We look at services that countries such as Japan, which has a much higher degree of broadband development than North America, are beginning to provide and how these motivate the development of the converged network. We examine the advantages of convergence and the barriers that are impeding its development. We look briefly at the Infranet, which is an initiative of several companies to develop an IP-based alternative to the Internet. The Infranet, if it is successful, may prove to be the enabling factor that removes most of the barriers. We conclude the chapter with an Applications section that discusses the ways in which the converged network is available today and considerations in applying it.

WHY CONVERGENCE?

Above the physical layer, the Internet is diametrically opposite the PSTN in most ways, so why is the push on for convergence? The answer lies in a complex brew of economics, politics, culture, and technology. The term convergence holds different meanings for many people, so let us begin this section with a definition.

Convergence Merging real-time applications such as voice, video, and instant messaging together with data onto a single broadband infrastructure that is based on IP.

The reasons for the trend toward convergence are many, not the least of which is an urgent need of manufacturers to find innovative new products to help them emerge from the technology slump that began in 2002. This is coupled with the zeal of the international Internet community that is persistently pursuing a vision of a unified network. Companies that see expanded opportunities for innovation in such online applications as gaming, music, television, education, and countless services that have scarcely been glimpsed support this vision.

In 2004 NTT DoCoMo introduced services in Tokyo that give us a preview of the kinds of applications that converged services make possible. Dubbed "Felicia," the service marries a 3G handset, a smartcard, and GPS to enable users to perform a variety of functions that are otherwise impractical or require separate devices. The smartcard, coupled to the target by wireless, is loaded with identification and financial information that enables users to make purchases from vending machines, payments at restaurants, and to board trains and buses without tickets. A user standing at a bus stop wondering when the next bus will arrive can link to the transit company to obtain the answer. The mobile handset provides its location from its GPS and the application compares it to the bus' location, also obtained from GPS. Users can purchase tickets to events, locate restaurants and shops in their locale, and link to RFID devices in the store to download more information about the products. Devices such as these will change the way we work and live, and the functions are impractical without a converged network.

The Business Case for Convergence

Convergence has a lot of appeal for the enterprise. If its promise is realized, operational costs will drop and better utilization will be made of an expensive resource. An all-IP network will allow service providers to create new value-added services. Much of the time-consuming provisioning process will become unnecessary as the network evolves to respond dynamically to changes in demand. Business models will change and new ones will emerge as inexpensive and ubiquitous bandwidth becomes available. In some countries with extensive broadband access, for example, the market for downloaded music exceeds that for CDs. Video rentals will fade in favor of video-on-demand, and presence engines that advertise personal preferences for reachability will drive personal communications. Barriers to entry of local phone markets will disappear as consumers move their telephone service to IP networks, and the phone will be a multimedia device more or less permanently associated with an individual, possibly with the address issued at birth. Let us look in more detail at these benefits.

Competitive Advantage

For some businesses, convergence is necessary to support business goals including reduced cost of doing business, increased productivity, and improved customer service. A converged network supports collaboration with customers, which results in improved customer retention. Businesses can tie customers to their internal processes through e-commerce. Multimedia contact centers can enable organizations to create an on-line experience that approximates face-to-face service delivery, enabling the customer to choose the preferred method of communication and shift to another mode in the middle of a session.

At first the converged network will catch on slowly with the early adopters gaining the most advantage, but gradually it will change the business model to the point that customers will expect the service. Today in Japan's Felicia service, for example, a customer can surf the Web from her 3G phone, inspect a photo of a product, order it, and pay for it, all from the same hand-held instrument.

Lower Cost of Ownership

The cost equation is difficult to factor into the decision to migrate to a converged network. To realize cost reductions, cultural changes and dedicated management effort are required. The following are some of the ways ownership costs can be reduced:

- With only one unified network infrastructure to design, manage, and support, labor costs should be reduced. The same staff that support servers, switches, routers, and LAN equipment can maintain the voice switching system, and reduce or eliminate the need for outsourced expertise.

- Users can do many of their own moves, adds, and changes without involving the IT staff.
- A converged network enables enterprises to make more effective use of access bandwidth, while simplifying administration and maintenance.
- Host-based services can eliminate the need to invest in a dedicated PBX or upgrade an obsolescent system, while still providing the productivity enhancements of feature telephones.
- Call processing can be centralized for multiple independent sites, eliminating the need for separate key systems, resulting in lower branch office costs.
- The quantity of station wiring drops can be reduced by combining both PCs and VoIP on the same Ethernet port.
- IP PBXs are more scalable than TDM PBXs, which have a maximum size and grow to that size by adding port cards and cabinets.

Support a Mobile and Geographically Independent Workforce

This advantage is important for telecommuters, sales people, and others who spend a great deal of time away from the office and to people who work from home at least part of the time. It is particularly difficult to provide conventional voice communications that are integrated with the office telephone system to such people. With VoIP the benefits of the PBX can be extended outside the bounds of the office. The need for office space is reduced and the organization is able to comply with government regulations that require employers to reduce traffic congestion. Part-time people can work from home to supplement a fixed call center workforce.

Flexibility

The TDM PBX is a proprietary and inflexible device, closed in every respect except for its CTI interface, which provides for limited call control. Although IP PBXs (with the exception of open-source systems) do not open their call-control programs, they provide more flexible interfaces such as SIP to permit development of server-based features. They also open options for branch and home offices that are expensive or difficult with traditional architectures.

Create Value-Added Services

Convergence opens an enormous variety of opportunities to provide new services that are infeasible with circuit switching. The user interface for telephone service has improved little over the years. This is not because of a lack of imagination on how to improve it or a lack of APIs for hooking new applications to proprietary systems. It is more because the applications must be customized for each type of CPE system, and no manufacturer has enough market share to generate a mass market. For example, unified messaging has been available for years, and there is little variance in the features that various products support, but it has not achieved enough market penetration to bring the cost down to the point of becoming popular. Open protocols such as SIP can separate the services from the call control, which by its nature must be closed. Carriers and third-party developers can create countless new services and make them operable across a variety of platforms.

Enriched User Experience

Ultimately, new services will change the way people work and communicate. Personal communication assistants can enable critical employees to be contacted while still screening unwanted calls without the need for human assistance. Functions that are difficult with a standard telephone interface, such as setting up conference calls and dialing by directory name, will become easier with an improved user interface. Productivity should improve through remote collaboration and shared access to documents or whiteboard. Just as the PC is a standard office tool today, these new applications will become such a way of office life that users will expect them to be available.

Rapid Deployment of New Applications

New and innovative applications can be deployed more rapidly with IP than with traditional fixed telephone systems, and the pace of improvement does not depend on the actions of a single vendor. Furthermore, in a geographically dispersed organization, new applications can be downloaded onto desktop clients without involving a generic program upgrade.

Barriers to Convergence

New telecommunications developments have always been over-hyped, and with notable exceptions such as fiber optics, many of these have flared briefly and fizzled out. After an initial period of exuberance, convergence is proceeding slowly for a variety of reasons, not the least of which is the difficulty in demonstrating a suitable return on the investment. Most of the advantages listed above require the organization to change and adapt to a new environment, and this often happens slowly.

The initial impetus was long distance cost saving, but that argument has largely disappeared now that long distance costs are so low. Savings from managing the converged network are difficult to prove unless the workforce shrinks with the new technology. When ROI is calculated, many of the savings are in soft dollars and demonstrating real cash saving is more difficult. Convergence will develop in time, but many issues listed in this section remain to be resolved.

Network Infrastructure

Carriers are converting portions of their network to VoIP, but islands of VoIP cannot support the performance and security that commercial-grade voice communication demands. For VoIP to be a viable alternative to the PSTN, it must support carrier interoperability. For services such as worldwide VPNs and telephone connections to be effective, they must transcend carrier boundaries because no carrier can fulfill all of the needs without relying on other providers. Carriers must be able to provide services to any corner of any rural area of any country, and no carrier has sufficient reach without relying on other service providers. Carriers therefore must be able to interconnect with appropriate levels of security and service definition and the sessions must be metered to compensate carriers for handling transit traffic.

Today, many outsiders expect the Internet to become the backbone for this multiservice, multinational, multiowner network, but those expectations are unrealistic. A network converged over IP does not mean the public Internet as it is now structured. Obviously, it can carry voice. It does it every day, but it cannot carry voice with the consistent quality that the world has enjoyed since the conversion to an all-digital network. The alternatives are developing an overlay network that has the stability isochronous applications need, or hardening the Internet. The latter means changing the basic design concepts that keep the Internet cheap and fast-paced.

Peering points on the Internet today do not meet any specified performance criteria and there is no incentive to support a guaranteed level of service. The converged network must provide an appropriate level of assured delivery in response to requests from the application. This is inconsistent with the intent of the Internet, which is to deliver inexpensive connections that are not sensitive to usage or distance. Either the Internet must be split to provide a separate network with reliability and security, or the cost of service must increase.

Flow Control

Another key issue is congestion control, which is a vital feature of any voice or data network. The difficulty is that voice and data behave differently when it comes to congestion. Both can throttle traffic back at the source, but the nature of the traffic flow is much different. Many data applications have peaks of high bandwidth demand for short intervals, but then demand drops to zero as the user operates on a downloaded file. If the network is congested, it is apt to be for only a brief interval, after which traffic begins to flow normally. During the heavy flow periods, TCP closes its window or routers discard traffic, but the process is transparent to the users, who see a slow response, but the session continues without interruption.

Unlike data with its heavy peaks, voice is a relatively even flow of half-duplex traffic that is predictable. Traffic engineers have mounds of data that enable them to predict voice loads by hour, day, and season until something unusual happens. Storms, disasters, significant news events, and other external events usually inspire an extraordinary number of customers to place telephone calls. These cause traffic to fall outside the normal range and the network has to protect itself while prioritizing service to essential customers. Voice networks shed load by a variety of techniques, the first of which is to delay dial tone. During heavy load periods the LEC can operate line-load control, but this is done only in extreme circumstances. Common-control equipment such as DTMF registers are engineered for normal peaks. In abnormal peaks, the registers may be tied up, so the caller does not receive dial tone. The caller can remain off the hook and dial tone will eventually be provided. If the congestion is in the trunking network, calls will not go outside the serving class 5 switch. The user hears reorder and must redial.

Flow control is a standard feature of TCP, but real-time packets work under UDP, which does not provide flow control. An IETF working group is working on Datagram Congestion Control Protocol, which is intended as an alternate transport protocol. DCCP offers functions that bridge the gap between TCP and UDP. These include packet acknowledgement, congestion notification and control, packet sequencing, and protection against denial-of-service attacks. This protocol may resolve flow control issues.

Security

This is the issue that is the most difficult to resolve on the Internet, while still keeping the service manageable. While users must get involved with encryption, tunneling, firewalls, and similar security provisions, VoIP will be confined to a narrow spectrum of users that are willing to put up with the complexity in exchange for the benefits. A major strength of the PSTN is the fact that uninvited guests cannot ride the coattails of a file or message, latch onto the telephone, and infect it with a virus. While trust is not an issue with the PSTN, eternal vigilance is required on the Internet to thwart a coterie of miscreants who, for whatever malevolent motivation are attempting to inflict damage.

Broadband Penetration

For VoIP to be successful, broadband must become nearly as ubiquitous as the PSTN. In 2004, the latest figures available as this book goes to press, broadband penetration in the U.S. is reported to be 42.5 percent of households with Internet access. About three-fourths of households have Internet access, which means broadband penetration is roughly one-third of the households. Other countries, notably Sweden, Japan, and South Korea, have penetration in the order of three-fourths of all households, significantly greater bandwidth, and at a cost that is more affordable than in North America. This issue must be resolved before the converged network can claim success.

Service Complexity

For VoIP to become universally accepted, it must be simplified. Today, customers need to know too much about VoIP to make it work. The strength of the PSTN is its simplicity. Customers do not need to know anything about addresses, E-911 access, NAT, and other such technical issues to make the telephone work. For VoIP to be ready for real time, users must be able to plug the service into a wall jack and have it work without the need to configure firewalls, install VoIP terminal adapters, or worry about IP addresses. Users must be able to connect to other VoIP users without using the PSTN as an intermediary and without the need to know the identity of the callee's service provider.

Lack of Carrier Agreements

IP networks are today a loose confederation of agreements among carriers. A carrier-class VoIP network cannot depend on a service provider's decision whether to carry transit traffic. This, in turn, means that international division of revenue processes must be developed. Competing carriers must deal promptly with impairments and failures and must cooperate to ensure that service is not affected by the structure of the interface points of diverse networks.

Carrier and Regulator Inertia

The existing PSTN infrastructure works well, represents a considerable investment, and is maintained and managed by workers that lack the skills to manage an all IP infrastructure. ILECs, in particular, will retain what they have until it is functionally obsolete or competitive pressures force them to change. Numerous other forces resist change. Congress is subject to numerous pressures that affect the shape telecommunications will take, and money is at the root of it. Cash-rich companies such as the ILECs have the ability to inject huge amounts of money to purchase influence and they have been successful in this endeavor.

The regulatory framework is based on conventional telephone technology as are the structure of taxes and fees and these tend to change slowly. Much of the economies of VoIP in the U.S. today result from Congress' reluctance to burden IP with the fees, restrictions, and taxes that it loads on the conventional telephone industry. Access charges, USF fees, excise taxes, and myriad state and local taxes, not to mention carriers' miscellaneous fees complicate the picture. Moreover, regulations are inconsistent. LEC cable pairs are subject to unbundled access regulations, but new fiber investments are not. Cable companies can prohibit other carriers from using their access facilities and have both the ability and the motivation to exclude other providers from using them for VoIP. All of this creates an atmosphere in which progress is impeded because there is little assurance that the rules will not change.

Reliability and Availability

IP networks are inherently robust with their ability to route around failures, but router convergence time is too lengthy for real-time applications. Operational changes such as hardware and software upgrades and configuration changes cannot be done in real time with many router platforms. The solution requires a new generation of routers and protocols, which will be slow to propagate themselves through the network.

Compliance Issues

Making VoIP comply with E-911 and CALEA is a major unresolved problem. Many countries censor Web content, block access to certain sites, or monitor access to parts of the Internet. These requirements are incompatible with using the Internet as a reliable communications channel.

Interworking between the PSTN and IP

Until the transition to an all-IP infrastructure is complete, communication between the two networks is required. This requires gateways, signaling, addressing, and numerous other complexities that must be transparent to the users. Much work remains to be done to make the interface with the PSTN seamless.

THE IPsphere INITIATIVE

Many of the changes necessary to make VoIP equivalent to the PSTN require voluntary or compulsory adherence to centralized authority, a concept that is anathema to the Internet community. If we conclude that the Internet in its present state is not an appropriate vehicle for isochronous traffic, then the solution may be an overlay network that is designed to provide the elements missing from the Internet: QoS, predictability, end-to-end management, security, and carrier interconnections with division of revenues or settlement process. Such a network would not be accessible from the public Internet and would provide the benefits of IP without subjecting its subscribers to the chaotic conditions that prevail on the Internet today.

An initiative known as the *IPsphere*, previously known as the Infranet is underway to create such a network. Equipment manufacturers and carriers are not unanimous in their support of the IPsphere and there is no assurance that such a network will happen. The goal of IPsphere is to deliver performance over a virtual network that is predictable, flexible, and secure so subscribers can entrust mission-critical information to it. At the edge of the IPsphere is a barrier that requires customers to authenticate themselves before admission.

The IPsphere requires communication between the subscriber's application and the network to enable the application to request the level of security, quality, and bandwidth it needs. Costs would be based on what the application needs, in contrast to the Internet where the cost is independent of the application. Since no single provider can guarantee worldwide connectivity, connections are needed between networks so providers can communicate levels of service and security when handing off traffic. In addition, accounting mechanisms are needed to enable carriers to bill each other for carrying traffic.

CONVERGENCE APPLICATION ISSUES

The motivation for the converged network in the enterprise network can be summed up in one word: productivity, both of personnel and capital. In the public network the motivation is also clear. For consumers it is saving money and getting enhanced services. For service providers it is the opportunity to make money. It will be many years before the technology has advanced to the point of replacing the PSTN for more than a narrow spectrum of users, but the obstacles will gradually be surmounted.

VoIP in the Public Network

At this stage of development, the ways of implementing VoIP are many and varied. From a residential or small business user's standpoint the ideal would be to connect a VoIP telephone into a wired or wireless LAN, assign it a telephone number, and use it as if it were a wired phone. The reality, however, is considerably different. Several companies have jumped onto the VoIP bandwagon and by the time this is published, many more would have joined the fray. IDC estimates that by 2008 some 14 million customers worldwide will subscribe to VoIP services. In this section we briefly discuss the alternative configurations and some of the considerations in selecting them. The services they offer are changing, however, so it is best to refer to the vendor's description on its Web page before relying on this discussion.

Computer-to-Computer Services

Representative services include Pulver, Skype, and Dialpad. These services generally cannot be reached from the PSTN because the user does not have an E.164 number. In the process of registering with the service provider, the user obtains an address that is valid within that network. Calls within the network can be made to others who have registered and are online.

The telephone instrument is usually a softphone and client, which can be downloaded without cost. These services usually do not carry a monthly fee, although some offer off-net prepaid packages. PC-to-PC calling is free. Calls that hop off to a wireline phone carry a charge that is usually much lower than small users can obtain, but may not be much of a saving for large users. The greatest savings are on international calls. Features include buddy lists, redial, and running account balance.

Firewalls and NAT generally do not bother these types of service because nothing identifies the call as a voice call. To place or receive calls the user logs onto the provider's Web service, so the session looks to the network like any other Web connection.

Computer to PSTN Services

Representative services include AT&T CallVantage, Vonage, Net2Phone, and Go2Call. A major difference between these and computer-to-computer services is the provision of an E.164 number, which enables the IP phone to be called as if it were a wireline phone. Since the service provider controls the design, the portion of the call that uses the Internet is under its control, so the service provider can control the quality. The architecture is invisible to the customer, so there is no way to evaluate the service in advance except to try it.

Some products permit or require the use of a VoIP terminal adapter, which connects an analog phone to the network. Some will work with downloaded softphone products. In most cases existing numbers can be ported to the IP service, and the IP phone is not tied to a physical location. This means it can be transported to another place and operate as it does from the primary location. Vonage offers a virtual number in certain local calling areas. This permits someone with a landline telephone to make a local call, which is transported to the destination across the Internet.

Most of these services carry a monthly rate, which may include unlimited domestic calling. The rates are generally lower than LEC phones, and may include a package of special features that the LEC charges for. Voice mail is typically available, with message retrieval from either telephone or browser. Voice mail messages can also be forwarded as e-mail attachments. Other typical features include caller ID, call forwarding, call logging, do-not-disturb, conferencing, and locate service.

The service has much in common with cell phone service. Most users will not give up their PSTN phone, but it is a good solution for a second line. The service has several downsides that must be considered, not the least of which is lack of compatibility with E-911. In addition, it does not work through power failures. If the line is to be used as an additional line throughout the house or business, the adapter must be wired in place, which defeats the easy portability. Finally, the service is not as simple to set up as buying a telephone and plugging it in. The author's experience with AT&T CallVantage is a case in point. After several hours of attempting to make it work through Comcast cable, the AT&T technician gave up concluding that Comcast's routers were blocking the service.

TDM over IP (TDMoIP) Multiplexers

Several manufacturers provide TDMoIP multiplexers for applications such as the one shown in Figure 39-1. Here, the company has an Ethernet connection between sites and T1/E1 compatible PBX and key systems. The IP multiplexers connect to the CPE devices with TDM and to the Ethernet switches with 100Base-T, sharing the bandwidth between sites with data. The multiplexers shown use T1 on both ends, but they could just as easily use analog trunks, in which case the connection would typically be FXS/FXO. The TDM frames are encapsulated into IP packets that are transported over the fast Ethernet ports. The ToS bits of the packets are set to classify the packets as high priority.

TDMoIP provides a circuit-emulation service, also called pseudowire, that is transparent to protocols and signaling. The multiplexer repeats the contents of each channel to the other end. The IETF PWE3 Working Group is working on protocol standards that are in draft form as this book is published, so products are likely to be proprietary. Typically, the payload of each channel connects to a 48-octet ATM cell, which does not have the 5-octet header. These are encapsulated into IP frames in some multiple. As the number of TDM octets per frame increases

FIGURE 39-1

TDM Over IP Application

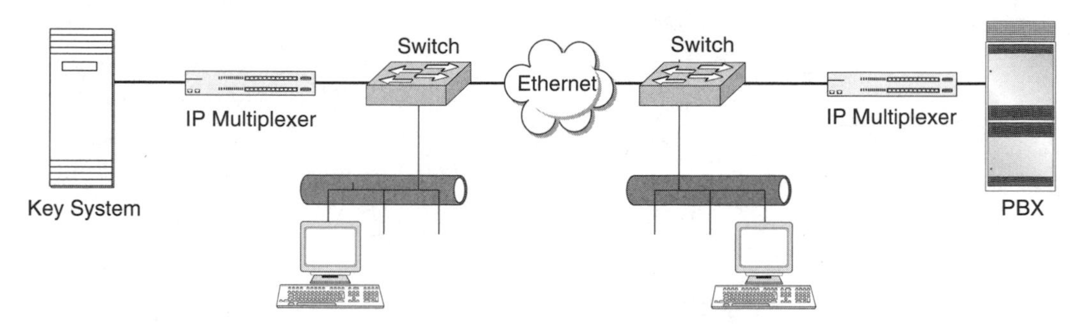

throughput increases because of lower packet overhead, but the effects of frame loss are more severe.

Compared to VoIP gateways, TDMoIP multiplexers have lower latency, so the circuit quality is likely to be better, provided packet loss is not excessive. As with VoIP gateways, the multiplexers compensate for packet loss by repeating the contents of the previous packet. The multiplexer transmits channel timeslots whether they are empty or not, so from this standpoint it is less bandwidth-efficient than VoIP. The impact of this is generally irrelevant, however, because a TDMoIP multiplexer is not used where bandwidth is restricted. For example, it would not be usable over the Internet. Some products support fractional T1/E1 so vacant channels are not transmitted.

VoIP Business Applications

VoIP has four principal categories of applications in the enterprise network:

- *Branch office*. Since the branch office is already equipped with a router or FRAD, adding voice may be a natural outgrowth of the existing network. For distant branch offices, the existing frame relay network may be used. For local branches, it may be feasible to replace off-premise extensions with VoIP.
- *Telecommuting*. Similar in architecture to the branch office, telecommuters may use VoIP to provide a voice and data connection to the main office.
- *Toll bypass*. Carrying long distance traffic for next to nothing is attractive, particularly when the distances are great. The ability to send large quantities of fax messages, which tend not to be time sensitive, over the Internet, can be particularly attractive.
- *Web-enabled call center*. Callers can browse the company's Web page by computer, and then click an icon to talk to a live agent. Costs are reduced and customer convenience is enhanced.

CHAPTER 40

Future Developments in Telecommunications

As we complete our journey through the varied and sometimes bewildering world of telecommunications, it is an appropriate time to peer further into the future to see what it might hold. One prediction we can make without fear of contradiction is that things will continue to change. The changes will not always be to our liking, and they will not necessarily pull in the same direction. In fact, the direction telecommunications takes is like a pack of undirected draft horses hitched to the same object, which moves as the vector sum of diverse forces. Pressures such as service providers, technology, regulators, governments, customers, embedded equipment, developers, and dozens of other factors are impinging on this technological colossus and it is moving; sometimes slowly, but it is always moving.

The next decade in telecommunications will be most interesting and replete with opportunities. We are approaching a decade of turmoil in which the traditional structures will have to change, but who has the foresight to lead their organizations in the right direction? The decade past has proved that financial strength and technical acumen are no guarantee of success in this industry. The giants and startups alike have assessed the industry, made their moves, and some have flopped spectacularly. Others have come from nowhere and risen to profitability because they guessed accurately what the market wanted. Unanimity of the experts is no guarantee of success. Not long ago the industry was solidly behind ATM as a universal end-to-end protocol. ATM is far from dead, but it will never fulfill the expectations the pundits then had. With the caveat that the predictions of this chapter may turn out to be wrong, we can still risk a few guesses on where technology is going.

The telecommunications industry has several unmistakable trends that will shape products and services in the future. IP is becoming a global language, wire is giving way to wireless, and broadband access unleashes a plethora of new services. The path ahead for these trends is far from clear, but judging by past performance, technologists will resolve the remaining issues. Make no mistake about it, though, the political issues are fully as important as the technical and more difficult to assess. A crucial issue to watch over the next few years is what happens to the Telecommunications Act of 1996. The first act served the nation for more than 60 years and now there is talk of another revision only a decade later. The issue lies with Congress and the FCC, neither of which is immune to external influence. Look for major battles over this legislation.

We are seeing evidence of the remonopolization of the telecommunications industry and this is sure to draw attention. This time the monopoly is not nationwide, but on a regional basis it bears more than a casual resemblance to the old Bell System.

Other trends that are already well underway include movement toward Ethernet and open source. Ethernet, is well entrenched in the LAN and in its 1G and 10G versions it will increasingly serve the metropolitan network. Open source protocols, operating systems, and software are a major disruptive force to the traditional developers. The trend is unmistakable, and will continue to provide users with attractive alternatives, even extending to PBXs.

DISRUPTIVE INFLUENCES

We are entering an era of major disruption for traditional providers and confusion for customers. The historical architectures are changing. In accordance with established custom vendors are proclaiming the obsolescence of the old and persuading customers of the superiority of new technologies that have deficiencies yet to be resolved. Any faults with the new, we are told, will be worked out. In any event, services will gradually migrate from the PSTN to IP networks and this causes major disruption. The IP migration penetrates to the core of the business models of the past and wrenches them apart. Barriers to entry into telecommunications markets are disappearing. Companies with software, a little hardware, and minimal capital investment can carve niches from the old-line markets. The impact is insignificant so far, but the new entrants often cut into the most profitable elements, leaving the remainder for companies that have a legal obligation to serve.

Long Distance Commoditization

One major influence is the fact that the IXCs' business models are evaporating. Consumers no longer think much about long distance prices or care which carrier they choose. Long distance comes bundled with most cell phone packages and local service promotions usually include a generous amount of domestic long distance. As a result, the distinction between local service and long distance is disappearing. Businesses are still major long distance consumers, but domestic toll prices have hit rock bottom and international toll is not far behind. The service that was once so costly has become a commodity and carriers cannot easily differentiate themselves on anything but price.

Profitability has been steadily declining along with prices, leaving the IXCs with little choice but to redefine their services or allow themselves to be acquired. For years the three dominant carriers were AT&T, MCI, and Sprint. Now all have been acquired or are acquisition targets. Sprint has a local service presence as an ILEC, but not in major markets. The other two offer local service as CLECs, but whether they are profitable is questionable. The IXCs survived until the RBOCs gained permission to sell long distance, at which point the competitive arena expanded. Moreover, WorldCom in particular, suffered from financial missteps that sent them into bankruptcy. WorldCom vanished and reverted to MCI, which they had absorbed a few years earlier.

The major value the IXCs have today is their networks and their international footprint. Network capacity can be leased, but as companies like SBC and Verizon advance into the long distance business, they must have an international presence. Alliances with international providers are not enough to satisfy large customers that prefer to deal with a minimum of service providers. These companies demand service providers with enough mass and stability to guarantee that they will survive.

Local Service Competition

The local market is served by three distinct categories: the ILECs in the core metropolitan areas, CLECs, whose serving areas overlap the ILECs, and the small rural exchanges that depend on universal service fund subsidies. The ILECs are somewhat better off than the IXCs, but their business models are also changing. The trend today is toward bundled service and flat-rate pricing, which disrupts the traditional pricing policies of many ILECs. In some cases long distance is cheaper than the zone pricing that many LECs provide. Competition may force them into flat-rate pricing and erode revenues.

ILECs are still profitable despite losing a significant portion of their business revenues to CLECs. They have held onto many of their largest business customers by making concessions to prevent them from defecting. Even when their customers subscribe to CLEC service, the ILECs retain much of the revenue because the CLECs use ILEC facilities to provide the service. This also is changing. Recent court and FCC decisions point to the need for CLECs to become facility based, i.e. own their switches if they are to survive. The ILECs have managed to stave off competition in the form of government-mandated fracturing of their networks but competition will not disappear. In contrast, cable companies are on the threshold of a deep sortie into ILEC territory and the phone companies have no way of preventing it. They have lost a few residential customers to this point because they are not profitable enough to attract competitive providers, but that is about to change. Although VoIP over cable has drawbacks as a lifeline service, the cable companies can offer additional lines at an attractive rate, often bundled with free long distance.

Cable providers serve the majority of households and have sufficient bandwidth to add voice, bypassing the ILECs by collocating a softswitch with their headend. Cable has the additional advantage of greater range than a copper loop. CATV providers are not required to provide equal access to their networks, however, and they can block VoIP traffic from other providers so cable is not an open path for competitors to bypass ILEC facilities. Quality is easier for cable to control than other VoIP providers. The portion of the call that rides the public Internet is apt to be short because it can immediately hop onto a private IP network or onto the PSTN through a gateway.

Not only does competition come from wireline carriers, but cell phones are having a comparable effect as many people forgo their wired telephones in favor of cellular/PCS. Cellular is rapidly becoming an essential service, to the point that many households and some businesses use it as their sole telephone. As a case in point, Ford Motor Co. in 2005 announced that it was replacing 8000 wired phones with Sprint wireless. Fortunately for the major ILECs, Verizon and SBC, they own a large portion of the cellular/PCS market, but allure of cannibalizing their core business is rather conflicted.

The ILECs are affected to some degree by the same problem that afflicts the IXCs. They are selling commodity services that other companies can provide at lower cost, particularly because they are not saddled with a huge investment in equipment that can be replaced at a much lower cost. The ILECs will hang onto their existing central offices as long as possible, but ultimately they must replace them. At that point, the shape of the local exchange network will have changed radically.

The ILECs have enjoyed a monopoly because they had something that was difficult to replace: copper wire to every household. That wire still gives them a major advantage, but cable companies are now positioned to displace that. This leaves the ILECs with the need for a broadband pipe of their own. DSL is sufficient today, but consumer demand for bandwidth seems almost insatiable. To stay competitive, the LECs have to build fiber optics to their customers. They have gained FCC concessions that exempt them from sharing new fiber with CLECs, but telephone service alone cannot support the cost. The most likely candidate for revenue enhancement is entertainment, which puts the LECs on a collision course with the cable companies. It also requires them to get into the content-delivery business, something for which they have no experience. They can hire the expertise to expand into entertainment, but the landscape is littered with companies that have wandered too far afield of their expertise.

ILEC fiber development plans also open another bitter contest—this time between the haves and the have-nots. The ILECs obviously cannot afford to convert all of their customers to fiber at once. They must space it out over years—perhaps decades. Left to their own devices, they would convert the affluent neighborhoods first because those are the most likely to yield the revenue to justify the investment. Poorer and rural neighborhoods would be converted last if at all. This further widens the digital divide, which incurs opposition from their regulators—the cities, counties, and state governments that control the right of way. The governmental bodies prefer that the benefits of broadband access go to the lesser developed neighborhoods. Some communities, unwilling to trust their broadband development to profit-making enterprises, are trying forays into publicly owned fiber or wireless. The ILECs' response is to head off this threat by appealing to legislators. A monumental clash seems inevitable.

Cable and Cellular Competitors

The cellular and cable competitors are likewise faced with major competition. Cellular providers enjoyed a comfortable oligopoly before PCS. They met that threat by acquiring PCS frequencies and competitors, but competition lurks in the wings from another quarter: wireless. Many people expect voice over WiFi to provide a viable alternative to cellular/PCS, but WiFi hotspots cannot offer a reliable alternative to cellular because the frequencies are too limited and the protocols have too many quality problems. Moreover, WiFi is designed as a private network, not a common carrier service. WiMax, on the other hand, is lurking in the future. Its protocols are far from complete, but by about 2007 WiMax should be ready for the market as a credible platform on which service providers can offer VoIP. WiMax will not only compete with cell phones (including 3G and 4G cellular), it is expected to be suitable for access, competing directly with cable and DSL. In some localities it may become the primary means of access. WiMAX can potentially do for metropolitan area wireless what WiFi did for wireless LANs. Prestandard products are on the market now, well ahead of the expected 2007-8 protocol approval.

Both the cable companies and the ILECs are studying the feasibility of IPTV: bringing video on demand to the customer over a broadband pipe. This displaces the old service model of passing all channels by every cable customer and replaces it with video that can be customized at the source. Instead of being restricted to a single view of live program material, the originator can send multiple views, which allows the viewers to select the ones they prefer. IPTV, in trials today, holds out the promise of personalizing video reception beyond the limited manipulations that are possible with video hard disk drives.

Another trend that is apt to occur is barely visible: a shift from broadcasting to narrowcasting. One of the hottest products on the market is the i-Pod, which gives users a choice of hours of their own musical selections. For video, devices such as Tivo, DVD recorders, and the old standby VCR give users a choice of what to watch and when they want to watch it without being bombarded with annoying commercials.

The cable companies are also vulnerable to FCC decisions that may require them to open their networks to competitors. Today they enjoy monopoly power over access services in most reception areas. They are also subject to competition from satellite providers, but these cannot compete favorably for access, which gives the cable companies a big advantage. They are falling behind DSL, however, in the race to be the dominant access provider.

Wireless

In the matter of broadband access, many people are relying on wireless to avoid the expense of digging up the streets. While SBC and Verizon have announced ambitious plans to build fiber, Qwest, already burdened with massive debt, has announced that it is looking to WiMax as its broadband vehicle. The major difficulty of wireless in all of its manifestations is a lack of frequency spectrum. This problem is not insoluble. Plenty of frequencies are underused, but the process of relocating spectrum is lengthy. Moreover, if mobility is unimportant and the cost can be supported, wired solutions are always preferable to wireless.

It is safe to predict that wireless will have a vigorous future. It will not, however, have the impact that many people expect. The spectrum is limited and competition for it is intense. Many people see WiFi as a key technology, but its limitations will prevent it from becoming more than is intended to be, which is a LAN alternative. Its spectrum and range are too limited to create a metropolitan network service. Hot spots in airports and other transportation centers have great potential. The airlines are preparing to offer WiFi in their cabins and this should prove to be enormously attractive. Much of the enthusiasm for WiFi will prove to be a little more than a hobbyist class-of-service. For commercial-grade service, a complete operating system is needed. A system that requires its users to change locations, fiddle with the controls, endure occasional dropouts will satisfy enthusiasts and those who are willing to put up with the inconvenience to save money, but it does not qualify as a business class of network. Those with a vision of ubiquitous and seamless WiFi are destined for disappointment.

Another group that is likely to have their expectations unfulfilled is the voice over WiFi advocates, particularly those who expect to mix voice and data on the same LAN and obtain service that remotely approaches the quality of the wired network. VoIP proponents state that you should not attempt to run it on a shared-media LAN, which is an inherent characteristic of 802.11. If plenty of bandwidth was available, nothing would prevent wireless from approximating the quality of the wired LAN, but that is not the case. With limited channels, limited range, and the potential of interference, WLANs are incapable of providing consistent voice quality. This is not to say the users will shun service because the convenience may offset the grade of service.

The rollout of 3G cellular is beginning, and with it come new phone models and features that obsolete the older models. 3G cellular should enjoy a respectable amount of demand. The desire for cellular service is nothing short of astounding, particularly when you watch the development in third-world nations, and consider the amount of money that people pay for trivial accessories such as customized ringing tones. New cell phones support video, albeit on a miniature screen, and content providers are beginning to offer custom video clips. How far the market for this will extend is anyone's guess, but society has demonstrated a boundless willingness to pay for entertainment.

Fourth generation wireless standards are under development, and an industry group is working on an interim service known as Super 3G that may be ready in 2007. Super 3G would have downstream speeds as high as 100 Mbps.

Increased Computing Power

Telecommunications continues to evolve under the effect of increasing computer power. In 1965, Gordon Moore of Intel predicted that the number of transistors per chip would double every couple of years. Moore's Law, as it is known, is holding true and Intel expects it to continue through the end of the decade. As a result, we will have even cheaper and faster computers. Computing power will reach into crannies that we can hardly imagine today. Voice recognition gets better with every new iteration. Smart appliances, climate control systems, security systems, highways—countless devices will communicate. Meter readers will become an endangered species as devices communicate their usage to the central authority and accept orders to turn themselves off and on. Increased computing power means more storage, which in turn requires telecommunications resources as organizations transmit massive files through storage area networks.

Users will have more choices than ever before as companies strive to differentiate their products and increase their appeal. New PCs, particularly laptops, are today inseparable from communications. Most travelers hate to carry a suitcase full of specialized devices so they purchase multifunction devices and put up with their constraints. Cellular phones and wireless PDAs are attempting to emulate the PC, and within their physical limitations they are succeeding. Wireless PDAs have full keyboards and cell phone functions. The keyboard is awkward and the cell phone is not shaped to fit the anatomy, but users love them. A decade from now, laptops will probably have a standard WiMax interface.

Unresolved Issues

These trends may be inevitable, but they will not be simple because they raise countless issues for which we have no answers. Simplicity is the first unresolved issue. Today users have to know too far much to configure devices. Everyone who has used technology has stories of technicians who had to make three trips to make it work, devices returned to the store because they could not be configured, mysterious and unexplained error messages, and long minutes or hours on hold to talk to a technician who spoke in an accent that was barely comprehensible. We enjoy the benefits these devices bring when they work, but in some ways we yearn for the old days of the simple telephone, which, if it did not work, the telephone company repaired.

A good model for configuration simplicity is the modern automobile. Today's vehicles have nearly 1000 times more computing power than the computers that guided the lunar and command modules to the moon, but its designers have done an admirable job of concealing the complexity from the driver, at least in the low-end models. Cars today work as they have for years. The only things that need to be configured, the music system, trip computer, and GPS device, can easily be ignored. The days of do-it-yourself repair have vanished, but the product has reliability that far exceeds the older models. Inside desktop computers, examples of simplicity are the Ethernet and USB adapters. A decade ago, installing and configuring Ethernet cards and serial ports was a frustrating chore. Now no one even thinks about the USB or Ethernet port: They are just there and they work. We need this kind of simplicity in telecommunications devices and computers.

Many other issues remain to be resolved. In addition to the aforementioned digital divide that separates computer uses into classes, we have the question of frequency spectrum. Users have announced unmistakably that they want to do away with wires. Wireless technology is becoming viable, but where will the spectrum come from? Carving up the bandwidth would be easy except that it awakens the guard dogs. Plenty of unused spectrum exists, particularly if we take into account the "Negroponte Flip." Several years ago Nicholas Negroponte, founding chairman of MIT's Media Laboratory, noted that most of today's wired applications are becoming wireless and vice versa. We can see the glimmerings of that in cell phones and cable television. We have reached the point in the United States, where cable could serve about 85 percent of the population, except cable is not free and it does not reach the fringe areas. Satellite TV covers the fringes, but it is not free.

Television is assigned an enormous amount of valuable spectrum. Of the lower 1 GHz of frequency spectrum, broadcast television is assigned 412 MHz—almost half. A large portion of the UHF television spectrum could potentially be released, but not without a fight. Part of the problem lies in the quest for inexpensive television receivers. Their front ends are not selective enough to reject adjacent channel interference, so the FCC makes a practice of leaving some channels vacant. Technology may provide the answer. Scientists and engineers are managing to cram more and more information into a narrow spectrum and we have faith that they will continue to do so. Digital televisions are readily available today, and the technology to make them sufficiently selective to receive off-the-air broadcasts without mutual interference is well known. Furthermore, plenty of bandwidth is available above 30 GHz and ways of using it effectively will undoubtedly be discovered.

THE CONVERGED NETWORK

The converged voice and data network will happen, but not as rapidly as many of its proponents hope. Many roadblocks stand in the way of the journey to an all-IP network with ubiquitous wireless and universal broadband access. Many of the solutions are impossible to predict because they have nothing to do with advances in technology. Will Congress revisit the Telecommunications Act of 1996? The act is deeply flawed, but whether they can improve it is questionable given the conflicting pressures, some with deep pockets. Will the old-line service providers adapt to the new network? History is replete with the carcasses of companies that failed to perceive the fundamental changes in their business. Companies that sat at the pinnacle of their industries have clung to their existing markets in the face of evidence that technology was passing them by. Western Union is a prime example, and the company that dislodged them, AT&T, suffered a similar fate. The top record companies of the past are bit players in the CD industry. Semiconductors companies such as Intel and Texas Instruments prospered because they did not have to cast off the baggage of manufacturing vacuum tubes, while the companies that did are minor participants today.

Broadband Development

Broadband has been increasing in the United States and the FCC has made it clear that it considers increased broadband development to be in the national interest. The main impediment is price. As long as broadband costs somewhere north of $50 per month a significant portion of the population will decide that dial-up or no Internet access is in their best interest. So far the FCC has not defined it as a critical service subject to subsidy ala the telephone, but that is probably coming.

The country's best hope for both broadband and video development is a price war between ILECs and cable companies, both of which have been successful in persuading the FCC to treat their monopolies benignly. As this edition is written, the FCC is under new management and some of its policies will undoubtedly change. If the ILECs begin to develop fiber to the home, delivering video on demand in direct competition with the cable providers, a price war could erupt and bring both video and broadband access down to more affordable levels. The outcome is difficult to predict because the FCC has chosen to exempt cable companies from a requirement to share their facilities and ILECs from sharing new fiber development.

VoIP Development

VoIP should experience sustained growth over the next decade while TDM slips down the slope of its life cycle curve. Voice over IP does not mean voice over the Internet, however, and in this respect the expectations many people have for the technology will not be realized. Telephone service of differing grades is likely to arise from the many alternatives that VoIP enables. A best-effort telephone service will continue to be available for minimal or no cost for those who wish to use their broadband connections to communicate with willing and similarly equipped correspondents. For the service to work, both parties must have compatible programs, computers turned on, a broadband connection, and software at the ready. Lacking any of these in the callee, a VoIP caller can initiate a call and terminate it over the PSTN for a price. As with metropolitan WiFi, the results fall short of commercial quality. If the conditions are right, VoIP connections over the Internet can be better than the PSTN and the price is free. Buddy sessions are no substitute for universal connectivity, however, which means the PSTN has a long, although diminishing future. Meanwhile, isolated islands of users will benefit from VoIP.

For VoIP on a public network to deliver voice service with a quality comparable to the PSTN, the IPsphere is an initiative to watch. Think of IPsphere as the Internet sans the World Wide Web, its free and open access, and resultant chaos. If IPsphere achieves its objectives, it can provide ubiquitous VoIP service with controlled quality, interconnection agreements, network management, and the other amenities users have come to associate with commercial telephone service. The probability of IPsphere's success is difficult to predict, however, because some major players are not participating. The market for and pricing of such a service has yet to be demonstrated, but if IPsphere is successful, it should have a major effect on network architectures.

ILECs are beginning to make forays into VoIP by installing softswitches and offering packaged services. This gives them the ability to range outside their defined territories. These ventures are experimental at this point. It is far too early to proclaim the death of the PSTN, but it is showing its age and upstarts are attempting to knock it off its perch with attractive packages of bundled services. No one is making money on VoIP yet, but it portends the shape of things to come. The industry has a long way to go to make VoIP services easy to use, but it will improve.

Unfortunately for the manufacturers and service providers, the strength of the VoIP industry depends on commoditization of local service. The early entrants hope to build a customer base large enough to make their business an acquisition target. Once the commoditization stage is reached, local telephone service will look much like it does today. It will not be a cottage industry; the service providers will look much like today's ILECs except their tentacles will extend outside their territories. The territories are likely to survive for a long time because someone must be the provider of last resort and that falls to today's franchised telephone companies. This vision does not bode well for rural areas.

Telephone Subsidization

A related issue arises out of this scenario: how subsidies will be administered. The VoIP providers are today exempt from taxes and surcharges that subsidize universal service. As VoIP erodes the revenues that are the basis of these subsidies, the fund will shrink. Congress will undoubtedly act to withdraw its exemptions for VoIP, but it will run up against the difficulty of distinguishing voice calls from Web browsing. Local taxing authorities have already begun licking their chops in anticipation of a new revenue stream. In the case of the State of Minnesota's attempt to tax Vonage's service as it does telephones, the FCC warned them to keep hands off, but the state appears not to have given up. The problem is not only political, it is also technical. When the voice stream flows through a gateway, it is feasible to identify and tax it, but peer-to-peer VoIP sessions are difficult to recognize, and it is almost impossible to tell whether they are intrastate. The IETF has yet to discover any VoIP requirement that they cannot solve with new protocols, but their enthusiasm wanes in the face of making it easier for authorities to tax the service. The government has so far been reluctant to stand in the way of these new technologies but when the new wears off and their revenues decline, they will discover ways to redefine the public interest.

Taxation and Regulation

The carriers are not the only ones whose business models are being disrupted. Taxing authorities and regulators are learning that IP technologies do not fit into the neat silos of the past. In the old days the providers organized themselves into wired, wireless, and cable and submitted themselves to regulation that was organized along the same lines. Services were segmented into inter- and intrastate categories and local authorities levy franchise fees for use of the public right of way. Local authorities can tax intrastate service, but the federal government reserves to itself the right to tax interstate commerce.

These convenient distinctions remain, but they are fading as surely as the service providers' traditional models are endangered. How can a taxing authority distinguish between a VoIP session and Web surfing? It is possible by deep packet inspection, but not without a lot of processing power and undermining the character of the Internet. Look for this to become another battleground.

Security

The overarching concern in the IP network is security. VoIP will not be successful until it is as simple to use as the PSTN. Making it simple is not as big a problem as making it secure. Equipping residences with broadband access can and will be accomplished. Telephone service can be brought to every room with an appliance that looks a lot like a key telephone system. It is even feasible to handle emergency calls with a server that routes the call to the appropriate agency. A considerable amount of work is needed, however, to make VoIP service as secure as the telephone is today, particularly in the case of wireless signals. It is not realistic to expect the broad base of consumers to configure firewalls and set up routers, but without the appropriate safeguards, the miscreants that populate the Internet will find ways to destroy it. This issue opens plenty of opportunity for creative innovation.

E-mail security is a case in point. E-mail is one of the most valuable services that has come out of the Internet experience. It quickly expanded from a service that was nice to have to the point that a world without e-mail is unthinkable. At the same time, however, criminals use it to propagate devastating viruses and fish for information that they can use for identity theft. Unwanted e-mail or spam is an expensive nuisance that clogs servers, consumes bandwidth, and wastes untold amounts of time.

Opportunities and Challenges

Over the past few decades telecommunications has been one of the most exciting and challenging fields of endeavor in the entire world. Fortunately for both consumers and those involved in the industry, nothing in sight will slow that down. Occasional blips will arise like the "telecom winter" of the early part of this century, but through this period consumer demand did not flag. Most of the wounds of the era were self-inflicted and many of the casualties have either recovered or revival is underway.

Nothing visible in the future portends anything but opportunity. The companies that prosper will be the ones that do the best job of determining what consumers want. Much of the result is impossible to foresee today, but it is clear that we are in for an explosion of innovation and at the heart of it will be computers and telecommunications.

APPENDIX A

Telecommunications Acronym Dictionary

The telecommunications industry is expansive in its use of acronyms. The following is a list of acronyms used in this book, without definitions. Protocols are defined in the text and can be located in the index. Definitions for most acronyms are indexed by their full name in the glossary.

3G	Third-Generation Cellular
AAA	Authentication, Accounting, and Authorization
AAC	Alternate Access Carrier
AAL	ATM Adaptation Layer
ABR	Available Bit Rate
AC	Alternating Current
ACD	Automatic Call Distributor
ACF/VTAM	Advanced Communications Function/Virtual Telecommunications Access Method
ACR	Attenuation-to-Crosstalk Ratio
ADM	Add-Drop Multiplexer
ADPCM	Adaptive Pulse-Code Modulation
ADSI	Analog Display Services Interface
ADSL	Asymmetric Digital Subscriber Line
AIN	Advanced Intelligent Network
AIS	Automatic Intercept System
ALI	Automatic Location Information
AM	Amplitude Modulation, Active Monitor
AMA	Automatic Message Accounting
AMI	Alternate Mark Inversion
AMIS	Audio Messaging Interchange Specification
AMPS	Advanced Mobile Phone System

AMR	Adaptive Multi-Rate
AMR-WB	Adaptive Multi-Rate-Wide Band
ANI	Automatic Number Identification
ANSI	American National Standards Institute
AO/DI	Always On/Dynamic ISDN
AP	Access Point
APD	Avalanche Photo Diode
API	Application Programming Interface
APPC	Advanced Peer-to-Peer Communications
APPN	Advanced Peer-to-Peer Networking
ARP	Address Resolution Protocol
ARPA	Advance Research Projects Agency
ARQ	Automatic Repeat Request
ARS	Automatic Route Selection
AS	Autonomous System
ASCII	American Standard Code for Information Interexchange
ASIC	Application Specific Integrated Circuit
ASN.1	Abstract Syntax Notation 1
ATIS	Alliance for Telecommunications Industry Solutions
ATM	Asynchronous Transfer Mode
ATSC	Advanced Television Systems Committee
AUI	Attachment Unit Interface
AWG	American Wire Gauge
B8ZS	Bipolar with 8-Zero Substitution
BACP	Bandwidth Allocation Control Protocol
BAN	Boundary Access Node
BBS	Bulletin Board System
BECN	Backward Explicit Congestion Notification
BER	Bit Error Rate
BERT	Bit Error Rate Test
BGAN	Broadband Global Area Network
BGP	Border Gateway Protocol
BHCA	Busy Hour Call Attempts
BHCC	Busy Hour Call Completions
BHCCS	Busy Hour Hundreds of Call Seconds
B-ICI	B-ISDN Inter-Carrier Interface
BIIS	Binary Interchange of Information and Signaling
B-ISDN	Broadband ISDN
BLER	Block Error Rate
BLF	Busy Lamp Field
BLSR	Bidirectional Line-Switched Ring
BNN	Boundary Network Node
BOC	Bell Operating Company

BOOTP	Boot Protocol
bps	Bits per Second
BPSK	Binary Phase Shift Keying
BRI	Basic Rate Interface
BSA	Basic Service Arrangement
BSE	Basic Service Element
BSRF	Basic Standard Reference Frequency
BSS	Basic Service Set
BTA	Basic Trading Area
CAC	Carrier Access Code, Call Admission Control
CALEA	Communications Assistance for Law Enforcement Act
CAMA	Centralized Automatic Message Accounting
CAMEL	Customized Application for Mobile Enhanced Logic
CAP	Control Access Protocol, Carrierless Amplitude Phase, Competitive Access Provider
CAS	Channel Associated Signaling, Centralized Attendant Service
CATV	Community Antenna Television
CBR	Constant Bit Rate
CCIS	Common Channel Interoffice Signaling
CCITT	Consultative Committee on International Telephone and Telegraph
CCS	Centum Call Seconds, Common Channel Signaling
CCTV	Closed Circuit Television
CDDI	Copper Distributed Data Interface
CDMA	Code Division Multiple Access
CDO	Community Dial Office
CDPD	Cellular Digital Packet Data
CDR	Call Detail Recorder
CED	Called Station Identification
CELP	Code Excited Linear Prediction
CEPT	Conference on European Post and Telecommunications
CES	Circuit Emulation Service
CGSA	Cellular Geographic Serving Area
CHAP	Challenge Handshake Authentication Protocol
CIDR	Classless Interdomain Routing
CIF	Common Intermediate Format
CIR	Committed Information Rate
CLASS	Custom Local Area Signaling Services
CLEC	Competitive Local Exchange Carrier
CLID	Calling Line Identification
CM	Cable Modem
CMIP	Common Management Information Protocol

CMTS	Cable Modem Termination System
CNLP	Connectionless Network Layer Protocol
CN	Common Name
CNG	Calling Tone
CNM	Customer Network Management
CO	Central Office
CO-ACD	Central Office Automatic Call Distributor
CODEC	Coder/Decoder
COS	Corporation on Open Systems
CP	Control Point
CPE	Customer Premises Equipment
CPU	Central Processing Unit
CRC	Cyclic Redundancy Check
CR-LDP	Constraint-Based Routed Label Distribution Protocol
CSMA/CD	Carrier Sense Multiple Access with Collision Detection
CSU	Channel Service Unit
CTI	Computer Telephony Integration
CTS	Clear To Send
CWDM	Coarse Wavelength-Division Multiplexing
DAMA	Demand Assignment Multiple Access
D-AMPS	Digital Advanced Mobile Phone System
DAP	Directory Access Protocol
DAS	Dual Attachment Stations
dB	Decibel
DBS	Direct Broadcast Satellite
DC	Direct Current, Domain Components
DCA	Document Content Architecture
DCE	Data Communications Equipment
DCS	Digital Crossconnect System
DCT	Discreet Cosine Transform
DDD	Direct Distance Dialing
DDS	Digital Data Service
DE	Discard Eligible
DECT	Digital Enhanced Cordless Telecommunications
DEN	Directory-Enabled Networks
DES	Data Encryption Standard
DHCP	Dynamic Host Configuration Protocol
DIA	Document Interchange Architecture
DID	Direct Inward Dialing
DiffServ	Differentiated Services
DISA	Direct Inward System Access
DIT	Directory Information Tree
DLC	Data Link Control

DLCI	Datalink Connection Identifier
DLSw	Datalink Switching
DMI	Digital Multiplexed Interface, Desktop Management Interface
DMT	Discrete Multitone
DN	Distinguished Name
DNIS	Dialed Number Identification System
DNS	Domain Name Service, Directory Naming Service
DOC	Dynamic Overload Control
DOCSIS	Data Over Cable System Interface Specification
DOD	Direct Outward Dialing
DoS	Denial of Service
DQDB	Distributed Queue Dual Bus
DSA	Directory System Agent
DSCP	DiffServ Code Point
DSE	DSA-Specific Entry
DSI	Digital Speech Interpolation
DSL	Digital Subscriber Line
DSLAM	DSL Access Multiplexer
DSMA/CD	Digital Sense Multiple Access with Collision Detection
DSP	Digital Signal Processor, Directory Service Protocol
DSS	Direct Station Selection
DSSS	Direct Sequence Spread Spectrum
DSU	Digital Service Unit
DSX	Digital Service Crossconnect
DTE	Data Terminal Equipment
DTMF	Dual Tone Multifrequency
DUA	Directory User Agent
DVC	Digital Video Compression
DWDM	Dense Wavelength-Division Multiplexing
E-911	Enhanced 911
EAP	Extensible Authentication Protocol
EAS	Extended Area Service
EBCDIC	Expanded Binary Coded Decimal Interexchange Code
EDFA	Erbium-Doped Fiber Amplifiers
EDI	Electronic Data Interchange
EDIFACT	EDI for Administration, Commerce, and Transport
EF	Entrance Facility
EFM	Ethernet in the First Mile
EFMC	EFM Copper
EFMF	EFM Fiber
EFMH	EFM Hybrid
EFMP	EFM PON

EFS	Error-Free Seconds
EGP	External Gateway Protocol
EIA	Electronic Industries Alliance
EIRP	Effective Isotropic Radiated Power
ELAN	Emulated LAN
E-LAN	Ethernet LAN
ELFEXT	Equal-Level FEXT
E-Line	Ethernet line service
EMF	Electromagnetic Force
EMI	Electromagnetic Interference
ENFIA	Exchange Network Facility Interconnecting Arrangement
ENUM	Electronic Number Mapping
EoS	Ethernet over SONET
EPON	Ethernet Passive Optical Network
ER	Equipment Room
ES	End System
ESF	Extended Super Frame
ESN	Electronic Serial Number, Emergency Service Number
ESS	Electronic Switching System, Extended Service Set
ETC	Enhanced Throughput Cellular
ETN	Electronic Tandem Network
ETSI	European Telecommunications Standardization Institute
EVC	Ethernet Virtual Connection
EV-DO	Evolution Data Optimized
FAS	Facility Associated Signaling
FCC	Federal Communications Commission
FDDI	Fiber Distributed Data Interface
FDL	Facility Datalink
FDM	Frequency Division Multiplexing
FDMA	Frequency Division Multiple Access
FEC	Forward Error Correction, Forwarding Equivalence Class
FECN	Forward Explicit Congestion Notification
FEX	Foreign Exchange
FEXT	Far-End Crosstalk
FITL	Fiber-in-the-Loop
FM	Frequency Modulation
FoD	Fax on Demand
FoIP	Fax over Internet Protocol
FR	Frame Relay
FRAD	Frame Relay Access Device
FTAM	File Transfer, Access, and Management
FTP	File Transfer Protocol
FTTC	Fiber-to-the-Curb

FTTH	Fiber-to-the-Home
FX	Foreign Exchange
FXO	Foreign Exchange Office
FXS	Foreign Exchange Subscriber
GbE	Gigabit Ethernet
Gbps	Gigabits per Second
GEF	Generic Extensibility Framework
GEO	Geosynchronous Earth Orbit (Satellite)
GF	Generic Function
GFC	Generic Flow Control
GGSN	Gateway GPRS Service Node
GHz	Gigahertz
GMLC	Gateway Mobile Location Center
GMSC	Gateway MSC
GPRS	General Packet Radio Service
GPS	Global Positioning Satellite
GSM	Global System for Mobile Communications
GUI	Graphical User Interface
HDLC	High-Level Data Link Control
HDSL	High-Speed Digital Subscriber Line
HDTV	High Definition Television
HE	Head-end
HEC	Header Error Control
HFC	Hybrid Fiber-Coax
HLR	Home Location Register
HSDPA	High-Speed Downlink Packet Access
HTML	Hypertext Markup Language
HTTP	Hypertext Transfer Protocol
Hz	Hertz
IAD	Integrated Access Device
IC	Intermediate Crossconnect
ICANN	Internet Corporation for Assigned Names and Numbers
ICMP	Internet Control Message Protocol
ICNAM	Internetwork Calling Name Service
ICPIF	Impairment/Calculated Planning Impairment Factor
ICSA	International Computer Security Association
IDF	Intermediate Distributing Frame
IDSL	ISDN Digital Subscriber Line
IE	Internet Exchange
IEEE	Institute of Electrical and Electronic Engineers
IETF	Internet Engineering Task Force
IF	Intermediate Frequency
IGMP	Internet Group Management Protocol

IGRP	Interior Gateway Routing Protocol
IKE	Internet Key Exchange
ILEC	Incumbent Local Exchange Carrier
ILMI	Integrated Link Management Interface
IM	Instant Messaging
IMAP	Internet Message Access Protocol
IMPP	Instant Messaging and Presence Protocol
IMT	Intermachine Trunk
IMTS	Improved Mobile Telephone Service
IN	Intelligent Network
INMARSAT	International Maritime Satellite Service
IntServ	Integrated Services
IOC	Interoffice Channel
IP	Internet Protocol
IPM	Interruptions per Minute
IPng	IP Next Generation
IPSec	IP Security
IPv6	IP Version 6
IS	Intermediate System
ISAKMP	Internet Security Association and Key Management Protocol
ISDN	Integrated Services Digital Network
ISG	Intelligent Services Gateway
ISI	Intersymbol Interference
ISM	Industrial, Scientific, Medical
ISO	International Standards Organization
ISP	Internet Service Provider
ISUP	ISDN Service User Part
ITFS	Instructional Television Fixed Service
ITU	International Telecommunications Union
IVR	Interactive Voice Response
IXC	Interexchange Carrier
JPEG	Joint Photographic Experts Group
Kbps	Kilobits per Second
KHz	Kilohertz
KVA	Kilo-Volt Amps
KSU	Key System Unit
KTS	Key Telephone System
L2F	Layer 2 Forwarding
L2TP	Layer 2 Tunneling Protocol
LAMA	Local Automatic Message Accounting
LAN	Local Area Network
LANE	LAN Emulation

LAPB	Balanced Link Access Procedure
LAPD	Distributed Link Access Procedure
LATA	Local Access Transport Area
LBO	Line Build-Out
LCR	Least Cost Routing
LDAP	Lightweight Directory Access Protocol
LDP	Label Distribution Protocol
LEC	Local Exchange Carrier, LAN Emulation Client
LECS	LAN Emulation Configuration Server
LED	Light Emitting Diode
LEO	Low Earth Orbit
LER	Label Edge Router
LERG	Line Equipment Routing Guide
LES	LAN Emulation Server
LIB	Label Information Base
LIT	Line Insulation Test
LIV	Link Integrity Verification
LLC	Logical Link Control
LMDS	Local Multipoint Distribution Service
LMI	Local Management Interface
LMS	Local Measured Service
LNA	Low Noise Amplifier
LNP	Local Number Portability
LOS	Line of Sight
LRN	Location Routing Number
LSP	Label-Switched Path
LSR	Label Switch Router
LTD	Local Test Desk
LU	Logical Unit
MAC	Media Access Control
MAN	Metropolitan Area Network
MAP	Manufacturing Automation Protocol
MAPI	Mail Application Programming Interface
MAT	Maintenance and Administration Terminal
MAU	Media Access Unit
Mbps	Megabits per Second
MC	Main Crossconnect
MCC	Maintenance and Control Center
MCU	Multipoint Control Unit
MCVD	Modified Chemical Vapor Deposit
MDBS	Mobile Data Base Station
MDF	Main Distributing Frame
MDIS	Mobile Data Intermediate Station

MDLP	Mobile Datalink Protocol
MDT	Mobile Data Terminal
MEGACO	MEdia GAteway COntrol protocol
MEO	Medium Earth Orbit
MFJ	Modified Final Judgment
MG	Media Gateway
MGC	Media Gateway Controller
MGCP	Media Gateway Control Protocol
MH	Modified Huffman
MHS	Message Handling Service
MHz	Megahertz
MIB	Management Information Base
MIF	Management Information Format
MIME	Multipurpose Internet Mail Extension
MIMO	Multiple Input–Multiple Output
MIN	Mobile Identification Number
MMDS	Multichannel Distribution Service
MMF	Multimode Fiber
MMR	Modified Modified Read
MOS	Mean Opinion Score
MP	Multipoint Processor
MPDS	Mobile Packet Data Service
MPEG	Motion Picture Experts Group
MPLS	Multiprotocol Label Switching
MPOA	Multiprotocol Over ATM
MR	Modified Read
MTA	Message Transfer Agent, Major Trading Area
MTBF	Mean Time Between Failures
MTP	Message Transfer Part
MTS	Message Telephone Service
MTSO	Mobile Telephone Switching Office
MTTR	Mean Time To Repair
MTU	Maximum Transmission Unit
MVIP	Multivendor Integration Protocol
NANPA	North American Numbering Plan Administration
NAP	Network Access Point
NAPTR	Naming Authority Pointer
NAT	Network Address Translation
NAU	Network Access Unit, Network Addressable Unit
NCP	Network Control Program, Network Control Point
NEBS	Network Equipment Building System
NEC	National Electrical Code
NECA	National Exchange Carrier Association

NEXT	Near-End Crosstalk
NFAS	Non-Facility Associated Signaling
NHRP	Next Hop Resolution Protocol
NIC	Network Interface Card
NLOS	Non-Line-of-Sight
NMCC	Network Management Control Center
NMS	Network Management System
NNI	Network-to-Network Interface, Network Node Interface
NOS	Network Operating System
NPA	Numbering Plan Area
NPAC	Number Portability Administration Center
NRT-VBR	Non-Real Time Variable Bit Rate
NRZ	Non-Return to Zero
NS/EP	National Security and Emergency Preparedness
NSAP	Network Service Access Point
NT1	Network Termination 1
NT12	Network Termination 1-2
NT2	Network Termination 2
NTN	Network Terminal Number
NTSC	National Television Systems Committee
OAM	Operations, Maintenance, and Administration
OAM&P	Operations, Administration, Maintenance, and Provisioning
OC	Optical Carrier
OCR	Optical Character Recognition
OCX	Optical Cross-Connect
OFDM	Orthogonal Frequency Division Multiplex
OLT	Optical Line Terminal
ONU	Optical Network Unit
OSI	Open Systems Interconnect
OSP	Outside Plant, Operator Service Provider
OSPF	Open Shortest Path First
OSS	Operations Support System
OTDR	Optical Time Domain Reflectometer
OU	Organizational Unit
OVD	Outside Vapor Deposition
PAD	Packet Assembler/Disassembler
PAL	Phase Alternate Line
PAP	Password Authentication Protocol
PBX	Private Branch exchange
PC	Personal Computer
PCA	Protective Coupling Arrangement
PCM	Pulse Code Modulation
PCS	Personal Communication Service

PCVD	Plasma-Activated Chemical-Vapor Deposition
PDA	Personal Digital Assistant
PDF	Portable Document Format
PDFFA	Praseodymium-Doped Fluoride Fiber Amplifier
PDN	Public Data Network
PDU	Protocol Data Unit
PEL	Picture Elements
PHY	PHYsical layer of a network model
PIC	Primary Interexchange Carrier
PICC	Presubscribed Interexchange Carrier Charge
PIN	Personal Identification Number
PISN	Private Integrated Services Network
PKI	Public Key Infrastructure
PNNI	Private Network-to-Network Interface
PoE	Power over Ethernet
POI	Point of Interface
PON	Passive Optical Network
POP	Point-of-Presence
POP3	Post Office Protocol Version 3
PoS	Packet over SONET
POTS	Plain Old Telephone Service
PPP	Point-to-Point Protocol
pps	Packets per Second
PPTP	Point-to-Point Tunneling Protocol
PRI	Primary Rate Interface
PS/ALI	Private Switch Automatic Line Identification
PSAP	Public Safety Answering Point
PSK	Phase Shift Keying
PSQM	Perceptual Speech Quality Measurement
PSS1	Private Signaling System No. 1
PSTN	Public Switched Telephone Network
PU	Physical Unit
PUC	Public Utilities Commission
PVC	Permanent Virtual Circuit, Polyvinyl Chloride
PWT	Personal Wireless Telecommunications
QAM	Quadrature Amplitude Modulation
QCIF	Quarter CIF
QoS	Quality of Service
QPSK	Quadrature Phase Shift Keying
RADIUS	Remote Authentication Dial-In User Service
RADSL	Rate-Adaptive DSL
RAID	Redundant Array of Independent Disks
RARP	Reverse Address Resolution Protocol

RAS	Registration, Admission, and Status
RF	Radio Frequency
RFC	Request for Comments
RIP	Routing Information Protocol
RLM	Remote Line Module
RMON	Remote Monitoring
RMS	Root Mean Square
ROTL	Remote Office Test Line
RPR	Resilient Packet Ring
RSMI	Removable Security Module Interface
RSU	Remote Switch Unit
RSVP	Resource Reservation Protocol
RTCP	Real-Time Control Protocol
RTMP	Routing Table Maintenance Protocol
RTP	Real-Time Transport Protocol
RTS	Ready-to-Send
RT-VBR	Real-Time Variable Bit Rate
SAN	Storage Area Network
SAP	Session Announcement Protocol
SAR	Segmentation and Reassembly
SAT	Supervisory Audio Tones
SCC	Satellite Communications Control
SCCP	Signaling Connection Control Part
SCN	Service Control Node
SCP	Service Control Point
SCSI	Small Computer System Interface
SCTP	Stream Control Transmission Protocol
SDH	Synchronous Digital Hierarchy
SDLC	Synchronous Datalink Control
SDN	Software Defined Network
SDP	Session Description Protocol
SDSL	Single-Pair Digital Subscriber Line
SECAM	Sequential Coleur Avec Memoire
SF	Superframe
SGCP	Simple Gateway Control Protocol
SID	Silence Insertion Descriptor, System Identification Code
SIMPLE	SIP for Instant Messaging and Presence Leveraging Extensions
SIP	Session Initiation Protocol
SIT	Special Identification Tones
SKIP	Simple Key Management for IP
SLA	Service Level Agreement
SMDI	Simplified Message Desk Interface

SMDR	Station Message Detail Recording
SMDS	Switched Multimegabit Data Service
SMF	Single-Mode Fiber
SMI	Structure of Management Information
SMR	Specialized Mobile Radio
SMS	Short Message Service
SMTP	Simple Mail Transfer Protocol
SNA	Systems Network Architecture
SNMP	Simple Network Management Protocol
SoHo	Small Office-Home Office
SONET	Synchronous Optical Network
SP	Signal Point
SPAG	Standards Promotion and Applications Group
SPC	Stored Program Control
SPID	Service Profile Identifier
SR	Selective Router
SS7	Signaling System 7
SSCP	Systems Services Control Point
SSI	Security System Interface
SSID	Service Set Identifier
SSL	Secure Sockets Layer
SSP	Service Switching Point
STP	Signal Transfer Point, Shielded Twisted-Pair
STS	Synchronous Transport Signal
SVC	Switched Virtual Circuit
TA	Terminal Adapter
TAPI	Telephony Applications Programming Interface
TAS	Telephony Application Server
Tbps	Terabits per Second
TC	Telecommunications Closet
TCAM	Telecommunications Communication Access Method
TCM	Trellis-Coded Modulation
TCP	Transmission Control Protocol
TCP/IP	Transport Control Protocol/Internet Protocol
TDD	Telecommunications Device for the Deaf
TDM	Time Division Multiplexing
TDMA	Time Division Multiple Access
TDMoIP	TDM over IP
TE1	Terminal Equipment Type 1
TE2	Terminal Equipment Type 2
TELNET	Terminal Emulation Protocol
TFTP	Trivial File Transfer Protocol
TIA	Telecommunications Industry Association

TIMS	Transmission Impairment Measuring Set
TKIP	Temporal Key Integrity Protocol
TLD	Top Level domain
TLP	Transmission Level Point
TLS	Transparent LAN Service
TMGB	Telecommunications Main Grounding Busbar
TO	Telecommunications Outlets
ToS	Type of Service
TP-PMD	Twisted Pair-Physical Medium Dependent
TRIP	Telephone Routing over IP
TSAPI	Telephony Systems Applications Programming Interface
TSP	Telecommunications Service Priority
TTL	Time-to-Live
TWT	Traveling Wave Tube
UA	User Agent
UBR	Unspecified Bit Rate
UCD	Uniform Call Distribution
UDP	User Datagram Protocol
UMTS	Universal Mobile Telecommunications System
UNE	Unbundled Network Elements
UNI	User-to-Network Interface
UNII	Unlicensed National Information Infrastructure
UPS	Uninterruptible Power Supply
UPSR	Unidirectional Path Switched Ring
URI	Uniform Resource Indicator
URL	Universal Resource Locater
USF	Universal Service Fund
USTA	United States Telephone Association
UTP	Unshielded Twisted Pair
UWB	Ultra-Wideband
V/FoIP	Voice and Fax over Internet Protocol
VAD	Voice Activity Detection, Vapor-Axial Deposition
VAN	Value-Added Network
VBD	Voice-Band Data
VC	Virtual Channel, virtual circuit
VCC	Virtual Control Channel, Virtual Channel Connection
VCI	Virtual Channel Indicator, Virtual Circuit Identifier
VDSL	Very High Bit-Rate DSL
VF	Voice Frequency
VLAN	Virtual LAN
VLR	Visiting Location Register
VoATM	Voice over ATM
VoD	Video on Demand

VoDSL	Voice over DSL
VoFR	Voice over Frame Relay
VoIP	Voice over IP
VPF	VPN Routing/Forwarding Table
VPI	Virtual Path Identifier
VPN	Virtual Private Network
VSAT	Very Small Aperture Terminal
VT	Virtual Terminal, Virtual Tributary
VTAM	Virtual Terminal Access Method
WA	Work Area
WADM	Wavelength Add-Drop Multiplexers
WAN	Wide Area Network
WAP	Wireless Application Protocol
WATS	Wide Area Telephone Service
WDM	Wavelength-Division Multiplexing
WEP	Wired Equivalent Privacy
WFQ	Weighted Fair Queuing
WLAN	Wireless LAN
WLL	Wireless Local Loop
WML	WAP Markup Language
WPA	WiFi Protected Access
WRED	Weighted Random Early Detection
WWDM	Wide Wave Division Multiplexing
XML	Extensible Markup Language
XMPP	Extensible Messaging and Presence Protocol

APPENDIX B

Glossary

This glossary defines many of the terms used in this book. The terminology is so vast and ambiguous, however, that space does not permit more than abbreviated definitions. Several Web pages have extensive glossaries that may be consulted for additional detail.

A bit: In T1 carrier, the signaling bit that is formed from the eighth bit of the sixth channel.

A-law: The coding law used in the European 30-channel PCM system.

Absorption: The attenuation of a lightwave signal by impurities or fiber core imperfections or of a microwave signal by oxygen or water vapor in the atmosphere.

Access control list (ACL): In a router, restrictions that prevent stations from reaching certain addresses or applications.

Access tandem: A LEC switching system that provides access for the IXCs to the local network. The access tandem provides the IXC with access to more than one end office within a LATA.

Adaptive differential pulse code modulation (ADPCM): A method approved by CCITT for coding voice channels at 32 kbps to increase the capacity of T1 to either 44 or 48 channels.

Adaptive equalizer: (1) Circuitry in a modem that allows the modem to compensate automatically for circuit conditions that impair high-speed data transmission; (2) a circuit installed in a microwave receiver to compensate for distortion caused by multipath fading.

Add-drop multiplexer (ADM): A device that extracts channels from a digital bit stream and inserts other channels into the bit stream.

Address resolution protocol (ARP): A protocol that translates between IP addresses and Ethernet addresses.

Addressing: The process of sending digits over a circuit to direct the switching equipment to the station address of the called number.

Advanced intelligent network (AIN): An interface between the LEC switching system and an external computer that can provide special and custom services to subscribers independently of features offered by the switch manufacturer.

Aerial cable: Any cable that is partially or completely run aerially between buildings or poles.

Agent: In ACD, an agent is a customer contact person, also known as a telephone service representative. In network management, an agent is a software that resides on the managed device and communicates with the network management workstation.

Alerting: The use of signals on a telecommunications circuit to alert the called party or equipment to an incoming call.

Alternate access carrier (AAC): See Competitive access carrier.

Alternate mark inversion (AMI): See Bipolar coding.

Alternate routing: The ability of a switching machine to establish a path to another machine over more than one circuit group.

Alternating current (AC): A current flow, usually in the form of a sine wave, that alternates from peak positive to peak negative.

Always on/dynamic ISDN (AO/DI): An ISDN service that keeps the D-channel actively connected to a service provider. B-channels are called in as needed.

American Standard Code for Information Interexchange (ASCII): A seven-bit (plus one parity bit) coding system used for encoding characters for transmission over a data network.

Amplitude: The magnitude of voltage variation from peak positive to peak negative in an AC signal.

Amplitude distortion: Any variance in the level of frequencies within the pass band of a communication channel.

Analog: A transmission mode in which information is transmitted by converting it to a continuously variable electrical signal.

Angle of acceptance: The angle of light rays striking an optical fiber aperture, within which light is guided through the fiber. Light outside the angle of acceptance escapes through the cladding.

Antenna gain: The increase in radiated power from an antenna compared to an isotropic antenna.

Armored cable: Multipair cable intended for direct burial that is armored with a metallic covering that serves to prevent damage from rodents and digging apparatus.

Aspect ratio: The ratio between the width and height of a video screen.

Asymmetric digital subscriber line (ADSL): A technology for multiplexing a high-speed data or compressed video signal above the voice channel in a subscriber loop.

Asynchronous transfer mode (ATM): A broadband connection-oriented switching service that carries data, voice, and video information in fixed-length 48-octet cells with a five-octet header.

Asynchronous transmission: A means of transmitting data over a network wherein each character contains a start and stop bit to keep the transmitting and receiving terminals in synchronism with each other.

Atmospheric loss: The attenuation of a radio signal because of absorption by oxygen molecules and water vapor in the atmosphere.

Attenuation to crosstalk ratio (ACR): In a structured cable, the ratio of attenuation divided by the amount of near-end crosstalk.

Audible ring: A tone returned from the called party's switching machine to inform the calling party that the called line is being rung.

Audio frequency (AF): A range of frequencies, nominally 20 Hz to 20 kHz, that the human ear can hear.

Audio Messaging Interchange Specification (AMIS): A standard that permits networking of voice mail systems from different manufacturers.

Audiotex: A voice mail service that prompts callers for the desired service and delivers information in audio form.

Automated attendant: A feature of voice mail and stand-alone systems that answers calls, prompts callers to enter DTMF digits in response to menu options, and routes the call to an extension or call distributor.

Automatic call distributor (ACD): A switching system that automatically distributes incoming calls to a group of answering positions without going through an attendant. If all answering positions are busy, the calls are held until one becomes available.

Automatic location information (ALI): Emergency equipment that enables a public safety answering point to determine the location of a caller.

Automatic message accounting (AMA): Equipment that registers the details of chargeable calls and enters them on a storage medium for processing by an offline center.

Automatic number identification (ANI): Identification of the calling line that is delivered from the calling station to the IXC for the purpose of billing the call. ANI is similar to CLID and comes from the same source, except that CLID may be blocked by the caller where ANI, which is used on long distance and 800 calls, cannot be blocked.

Automatic repeat request (ARQ): A data communications protocol that automatically initiates a request to repeat the last transmission if an error is received.

Automatic route selection (ARS): A software feature of PBXs and hybrids that selects the appropriate trunk route for a call to take based on digits dialed and the caller's class of service.

Autonomous system (AS): See domain.

Avalanche photo diode (APD): A light detector that generates an output current many times greater than the light energy striking its face.

B bit: In T1 carrier, the signaling bit that is formed from the eighth bit of the twelfth channel.

B-channel: The 64-kbps "bearer" channel that is the basic building block of ISDN. The B-channel is used for voice and circuit-switched or packet-switched data.

Backbone cable: Cabling connecting an MDF to intermediate distributing frames located in telecommunications closets.

Backoff algorithm: The process built into the media access control of contention network to determine when to reattempt to send a frame.

Backplane: The wiring connecting the sockets of an equipment cabinet.

Balance: The degree of electrical match between the two sides of a cable pair or between a two-wire circuit and the matching network in a four-wire terminating set.

Balancing network: A network used in a four-wire terminating set to match the impedance of the two-wire circuit.

Balun: A device that converts the unbalanced wiring of a coaxial terminal system to a balanced twisted pair system.

Bandwidth: The range of frequencies a communications channel is capable of carrying without excessive attenuation.

Banyan switch: A high-speed switching system that takes its name from the many branches of a banyan tree. A banyan switch is bi-directional and chooses its path from the address contained in the header of an incoming packet. If an address bit is a 1, the upper path of the switch is taken. Otherwise the switch takes the lower, or 0, path.

Baseband: A form of modulation in which data signals are pulsed directly on the transmission medium without frequency division.

Basic rate interface (BRI): The basic ISDN service consisting of two 64-kbps information or bearer channels and one 16-kbps data or signaling channel.

Battery: A direct current voltage supply that powers telephones and telecommunications apparatus.

Baud: The number of data signal elements per second a data channel is capable of carrying.

Beacon: In a token ring network, beacons are signals sent by stations to isolate and bypass failures.

Bearer channel: A 64-kbps information-carrying channel that furnishes ISDN services to end-users.

Bell Operating Company (BOC): One of the 22 local exchange companies (LECs) that were previously part of the Bell System.

Binary synchronous communications (BSC or Bisync): An IBM byte-controlled half-duplex protocol using a defined set of control characters and sequences for data transmission.

Binary: A numbering system consisting of two digits, 0 and 1.

Bipolar coding: The T carrier line-coding system that inverts the polarity of alternate 1 bits. Also called alternate mark inversion (AMI).

Bipolar violation: The presence of two consecutive 1 bits of the same polarity on a T carrier line.

Bipolar with 8-zero substitution (B8ZS): A line-coding scheme used with T1 clear channel to send a string of eight 0s with a deliberate bipolar violation. The 1 bits in the bipolar violation maintain line synchronization.

Bit error rate (BER): The ratio of bits transmitted in error to the total bits transmitted on the line.

Bit rate: The speed at which bits are transmitted on a circuit; usually expressed in bits per second.

Bit robbing: The use of the least-significant bit per channel in every sixth frame of a T1 carrier system for signaling.

Bit stream: A continuous string of bits transmitted serially in time.

Bit stuffing: Adding bits to a digital frame for synchronizing and control. Used in T carrier to prevent loss of synchronization from 15 or more consecutive 0 bits.

Bit: The smallest unit of binary information; a contraction formed from the words BInary digIT.

Block error rate (BLER): In a given unit of time, BLER measures the number of blocks that must be retransmitted because of error.

Blocking: A switching system condition in which no circuits are available to complete a call and a busy signal is returned to the caller.

Bonding: The permanent connecting of metallic conductors to equalize potential between the conductors and carry any current that is likely to be imposed. Also combining two ISDN B-channels to increase the bandwidth.

Bootp: A boot protocol that dynamically assigns IP addresses to nodes.

Branch feeder: A cable between distribution cable and the main feeder cable that connects users to the central office.

Branching filter: A device inserted in a waveguide to separate or combine different microwave frequency bands.

Breakdown voltage: The voltage at which electricity will flow across an insulating substance between two conductors.

Bridge: Circuitry used to interconnect networks with a common set of higher level protocols.

Bridged tap: Any section of a cable pair that is not on the direct electrical path between the central office and the users' premises, but which is bridged onto the path.

Bridger amplifier: An amplifier installed on a CATV trunk cable to feed branching cables.

Broadband ISDN (B-ISDN): A broadband service based on the use of ATM and SONET.

Broadband: A term used to describe always-on access to the Internet by cable, DSL, or satellite. Also, a form of LAN modulation in which multiple channels are formed by dividing the transmission medium into discrete frequency segments. Also, a term used to describe high bandwidth transmission of data signals.

Broadcast: A transmission to all stations on a network.

Bus: A group of conductors that connects two or more circuit elements, usually at a high speed for a short distance.

Byte: A set of eight bits of information equivalent to a character. Also called an octet.

Cable modem termination service (CMTS): In a DOCSIS cable system, the application that manages admission to the network.

Cable racking: Framework fastened to bays to support interbay cabling.

Caching: The use of memory in a file server to read more information than is requested, storing it so the next information request can be served from memory instead of from the disk. Also, the storage of frequently accessed Web pages in a server.

Call detail recorder (CDR): An auxiliary device attached to a PBX to capture and record call details such as called number, time of day, duration, etc.

Call progress tones: Tones returned from switching systems to inform the calling party of the progress of the call. Examples are audible ring, reorder, and busy.

Call sequencer: An electronic device similar to an ACD that can answer calls, inform agent positions of which call arrived first, hold callers in queue, and provide limited statistical information.

Call store: The temporary memory used in an SPC switching system to hold records of calls in progress and pending changes to permanent memory.

Call-by-call service selection: An ISDN feature that lets more than one service be assigned to a single channel. ISDN-compatible switching systems communicate across the D-channel to select the appropriate service.

Called party control: The provision in a 911 system for the called party to supervise a call and hold it up for tracing.

Calling line identification (CLID): A service offered by LECs in which the calling line number is delivered with the call.

Carbon block protector: A form of electrical protector that uses a pair of carbon blocks separated from ground by a narrow gap. When the voltage from the block to ground exceeds a specified value, the blocks arc across to ground the circuit.

Carrier access code (CAC): A seven-digit code consisting of the digits 101 plus a four-digit carrier identification code. For example, the CAC for MCI is 1010222.

Carrier sense multiple access with collision detection (CSMA/CD): A system used in contention networks where the network interface unit listens for the presence of a carrier before attempting to send and detects the presence of a collision by monitoring for a distorted pulse.

Carrier-to-noise ratio: The ratio of the received carrier to the noise level in a satellite link.

Carrier: A type of multiplexing equipment used to derive several channels from one communications link by combining signals on the basis of time or frequency division. Also, a card cage used in an apparatus cabinet to contain multiple circuit packs. Also, a company that carries telecommunications messages and private channels for a fee.

Carrierless amplitude phase (CAP): A DSL multiplexing method that divides the data signal into two parts and combines them with QAM.

Cell relay: A data communications technology based on fixed-length cells.

Cell: A hexagonal subdivision of a mobile telephone service area containing a cell-site controller and radio frequency transceivers. Also, a group of octets conditioned for transmission across a network.

Cell-site controller: The cellular radio unit that manages radio channels within a cell.

Cellular geographic serving area (CGSA): A metropolitan area in which the FCC grants cellular radio licenses.

Central office (CO): A switching center that terminates and interconnects lines and trunks from users.

Central processing unit (CPU): The control logic element used to execute instructions in a computer.

Centralized attendant service (CAS): A PBX feature that allows the using organization to route all calls from a multi-PBX system to a central answering location where attendants have access to features as if they were collocated with the PBX.

Centralized automatic message accounting (CAMA): A LEC message accounting option in which call details are sent from the serving central office to a central location for recording.

Centrex: A class of central office service that provides the equivalent of PBX service from a LEC switching machine. Incoming calls can be dialed directly to extensions without operator intervention.

Centum call seconds (CCS): See Hundred call seconds.

Channel bank: Apparatus that converts multiple voice frequency signals to frequency or time division multiplexed signals for transmitting over a transmission medium.

Channel service unit (CSU): Apparatus that terminates a T1 line providing various interfacing, maintenance, and testing functions.

Channel: A path in a communications system between two or more points, furnished by a wire, radio, lightwave, satellite, or a combination of media.

Chrominance: The portion of a television signal that carries color encoding information to the receiver.

Circuit pack: A plug-in electronic device that contains the circuitry to perform a specific function. A circuit pack is not capable of stand-alone operation but functions only as an element of the parent device.

Circuit switching: A method of network access in which terminals are connected by switching together the circuits to which they are attached. In a circuit-switched network, the terminals have full real-time access to each other up to the bandwidth of the circuit.

Circuit: A transmission path between two points in a telecommunications system.

Cladding: The outer coating of glass surrounding the core in fiber optics.

Class 5 office: See End office.

Class of service: In a voice network, the service classification within a telecommunications system that controls the features, calling privileges, and restrictions the user is assigned. In a data network the differentiation between requirements of the application for parameters such as delay, jitter, and packet loss.

Clear channel: A 64-kbps digital channel that uses external signaling and therefore permits all 64 kbps to be used for data transmission.

Clock: A device that generates a signal for controlling network synchronization.

Closed circuit television (CCTV): A privately operated television system not connected to a public distribution network.

Cluster controller: A device that controls access of a group of terminals to a higher level computer.

Coarse wavelength division multiplexing (DWDM): The process of multiplexing fiber optics with multiple (usually about 16) wavelengths.

Coaxial cable: A single-wire conductor surrounded by an insulating medium and a metallic shield that is used for carrying a telecommunications signal.

Code-excited linear prediction (CELP): A speech encoding algorithm that enables speech to be digitized at 8.0 kbps.

Coder/decoder (codec): The analog-to-digital conversion circuitry in the line equipment of a digital CO. Also, a device in television transmission that compresses a video signal into a narrow digital channel.

Coherence bandwidth: The bandwidth of a range of frequencies that are subjected to the same degree of frequency-selective fading.

Collimate: The condition of parallel light rays.

Collision window: The time it takes for a data pulse to travel the length of a contention network. During this interval, the network is vulnerable to collision.

Collision domain: Ethernet stations that are subject to collisions because they connect to the same hub.

Collision: A condition that occurs when two or more terminals on a contention network attempt to acquire access to the network simultaneously.

Committed information rate (CIR): In a frame relay network the CIR is the speed the carrier guarantees to provide. Frames above the CIR are carried on a permissive basis up to the port speed, but are marked discard eligible.

Common carrier: A company that carries communications services for the general public within an assigned territory.

Common channel signaling (CCS): A separate data network used to route signals between switching systems.

Common control switching: A switching system that uses shared equipment to establish, monitor, and disconnect paths through the network. The equipment is called into the connection to perform a function and then released to serve other users.

Communications controller: see Front end processor.

Community antenna television (CATV): A network for distributing television signals over coaxial cable throughout a community. Also called cable television.

Community dial office (CDO): A small CO designed for unattended operation in a community, usually limited to about 10,000 lines.

Compandor: A device that compresses high-level voice signals in the transmitting direction and expands them in the receiving direction with respect to lower level signals. Its purpose is to improve noise performance in a circuit.

Competitive access carrier (CAP): A common carrier that builds a local access network, usually of fiber optics, to provide access service to the IXCs in competition to the LECs.

Competitive local exchange carrier (CLEC): A company offering local service in competition with the incumbent local exchange carrier.

Complement: A group of 50 cable pairs (25 pairs in small cable sizes) that are bound together and identified as a unit.

Computer-telephony integration (CTI): Marriage of the PBX with a host computer or file server. The PBX provides call information to the computer and accepts call-handling instructions from the computer.

Concentration ratio: As applied to CO line equipment, it is the ratio between the number of lines in an equipment group and the number of links or trunks that can be accessed from the lines.

Concentration: The process of connecting a group of inputs to a smaller number of outputs in a network. If there are more inputs than outputs, the network has concentration.

Concentrator: A data communications device that subdivides a channel into a larger number of data channels. Asynchronous channels are fed into a high-speed synchronous channel via a concentrator to derive several lower speed channels.

Conditional routing: An ACD feature that routes calls based on variables such as number of agents logged on, length of oldest waiting call, etc.

Conditioning: Special treatment given to a transmission facility to make it acceptable for high-speed data communication.

Conference on European Post and Telecommunications (CEPT): The European telecommunications standards-setting body.

Connectionless: A data transmission method in which packets are launched into the network with the sending and receiving addresses, but without a defined path. For example, LANs use connectionless transmissions.

Connection-oriented: A circuit that is set up over a network so that the originating and terminating stations share a defined path, either real or virtual. For example, a telephone call uses a connection-oriented circuit.

Connectivity: The ability to connect a device to a network.

Consultative Committee on International Telephone and Telegraph (CCITT): An international committee that sets telephone, telegraph, and data communications standards. Now known as the International Telecommunications Union (ITU-T).

Contention: A form of multiple access to a network in which the network capacity is allocated on a "first come first served" basis.

Control equipment: Equipment used to transmit orders from an alarm center to a remote site to perform operations by remote control.

Convergence: Merging real-time applications such as voice, video, and instant messaging together with data onto a single broadband infrastructure that is based on IP. Also, the process of routers exchanging routing table updates in response to a change.

Converter: A device for changing central office voltage to another DC voltage for powering equipment.

Core: The inner glass element that guides the light rays in an optical fiber. Also, the internal portion of a network.

Country code: The highest tier in the E.164 telephone numbering system. It is a set of digits that defines the country in the international numbering plan.

Coverage path: In a PBX the coverage path determines where the call will be routed if the called telephone is busy or does not answer.

Critical rain rate: The amount of rainfall where the drops are of sufficient size and intensity to cause fading in a microwave signal.

Cross-polarization: The relationship between two radio waves when one is polarized vertically and the other horizontally.

Cross: A circuit impairment where two separate circuits are unintentionally interconnected.

Crossconnect: A wired connection between two or more elements of a telecommunications circuit.

Cross-polarization discrimination (XPD): The amount of decoupling between radio waves that exists when they are cross-polarized.

Crosstalk: The unwanted coupling of a signal from one transmission path into another.

Custom local area signaling service (CLASS): A suite of telephone services LECs offer. Examples are CLID, distinctive ringing, automatic callback, etc.

Customer premise equipment (CPE): An apparatus mounted on the user's premises and connected to the telecommunications network.

Cut-through: A LAN switching method that makes a connection immediately on reading the destination address.

Cyclical redundancy checking (CRC): A data error-detecting system wherein an information block is subjected to a mathematical process designed to ensure that errors cannot occur undetected.

D-channel: The ISDN 16-kbps data channel that is used for out-of-band signaling functions such as call setup.

Daisy chain: A LAN configuration in which nodes are directly connected in series.

Data communications equipment (DCE): Equipment designed to establish a connection to a network, condition the input and output of DTE for transmission over the network, and terminate the connection when completed.

Data compression: A data transmission system that replaces a bit stream with another bit stream having fewer bits.

Datalink: A circuit capable of carrying digitized information. Usually refers to layer 2 of the OSI protocol stack.

Data network identification code (DNIC): A 14-digit number used for worldwide numbering of data networks.

Data service unit (DSU): Apparatus that interfaces DTE to a line. Used with CSU when DTE lacks complete digital line interface capability or alone when DTE includes digital line interface capability.

Data terminal equipment (DTE): Any form of computer, peripheral, or terminal that can be used for originating or receiving data over a communication channel.

Data: Digitized information in a form suitable for storage or communication over electronic means.

Datagram: An unacknowledged packet sent over a network as an individual unit without regard to previous or subsequent packets.

Datalink connection identifier (DLCI): The numerical designation in a frame relay circuit that defines the PVC.

dBm: A measure of signal power as compared to 1 mW (1/1000 watt) of power. It is used to express power levels. For example, a signal power of –10 dBm is 10 dB lower than 1 mW.

Decibel (dB): A measure of relative power level between two points in a circuit.

Dedicated access: The interconnection of a station to an IXC through a dedicated line.

Dedicated circuit: A communications channel assigned for the exclusive use of an organization. Also known as a private line.

Delay: The time required for a signal to transit the communications facility; also known as latency.

Delta modulation: A system of converting analog to digital signals by transmitting a single bit indicating the direction of change in amplitude from the previous sample.

Demand-assigned multiple access (DAMA): A method of sharing the capacity of a communications satellite by assigning capacity on demand to an idle channel or time slot from a pool.

Demarcation point: The point at which the customer-owned wiring and equipment interfaces with the telephone company.

Demodulation: The process of extracting intelligence from a carrier signal.

Dense wavelength division multiplexing (DWDM): The process of multiplexing fiber optics with multiple (usually 40 or more) wavelengths.

Dialed number identification service (DNIS): A service offered over T1/E1 lines by most 800 carriers that reports to the PBX which 800 number was dialed. The PBX can then use the 800 number to route the call. DNIS on 800 circuits is equivalent to DID on central office trunks.

Dial-up: A data communications session that is initiated by dialing a switched telephone circuit.

Digital crossconnect system (DCS): A specialized digital switch that enables cross-connection of channels at the digital line rate.

Digital service crossconnect (DSX): A physically wired crossconnect frame to enable connecting digital transmission equipment at a standard bit rate.

Digital switching: A process for connecting ports on a switching system by routing digital signals without converting them to analog.

Digital: A mode of transmission in which information is coded in binary form for transmission on a network.

Digroup: Two groups of 12 digital channels integrated to form a single 24-channel system.

Diplexer: A device that couples a radio transmitter and receiver to the same antenna.

Dipole: An antenna that has two radiating elements fed from a central point.

Direct broadcast satellite (DBS): A television broadcast service that provides television programming services throughout a country from a single source through a satellite.

Direct control switching: A system in which the switching path is established directly through the network by dial pulses without central control.

Direct current (DC): Current that always flows in only one direction.

Direct distance dialing (DDD): A long-distance calling system that enables a user to place a call without operator assistance.

Direct inward dialing (DID): A method of enabling callers from outside a PBX to reach an extension by dialing the access code plus the extension number.

Direct trunks: Trunks dedicated exclusively to traffic between the terminating offices.

Directional coupler: A device inserted in a waveguide to couple a transmitter and a receiver to the same antenna. Also a passive device installed on a CATV cable to isolate the feeder cable from another branch.

Discard eligible: In a frame relay network, frames above the CIR are marked discard eligible. In case of congestion the carrier can discard such frames to preserve network integrity.

Discrete multitone: A DSL modulation method that separates the spectrum into multiple channels and modulates each channel with part of the data signal.

Dispersion: The rounding and overlapping of a light pulse that occurs to different wavelengths because of reflected rays or the different refractive indexes of the core material.

Dispersive fade margin: A property of a digital microwave signal that expresses the amount of fade margin under conditions of distortion caused by multipath fading.

Distance vector: A routing algorithm that is based on selecting circuits based on cost.

Distortion: An unwanted change in a waveform.

Distributed control: A switching system architecture in which more than one processor controls certain groups of line ports.

Distributed processing: The distribution of call processing functions among numerous small processors rather than concentrating all functions in a single central processor.

Distributed queue dual bus (DQDB): The protocol used in the 802.6 metropolitan area network.

Distributed switching: The capability to install CO line circuits close to the served subscribers and connect them over a smaller group of links or trunks to a CO that directly controls the operation of the remote unit.

Distributing frame: A framework holding terminal blocks that interconnect cable and equipment and provide test access.

Distribution cable: Cable that connects the user's serving terminal to an interface with a branch feeder cable.

Diversity: A method of protecting a radio signal from failure of equipment or the radio path by providing standby equipment.

Domain: A network that has its own numbering system and the capability of autonomous operation. Also known as an autonomous system.

Domain name service (DNS): Translates host names to IP addresses.

Downlink: The radio path from a satellite to an earth station.

Download: To send information from a host computer to a remote terminal.

Downstream channel: The frequency band in a CATV system that distributes signals from the headend to the users.

Drop wire: Wire leading from the user's serving terminal to the station protector.

DSL access multiplexer (DSLAM): A device that connects DSL lines to terminating equipment in the central office.

Dual tone multifrequency (DTMF): A signaling system that uses pairs of audio frequencies to represent a digit. Usually synonymous with the Lucent Technologies trademark Touch-tone.

Dumb terminal: A terminal that has no processing capability. It is functional only when connected to a host.

Dwell time: In frequency-hopping spread spectrum radio, dwell time is the time the system stays on one frequency before changing to another.

Dynamic Host Configuration Protocol (DHCP): A protocol that runs on a server and provides IP addresses to stations as requested.

E & M signaling: A method of signaling between offices by voltage states on the transmit and receive leads of signaling equipment at the point of interface.

Earth station: The radio equipment, antenna, and satellite communication control circuitry that is used to provide access from terrestrial circuits to satellite capacity.

Echo canceller: An electronic device that processes the echo signal and cancels it out to prevent annoyance to the talker.

Echo cancellation: A protocol used to obtain full-duplex data communication over a two-wire line.

Echo checking: A method of error checking in which the receiving end echoes received characters to the transmitting end.

Echo return loss (ERL): The weighted return loss of a circuit across a band of frequencies from 500 to 2500 Hz.

Echo suppressor: A device that opens the receive path of a circuit when talking power is present in the transmit path.

Echo: The reflection of a portion of a signal back to its source.

Effective isotropic radiated power (EIRP): Power radiated by a transmitter compared to the power of an isotropic antenna, which is one that radiates equally in all directions.

Electromagnetic interference (EMI): An interfering signal that is radiated from a source and picked up by a telecommunications circuit.

Electronic data interchange (EDI): The intercompany exchange of legally binding trade documents over a telecommunications network.

Electronic mail: A service that enables messages and file attachments to be transferred across a communication network.

Electronic tandem network (ETN): A private telecommunications network that consists of switching nodes and interconnecting trunks.

Emergency ringback: A feature that enables a 911 center to connect a caller to the appropriate public service answering point (PSAP) and to enable the PSAP to re-ring the circuit if the caller disconnects. This enables the PSAP to connect to the caller to get additional emergency information.

End office: The central office in the LEC's network that directly serves subscriber lines. Also known as a class 5 office.

End system (ES): In a data network, the system that directly serves subscriber locations.

Endpoint: The originating and terminating points in a telecommunications session.

End-to-end signaling: A method of connecting signaling equipment so that it transmits signals between the two ends of a circuit with no intermediate appearances of the signaling leads.

Enhanced 911 (E-911): An emergency service in which information relative to the specific caller, such as identity and location, is forwarded to the 911 agency by the serving central office.

Enterprise network: A private network of both switched and dedicated facilities that enables users to connect to services wherever they are located without concern about how to establish the session.

Entrance facility (EF): The physical structure and cable connecting the main equipment room to the common carrier's facilities.

Entrance link: A coaxial or fiber optic facility used to connect the last terminal in a microwave signal to multiplex or video-terminating equipment.

Envelope delay: The difference in propagation speed of different frequencies within the pass band of a voice channel.

Equal access: A central office feature that allows all interexchange carriers to have access to the trunk side of the switching network in an end office.

Equipment rooms (ER): Building areas intended to house telecommunications equipment. Equipment rooms also may fill the functions of a telecommunications closet.

Error-free seconds: The number of seconds per unit of time that a circuit vendor guarantees the circuit will be free of errors.

Error: Any discrepancy between a received data signal from the signal as it was transmitted.

Ethernet: A proprietary contention bus network developed by Xerox, Digital Equipment Corporation, and Intel. Ethernet formed the basis for the IEEE 802.3 standard.

Exchange: A geographical area approximating a metropolitan area in which the LEC provides a uniform collection of services and rates.

Extended area service: Service from a central office to central offices in outlying exchanges.

Extended super frame (ESF): T1 carrier framing format that provides 64-kbps clear channel capability, error checking, 16 state signaling, and other data transmission features.

Extranet: A virtual data network tying a company to outside organizations, often through the Internet, but also through frame relay and other such network services.

Facility: Any set of transmission paths that can be used to transport voice or data. Facilities can range from a cable to a carrier system or a microwave radio system.

Facsimile: A system for scanning a document, encoding it, transmitting it over a telecommunications circuit, and reproducing it in its original form at the receiving end.

Far end crosstalk (FEXT): The amount of crosstalk measured at the distant end of a receive circuit when a signal is applied at the near end.

Fax on demand (FOD): Equipment that prompts a caller to enter digits from the dial pad identifying the information needed and a fax number. FOD equipment retrieves the information from a database and automatically faxes it to the caller.

Feeder cables: Cables that route from the central office to neighborhood serving areas.

Fibre Channel: An ANSI standard protocol for connecting high-speed devices that are located too distant to use a protocol such as small computer system interface (SCSI).

File transfer protocol (FTP): A protocol used by TCP/IP networks to transfer files from one system to another.

Filtering: The process of reading packets or frames and forwarding them or not depending on setup definitions.

Final trunk: In a telephone network final trunks are engineered to provide a low degree of blockage.

Firewall: A device that protects the connection between a network and an untrusted connecting network such as Internet. The firewall blocks unwanted traffic from entering the network and allows only authorized traffic to leave.

Flow control: The process of protecting network service by slowing or denying access to additional traffic that would add further to congestion.

Foreign exchange (FEX): A special service that connects station equipment located in one telephone exchange with switching equipment located in another.

Forward error correction (FEC): A process for correcting errors in a data stream by processing redundant code bits.

Forwarding equivalence class (FEC): In MPLS, FEC is a group of sessions having the same destination and service requirements.

Four-wire circuit: A circuit that uses separate paths for each direction of transmission.

Fragmentation: The process of subdividing long packets for transport across a network.

Frame relay: A data communications service that transports frames of information across a network to one or more points. Cost is based on three elements: CIR, access circuit, and port speed.

Frequency: The rapidity, measured in cycles per second or Hertz (Hz), with which an alternating current varies from peak to peak.

Front-end processor: An auxiliary computer attached to a network to perform control operations and relieve the host computer for data processing.

Full-duplex: A data communication circuit over which data can be sent in both directions simultaneously.

Gas tube protector: A protector containing an ionizing gas that conducts external voltages to ground when they exceed a designed threshold level.

Gateway: Circuitry used to interconnect networks by converting the protocols of each network to that used by the other.

Gauge: The physical size of an electrical conductor, specified by American Wire Gauge (AWG) standards.

Generic program: The operating system in an SPC central office that contains logic for call processing functions and controls the overall machine operation.

Glare: A condition that exists when both ends of a circuit are simultaneously seized.

Go-back-n: A transport protocol process in which the receiving end of the circuit requests the sender to go back to the *n*th previous packet and resend all following that.

Grade of service: The percentage of time or probability that a call will be blocked in a network. Also, a quality indicator used in transmission measurements to specify the quality of a circuit based on both noise and loss.

Grooming: The process of combining and configuring signals to optimize the use of bandwidth.

Ground start: A method of circuit seizure between a central office and a PBX that transmits an immediate signal by grounding the tip of the line.

Half-duplex: A data communications circuit over which data can be sent in only one direction at a time.

Handshaking: Signaling between two DCE devices on a link to set up communications between them.

Headend: In a CATV system, the equipment at the head of the network that receives and generates signals and applies them to the cable.

Heat coil: A protection device that opens a circuit and grounds a cable pair when operated by stray currents.

Hexadecimal: A base-16 numbering system. Hexadecimal digits use 0 to 9 plus A to F to represent groups of four binary bits. For example, binary 1111 equals hexadecimal F.

High-usage trunk: Trunk groups established between two switching machines to serve as the first choice path between the machines and handle the bulk of the traffic.

Holding time: The average length of time per call that calls in a group of circuits are off hook.

Horizontal wiring: The wiring from the equipment rooms or telecommunications closets to the work area.

Hundred call seconds (CCS): A measure of network load. Thirty-six CCS represents 100 percent occupancy of a circuit or piece of equipment.

Hybrid: A key telephone system that has many of the features of a PBX. Such features as pooled trunk access characterize a hybrid. Also, a multiwinding coil or electronic circuit used in a four-wire terminating set or switching system line circuits to separate the four-wire and two-wire paths.

Impedance: The ratio of voltage to current in an alternating current electrical circuit.

Impulse noise: Short bursts of high-amplitude interference.

Incumbent local exchange company (ILEC): The traditional telephone company that serves a particular franchised area.

Independent telephone company (IC): A non-Bell ILEC.

Inside wiring: The wiring on the customers' premises between the telephone set and the telephone company's demarcation point.

Integrated services digital network (ISDN): A set of standards promulgated by ITU-T to prescribe standard interfaces to a switched digital network.

Integrated voice/data: The combination of voice and data signals from a workstation over a communication path to the PBX.

Interactive voice response (IVR): Also known as voice response unit (VRU). Equipment that acts as an automatic front-end for a computer system, enabling callers to conduct their own transactions. The IVR prompts the caller to dial identification digits plus digits to complete transactions such as checking account balance, transferring funds, etc.

Interexchange carrier (IXC): A common carrier that provides long-distance service between LATAs.

Interface: The connection between two systems. Usually hardware and software connecting a computer terminal with peripherals such as DCE, printers, etc.

Intermediate distributing frame (IDF): A crossconnection point between the MDF and station wiring.

Intermediate system (IS): A network system that interconnects end systems.

Intermodulation distortion: Distortion or noise generated in electronic circuits when the power carried is great enough to cause nonlinear operation.

International Telecommunications Union (ITU): An agency of the United Nations that is responsible for setting telecommunications standards.

Internet protocol (IP): A connectionless protocol used for delivering data packets from host to host across an internetwork.

Interoffice channel (IOC): In a telecommunications circuit, the IOC is the circuit interconnecting the local channels, which serve the subscribers directly.

Intranet: A network linking a company's offices together and making information accessible, usually with a Web browser.

Inverse multiplexer: A device that combines multiple 64-kbps or 56-kbps channels into a higher speed bit stream. It is often used to combine multiple switched 56 or ISDN channels for video conferencing.

IPsphere: An initiative to develop an overlay IP network that provides the security and stability that commercial-class communications can be carried over the network.

Isochronous: Time dependent, referring to processes such as voice and video where media streams must be delivered in a specified interval.

Jabber: A single or steady stream of Ethernet frames longer than the maximum.

Jitter: The variation in arrival intervals of a stream of packets. Also, the phase shift of digital pulses over a transmission medium.

Jumper: Wire used to interconnect equipment and cable on a distributing frame.

Key telephone system (KTS): A method of allowing several central office lines to be accessed from multiple telephone sets.

Lambda: Another name for a wavelength in a fiber-optic DWDM system.

Latency: The time it takes for a bit to pass from origin to destination through network; also known as delay.

Leased line: A nonswitched telecommunications channel leased to an organization for its exclusive use.

Least cost routing (LCR): A PBX service feature that chooses the most economical route to a destination based on cost of the terminated services and time of delay.

Level: The signal power at a given point in a circuit.

Line conditioning: A service offered by common carriers to reduce envelope delay, noise, and amplitude distortion to enable transmission of higher speed data.

Link: A circuit or path joining two communications channels in a network.

Link-state: A routing algorithm that periodically broadcasts the state of its attached links to other routers on the network.

Loading: The process of inserting fixed inductors in series with both wires of a cable pair to reduce voice frequency loss.

Local access transport area (LATA): The geographical boundaries within which Bell Operating Companies are permitted to offer long-distance traffic.

Local area network (LAN): A narrow-range data network using one of the nonswitched multiple access technologies.

Local automatic message accounting (LAMA): A LEC message accounting option in which call details are recorded in the serving central office.

Local exchange company (LEC): Any local exchange telephone company that serves a particular area. A LEC may be an incumbent LEC (ILEC) or a competitive LEC (CLEC).

Local loop: See Subscriber loop.

Local multipoint distribution service (LMDS): A microwave-based service that serves multiple voice and data users from a central base station hub.

Loop start: A method of circuit seizure between a central office and station equipment that operates by bridging the tip and ring of the line through a resistance.

Local trunk: A trunk interconnecting two local switching systems.

Loopback: The process of connecting the receive leads of a circuit back to the transmit leads so they are reflected back to the source.

Loss: The drop in signal level between points on a circuit.

Low earth orbiting satellite (LEOS): A global PCS technology using a constellation of satellites orbiting the earth at a few hundred miles for communications with hand-held units.

Main distributing frame (MDF): The cable rack used to terminate all distribution and trunk cables in a central office or PBX.

Management information base (MIB): A database contained in an SNMP-compatible device that defines the object that is managed.

Maximum transfer unit (MTU): The largest packet size that a network can transport.

Mean time between failures (MTBF): The average time a device or system operates without failing.

Mean time to repair (MTTR): The average time required for a qualified technician to repair a failed device or system.

Message switching: A form of network access in which a message is forwarded from a terminal to a central switch where it is stored and forwarded to the addressee after some delay.

Message telephone service (MTS): A generic name for the switched long-distance telephone service offered by all interexchange carriers.

Message transfer agent (MTA): In a messaging system the MTA transfers messages between the user agents and other MTAs.

Messaging: The use of computer systems to exchange messages among people, applications, systems, and organizations.

Messenger: A metallic strand attached to a pole line to support aerial cable.

Microwave: A high-frequency, high-capacity radio system, usually used to carry multiple voice channels.

Milliwatt: One-thousandths of a watt. Used as a reference power for signal levels in telecommunications circuits.

Mobile data terminal (MDT): A wireless terminal that permits one-way or two-way data communications. The MDT may be vehicular or hand-held.

Mobile telephone switching office (MTSO): The electronic switching system that switches calls between mobile and wireline telephones, controls handoff between cells, and monitors usage. This equipment is known by various trade names.

Modeling: The process of designing a network from a series of mathematical formulas that describe the behavior of network elements.

Modem pool: A centralized pool of modems accessed through a PBX or LAN to provide off-net data transmission from modemless terminals.

Modem: A contraction of the terms MOdulator/DEModulator. A modem is used to convert analog signals to digital form and vice versa.

Modulation: The process by which some characteristic of a carrier signal, such as frequency, amplitude, or phase is varied by a low-frequency information signal.

Multicast: A transmission that includes multiple selected stations on a network.

Multidrop: A circuit dedicated to communication between multiple terminals that are connected to the same circuit.

Multiline hunt: The ability of a switching machine to connect calls to another number in a group when other numbers in the group are busy.

Multiple access: The capability of multiple terminals connected to the same network to access one another by means of a common addressing scheme and protocol.

Multiplexer: A device used for combining several lower speed channels into a higher speed channel.

Near end crosstalk (NEXT): The amount of signal received at the near end of a circuit when a transmit signal is applied at the same end of the link.

Network access restriction (NAR): A software restriction built into Centrex systems to limit the number of simultaneous trunk calls the subscriber can place.

Network address translation (NAT): A process by which a network maintains a private addressing system behind a server that relays sessions to the public Internet.

Network administration: The process of monitoring network loads and service results and making adjustments needed to maintain service and costs at the design objective level.

Network channel terminating equipment (NCTE): Apparatus mounted on the user's premises that is used to amplify, match impedance, or match network signaling to the interconnected equipment.

Network design: The process of determining quantities and architecture of circuit and equipment to achieve a cost/service balance.

Network: A set of communications nodes connected by channels.

Node: A major point in a network where lines from many sources meet and may be switched.

Noise: Any unwanted signal in a transmission path.

Nonfacility associated signaling (NFAS): The capability in a primary rate ISDN network of using one 64-kbps D-channel to control the signaling of multiple PRIs.

Numbering plan area (NPA): A three-digit (in North America) number, sometimes called an area code, that identifies telephone service within a geographical area.

Octet: A group of eight bits. Often used interchangeably with byte, although a byte can have other than eight bits.

Off-premise extension (OPX): An extension telephone that uses LEC facilities to connect to the main telephone service.

Off-hook: A signaling state in a line or trunk when it is working or busy.

On-hook: A signaling state in a line or trunk when it is nonworking or idle.

On-net: In a virtual network, an on-net call is one that uses a dedicated access line (DAL)

Open network architecture (ONA): A telephone architecture that provides the interfaces to enable service providers to connect to the PSTN.

Open systems interconnect (OSI): A seven-layer data communications protocol model that specifies standard interfaces which all vendors can adapt to their own designs.

Overflow: The ACD process in which a call is routed to a queue other than the original one.

Overhead: Any noninformation bits such as headers, error-checking bits, start and stop bits, etc., used for controlling a network.

Overlay area code: Two or more NPAs that cover the same geographical area. Overlay area codes require callers to dial the area code to make a local call.

Packet assembler/disassembler (PAD): A device used on a packet-switched network to assemble information into packets and to convert received packets into a continuous data stream.

Packet switching: A method of allocating network time by forming data into packets and relaying it to the destination under control of processors at each major node. The network determines packet routing during transport of the packet.

Packet: A unit of data information consisting of header, information, error detection, and trailer records.

Pair gain: The degree to which digital loop carrier increases the carrying capacity of a cable pair.

Parity: A bit or series of bits appended to a character or block of characters to ensure that either an odd or even number of bits are transmitted. Parity is used for error detection.

Patch: The temporary interconnection of transmission and signaling paths; used for temporary rerouting and restoral of failed facilities or equipment. Also, changes to a computer program designed to correct a coding deficiency.

Peg count: The number of times a specified event occurs. Derived from an early method of counting the number of busy lines in a manual switchboard.

Permanent virtual circuit (PVC): In a data network a PVC is defined in software. The circuit functions as if a hardware path was in place, but the path is shared with other users.

Personal communications service (PCS): A radio-based service that allows subscribers to roam anywhere, using a telephone number that is not associated with a fixed location.

Personal identification number (PIN): A billing identification number dialed by the user to enable the switching machine to identify the calling party.

Phase: A means of describing by compass degrees, the relative azimuth of different AC signals or different points on the same AC signal.

Phishing: The practice of obtaining confidential information such as credit card numbers by masquerading as a trusted source.

Point-of-presence (POP): The point at which a carrier meets the customer-provided portion of a circuit. For LECs the POP is usually on the customers' premises. For IXCs the POP is usually on the carriers' premises.

Point-to-point circuit: A telecommunications circuit that is exclusively assigned to the use of two devices.

Pointer: A process used by SONET/SDH to align the payload and minimize jitter.

Poll cycle time: The amount of time required for a multidrop data communications controller to make one complete polling cycle through all devices on the network.

Polling: A network sharing method in which remote terminals send traffic upon receipt of a polling message from the host. The host accesses the terminal, determines if it has traffic to send, and causes traffic to be uploaded to the host.

Port: Either a physical piece of hardware that provides access to a device or a logical definition, such as a TCP port, that defines the application using the transport mechanism.

Power fail transfer: A unit in KTS or a PBX that transfers one or more telephone instruments to central office lines during a power failure.

Power sum: In multipair structured cable, a process that looks at interference across all pair combinations.

Prefix: A three-digit (in North America) code that is the third tier in the E.164 numbering plan, after country code and area code.

Primary rate interface (PRI): In North America a 1.544-mbps information-carrying channel that furnishes ISDN services to end users. Consists of 23 bearer channels and one signaling channel. In Europe a 2.048-mbps channel consisting of 30 bearer and two signaling channels.

Private automatic branch exchange (PABX): A term often used synonymously for PBX. A PABX is always automatic, whereas switching is manual in some PBXs.

Private branch exchange (PBX): A switching system dedicated to telephone and data use in a private communication network.

Propagation delay: The absolute time delay of a signal from the sending to the receiving terminal.

Propagation speed: The speed at which a signal travels over a transmission medium.

Property management system (PMS): A computer application used in the hospitality industry to handle functions such as check-in/check-out, room status, etc. The PMS is often linked to the telephone system to provide such functions from the PMS terminal.

Protector: A device that prevents hazardous voltages or currents from injuring a user or damaging equipment connected to a cable pair.

Protocol analyzer: A data communications test set that enables an operator to observe bit patterns in a data transmission, trap specific patterns, and simulate network elements.

Protocol converter: A device that converts one communications protocol to another.

Protocol data unit (PDU): A formatted data element, the size and structure of which depends on its position in the protocol stack. For example, a frame is a data-link layer PDU, a packet a network layer PDU, and an X.400 message an application layer PDU.

Protocol: The conventions used in a network for establishing communications compatibility between terminals and for maintaining the line discipline while they are connected to the network.

Provisioning: The process of providing telecommunications service such as equipment, wiring, and transmission.

Pseudowire: A point-to-point connection between service provider edge routers that emulates a wired connection through a core MPLS network.

Public data network (PDN): A data transmission network operated by a private telecommunications company for public subscription and use.

Public switched telephone network (PSTN): A generic term for the interconnected networks of operating telephone companies.

Pulse code modulation (PCM): A digital modulation method that encodes a voice signal into an eight-bit digital word representing the amplitude of each pulse.

QSIG: A protocol for networking PBXs of different manufacture.

Quadrature amplitude modulation (QAM): A modem modulation method in which two carrier tones combine in quadrature to produce the output signal.

Quality of service (QoS): A measure of parameters such as delay, jitter, and packet loss that affect isochronous transmissions.

Quantizing: In a PCM system, the process of scaling analog samples into digital octets.

Queuing: The holding of calls in queue when a trunk group is busy and completing them in turn when an idle circuit is available.

Redundancy: The provision of more than one circuit element to assume call processing when the primary element fails.

Register: A device in a switching system that provides dial tone and receives and registers dialed digits.

Remote access: A family of products that allows users who are away from the office to dial into the LAN and access its resources.

Remote MONitoring (RMON): A network management function that enables the manager to monitor network functions from a remote management workstation.

Reorder: A fast busy tone used to indicate equipment or circuit blockage.

Repeater: A bi-directional signal regenerator (digital) or amplifier (analog). Repeaters are available to work on analog or digital signals from audio to radio frequency.

Resolving: The process of propagating DNS information among servers.

Response time: The interval between the user's sending time of the last character of a message and the time the first character of the response from the host arrives at the terminal.

Restriction: Limitations to a station on the use of PBX features or trunks on the basis of service classification.

Return loss: The degree of isolation, expressed in dB, between the transmit and the receive ports of a four-wire terminating set.

Ring: The designation of the side of a telephone line that carries talking battery to the users' premises.

Riser cable: In a building, riser cable runs between the MDF and telecommunications closets on other floors. In a campus environment riser connects the MDF to TCs in other buildings.

RJ-11: A standard four-conductor jack and plug arrangement typically used for connecting a standard telephone to inside wiring.

RJ-45: A standard eight-conductor jack and plug arrangement typically used for connecting a telephone or data terminal to inside wiring.

Route node: The top level in the DNS hierarchy.

Router: A device that operates at layer 3 in the OSI hierarchy to forward packets between devices across a network.

Routing: The path selection made for a telecommunications signal through the network to its destination.

Rules-based messaging: An e-mail feature that enables a user to screen messages based on criteria that the user specifies.

Runt: An Ethernet frame shorter than the 64-byte minimum.

Segment: See Subnet.

Selective repeat: A process in which the transport protocol in a data circuit asks the sender to repeat certain packets.

Serial interface: Circuitry used in DTE to convert parallel data to serial data for transmission on a network.

Server: In a telecommunications network, servers are the trunks or the service processes, such as call center agents, that fulfill the users' service requests. In a LAN, servers are devices that provide specialized services such as file, print, and modem or fax pool services.

Service level: In a call center, service level is defined as the percentage of calls answered within an objective time interval.

Service profile identifier (SPID): A unique number that ISDN equipment uses to identify itself to the central office.

Short: A circuit impairment that exists when two conductors of the same pair are connected at an unintended point.

Sidetone: The sound of a talker's voice audible in the handset of the telephone instrument.

Signal constellation: A graphic representation of combinations of amplitude and phase points in a QAM signal.

Signaling System #7 (SS7): An out-of-band signaling protocol between public switching systems.

Signal-to-noise ratio: The ratio between signal power and noise power in a circuit.

Simple Mail Transfer Protocol (SMTP): A protocol for delivering messages across a TCP/IP network.

Simple Network Management Protocol (SNMP): A management protocol for monitoring and controlling network devices.

Simulation: The process of designing a network by simulating the events and facilities that represent network load and capacity.

Skill-based routing: An ACD function that routes calls based on a table containing the skills of each agent.

Slip: In T carrier, a momentary loss of framing.

Socket: A software object that connects an application to a network protocol.

Spanning tree: An algorithm that prevents redundant paths in a layer 2 network.

Spatial reuse: The capability of a node on a ring network to send traffic directly to the destination node on either ring and in either direction.

Split: A designated group of answering stations in an ACD.

Spoofing: The creation of TCP/IP packets using a fictitious IP address. Hackers use spoofing to hide their true IP address. Satellite carriers can use spoofing to acknowledge packets in a manner that avoids the delay problems of the satellite link.

Spread spectrum: A radio modulation method that transmits signal over a broad range of frequencies (direct sequence method) or rapidly jumps from one frequency to another (frequency hopping). Spread spectrum provides excellent security and resists interference.

Split-channel modem: A modem that achieves four-wire transmission by dividing the analog frequency band into two segments.

Stateful firewall: A firewall that maintains knowledge of processes under operation.

Station equipment: The equipment in a telecommunications network that the subscriber uses directly.

Station message detail recording (SMDR): The port in a PBX that provides information such as called and calling station, time of day, and duration on long-distance calls. The SMDR port is usually connected to a call accounting system to produce the necessary reports.

Station range: The number of feet or ohms over which a telephone instrument can signal and transmit voice and data.

Statistical multiplexing: A form of data multiplexing in which the time on a communications channel is assigned to terminals only when they have data to transport.

Store-and-forward: A method of switching messages in which a message or packet is sent from the originating terminal to a central unit where it is held for retransmission to the receiving terminal. In LAN switches, a switch that reads the entire frame before forwarding it.

Striping: In a RAID disk, striping is the process of parceling a file across several disks. In Fibre Channel it is the method of aggregating multiple N_ports to obtain greater bandwidth.

Structure of management information (SMI): The framework that defines network management variables in a network management system.

Subnet: A group of workstations in a LAN that is separated from other groups by a bridge or, more frequently, a router.

Subscriber carrier: A multichannel device that enables several subscribers to share a single facility in the local loop.

Subscriber loop: The circuit that connects a user's premises to the telephone central office.

Superframe: In T carrier, a repetitive sequence of two groups of 12 frames.

Supervision: The process of monitoring the busy/idle status of a circuit to detect changes of state.

Sustained information rate (SIR): The maximum guaranteed rate at which data can transit an SMDS network. SIR is similar to CIR in a frame relay network.

Switched access: Access to a long-distance carrier by switching through the LEC—provided as an alternative to dedicated access.

Switched multimegabit data service (SMDS): A high-speed connectionless data transport service offered by LECs and some IXCs.

Switched virtual circuit (SVC): A logical link between points on a carrier network that is set up and disconnected with each session.

Synchronous: A method of transmitting data over a network wherein the sending and receiving terminals are kept in synchronism with each other by a clock signal embedded in the data.

System integration: The process of bringing software and equipment from different manufacturers together to form an operational unit.

Systems Network Architecture (SNA): An IBM data communications architecture that includes structure, formats, protocols, and operating sequences.

T1 multiplexer: An intelligent device that divides a 1.544-Mbps facility into multiple voice and data channels.

Tandem switch: A switching system that terminates trunks from multiple connecting switches and relays calls among them.

Telecommunications bonding backbone: A backbone cable run between equipment rooms and telecommunications closets and the building's TMGB.

Telecommunications closet (TC): In a telecommunications wiring plan, a satellite closet containing a junction between backbone and horizontal cable.

Telecommunications grounding busbar: A grounding point for telecommunications services and equipment in each telecommunications equipment room and closet.

Telecommunications main grounding busbar (TMGB): The central grounding point for telecommunications equipment rooms and closets. The TMGB is bonded to the electrical ground and to the building's metal framework.

Telecommunications: The electronic movement of information.

Telephony applications programming interface (TAPI): A programming interface developed by Microsoft that connects PCs to telephone instruments to enable computer-telephony applications.

Telephony services applications programming interface (TSAPI): A programming interface for Novell servers that enables developers to write computer-telephony software for applications that work on Novell servers.

Terminal: A fixture attached to distribution cable to provide access for making connections to cable pairs. Also, any device meant for direct operation over a telecommunications circuit by an end user.

Text messaging: The use of a computer-based network of terminals to store and transmit messages among users.

Thermal noise: Noise created in an electronic circuit by the movement and collisions of electrons.

Throughput: Information bits correctly transported over a data network per unit of time.

Tie trunk: A privately owned or leased trunk used to interconnect PBXs in a private switching network.

Time division multiplexing (TDM): A method of combining several communications channels by dividing a channel into time increments and assigning each channel to a time slot. Multiple channels are interleaved when each channel is assigned the entire bandwidth of the backbone channel for a short period of time.

Time slot: In a TDM system each byte is assigned a unit of time sufficient for the transmission of eight bits. This unit of time recurs at the same instant of time in a transmission frame.

Timed overflow: In an ACD this feature allows a call to overflow to an alternate queue after it has waited in the original queue for a specified period of time.

Tip: The designation of the side of a telephone line that serves as the return path to the central office.

Token passing: A method of allocating network access wherein a terminal can send traffic only after it has acquired the network's token.

Token: A software mark or packet that circulates among network nodes.

Top-level domain (TLD): The servers in the DNS system that handle generic, country code, and chartered domains.

Topology: the architecture of a network or the way circuits are connected to link the network nodes.

Traffic usage recorder (TUR): Hardware or software that monitors traffic-sensitive circuits or apparatus and records usage, usually in terms of CCS and peg count.

Transceiver: A device that has the capability of both transmitting and receiving information.

Transducer: Any device that changes energy from one state to another. Examples are microphones, speakers, and telephone handsets.

Translations: Software in a switching system that establishes the characteristics and features of lines and trunks.

Transmission Control Protocol (TCP): A protocol for providing reliable end-to-end delivery of data across an internetwork, usually used with IP.

Transmission level point (TLP): A designated measurement point in a circuit where the designer has specified the transmission level.

Transmission: The process of transporting voice or data over a network or facility from one point to another.

Trap: A message sent by an SNMP-managed device to indicate that a threshold or alarm has been reached.

Traveling class mark: A feature of switches used in ETNs to carry the class of service of the user with the call so downstream tandems know what features and restrictions to apply.

Trellis-coded modulation (TCM): A modem coding method in which the definition of each symbol is dependent on adjacent signals.

Trunk: A communications channel between two switching systems equipped with terminating and signaling equipment.

Tunneling: The process of encapsulating one protocol in another for transport across a network.

Unbundled network element (UNE): An element defined in the Telecommunications Act of 1996 as a service that CLECs can obtain from ILECs and resell or lease.

Uplink: The radio path from an earth station to a satellite.

User agent (UA): In a messaging system the UA is the user's interface to the system. The UA provides for message composition, receipt, sending, and handling.

Universal service: The concept that telecommunications service is physically available and affordable for anyone.

Utilization: The ratio of the time a resource is used to the total time it is available. For example, if a circuit carries 25 CCS of traffic during the busy hour, its utilization is 25/36 = 69.4 percent.

Value added network: A data communication network that adds processing services such as error correction and storage to the basic function of transporting data.

Vector: A series of routing and call-handling instructions in an ACD.

Video on demand (VOD): The delivery of video services to customers in response to their specific request. VOD is contrasted to conventional cable television where all channels are delivered over the medium.

Videotex: An interactive information retrieval service that usually employs the telephone network as the transmission medium to provide information with text and color graphics.

Virtual circuit: A circuit that is established between two terminals by assigning a logical path over which data can flow. A virtual circuit can either be permanent, in which terminals are assigned a permanent path, or switched, in which the circuit is reestablished each time a terminal has data to send.

Virtual LAN (VLAN): A LAN composed of users that are attached to different hubs. Users can be assigned to a LAN segment regardless of their physical location.

Virtual network: A switched voice network offered by interexchange carriers that provides service similar to a private voice network. Virtual networks offer reduced rates for on-net calling, which is available from stations with a direct T1 connection from the user to the IXC.

Virtual private network (VPN): An IP network that is defined over a carrier's IP network, providing security by authentication and encryption.

Virtual tributary: In SONET, a synchronous format below STS-1 into which T1/E1 signals can be mapped.

Voice grade: Analog service that has the transmission characteristics of a voice circuit, i.e., bandwidth of approximately 300 to 3000 Hz.

Voice mail: A service that allows voice messages to be stored digitally in secondary storage and retrieved remotely by dialing access and identification codes.

Voice response unit (VRU): See Interactive voice response.

Wide area telephone service (WATS): A bulk-rated long-distance telephone service that carries calls at a cost based on usage and the state in which the calls terminate.

Wire center: A point, usually in a central office building, where cable to telephone subscribers comes together.

Wireless: A radio or infrared-based service that enables telephone or LAN users to connect to the telecommunications network without wires.

Wireless local loop (WLL): An access service that uses wireless to distribute signals to the user.

Wiring closet: See Telecommunications closet.

Work area: In the EIA/TIA 568 specifications work area is the area in which the station jack is located to feed the terminating devices.

BIBLIOGRAPHY

Bates, Regis J. and Donald Gregory. *Voice and Data Communications Handbook (Standards & Protocols)*. Berkeley, CA: Osborne/McGraw-Hill, 2001.

Bates, Regis J. *Broadband Telecommunications Handbook*. New York: McGraw-Hill, 2002.

Buckley, John. *Telecommunications Regulation*. London: IEEE Press, 2003.

Camarillo, Gonzalo, Miguel-Angel, and Garcia-Martin. *The 3G IP Multimedia Subsystem (IMS): Merging the Internet and the Cellular Worlds*. New York: Wiley, 2004.

Camarillo, Gonzalo. *SIP Demystified*. New York: McGraw-Hill, 2002.

Camp, Ken. *IP Telephony Demystified*. New York: McGraw-Hill, 2003.

Collins, Daniel. *Carrier Grade Voice over IP*. New York: McGraw-Hill, 2003.

Collins, Daniel and Clint Smith. *Wireless Networks*. New York: McGraw-Hill, 2002.

Davidson, Jonathan, James Peters, and Brian Gracely. *Voice over IP Fundamentals*. Indianapolis: Cisco Press, 2000.

Douskalis, Bill. *Putting VoIP to Work: Softswitch Network Design and Testing*. Upper Saddle River, NJ: Prentice Hall, 2002.

Dunsmore, Brad and Toby Skandier. *Telecommunications Technologies Reference*. Indianapolis: Cisco Press, 2003.

Green, James H. *The Irwin Handbook of Telecommunications Management*, 3rd ed. New York: McGraw-Hill, 2001.

Johnston, Alan. *SIP: Understanding the Session Initiation Protocol*, 2nd ed. Norwood, MA: Artech House, Inc., 2004.

Keagy, Scott. *Integrating Voice and Data Networks*. Indianapolis: Cisco Press, 2000.

Kennedy, Charles H. *An Introduction to U.S. Telecommunications Law*. Norwood, MA: Artech House, Inc., 2001.

Korhonen, Juha. *Introduction to 3G Mobile Communications*, 2nd ed. Norwood, MA: Artech House, Inc., 2003.

Louis, P. J. *Telecommunications Internetworking: Delivering Services across the Networks*. New York: McGraw-Hill, 2000.

Miller, Mark A. *Voice over IP Technologies: Building the Converged Network*. New York: M&T Books, 2002.

Muller, Nathan J. *Desktop Encyclopedia of Telecommunications*. New York: McGraw-Hill, 2002.

Newton, Harry. *Newton's Telecom Dictionary*, 21st ed. San Francisco: CMP Books, 2004.

Ohrtman, Frank, Jr. *Softswitch: Architecture for VoIP*. New York: McGraw-Hill, 2003.

Oodan, Antony, et al. *Telecommunications Quality of Service Management: From Legacy to Emerging Services*. London: IEEE Press, 2003.

Panko, Ray. *Business Data Networks and Telecommunications*, 5th ed. Upper Saddle River, NJ: Prentice Hall, 2004.

Schoening, Heinrich M. *Business Management of Telecommunications*. Upper Saddle River, NJ: Prentice Hall, 2004.

Sheldon, Tom. *McGraw-Hill's Encyclopedia of Networking & Telecommunications*. New York: McGraw-Hill, 2001.

Sinnreich, Henry and Alan B. Johnston. *Internet Communications Using SIP*. New York: Wiley, 2001.

Sulkin, Allan. *PBX Systems for IP Telephony*. New York: McGraw-Hill, 2002.

Wisely, Dave, Philip Eardley, and Louise Burness. *IP for 3G: Networking Technologies for Mobile Communications*. New York: Wiley, 2002.

INDEX